Lecture Notes in Computer Science 1295

Edited by G. Goos, J. Hartmanis and J. van Leeuwen

Advisory Board: W. Brauer   D. Gries   J. Stoer

T0239193

Springer
Berlin
Heidelberg
New York
Barcelona
Budapest
Hong Kong
London
Milan
Paris
Santa Clara
Singapore
Tokyo

Igor Prívara   Peter Ružička (Eds.)

Mathematical Foundations of Computer Science 1997

22nd International Symposium, MFCS '97
Bratislava, Slovakia, August 25-29, 1997
Proceedings

 Springer

Series Editors

Gerhard Goos, Karlsruhe University, Germany

Juris Hartmanis, Cornell University, NY, USA

Jan van Leeuwen, Utrecht University, The Netherlands

Volume Editors

Igor Prívara
INFOSTAT
Dúbravská cesta 3, 842 21 Bratislava, Slovakia
E-mail: privara@vuseiar.sk

Peter Ružička
Comenius University
Faculty of Mathematics and Physics, Institute of Informatics
842 15 Bratislava, Slovakia
E-mail: ruzicka@dcs.fmph.uniba.sk

Cataloging-in-Publication data applied for

Die Deutsche Bibliothek - CIP-Einheitsaufnahme

Mathematical foundations of computer science 1997 : 22nd
international symposium ; proceedings / MFCS '97, Bratislava,
Slovakia, August 25 - 29, 1997. Igor Prívara ; Peter Ružička (ed.). -
Berlin ; Heidelberg ; New York ; Barcelona ; Budapest ; Hong Kong
; London ; Milan ; Paris ; Santa Clara ; Singapore ; Tokyo : Springer,
1997
  (Lecture notes in computer science ; Vol. 1295)
  ISBN 3-540-63437-1

CR Subject Classification (1991): F.1-4, D.2-3, G.2

ISSN 0302-9743
ISBN 3-540-63437-1 Springer-Verlag Berlin Heidelberg New York

© Springer-Verlag Berlin Heidelberg 1997
Printed in Germany

Typesetting: Camera-ready by author
SPIN 10546367   06/3142 – 5 4 3 2 1 0   Printed on acid-free paper

Preface

The 22nd Symposium on Mathematical Foundations of Computer Science (MFCS'97) was held in Bratislava, Slovakia, August 25–29, 1997. This volume contains all contributed papers as well as 11 invited papers presented at the symposium.

The MFCS series was established in 1972 as an annual event for researchers in mathematical foundations of computer science. The aim of these symposia is to bring together specialists in theoretical fields of computer science from various countries and to stimulate mathematical research in theoretical computer science. The previous meetings took place in Jablona, 1972; Štrbské Pleso, 1973; Jadwisin, 1974; Mariánské Lázně, 1975; Gdańsk, 1976; Tatranská Lomnica, 1977; Zakopane, 1978; Olomouc, 1979; Rydzina, 1980; Štrbské Pleso, 1981; Prague, 1984; Bratislava, 1986; Carlsbad, 1988; Porąbka-Kozubnik, 1989; Banská Bystrica, 1990; Kazimierz Dolny, 1991; Prague, 1992; Gdańsk, 1993; Košice, 1994; Prague, 1995; and Cracow, 1996.

The program committee for MFCS'97 consisted of: G. Ausiello (Italy), J. Díaz (Spain), P. Ďuriš (Slovakia), T. Eiter (Austria), R. Freivalds (Latvia), F. Gécseg (Hungary), J. Karhumäki (Finland), H. Kirchner (France), H.-J. Kreowski (Germany), M. Nielsen (Denmark), W. Penczek (Poland), V. Pratt (USA), I. Prívara (co-chair, Slovakia), B. Rovan (Slovakia), P. Ružička (chair, Slovakia), A. Salwicki (Poland), D. Sotteau (France), J. van Leeuwen (The Netherlands), J. Wiedermann (Czech Republic), J. Zlatuška (Czech Republic).

The program committee met on May 10–11, 1997, in Bratislava and out of 94 submitted papers it selected 40 papers for inclusion in the scientific program of MFCS'97. The selection was based on originality, quality, and relevance to theoretical computer science. We would like to thank the program committee members for their meritorious work on evaluating and selecting the papers. We would also like to thank all those who submitted papers for consideration, and all referees and colleagues who helped in the evaluation process.

The symposium was organized by the Slovak Society for Computer Science and the Comenius University in Bratislava in cooperation with other institutions in Slovakia, and it is supported by the Europian Association for Theoretical Computer Science. Many thanks are due to the organizing committee of MFCS'97 for making the symposium happen. The organizing committee consisted of: R. Červenka, Š. Dobrev, J. Džubas, P. Ďuriš (chair), V. Hambálková, M. Nehéz, D. Olejár, D. Pardubská, J. Staudek, A. Zavarský. The program committee profited a lot from the electronic support system designed and managed by R. Červenka.

Last but not least, we would like to thank Springer-Verlag for excellent co-operation in publication of this volume.

Bratislava, June 1997   Igor Prívara, Peter Ružička

TABLE OF CONTENTS

INVITED PAPERS

Game Semantics for Programming Languages

Samson Abramsky
Laboratory for the Foundations of Computer Science
University of Edinburgh

Game semantics models types as two-person games, and programs as strategies for playing such games. The key feature of this semantic paradigm is that interaction between the players, who should be thought of as representing a system and its environment is represented explicitly and intrinsically, at a very basic level [Abr96b]. This brings to light a wealth of structure which is not apparent in more traditional behavioural models. In fact, there is enough structure for game semantics to be seen as an intensional version of domain theory.

In 1993, game semantics was used as the basis for the first syntax-independent constructions of fully abstract models for PCF, thus addressing a long-stading problem [AJM96,HO96]. Since then, there has been an extensive further development and application of game semantics to capture many of the more subtle features of programming languages, including local references and non-local control operators [AM97a,AM97b,AM97c,HY97,Lai97]. A picture has begun to develop, based on "semantic cube". Pure functional languages are characerized semantically by a highly constrained class of strategies; there are several, essentially orthogonal constraints. Non-functional features such as local state, control operators on non-determinism can be modelled by relaxing one or another of these constraints. There is then a "factorization theorem" which reduces definability for the extended language to that for the pure functional language. In this way, a strikingly uniform and coherent map of the space of programming language features is beginning to emerge. Moreover, the combinations of features which can be captured precisely, in the form of fully abstract models, with these means are now approaching those found in "real" programming languages such as Standard ML.

References

[AJM96] S. Abramsky, R. Jagadeesan, P. Malacaria: *Full abstraction for PCF*, (submitted for publication, 1996)

[Abr96b] S. Abramsky: *Retracing some paths in process algebra*, Springer Lecture Notes in Computer Science 1119, 1-17, 1996

[AM97a] S. Abramsky, G. McCusker: *Linearity, Sharing and State*, In Algol-like languages, ed. P.O'Hearn and R.D. Tennent, 317-348, Birkhauser, 1997

[AM97b] S. Abramsky, G. McCusker: *Full Abstraction for Idealized Algol with Passive Expressions*, (submitted for publication, 1997)

[AM97c] S. Abramsky, G. McCusker: *Call-by-value games* (submitted for publication, 1997)

[HY97] K. Honda, N. Yoshida: *Game-theoretic analysis of call-by-value computation* To appear in Proceedings of ICALP'97

[HO96] K. Hyland and C.H.L. Ong: *On full abstraction for PCF* (submitted for publication, 1996)

[Lai97] J.Laird: *Full abstraction for functional languages with control*, To appear in Proceedings of LiCS'97

Communication Complexity

László Babai

University of Chicago, 1100 E. 58th St
Chicago, IL 60637, USA
E-mail: laci@cs.uchicago.edu

Abstract. We discuss some aspects of two-party and multi-party communication complexity theory. The topics include a sample from the long list of connections of communication complexity to other models of computation which provide strong motivation to the study of this subject; separation results for restricted models such as simultaneous and one-way communication; some counter-intuitive upper bounds in these models; a new model called "communication with help," and a lower bound technique in this model, based on discrete Fourier analysis and multi-color discrepancy.

Most of the recent results surveyed are joint work with my former and current students Anna Gál, Tom Hayes, Peter Kimmel, Satya V. Lokam.

1 Introduction

Communication complexity is an abstract model of computation which focuses on inter-processor communication and ignores all other costs. This model has gained increasing significance in the past decade because of its links to a variety of other models of computation. It appears to be at the heart of complexity questions especially in highly parallel models such as shallow circuits, and in models of space-bounded computation such as space-bounded Turing machines, decision trees and branching programs. The lower bounds obtained allow in several cases the construction of pseudorandom generators which fool the models in question.

Lovász's beautiful survey [L] discusses lower bound techniques used in the two-party model, with special attention to linear algebra methods. The excellent recent monograph by Kushilevitz and Nisan [KN] gives an in-depth treatment of the entire subject, including many of its applications to other models of computation, and provides a long list of up-to-date references.

In the present paper we briefly review applications that motivate the subject and focus on some variations of the basic models, especially on the simultaneous and the one-way communication models. Our main subject is lower bound techniques leading to the separation of these models for less than a logarithmic number of players.

1.1 The two-party model

We first describe the model introduced by Yao in 1979 [Y79]. Let us consider a function $f : X \times Y \to G$ where X, Y, G are arbitrary nonempty sets. (In applications we often set $X = Y = \{0,1\}^n$). We say f is a *boolean function* if $G = \{0,1\}$. Two players, Alice and Bob, wish to collaboratively evaluate f on input (x,y) ($x \in X, y \in Y$). Each player has unlimited computational power and full knowledge of the function f. However, each has only partial information about the input: Alice sees only x, and Bob sees only y. The players take turns sending messages to each other according to an agreed protocol $\mathcal{P}$. (The protocol depends on f but not on the particular input (x,y), like a bidding scheme in the game of bridge.) The protocol determines, as a function of the communication history (messages already sent), whether or not the communication has ended; if it has not, it also determines who speaks next and how long the next message will be. Moreover, it determines, as a function of the communication history and the part of the input known to the next speaker, what the next message should be. Finally, we say that $\mathcal{P}$ *computes* f if for all inputs (x,y), the the value $f(x,y)$ is uniquely determined by the total communication string. (In particular, at the end, all players will know the value $f(x,y)$.)

The *cost* of protocol $\mathcal{P}$ is the length of the communication string triggered by the worst input. The *communication complexity* $C(f)$ of the function f is the minimum cost of a protocol computing f.

Under a *randomized* protocol, the players may use random bits in addition to the information available to them to determine each message. For each input, the result may be in error with small probability. The minimum cost of a randomized protocol with, say, $\leq 1/3$ probability of error, is the *randomized communication complexity*, denoted by $R(f)$. There are two subcases to this model, depending on whether the coins used are public or private; we reserve the notation $R(f)$ for the *public coin* model.

1.2 The multi-party model

An immediate generalization of this model would divide up the input between several players so that each player sees a separate piece. This model, however, seems to present few mathematical challenges beyond those encountered in the 2-party model and is therefore of limited use. One open problem in this model is the following: find a non-constant lower bound on the randomized complexity of the question $x + y + z = 0$ where x, y, z are n-digit integers, known to players 1, 2, and 3, resp. [Ni].

Henceforth we shall only consider the more profound generalization introduced by Chandra, Furst, and Lipton in 1979 [CFL]. In this model, players $A_1, \ldots, A_k$ wish to collaboratively evaluate the function $f : X_1 \times \cdots \times X_k \to G$ on input $x = (x_1, \ldots, x_k)$ ($x_i \in X_i$). Again, each player has unlimited computational power and full knowledge of the function f. In this model, however, each player has access to a large fraction of the input: player A_i sees all pieces of the input *except* x_i (x_i is "written on the forehead" of A_i). We say that A_i *misses* x_i. The players communicate by broadcasting messages to all other players according to a preagreed protocol as in the two-party model; the cost is similarly defined.

Randomized communication complexity is defined analogously as above.

1.3 Links to other models: two-party

A multitude of connections to other models of computation have been established. The list below is just a small sample.

The goal is to prove lower bounds in other models of computation using lower bounds on communication complexity.

First we mention some applications to the well-studied two-party model. C. D. Thompson's method of proving area-time tradeoffs for *VLSI* [Th] was the original motivation for the model. Applications to *slightly random sources* followed (Chor, Goldreich [CG], U. Vazirani [Vaz]). Lower bounds for linear *decision-trees* were obtained by Gröger and Turán [GT] and Nisan [Ni]. Lower bounds for threshold circuits were given, among others, by Hajnal et al. [HM+], Roychowdhuri et al. [ROS], Vatan [Vat], and Nisan [Ni].

A variant of the two-party model has been used very successfully to establish strong lower bounds on the *depth of monotone circuits* (Karchmer, Raz, Wigderson, McKenzie [KaW, RW, KRW, RM]). The same method yields an approach to the open problem of superlogarithmic lower bounds for the depth of a not necessarily monotone circuit (cf. [KaW]).

1.4 Links to other models: multi-party

In contrast to the two-party models where a great variety of lower bound techniques are known, for three or more players the only strong lower bounds known (for explicit functions) are the ones in [BNS]. These lower bounds are of the form $\Omega(n/c^k)$ where n is the bit-length of the input and k is the number of players. (c is a positive constant; $c \in \{2, 4\}$ in the [BNS] examples.) While these bounds are optimal for a constant number of players, they degrade as k approaches $\log n$; no lower bounds are known for $\log n$ or more players.

While lower bound techniques in this area are scarce, the existing and potential applications of multi-party communication are numerous and significant. Multi-party communication seems particularly relevant to lower bounds for space-bounded computation and for models of highly parallel computation (shallow circuits).

The applications given in [BNS] include several problems of space-bounded computation: an optimal time-space tradeoff for k vs. $(k + 1)$-head Turing machines, a pseudorandom generator for logspace Turing machines, and a lower bound for branching programs (a combinatorial model of space-bounded computation).

Applications to shallow circuits were originated by Håstad and Goldmann [HG]. They recognized that a function defined by a depth-2 circuit with a symmetric function gate on the top and S arbitrary gates of fan-in $\leq k$ on the bottom has communication complexity $\leq \log S$ if the input is divided in an arbitrary way among the (foreheads of) $k + 1$ players.

Further applications of this idea can be found in [LVW]. That paper constructs a pseudorandom generator which fools certain depth-2 circuits; this generator is in turn used for efficient approximate counting.

An important potential application of the Håstad–Goldmann approach follows from the work of Yao [Y90] and Beigel and Tarui [BT]: they show that ACC (bounded depth, polynomial size circuits with boolean and MOD_m gates) can be simulated by depth-two circuits of the Håstad–Goldmann variety, with both $\log S$ and k bounded by $(\log n)^{O(1)}$. Therefore, if the communication complexity of a function f for more than polylog players is more than polylog then f does not belong to ACC. The hope is that such an explicit function can be constructed. Currently it is not known whether ACC=P, or even ACC=NP.

In [Ni], Nisan gives very elegant applications of two-party and multiparty communication lower bounds to obtain lower bounds for models involving low-degree-polynomial threshold gates. He obtains a lower bound of $\Omega(n/\log^2 n)$ on the size of a threshold circuit with polynomial threshold gates of bounded degree, and the same lower bound for the depth of decision trees of bounded-degree-polynomial threshold gates for the "generalized inner product" (GIP) function, using an $\Omega(n/c^k)$ lower bound on the k-party communication complexity of GIP. The idea is to show that each threshold gate has low communication complexity under any partition of the inputs; therefore, if a combination of such gates computes a function of high communication complexity, then there must be a large number of gates.

2 Simultaneous messages

2.1 The model

Since lower bounds in the multi-party model are notoriously hard to come by, it is desirable to look for more tractable substitutes. An important restriction of the model arises when the communication is required to be oblivious: the players do not communicate with each other but simultaneously send messages to a "referee" who does not see any of the input. From the messages received, the referee must be able to produce the value of f. We call this the *"simultaneous messages"* (SM) model; the associated cost is the *SM-complexity* or *simultaneous complexity* of f, denoted by $C_{SM}(f)$.

For $k = 2$ players, the SM complexity is just the log of the number of different rows plus the log of the number of different columns of the $|X| \times |Y|$ matrix $M_f := (f(x,y))$. However, the private coin randomized version holds surprises even in the 2-player case. The question of the private-coin randomized SM complexity of the predicate "$x = y$" was raised in Yao's 1979 paper and was only resolved 17 years later. While the public-coin SM-complexity of "$x = y$" is easily seen to be $O(1)$ (even with only $O(\log n)$ public random bits, using Justesen's code, for instance), it turns out that the private-coin SM complexity of "$x = y$" is $\Theta(\sqrt{n})$. (Here $X = Y = \{0,1\}^n$.) The upper bound was established by A. Ambainis [A2], Moni Naor [Na], and Ilan Newman [Ne]; the lower bound by Newman and Szegedy [NS], Bourgain and Wigderson [BW], and [BK]. The lower bound proofs reveal a curious connection with the following problem in extremal graph theory. Let us say that two subsets A and B of the vertex set of a graph G are *densely connected* if the number of edges $\{\{a,b\} : a \in A, b \in B\}$ is $\geq |A||B|/2$. The question is the maximum number of pairwise densely connected independent sets in graphs on V vertices. The answer is $V^{\Theta(\log V)}$, and this answer is essentially equivalent to the $\Theta(\sqrt{n})$ private-coin SM-complexity of "$x = y$". The exact constant in the exponent of the $V^{\Theta(\log V)}$ expression is not known. – The [BW] and [BK] proofs also show that for any boolean function f, the private coin SM-complexity is $\Omega(\sqrt{C_{SM}(f)})$.

For $k = 3$ players, the SM model was first considered by Nisan and Wigderson [NW]. The general SM model was introduced Pudlák, Rödl, and Sgall [PRS] and by [BKL]; these papers also established an exponential separation of the SM model from the general communication model for $k \leq (\log n)^{1-\epsilon}$ players. We return to this question in Section 2.3.

2.2 Potential applications

Strong motivation for considering the SM model is provided by the fact that *simultaneous messages suffice* for the Håstad–Goldmann simulation; therefore a super-polylog SM lower bound for super-polylog players would suffice for the ACC separation.

Unfortunately, even in the SM model, no lower bounds (for explicit functions) are known for more than $\log n$ players.

One of the long recognized open problems in the theory of computing is to give a *superlinear lower bound on the size of log-depth boolean circuits,* even allowing n-bit output.

A potential application of the 3-party SM model to this problem originates from Valiant's separator approach. Building on a separator theorem for DAGs by Erdős, Graham, and Szemerédi [EGS], Valiant gave a remarkable reduction of the circuits in question [V]. This reduction was in turn used by Nisan and Wigderson [NW] to give a 3-player SM simulation of such functions in the following sense. Let $f : \{0,1\}^n \times \{0,1\}^{\log n} \to \{0,1\}^n$ be a function computed by a (fanin-2) boolean circuit of depth $O(\log n)$ and size $O(n)$. Let $g(x, j, i)$ denote the i-th bit of $f(x,j)$ ($x \in \{0,1\}^n$, $j \in \{0,1\}^{\log n}$). Then $g(x,i,j)$ *can be computed by a 3-party SM protocol where the player missing x sends $O(\log n)$ bits, the player missing j sends $O(n^{\epsilon})$ bits (for any fixed $\epsilon > 0$) and the player missing i sends $O(n/\log\log n)$ bits to the referee* (cf. [KN, p. 135]). An SM lower bound beating these communication constraints would yield the desired circuit size-depth tradeoff.

2.3 Simultaneous messages: surprising upper bounds

The *generalized addressing function* (GAF) for 3 players is defined as follows. Let H be a group of n elements. Player 1 misses a function $h : H \to \{0,1\}$; players 2 and 3 miss elements $a, b \in H$, resp. The desired output is $f(h, a, b) = h(ab)$. So the inputs missed by players 2 and 3 are $\log n$ bits each. For k players, the function is defined by putting an element of H on the forehead of each additional player; the otput is $f(h, a_2, \ldots, a_k) = h(\prod_{i=2}^k a_i)$. (For the case of the cyclic group, this function has been referred to as "shift.")

Clearly, $C(\text{GAF}) \leq 1 + \log n$ (player 1 announces $\prod_{i=2}^k a_i$ and player 2 announces the output). However, a simple information theory argument shows that for 3 players, the simultaneous complexity of GAF is $\Omega(\sqrt{n})$ (for any group), and more generally for k players the simultaneous complexity is $\Omega(n^{1/(k-1)}/k)$ [BKL, PRS]. This observation establishes the exponential separation of $C(f)$ from $C_{SM}(f)$ for $f = $GAF.

Intuition would suggest, however, that even the $\sqrt{n}$ SM lower bound for the 3-player GAF is quite weak; it is not clear how to improve upon the trivial $\Omega(n)$-bit protocol at all. Indeed, for most groups, we don't have any improvement. There are, however, two intriguing classes of exceptions: the *cyclic groups* $H = \mathbf{Z}_n$ and the elementary abelian 2-groups $H = \mathbf{Z}_2^t$ (vector spaces over $GF(2)$; $n = 2^t$).

For cyclic groups, an $O(n \log \log n / \log n)$ upper bound was found by [PRS], which was subsequently improved to $O\left(n(\log n)^{1/4} 2^{-\sqrt{\log n}}\right)$ by Ambainis [A1]. No $n^{1-\epsilon}$ upper bound is known for this case.

In contrast, for the case $H = \mathbf{Z}_2^t$, an $O(n^{0.92})$ upper bound was found in [BKL]; the exponent for k players is $H(1/k)$ where $H(x)$ is the binary entropy function. The exponent for 3 players was subsequently improved to 0.73 by Ambainis and Lokam [AL].

Even stronger, highly counterintuitive upper bounds are known for more than $\log n$ players, indicating the need for caution in appealing to "intuition" in this area. For the cyclic groups, Ambainis gave an n^ϵ upper bound for GAF when $k > (\log n)^c$ for some constant c [A1]. For $H = \mathbf{Z}_2^t$, [BKL] give an upper bound of $3 + \log n$ bits per player, assuming $k > \log n$.

Anna Gál contributed a large class of functions which show suprisingly low SM complexity for more than $\log n$ players [BGKL]. The class consists of functions defined by the following type of depth-2 circuits: the top gate is an arbitrary symmetric function of n boolean variables; the bottom gates are identical symmetric functions of k boolean variables each; each of the k players misses exactly one input bit from each bottom gate. We require the bottom gates to be *compressible* in the sense to be defined below; examples of compressible symmetric functions include *parity, threshold, MOD_m*.

We say that a boolean function f is *compressible* if for any partition $A \dot\cup B$ of its inputs, its two-player one-way communication complexity is $O(\log|B|)$, where the player who sees A speaks first.

(An example of a symmetric function that is *not* compressible is the quadratic character of the sum of the boolean variables; this can be shown using Weil's character sum estimate [BGKL].) Gál's result [BGKL] asserts the following.

Theorem 1. *If f is a function defined by a depth-2 circuit as above with compressible symmetric function gates of k variables at the bottom, and $k \geq 3 + \log n$ then the k-player simultaneous complexity of f is $\leq (\log n)^c$.*

Note that the functions to which the theorem applies includes "majority of majorities," "generalized inner product," "parity of thresholds."

3 Discrepancy: a lower bound technique revisited

The *discrepancy* of a 2-coloring of a set is defined as the difference between the number of elements of each color. Discrepancy bounds for subsets of carte-sian products called "cylinder intersections" were at the heart of the [BNS] lower bounds for multiparty communication complexity. In this section we in-dicate two generalizations of this concept to multi-colored sets ("strong" and "weak" discrepancies), state upper bounds on them for classes of explicit func-tions, and indicate how such bounds imply lower bounds in a multi-party model called "communication with help." Finally, we apply these bounds to resolve the question of separating the one-way and general communication complexities of boolean functions for several players. The material of this section is joint work with Tom Hayes and Peter Kimmel.

3.1 Communication with help

The significance of this variant of the multi-party model will be explained in Section 3.6. We believe, however, that the mathematical beauty and the strength of the results alone justify this model. The model is of interest when the range G is not boolean ($|G| > 2$), although it includes the boolean model as a special case and the lower bounds obtained remain strong when specialized to the boolean case (cf. Corollary 8).

In the "communication with help" model, our players are aided by a "Helper" who can see the entire input and broadcasts a message to the players before they begin to communicate. The Helper's message is from the set $\{1, \ldots, R\}$. We assume $R \leq |G|(1 - c)$ for some constant $c > 0$ (at least a positive constant amount of information should be missing after the Helper's message). The value $f(x_1, \ldots, x_k)$ is again a function of the communication, which now includes the Helper's message. The cost of the protocol includes the number of help bits; the worst-case cost of the best protocol is denoted by $C^{\text{help}}(f)$. This model is introduced in [BHK].

3.2 Cylinder intersections

Let $X = X_1 \times \cdots \times X_k$. A subset $S \subseteq X$ is called a *cylinder in the ith dimension* if membership in S does not depend on the ith coordinate. A subset of X is called a *cylinder intersection* if it is an intersection of cylinders.

In the k-party model (without help), the set $X(s)$ of inputs triggering a given communication string s is a cylinder intersection [BNS]. Each $X(s)$ is *homogeneous* with respect to f, i.e. f restricted to $X(s)$ is constant. Therefore X is the disjoint union of T homogenous cylinder intersections, where $C(f) = \log T$. Now if the size of every homogeneous cylinder-intersection is $\leq V$ then $T \geq |X|/V$. If, in the boolean case, we can show that every cylinder intersection has discrepancy $\leq \Delta(f)$ with respect to some function f then it follows that $V \leq \Delta$ and therefore

$$C(f) \geq \log(1/\Gamma(f)), \tag{1}$$

where Γ is the *normalized discrepancy*: $\Gamma(f) = \Delta(f)/|X|$. In fact, the discrepancy bound yields a lower bound even in the case of randomized protocols: $R(f) \geq \log(1/\Gamma(f)) - O(1)$.

With the appropriate generalization of discrepancy to multi-colored sets, we shall see that inequality (1) remains in force under the "communication with help" model.

3.3 Strong multi-color discrepancy

In this section we define the concept of *strong discrepancy* of multi-colored sets. We drop the adjective "strong;" simply the absence of the adjective "weak" will indicate that we have the "strong" discrepancy in mind.

Let f be a function $f : X \to G$ where G is any set with $|G| \geq 2$. For a subset $S \subseteq X$ and for $\zeta \in G$, we define the *absolute ζ-discrepancy of f in S* by

$$\Delta_\zeta(f, S) := \big| |f^{-1}(\zeta) \cap S| - |S|/|G| \big|.$$

The *absolute discrepancy of f in S* is defined by

$$\Delta(f, S) := \max_{\zeta \in G}\{\Delta_\zeta(f, S)\}.$$

The *discrepancy of f in S* is the *normalized* version of this quantity (the adjective "normalized" will always be omitted):

$$\Gamma(f, S) = \Delta(f, S)/|X|.$$

The discrepancy of f over a family $\mathcal{F}$ of subsets of X is defined as

$$\Gamma(f, \mathcal{F}) := \max_{S \in \mathcal{F}}\{\Gamma(f, S)\}.$$

When X is a cartesian product $X = X_1 \times \cdots \times X_k$, we use $\Gamma(f)$ to denote the discrepancy of f over the family of cylinder intersections. The analysis of protocols with help will be based on the following extension of inequality (1):

$$C^{\text{help}}(f) \geq \log(1/\Gamma(f)) - O(1). \tag{2}$$

The main technical contributions of [BHK] are inequalities of the form $\Gamma(f) < \exp(-n/c^k)$ for several classes of explicit functions f with possibly very large range G (up to $|G| = 2^{(1-\epsilon)n}$; the typical range in applications to one-way complexity would have $|G| \approx 2^{\sqrt{n}}$). In these cases the Helper can reveal nearly full information about the output, yet, by inequality (2), the missing information (fraction of a bit) forces our players to communicate $\Omega(n/c^k)$ bits.

3.4 Weak multi-color discrepancy

Our main tool in bounding the strong discrepancy will be the concept of *weak discrepancies*, available in the case when the range G of f is an abelian group. The notion of weak discrepancy generalizes the observation that for $G = \{\pm 1\}$ we have $\Delta(f, S) = \frac{1}{2} |\sum_{x \in S} f(x)|$. In this (boolean) case, weak and strong discrepancies coincide. In general, we shall be able to bound the strong discrepancy in terms of weak discrepancies using the elements of *discrete Fourier analysis* (cf. [Ba]).

Let G be a finite abelian group of order m. The group of characters of G is denoted by $\widehat{G}$, and the principal character is denoted by χ_0.

As before, let $f : X \to G$ be a function. For a subset $S \subseteq X$ and for $\chi \in \widehat{G}$, we define the *weak, absolute χ-discrepancy of f in S* by

$$\Delta_\chi^{\mathrm{weak}}(f, S) := \left| \sum_{x \in S} \chi(f(x)) \right|.$$

The normalized version is the *weak χ-discrepancy of f in S*, defined by

$$\Gamma_\chi^{\mathrm{weak}}(f, S) = \Delta_\chi^{\mathrm{weak}}(f, S) / |X|.$$

We define the *weak discrepancy of f in S* by averaging over all non-principal characters:

$$\Gamma^{\mathrm{weak}}(f, S) := \frac{\Delta^{\mathrm{weak}}(f, S)}{|X|} = \frac{1}{m - 1} \sum_{\substack{\chi \in \widehat{G} \\ \chi \neq \chi_0}} \Gamma_\chi^{\mathrm{weak}}(f, S).$$

Finally, for a family $\mathcal{F}$ of subsets of X, we define the *weak discrepancy of f over $\mathcal{F}$* by

$$\Gamma^{\mathrm{weak}}(f, \mathcal{F}) := \max_{S \in \mathcal{F}} \{ \Gamma^{\mathrm{weak}}(f, S) \}.$$

The connection between the strong and weak discrepancies is established by the inequality

$$\Gamma(f, S) \leq (1 - 1/m) \Gamma^{\mathrm{weak}}(f, S). \tag{3}$$

Therefore inequality (2) remains valid with $\Gamma^{\mathrm{weak}}(f)$ in the place of $\Gamma(f)$ (where $\Gamma^{\mathrm{weak}}(f)$ is defined as the weak discrepancy over the family of cylinder intersections).

3.5 Bounding the weak discrepancy

Next we consider the problem of bounding the weak discrepancies for a general class of functions. In addition to obtaining strong "communication with help" lower bounds which extend the bounds for the "generalized inner product" function proved in [BNS], we also obtain similarly strong lower bounds for new classes of functions (yielding new results even in the boolean case). The letter n denotes the common length (in binary) of the input strings on each player's forehead. All discrepancies are over the family of cylinder intersections.

Theorem 2. *Let q be a prime power, $X_1, \ldots, X_k$ finite dimensional vector spaces over $\mathbf{F}_q$. Let $X = X_1 \times \ldots \times X_k$, and let $f : X \to \mathbf{F}_q$ be a k-linear map. Let χ be a non-principal additive character of $\mathbf{F}_q$. Then*

$$
\Gamma_\chi^{\text{weak}}(f) \leq \left(\mathop{\mathrm{E}}_{x \in X} \chi(f(x)) \right)^{2^{1-k}} = \left(\mathop{\Pr}_{u \in \widetilde{X}} [\forall x \in X_1 \ f(x, u) = 0] \right)^{2^{1-k}},
$$

where $\widetilde{X} := X_2 \times \cdots \times X_k$; expected values and probabilities are with respect to the uniform distribution over X and $\widetilde{X}$, resp.

As a corollary, we obtain an extension of the GIP lower bound of [BNS].

Definition 3. For a prime power q, positive integers s and k, we define the *generalized inner product*

$$
\mathrm{GIP}_{q,s,k} : (\mathbf{F}_q^s)^k \to \mathbf{F}_q
$$

by

$$
\mathrm{GIP}_{q,s,k}(x_1, \ldots, x_k) = \sum_{i=1}^{s} x_{1,i} x_{2,i} \cdots x_{k,i}.
$$

Corollary 4.
$$
C^{\text{help}}(\mathrm{GIP}_{q,s,k}) \geq \Omega\left(n/c^k\right). \tag{4}
$$

Here the constant c satisfies $2 \leq c \leq 4$. The boolean GIP function considered in [BNS] is the special case when $q = 2$; inequality (4) then reduces to the [BNS] bound.

We now mention another class of functions to which Theorem 2 applies, albeit with a rather different analysis.

Definition 5. Let q be a prime power, and let t be a positive integer. Let X_1 be the space of $1 \times t$ row vectors over $\mathbf{F}_q$, let $X_2 = \cdots = X_{k-1}$ all equal the space of $t \times t$ matrices over $\mathbf{F}_q$, and let X_k be the space of $t \times 1$ column vectors over $\mathbf{F}_q$. The function $\mathrm{MP}_{q,t} : X_1 \times \cdots \times X_k \to \mathbf{F}_q$ is defined by $\mathrm{MP}_{q,t}(A_1, A_2, \ldots, A_k) := A_1 A_2 \cdots A_k$ and is called the *matrix product function*.

Corollary 6. *The discrepancy of $\mathrm{MP}_{q,t}$ is at most*

$$
\left(kq^{-t}\right)^{2^{1-k}}.
$$

Corollary 7. $C^{\text{help}}(\mathrm{MP}_{q,t}) \geq \Omega((t \log q)/2^k)$.

Even the no-help boolean case of this result is new. We state it separately.

Corollary 8. *In the special case $q = 2$, $t = \sqrt{n}$, we have $C(\mathrm{MP}_{2,\sqrt{n}}) \geq \Omega(\sqrt{n}/2^k)$.*

We observe that for fixed k this lower bound is tight apart from a constant factor since two of the players have inputs of size $O(\sqrt{n})$ on their foreheads, so $C(MP_{2,\sqrt{n}}) = O(\sqrt{n})$. We note that MP is the first new class of explicit boolean functions for which such a multi-party lower bound has been found since [BNS].

3.6 Application: one-way communication

An intermediate model of interest between the simultaneous and the unrestricted communication models is the *one-way communication* model in which each player may speak only once, and they proceed in a specified order. The separation of simultaneous and one-way complexities is accomplished by the GAF discussed in Section 2.3. The separation of one-way complexity from unrestricted communication complexity has proved more difficult. Nisan and Wigderson [NW] gave an exponential separation of these two complexity measures for $k = 3$ players. The study of their approach motivated our definition of "communication with help" which has led to the desired separation for k players, $3 \leq k \leq (1 - \epsilon) \log n$.

We accomplish this separation in the following stronger sense. We use $^iC_1(f)$ to denote the communication complexity of f when *player A_i speaks first and is followed by an unrestricted (multi-round) communication between the other players* (but player A_i is not allowed to speak again). What we show in fact is that for certain classes of explicit functions, $^1C_1(f) = \Omega(\sqrt{n}/c^k)$ while $O(\log n)$ bits of communication suffice if any other player speaks first (then player A_1 can immediately announce the result and no one else needs to speak) [BHK].

The basic idea, due to [NW], is the following. Consider a function $g : X \to \{0,1\}^b$ where $X = X_2 \times \cdots \times X_k$. Let $X_1 = \{1, \ldots, b\}$ and for $x \in X$ define the *boolean* function $f_g : X_1 \times \ldots \times X_k \to \{0,1\}$ by setting $f_g(i, x)$ to be the ith bit of $g(x)$.

Proposition 9. $^1C_1(f_g) \geq \min\left\{b, \left(C^{\text{help}}(g)/b\right) - 1\right\}.$

The idea, implicit in [NW], is that either player A_1 provides b bits of information and we are done, or A_i can be viewed as the Helper for a protocol to evaluate $g(x)$ in b rounds as follows. We use an optimal protocol for $^1C_1(f_g)$ successively with inputs (i, x), $i = 1, \ldots, b$. Since A_1 does not see i, the message of A_1 will not vary, so the total number of bits communicated will be $\leq b + b \cdot {}^1C_1(f_g)$. This being an upper bound on $C^{\text{help}}(g)$, the claim follows. $\qquad\Box$

In our examples one can choose b to be $b = \Theta\left(\sqrt{C^{\text{help}}(g)}\right)$; hence we obtain the desired *one-way lower bounds*

$$^1C_1(f_g) = \Theta\left(\sqrt{C^{\text{help}}(g)}\right). \tag{5}$$

Acknowledgment

I wish to thank Tom Hayes and Peter Kimmel for their help and advice during the preparation of this paper.

References

[A1] A. Ambainis: Upper Bounds on Multiparty Communication Complexity of Shifts. *13th Symp. on Theoretical Aspects of Comp. Sci.*, Springer Lecture Notes In Comp. Sci. **1046** (1996) 631–642.

[A2] A. Ambainis: Communication Complexity in a 3-Computer Model. *Algorithmica* **16** (1996), 298–301.

[AL] A. Ambainis, S. V. Lokam: Improved bounds for 3-player generalized addressing. Private communicatioin, 1996.

[AS] N. Alon, J. H. Spencer: *The Probabilistic Method*. John Wiley & Sons, Inc, 1992.

[Ba] L. Babai: *The Fourier Transform and Equations Over Finite Abelian Groups*, Lecture Notes, version 1.2., Univ. of Chicago, December 1989

[BGKL] L. Babai, A. Gál, P. Kimmel, S. V. Lokam: Simultaneous Messages vs. Communication. Updated version of [BKL], Univ. Chicago Tech. Rep. 96-23, November 1996.

[BHK] L. Babai, T. Hayes, P. Kimmel: The cost of the missing bit: communication complexity with help. Manuscript, June 1997.

[BK] L. Babai, P. Kimmel: Randomized simultaneous messages: solution of a problem of Yao in communication complexity. Univ. Chicago Tech Report, March 1996. *12th IEEE Symp. on Structure in Complexity Theory* (1997), to appear.

[BKL] L. Babai, P. Kimmel, S. V. Lokam: Simultaneous Messages vs. Communication. *12th Symposium on Theoretical Aspects of Comp. Sci.*, Springer Lecture Notes in Comp. Sci. **900** (1995) 361–372.

[BNS] L. Babai, N. Nisan, M. Szegedy: Multiparty Protocols, Pseudorandom Generators for Logspace and Time-Space Trade-offs. *Journal of Comp. and Sys. Sci.* **45** (1992) 204–232. (Prelim. version *21st ACM STOC* (1989) 1–11.)

[BT] R. Beigel, J. Tarui: On ACC. *32nd IEEE FOCS* (1991) 783-792.

[BW] J. Bourgain, A. Wigderson. Private communication by Avi Wigderson, Feb. 1996. Cf. [BK] for details.

[CFL] A. K. Chandra, M. L. Furst, R. J. Lipton: Multiparty protocols. *15th ACM STOC* (1983) 94–99.

[CG] B. Chor, O. Goldreich: Unbiased bits from sources of weak randomness and probabilistic communication complexity. *SIAM J. Comp.* **17** (1988) 230–261. (Prelim. version: *26th FOCS* (1985) 429–442.)

[CGKS] B. Chor, O. Goldreich, E. Kushilevitz, M. Sudan: Private Information Retrieval. *36th IEEE FOCS* (1995) 523–531.

[EGS] P. Erdős, R. L. Graham, E. Szemerédi: On sparse graphs with dense long paths. *Computers and Math. with Appl.* **1** (1975) 365–369.

[GT] H. D. Gröger, G. Turán: On linear decision trees computing boolean functions. *18th ICALP*, Springer Lecture Notes in Comp. Sci. **510** (1991) 707–719.

[G1] V. Grolmusz: The BNS Lower Bound for Multi-Party Protocols is Nearly Optimal. *Information and Computation* **112** (1994) 51–54.

[G2] V. Grolmusz. *Harmonic Analysis, Real Approximation, and the Communication Complexity of Boolean Functions*. Proc. *COCOON* 1996.

[G3] V. Grolmusz. *Separating the Communication Complexities of MOD m and MOD p Circuits*. *33rd IEEE FOCS* (1992) 278–287.

[G4] V. Grolmusz: A Weight-Size Trade-Off for Circuits with MOD m Gates. *26th ACM STOC* (1994) 68–74.

[HM+] A. Hajnal, W. Maass, P. Pudlák, M. Szegedy, G. Turán: Threshold circuits of bounded depth. *28th IEEE FOCS* (1987) 99–110.

[HG] J. Håstad, M. Goldmann: On the Power of Small-Depth Threshold Circuits. *Computational Complexity* 1 (1991) 113–129.

[KRW] M. Karchmer, R. Raz, A. Wigderson: On proving super-logarithmic depth lower bounds via the direct sum in communication complexity. *6th IEEE Symp. on Structure in Complexity Theory* (1991) 299–304.

[KaW] M. Karchmer and A. Wigderson. Monotone circuits for connectivity require super-logarithmic depth. *SIAM J. on Disc. Math.* 3 (1990) 255-265. (Prelim. version: *20th ACM STOC* (1988) 539–550.)

[KN] E. Kushilevitz, N. Nisan: *Communication Complexity*. Cambridge University Press, 1997.

[KNR] I. Kremer, N. Nisan, D. Ron: On Randomized One-Round Communication Complexity. *27th ACM STOC* (1995) 596–605.

[KrW] M. Krause, S. Waack: Variation ranks of communication matrices and lower bounds for depth-two circuits having symmetric gates with unbounded fan-in. *Math. Sys. Theory* 28 (1995) 553–564. (Prelim. version: *32nd IEEE FOCS* (1991) 777–782.)

[L] L. Lovász: Communication Complexity: A Survey. In: *Paths, flows, and VLSI-layout*. (B. Korte, ed.) Springer-Verlag, 1990, pp. 235–265.

[LV] S. V. Lokam: *Algebraic Methods in Computational Complexity: Arithmetic Circuits, Communication Complexity, and Interactive Proof Systems*. Ph.D. Thesis, University of Chicago, August 1996

[LVW] M. Luby, B. Veličković, A. Wigderson: Deterministic approximate counting of depth-2 circuits. *Proc. 2nd Israel Symp. on Theory and Computing Systems*, 1993, pp. 18–24.

[MS] K. Mehlhorn, E.M. Schmidt: Las Vegas is better than determinism in VLSI and distributed computing. *14th ACM STOC* (1982) 330–337.

[Na] M. Naor. Private communication, 1994, cited in [KNR].

[Ne] I. Newman. Private communication, 1994, cited in [KNR].

[Ne2] I. Newman: Private vs. common random bits in communication complexity. *Info. Proc. Letters* 39 (1991) 67–71.

[NS] I. Newman, M. Szegedy: Public vs. Private Coin Flips in One Round Communication Games. *28th ACM STOC* (1996) 561–570.

[Ni] N. Nisan: The communication complexity of threshold gates. In: *Combinatorics, Paul Erdős is Eighty* (D. Miklós, T. Szőnyi, V. T. Sós, eds.), Vol. 1. Bolyai Society Mathematical Studies 1, Budapest 1993 (distributed by the A. M. S.), pp. 301–315.

[NW] N. Nisan, A. Wigderson: Rounds in Communication Complexity Revisited. *SIAM J. Comp.* 22 (1993) 211–219. (Prelim. version: 23rd ACM STOC (1991) 419–429.)

[PS] R. Paturi and J. Simon: Probabilistic communication complexity. *J. Comp. Sys. Sci.* 33 (1986) 106–123. (Prelim. version: 25th IEEE FOCS (1984) 118–126.)

[P] P. Pudlák: Large communication in constant depth circuits. *Combinatorica* 14 (1994) 203–216.

[PRS] P. Pudlák, V. Rödl, J. Sgall: Boolean circuits, tensor ranks and communication complexity. *SIAM J. Comp.*, to appear.

[RM] R. Raz, P. McKenzie: Separation of the Monotone NC Hierarchy. Manuscript, 1997.

[RW] R. Raz, A. Wigderson: Monotone circuits for matching require linear depth. *J. ACM* 39 (1992) 736–744. (Prelim. version: *22nd ACM STOC* (1990) 287–292.)

[ROS] V. P. Roychowdhury, A. Orlitsky, K. Y. Siu: Lower bounds on threshold and related circuits via communication complexity. *IEEE Trans. Info. Theory* 40 (1994), 467–474.

[Th] C. D. Thompson: Area-time complexity for VLSI. *11th ACM STOC* (1979), 81–88.

[V] L. G. Valiant: Graph theoretic arguments in low-level complexity. *6th Symp. MFCS* (1977), 162–176.

[Vat] F. Vatan: Some lower and upper bounds for algebraic decision trees and the separation problem. *7th IEEE Symp. Structure in Complexity Theory* (1992) 295–304.

[Vaz] U. V. Vazirani: Strong communication complexity or generating quasirandom sequences from two communicating semirandom sources. *Combinatorica* 7 (1987) 375–392.

[Y79] A. C.-C. Yao. *Some Complexity Questions Related to Distributed Computing.* Proc. of the 11th ACM STOC, 1979, pp. 209-213.

[Y83] A. C.-C. Yao. *Lower Bounds by Probabilistic Arguments.* Proc. of the 24th IEEE FOCS, 1983, pp. 420-428.

[Y90] A. C-C. Yao. *On ACC and Threshold Circuits. 31st IEEE FOCS* (1990) 619–627.

Treewidth: Algorithmic Techniques and Results

Hans L. Bodlaender

Department of Computer Science, Utrecht University
P.O. Box 80.089, 3508 TB Utrecht, the Netherlands

Abstract. This paper gives an overview of several results and techniques for graphs algorithms that compute the treewidth of a graph or that solve otherwise intractable problems when restricted graphs with bounded treewidth more efficiently. Also, several results on graph minors are reviewed.

1 Introduction

The notion of treewidth is playing a central role in many recent investigations in algorithmic graph theory. There are several reasons for the interest in this, at first sight perhaps somewhat unnatural notion. One of these reasons is the central role that the notion plays in the theory on graph minors by Robertson and Seymour (see Section 5); another reason is that many problems that are otherwise intractable become polynomial time solvable when restricted to graphs of bounded treewidth (see Section 4).

There are several 'real world' applications of the notion of treewidth, amongst others in expert systems [93], telecommunication network design ([46]), VLSI-design, Choleski factorization, natural language processing [91] (see e.g. [21] for a brief overview.) An interesting recent application has been found by Thorup [131]. He shows that for many well known programming languages (like C, Pascal, Modula-2), the control-flow graph of goto-free programs has treewidth bounded by a small constant (e.g., 3 for Pascal, 6 for C). Thus, certain optimization problems arising in compiling can be solved using techniques relying on small treewidth.

2 Definitions

The notion of treewidth was introduced by Robertson and Seymour in their work on graph minors [104].

Definition 1. A *tree decomposition* of a graph $G = (V, E)$ is a pair $(\mathcal{X}, T)$ with $T = (I, F)$ a tree, and $\mathcal{X} = \{X_i \mid i \in I\}$ a family of subsets of V, one for each node of T, such that

- $\bigcup_{i \in I} X_i = V$,
- for all edges $\{v, w\} \in E$ there exists an $i \in I$ with $v \in X_i$ and $w \in X_i$, and
- for all $i, j, k \in I$: if j is on the path from i to k in T, then $X_i \cap X_k \subseteq X_j$.

The *width* of a tree decomposition $((I, F), \{X_i \mid i \in I\})$ is $\max_{i \in I} |X_i| - 1$. The *treewidth* of a graph G is the minimum width over all tree decompositions of G.

There are several equivalent notions to treewidth (for an overview, also of classes of graphs that have a uniform upper bound on the treewidth, see [25]); amongst others, graphs of treewidth at most k are also known as *partial k-trees*. A notion related to treewidth is *pathwidth*, defined first in [102]. A tree decomposition $(\mathcal{X}, T)$ is a path decomposition if T is a path; the pathwidth of a graph G is the minimum width over all path decompositions of G. A survey giving relations to notions of graph searching has been written by Bienstock [14].

Another notion that is related to treewidth and that might be more suitable in some cases for implementation purposes is *branchwidth* [118].

A tree decomposition can easily be converted (in linear time) in a *nice* tree decomposition of the same width (and with a linear size of T): here the tree T is rooted and binary, and nodes are of four types:

- Leaf nodes i are leaves of T and have $|X_i| = 1$.
- Introduce nodes i have one child j with $X_i = X_j \cup \{v\}$ for some vertex $v \in V$.
- Forget nodes i have one child j with $X_i = X_j - \{v\}$ for some vertex $v \in V$.
- Join nodes i have two children j with $X_i = X_{j_1} = X_{j_2}$.

Using nice tree decompositions instead of normal ones does in general not give additional algorithmic possibilities, but it considerably eases the design of algorithms, and one can also expect in several cases to have better constant factors in the running times of algorithms that use nice instead of normal tree decompositions.

3 Determining treewidth

In this section we review a number of results on the problem, to determine the treewidth of a given graph.

The problem to determine, when given a graph G and an integer k, whether the treewidth of G is at most k is NP-complete [5], even for graphs of maximum degree at most 9 [36], bipartite graphs, or cocomparability graphs. For several special graph classes, there exist polynomial time algorithms to determine the treewidth of graphs in the class, e.g., for chordal graphs, permutation graphs [33], circular arc graphs [127], circle graphs [87], and distance hereditary graphs [40]. See also [34, 60, 72, 78, 79, 89]. One of the most interesting open problems here is the complexity of treewidth when restricted to planar graphs. As branchwidth can be solved in polynomial time on planar graphs [126], and branchwidth differs at most a factor 1.5 from treewidth, we have a polynomial time approximation algorithm for treewidth on planar graphs with performance ratio 1.5. For arbitrary graphs, there is a polynomial time approximation algorithm for treewidth with performance ration $O(\log n)$ [30]; it is an interesting (but probably hard) open problem whether treewidth can be approximated with constant performance ratio.

A well studied case is when the parameter k is a fixed constant. We distinguish here results for two versions of the problem: the decision problem, where we only must decide whether the treewidth of the input graph is at most k, and the construction problem, where also a tree decomposition of width at most k must be given, when existing.

The first polynomial time algorithm for the (construction and decision) problem used $O(n^{k+2})$ time and was found by Arnborg, Corneil, and Proskurowski [5]. Using deep results on graph minors, Robertson and Seymour then gave a non-constructive proof of the existence of a decision algorithm that uses $O(n^2)$ time [118]. This algorithm has the following structure. First, in $O(n^2)$ time, we can find a tree decomposition of the input graph G with width at most $4k + 3$, or decide that the treewidth of G is at most k. (To be precise, Robertson and Seymour use branchwidth to give a similar result; the technical difference is not important.) Then, this tree decomposition is used to check in linear time whether G contains an element of the obstruction set of graphs of treewidth at most k (see section 5.)

The first step of this algorithm was improved by Matousek and Thomas [96], who gave a faster randomized algorithm, by Lagergren [92], who gave a parallel algorithm using $O(log^3 n)$ time and $O(n)$ processors on a CRCW PRAM, or a sequential $O(n log^2 n)$ time algorithm, and Reed, who gave an algorithm running in $O(n \log n)$ time. Each of these algorithms either determines that the treewidth of input graph is more than G or finds a tree decomposition of width at most $f(k)$ for some linear function f. (See [42] for a simple linear time algorithm for the pathwidth variant of this problem.) In [32, 88], Bodlaender and Kloks address the second step of the algorithm of Robertson and Seymour: they give an algorithm for the second step that solves the construction problem in linear time (i.e., provided a tree decomposition of bounded but perhaps not optimal width has been found). Using the algorithm from [32], in [24] for each fixed k, a *linear time* algorithm is given for both the decision and construction problem. Each of the algorithms mentioned in this paragraph has a hidden constant factor that is at least exponential in k — in all cases, the factors seem too large for practical purposes, but little experimental work has been done so far.

An interesting different approach was taken by Arnborg et al. [6]. They use graph reduction to obtain algorithms that use linear time, but more than linear memory. More on graph reduction can be found in Section 4.

Work has also been done on parallel algorithms for the fixed parameter case of the treewidth problem. Older algorithms by Bodlaender [17] and Chandrasekharan and Hedetniemi [43] need large numbers of processors. The first algorithm with work (product of time and number of processors) only a polylogarithmic factor more than linear was the algorithm by Lagergren [92], discussed above. This result was improved by Bodlaender and Hagerup [31], who, combining parallizations of the sequential algorithms of [24] and [6] with new techniques, obtained the following results. On an EREW PRAM, the construction problem can be solved in $O(log^2 n)$ time; the decision problem can be solved in $O(\log n \log^* n)$ time on a EREW PRAM and $O(\log n)$ time on a CRCW PRAM. Each of the algorithm has optimal speedup, i.e., the product of time and number of processors is linear.

In the case that k is 2, 3, or 4, better algorithms have been found. Practically efficient linear time algorithm exist, based on graph reduction [8, 96, 122]. The parallel algorithms for $k = 2, 3$ by [75] were improved by the results in [31], mentioned above. For $k = 2$, a parallel algorithm for the construction problem that uses $O(\log n \log^* n)$ time on a EREW PRAM and $O(\log n)$ time on a CRCW PRAM and has optimal speedup has been found by de Fluiter and Bodlaender [54, 55].

See also [37] for a closely related problem.

4 Finding algorithms for problems on graphs of small treewidth

For large numbers of graph problems, it has been shown that they are solvable in linear time, polynomial time, or become a member of NC, when the inputs are restricted to graphs of treewidth at most k for some constant k. Underlying many of these results, there are a few common techniques. In this section, these techniques are reviewed.

A technique that is very often applicable for solving problems on graphs of bounded treewidth is the one, discussed below. It can be characterized as: 'computing tables of characterizations of partial solutions' for each node $i \in I$ in a tree decomposition of bounded width – in bottom-up order. The technique first appeared (in 1992) in the context of graphs of bounded treewidth in a paper by Arnborg and Proskurowski [9]; another paper founding this technique was Bern et al. [13].

The algorithm has the following structure (k is assumed the assumed constant upper bound on the treewidth of input graph $G = (V, E)$):

- Find a tree decomposition of G of width at most k. (This can be done in linear time, as discussed above.)
- Transform it into a nice tree decomposition, say $(\{X_i \mid i \in I\}, T = (I, F))$ of width at most k, $|I| = O(|V|)$, r the root of T.
- Compute for each node $i \in I$ a certain table. To compute a table for a node i, one only uses tables already computed for the children of i, the type of node i (leaf, forget, introduce, or join), and the information about G, restricted to X_i. Thus, these tables are computed in bottom-up order.
- The answer to the problem can be found by inspecting the table of the root r.
- Construction versions of problems usually need another phase, where tables are used again to construct a solution (when one exists). We will not go into detail for this step.

To describe what type of tables are needed, we first introduce some additional notions.

A terminal graph is a triple $H = (V, E, X)$, where (V, E) is a graph with vertex set V and edge set E, and X is an ordered subset of the vertices in V, called the *terminals* of G. A terminal graph with l terminals is also called an l-terminal graph. The operation $\oplus$ is defined on pairs of terminal graphs with the same number l of terminals: $H \oplus H'$ is obtained by taking the disjoint union of H and H' and then identifying the ith terminal of G with the ith terminal of H' for all i, $1 \le i \le l$. A terminal graph H is a terminal subgraph of a graph G, iff there exists a terminal graph H' with $G = H \oplus H'$.

To every node i in a nice tree decomposition $(\{X_i \mid i \in I\}, T = (I, F))$ we associate the terminal graph $G_i = (V_i, E_i, X_i)$, where V_i is the set $\{v \mid v \in X_j$ and $j = i$ or j is a descendant of i in $T\}$; $E_i = E[V_i] = \{\{v, w\} \in E \mid v, w \in V_i\}$, or, in other words: the subgraph induced by vertices in the sets of the node j and all nodes below j in T, with X_i as the set of terminals. (The ordering of X_i is not important.)

At this point, we are ready to describe the 'build tables of partial solutions technique' more precisely. Suppose we want to solve a certain problem X, where for the moment we assume that X is a graph property. The algorithm design follows the following steps.

1. Define a notion of *solution*. For instance, if X is the Hamiltonian circuit problem, then a solution for a graph G is an actual Hamiltonian circuit in G.

2. Define a notion of *partial solution*. A partial solution is an object that can be associated with a terminal graph. When this terminal graph H is a terminal subgraph of a graph G, then the partial solution should describe possible behavior of a solution on G, when we 'only look to what happens on H'. For instance, a partial solution for Graph Coloring is a coloring of the vertices of the terminal graph, a partial solution for Hamiltonian circuit is a set of paths between the terminals in the terminal graph, disjoint except possibly for their endpoints and covering all vertices in the terminal graph.

3. Define a notion of *extension of partial solution*. We must specify what it means that a solution is an extension of a partial solution. This is usually very natural, for instance, for graph coloring, a solution for $G = (V, E)$, i.e., a coloring f of G, is an extension of a partial solution (coloring g) for terminal subgraph $H = (V', E', X)$ iff g is the restriction of f to W.

4. Define a notion of *characteristic* of a partial solution. It is meant to describe 'what is needed to know about the partial solution to see whether it can be extended to a solution', i.e., if two partial solutions have the same characteristic, then one can be extended to a solution if and only if the other can be extended to a solution. See below for examples.

5. A *full set of characteristics* for a terminal graph G is the set of all characteristics of partial solutions on G. The full set of characteristics of a graph G_i for a node i in a nice tree decomposition is also called the full set for i. Show for each of the four types of nodes (leaf, introduce, forget, join), that for node i, a full set for i can be computed efficiently (in constant or polynomial time), given the full sets for all children of i, assuming that there are at most $k + 1$ terminals in each of the involved terminal graphs.

6. Show that the problem can be decided efficiently (in constant or polynomial time), given a full set for the root node r of the nice tree decomposition.

More formally: we have relations sol_X, $psol_X$, ex_X, and a function ch_X, capturing respectively the notions 'solution', 'partial solution', 'extension', and 'characteristic'. sol_X is a relation with two arguments (G, s), G a graph and s a 'solution string', such that for all graphs G: $X(G) \Leftrightarrow \exists s : sol_X(G, s)$. (This is comparable to a definition of NP.) $psol_X$ is a relation with two arguments (H, s), H a terminal graph, and s a 'partial solution string'. ex_X is a relation with four arguments (G, s, H, s'), G a graph, s a solution string, H a terminal graph, s' a terminal solution string. The following must hold, for all G, s, H, s'

$$ex_X(G, s, H, s') \rightarrow \exists H' : G = H \oplus H' \wedge sol_X(G, s) \wedge sol_X(H, s')$$

Also, the following must hold, expressing that every solution has a partial solution on any terminal subgraph:

$$\forall G, s, H, H' : (sol_X(G, s) \wedge G = H \oplus H') \rightarrow \exists s' : psol_X(H, s') \wedge ex_X(G, s, H, s')$$

ch_X is a function on pairs (H, s), H a terminal subgraph and s a partial solution string, defined when $psol_X(H, s)$ is true. It must fulfill, for all terminal graphs H, H', H'', partial solution strings s, s':

$$ch_X(H, s) = ch_X(H', s') \Rightarrow (\exists s'' : ex_X(H \oplus H'', s'', H, s)) \Leftrightarrow (\exists s''' : ex_X(H' \oplus H'', s''', H', s')) \quad (1)$$

The full set of characteristics of terminal graph H is $full_X(H) = \{ch_X(H, s) \mid psol_X(H, s)\}$.

As example, we look to the problem to decide whether a graph $G = (V, E)$ has chromatic number at most three, i.e., is there an $f : V \rightarrow \{1, 2, 3\}$, such that $\forall \{v, w\} \in E : f(v) \neq f(w)$?

The notions 'solutions', 'partial solution', and 'extension' are obvious, and discussed above. The characteristic of a partial solution $f : W \rightarrow \{1, 2, 3\}$ of terminal graph $H = (W, F, X)$ is the restriction of f to X, $f|_X$. The following lemma shows the correctness of this notion of characteristic.

Lemma 2. *Let $H = (V, E, (v_1, \ldots, v_l))$, $H' = (V', E', (v'_1, \ldots, v'_l))$ be l-terminal graphs. Let c, c' be 3-colorings of the vertices of H and H', such that for $1 \leq i \leq l$, $c(v_i) = c'(v'_i)$. Then for any l-terminal graph H'', there exists a 3-coloring of $H \oplus H''$ that extends c, if and only if there exists a 3-coloring of $H' \oplus H''$ that extends c''.*

The proof of the lemma is easy: color the new vertices in H'' in both cases is the same way. An important element in the proof is the following observation: there are no edges between a vertex that belongs to H but not to X, and a vertex that belongs to H' but not to X in the graph $H \oplus H'$.

Next, notice that for any node i in the nice tree decomposition $(\{X_i \mid i \in I\}, T = (I, F))$ of $G = (V, E)$, G is of the form $G = G_i \oplus H$ for some terminal graph H. This follows from the definition of tree decomposition: the only vertices in V_i, adjacent to vertices in $V - V_i$ are those that belong to X_i.

What this all amounts to is that the full set of characteristics of G_i is actually all one needs to know about the terminal subgraph G_i when solving the 3-coloring problem. Additionally, as for each i, $|X_i| \leq k + 1$, if G is a graph of treewidth at most k, each full set contains at most 3^{k+1} elements, which is constant when k is a constant. It is not hard to show that full sets can be computed in constant time, given the full sets of the children of a node. This has to be shown for each of the four cases: leaf node, introduce node, forget node, join node. We only look at the case of introduce node here.

Lemma 3. *Let $\chi 3$ be the 3-coloring problem, $(\{X_i \mid i \in I\}, T = (I, F))$ a nice tree decomposition of $G = (V, E)$, i an introduce node with child j, and $X_i = X_j \cup \{v\}$. A function $f : X_i \to \{1, 2, 3\}$ belongs to $full_{\chi 3}(i)$, if and only if $f|_{X_j} \in full_{\chi 3}(j)$ and for all $w \in X_j$, if $\{v, w\} \in E$, then $f(v) \neq f(w)$.*

The proof relies on the fact that v can only be adjacent in G_i to vertices in X_j — this follows from the definition of tree decomposition. The lemma shows that we can compute full sets for introduce nodes, given a full set of the child, and only very local information. Similar lemmas exist for the other types of nodes.

Computing full sets in bottom up order, we finally have a full set for the root node. Now, as G is 3-colorable, if and only if the full set for the root node is non-empty, we can directly decide the problem.

When we used a tree decomposition of width, bounded by a constant, each computation of a full set took time, only exponential in that constant, hence the entire algorithm uses linear time.

Similar approaches work also for many other problems. In other cases, notions of partial solution, extension, and especially characteristic are less obvious or hard to find. For instance, look at the problem to decide, whether on a graph $G = (V, E)$, given with a number of pairs (v_i, w_i), $1 \leq i \leq r$ of vertices, there are paths from each v_i to w_i that are mutually disjoint. Clearly, a solution here is the desired set of paths. A partial solution for a terminal subgraph H is a collection of disjoint paths, some of which end in a terminal, such that certain properties hold: e.g., if v_i and w_i both belong to H, there either must be a path between v_i and w_i in the subgraph, or both v_i and w_i must have a path ending in a terminal (there then must be another path joining these terminals in the solution). Note there can also be paths between terminals. The characteristic of such a partial solution then describes which terminals are joined with which vertices from pairs; one can actually show there are at most a constant number of possibilities when $|X_i|$ is bounded by a constant, and that for each type of node, a full set can be computed in constant time. See [125].

When designing these types of algorithms, the most important step is the right choice of characteristic. First, it should fulfill property (1). Secondly, one should aim for characteristics, such that full sets of l-terminal graph (or l-terminal graphs of bounded treewidth) have bounded size, i.e., size only depending on l. Experience shows, that once the right choices for solution, partial solution, extension, and characteristic are made, the design of the algorithm (i.e., procedures how to compute full sets for the four types of nodes, and deciding the property given the full set of the root) usually succeeds, although it often is a lot of detailed work.

A similar technique works for optimization problems. We omit the more formal framework here, and give only a sketch. For characteristics, we take pairs (s, r), with r a member of a totally ordered set, usually an integer or real number — the value of the partial solution. If we have partial solutions with characteristics (s, r_1) and (s, r_2), then we put only that characteristic of these in the full set with the best value r_1 or r_2.

For instance, suppose we want to solve the problem to find a mapping f of the vertices of a graph $G = (V, E)$ to colors 1, 2, 3, such that the number of edges $\{v, w\} \in E$ with $f(v) = f(w)$ is as small as possible. A partial solution for a terminal graph $H = (W, F, X)$ is a coloring f of the vertices in W; the characteristic of this partial solution is the pair $(f|_X, r)$, where $f|_X$ is the restriction of f to X, and r is the number of edges $\{v, w\} \in F$ with $f(v) = f(w)$. For each possible function $g : X \to \{1, 2, 3\}$, we have at most one pair (g, r) in the full set of H, namely the pair with the smallest possible value of r. Again, we can compute the full set for a node when given the full sets for its children in a tree decomposition in constant time, giving a linear time algorithm. (For similar algorithms, see e.g., [82].)

Actually, results exploiting analogues to Myhill-Nerode theory (for finite state automata) can be used to show existence of an algorithm at an earlier stage of the design process, when dealing with certain types of decision problems.

Let P be a graph property. Define the relation $\sim_{P,k}$ on k-terminal graphs as follows:

$$G \sim_{P,k} H \Leftrightarrow (\forall K : P(G \oplus K) \leftrightarrow P(H \oplus K))$$

We say that P is of *finite index*, when for every k, the equivalence relation $\sim_{P,k}$ has a finite number of equivalence classes.

One can show that every finite index problem can be solved in linear time on graphs of bounded treewidth (see e.g., [2]). Now, as soon as we have characteristics which need $O(1)$ bits to describe, we know that the problem is finite state: if k-terminal graphs G and H have the same full set, then $G \sim_{P,k} H$, and there are only a constant number of different possible full sets.

Graph reduction Another interesting algorithmic method is based on graph reduction. Here, we observe that when $H \sim_{P,k} H'$, then when we have a graph of the type $H \oplus K$, we can replace it by $H' \oplus K$ and not change the answer of the problem. When H' is smaller than H, we have reduced the problem to an equivalent one of equal size. If P is of finite index, one can show that there exists a set of such 'safe' graph reduction rules for P, that can be used for a linear time algorithm of the following form: repeatedly apply a reduction rule to G. When no rule can be applied, we have a graph of size at most some constant, or for which P does not hold. This method was introduced to the setting of treewidth in [6]. More on graph reduction can be found in [26, 54].

Monadic second order logic An interesting general framework to quickly establish that a problem can be solved in linear time on graphs of bounded treewidth has been established by Courcelle [49, 48, 47, 51], and extended by Borie et al. [39], Arnborg et al. [7], and Courcelle and Mosbah [53]. Courcelle results states that each problem that can be stated in *monadic second order logic* can be solved in linear time on graphs of bounded treewidth. Monadic second order logic is a language to describe graph properties, using the following constructions: quantifications over vertices, edges, sets of vertices, sets of edges, ($\exists v \in V$, $\forall F \subseteq E$, ...); membership tests ($v \in W$, $e \in F$), adjacency tests (v is endpoint of e), and logic operations ($\lor$, $\neg$, ...). Extensions allow e.g., to optimize over the size of a free set variable. For instance, the problem whether a graph has a partition of its vertices in triangles (i.e., we want to partition V into $V_1, \ldots, V_r$, such that each V_i has three vertices and induces a triangle in G) can be expressed as:

$$\exists F \subseteq E \colon \forall v \in V \colon \exists w \in V \colon \exists x \in V \colon \{v, w\} \in F \land \{v, x\} \in F \land$$
$$\{w, x\} \in F \land \neg (\exists y \in V : y \neq w \land y \neq x \land \{v, y\} \in F).$$

(To be precise, instead of $\{v, w\} \in F$, we should write: $\exists e : e \in F \land v \in e \land w \in e$.) The maximum independent set problem can be formulated as:

$$\max_{W \subseteq V} |W| : \forall v : \forall w : (v \in W \land w \in W) \rightarrow \neg(\{v, w\} \in E)$$

Especially the paper by Borie et al. [39] is very helpfull to see what kind of constructions can be used to express problems in (extensions of) monadic second order logic. Problems in monadic second order logic are finite index.

An interesting question is whether language constructions can be added to monadic second order logic, such that its expressive power becomes sufficient to describe all problems that are finite index. See e.g., [50, 52, 83].

Additional remarks Some problems whose decision versions are not in NP can also be solved in linear time on graphs of bounded treewidth. See e.g. [4, 3, 19].

For more algorithms that exploit the small treewidth of graphs, see also (amongst others) [10, 18, 38, 44, 61, 76, 81, 94, 95, 96, 129, 130, 133, 134].

Dynamic algorithms for graphs of bounded treewidth have been considered amongst others in [22, 45, 70, 80]. Parametric problems can also be solved efficiently on graphs of bounded treewidth in many cases [68, 69].

5 Graph minors

In a long series of papers, [102, 104, 103, 108, 105, 106, 107, 110, 109, 111, 116, 117, 118, 119, 120, 112, 113, 101, 114, 115] (and others), Robertson and Seymour showed many deep results on graph minors. Some of these results will be discussed here.

A graph $G = (V, E)$ is a *minor* of a graph $H = (W, F)$, if G can be obtained from H by a series of vertex deletions, edge deletions, and edge contractions, where an edge contraction is the operation that replaces two adjacent vertices v, w by a new vertex that is adjacent to all vertices that were adjacent to v or w.

Theorem 4. *If $\mathcal{G}$ is a class of graphs that is closed under taking of minors, then there exists a finite set of graphs $ob(\mathcal{G})$, the obstruction set of $\mathcal{G}$, such for every graph H: $H \in \mathcal{G}$, if and only if no element of $ob(\mathcal{G})$ is a minor of H.*

The theorem, formerly known as Wagner's conjecture, is equivalent to stating that every class of graphs has a finite number of minor-minimal elements. An example of an obstruction set is the set $\{K_5, K_{3,3}\}$ for the minor closed class of planar graphs, or the set $\{K_3\}$ for the minor closed class of forests. There are several results giving the obstruction sets of specific minor closed classes of graphs, e.g., the obstruction set of graphs of treewidth two is $\{K_4\}$; see [11, 123] for the obstruction set of graphs of treewidth 3, and [86] for the obstruction sets of graphs of pathwidth 1, respectively 2. The size of the obstruction sets can grow very fast: for instance, the obstruction set of the graphs with pathwidth at most k contains at least $k!^2$ trees, each containing $\frac{5 \cdot 3^k - 1}{2}$ vertices [128]. Ramachandramurthi [99, 100] investigated the graphs with $k+1$, $k+2$ and $k+3$ vertices that belong to the obstruction sets for graphs of treewidth or pathwidth k. See, e.g., also [41, 56].

Some additional results make that Theorem 4 has surprising algorithmic implications. Robertson and Seymour [118] have shown that for every fixed graph H, there exists an algorithm that decides in $O(n^3)$ time whether H is a minor of a given graph G. Combining this algorithm with Theorem 4, we have the following result:

Theorem 5 Robertson, Seymour. *If $\mathcal{G}$ is a class of graphs, closed under taking of minors, then there exists an algorithm that decides membership in $\mathcal{G}$ in $O(n^3)$ time.*

(The algorithm checks for each element in the obstruction set of $\mathcal{G}$ whether it is a minor of the input graph.) Note that the result is non-constructive in two ways: only existence of the algorithm is shown, and the algorithm only decides but does not construct solutions.

If the minor closed class of graphs $\mathcal{G}$ does not contain all planar graphs, then a linear time algorithm is possible.

Theorem 6 Robertson, Seymour [104]. *For any planar graph H, there is a constant c_H, such that every graph with no minor isomorphic to H has treewidth at most c_H.*

There are planar graphs with arbitrary large treewidth, and planarity is preserved under minor taking, thus it is not possible to prove a variant of Theorem 6 for non-planar graphs H. If H is a forest, then there exists a similar upper bound c_H on the *pathwidth* of graphs that do not contain H as a minor (see [16, 42]). In [121], it is shown that one can take $c_H = 20^{2(2|V_H|+4|E_H|)^5}$. A similar type of bound was proved by Gorbunov [74]. In some special cases, one can prove better bounds. For instance, for $H = C_k$, the cycle with k vertices, then one can take $c_H = k - 1$ [64]. If $H = K_{2,k}$, then one can take $c_H = 2k - 2$ [35]. Other special cases are discussed in [20, 23, 35].

As on graphs of bounded treewidth, one can check for any fixed graph H whether it is a minor of input graph G in linear time (e.g., for fixed H, the property can be formulated in monadic second order logic, see e.g., [12]), we have the following result.

Theorem 7. *If $\mathcal{G}$ is a class of graphs, closed under taking of minors and that does not contain all planar graphs, then there exists an algorithm that decides membership in $\mathcal{G}$ in $O(n)$ time.*

(Let planar graph $H \notin \mathcal{G}$. Check whether the treewidth of G is at most c_H (linear time by the algorithm of [24]). If not, answer no. If so, test minorship of all graphs in the obstruction set — this can now be done in linear time.)

In some cases, self-reduction can help to overcome the non-constructive aspects of this theory. A general technique has been established by Fellows and Langston [67].

An application of the theorem recently arose in the area of distributed computing, specifically interval routing. The class of graphs for which k-interval routing schemes exist under varying edge lengths, k-IRS is closed under taking of minors, hence there exists a linear time checkable characterization for each fixed k. (See [35] for precise definitions and more results.) Still, no actual characterization is known.

Several applications for problems from graph layout, VLSI-design, and graph theory have been found by Fellows and Langston. See e.g., [62, 63, 66, 65, 15].

6   Fixed parameter complexity

Some problems are not (known to be) linear time solvable when restricted to graphs of bounded treewidth. The following behaviors can often be observed: the problem is NP-complete; the problem can be solved in $O(f(k)n^{g(k)})$ time (k the treewidth, f, g some functions growing with k); the problem can be solved in time $O(f(k)n^c)$ time (c a constant, f a function). More generally, this type of behavior can be seen in *parameterized problems*: part of the instance is distinguished as the 'parameter' often an integer, which might be small in practice. To distinguish between the second and third type of behavior, Downey and Fellows introduced the theory of *fixed parameter complexity* [58, 59, 57, 1]. Hereto, they introduced the notion of parameterized language (or problem): a subset $L \subseteq \Sigma^* \times \Sigma^*$ for some fixed alphabet Σ. The second part of the input is called the parameter; we are interested in what happens if this parameter is 'small'. Downey and Fellows also define a notion of reduction between parameterized languages, the class of *fixed parameterized tractable problems* FPT (the class of parameterized languages L for which there exists an algorithm that decides whether $< x, k > \in L$ in $f(k)|x|^c$ time, f a function and c a constant), and a notion of reduction between parameterized languages (that preserves fixed parameterized tractability). Then they introduce a hierarchy of complexity classes $FTP \subseteq W[1] \subseteq W[2] \subseteq \cdots \subseteq W[i] \subseteq \cdots W[P]$ of parameterized problems. Classes $W[i]$, $W[P]$ are defined in terms of reductions to certain parameterized problems on Boolean circuits. It is conjectured that the hierarchy is proper. So, hardness for $W[1]$ or any larger class means for a problem that it is unlikely that there exists an algorithm for it with time complexity of the form $O(f(k)n^c)$.

As an example: the treewidth of a graph is never larger than its bandwidth. Bandwidth is solvable in $O(f(K)n^K)$ time, K the bandwidth to obtain [77]. Bandwidth is hard for all $W[i]$, $i \in \mathbf{N}$ [28], and hence it is unlikely that the bounded treewidth of yes-instances will help to get an e.g., a linear time algorithm for bandwidth for fixed k, even with the help of tree decompositions. Other graph problems where yes-instances have bounded treewidth, and which are hard for $W[1]$ or a larger class, can be found in [28, 27, 29, 84, 85].

Postscript

I want to thank all who cooperated with me and informed me on all kinds of treewidth related topics in the past years, and to apologize to those whose work I forgot to mention here.

References

1. K. A. Abrahamson, R. G. Downey, and M. R. Fellows. Fixed-parameter tractability and completeness IV: On completeness for $W[P]$ and PSPACE analogues. *Annals of Pure and Applied Logic*, 73:235–276, 1995.
2. K. R. Abrahamson and M. R. Fellows. Finite automata, bounded treewidth and well-quasiordering. In N. Robertson and P. Seymour, editors, *Proceedings of the AMS Summer Workshop on Graph Minors, Graph Structure Theory, Contemporary Mathematics vol. 147*, pages 539–564. American Mathematical Society, 1993.
3. S. Arnborg. Some PSPACE-complete logic decision problems that become linear time solvable on formula graphs of bounded treewidth. Manuscript, 1991.
4. S. Arnborg. Graph decompositions and tree automata in reasoning with uncertainty. *J. Expt. Theor. Artif. Intell.*, 5:335–357, 1993.
5. S. Arnborg, D. G. Corneil, and A. Proskurowski. Complexity of finding embeddings in a k-tree. *SIAM J. Alg. Disc. Meth.*, 8:277–284, 1987.
6. S. Arnborg, B. Courcelle, A. Proskurowski, and D. Seese. An algebraic theory of graph reduction. *J. ACM*, 40:1134–1164, 1993.
7. S. Arnborg, J. Lagergren, and D. Seese. Easy problems for tree-decomposable graphs. *J. Algorithms*, 12:308–340, 1991.
8. S. Arnborg and A. Proskurowski. Characterization and recognition of partial 3-trees. *SIAM J. Alg. Disc. Meth.*, 7:305–314, 1986.
9. S. Arnborg and A. Proskurowski. Linear time algorithms for NP-hard problems restricted to partial k-trees. *Disc. Appl. Math.*, 23:11–24, 1989.
10. S. Arnborg and A. Proskurowski. Canonical representations of partial 2- and 3-trees. *BIT*, 32:197–214, 1992.

11. S. Arnborg, A. Proskurowski, and D. G. Corneil. Forbidden minors characterization of partial 3-trees. *Disc. Math.*, 80:1–19, 1990.

12. S. Arnborg, A. Proskurowski, and D. Seese. Monadic second order logic, tree automata and forbidden minors. In E. Börger, H. Kleine Büning, M. M. Richter, and W. Schönfeld, editors, *Proceedings 4th Workshop on Computer Science Logic, CSL'90*, pages 1–16. Springer Verlag, LNCS, vol. 533, 1991.

13. M. W. Bern, E. L. Lawler, and A. L. Wong. Linear time computation of optimal subgraphs of decomposable graphs. *J. Algorithms*, 8:216–235, 1987.

14. D. Bienstock. Graph searching, path-width, tree-width and related problems (a survey). *DIMACS Ser. in Discrete Mathematics and Theoretical Computer Science*, 5:33–49, 1991.

15. D. Bienstock and M. A. Langston. Algorithmic implications of the graph minor theorem. To appear as chapter in: Handbook of Operations Research and Management Science: Volume on Networks and Distribution, 1993.

16. D. Bienstock, N. Robertson, P. D. Seymour, and R. Thomas. Quickly excluding a forest. *J. Comb. Theory Series B*, 52:274–283, 1991.

17. H. L. Bodlaender. NC-algorithms for graphs with small treewidth. In J. van Leeuwen, editor, *Proceedings 14th International Workshop on Graph-Theoretic Concepts in Computer Science WG'88*, pages 1–10. Springer Verlag, LNCS, vol. 344, 1988.

18. H. L. Bodlaender. Polynomial algorithms for graph isomorphism and chromatic index on partial k-trees. *J. Algorithms*, 11:631–643, 1990.

19. H. L. Bodlaender. Complexity of path-forming games. *Theor. Comp. Sc.*, 110:215–245, 1993.

20. H. L. Bodlaender. On linear time minor tests with depth first search. *J. Algorithms*, 14:1–23, 1993.

21. H. L. Bodlaender. A tourist guide through treewidth. *Acta Cybernetica*, 11:1–23, 1993.

22. H. L. Bodlaender. Dynamic algorithms for graphs with treewidth 2. In *Proceedings 19th International Workshop on Graph-Theoretic Concepts in Computer Science WG'93*, pages 112–124, Berlin, 1994. Springer Verlag.

23. H. L. Bodlaender. On disjoint cycles. *Int. J. Found. Computer Science*, 5(1):59–68, 1994.

24. H. L. Bodlaender. A linear time algorithm for finding tree-decompositions of small treewidth. *SIAM J. Comput.*, 25:1305–1317, 1996.

25. H. L. Bodlaender. A partial k-arboretum of graphs with bounded treewidth. Tech. Rep. UU-CS-1996-02, Dep. of Comp. Sc., Utrecht Univ., Utrecht, 1996.

26. H. L. Bodlaender and B. de Fluiter. Reduction algorithms for constructing solutions in graphs with small treewidth. In J.-Y. Cai and C. K. Wong, editors, *Proceedings 2nd Annual International Conference on Computing and Combinatorics, COCOON'96*, pages 199–208. Springer Verlag, LNCS, vol. 1090, 1996.

27. H. L. Bodlaender and J. Engelfriet. Domino treewidth. To appear in *J. Algorithms*, 1997.

28. H. L. Bodlaender, M. R. Fellows, and M. Hallett. Beyond *NP*-completeness for problems of bounded width: Hardness for the *W* hierarchy. In *Proceedings of the 26th Annual Symposium on Theory of Computing*, pages 449–458, New York, 1994. ACM Press.

29. H. L. Bodlaender, M. R. Fellows, M. T. Hallett, H. T. Wareham, and T. J. Warnow. The hardness of problems on thin colored graphs. Tech. Rep. UU-CS-1995-36, Dep. of Comp. Sc., Utrecht Univ., Utrecht, 1995.

30. H. L. Bodlaender, J. R. Gilbert, H. Hafsteinsson, and T. Kloks. Approximating treewidth, pathwidth, and minimum elimination tree height. *J. Algorithms*, 18:238–255, 1995.

31. H. L. Bodlaender and T. Hagerup. Parallel algorithms with optimal speedup for bounded treewidth. In Z. Fülöp and F. Gécseg, editors, *Proceedings 22nd International Colloquium on Automata, Languages and Programming*, pages 268–279, Berlin, 1995. Springer-Verlag, LNCS 944.

32. H. L. Bodlaender and T. Kloks. Efficient and constructive algorithms for the pathwidth and treewidth of graphs. *J. Algorithms*, 21:358–402, 1996.

33. H. L. Bodlaender, T. Kloks, and D. Kratsch. Treewidth and pathwidth of permutation graphs. *SIAM J. Disc. Meth.*, 8(4):606–616, 1995.

34. H. L. Bodlaender and R. H. Möhring. The pathwidth and treewidth of cographs. *SIAM J. Disc. Meth.*, 6:181–188, 1993.

35. H. L. Bodlaender, R. B. Tan, D. M. Thilikos, and J. van Leeuwen. On interval routing schemes and treewidth. In M. Nagl, editor, *Proceedings 21th International Workshop on Graph Theoretic Concepts in Computer Science WG'95*, pages 181–186. Springer Verlag, LNCS, vol. 1017, 1995.

36. H. L. Bodlaender and D. M. Thilikos. Treewidth and small separators for graphs with small chordality. Tech. Rep. UU-CS-1995-02, Dep. of Comp. Sc., Utrecht Univ., Utrecht, 1995. To appear in Disc. Appl. Math.

37. H. L. Bodlaender and D. M. Thilikos. Constructive linear time algorithms for branchwidth. To appear in *Proceedings ICALP'97*, 1997.

38. R. B. Borie. Generation of polynomial-time algorithms for some optimization problems on tree-decomposable graphs. *Algorithmica*, 14:123–137, 1995.

39. R. B. Borie, R. G. Parker, and C. A. Tovey. Deterministic decomposition of recursive graph classes. *SIAM J. Disc. Meth.*, 4:481 – 501, 1991.

40. H. Broersma, E. Dahlhaus, and T. Kloks. A linear time algorithm for minimum fill in and tree width for distance hereditary graphs. In U. Faigle and C. Hoede, editors, *Scientific program 5th Twente Workshop on Graphs and Combinatorial Optimization*, pages 48–50, 1997.

41. K. Cattell, M. J. Dinneen, and M. R. Fellows. Obstructions to within a few vertices or edges of acyclic. In *Proceedings 4th Int. Workshop on Algorithms and Data Structures*. Springer Verlag, LNCS, vol. 955, 1995.

42. K. Cattell, M. J. Dinneen, and M. R. Fellows. A simple linear-time algorithm for finding path-decompositions of small width. *Inform. Proc. Letters*, 57:197–204, 1996.

43. N. Chandrasekharan and S. T. Hedetniemi. Fast parallel algorithms for tree decomposing and parsing partial k-trees. In *Proc. 26th Annual Allerton Conference on Communication, Control, and Computing*, Urbana-Champaign, Illinois, 1988.

44. S. Chaudhuri and C. D. Zaroliagis. Optimal parallel shortest paths in small treewidth digraphs. In P. Spirakis, editor, *Proceedings 3rd Annual European Symposium on Algorithms ESA'95*, pages 31–45. lncs979, 1995.

45. R. F. Cohen, S. Sairam, R. Tamassia, and J. S. Vitter. Dynamic algorithms for optimization problems in bounded tree-width graphs. In *Proceedings of the 3rd Conference on Integer Programming and Combinatorial Optimization*, pages 99–112, 1993.

46. W. Cook and P. D. Seymour. An algorithm for the ring-routing problem. Bellcore technical memorandum, Bellcore, 1993.

47. B. Courcelle. The monadic second-order logic of graphs II: Infinite graphs of bounded width. *Mathematical Systems Theory*, 21:187–221, 1989.

48. B. Courcelle. Graph rewriting: an algebraic and logical approach. In J. van Leeuwen, editor, *Handbook of Theoretical Computer Science, volume B*, pages 192–242, Amsterdam, 1990. North Holland Publ. Comp.

49. B. Courcelle. The monadic second-order logic of graphs I: Recognizable sets of finite graphs. *Information and Computation*, 85:12–75, 1990.

50. B. Courcelle. The monadic second-order logic of graphs V: On closing the gap between definability and recognizability. *Theor. Comp. Sc.*, 80:153–202, 1991.

51. B. Courcelle. The monadic second-order logic of graphs III: Treewidth, forbidden minors and complexity issues. *Informatique Théorique*, 26:257–286, 1992.

52. B. Courcelle and J. Lagergren. Equivalent definitions of recognizability for sets of graphs of bounded tree-width. *Mathematical Structures in Computer Science*, 6:141–166, 1996.

53. B. Courcelle and M. Mosbah. Monadic second-order evaluations on tree-decomposable graphs. *Theor. Comp. Sc.*, 109:49–82, 1993.

54. B. de Fluiter. *Algorithms for Graphs of Small Treewidth*. PhD thesis, Utrecht Univ., 1997.

55. B. de Fluiter and H. L. Bodlaender. Parallel algorithms for treewidth two, 1997. To appear in Proceedings WG'97.

56. M. J. Dinneen. *Bounded Combinatorial Width and Forbidden Substructures*. PhD thesis, Univ. of Victoria, 1995.

57. R. G. Downey and M. R. Fellows. Fixed-parameter tractability and completeness III: Some structural aspects of the W hierarchy. In K. Ambos-Spies, S. Homer, and U. Schöning, editors, *Complexity Theory*, pages 191–226. Cambridge Univ. Press, 1993.

58. R. G. Downey and M. R. Fellows. Fixed-parameter tractability and completeness I: Basic results. *SIAM J. Comput.*, 24:873–921, 1995.

59. R. G. Downey and M. R. Fellows. Fixed-parameter tractability and completeness II: On completeness for $W[1]$. *Theor. Comp. Sc.*, 141:109–131, 1995.

60. J. A. Ellis, I. H. Sudborough, and J. Turner. The vertex separation and search number of a graph. *Information and Computation*, 113:50–79, 1994.

61. D. Eppstein. Subgraph isomorphism in planar graphs and related problems. In Proceedings SODA'95, 1995.

62. M. R. Fellows and M. A. Langston. Nonconstructive advances in polynomial-time complexity. *Inform. Proc. Letters*, 26:157–162, 1987.

63. M. R. Fellows and M. A. Langston. Nonconstructive tools for proving polynomial-time decidability. *J. ACM*, 35:727–739, 1988.

64. M. R. Fellows and M. A. Langston. On search, decision and the efficiency of polynomial-time algorithms. In *Proceedings of the 21rd Annual Symposium on Theory of Computing*, pages 501–512, 1989.

65. M. R. Fellows and M. A. Langston. Fast search algorithms for layout permutation problems. *Int. J. on Computer Aided VLSI Design*, 3:325–340, 1991.

66. M. R. Fellows and M. A. Langston. On well-partial-order theory and its application to combinatorial problems of VLSI design. *SIAM J. Disc. Meth.*, 5:117–126, 1992.

67. M. R. Fellows and M. A. Langston. On search, decision and the efficiency of polynomial-time algorithms. *J. Comp. Syst. Sc.*, 49:769–779, 1994.

68. D. Fernández-Baca and A. Medipalli. Parametric module allocation on partial k-trees. *IEEE Trans. on Computers*, 42:738–742, 1993.

69. D. Fernández-Baca and G. Slutzki. Parametic problems on graphs of bounded treewidth. *J. Algorithms*, 16:408–430, 1994.

70. G. N. Frederickson. Maintaining regular properties dynamically in k-terminal graphs. Manuscript. To appear in *Algorithmica.*, 1993.

71. G. N. Frederickson and R. Janardan. Designing networks with compact routing tables. *Algorithmica*, 3:171–190, 1988.

72. R. Garbe. Tree-width and path-width of comparability graphs of interval orders. In E. W. Mayr, G. Schmidt, and G. Tinhofer, editors, *Proceedings 20th International Workshop on Graph Theoretic Concepts in Computer Science WG'94*, pages 26–37. Springer Verlag, LNCS, vol. 903, 1995.

73. C. Gavoille. On the dilation of interval routing. In: Proceedings MFCS'97, 1997.

74. K. Y. Gorbunov. An estimate of the tree-width of a graph which has not a given planar grid as a minor. Manuscript, 1993.

75. D. Granot and D. Skorin-Kapov. NC algorithms for recognizing partial 2-trees and 3-trees. *SIAM J. Disc. Meth.*, 4(3):342–354, 1991.

76. A. Gupta and N. Nishimura. The complexity of subgraph isomorphism: Duality results for graphs of bounded path- and tree-width. Tech. Rep. CS-95-14, Univ. of Waterloo, Comp. Sc. Dep., Waterloo, Ontario, Canada, 1995.

77. E. M. Gurari and I. H. Sudborough. Improved dynamic programming algorithms for bandwidth minimization and the mincut linear arrangement problem. *J. Algorithms*, 5:531–546, 1984.

78. J. Gustedt. On the pathwidth of chordal graphs. *Disc. Appl. Math.*, 45(3):233–248, 1993.

79. M. Habib and R. H. Möhring. Treewidth of cocomparability graphs and a new order-theoretic parameter. *ORDER*, 1:47–60, 1994.

80. T. Hagerup. Dynamic algorithms for graphs of bounded treewidth. To appear in Proceedings ICALP'97, 1997.

81. T. Hagerup, J. Katajainen, N. Nishimura, and P. Ragde. Characterizations of k-terminal flow networks and computing network flows in partial k-trees. In *Proceedings SODA'95*, 1995.

82. K. Jansen and P. Scheffler. Generalized coloring for tree-like graphs. In *Proceedings 18th International Workshop on Graph-Theoretic Concepts in Computer Science WG'92*, pages 50–59, Berlin, 1993. Springer Verlag, LNCS, vol. 657.

83. D. Kaller. Definability equals recognizability of partial 3-trees. In *Proceedings 22nd International Workshop on Graph-Theoretic Concepts in Computer Science WG'96*, pages 239–253. Springer Verlag, LNCS, vol. 1197, 1997.

84. H. Kaplan and R. Shamir. Pathwidth, bandwidth and completion problems to proper interval graphs with small cliques. *SIAM J. Comput.*, 25:540–561, 1996.

85. H. Kaplan, R. Shamir, and R. E. Tarjan. Tractability of parameterized completion problems on chordal and interval graphs: Minimum fill-in and physical mapping. In *Proceedings of the 35th annual symposium on Foundations of Computer Science (FOCS)*, pages 780–791. IEEE Computer Science Press, 1994.

86. N. G. Kinnersley and M. A. Langston. Obstruction set isolation for the gate matrix layout problem. *Disc. Appl. Math.*, 54:169–213, 1994.

87. T. Kloks. Treewidth of circle graphs. In *Proceedings Forth International Symposium on Algorithms and Computation, ISAAC'93*, pages 108–117, Berlin, 1993. Springer Verlag, LNCS, vol. 762.

88. T. Kloks. *Treewidth. Computations and Approximations.* LNCS, Vol. 842. Springer-Verlag, Berlin, 1994.

89. T. Kloks and D. Kratsch. Treewidth of chordal bipartite graphs. *J. Algorithms*, 19:266–281, 1995.

90. E. Korach and N. Solel. Linear time algorithm for minimum weight Steiner tree in graphs with bounded treewidth. Manuscript, 1990.

91. A. Kornai and Z. Tuza. Narrowness, pathwidth, and their application in natural language processing. *Disc. Appl. Math.*, 36:87–92, 1992.

92. J. Lagergren. Efficient parallel algorithms for graphs of bounded tree-width. *J. Algorithms*, 20:20–44, 1996.

93. S. J. Lauritzen and D. J. Spiegelhalter. Local computations with probabilities on graphical structures and their application to expert systems. *J. of the Royal Statistical Society. Series B (Methodological)*, 50:157–224, 1988.

94. S. Mahajan and J. G. Peters. Regularity and locality in k-terminal graphs. *Disc. Appl. Math.*, 54:229–250, 1994.

95. E. Mata-Montero. Resilience of partial k-tree networks with edge and node failures. *Networks*, 21:321–344, 1991.

96. J. Matoušek and R. Thomas. On the complexity of finding iso- and other morphisms for partial k-trees. *Disc. Math.*, 108:343–364, 1992.

97. B. Monien. The bandwidth minimization problem for caterpillars with hair length 3 is NP-complete. *SIAM J. Alg. Disc. Meth.*, 7:505–512, 1986.

98. A. Parra. *Structural and Algorithmic Aspects of Chordal Graph Embeddings*. PhD thesis, Tech. Univ. Berlin, 1996.

99. S. Ramachandramurthi. *Algorithms for VLSI Layout Based on Graph Width Metrics*. PhD thesis, Comp. Sc. Dep., Univ. of Tennessee, Knoxville, Tennessee, USA, 1994.

100. S. Ramachandramurthi. The structure and number of obstructions to treewidth. *SIAM J. Disc. Meth.*, 10:146–157, 1997.

101. N. Robertson and P. D. Seymour. Graph minors. XX. Wagner's conjecture. In prepartion.

102. N. Robertson and P. D. Seymour. Graph minors. I. Excluding a forest. *J. Comb. Theory Series B*, 35:39–61, 1983.

103. N. Robertson and P. D. Seymour. Graph minors. III. Planar tree-width. *J. Comb. Theory Series B*, 36:49–64, 1984.

104. N. Robertson and P. D. Seymour. Graph minors. II. Algorithmic aspects of tree-width. *J. Algorithms*, 7:309–322, 1986.

105. N. Robertson and P. D. Seymour. Graph minors. V. Excluding a planar graph. *J. Comb. Theory Series B*, 41:92–114, 1986.

106. N. Robertson and P. D. Seymour. Graph minors. VI. Disjoint paths across a disc. *J. Comb. Theory Series B*, 41:115–138, 1986.

107. N. Robertson and P. D. Seymour. Graph minors. VII. Disjoint paths on a surface. *J. Comb. Theory Series B*, 45:212–254, 1988.

108. N. Robertson and P. D. Seymour. Graph minors. IV. Tree-width and well-quasi-ordering. *J. Comb. Theory Series B*, 48:227–254, 1990.

109. N. Robertson and P. D. Seymour. Graph minors. IX. Disjoint crossed paths. *J. Comb. Theory Series B*, 49:40–77, 1990.

110. N. Robertson and P. D. Seymour. Graph minors. VIII. A Kuratowski theorem for general surfaces. *J. Comb. Theory Series B*, 48:255–288, 1990.

111. N. Robertson and P. D. Seymour. Graph minors. X. Obstructions to tree-decomposition. *J. Comb. Theory Series B*, 52:153–190, 1991.

112. N. Robertson and P. D. Seymour. Graph minors. XVI. Excluding a non-planar graph. Manuscript, 1991.

113. N. Robertson and P. D. Seymour. Graph minors. XVII. Taming a vortex. Manuscript, 1991.

114. N. Robertson and P. D. Seymour. Graph minors. XXI. Graphs woth unique linkages. Manuscript, 1992.

115. N. Robertson and P. D. Seymour. Graph minors. XXII. Irrelevant vertices in linkage problems. Manuscript, 1992.

116. N. Robertson and P. D. Seymour. Graph minors. XI. Distance on a surface. *J. Comb. Theory Series B*, 60:72–106, 1994.

117. N. Robertson and P. D. Seymour. Graph minors. XII. Excluding a non-planar graph. *J. Comb. Theory Series B*, 64:240–272, 1995.

118. N. Robertson and P. D. Seymour. Graph minors. XIII. The disjoint paths problem. *J. Comb. Theory Series B*, 63:65–110, 1995.

119. N. Robertson and P. D. Seymour. Graph minors. XIV. Extending an embedding. *J. Comb. Theory Series B*, 65:23–50, 1995.

120. N. Robertson and P. D. Seymour. Graph minors. XV. Giant steps. *J. Comb. Theory Series B*, 68:112–148, 1996.

121. N. Robertson, P. D. Seymour, and R. Thomas. Quickly excluding a planar graph. *J. Comb. Theory Series B*, 62:323–348, 1994.

122. D. P. Sanders. On linear recognition of tree-width at most four. *SIAM J. Disc. Meth.*, 9(1):101–117, 1996.

123. A. Satyanarayana and L. Tung. A characterization of partial 3-trees. *Networks*, 20:299–322, 1990.

124. J. B. Saxe. Dynamic programming algorithms for recognizing small-bandwidth graphs in polynomial time. *SIAM J. Alg. Disc. Meth.*, 1:363–369, 1980.

125. P. Scheffler. A practical linear time algorithm for disjoint paths in graphs with bounded tree-width. Report 396/1994, TU Berlin, Fachbereich Mathematik, Berlin, 1994.

126. P. D. Seymour and R. Thomas. Call routing and the ratcatcher. *Combinatorica*, 14(2):217–241, 1994.

127. R. Sundaram, K. Sher Singh, and C. Pandu Rangan. Treewidth of circular-arc graphs. *SIAM J. Disc. Meth.*, 7:647–655, 1994.

128. A. Takahashi, S. Ueno, and Y. Kajitani. Minimal acyclic forbidden minors for the family of graphs with bounded path-width. *Disc. Math.*, 127(1/3):293 – 304, 1994.

129. J. Telle and A. Proskurowski. Efficient sets in partial k-trees. *Disc. Appl. Math.*, 44:109–117, 1993.

130. J. Telle and A. Proskurowski. Practical algorithms on partial k-trees with an application to domination-like problems. In *Proceedings of Workshop on Algorithms and Data Structures WADS'93*, pages 610–621. Springer Verlag, LNCS, vol. 700, 1993.

131. M. Thorup. Structured programs have small tree-width and good register allocation. Tech. Rep. DIKU-TR-95/18, Dep. of Comp. Sc., Univ. of Copenhagen, Denmark, 1995. To appear in: Proceedings WG'97.

132. E. Wanke. Bounded tree-width and LOGCFL. *J. Algorithms*, 16:470–491, 1994.

133. T. V. Wimer. *Linear Algorithms on k-Terminal Graphs*. PhD thesis, Dept. of Comp. Sc., Clemson Univ., 1987.

134. X. Zhou, S. Nakano, and T. Nishizeki. Edge-coloring partial k-trees. *J. Algorithms*, 21:598–617, 1996.

When are Two Rewrite Systems
More than None?[*]

Nachum Dershowitz

Department of Computer Science, University of Illinois
Urbana, IL 61801, USA
nachum@uiuc.edu

Abstract. It is important for programs to have modular correctness properties. We look at non-deterministic programs expressed as term-rewriting systems (which compute normal forms of input terms) and consider the case where individual systems share constructors, but not defined symbols. We present some old and new sufficient conditions under which termination (existence of normal forms, regardless of computation strategy) and confluence (uniqueness) are preserved by such combinations.

1   Introduction

Rewriting is an important model of computation, with its clean syntax and simple semantics. Rewriting is also an important tool for equational reasoning in automated theorem proving and symbolic computation systems. Recent surveys of rewriting include [Avenhaus and Madlener, 1990; Dershowitz and Jouannaud, 1990; Klop, 1992; Plaisted, 1993].

A *rewrite system* is a set of oriented equations, called *(rewrite) rules*. We use an arrow instead of an equal sign, as in $append(nil, x) \rightarrow x$, to distinguish the left side, $append(nil, x)$, from the right side, x. A rule $l \rightarrow r$ is applied to a term t by finding a subterm s of t that matches the left side l (meaning that there exists a substitution σ of terms for variables in l such that $s = l\sigma$) and replacing s with the corresponding instance ($r\sigma$) of the rule's right side. We write $t \rightarrow t'$ to indicate that the result of the replacement is t'. One computes with rewrite systems by repeatedly, and nondeterministically, applying rules to rewrite an input term until a *normal form* (unrewritable term) is obtained. When the normal form is unique, it can be taken as the value of the initial term.

Two of the most central properties of relevance for rewrite systems are confluence (the Church-Rosser property; see Section 2)—which implies that there can be at most one normal form for any term, and termination (strong normalization in lambda calculus parlance; see Section 3)—which implies the existence of at least one normal form. A confluent and terminating system is called *convergent*

---

[*] This research was supported in part by the National Science Foundation under grants INT-95-07248 and CCR-97-00070 and was performed while on leave at the Hebrew University, Jerusalem, Israel and at École Normale Supérieure de Cachan, France.

(or *complete* or *canonical*) and defines exactly one normal form for each input term (see Section 4).

If rewriting is to be recommended as a practical programming paradigm, then it is important that one at least be able to combine two independent rewrite systems into one, and still maintain the desired properties for the combined system. Unfortunately, this is not always the case, but—as we will see—in certain more or less reasonable situations one can obtain such modularity.

For example, suppose one has a red system (over a red alphabet consisting of the defined symbol +)

$$
\begin{aligned}
x + 0 &\rightarrow x \\
x + s(y) &\rightarrow s(x + y)
\end{aligned}
$$

for adding two numbers (in successor notation, with constructors s and 0) and a blue system (with blue defined symbol *append*)

$$
\begin{aligned}
append(nil, x) &\rightarrow x \\
append(cons(x, y), z) &\rightarrow cons(x, append(y, z))
\end{aligned}
$$

for appending two lists (using the list constructors *cons* and *nil*). We would like to be certain that the union of these two unrelated programs is terminating and confluent, just as its constituent systems are. That way, we could be certain that terms containing a mixture of red and blue symbols, such as

$$
append(cons(s(0) + s(0), nil), cons(s(s(0)) + s(0), nil))
$$

have unique normal forms. (For the purposes of this exposition, a *defined symbol* is any function symbol or constant that appears at the head of a left side and a *constructor* is any other non-variable symbol appearing in the rules.) We would like modularity to hold even in the presence of additional rules, like

$$
\begin{aligned}
0 + x &\rightarrow x \\
append(append(x, y), z) &\rightarrow append(x, append(y, z))
\end{aligned}
$$

The above red and blue systems have no symbols at all in common. In most practical situations, one would want to be able to combine the blue system with a system like:

$$
\begin{aligned}
interleave(nil, x) &\rightarrow x \\
interleave(cons(x, y), z) &\rightarrow cons(y, interleave(x, z))
\end{aligned}
$$

that interleaves, rather than concatenates, two lists. Here the two list constructors appear in both programs.

In our definition of a rewrite rule we imposed no restrictions on the appearance of variables: Both $x \times 0 \rightarrow 0$ and $0 \rightarrow x \times 0$ are legitimate rewrite rules. Applying the latter to a term containing the constant 0 results in the replacement of that occurrence of 0 with any term of the form $u \times 0$ (u can be any term at all). A system having a rule with a variable on the right that is not also on the left, is nonterminating and likely nonconfluent. Similarly, a priori a rule could have just a variable on the left (for example, $x \to x \times 1$), in which case it is nonterminating. Since we are interested here in combinations of conceptually independent programs, we must rule out such cases from our discussions (as is indeed the convention of some authors, including [Huet, 1980]): a rule with a new variable on the right could introduce arbitrary nesting of variegated symbols; a rule with a variable for left side would apply at all positions of all terms and interfere with any other intended computation step. Accordingly, we define *constructor-sharing* pairs of rewrite systems as including only rules with nonvariable left sides and no new right side variables and for which all function symbols that appear at the top of the left side of a rule of one system are prohibited from also appearing at the top left of a rule in the conjoined system.

In the following sections, we summarize some of what is known about constructor-sharing combinations, and sketch some new results. Properties other than confluence and termination, as well as (hierarchical) combinations that share more than constructors, lie beyond the scope of this paper.

2   Confluence

The *rewrite relation* on terms, for a given system, is denoted by $\to$, its reflexive-transitive closure, called *derivability*, is $\to^*$, and $\leftrightarrow^*$ is its reflexive-symmetric-transitive closure, called *convertibility*. A system (or indeed any binary relation) is *confluent* if $s, t \to^* v$ for some v, whenever if $u \to^* s, t$. Confluence is equivalent to the *Church-Rosser* property: $s, t \to^* v$ whenever $s \leftrightarrow^* t$.

The confluence of unions of confluent relations was considered early on in [Hindley, 1964; Rosen, 1973; Staples, 1975].

In the following circumstances, it is known that the union of two confluent systems is confluent:

(a) The systems are both *left-linear* (that is, no variable appears more than once on the left side) [Raoult and Vuillemin, 1980].
(b) There are no shared constructors [Toyama, 1987b].
(c) Both systems are *bright* (meaning that the right-hand side of each rule is a defined symbol, not a variable or constructor) [Ohlebusch, 1994a].
(d) Each system is *normalizing* (in the sense that every term has at least one normal form) [Ohlebusch, 1994a].
(e) One system is terminating and left-linear and the other is bright [Dershowitz, 1997].

(This list and those in the sequel omit some known conditions that involve undecidable properties of the union.)

The necessity of these conditions may be seen from the following standard example [Huet, 1980]:

$$\begin{array}{rcl} g(x, x) & \to & 0 \\ g(x, c(x)) & \to & 1 \\ \hline a & \to & c(a) \end{array} \qquad \text{(A)}$$

The upper part is not left-linear; the lower is not normalizing; c is a shared constructor; neither is bright.

A careful analysis of why modularity fails [Dershowitz *et al.*, 1997] shows that at the crux of the problem lie certain instances $s\sigma$ and $t\tau$ of terms s and t appearing in left sides of one system such that $t\tau$ contains $s\sigma$ as a subterm, but no other defined symbols. If $s\sigma \leftrightarrow^* t\tau$ holds in the union, but not in the one system alone, then confluence is not guaranteed. The above results follow from this observation.

3   Termination

A rewrite system (or any binary relation) is *terminating* if there are no infinite derivations $t_1 \to t_2 \to \cdots$.

Modularity of termination was considered in [Dershowitz, 1981].

In the following circumstances, it is known that the union of two constructor-sharing terminating systems is terminating:

(a) One system is left-linear; the other is right linear (no variable appears more than once on the right side) and bright [Bachmair and Dershowitz, 1986].

(b) The systems are each *finitely-branching* (no term rewrites in one step to infinitely many terms) and remain terminating when combined with the (non-confluent, nonbright) system $\{h(x, y) \to x, h(x, y) \to y\}$ (for new function symbol h) [Gramlich, 1994].

(c) The systems do not share constructors and each remains terminating when combined with $\{h(x, y) \to x, h(x, y) \to y\}$ (for new function symbol h) [Ohlebusch, 1994b].

(d) Both systems bright [Gramlich, 1994; Ohlebusch, 1994b].

(e) The systems are both *non-duplicating* (that is, each rule's right side contains no more occurrences of any variable than does the left) [Dershowitz, 1995; Ohlebusch, 1994b].

(f) One of the systems is both bright and non-duplicating [Dershowitz, 1995; Ohlebusch, 1994b].

The necessity of most of these conditions can be seen from the following nonterminating union [Toyama, 1987a]:

$$
\begin{array}{rcl}
g(x, y) & \to & x \\
g(x, y) & \to & y \\
\hline
f(0, 1, x) & \to & f(x, x, x)
\end{array}
\qquad\text{(B)}
$$

Its upper half is not bright; its lower half duplicates x, is not right linear, and is nonterminating when conjoined with the rules for h.

4 Convergence

A convergent system is one that is both terminating and confluent. Confluence of the union follows from termination of the union by Knuth's Critical Pair Lemma [Knuth and Bendix, 1970], so one needs to find conditions under which termination is preserved for confluent systems. Modularity of convergence was investigated in [Bidoit, 1981].

In the following circumstances, it is known that the union of two constructor-sharing convergent systems is convergent:

(a) For each system no left side unifies with a proper subterm of any left side (with variables of the two sides considered disjoint) [Gramlich, 1992; Dershowitz, 1995].

(b) They have no shared constructors and both are left-linear [Toyama et al., 1995].

(c) One is *constructor-based* (proper subterms of left sides do not contain defined symbols) and left-linear [Dershowitz, 1997].

The case when both are constructor-based [Middeldorp and Toyama, 1993] follows from (a).

Even without shared constructors, modularity fails in general (as seen, for example, from the following nonterminating combination due to [Drosten, 1989]):

$$
\begin{array}{rcl}
g(x, x, y) & \to & y \\
g(x, y, y) & \to & x \\
\hline
f(a, b, x) & \to & f(x, x, x) \\
f(x, y, z) & \to & 0 \\
a & \to & 0 \\
b & \to & 0
\end{array}
\tag{C}
$$

The upper part is not left-linear; the lower part is not constructor-based and a and b appear as proper subterms on its left.

If the union is nonterminating, then there is an infinite derivation with minimal *rank* (alternation of colors of symbols along a path from root to leaf) with infinitely many rewrites in the *cap* (topmost maximal monochrome context). Thus, subterms of lesser rank are terminating. To show termination of the union, we need to find a transformation of the *alien* terms (subterms below the cap) such that a rewrite in the cap can be mirrored by a rewrite of transformed terms and such that a rewrite below the cap does not affect the transformation. Variations on this approach lead to the above results. Using the idea of [Marchiori, 1995] for proving (b), one can extend the modularity of confluence to some constructor-sharing unions of left-linear systems.

References

[Avenhaus and Madlener, 1990] Jürgen Avenhaus and Klaus Madlener. Term rewriting and equational reasoning. In R. B. Banerji, editor, *Formal Techniques in Artificial Intelligence: A Sourcebook*, pages 1–41. Elsevier, Amsterdam, 1990.

[Bachmair and Dershowitz, 1986] Leo Bachmair and Nachum Dershowitz. Commutation, transformation, and termination. In J. H. Siekmann, editor, *Proceedings of the Eighth International Conference on Automated Deduction (Oxford, England)*, volume 230 of *Lecture Notes in Computer Science*, pages 5–20, Berlin, July 1986. Springer-Verlag.

[Bidoit, 1981] Michel Bidoit. *Une méthode de présentation de types abstraits: Applications*. PhD thesis, Université de Paris-Sud, Orsay, France, June 1981. Rapport 3045.

[Dershowitz and Jouannaud, 1990] Nachum Dershowitz and Jean-Pierre Jouannaud. Rewrite systems. In J. van Leeuwen, editor, *Handbook of Theoretical Computer Science*, volume B: Formal Methods and Semantics, chapter 6, pages 243–320. North-Holland, Amsterdam, 1990.

[Dershowitz et al., 1997] Nachum Dershowitz, Maribel Fernández, and Jean-Pierre Jouannaud. Modular confluence revisited: The constructor-sharing case, 1997. In preparation.

[Dershowitz, 1981] Nachum Dershowitz. Termination of linear rewriting systems (preliminary version). In *Proceedings of the Eighth International Colloquium on Automata, Languages and Programming (Acre, Israel)*, volume 115 of *Lecture Notes in Computer Science*, pages 448–458, Berlin, July 1981. European Association of Theoretical Computer Science, Springer-Verlag.

[Dershowitz, 1995] Nachum Dershowitz. Hierarchical termination. In N. Dershowitz and N. Lindenstrauss, editors, *Proceedings of the Fourth International Workshop on Conditional and Typed Rewriting Systems (Jerusalem, Israel, July 1994)*, volume 968 of *Lecture Notes in Computer Science*, pages 89–105, Berlin, 1995. Springer-Verlag.

[Dershowitz, 1997] Nachum Dershowitz. Innocuous constructor-sharing combinations. In H. Comon, editor, *Proceedings of the Eighth International Conference on Rewriting Techniques and Applications (Sitges, Spain)*, number 1232 in Lecture Notes in Computer Science, pages 203–216, Berlin, June 1997. Springer-Verlag.

[Drosten, 1989] K. Drosten. *Termersetzungssysteme*. PhD thesis, Universitat Passau, Passau, Germany, 1989. Informatik Fachberichte 210, Springer-Verlag (Berlin).

[Gramlich, 1992] Bernhard Gramlich. Relating innermost, weak, uniform and modular termination of term rewriting systems. In A. Voronkov, editor, *Proceedings of the Conference on Logic Programming and Automated Reasoning (St. Petersburg, Russia)*, volume 624 of *Lecture Notes in Artificial Intelligence*, pages 285–296, Berlin, July 1992. Springer-Verlag.

[Gramlich, 1994] Bernhard Gramlich. Generalized sufficient conditions for modular termination of rewriting. *Applicable Algebra in Engineering, Communication and Computing*, 5:131–158, 1994.

[Hindley, 1964] J. Roger Hindley. *The Church-Rosser Property and a Result in Combinatory Logic*. PhD thesis, 1964.

[Huet, 1980] Gérard Huet. Confluent reductions: Abstract properties and applications to term rewriting systems. *J. of the Association for Computing Machinery*, 27(4):797–821, October 1980.

[Klop, 1992] Jan Willem Klop. Term rewriting systems. In S. Abramsky, D. M. Gabbay, and T. S. E. Maibaum, editors, *Handbook of Logic in Computer Science*, volume 2, chapter 1, pages 1–117. Oxford University Press, Oxford, 1992.

[Knuth and Bendix, 1970] Donald E. Knuth and P. B. Bendix. Simple word problems in universal algebras. In J. Leech, editor, *Computational Problems in Abstract Algebra*, pages 263–297. Pergamon Press, Oxford, U. K., 1970. Reprinted in *Automation of Reasoning 2*, Springer-Verlag, Berlin, pp. 342–376 (1983).

[Marchiori, 1995] Massimo Marchiori. Modularity of completeness revisited. In *Proceedings of the Sixth International Conference on Rewriting Techniques and Applications (Kaiserslautern, Germany)*, volume 914 of *Lecture Notes in Computer Science*, pages 2–10, Berlin, April 1995. Springer-Verlag.

[Middeldorp and Toyama, 1993] Aart Middeldorp and Yoshihito Toyama. Completeness of combinations of constructor systems. *J. Symbolic Computation*, 15:331–348, 1993.

[Ohlebusch, 1994a] Enno Ohlebusch. On the modularity of confluence of constructor-sharing term rewriting systems. In S. Tison, editor, *Proceedings of the Nineteenth International Colloquium on Trees in Algebra and Programming (Edinburgh, UK)*, volume 787 of *Lecture Notes in Computer Science*, pages 262–275, Berlin, April 1994. Springer-Verlag.

[Ohlebusch, 1994b] Enno Ohlebusch. On the modularity of termination of term rewriting systems. *Theoretical Computer Science*, 136(2):333–360, December 1994.

[Plaisted, 1993] David A. Plaisted. Term rewriting systems. In D. M. Gabbay, C. J. Hogger, and J. A. Robinson, editors, *Handbook of Logic in Artificial Intelligence and Logic Programming*, volume 4, chapter 2. Oxford University Press, Oxford, 1993. To appear.

[Raoult and Vuillemin, 1980] Jean-Claude Raoult and Jean Vuillemin. Operational and semantic equivalence between recursive programs. *J. of the Association for Computing Machinery*, 27(4):772–796, October 1980.

[Rosen, 1973] Barry K. Rosen. Tree-manipulating systems and Church-Rosser theorems. *J. of the Association for Computing Machinery*, 20(1):160–187, January 1973.

[Staples, 1975] John Staples. Church-Rosser theorem for replacement systems. In J. N. Crossley, editor, *Algebra and Logic: 1974 Summer Research Institute of the Australian Mathematical Society*, volume 450 of *Lecture Notes in Mathematics*, pages 291–307, Berlin, West Germany, 1975. Springer-Verlag.

[Toyama et al., 1995] Yoshihito Toyama, Jan Willem Klop, and Hendrik Pieter Barendregt. Termination for direct sums of left-linear complete term rewriting systems. *J. of the Association for Computing Machinery*, 42(6):1275–1304, November 1995.

[Toyama, 1987a] Yoshihito Toyama. Counterexamples to termination for the direct sum of term rewriting systems. *Information Processing Letters*, 25:141–143, 1987.

[Toyama, 1987b] Yoshihito Toyama. On the Church-Rosser property for the direct sum of term rewriting systems. *J. of the Association for Computing Machinery*, 34(1):128–143, January 1987.

Positive Applications of Lattices to Cryptography

Cynthia Dwork

IBM Almaden Research Center

Abstract. We describe constructions of several cryptographic primitives, including hash functions, public key cryptosystems, pseudo-random bit generators, and digital signatures, whose security depends on the assumed worst-case or average-case hardness of problems involving lattices.

1 Introduction

Initiated by Ajtai's paper "Generating Hard Instances of Lattice Problems," a burgeoning effort to build cryptographic primitives based on the assumed hardness of worst-case or random instances of problems involving lattices has proved extremely fruitful. Prior to Ajtai's work, lattices, and in particular, the lattice basis reduction algorithm of Lenstra, Lenstra, and Lovász, were used in cryptography principally to prove cryptographic *insecurity* [1, 9, 10, 20, 22, 25]. We describe more positive applications of lattices: constructions for public key cryptosystems, cryptographically strong hash functions, and pseudo-random bit generators whose security depends only on the worst-case hardness of the underlying lattice problem; a digital signature scheme whose security depends on the average hardness of the underlying problem.

2 Definitions

Many of the definitions included here are *extremely* informal. References for precise definitions are included in every case.

2.1 Cryptography

A *one-way* function is easy to compute and hard to invert. A *trapdoor* function is a one-way function for which there exists some special "trapdoor" information, so that given the trapdoor information the function is easy to invert, but without the trapdoor information the function is hard to invert (see [12]). A *public key cryptosystem* is a method of encrypting messages using publicly known information called the *public key*, in such a way that only the party knowing the corresponding *private key* can decrypt the ciphertext. Thus, encryption has a trapdoor nature: without the trapdoor information (the private key) decryption is hard, but decryption is easy given the private key (see [16] and [11]).

A *digital signature scheme* is a method of generating a (public key, private key) pair, together with a pair of procedures SIGN, and VERIFY. SIGN requires as input the message to be signed and the private key of the signer, while VERIFY, requires as input the message, its purported signature, and the public key of the claimed signer. Let (K, s) be a (public key, private key) pair. Let (m, α) be a claimed (message, signature) pair. Given (m, α, K) the VERIFY procedure, without knowing the secret s, verifies that $\alpha = \text{SIGN}(m, s)$ (see [17]).

A *one-way hash function* is a one-way function h mapping long strings to short strings, say, $h : \{0,1\}^n \to \{0,1\}^\ell$ for $n > \ell$. One-way hash functions have many uses in cryptography. In particular they are used to "shrink" long messages before signing (see [24]). Thus, what is actually signed is $h(m)$ rather than m ($h(m)$ is sometimes called a *message digest*). In this case the VERIFY procedure checks that $\alpha = \text{SIGN}(h(m), s)$. For this application it is essential that, given $h(m)$, it is hard to find a different message $m' \neq m$, for which $h(m') = h(m)$. A little more formally, a family of *universal one-way hash functions* is a collection $\mathcal{F}$ of functions $f : \{0,1\}^m \to \{0,1\}^{i(m)}$ with the property that for any element $x \in \{0,1\}^m$, if f is chosen at random from the collection $\mathcal{F}$, then it is hard to find an element $y \neq x$ such that

$f(y) = f(x)$. Each choice of $l(m)$ yields a class of hash functions. A slightly stronger notion is *collision-intractability*: for a randomly selected function $f \in \mathcal{F}$, it is hard to find x, y such that $x \neq y$ and $f(x) = f(y)$.

A *pseudorandom bit generator* is a (deterministic) function that takes as input a string $s \in \{0,1\}^n$ and produces as output a string $p \in \{0,1\}^m$ where $m > n$. Moreover, the strings produced in this way when the inputs s are random should be polynomial-time indistinguishable from truly random strings of length m. Thus these functions appear to manufacture some additional bits of randomness (see [6, 26]; extensive treatment appears in [23]).

The *subset sum problem* of dimensions m and l is: given m numbers $\mathbf{a} = (a_1, \ldots, a_m)$, each of length l, and a number T, find a subset $S \subset \{1, \ldots, m\}$ such that $\sum_{i \in S} a_i = T \bmod 2^l$. The subset sum problem can be viewed as that of inverting the function $f(\mathbf{a}, S) = \mathbf{a}, \sum_{i \in S} a_i \bmod 2^{l(n)}$.

2.2 Lattices

The fundamental concepts concerning lattices can be found in [8, 18, 19].

If $a_1, \ldots, a_n$ are linearly independent vectors in $\mathbb{R}^n$, then we say that the set $\{\sum_{i=1}^n k_i a_i \,|\, k_1, \ldots, k_n \in \mathbb{Z}\}$ is a lattice in $\mathbb{R}^n$. We will denote this lattice by $L(a_1, \ldots, a_n)$. The set $a_1, \ldots, a_n$ is called a basis of the lattice; its length is $\max_{1 \leq i \leq n} \|a_i\|$. The determinant of a lattice L will be the absolute value of the determinant of the matrix whose columns are the vectors $a_1, \ldots, a_n$. We let $\mathrm{bl}(L)$ denote the length of the shortest basis for L.

The *dual* lattice of L, denoted L^*, is defined as

$$L^* = \{x \in \mathbb{R}^n \mid x^T y \in \mathbb{Z} \text{ for all } y \in L\}.$$

If $(b_1, \ldots, b_n)$ is a basis of L then $(c_1, \ldots, c_n)$ is a basis for L^*, where

$$c_i^T b_j = \begin{cases} 1 \text{ if } i = j \\ 0 \text{ if } i \neq j \end{cases}$$

Thus, if we represent the lattice $L = L(b_1, \ldots, b_n)$ by a matrix B with columns $b_1, \ldots, b_n$, then the dual of L is the lattice spanned by the rows of B^{-1}. Each basis vector b_i in $L = L(b_1, \ldots, b_n)$ induces a collection of mutually parallel $(n-1)$-dimensional hyperplanes, where, for $k \in \mathbb{Z}$, the kth hyperplane in the collection is the set of all points whose inner product with b_i is equal to k. The distance between adjacent hyperplanes in the collection is $\|b_i\|^{-1}$. Thus, if $\|b_i\| < \|b_j\|$, then adjacent hyperplanes in the ith collection are farther apart than adjacent hyperplanes in the jth collection. As the formula for computing the basis for the dual makes clear, the dual lattice is the set of points that are intersections of n hyperplanes, one from each of the n collections.

Assume n is a positive integer, $M > 0$, $d > 0$ are real numbers, and $L \subseteq \mathbb{Z}^n$ is a lattice which has an $n-1$ dimensional sublattice L' with the following properties:

1. L' has a basis of length at most M;
2. if H is the $n-1$ dimensional subspace of $\mathbb{R}^n$ containing L' and $H' \neq H$ is a coset of H intersecting L, then the distance of H and H' is at least d.

We say that L is a (d, M)-lattice. If $d > M$, then L' is unique. In this case L' will be denoted by $L^{(d,M)}$. If $a_1, \ldots, a_n \in \mathbb{R}^n$ are linearly independent vectors, then $\mathcal{P}^-(a, \ldots, a_n)$ denotes the half-closed parallelepiped $\{\sum_{i=1}^n \gamma_i a_i \,|\, 0 \leq \gamma_i < 1, i = 1, \ldots, n\}$. By "$x \bmod \mathcal{P}$" we mean the unique vector $x' \in \mathcal{P}^-(a_1, \ldots, a_n)$ so that $x - x'$ is an integer linear combination of the vectors $a_1, \ldots, a_n$.

The *orthogonality defect* of an $n \times n$ matrix B is the quantity $\frac{1}{\det(B)} \prod_{i=1}^n \|b_i\|$. The *dual orthogonality defect* of B is the quantity $\frac{1}{\det(B^{-1})} \prod_{i=1}^n \|\dot{b}_i\|$, where for $1 \leq i \leq n$, $\dot{b}_i$ is the ith row of B^{-1}.

3 Generating Hard Instances of Lattice Problems

Cryptographic constructions necessarily require random choices: if, for example, the choice of a key were deterministic, then the key could not be secret. Thus, the security of the construction relies on the intractability of a *random* instance of the problem on which the construction is based. It has therefore been a longstanding goal in cryptography to find a "hard" problem for which one can establish an explicit connection between the hardness of random instances and the hardness of the hardest, or worst-case, instances.

Such a connection is the contribution of the celebrated paper of Ajtai, "Generating Hard Instances of Lattice Problems" [2]. Specifically, the paper presents a random problem whose solution would imply the solution of three famous worst-case problems:

1. Find the length of a shortest nonzero vector in an n-dimensional lattice approximately, up to a polynomial factor.

2. Find the shortest nonzero vector in an n-dimensional lattice L where the shortest vector v is unique in the sense than any other vector whose length is at most $n^c\|v\|$ is parallel to v, where c is a sufficiently large absolute constant.

3. Find a basis $b_1, \ldots, b_n$ in the n-dimensional lattice L whose length, defined as $\max_{i=1}^n \|b_i\|$, is the smallest possible up to a polynomial factor.

Ajtai's Random Lattice Problem. For $n, m, q \in \mathcal{N}$ such that $n \log q < m \leq \frac{q}{2n^4}$ and $q = O(n^c)$ for a fixed $c > 0$, given a matrix $M \in \mathbb{Z}_q^{n \times m}$ (that is, an $n \times m$ matrix of integers in [0,q-1] of a certain form described below), find a vector $x \neq 0 \in \mathbb{Z}_q^m$ so that $Mx \equiv 0 \bmod q$ and $\|x\| < n$. The lattices are defined modulo q, in the sense that if two vectors are congruent modulo q then either both are in the lattice or neither is in the lattice. Thus the matrix M and the integer q define the lattice: $x \in \Lambda(M, q)$ iff $Mx \equiv 0 \pmod q$.

The matrix M is obtained as follows. Randomize vectors $v_1, \ldots, v_{m-1}$ independently and with uniform distribution on the set of all vectors $\langle x_1, \ldots, x_n \rangle \in \mathbb{Z}_q^n$. Independently randomize a $0, 1$ sequence $\delta_1, \ldots, \delta_{m-1}$, where the numbers δ_i are chosen independently and uniformly. Then define $v_m = -\sum_{i=1}^{m-1} \delta_i v_i \bmod q$ with the additional constraint that each component of v_m is an integer in $[0, q-1]$. The matrix M has columns $v_1, \ldots, v_m$. The class of lattices $\Lambda(M, q)$ defined by matrices of this type will be called λ. The random problem is to find a vector in $\Lambda(M, q)$ of length less than n. Note that $(\delta_1, \ldots, \delta_{m-1}, 1) \in \Lambda(M, q)$ and its length is $O(\sqrt{m})$, so this vector is a solution when $m < n^2$.

Let L be an n-dimensional lattice, let $a_1, \ldots, a_n$ be a set of linearly independent vectors in L and let $M = \max_{i=1}^n \|a_i\|$. The heart of Ajtai's work is a procedure which, if $M > n^c \mathrm{bl}(L)$ for a fixed consant c, uses an oracle for the random lattice problem just defined to obtain another set of linearly independent elements in L whose maximum length is at most $\frac{1}{2} \max_{1 \leq i \leq n} \|a_i\|$.

In rough outline the procedure works as follows. Starting from $a_1, \ldots, a_n$, construct a set of linearly independent lattice vectors $f_1, \ldots, f_n$ such that $\max_{i=1}^n \|f_i\| \leq n^3 M$ and $W = \mathcal{P}(f_1, \ldots, f_n)$ is close to a cube, in the sense that each vertex of W will be at most distance nM from a fixed cube. If the space is covered with the cells of a lattice determined by a short basis, then most of the cells intersecting W lie completely in the interior of W. This implies that every parallelepiped of the form $u + W$, $u \in \mathbb{R}^n$, has roughly the same number of lattice points. Moreover, this also holds for parallelepipeds of the form $u + \frac{1}{q}W$ for $q = [n^{c_2}]$, where c_2 is sufficiently small with respect to c. Thus, if we pick a lattice point v at random from a set D of parallelepipeds of the form $u + \frac{1}{q}W$ with non-overlapping interiors, then the distribution induced on D – that is, the choice of which element in D contains v – is very close to the uniform distribution.

The set D of parallelepipeds $u + \frac{1}{q}W$ that is of interest to us is that obtained by cutting W into q^n small parallelepipeds by dividing each of the vectors f_i into q pieces of equal length. Thus each of the small parallelepipeds is of the form $(\sum_{i=1}^n t_i \frac{f_i}{q}) + \frac{1}{q}W$, where $0 \leq t_i < q$, $i = 1, \ldots, n$ is a sequence of integers; that is, $\langle t_1, \ldots t_n \rangle \in \mathbb{Z}_q^n$. Let us call the vector $o = \sum_{i=1}^n t_i \frac{f_i}{q}$ the *origin* of the parallelepiped. We will name an element of D by the vector $t(o) = \langle t_1, \ldots, t_n \rangle$ of coefficients of the $\frac{f_i}{q}$ defining its origin. If we choose a random set of lattice points $\xi_1 \ldots, \xi_m$ in W and look at, for each ξ_j, the name $t(o_j)$ of the parallelepiped containing ξ_j, then we get a sequence $t(o_1), \ldots, t(o_m)$ of elements chosen almost uniformly from D. Express each ξ_j as the sum of the origin o_j and an offset $\delta_j \in \frac{1}{q}W$. Note that the offset is relatively short: since δ_j is contained in $\frac{1}{q}W$, $\|\delta_j\|$ is bounded by n times the length of the longest side of W. That is, $\max_{1 \leq j \leq m} \|\delta_j\| \leq n(\frac{1}{q}n^3 M)$.

By definition of D and the fact that the distribution induced on D by the choice of ξ is almost uniform, each $t(o_j)$ is distributed almost uniformly in $\mathbb{Z}_q^n$. Let $m = [c_1 n \log n]$. Consider the sequence $t(o_1), \ldots, t(o_m)$ as a value of the random variable λ (it is shown in [2] that the distribution of $t(o_1), \ldots, t(o_m)$ is extremely close to that of λ). If there exists an algorithm $\mathcal{A}$ that can solve Ajtai's random lattice problem, then using $\mathcal{A}$ we can find a short (length at most n) vector $h = \langle h_1, \ldots, h_m \rangle \in \mathbf{Z}^m$ satisfying $\sum_{j=1}^{m} h_j t(o_j) \equiv 0 \bmod q$.

Writing the lattice vector $\sum_{j=1}^{m} h_j \xi_j$ as the weighted sum of origins and offsets, we get

$$w = \sum_{j=1}^{m} h_j \xi_j = \sum_{j=1}^{m} h_j o_j + \sum_{j=1}^{m} h_j \delta_j .$$

Critically, since $\sum_j h_j t(o_j) \equiv 0 \bmod q$, we have that $\sum_j h_j o_j$ is an integer linear combination of the vectors $(f_1, \ldots, f_n)$. Since the f_i are lattice vectors, so is $\sum_j h_j o_j$. Since w is also in L the difference $w - \sum_j h_j o_j = \sum_j h_j \delta_j \in L$. Finally, since $|\sum_{j=1}^{m} h_j^2| \leq n^2$ and, as noted above, each of the offsets is also relatively short, the lattice vector $\sum_{1 \leq j \leq n} h_j \delta_j$ is relatively short: $\| \sum_{1 \leq j \leq n} h_j \delta_j \| \leq n^2(n^4 M \frac{1}{q})$, which is less than $\frac{M}{2}$ if q is sufficiently large (say, $q \geq n^7$).

Recently, Ajtai's results have been tightened by Cai and Nerurkar [7]. Through a number of technical steps, Cai and Nerurkar are able to shrink the constant c in Ajtai's reduction, slightly better than halving it.

Based solely on the results in [2], it is possible to design a number of *interactive* cryptographic procedures, including schemes for identification, bit commitment, and coin flipping [3].

4 Hashing

The reduction described in the previous section has implications for the security of the following family of hash functions, studied by Impagliazzo and Naor [21]:

Let $l(m) = (1 - c)m$ for $c > 0$. For $a_1, \ldots, a_m \in \{0, 1\}^{l(m)}$ the function $f_\mathbf{a} = f_{a_1, \ldots, a_m} : \{0, 1\}^m \rightarrow \{0, 1\}^{l(m)}$ is defined as follows. Let the m-bit number x be written $x = x_1 x_2 \ldots x_m$ where each $x_i \in \{0, 1\}$. Then $f_\mathbf{a}(x) = \sum_{i=1}^{m} x_i a_i \bmod 2^{l(m)}$.

The bits of x act as selectors to determine which of the a_i are summed. We can represent the function as a $1 \times m$ matrix M with columns $a_1, \ldots, a_m$. Given $x \in \{0, 1\}^m$, the value of the function is $Mx \bmod 2^{l(m)}$. As we next explain, Ajtai's proof shows that the ability to solve a random instance of the subset sum problem implies the ability to solve the worst-case lattice problems listed in Section 3 (additional details appear in [2]). So if we assume that these worst-case problems are hard for dimension n, then these randomized subset sum problems will be hard as well. To illustrate this connection, let $q = [n^{c_2}]$ and $m = [c_1 n \log n]$ as in the discussion of Ajtai's reduction in Section 3. Let $N = q_1 q_2 \ldots q_n$ where each q_i is a distinct prime in $[q, 2q]$. Let $a_1, \ldots, a_m, b$ be random integers modulo N. Consider the subset sum problem of finding $x \in \{0, 1\}^m$ such that $\sum_{i=1}^{m} x_i a_i \equiv b \bmod N$.

Remarks.

(1) The numbers a_i are of length $l(m) \approx n \log q = (1 - c)m$ for some $c > 0$ if $c_1 > c_2$. So subset sum problems of this type are essentially those in the Impagliazzo-Naor family of hash functions.

(2) If $x \in \{0, 1\}^m$ then $\|x\| \leq \sqrt{m} < n$.

We may express each a_i as a vector of remainders modulo the primes $q_1, \ldots, q_n$: $a_i' = (a_i^1, \ldots, a_i^n)$, where $a_j^i \in \mathbb{Z}_{q_j}$, for $1 \leq i \leq m$ and $1 \leq j \leq n$. Note that if a_i is chosen uniformly from $\mathbb{Z}_N$ then a_i' is implicitly chosen uniformly from $\mathbb{Z}_{q_1} \times \ldots \times \mathbb{Z}_{q_n}$. Similarly, let b' be the Chinese remainder decomposition of b. Let M be the $n \times m$ matrix with columns $a_1', \ldots, a_n'$. If we can find $x \in \{0, 1\}^m$ satisfying $\sum_{i=1}^{m} x_i a_i \equiv b \bmod N$, then $Mx \equiv b'$ (where the jth component of the product is reduced modulo q_j, $1 \leq j \leq n$).

The hardness of this problem follows from Ajtai's proof. The key modification is as follows. Recall that $W = \mathcal{P}(f_1, \ldots, f_n)$. Rather than cutting each vector f_i, $1 \leq i \leq n$, into q equal pieces (for a fixed q), instead for each $1 \leq i \leq n$, cut f_i into q_i pieces. Thus, instead of having q^n little parallelepipeds we will have $N = q_1 \ldots q_n$ of them. Any solution x plays the role of the solution $h = \langle h_1, \ldots, h_m \rangle$ in the original proof. See [2] for more details and extensions of these results.

Impagliazzo and Naor proved that if the subset sum function for length $(1 - c)m$, $c > 0$, is one-way in the sense that no polynomial time algorithm can invert the function on a random input, then it is also a family of universal one-way hash functions [21]. Since this class of subset sum problem is hard on average (assuming the worst-case lattice problems are difficult for dimension n), the Impagliazzo and Naor construction yields a family of universal one-way hash functions.

In a related note, Goldreich, Goldwasser, and Halevi [13] observed that these hash functions are actually *collision-intractable*. Specifically, they show that if M is a random matrix in $Z_q^{n \times m}$, then finding collisions of the function $h(x) = Mx \bmod q$ is hard provided a slight modification of Ajtai's random lattice problem is hard. The modification is to only require that the vector x have coefficients in $\{-1, 0, 1\}$ (rather than to require $x \in \mathbb{Z}_q^m$ and $\|x\| < n$), and the proof of collision-intractability relies on the fact that Ajtai's results hold even if the random lattice problem is relaxed so that $\|x\|$ is bounded by a polynomial in n. (The more relaxed version incurs a cost in the quality of the approximation obtained in Ajtai's reduction.) Collision-intractability follows from the fact that if it were easy to find $x, y \in \{0, 1\}^m$ such that $Mx \equiv My \bmod q$ then $M(x - y) \equiv 0 \bmod q$. Since $x - y \in \{-1, 0, 1\}^m$, finding such a pair x, y is difficult.

5 Public Key Cryptography

Ajtai and Dwork constructed a public key cryptosystem generator with the property that if a random instance of the cryptosystem can be broken, that is, if for a random instance the probability that an encryption of a zero can be distinguished from an encryption of a one (without the private key) in polynomial time is at least $\frac{1}{2} + n^{-c_1}$ for some absolute constant $c_1 > 0$, then the worst-case unique shortest vector problem has a probabilistic polynomial time solution. Intuitively, this worst-case/average-case equivalence means that there are essentially no "bad" instances of the cryptosystem. In this discussion we will work with real numbers, ignoring issues of finite precision. The private key is a vector $u \in \mathbb{R}^n$ chosen uniformly at random from the n-dimensional unit ball. u induces a collection of $(n - 1)$-dimensional hyperplanes, where for $i \in \mathbb{Z}$ the ith hyperplane is the set of vectors v whose inner product satisfy $u \cdot v = i$. Very roughly speaking, the public key is a method of generating a point guaranteed to be near one of the hyperplanes in the collection. The public key is chosen so as not to reveal the collection of hyperplanes – indeed, Ajtai and Dwork prove that any ability, given only the public key, to discover the collection implies the ability to solve the worst-case unique shortest vector problem. Encryption is bit-by-bit: zero is encrypted by using the public key to find a random vector $v \in \mathbb{R}^n$ near one of the hyperplanes – the ciphertext is v; one is encrypted by choosing a random vector u uniformly from $\mathbb{R}^n$ – the ciphertext is simply u. Decryption of a ciphertext x is simple using the private key u: if $u \cdot x$ is close to an integer then x is by definition near one of the hidden hyperplanes, and so x is interpreted as zero; otherwise x is interpreted as one.

If a lattice Λ has an n^c-unique shortest vector v, then $L = \Lambda^*$ is a $(\|v\|^{-1}, n^{-c'}\|v\|^{-1})$ lattice, where c' is roughly $c - 2$ (a proof appears in [2]). Moreover, v is orthogonal to the $(n - 1)$-dimensional space containing $L' = L^{(\|v\|^{-1}, n^{-c'}\|v\|^{-1})}$, and if H is the $(n - 1)$-dimensional subspace of $\mathbb{R}^n$ containing L', then the hyperplanes induced by v are the cosets of H intersecting L (recall the discussion of the dual in Section 2).

Define pert(R) to be a random variable that, roughly speaking, is normally distributed about the origin in a ball of radius R. Let $\mathcal{K}$ be a very large cube, and let $R = n^c$. It is first shown that if the n^{c_1}-unique shortest vector problem is hard, for c_1 sufficiently larger than c, then the distribution obtained by choosing a random lattice point in $\mathcal{K}$ and perturbing it by adding a value of pert(R) (for sufficiently large R) is polynomially indistinguishable from the distribution obtained by choosing a vector uniformly at random from $\mathcal{K}$.

To see this, suppose we are given a *random* lattice Λ with an n^{c_1}-unique shortest vector v, and let $L = \Lambda^*$. Let $d = \|v\|^{-1}$ and $M = n^{-c_1'}\|v\|^{-1}$. where c_1' is roughly $c_1 - 2$. Then L is a (d, M) lattice. Let $L' = L^{(d,M)}$ have basis $b_1, \ldots, b_{n-1}$. Let $H = H_0$ be the $(n - 1)$-dimensional hyperplane containing L'. If R is sufficiently large with respect to $b_1, \ldots, b_{n-1}$. then the random variable obtained by sampling pert(R), projecting the result onto the $(n - 1)$-dimensional hyperplane containing L', and taking the projection modulo $\mathcal{P}^-(b_1, \ldots, b_{n-1})$ is extremely close to the value obtained by choosing a point uniformly in $\mathcal{P}^-(b_1, \ldots, b_{n-1})$.

Intuitively, this means that any algorithm distinguishing between "lattice point + pert(R)" and the uniform distribution is really distinguishing between points close to the cosets of H intersecting L and random points. From this it is possible (with some effort – see [4]) to find H. Finally, given H we can recover v, the unique shortest vector in $\Lambda = L^*$ as follows. As noted above, v is perpendicular to H. By definition $\|v\| = d^{-1}$; given a basis for L (computable from the given basis for Λ), we can sample points from L and compute for each its distance from H. By taking the gcd of many random such distances we can find d.

The next step is to dispense with the lattice L. Let u be chosen uniformly at random from the n-dimensional unit ball and let $\mathcal{H}_u$ be the collection of hyperplanes induced by u. The distribution obtained by choosing a random point in $\mathcal{H}_u \cap \mathcal{K}$ and then sufficiently perturbing the chosen point, is indistinguishable from the uniform distribution in $\mathcal{K}$ – otherwise there would be a way of distinguishing points close to the hyperplanes from random points. The scheme is therefore as follows.

Private Key: vector u chosen at random from the n-dimensional unit ball

Public Key: $v_1, \ldots, v_m$: a collection of perturbations of points chosen uniformly from $\mathcal{H}_u \cap \mathcal{K}$, and a parallelepiped $\mathcal{P}$

Encryption: To encrypt zero, choose $\delta_1, \ldots, \delta_m$, each $\delta_i \in_R \{0, 1\}$. The ciphertext is $\sum_{i=1}^{m} \delta_i \bmod \mathcal{P}$. To encrypt one, choose a random point in $\mathcal{P}^-$.

Decryption: given ciphertext x, compute $x \cdot u$. If the result is sufficiently close (as a function of R) to an integer, then decrypt x as zero; else decrypt x as one.

There is some chance of a decryption error. This can be avoided by including in the public key a point B obtained by averaging two encryptions of zero lying on hyperplanes of different parity. (A related solution appears in [15].) The procedure for encrypting one becomes: follow the procedure for encrypting zero but add B before modding out by $\mathcal{P}$.

Very roughly, worst-case/average-case equivalence is shown as follows. Suppose we have an algorithm $\mathcal{A}$ that can break random instances of the cryptosystem with non-negligible probability over the choice of u. Given any instance L of the unique shortest vector problem, we convert it to an instance of the cryptosystem by choosing a number of random linear transformations $U = \theta\nu$ where $\theta \in \mathbb{R}$ and ν is an orthogonal linear transformation. Intuitively, ν rotates the lattice L leaving the lengths of the basis vectors unchanged, while θ scales the rotated basis. If v is the unique shortest vector and we choose enough transformations, then for one of them $\|Uv\| < 1$ and $\mathcal{A}$ can crack the instance of the cryptosystem defined by u. Note that v is the n^c-unique shortest vector of L if and only if Uv is the n^c-unique shortest vector of UL. It follows that J, the dual lattice of UL, is a $(1, n^{-c'})$ lattice, where $c' \sim c - 2$. Moreover, the distribution obtained by perturbing points of J is exponentially close to the distribution obtained by perturbing points in the hyperplanes induced by Uv. But Uv describes (the private key of) a *random* instance of the Ajtai-Dwork cryptosystem: it is random because U is random. Moreover, the ability to distinguish zeros – points close to the hyperplanes induced by Uv – from ones – random points– would imply the ability to distinguish perturbations of lattice points in J from random points. As argued above, this ability would yield Uv, the unique shortest vector in J^*, and hence, by the invertibility of U, v.

6 Pseudorandom Bit Generators

The Ajtai-Dwork construction suggests a pseudorandom bit generator with a very natural geometrical interpretation. Note that, given the secret information u, it requires fewer bits to describe a point that is close to one of the hyperplanes induced by u than to describe a point chosen at random from $\mathbb{R}^n$. To see this, consider a basis $b_1, \ldots, b_n$ for $\mathbb{R}^n$ in which the first $n - 1$ vectors lie in H_0, the $(n - 1)$-dimensional space orthogonal to u, and b_n is parallel to u. Using this basis it is easy to see that to describe a random point requires more bits than to describe a point close to one of the hyperplanes because, intuitively, there are more choices for the random point (the distance of a random point to the nearest hyperplane can be any value in $[0, \|u\|/2]$, while the distance of a point close to the hyperplane is in $[0, n^{-c}]$ for a fixed constant $c > 0$).

7 Digital Signatures

Goldreich, Goldwasser, and Halevi have suggested a digital signature scheme based on a trapdoor function related to the problem of finding the lattice vector closest to a given vector v [14]. Their approach, which also yields a public-key cryptosystem, depends on the hardness of random instances of the underlying problem (rather than worst-case instances). Naor and Yung have shown how to obtain a digital signature scheme from any one-way function [24]. Other than schemes obtained by applying this general construction to the one-way functions of [2, 4], we know of no proposed digital signature scheme with worst-case/average-case equivalence.

The trapdoor function proposed by Goldreich, Goldwasser, and Halevi relies on the difficulty, given a basis B for a lattice L, of finding a basis for L with small dual orthogonality defect. Call such a basis *reduced*.

The trapdoor information is a reduced basis R for an n-dimensional lattice (defined implicitly by R). Given R, it is possible to generate a second basis B for $L = L(R)$ so that B has high dual orthogonality defect. The trapdoor function is specified by B and a real parameter $\sigma \in \mathbb{R}$. Given vectors $v, e \in \mathbb{R}^n$, the function $f_{(B,\sigma)}(v, e) = Bv + e$. Note that the value σ does not appear in the definition of the function. Rather, σ governs the selection of e: each entry in e is chosen at random according to a distribution with zero mean and variance σ^2. For example, each entry in e can be chosen uniformly from $\{\sigma, -\sigma\}$.

Assume e is chosen as described and each component of v is chosen uniformly from, say, $\{-n^2, -n^2 + 1, \ldots, n^2 - 1, n^2\}$. Let $c = f_{(B,\sigma)}(v, e) = Bv + e$. If σ is chosen carefully, the function can be inverted using R by applying Babai's rounding technique [5]: represent c as a linear combination of the columns of R and then round the coefficients in the linear combination to the nearest integers to obtain a lattice point (integer linear combination of the columns of R). Once v is recovered we find $e = c - Bv$.

In the Goldreich, Goldwasser, and Halevi digital signature scheme, the private key is a reduced basis R and the public key is a non-reduced basis B. To sign a message m encoded as a vector $v \in \mathbb{R}^n$, the signer computes, using the reduced basis, a lattice vector w close to v. The public verification key is a threshold τ and the non-reduced basis B; the signature is verified by checking that $\|v - w\| \leq \tau$. As the authors point out, if $u, u' \in \mathbb{R}^n$ are sufficiently close, then a signature on u is likely also to be a signature on u'; it is therefore important to use a "good hash function" to hash a message before interpreting it as a vector in $\mathbb{R}^n$ [14].

References

1. L. Adleman, On Breaking Generalized Knapsack Public Key Cryptosystems, Proceedings 15th Annual ACM Symposium on Theory of Computing, 1983, pp. 402–412
2. M. Ajtai, Generating Hard Instances of Lattice Problems, Proceedings 28th Annual ACM Symposium on Theory of Computing, 1996, pp. 99–108 *Electronic Colloquium on Computational Complexity TR96-007*, http://www.eccc.uni-trier.de/eccc-local/Lists/TR-1996.html
3. M. Ajtai, *discussion with the author*, 1996
4. M. Ajtai, C. Dwork, A Public-Key Cryptosystem with Average-Case/Worst-Case Equivalence, Proceedings 29th Annual ACM Symposium on Theory of Computing, 1997; see also *Electronic Colloquium on Computational Complexity TR96-065*, http://www.eccc.uni-trier.de/eccc-local/Lists/TR-1996.html
5. L. Babai, On Lovász' Lattice Reduction and the Nearest Lattice Point Problem, *Combinatorica* 6(1), 1986, pp. 1–13
6. M. Blum and S. Micali, How to Generate Cryptographically Strong Sequences of Pseudo-Random Bits, *SIAM J. Computing* 13, 1984, pp. 850–864
7. J.-Y. Cai and A. P. Nerurkar, An Improved Worst-Case to Average-Case Connection for Lattice Problems, *private communication*, 1997
8. J.W.S. Cassels, *An Introduction to the Geometry of Numbers*, Springer, 1959
9. D. Coppersmith, Finding a Small Root of a Univariate Modular Equation, *Proc. EUROCRYPT'96*
10. D. Coppersmith, M. Franklin, J. Patarin, and M. Reiter, Low Exponent RSA with Related Messages, *Proc. EUROCRYPT'96*
11. D. Dolev, C. Dwork, and M. Naor, Non-Malleable Cryptography, Proceedings 23th Annual ACM Symposium on Theory of Computing, 1991, pp. 542–550
12. O. Goldreich, *Foundations of Cryptography (Fragments of a Book)*, http://www.wisdom.weizmann.ac.il/people/homepages/oded/frag.html

13. O. Goldreich, S. Goldwasser, and S. Halevi, Collision-Free Hashing from Lattice Problems, *Electronic Colloquium on Computational Complexity TR96-042*, http://www.eccc.uni-trier.de/eccc-local/Lists/TR-1996.html

14. O. Goldreich, S. Goldwasser, and S. Halevi, Public-Key Cryptosystems from Lattice Reduction Problems, *Electronic Colloquium on Computational Complexity TR96-056*, http://www.eccc.uni-trier.de/eccc-local/Lists/TR-1996.html

15. O. Goldreich, S. Goldwasser, and S. Halevi, Eliminating the Decryption Error in the Ajtai-Dwork Cryptosystem, *to appear. Proc. CRYPTO'97*

16. S. Goldwasser and S. Micali, Probabilistic Encryption, *J. Comput. System Sci. 28*, 1984, pp. 270-299

17. S. Goldwasser, S. Micali, and R. Rivest, A "Paradoxical" Solution to the Signature Problem, *SIAM J. Computing 17*, 1988, pp. 281-308

18. M. Grötschel, Lovász, A. Schrijver, *Geometric Algorithms and Combinatorial Optimization*, Springer, Algorithms and Combinatorics 2, 1988

19. P.M. Gruber, C.G. Lekkerkerker, *Geometry of Numbers*, North-Holland, 1987

20. J. Hastad, Solving Simultaneous Modular Equations of Low Degree, *SIAM J. Computing 17(2)*, pp.336-341, 1988

21. R. Impagliazzo and M. Naor, Efficient Cryptographic Schemes Provably as Secure as Subset Sum, *J. Cryptology 9*, pp. 199-216, 1996

22. J.C. Lagarias, A.M. Odlyzko, Solving low-density subset sum problems, *Journal of the Association for Computing Machinery 32* pp. 229-246, 1985. An earlier version appeared in *Proc. 24th Annual Symposium on Foundations of Computer Science*, 1983

23. M. Luby, **Pseudo-randomness and applications**, Princeton University Press, 1996.

24. M. Naor and M. Yung, Universal One-Way Hash Functions and Their Cryptographic Applications, Proceedings 21th Annual ACM Symposium on Theory of Computing, 1989, pp. 33-43

25. A. Shamir, A Polynomial-Time Algorithm for Breaking the Basic Merkle-Hellman Cryptosystem, *Proc. 23rd Annual Symposium on Foundations of Computer Science*, 1982, pp. 145-152

26. A. C. Yao, Theory and Applications of Trapdoor Functions, *Proc. 23rd Annual Symposium on Foundations of Computer Science*, 1982, pp. 80-91

A Tile-Based Coordination View of Asynchronous π-calculus*

GianLuigi Ferrari[1] and Ugo Montanari[2]

[1] Dipartimento di Informatica, Università di Pisa, `giangi@di.unipi.it`
[2] Computer Science Laboratory, SRI International, Menlo Park, `ugo@csl.sri.com`

Abstract. Tiles are rewrite rules with side effects, reminiscent of both Plotkin SOS and Meseguer rewriting logic rules. They are well suited for modeling coordination languages, since they can be composed both statically and dynamically via possibly complex synchronization and workflow mechanisms. In this paper, we give a tile-based bisimilarity semantics for the asynchronous π-calculus of Honda and Tokoro and prove it equivalent to the ordinary semantics. Two kinds of tiles are provided: *activity tiles* and *coordination tiles*. Activity tiles specify the basic interactions sequential processes are able to perform, without considering the operational environment where they live. Instead, coordination tiles control the global evolution of programs.

1 Introduction

Coordination is considered as a key concept for modeling and designing heterogeneous, distributed, open ended systems. It applies typically to systems consisting of a large number of software components, independently programmed in different programming languages, which may change their configuration during execution. Changes may be due to the external addition/deletion of software components and sites, or to the transmission of programs and resource references made possible by operating systems and languages supporting mobility.

Coordination languages (see e.g. [2, 7]) provide the constructs to define configurations and interaction protocols, but they assume as given the software agents being coordinated. Hence, systems are designed and developed in a structured way, starting from the basic computational components and adding suitable software modules called *coordinators*. This approach has several advantages. In fact, it increases the potential reuse of both software agents and coordinators with usually acceptable overheads. Moreover, it provides a simple but powerful technique to reason about and prove properties of systems: induction on their structure.

The main aim of this paper is to present an approach to coordination based on the *tile model* [11, 12, 13]. Tiles have been used for coordination formalisms

---

* Research supported in part by CNR Integrated Project *Metodi e Strumenti per la Progettazione e la Verifica di Sistemi Eterogenei Connessi mediante Reti di Comunicazione*; and Esprit Working Groups *CONFER2* and *COORDINA*. The second author is on leave from Dipartimento di Informatica, Pisa, Italy.

equipped with flexible synchronization primitives [20, 4]. Advantages of the tile model with respect to standard operational models of coordination languages are generality and formal foundations. Also it enables to compose systems both statically and dynamically, specifying possibly complex synchronization and workflow mechanisms.

In this paper, as a case study, we consider the asynchronous π-calculus [6, 14]. The π-calculus [19] is one of the best studied examples of *mobile* process calculi, namely calculi in which the communication topology among processes can dynamically evolve when computation progresses. In the π-calculus, mobility is achieved via the communication of *names* rather than of *processes*, as happens in so called *higher order* process calculi (e.g. [24]). The π-calculus has been used to model objects [25], higher order communication [23] and also higher order configuration languages [22]. The asynchronous π-calculus is a variant of the π-calculus where emission of messages is non-blocking.

We now shortly introduce the tile model. It is based on certain rewrite rules with side effects, called *tiles*, reminiscent of both SOS rules [21] and rewrite logic rules [18]. Also related are SOS contexts [16]. To match SOS notation as much as possible, here a tile has the form

$$\frac{a}{s \xrightarrow{b} s'}$$

and states that the *initial configuration* s of the system evolves to the *final configuration* s' producing an *effect* b. However the rewrite step is actually possible only if the subcomponents of s also evolve producing the *trigger* a. Both trigger and effect are called *observations*, and model the interaction, during a computation, of the system being described with its environment. More precisely, both system configurations are equipped with an *input* and an *output interface*, and the trigger just describes the evolution of the input interface from its initial to its final configuration. Similarly for the effect. It is convenient to visualize a tile as a two-dimensional structure (see Fig.1), where the horizontal dimension corresponds to the extension of the system, while the vertical dimension corresponds to the extension of the computation. Actually, we should also imagine a third dimension (the thickness of the tile), which models parallelism: configurations, observations, interfaces and tiles themselves are all supposed to consist of several subcomponents in parallel.

Both configurations and observations are assumed to be equipped with operations of parallel and sequential composition (represented by infix operators $\otimes$ and ; respectively) which allow us to build more parallel and larger components, extended horizontally for the configurations and vertically for the observations. Similarly, tiles themselves possess three operations of composition[3]:

---

[3] In general, tiles are also equipped with proof terms which distinguish between sequents with the same configurations and observations, but derived in different ways. Suitable axioms for normalizing proof terms are also provided [11, 12, 13]. Normalized proof terms actually model concurrent computations. Here we consider only *flat* tiles, i.e. without proof terms, since we are only interested in interleaving semantics.

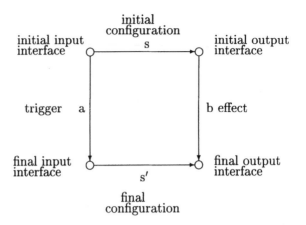

Fig. 1. A tile.

parallel ($\_\otimes\_$), horizontal or *coordinate* ($\_*\_$), and vertical or *sequential* ($\_\cdot\_$) composition. If we consider tiles as logical sequents, it is natural to define the three operations via inference rules called *composition* rules (see Definition 6).

The operations of parallel and sequential composition are self explanatory. More interesting is the operation of coordination: the effect of the first tile acts as trigger of the second tile, and the resulting tile expresses the synchronized behavior of both. Thus horizontal composition models in a sense the dynamic creation of coordinators[4]. Computing in a *tile logic* consists of starting from a set of basic tiles called *rewrite rules* (and from a set of *auxiliary* tiles which depend on the version of the tile model at hand), and of applying the composition rules in Definition 6 in all possible ways.

A tile logic also can be seen as a double category [10] and tiles themselves as double cells. The categorical interpretation [11, 12] is useful since it makes the model more general (configurations and observations can be arrows of any category), allows for universal constructions (e.g. a tile logic is the double category freely generated by its rewrite rules) and suggests analogies with fruitful concepts of algebraic semantics, like institutions. However, the tile model is presented here in a purely logical form.

In this paper, configurations and observations are certain acyclic graphs called *agent graphs*. They are a small extension of term graphs [5], in that they may lack roots. Term graphs are like terms, but two term graphs may explicitly share some of their subterms. Thus in a term graph it is in general not allowed to copy the shared subterms to make the two terms disjoint, since this would yield a different term graph. An axiomatization of term graphs (which can be easily extended to agent graphs) in terms of ps-monoidal theories has been recently proposed by Corradini and Gadducci [8, 9], and it is reported in the Appendix.

---

[4] While tiles can be considered as a generalization of SOS inference rules, their algebraic structure is new. Larsen contexts [16] are analogous, but their algebraic structure is limited to ordinary terms and not axiomatized.

Agent graphs have important advantages over ordinary syntactic terms for representing configurations with shared, possibly bound names. In fact, bound names can be simply modelled as different, shared subterms which consist of the same constant. For instance, the two different CCS agents $(a.nil)\backslash a \mid (\overline{a}.nil)\backslash a$ and $(a.nil \mid \overline{a}.nil)\backslash a$ could be represented by the two term graphs $a_1.nil \mid \overline{a_2}.nil, a_1 := \nu, a_2 := \nu$ and $a.nil \mid \overline{a}.nil, a := \nu$, while, as terms, they would be identified. Another way of expressing this remark is that in agent graphs the handling of bound names (with the associated mechanisms for alpha conversion and for generating new names in mobile calculi) can be fully delegated to the underlying logic, while explicit mechanisms are needed in the case of terms. As we think it is clearly shown by the π-calculus case study in this paper, the combined use of tiles and agent graphs frees mobile calculi from complex name bookkeeping and its unpleasant consequences, like infinitely many names, infinite branching and the need for specialized versions of bisimilarity.

Tiles and agent graphs are also particularly well suited for the coordination approach. In fact the separation between software agents and coordinators can be reflected in the structure of rewrite rules. In this paper we distinguish two kinds of rules: activity rules and coordination rules. Activity rules describe the evolution of software agents, and they do not consider the operational context where agents live. As a consequence, they have no (i.e. empty) trigger, but only effect. Instead, coordination tiles define the behaviour of coordinators: they determine the evolution of software agents inside their operational environment by taking coordinating actions in response to actions of software agents.

As an example of an activity rule for the asynchronous π-calculus, let us consider the Input rule which models a message entering the system.

$$(\text{Input}) \qquad !(e) \xrightarrow{e := input(e', a, b)} M(e', a, b)$$

The same rule is shown in graphical form (wires-and-boxes notation) on the left side of Fig2.

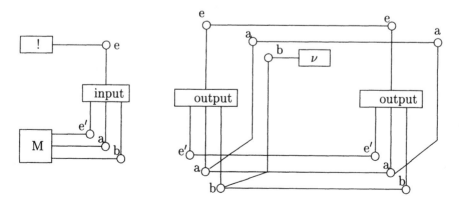

Fig. 2. Graphical representation of the rules Input and Open.

In this rule the trigger is the empty graph. The initial configuration is the agent graph with one variable and no root, which just discharges (creates?) the variable. The effect is the graph $e := input(e'a, b)$ and the final configuration is the graph $M(e', a, b)$. The initial interface of the effect is variable e, which models the sequential dependence of the present rewriting (event) from the previous events. In fact the effect replaces e in the output interface with a new event e'. Messages (resp. input observations) are represented by means of a symbol M (resp $input$) of the signature, with three variables: the current event, the subject and the object.

This rule models the case in which the incoming message has *new* names as subject and object. In fact, the input constructor creates variables a and b. Other (coordination) rules are used to instantiate subject, object or both. Notice that identifiers e, e', a and b we use in the description above have a purely local scope. The only name sharing mechanism between two configurations or tiles is via explicit composition. For instance, the fact that subject and object are new can be determined on the above rule purely by inspection.

As an example of a coordination rule, let us consider the Open rule which models the extrusion of a name while a message is leaving the system.

$$(\text{Open}) \quad \frac{e := input(e', a, b), a, b}{e, a, b := \nu \xrightarrow{e:=input(e',a,b),a} e', a, b}$$

In this rule the precondition (trigger) is $e := output(e', a, b), a, b$, an agent graph with roots $\{e, a, b\}$ and variables $\{e', a, b\}$. The application of the rule causes the transformation of the initial configuration $e, a, b := \nu$ into the final configuration e', a, b (i.e. the restriction is removed) and the creation of effect $e := output(e', a, b), a$. Notice that while the trigger of the rule has as both source and target an event and two actions, the effect has only an event and *one* action as source. Thus the effect creates a new name, which is clearly "the same" name (following the wires) which occurred restricted in the initial configuration.

The paper is organized as follows. Section 2 introduces agent graphs, and the operations of parallel and sequential composition on them. Section 3 presents the tile model, in the simplified version needed in the paper, while Section 4 describes the asynchronous π-calculus. Section 5 defines a tile rewrite system for it, and outlines its equivalence with the ordinary semantics of Section 4.

2 Agent Graphs

In this section we introduce *agent graphs*, on which is based the data stucture we use for modeling configurations and observations. Agent graphs are similar to term graphs [5, 3, 8, 9], have a nice algebraic structure and can be finitely axiomatized essentially as gs-monoidal theories [11, 12, 8, 9]. We follow a style of presentation similar to [3].

Definition 1. *(many-sorted agent graphs)* Let us consider a set K of sorts, a many-sorted signature Σ containing symbols f of type $f : s \to w$ and p of type

$p : \epsilon \to w$, being $s \in K$ and $w, \epsilon \in K^*$, with ϵ the empty string. Furthermore let V be a totally ordered infinite set of sorted *names*, a name being denoted by n and similar letters. An *agent graph* is a triple $G = (S, rt, var)$ where:

- S is a finite set of sentences, which are assignments of the form $n :=$ $f(n_1, \ldots, n_k)$ or of the form $n := n'$, or terms of the form $p(n_1, \ldots, n_k)$, with the obvious restrictions among symbol types and name sorts. In addition, every name must be assigned at most once and no cycles (with the obvious meaning) must be present;
- rt is a list, without repetitions, of all the *roots* of G, i.e. the names which appear as left members of assignments of the form $n := n'$. Roots are usually denoted by r or similar letters;
- var is a list, without repetitions, of the *variables* of G, variables being both all the names which are not assigned in S and possibly other names which do not appear in S. Variables are usually denoted by v or similar letters.

The ordering of roots and variables must respect the ordering in V. Furthermore, agent graphs are defined up to isomorphic renaming.

Given an agent graph G, let u (resp. w) be the string of sorts of the roots (resp. variables). Then G can be seen as an arrow of *type* $u \to w$. We write $G : u \to w$ and call u and w the source and target of G respectively. □

We need a concise term-like notation for representing agent graphs. Given an agent graph $G = (S, rt, var)$, we actually replace all names which are left members of an assignment and which also occur exactly another time in S. For instance, given the set $K = \{a, b\}$ of sorts, the signature $\Sigma = \{f : a \to b, g : b \to a, h : a \to ba, p : \epsilon \to a\}$ and the typed names $n_1, n_3, n_4, v_3 : a$ and $r_1, n_2, v_1, v_2 : b$, let $G : b \to bba = (S, rt, var)$ with $S = \{r_1 := v_1, p(n_1), n_1 := h(n_2, n_3), n_2 := g(n_3), n_3 := f(v_1), n_4 := f(v_2)\}$, $rt = (r_1)$ and $var = (v_1, v_2, v_3)$. Notice that, given the signature and the type of G, all the names of G are univocally typed. The term-like notation for G is as follows: $G : b \to bba = \{r_1 := v_1, p(h(g(n_3), n_3)), n_3 := f(v_1), n_4 := f(v_2)\}$.

Notice that if we consider as standard the names in the lists of roots and variables, isomorphic renaming is restricted to the names which are neither roots nor variables, in the example only n_3 and n_4. Furthermore, in this case an agent graph is fully specified when its type and its set of sentences are given, since the length of its list of variables can be recovered from its type.

If we restrict agent graphs to signatures with symbols $f : s \to w$ only, we obtain term graphs. Notice that in this case our notation is consistent with the ordinary record notation for terms and substitutions, and coincides with it when no names are left besides roots and variables. If this is not the case, the term corresponding to a term graph can be simply obtained by anyway executing the replacements specified by the assignments. For instance given the term graph $G' : ba \to bba = \{r_1 := v_1, r_2 := h(g(n_1), n_1), n_1 := f(v_1), n_2 := f(v_2)\}$ we obtain the tuple of terms $T : ba \to bba = \{r_1 := v_1, r_2 := h(g(f(v_1)), f(v_1))\}$. Notice however that the same tuple of terms corresponds to different term

graphs. For instance the term graphs
$\{r_1 := v_1, r_2 := h(g(n_1), n_1), n_1 := f(v_1)\}$ and
$\{r_1 := v_1, r_2 := h(g(f(v_1)), f(v_1)), n_1 := f(v_2)\}$ both correspond to the same
term as G'.

Particularly interesting are the following agent graphs which are called *atomic*.
Assume $|u| = k$ and $|w| = i$, then:

generators: for every $f : s \rightarrow w$ and $p : \epsilon \rightarrow w$ in Σ

$$f : s \rightarrow w = \{r_1 := f(v_1, \ldots, v_i)\} \text{ and } p : \epsilon \rightarrow w = \{p(v_1, \ldots, v_i)\};$$

identities: id_u

$$id_u : u \rightarrow u = \{r_1 := v_1, \ldots, r_k := v_k\};$$

permutations: $\rho_{u,w}$

$$\rho_{u,w} : uw \rightarrow wu = \{r_1 := v_{k+1}, \ldots, r_i := v_{k+i}, r_{1+i} := v_1, \ldots, r_{k+i} := v_k\};$$

duplicators: ∇_u

$$\nabla_u : uu \rightarrow u = \{r_1 := v_1, \ldots, r_k := v_k, r_{k+1} := v_1, \ldots, r_{2k} := v_k\};$$

dischargers: $!_u$

$$!_u : \epsilon \rightarrow u = \{\}.$$

We now introduce two operations on agent graphs. The *sequential composition* of two agent graphs is obtained by gluing the list of variables of the first graph with the list of roots of the second, and it is defined only if their types are equal. The *parallel composition* instead is always defined, and it is a sort of disjoint union where root and variable lists are concatenated.

Definition 2. *(sequential and parallel composition of agent graphs)* Given two agent graphs $G_1 : u \rightarrow w = (S_1, rt_1, var_1)$ and $G_2 : w \rightarrow z = (S_2, rt_2, var_2)$, let us take two instances in their isomorphism classes such that $var_1 = rt_2$ and that no other names are shared between G_1 and G_2. Furthermore, let S be the set of clauses in S_2 of the form $r := n$ and let σ be the corresponding name substitution. The *sequential composition* of G_1 and G_2 is the agent graph $G_1; G_2 : u \rightarrow z = (S_1\sigma \cup S_2 \setminus S, rt_1, var_2)$.

Given two agent graphs $G_1 : u_1 \rightarrow w_1 = (S_1, rt_1, var_1)$ and $G_2 : u_2 \rightarrow w_2 = (S_2, rt_2, var_2)$, let us take two instances in their isomorphism classes such that no names are shared between G_1 and G_2. The *parallel composition* of G_1 and G_2 is the agent graph $G_1 : u_1u_2 \rightarrow w_1w_2 = (S_1 \cup S_2, rt_1 \, rt_2, var_1 \, var_2)$. □

The following theorem is an easy extension of a similar result proved in [8, 9] for term graphs.

Theorem 3. (decomposition of agent graphs) *Every agent graph can be obtained by evaluating an expression containing only atomic agents as constants, and sequential and parallel composition as operators.* □

For instance the agent graph G of our previous example can be represented as:

$$G : b \to bba = \{r_1 := v_1; p(h(g(n_3), n_3)), n_3 := f(v_1), n_4 := f(v_2)\} =$$
$$((id_b \otimes (p; h; (g \otimes id_a); \nabla_a; f)); \nabla_b) \otimes (!_a; f) \otimes !_a.$$

The following theorem, which is also an easy generalization of a result by Corradini and Gadducci [8, 9], gives a characterization of agent graphs in terms of *gs-monoidal theories*. A gs-monoidal theory is a logical theory similar to, but weaker then, the algebraic theory of terms and substitutions.

Theorem 4. (characterization of agent graphs) *The agent graphs on the signature Σ are the arrows of the opposite* gs-monoidal theory $\mathbf{GS^{OP}}(\Sigma^{op})$ *generated by Σ^{op} (i.e. Σ with the opposite types for each symbol).* □

We call *agent theory on Σ* (in short $\mathbf{AG}(\Sigma)$) the theory of agent graphs on the signature Σ. In the Appendix we give the finitary axiomatization of gs-monoidal theories presented (in the one-sorted case) in [11, 12, 8, 9].

3 The Tile Model

We now describe the basic features of the *tile model*, in the version where both configurations and observations are agent graphs. Our presentation follows [12], but is made simpler, since the tile sequents we have here (which could be called *flat* sequents) have no associated proof term. However in the following we will call them simply tile sequents.

Definition 5. *(tile sequent, tile rewrite system)* Let Σ_h and Σ_v be two (many sorted) agent signatures, called the *horizontal* and the *vertical* signature respectively, on the same set of sorts K.

A Σ_h-Σ_v *tile sequent* is a quadruple $s \xrightarrow{a}_{b} t$, where $s : u \to w$ and $t : y \to z$ are agent graphs on Σ_h, while $a : u \to y$ and $b : w \to z$ are agent graphs on Σ_v. Graphs s, t, a and b are called respectively the *initial configuration*, the *final configuration*, the *trigger* and the *effect* of the tile. Trigger and effect are called *observations*. Sort strings u, w, y and z are called respectively the *initial input interface*, the *initial output interface*, the *final input interface* and the *final output interface*.

A *tile rewrite system* (TRS) $\mathcal{R}$ is a triple $\langle \Sigma_h, \Sigma_v, R \rangle$, where R is a set of Σ_h-Σ_v sequents called *rewrite rules*. □

A TRS $\mathcal{R}$ can be considered as a logical theory, and new sequents can be derived from it via certain inference rules.

Definition 6. *(tile logic)* Let $\mathcal{R} = \langle \Sigma_h, \Sigma_v, R \rangle$ be a TRS. Then we say that $\mathcal{R}$ *entails* the class $\mathbf{R}$ of the *tile sequents* $s \xrightarrow{a}_{b} t$ obtained by finitely many applications of the following inference rules *basic rules:*

$$(\textit{generators}) \quad \frac{s \xrightarrow[b]{a} t \in R}{s \xrightarrow[b]{a} t \in \mathbf{R}}$$

$$(\textit{h-refl}) \quad \frac{s : u \to w \in \mathbf{AG}(\Sigma_h)}{id_s = s \xrightarrow[id_w]{id_u} s \in \mathbf{R}} \qquad (\textit{v-refl}) \quad \frac{a : u \to w \in \mathbf{AG}(\Sigma_v)}{id_a = id_u \xrightarrow{a} id_w \in \mathbf{R}};$$

composition rules:

$$(\textit{p-comp}) \quad \frac{\alpha = s \xrightarrow[b]{a} t, \alpha' = s' \xrightarrow[b']{a'} t' \in \mathbf{R}}{\alpha \otimes \alpha' = s \otimes s' \xrightarrow[b \otimes b']{a \otimes a'} t \otimes t' \in \mathbf{R}}$$

$$(\textit{h-comp}) \quad \frac{\alpha = s \xrightarrow[c]{a} t, \alpha' = s' \xrightarrow[b]{c} t' \in \mathbf{R}}{\alpha * \alpha' = s; s' \xrightarrow[b]{a} t; t' \in \mathbf{R}}$$

$$(\textit{v-comp}) \quad \frac{\alpha = s \xrightarrow[b]{a} r, \alpha' = r \xrightarrow[b']{a'} t \in \mathbf{R}}{\alpha \cdot \alpha' = s \xrightarrow[b;b']{a;a'} t \in \mathbf{R}};$$

auxiliary rules: (permutations)

$$\rho_{u,w}^{0,0} = \rho_{u,w} \xrightarrow[id_{wu}]{\rho_{u,w}} id_{wu} \in \mathbf{R} \qquad \rho_{u,w}^{0,1} = \rho_{w,u} \xrightarrow[\rho_{u,w}]{id_{wu}} id_{wu} \in \mathbf{R}$$

$$\rho_{u,w}^{1,0} = id_{wu} \xrightarrow[id_{wu}]{\rho_{w,u}} \rho_{u,w} \in \mathbf{R} \qquad \rho_{u,w}^{1,1} = id_{wu} \xrightarrow[\rho_{w,u}]{id_{wu}} \rho_{w,u} \in \mathbf{R}.$$

$\square$

Basic rules provide the sequents corresponding to rewrite rules, together with suitable identity tiles, whose intuitive meaning is that an element of $\mathbf{AG}(\Sigma_h)$ can be rewritten to itself using only trivial trigger and effect. Similarly for $\mathbf{AG}(\Sigma_v)$. Composition rules provide all the possible ways in which sequents can be composed, while auxiliary rules are the counterpart of the atomic permutation graphs discussed above for agent theories.

For instance a tile denoted by $\rho_{u,w}^{0,1}$ consists of a horizontal permutation on the *initial configuration* (notice the character 0 as the *first* upper index) of the tile, and of the inverse permutation on the *effect observation* (notice the character 1 as the *second* upper index). The remaining sides are identities, and similarly for the other permutation tiles.

The role of permutation tiles is to permute the names on one vertex of the tile (the initial output interface in the example), still mantaining the same connections between the adjacent agent graphs. For instance, given any tile $\alpha = s \xrightarrow[b]{a} t$ with s and b having uw as target and source respectively, the composition

$(id_s * \rho_{w,u}^{0,1}) \cdot (\alpha * id_b)$ produces the tile $\alpha' = s; \rho_{u,w} \xrightarrow[\rho_{w,u};b]{a} t$. Here two permutations have been introduced, but the connections between the two involved agent graphs, represented by the composition $s; \rho_{u,w}; \rho_{w,u}; b$, are still the same as $s; b$, since $\rho_{u,w}; \rho_{w,u} = id_{uw}$.

It is easy to see that, by horizontal and vertical composition of the basic permutation tiles, it is possible to obtain permutation tiles $\rho_1 \xrightarrow[\rho_2]{\rho_3} \rho_4$ with arbitrary permutations ρ_1, ρ_2, ρ_3 and ρ_4 on the four sides, provided that $\rho_1; \rho_2 = \rho_3; \rho_4$.

It is straightforward to extend the notion of bisimilarity to deal with our framework.

Definition 7. *(tile bisimilarity)*. Let $\mathcal{R} = \langle \Sigma_h, \Sigma_v, R \rangle$ be a TRS. A symmetric equivalence relation $\equiv_b \subseteq \mathbf{AG}(\Sigma_h) \times \mathbf{AG}(\Sigma_h)$ is a *tile bisimulation* for $\mathcal{R}$ if, whenever $s \equiv_b t$ for generic s, t elements of $\mathbf{AG}(\Sigma_h)$, then for any sequent $\alpha = s \xrightarrow{a}_b s'$ entailed by $\mathcal{R}$ there exists a corresponding one $\beta = t \xrightarrow{a}_b t'$ with $s' \equiv_b t'$. The maximal tile bisimulation equivalence is called *tile bisimilarity*, and denoted by $\sim_t$. $\square$

Notice that this notion of bisimilarity is more general than the ordinary one, since it applies to pairs of system components which are incomplete, while the ordinary notion applies only to closed agents.

4   The Asynchronous π-calculus

The asynchronous π-calculus [14] is a variant of the π-calculus [19] where there is no output prefixing. We assume an infinite set of channel names $a, b, c \ldots$ and, following [1], we distinguish between processes (or agents) $P, Q, \ldots$, sequential processes or guards $G, H, \ldots$ and messages M. The syntax of the calculus is specified by the following grammar:

$$P ::= M \quad \big| \quad G \quad \big| \quad P \mid P \quad \big| \quad \nu a P$$
$$M ::= \overline{a}b$$
$$G ::= 0 \quad \big| \quad a(b).P \quad \big| \quad \tau.P \quad \big| \quad !G.$$

In $a(b).P$ and $(\nu b)P$ the occurrences of b are binding with scope P. We assume the standard notions of free names, bound names and substitution and we denote with $\mathtt{fn}(P)$ (resp. $\mathtt{bn}(P)$) the names free (resp. bound) in P.

Processes are defined up to a *structural congruence* $\equiv$. This is the smallest congruence such that:

1. processes which differ by α-conversion are equivalent;
2. $\mid$ is associative and commutative and 0 is its identity;

3. if $a \notin \mathbf{fn}(P)$ then $P \mid (\nu a Q) \equiv \nu a(P \mid Q)$;
4. $\nu a \, \nu b \, P \equiv \nu b \, \nu a \, P$;
5. $!G \equiv G \mid !G$.

Table 1 illustrates the structural rules of the operational semantics with early instantiation. The actions α are specified as follows

$$\alpha ::= \tau \quad \Big| \quad \bar{a}b \quad \Big| \quad \bar{a}(b) \quad \Big| \quad ab$$

The notions of free and bound names are naturally extended to actions α.

$(\tau)\qquad \tau.P \xrightarrow{\tau} P$ $\qquad\qquad (synch)\ \ \bar{a}c \mid a(b).P \xrightarrow{\tau} P[c/b]$

$(in)\qquad 0 \xrightarrow{ab} \bar{a}b$ $\qquad\qquad (out)\qquad \bar{a}b \xrightarrow{\bar{a}b} 0$

$(open)\ \ \dfrac{P \xrightarrow{\bar{a}b} P' a \neq b}{\nu b P \xrightarrow{\bar{a}(b)} P'}$ $\qquad (\nu)\qquad \dfrac{P \xrightarrow{\alpha} P' a \notin n(\alpha)}{\nu a P \xrightarrow{\alpha} \nu a P'}$

$(comp)\ \ \dfrac{P \xrightarrow{\alpha} P', \mathbf{bn}(\alpha) \cap \mathbf{fn}(Q) = \emptyset}{P \mid Q \xrightarrow{\alpha} P' \mid Q}$ $\qquad (cong)\ \ \dfrac{P \equiv Q \ Q \xrightarrow{\alpha} Q' \ Q' \equiv P'}{P \xrightarrow{\alpha} P'}$

Table 1. Labelled transition system with early instantiation

We now present the bisimilarity semantics for the asynchronous π-calculus.

Definition 8. A symmetric relation S on processes is a bisimulation if PSQ, $P \xrightarrow{\alpha} P'$, and $\mathbf{bn}(\alpha) \cap \mathbf{fn}(Q) = \emptyset$ implies $Q \xrightarrow{\alpha} Q'$ and $P'SQ'$. Let the *asynchronous bisimilarity* $\sim_{HT}$ be the maximal bisimulation.

$\square$

Recently, Amadio, Castellani and Sangiorgi [1] provided several alternative definitions of bisimilarity for the asynchronous π-calculus, which differ in the formulation of the input clause, and proved them all equivalent to the asynchronous bisimilarity $\sim_{HT}$.

Coordination and Names. Sequential processes G are autonomous entities and are at the source of any activity. Hence, sequential processes can be seen as software components possibly distributed over a network. In this perspective, process combinators define the *coordination language*. However, in the π-calculus (both synchronous and asynchronous), syntactic process combinators (i.e. parallel composition and restriction) are not the only language features which are relevant for defining configurations. In fact *free names* are used to this purpose, since they specify the *interfaces* of processes. Therefore the result of the operation of parallel composition heavily depends on which are the homonymous ports of the components (since $\mathtt{fn}(P \mid Q) = \mathtt{fn}(P) \cup \mathtt{fn}(Q)$).

This idea of viewing the free names as the specification of the communication interface leads to a more abstract representation of π-processes, and in particular of sequential processes.

Given any sequential process G, we assume the existence of a function *std* which returns an *abstract* sequential process $std(G)$ such that

$$G \equiv std(G)(a_1, \ldots, a_n)$$

for a suitable ordering of the free names $a_1, \ldots, a_n$ of G. We want *std* to be invariant under substitution (i.e. $std(G) = std(G\sigma)$ for any free name substitution σ) and under structural equivalence (i.e. if $G \equiv G'$ then $std(G) = std(G')$). For instance *std* could be defined by first selecting a standard representative G' of the equivalence class $[G]_{\equiv}$ and then writing $std(G) = \lambda a_1. \ldots \lambda a_n.G'$, where $a_1, \ldots, a_n$ are the free names G' (and of G) listed according to the order of their first occurrence in the *preorder tree walk* of the abstract syntax tree of the sequential process G'.

It is now easy to see that for any process P we have an equivalent process of the form
$P \equiv \nu a_1 \ldots \nu a_k(G_1(\mathbf{a_1}) \mid \ldots \mid G_n(\mathbf{a_n}) \mid \overline{b_1}c_1 \mid \ldots \mid \overline{b_m}c_m)$
where $G_1, \ldots, G_n$ are abstract sequential processes and $\mathbf{a_1}, \ldots, \mathbf{a_n}$ are suitable tuples of names.

5 A Tile Rewrite System for the Asynchronous π-calculus

The aim of this section is to show how the framework provided by the tile model can be applied to study coordination disciplines for distributed mobile systems specified in the asynchronous π-calculus.

In what follows we introduce the components of the tile rewrite system, i.e. sorts, horizontal signature, vertical signature and rewrite rules. While the total number of rules will be infinite, finitely many rules will suffice to describe the evolution of any process.

We will then show the coincidence of our notion of bisimilarity $\sim_t$ with the bisimilarity semantics $\sim_{HT}$ of [14]. To allow for a concise statement of the property, we will constrain the initial configurations to be sequential processes. However any process becomes a sequential process when prefixed by a "start" action.

Sorts. The first sort, e, labels a single name in every agent graph representing a configuration, and this name is shared by all sequential processes and messages. Its role is to sequentialize all the activities, since here we are modelling the interleaving version of the calculus. The symbol we choose, e, is reminiscent of the word *event*, since the only name of sort e can be considered a name generated by the last transition and corresponding to the last event. The second sort, a, is reminiscent of *action*, and the names of this sort will correspond to names of the calculus. Thus, we remind, symbols of a many sorted signature for agent graphs on our two sorts are arrows from ϵ, e or a into strings of e's and a's.

Horizontal Signature. The symbols of Σ_h are as follows.

$$A_G : \epsilon \to e\, a^{|\mathtt{fn}(G)|}\ (Agent)$$

$$M : \epsilon \to e\, a\, a \qquad (Message)$$

$$\nu : a \to \epsilon \qquad\qquad (Restriction)$$

where we have symbols A_G for all abstract sequential processes G.

Vertical Signature. The symbols of Σ_v are as follows.

$$input : e \to e\, a\, a\ (Input)$$

$$output : e \to e\, a\, a\ (Output)$$

$$tau : e \to e \qquad (Tau)$$

Rewrite Rules. To match the SOS notation as closely as possible, from now on a tile $s \xrightarrow[b]{a} t$ will be represented as

$$\frac{a}{s \xrightarrow{b} t}$$

Following the SOS convention, the antecedent (trigger) a will be omitted when it is the empty agent graph.

Since configurations and observations of rules are agent graphs, we will use for them the notation developed in Section 2. We denote a name of type e just as e, rather than as n_e. Moreover we do not want to specify explicitly root and variable names. To this purpose, we add a term $!(n)$ for every name n of the graph which is not a root, but which is not used in any assignment. If n is a variable, we sometimes omit it completely. In this way the set of assignments just specifies a partial ordering, whose maxima (resp. minima) are roots (resp. variables). Moreover, when a variable is assigned to a root, we often use the same name as both root and variable name, and we just write this name instead of the assignment. Also we assume some total ordering of names, but we do not really need to specify it. In fact, due to the presence of permutation tiles in the logic (see Def.6), tiles with all the possible orderings will be generated anyway, no matter what is the ordering of names in rewrite rules, and thus the set of generated tiles will be the same.

There are two kinds of rewrite rules: *activity* rules and *coordination* rules. Activity rules describe the evolution of a sequential process and they depend on the specific structure of the process. Instead, coordination rules coordinate the evolution of sequential processes inside the operational environment.

Activity Rules. We first present the Input and Output rules for messages. They model messages entering and leaving the system.

$$\text{(Input)} \quad !(e) \xrightarrow{e:=input(e',a,a')} M(e',a,a')$$

$$\text{(Output)} \; M(e,a,b) \xrightarrow{e:=output(e',a,b),a,b} !(e'), !(a), !(b)$$

We now introduce the activity rules for sequential processes. We have a rule (without trigger) for every possible τ move of a sequential process, and for every possible synchronization of a sequential process and of a message. We employ two inference rules.

The first inference rule, called Tau, creates a tile for every possible τ move of a sequential process G.

$$G(\mathbf{a}) \xrightarrow{\tau} \nu a_1 \ldots \nu a_k (G_1(\mathbf{a_1}) \mid \ldots \mid G_n(\mathbf{a_n}) \mid \overline{b_1}c_1 \mid \ldots \mid \overline{b_m}c_m)$$

(Tau) implies

$$A_{G}(e,\mathbf{a}) \xrightarrow{e:=tau(e'),\mathbf{a}} \begin{array}{l} A_{G_1}(e',\mathbf{a_1}), \ldots, A_{G_n}(e',\mathbf{a_n}), \\ M(e',b_1,c_1), \ldots, M(e',b_m,c_m), \\ a_1 := \nu, \ldots, a_k := \nu. \end{array}$$

The second activity tile, called Synch, describes the reception of messages.

$$G(\mathbf{a}) \mid \overline{b}c \xrightarrow{\tau} \nu a_1 \ldots \nu a_k (G_1(\mathbf{a_1}) \mid \ldots \mid G_n(\mathbf{a_n}) \mid \overline{b_1}c_1 \mid \ldots \mid \overline{b_m}c_m)$$

(Synch) implies

$$A_{G}(e,\mathbf{a}), M(e,b,c) \xrightarrow{e:=tau(e'),\mathbf{a},b,c} \begin{array}{l} A_{G_1}(e',\mathbf{a_1}), \ldots, A_{G_n}(e',\mathbf{a_n}), \\ M(e',b_1,c_1), \ldots, M(e',b_m,c_m), \\ a_1 := \nu, \ldots, a_k := \nu. \end{array}$$

As an example let us consider the sequential process

$$G(a) = a(b).\nu c(c(d).0 \mid \overline{b}c)$$

and the transition

$$G(a) \mid \overline{a}a' \xrightarrow{\tau} \nu c(c(d).0 \mid \overline{a'}c).$$

The corresponding tile is:

$$A_G(e,a), M(e,a,a') \xrightarrow{e:=tau(e'),a,a'} A_{G'}(e',c), M(e',a',c), c := \nu, !_a(a)$$

with $G'(c) = c(d).0$.

Notice that while the transition and the rule above look similar, they are actually very different, in that the names occurring in the transition are global, while those in the rule are just used in the linear notation for it. Thus if we rename the free names a and a' in the transition, we obtain a different transition, while if we rename them in the rule, provided we keep the same ordering between them, we obtain the same rule. Moreover, if we change the order, we obtain a tile which differs from the previous one only by composition with permutation tiles.

Coordination Rules. We have the following rewrite rules.

(Input Obj. Inst.)
$$\dfrac{e := input(e', a, b)}{e, !(b) \xrightarrow{e:=input(e',a,b),b} e', a, b}$$

(Twin Input)
$$\dfrac{e := input(e', a, b)}{e \xrightarrow{e:=input(e',a,a)} e', a, b := a}$$

(Open)
$$\dfrac{e := output(e', a, b), a, b}{e, a, b := \nu \xrightarrow{e:=output(e',a,b),a} e', a, b}$$

(Event Output Sharing)
$$\dfrac{e := output(e', a, b), e'', a, b}{e, a, b, e'' := e \xrightarrow{e:=output(e',a,b),a,b} e', a, b, e'' := e'}$$

The first rule, Input Object Instantiation, having an *input* trigger, must be synchronized with an Input rule. It links the object of the message created by the Input rule to a name existing in the system. Similarly, the second tile Twin Input identifies subject and object of a created message. Besides those two, we need more rules to cover all the cases: Input Subject Instantiation, Input Object Instantiation II (when the subject is already instantiated), Input Twin Instantiation. We have also a rule Twin Output. We do not show these rules since they are very similar to the first two.

The Open rule has already been commented in the Introduction. The last rule, Event Output Sharing, has an extra event name e'' in the trigger, which is identified with e in the initial and with e' in the final configuration. This rule corresponds directly to the *comp* rule in Table 1. Here we have several other sharing tiles and all of them allow for some trigger to be transmitted unmodified as effect, but sharing some name. More precisely, we have x Output Sharing, where x besides Event can be Subject, Object, Event Twin or Event Extruding, Action Twin or Action Extruding.

Similarly we have x y Input Sharing, where x can be empty, Subject Instantiated, Object Instantiated, Subject/Object Instantiated, Twin and also Twin Instantiated; while y can be Event, Subject or Object. Finally we have Event Tau Sharing. We do not show all these rules.

While there is an infinite number of activity tiles, it is possible to show the following finiteness result.

Proposition 9. (finiteness) *For any given sequential process G, the rewrite rules needed to entail all the tile sequents having $A_{std(G)}$ as initial configuration are finitely many.*

The proof relies on the fact that in every reachable configuration at most those abstract sequential processes can appear, which correspond to syntactic subagents of G. Thus only a finite number of activity tiles can apply.

We now prove the coincidence of our notion of bisimilarity $\sim_t$ with the bisimilarity of [14].

Proposition 10. (tile semantics vs. SOS semantics) *Let* $\{a_1, \ldots, a_m\} = \mathtt{fn}(G_1)$, $\{b_1, \ldots, b_n\} = \mathtt{fn}(G_2)$, $F = \mathtt{fn}(G_1) \cup \mathtt{fn}(G_2)$, $\{c_1, \ldots, c_h\} = F \setminus \mathtt{fn}(G_1)$ *and* $\{d_1, \ldots, d_k\} = F \setminus \mathtt{fn}(G_2)$. *Then*

$$G_1(a_1, \ldots, a_m) \sim_{HT} G_2(b_1, \ldots, b_m)$$
if and only if
$$\{A_{G_1}(e, a_1, \ldots, a_m), !_a(c_1), \ldots, !_a(c_h)\} \sim_t \{A_{G_2}(e, b_1, \ldots, b_n), !_a(d_1), \ldots, !_a(d_k)\}$$

where G_1 *and* G_2 *are abstract sequential processes.* □

We now briefly outline the proof of Proposition 10.

1. We define a "fully typed" version of the transition relation $P \xrightarrow{\alpha} P'$ where P' remembers all the free variables of P and where an extended label α contains all these variables. bisimilarity in the fully typed LTS coincides with bisimilarity in the ordinary LTS.
2. We establish a direct correspondence $\Leftrightarrow$ between processes P and those agent graphs H on Σ_h having type $H : \epsilon \to u$, where u contains only an occurrence of sort e. We have $P \Leftrightarrow H$ iff
   $P \equiv \nu a_1 \ldots \nu a_k (G_1(\mathbf{a_1}) \mid \ldots \mid G_n(\mathbf{a_n}) \mid \overline{b_1}c_1 \mid \ldots \mid \overline{b_m}c_m)$ and
   $H = \{A_{G_1}(e, \mathbf{a_1}), \ldots, A_{G_n}(e, \mathbf{a_n}), M(e, b_1, c_1), \ldots, M(e, b_m, c_m),$
   $a_1 := \nu, \ldots, a_k := \nu\}$.
   We show that $P \Leftrightarrow H$ implies $P\sigma \Leftrightarrow H; \rho$ for every injective substitution σ, where permutation ρ is the projection of σ on the ordered set of names actually present in P. We establish also a similar correspondence $\Updownarrow$ between extended labels α and certain agent graphs O on Σ_v.
3. We show that if $P \Leftrightarrow H$ then we have $P \xrightarrow{\alpha} P'$ in the fully typed LTS iff for every permutation ρ we can entail the tile $H \xrightarrow{O;\rho} H'; \rho$, with $\alpha \Updownarrow O$ and $P' \Leftrightarrow H'$. The two equal permutations ρ are created by the permutation tiles. Permutation tiles are needed in order to "twist the wires" while simulating SOS proofs in the tile model.
4. We show that if $P_1 \Leftrightarrow H_1$ and $P_2 \Leftrightarrow H_2$, then we have that $P_1 \sim_{HT} P_2$ iff $H_1 \sim_t H_2$.

6   Conclusion

In the paper we have shown a version of the tile model particularly well suited for defining coordination models of systems with bound names, extrusion and generation of new names. As a case study, we have also presented a tile-based semantics for the asynchronous π-calculus of Honda and Tokoro and proved it equivalent to the ordinary bisimilarity semantics. An advantage of the tile approach is the full compositionality of the underlying logic, which is able to handle computations of (nonclosed) system components as they were new rewrite rules specifying complex coordinators. Another innovative aspect is related to the use of agent graphs, a simple extension of term graphs, for representing configurations and observations: no global names exist in the system, but just operators for specifying name sharing among contiguous sequential processes, and rules for controlling the behavior accordingly. All the name handling issues, including alpha conversion and new name generation are thus delegated to the underlying logic.

The case study in the paper concerns the asynchronous π-calculus, but, due to our coordination viewpoint, its only constructs relevant for our model are parallel composition, restriction and name sharing. Thus we are confident that the same approach (in particular the notion of abstract sequential process) applies to all the calculi with similar configuration structure.

Since we wanted to adhere as much as possible to the Honda-Tokoro presentation of the asynchronous π-calculus, we used very simple coordination mechanisms. In particular we delegated to activity rules also the synchronization of sequential processes. Thus the observations of our rules always contain at most one constructor (symbol) of the vertical signature. However in general triggers and effects of coordinators can be arbitrarily complicated, and their agent graphs can specify complex causal dependencies (workflows) between tasks to be executed by sequential processes or by other coordinators.

7   Acknowledgments

We would like to thank Andrea Corradini and Fabio Gadducci for their help with the basic definitions of graphs and tiles, and Carolyn Talcott for several interesting discussions on related issues.

References

1. R. Amadio, I. Castellani; D. Sangiorgi, *On Bisimulations for the Asynchronous π-calculus.* CONCUR'96, LNCS, 1996.
2. J. Andreoli, C. Hankin, D. Le Metayer (Eds), *Coordination Programming: Mechanisms, Models and Semantics*, Imperial College Press, 1996.
3. Z.M. Ariola, J.W. Klop, *Equational Term Graph Rewriting*, Fundamenta Informaticae 26, 207–240, 1996.
4. Bruni, R. and Montanari, U., *Zero-Safe Nets, or Transition Synchronization Made Simple*, to appear in Proc. Express'97, Santa Margherita, September 1997.
5. H.P. Barendregt, M.C.J.D. van Eekelrn, J.R.W. Glauert. J.R. Kennaway. M.J. Plasmeijer, M.R. Sleep, Term Graph Reduction, *Proc. PARLE*, Springer LNCS 259, 141–158, 1987.
6. G. Boudol, Asynchrony and the π-calculus (note), *Rapport de Recherche 1702*, INRIA Sophia-Antipolis, May 1992.
7. N. Carriero, D. Gelenter, *Coordination Languages and Their Significance. Communications of the ACM*, 35(2), 97–107, 1992,
8. A. Corradini, F. Gadducci, *An Algebraic Presentation of Term Graphs via Gs-Monoidal Categories*, submitted for publication. Available at http://www.di.unipi.it/ gadducci/papers/aptg.ps, 1997.

9. A. Corradini, F. Gadducci, *A 2-Categorical Presentation of Term Graph Rewriting*, Proc. CTCS'97, Springer LNCS, to appear, 1997.

10. C. Ehresmann, *Catégories Structurées*: I and II, Ann. Éc. Norm. Sup. 80, Paris (1963), 349-426; III, Topo. et Géo. diff. V, Paris (1963).

11. F. Gadducci, *On the Algebraic Approach to Concurrent Term Rewriting*, PhD Thesis, Università di Pisa, Pisa. Technical Report TD-96-02, Department of Computer Science, University of Pisa, 1996.

12. F. Gadducci, U. Montanari, *The Tile Model*, Technical Report TR-96-27, Department of Computer Science, University of Pisa, 1996.

13. F. Gadducci, U. Montanari, *Tiles, Rewriting Rules and CCS*, in Proc. 1st international workshop on Rewriting Logic and Applications, J. Meseguer Ed., ENTCS 4 (1996), pp.1-19.

14. K. Honda, M. Tokoro, *An Object Calculus for Asynchronous Communication*, In: M. Tokoro, O. Nierstrasz, P. Wegner, Eds., Object-Based Concurrent Computing, Springer LNCS 612, 21–51, 1992.

15. G. Kelly, R. Street, *Review of the Elements of 2-categories*, Lecture Notes in Mathematics 420, 75–103, 1974.

16. K.G. Larsen, L. Xinxin, *Compositionality Through an Operational Semantics of Contexts*, in Journal of Logic and Computation, vol.1, n.6, 1991 (conference version in Proc. ICALP'90, Springer-Verlag, LNCS 443, 1990).

17. F. W. Lawvere, *Functorial Semantics of Algebraic Theories*, Proc. National Academy of Science **50**, 1963, pp. 869-872.

18. J. Meseguer, *Conditional Rewriting Logic as a Unified Model of Concurrency*, Theoretical Computer Science **96**, 1992, pp. 73-155.

19. R. Milner, J. Parrow and D. Walker. *A calculus of mobile processes* (parts I and II). *Information and Computation*, 100:1–77, 1992.

20. U. Montanari and F. Rossi, Graph Rewriting and Constraint Solving for Modelling Distributed Systems with Synchronization, in: Paolo Ciancarini and Chris Hankin, Eds., Coordination Languages and Models, LNCS 1061, 1996, pp. 12-27. Full paper submitted for publication.

21. G. Plotkin, *A Structural Approach to Operational Semantics*, Technical Report DAIMI FN-19, Computer Science Department, Aarhus University, 1981.

22. M. Radestock, S. Eisenbach, *Semantics of Higher Order Coordination Languages*, in *Proc. COORDINATION'96*, Springer LNCS 1061, 339-356, 1996.

23. D. Sangiorgi. *Expressing mobility in process algebras: first-order and higher-order paradigms.* PhD Thesis CST-99-93, University of Edinburgh, 1992.

24. B. Thomsen. *Plain Chocs, Acta Informatica*, 30, 1993.

25. D. Walker. *π-calculus semantics for object-oriented programming languages.* In *Proc. TACS'91.* Springer Verlag, 1995.

Appendix: Gs-Monoidal Theories

Here the interested reader may find the axiomatic definition (taken from [11, 12, 8, 9]) of gs-monoidal theories. They are similar to the ordinary algebraic (Lawvere) theories [17], the differences being the extended signature and the missing naturality axioms for duplicators and dischargers. Gs-monoidal theories are monoidal theories, since the naturality axiom of permutations holds instead.

Definition 11. *(graphs)* A *graph* G is a 4-tuple $\langle O_G, A_G, \delta_0, \delta_1 \rangle$: O_G, A_G are sets whose elements are called respectively *objects* and *arrows* (ranged over by $a, b, \ldots$ and $f, g, \ldots$), and $\delta_0, \delta_1 : A_G \to O_G$ are functions, called respectively *source* and *target*. A graph G is *reflexive* if there exists an *identity* function $id : O_G \to A_G$ such that $\delta_0(id(a)) = \delta_1(id(a)) = a$ for all $a \in O_G$; it is *with pairing* if its class O_G of objects forms a monoid; it is *monoidal* if it is reflexive with pairing and also its class of arrows forms a monoid, such that id, δ_0 and δ_1 respect the neutral element and the monoidal operation. □

Definition 12. *(hyper-signatures)* A *hyper-signature* $\Sigma_S = \langle S, \delta \rangle$ is a set of operators S, and a function $\delta = \langle \delta_0, \delta_1 \rangle : S \to \mathbf{N} \times \mathbf{N}$, where $\mathbf{N}$ is the set of underlined natural numbers. □

Definition 13. *(one-sorted gs-monoidal theories)* Given a hyper-signature Σ_S, the associated *gs-monoidal theory* $\mathbf{GS}(\Sigma_S)$ is the monoidal graph with objects the elements of the commutative monoid $(\mathbf{N}, \otimes, \underline{0})$ (where $\underline{0}$ is the neutral object and the sum is defined as $\underline{n} \otimes \underline{m} = \underline{n+m}$); and arrows those generated by the following inference rules:

$$(generators) \quad \frac{f : \underline{n} \to \underline{m} \in S}{f : \underline{n} \to \underline{m} \in \mathbf{GS}(\Sigma_S)} \qquad (sum) \quad \frac{s : \underline{n} \to \underline{m}, t : \underline{n'} \to \underline{m'}}{s \otimes t : \underline{n} \otimes \underline{n'} \to \underline{m} \otimes \underline{m'}}$$

$$(identities) \quad \frac{\underline{n} \in \mathbf{N}}{id_{\underline{n}} : \underline{n} \to \underline{n}} \qquad (composition) \quad \frac{s : \underline{n} \to \underline{m}, t : \underline{m} \to \underline{k}}{s; t : \underline{n} \to \underline{k}}$$

$$(duplicators) \quad \frac{\underline{n} \in \mathbf{N}}{\nabla_{\underline{n}} : \underline{n} \to \underline{n} \otimes \underline{n}} \qquad (dischargers) \quad \frac{\underline{n} \in \mathbf{N}}{!_{\underline{n}} : \underline{n} \to \underline{0}}$$

$$(permutations) \quad \frac{n, m \in \mathbf{N}}{\rho_{\underline{n}, \underline{m}} : \underline{n} \otimes \underline{m} \to \underline{m} \otimes \underline{n}}$$

Moreover, the composition operator ; is associative, and the monoid of arrows satisfies the functoriality axiom

$$(s \otimes t); (s' \otimes t') = (s; s') \otimes (t; t')$$

whenever both sides are defined; the identity axiom $id_{\underline{n}}; s = s = s; id_{\underline{m}}$ for all $s : \underline{n} \to \underline{m}$; the monoidality axioms

$$id_{\underline{n} \otimes \underline{m}} = id_{\underline{n}} \otimes id_{\underline{m}} \qquad \rho_{\underline{n} \otimes \underline{m}, \underline{p}} = (id_{\underline{n}} \otimes \rho_{\underline{m}, \underline{p}}); (\rho_{\underline{n}, \underline{p}} \otimes id_{\underline{m}})$$

$$!_{\underline{n} \otimes \underline{m}} = !_{\underline{n}} \otimes !_{\underline{m}} \qquad \nabla_{\underline{n} \otimes \underline{m}} = (\nabla_{\underline{n}} \otimes \nabla_{\underline{m}}); (id_{\underline{n}} \otimes \rho_{\underline{n}, \underline{m}} \otimes id_{\underline{n}})$$

$$!_{\underline{0}} = \nabla_{\underline{0}} = \rho_{\underline{0}, \underline{0}} = id_{\underline{0}} \qquad \rho_{\underline{0}, \underline{n}} = \rho_{\underline{n}, \underline{0}} = id_{\underline{n}}$$

for all $\underline{n}, \underline{m}, \underline{p} \in \mathbf{N}$; the *coherence* axioms

$$\nabla_{\underline{n}}; (id_{\underline{n}} \otimes \nabla_{\underline{n}}) = \nabla_{\underline{n}}; (\nabla_{\underline{n}} \otimes id_{\underline{n}}) \qquad\qquad \nabla_{\underline{n}}; \rho_{\underline{n}, \underline{n}} = \nabla_{\underline{n}}$$

$$\nabla_{\underline{n}}; (id_{\underline{n}} \otimes !_{\underline{n}}) = id_{\underline{n}} \qquad\qquad \rho_{\underline{n}, \underline{m}}; \rho_{\underline{m}, \underline{n}} = id_{\underline{n}} \otimes id_{\underline{m}}$$

for all $\underline{n}, \underline{m} \in \mathbf{N}$; and the *naturality* axiom

$$(s \otimes t); \rho_{\underline{m}, \underline{q}} = \rho_{\underline{n}, \underline{p}}; (t \otimes s)$$

for all $s : \underline{n} \to \underline{m}, t : \underline{p} \to \underline{q} \in S$. □

Communication Complexity
and Sequential Computation

Juraj Hromkovič[1,*] and Georg Schnitger[2]

[1] Lehrstuhl für Informatik I, RWTH Aachen, 52056 Aachen, Germany
[2] Fachbereich Informatik, Johann Wolfgang Goethe–Universität Frankfurt, Robert Mayer Strasse 11–15, 60054 Frankfurt am Main, Germany

Abstract. The communication complexity of two-party protocols introduced by Abelson and Yao is one of the most intensively studied complexity measures for computing problems. This is a consequence of the relation of communication complexity to many fundamental (mainly parallel) complexity measures. This paper focuses on the relation between communication complexity and the following three complexity measures of sequential computation:
- the size of finite automata,
- the time- and space-complexity measures of Turing machines and
- the time- and space-complexity for data structure problems.

We present a survey of the known relations between communication complexity and these three problem areas and formulate several open problems for further research.

Keywords: computational and structural complexity, Las Vegas, determinism, communication complexity, automata.

1 Introduction

The communication complexity of two-party protocols has been introduced by Abelson [1] and Yao [43] in 1978-1979. The initial goal was to develop a method for proving lower bounds on the complexity of distributed and parallel computations, with a special emphasis on VLSI computations.

In the last 15 years the study of communication complexity has brought much more than was expected in the early eighties. Analogous to Kolmogorov complexity in the theory of sequential computations, communication complexity has been developed as a method for the study of the complexity of concrete computing problems especially (but not only) in parallel information processing. Mainly, it has been applied to prove lower bounds on required computer resources (i.e., time, hardware, memory size, etc.) in order to compute a given task. The following, not exhaustive list shows fundamental complexity measures for which communication complexity has been used to prove lower bounds. This list illustrates the broad applicability of communication complexity [1]:

---

[*] The work of this author has been supported by DFG Project HR 14/3-1.

[1] An overview on relations between two-party communication and the computing models listed above can be found in Hromkovič [15].

- VLSI circuits (see for instance Thompson [38], Lengauer [21], Hromkovič [15], Lipton and Tarjan [25], and Ullman [40])
  - trade-offs of area and time (for two- and three-dimensional VLSI circuits),
  - area complexity,
  - complexity measures of multilective VLSI circuits.
- Boolean circuits
  - depth of Boolean circuits and monotone Boolean circuits [18], [34],
  - combinational complexity of planar Boolean circuits and circuits with sublinear separators [39],
  - area complexity of Boolean circuits,
  - combinational complexity of unbounded fan-in circuits [31], [36],
  - length of Boolean formulae.
- Size of finite automata [14], [9].
- Time- and space complexity of deterministic and nondeterministic Turing machines [6].
- Data structure problems [29], [30], [17].
- Size of linear programs [42].
- Size of branching programs [3].
- Depth of decision trees [11].
- Complexity trade-offs for interconnection networks with different topologies.

Communication complexity has established itself as a subarea of complexity theory due to the developed mathematical machinery to approximately determine the communication complexity of concrete computing problems (see, for instance, Aho, Ullman, Yanakakis [2], Dietzfelbinger, Hromkovič, Schnitger [7], Hromkovič [15], Kushilevitz and Nisan [20], Lovász [26], Nisan and Wigderson [32]). In addition to strong relations to several fundamental complexity measures, communication complexity has contributed to the study of determinism, nondeterminism and randomness in algorithms. For instance, for two-party communication,

- deterministic communication can be bounded by at most twice the product of nondeterministic communication for the language and its complement (Aho, Ullman and Yannakakis [2]). This implies an at most quadratic gap between deterministic and Las Vegas communication.
- There is an exponential gap between
  - determinism and Monte Carlo communication
  - nondeterminism and bounded error probabilism (Kalyanasundaram and Schnitger [16], Razborov [35]).

Thus, the investigation of two-party communication complexity may be divided into three main streams:

(1) The study of the relation between communication complexity and other complexity measures.

(2) The development of mathematical machinery for proving lower bounds on the communication complexity of concrete computing problems.

(3) The study of the computation modes determinism, nondeterminism, and (Las Vegas, Monte Carlo, bounded and unbounded error) randomness as well as further fundamental theoretical properties of the communication complexity measure.

Overviews on communication complexity may be found in Lovász [26], Hromkovič [15], Kushilevitz and Nisan [20], Orlitsky and El Gamal [33] and Wigderson [41]. The text of Kushilevitz and Nisan also discusses the important areas of multiparty communication and communication games that we have omitted in our discussion.

This paper is mainly concerned with directions (1), and (3). We would like to point out three problem areas in sequential computation which can be attacked with two party communication and its variants. The goal is not only to survey the state of the art, but to formulate questions for further research. The three chosen problem areas are as follows:

(1) To estimate the size of minimal deterministic, nondeterministic or probabilistic finite automata and to study the relation between them.

(2) To prove lower bounds on the complexity measures of time and space for Turing machines.

(3) To prove lower bounds on the time and space to solve data structure problems.

The paper is organized as follows. Section 2 introduces the basic definitions. Section 3 discusses communication complexity and finite automata, whereas Section 4 is devoted to Turing machines. Section 5 deals with the complexity of data structure problems.

2   Preliminaries

In this section we state our notation and give some informal definitions. For formal definitions of notions considered see, for instance, [43, 15].

N denotes the set of nonnegative integers. For every language $L \subseteq \{0,1\}^*$ and every $n \in N$, the Boolean function $h_n : \{0,1\}^n \to \{0,1\}$ is defined by $h_n(x) = 1$ iff $x \in L \cap \{0,1\}^n$.

A (two-party) communication model consists of two computers C_I and C_{II} and a protocol to compute a function $f : X \times Y \to \{0,1\}$ in the following way. C_I (with input $x \in X$) and C_{II} (with input $y \in Y$) communicate according to the protocol by exchanging binary messages until one of them knows the result $f(x,y)$. The complexity of the computation on input (x,y) is the sum of the lengths of messages exchanged. A protocol is one-way if, for every input, C_I sends only one message and C_{II} determines the result based on its input and the received message.

Usually f is considered to be a Boolean function of n variables for some $n \in N$ and $X = \{0,1\}^{\lceil n/2 \rceil}$ and $Y = \{0,1\}^{\lfloor n/2 \rfloor}$ are the respective input domains.

The **communication complexity of f, $cc(f)$**, is the complexity of the best protocol for f.

The **one-way communication complexity of f, $cc_1(f)$**, is the complexity of the best one-way protocol for f. The two-party protocol model can be straightforwardly extended to nondeterministic [Las Vegas and Monte Carlo] protocols. Correspondingly, **$ncc(f)$ [$lvcc(f)$ and $mccc(f)$)] denote the nondeterministic [Las Vegas and Monte Carlo] communication complexity of a function f**.

3 Finite Automata and One-Way Uniform Communication Complexity

We want to consider the two following fundamental problems on finite automata and to show how communication complexity can be helpful to provide at least partial solutions.

(i) Estimate the size of a minimal deterministic (nondeterministic, randomized) finite automaton for a given regular language.

(ii) Compare the power of nondeterminism, determinism, and randomization for finite automata with respect to their size.

In what follows we consider the standard models of deterministic and non-deterministic finite automata (DFA, NFA resp.) as 5-tuples (Q, Σ, δ, q_0, F) (for a formal definition see [13, 15]). Additionally we consider selfverifying non-deterministic finite automata (SNFA) and Las Vegas finite automata (LVFA) introduced in [9].

A SNFA can be considered as an NFA whose states are partitioned into three disjoint groups: accepting states, rejecting states, and neutral states. An input word w is accepted (rejected) by an SNFA A if there exists a computation of A on w finishing in an accepting (rejecting) state. Moreover, for no input there exist computations finishing in accepting states and computations finishing in rejecting states and for each input word at least one computation is accepting or rejecting.

A LVFA is a SNFA A which, for every $x \in L(A)$, reaches an accepting state with probability[2] at least $1/2$ and which, for every $x \notin L(A)$, reaches a rejecting state with probability at least $1/2$.

For a finite automaton A, **state(A)** denotes the number of states of A.

$$s(L) = \min\{\text{state}(A) \mid A \text{ is a DFA and } L = L(A)\}$$

is the minimal number of states required to recognize a regular languae L. Analoguously, **$ns(L)$, $svns(L)$, and $lvs(L)$** resp. denote the number of states of a minimal NFA, SNFA, and LVFA resp. for L.

---

[2] The probability of a computation of a LVFA is defined through the transition probabilities of the automaton.

While we have efficient algorithms constructing the minimal DFA for a given regular language L (and so determining $s(L)$), we do not know any efficient procedure estimating $ns(L)$, $svns(L)$ or $lvs(L)$. Moreover, we did not have general techniques for proving reasonable lower bounds on $ns(L)$ ($svns(L)$, $lvs(L)$) for a long time. The first general method which is often succesful in proving lower bounds on $ns(L)$ is based on one-way communication complexity. The approach is based on the simple observation [14] that

$$s(L) \geq 2^{cc_1(h_n(L))-1} \text{ and } ns(L) \geq 2^{ncc_1(h_n(L))-1}$$

for every regular language L and every $n \in N$. This follows because a one-way protocol computing $h_n(L)$ can act as follows. If C_I (C_{II}) has input x_1 (x_2), then C_I submits the binary code of the state $p = \hat{\delta}(q_0, x_1)$ reached after reading x_1 by the minimal deterministic automaton A for L. Then C_{II} accepts if $\hat{\delta}(p, x_2) \in F_2$ (if A accepts x_2 starting from state p). Obviously, this simulation works for all kinds of nondeterministic and randomized protocols and finite automata, too[3].

Unfortunately, this simple method may be weak for some regular languages. A good example is that of regular languages over a one-letter alphabet, for which the one-way communication complexity is zero. Clearly, the trouble is caused by the fact that communication protocols are a nonuniform computing model, whereas finite automata are a uniform computing model. In order to establish a closer relation between one-way communication complexity and finite automata, Ďuriš, Hromkovič, Schnitger and Rolim [9] introduced uniform communication protocols as follows.

Definition 1. Let Σ be an alphabet and let $L \subseteq \Sigma^*$. A **one-way uniform protocol** over Σ is a pair $D = \langle \Phi, \varphi \rangle$, where:

- $\Phi : \Sigma^* \to \{0,1\}^*$ is a function with the prefix freeness property[4], and
- $\varphi : \Sigma^* \times \{0,1\}^* \to \{\bar{0}, \bar{1}\}$ is a function.

We say that $D = \langle \Phi, \varphi \rangle$ accepts L (i.e., $L(D) = L$), if for all $x, y \in \Sigma^*$,

$$\varphi(y, \Phi(x)) = \bar{1} \text{ iff } xy \in L.$$

The **message complexity of D** is

$$\mathbf{mc}(D) = |\{\Phi(x) \mid x \in \Sigma^*\}|.$$

Let L be a regular language over Σ. Then the **message complexity of L** is

$$\mathbf{mc}(L) = \min\{mc(D) \mid D \text{ is a one-way uniform protocol and } L = L(D)\}.$$

The following result shows that deterministic uniform protocols and deterministic finite automata are equivalent if we compare message complexity and the number of states.

---

[3] For more details see [15].
[4] $\Phi(x)$ is no proper prefix of $\Phi(y)$ for any $x, y \in \Sigma^*$.

Theorem 2. *For every regular language L,*

$$s(L) = mc(L).$$

Proof. As is usual in communication complexity theory we represent the computing problem by its communication matrix M_L. The rows (columns) of M_L correspond to the inputs of C_I (C_{II}). For inputs $x \in \Sigma^*$ of C_I and $y \in \Sigma^*$ of C_{II}, the entry $M_L[x,y]$ is one iff $xy \in L$ (and zero otherwise).

In our uniform case, the matrix M_L is an infinite Boolean matrix, and one can easily observe that $mc(L)$ is exactly the number of different rows of M_L[5].

On the other hand $s(L) = i(R_L)$[6], where $i(R_L)$ is the number of equivalence classes of the right invariant equivalence relation R_L on Σ^*. Now, it suffices to observe that two words $x, y \in \Sigma^*$ are in the same equivalence class of R_L iff the rows of M_L corresponding to x and y are identical. Thus, $s(L) = i(R_L)$. $\square$

What happens if we consider the Las Vegas, Monte-Carlo or nondeterministic versions of this relation between automata and one-way communication complexity? The extension of one-way uniform protocols to one-way uniform Las Vegas, Monte-Carlo or nondeterministic computation is straightforward[7]. The fact that the message complexity of these extended protocols provides lower bounds on the size of the corresponding finite automata is obvious as well.

Unfortunaly, the opposite direction does not work and there are already exponential gaps between the minimal number of states of Monte-Carlo one-way automata and uniform Monte-Carlo one-way communication complexity: Consider for an arbitrary language $L \subseteq \{0,1\}^n$ the language

$$K(L) = \{xy \mid |x| = |y| = n \text{ and } (x \neq y \text{ or } x \in L)\}.$$

A uniform Monte-Carlo one-way protocol is capable of recognizing $K(L)$ with polynomially many messages, if the processor with the larger input part determines whether its first n bits (resp. its last n bits) belong to L and if both processors run an efficient inequality check simultaneously. Of course Monte-Carlo one-way automata will in general require exponentially many states. Below we will see however that there is only a quadratic gap for Las Vegas computation.

So, on one side, we have the advantage to have a powerful general method for proving lower bounds on $ns(L)$, but have to face the drawback that this method has no chance to prove reasonable lower bounds for certain regular languages L[8].

Following the considerations above we would like to formulate the following research problems.

---

[5] This is a well-known fact for one-way communication complexity. For a detailed proof see Hromkovič [15].

[6] This is the assertion of the Myhill-Nerode theorem.

[7] For more details see Ďuriš et. al. [9], and Hromkovič [15].

[8] We are not aware of a method of any method which provides better lower bounds on $ns(L)$ than the communication method.

Research problem 3.1 Search for a lower bound method for ns(L) which may be useful even in those cases where uniform nondeterministic communication does not help. □

Research problem 3.2 Search for an efficient approximation algorithm for the estimation of ns(L) [svns(L), lvs(L)] for regular languages L. □

Both problems are of course also of interest for Monte Carlo and bounded error probabilistic automata.

Next, we show how two-party communication can help to study the relations between s(L), ns(L), svns(L) and lvs(L). The regular language $L_k = \{w \in \{0,1\}^* \mid w = u1v \text{ and } |v| = k - 1\}$ provides an example of an exponential gap between s(L) and ns(L). The following result provides the first polynomial relation between determinism and Las Vegas computation for a uniform computing model and also provides a polynomial approximation of the minimal number of states of Las Vegas one-way automata.

Theorem 3. *For every regular language L,*

$$\text{lvs}(L) \geq \sqrt{\text{s}(L)}.$$

Sketch of the proof. In Theorem 2 we have shown that s(L) = mc(L). Since the message complexity **lvmc(L)** of one-way uniform Las Vegas protocols for L provides a lower bound on lvs(L), it is sufficient to prove lvmc(L) $\geq \sqrt{\text{mc}(L)}$. But this is shown in Ďuriš, Hromkovič, Schnitger and Rolim [9]. □

Research problem 3.3 How large is the gap between lvs(L) and the message complexity of one-way uniform Las Vegas protocols for L? Note, that the gap is at most quadratic with theorem 3. □

To prove the asymptotical optimality of the lower bound of Theorem 3 we again consider the language L_k. For every positive integer k, s(L_k) = 2^k and lvs(L_k) $\leq 4 \cdot 2^{\lceil k/2 \rceil} + 1 = 0(\sqrt{\text{s}(L_k)})^9$. One can easily observe that svns(L_k) $\leq 2k + 3$ which proves an exponential gap between self-verifying nondeterministic and Las Vegas automata. The language $\text{INEQ}_n = \{uv \mid u, v \in \{0,1\}^n, u \neq v\}$ on the other hand provides an exponential gap between nondeterminism and self-verifying nondeterminism [9].

The main open questions are related to two-way finite automata. In the two-way case one does not even know whether nondeterminism may be exponentially more powerful than determinism[10].

Research problem 3.4 Is there an exponential gap between nondeterminism and determinism for two-way finite automata? What is the power of Las Vegas and self-verifying nondeterminism for two-way automata[11]? □

---

[9] An LVFA for L_k tosses a fair coin to decide whether the important bit of an input is on an odd or even position.

[10] This is a classical open problem of formal language theory. However expoential lower bounds for sweeping automata are known[28].

[11] Note, that two-sided error Monte Carlo two-way finite automata may even recognize non-regular languages.

4 Turing Machines

The goal of this section is to show that communication complexity may be successfully used to prove lower bounds on time, space, and time-space trade-offs of various Turing machine models. First, we consider the original model of one-head Turing machines (TM)[12]. Communication complexity is tailormade to express the crossing sequence argument introduced by Cobham [5].

Let, for any machine A, $T_A(n)$ [$S_A(n)$] be the time [space] complexity of A, and $L(A)$ be the language accepted by A.

Theorem 4. Let $L \subseteq \{0,1\}^*$. For any nondeterministic TM A such that $L = L(A)$,

$$T_A(n) = \Omega((\mathrm{ncc}(h_n(L)))^2).$$

Proof sketch. Obviously, for every input $w = w_1 w_2 \ldots w_n \in L(A)$, $T_A(w)$ is at least the sum of the lengths of the crossing sequences of the shortest accepting computation B_w of A on x. Let C_i be the i-th crossing sequence of B_w[13] for some i.

Let $c_n = \mathrm{ncc}(h_n(L))$. We simulate the nondeterministic machine A as follows. C_I nondeterministically guesses $i \in \{\lceil n/2 \rceil - \frac{c_n}{2}, \ldots, \lceil n/2 \rceil\}$ and a sequence C of states. If C is a possible crossing sequence for C_I, then C_I submits the binary codes of $\lceil n/2 \rceil - i$ and C as well as $w_{i+1}, \ldots, w_{\lceil n/2 \rceil}$ to C_{II}. Now, C_{II} checks whether C is indeed an i-th crossing sequence and accepts if there is an accepting computation on w with C as the i-th crossing sequence.

Thus we communicate at most

$$\lceil n/2 \rceil - i + q \cdot |C_i| + \log_2(\lceil n/2 \rceil - i)\}$$

bits, where q is the logarithm of the number of states of A. Hence there will be some input such that no crossing sequence C_j may be short for $j \in \{\lceil n/2 \rceil - \frac{c_n}{2}, \ldots, \lceil n/2 \rceil\}$. Using simple combinatorics one obtains the result of the theorem. □

We observe that the proof of Theorem 4 crucially employs nondeterminism in the construction of the protocol. Recently Dietzfelbinger [6] settled the open problem of establishing a deterministic version of Theorem 4:

Theorem 5. Let $L \subseteq \{0,1\}^*$. For every TM A such that $L = L(A)$,

$$T_A(n) = \Omega((\mathrm{cc}(h_n(L)))^2).$$

Idea of the proof. Let d_n be a lower bound on the number of messages of any protocol computing $h_n(L)$ (according to the standard partition of the input into two halves). In particular, any protocol with C_I knowing the first i input bits

---

[12] A Turing machine consists of a finite state control and an infinite tape which is simultaneously the input and working tape.

[13] The i-th crossing sequence of B_w is the sequence of states of A in which A moves its head to position i of the tape in computation B_w.

and C_{II} knowing the remaining $n-i$ bits must use at least $d_n/2^{|n/2-i|}$ messages to compute $h_n(L)$. Therefore, any TM A accepting $L \cap \{0,1\}^n$ has to produce at least $d_n/2^{|n/2-i|}$ different i-th crossing sequences in all its computations on words in $\{0,1\}^n$, i.e., the average length of the i-th crossing sequence over all $w \in \{0,1\}^n$ is at least

$$\log_q(d_n/2^{|n/2-i|}) = \log_q d_n - (\log_q 2) \cdot |n/2 - i|.$$

Dietzfelbinger uses these facts to prove that one can find a word w such that

$$T_A(w) \geq c \cdot \sum_{j=\frac{n}{2}-(\log_q d_n)/2}^{\frac{n}{2}+(\log_q d_n)/2} \log_q d_n$$

for some suitable constant $c > 0$. Note that this fact is far from obvious. □

Next we discuss one-tape on-line Turing machines, i.e, one-tape Turing machines with an additional read-only one-way input tape. Maass [23] and Li, Longpre and Vitanyi [22] give quadratic lower bounds for the language recognition problem via Kolmogorov complexity arguments. [17] compares the power of non-deterministic and probabilistic bounded error one-tape on-line Turing machines and shows an almost quadratic speedup by nondeterminism. The argument there proceeds via communication complexity where the corresponding speedup result for a version of set disjointness ([16], [35]) is utilized. Moreover the lower bound arguments carry over to a large class of functions definable with communication concepts.

One-tape off-line Turing machines (i.e., one-tape Turing machines with an additional read-only two-way input tape) are quite a bit more powerful. For the problem of transposing matrices a lower bound of almost $\Omega(n^{3/2})$ is shown in [8] again via Kolmogorov complexity arguments. The problem of language recognition seems quite difficult and only a lower bound of $\Omega(n \log n)$ [24] is known. The argument of [24] is based on communication complexity.

Research problem 4.1 For some constant $c > 0$ give a lower bound of $\Omega(n^{1+c})$ for a language recognition problem on one-tape off-line Turing machines. □

We now turn to time-space trade-offs. Considering internal configurations instead of states in the proofs of theorems 4 and 5 one can obtain lower bounds on the time-space trade-off of off-line sequential machines[14], a general model of sequential computation. Let $S_A(n)$ denote the space complexity of the machine A.

---

[14] An off-line sequential machine A is a machine consisting of a two-way input tape with one read-only head and an infinite memory consisting of binary cells. In one step A reads one symbol from the input tape and, depending on the symbol read and the content of the memory, modifies a constant number of cells and moves the head to a neighboring position on the input tape.

Theorem 6. *Let $L \subseteq \{0,1\}^*$. For every [nondeterministic] off-line sequential machine A accepting L,*

$$T_A(n) \cdot S_A(n) = \Omega((cc(h_n(L)))^2) \; [\Omega((ncc(h_n(L)))^2)].$$

Of course the major open problem is to provide time-space trade-off's for languages in the model of branching programs. To date non-trivial trade-offs are only known for functions [4].

Research problem 4.2 For some constant $c > 0$ show a lower bound of

$$\text{length} \cdot \log_2 \text{ size} = \Omega(n^{1+c})$$

for branching programs. □

Communication complexity is certainly no universal remedy and the problem of time-space trade-offs for languages may very well be out of scope. On the other hand, communication complexity is appropriate to prove lower bounds on the space complexity of general on-line[15] sequential machine models. The following observation is straightforward.

Observation 7 *For every on-line sequential machine A accepting L*

$$S_A(n) \geq \log_2 mc_1(h_n(L)) - O(1) \geq cc_1(h_n(L)) - O(1).$$

5 Data Structures

In this section we discuss asymmetric communication complexity and its relation to Yao's cell probe model [44]. We conclude by describing a non-standard communication model to investigate a real-time comparison between the machine model of Kolmogorov and Uspenskii (KUM) [19] and the storage modification machines of Sch"onhage (SMM) [37].

Yao's cell probe model is designed to study the time- and space-complexity of static data structure problems:

determine for a data domain D and a set Q of queries the result $f(q,d)$ of query $q \in Q$ relative to the data $d \in D$.

Consider for instance a two-dimensional orthogonal range query problem. We set D to be the set of (say) pixel matrices and take rectangles as queries. For a given pixel matrix d and a given query q we would like to determine whether the rectangle q contains a black pixel of d (i.e., $f(q,d) = 1 \leftrightarrow q$ contains a black pixel of d).

An (s,b,t)-solution for the data structure problem (D,Q,f) is a random access machine algorithm that stores $d \in D$ as a data structure in s memory cells containing at most b bits each. Moreover $f(q,d)$ is to be evaluated by accessing at most t memory cells. Miltersen [29] simulated an (s,b,t)-solution by a communication model, where processor C_I receives the data d and C_{II} receives the query

---

[15] On-line access allows to read the input tape from left to right only.

q. (Observe that for natural data structure problems $\log_2 Q \ll \log_2 D$ and hence we are led to asymmetric communication problems.) The two processors communicate over $2t$ "rounds" of communication where in each round processor C_{II} sends a message of length at most $\log_2 s$ which is "answered" by a message of C_I of length at most b. Thus intuitively in each round the query-processor C_{II} may request one of the s memory cells and the data-processor C_I fullfills this request by sending the b-bit contents. Let us say such a communication protocol is of type $(\log_2 s, b, 2t)$.

Miltersen, Nisan, Safra and Wigdersen [30] observe that the converse of Miltersen's result holds, whenever a constant number of rounds is sufficient:

Theorem 8. *Let $P = (D, Q, f)$ be a data structure problem.*
(a) *If P has a (s, b, t)-solution, then there is a communication protocol of type $(\log_2 s, b, 2t)$ for P [29].*
(b) *If there is a communication protocol for P of type $(\log_2 s, b, O(1))$, then P has a $(s^{O(1)}, b, O(1))$-solution [30].*

In [30] several lower und upper bounds for data structure problems are given. For instance let us return to the two-dimensional orthogonal range query problem where we assume now that the data d corresponds to a subset $d \subseteq \{1, \ldots, 2^n - 1\}^2$ of size at most $2^{(\log_2 n)^2}$. For a query rectangle $q = [1, x] \times [1, y]$ we would like to know whether $d \cap q = \emptyset$. Using Theorem 8(a) query time $\Omega(\sqrt{\log_2 n})$ is shown, provided the data structure has size $s = O(2^{(\log_2 n)^3})$ and the cells contain at most poly(n) bits each [30].

We now turn our attention to an on-line comparison of KUM's and SMM's, where a quite different communication approach is used. We briefly sketch the definition of the respective machine models and refer the reader to [10] for a more detailed discussion. The storage structure of a KUM corresponds to an undirected graph of bounded degree. The KUM possesses a head which in one step moves from its current vertex to a neighbor or modifies the neighborhood of its vertex. Information is retrieved by asking whether two paths originating at the current vertex end in the same vertex. SMM's proceed analogously, but work on a directed graph with bounded fanout and potentially unbounded fanin. SMM's (as well as KUM's) are tailormade to tackle data structure problems. For instance, SMM's can solve the Union-Find problem in almost linear time.

A real-time simulation of KUM's by SMM's is immediate, whereas the SMM *versus* KUM problem, namely the existence of a real-time simulation of SMM's by KUM's remains unresolved for now over fifteen years. In [17] the KUM *versus* SMM problem is formulated as a problem of non-standard two-party communication:

- Functions $f_1, \ldots, f_m : \{0, 1\}^r \times \{0, 1\}^s \to \{0, 1\}^t$ are to be computed.
- At the beginning of the computation processor C_{II} is allowed to see the entire input of its partner, but has to condense this input into relatively few (say $s^{1+\varepsilon}$) bits and can use only the condensed string during the computation.
- The two processors receive a request $i \in \{1, \ldots m\}$ and communicate to compute f_i.

– The communication complexity of a protocol is then defined as the maximal number of exchanged bits (over all inputs and all requests).

This communication problem is tailormade to tackle data structures problems, where a large data structure (represented by r bits) is built in a first phase. A second, far shorter phase is described by s bits and modifies the data structure, whereas the third even shorter phase corresponds to testing the resulting data structure by evaluating one of the functions f_i. Thus, in contrast to Miltersen's approach, this non-standard model partitions the data structure information d and makes the query information q available to both processors.

Of particular interest for the KUM *versus* SMM problem are the parameters $r = n^2 \cdot \log_2 n, s = n$ and $t = \log_2 n$. Imagine that the input $x \in \{0,1\}^r$ of C_I specifies a function $A_x : \{0,1\}^{n^2} \to \{0,1\}^s$ and that the input $y \in \{0,1\}^s$ of C_{II} specifies a function $B_y : \{0,1\}^s \to \{0,1\}$. Then we define, for $1 \le i \le n - t + 1$, $f_i : \{0,1\}^r \times \{0,1\}^s \to \{0,1\}^t$ by

$$f_i(x,y) = (B_y \circ A_x(i), \ldots, B_y \circ A_x(i + t - 1)).$$

In [17] the following connection between non-standard communication and the SMM *versus* KUM problem is shown:

Theorem 9. *Let ε be an arbitrarily small positive constant and assume that C_{II} has to condense its (stolen) information into at most $s^{1+\varepsilon}$ bits. If any nonstandard communication model has to exchange c_ε bits to compute $f_1, \ldots, f_n$, then a real-time simulation of SMM's by KUM's requires a delay of at least*

$$\Omega\left(\frac{c_\varepsilon}{\log_2 n \cdot \log_2 \log_2 n}\right) \quad steps.$$

Observe that KUM's are able to simulate SMM's with logarithmic delay. On the other hand, it is not hard to show that $\Theta(\log_2^2 n)$ bits have to be exchanged, if C_{II} is not allowed to see the input of C_I. Thus at best a delay of $\Omega(\log_2 n / \log_2 \log_2 n)$ can be derived.

Research problem 5.1 Show that $c_\varepsilon = \omega(\log_2 n \cdot \log_2 \log_2 n)$. $\square$

References

1. Abelson, H., Lower bounds on information transfer in distributed computations. *Proc. 19th Annual Symp. on Foundations of Computer Science*, 151-158, 1978.
2. Aho,A.V., Ullman, J.D., Yannakakis, M., On notions of informations transfer in VLSI circuits. *Proc. 15th Annual ACM Symp. on Theory of Computing*, 133-139, 1983.
3. Babai, L., Nisan, N., Szegedy, M., Multiparty protocols and logspace-hard pseudorandom sequences. *Proc. 21st Annual ACM Symp. on Theory of Computing*, 1-11, 1989.
4. Beame, P., A general sequential time-space trade-off for finding unique elements. *Proc. 21st Annual ACM Symp. on Theory of Computing*, 197-203, 1989.

5. Cobham, A., The intrinsic computational difficulty of functions. Proc. *1964 Congress for Logic, Mathematics and Philosophy of Science*, North Holland, 24-30, 1964.

6. Dietzfelbinger, M., The linear-array problem in communication complexity resolved. To appear in *Proc. 29th Annual ACM Symp. on Theory of Computing*, 1997.

7. Dietzfelbinger, M., Hromkovič, J., Schnitger, G., A comparison of two lower bounds methods for communication complexity. *Theo. Comp. Sci.* 168 (1), 39-51,1996.

8. Dietzfelbinger, M., Maass, W., Schnitger, G., The complexity of matrix transposition on one-tape Turing machines. *Theoretical Computer Science*, 82(1), 113-129, 1991.

9. Ďuriš P., Hromkovič, J., Rolim, J., Schnitger, G., Las Vegas versus determinism for one-way communication complexity, finite automata, and polynomial-time computations. *Proc. Symp. on Theoretical Aspects of Computer Science*, Lecture Notes in Computer Science 1200, Springer-Verlag,117-128, 1997.

10. Gurevich, Y., On Kolmogorov machines and related issues. *Bulletin of the European Association for Theoretical Computer Science*, 1988.

11. Gr"oger, H.D., Turan, G., On linear decision trees computing Boolean Functions. *Proc. 18th International Colloquium on Automata, Languages and Programming*, 707-718, 1991.

12. Gr"oger, H.D., Turan, G., A linear lower bound for the size of threshold circuits. *Bulletin of the European Association for Theoretical Computer Science.* 1993.

13. Hopkroft, J.E., Ullman, J.D., *Introduction to Automata Theory, Languages, and Computation.* Addison-Wesley, 1979.

14. Hromkovič, J., Relation between Chomsky hierarchy and communication complexity hierarchy. *Acta Math. Univ. Com.* (48-49), 311-317, 1986.

15. Hromkovič, J.: *Communication Complexity and Parallel Computing.* Springer-Verlag 1997.

16. Kalyanasundaram, B., Schnitger, G., The probabilistic communication complexity of set intersection. *SIAM J. on Discrete Math.* 5(4), 1992.

17. Kalyanasundaram, B., Schnitger, G., Communication complexity and lower bounds for sequential machines. In: *Festschrift zum 60. Geburtstag von Günter Hotz*, Teubner Verlag, 840-849, 1992.

18. Karchmer, M., Wigderson, A., Monotone circuits for connectivity require super-logarithmic depth, *SIAM J. on Disc. Math.*, 718-727, 1990.

19. Kolmogorov, A. N., Uspenskij, V. A., On the definition of an algorithm. *Russian Math. Surveys* 30, 217-245, 1963.

20. Kushilevitz E., Nisan N., *Communication Complexity.* Cambridge University Press 1997.

21. Lengauer, Th., VLSI Theory. In: *Handbook of Theoretical Computer Science, Vol. A, Algorithms and Complexity*, Elsevier, 835-868, 1990.

22. Li, M., Longpre, L., Vitanyi, P.M.B., On the Power of the Queue. Structure in Complexity Theory, *Lecture Notes in Computer Science*, vol. 223, 219-223, 1986.

23. Maass, W., Quadratic lower bounds for deterministic and nondeterministic one-Tape Turing machines. *Proc. 16th Annual ACM Symp. on Theory of Computing*, 401-408, 1984.

24. Maass, W., Schnitger, G., Szemeredi, E., Turan, G., Two tapes are better than one for off-line Turing machines. *Computational Complexity* 3, 392-401, 1993.

25. Lipton, R. J., Tarjan, R. E., Applications of a planar separator theorem. *SIAM J. Computing* 9, 615-627, 1980.

26. Lovász, L., Communication Complexity: A Survey. In: *Paths, Flow and VLSI Layout* (Korte, Lovász, Pr"omel, and Schrijver, eds.), Springer-Verlag, 235-266, Berlin 1990.

27. Mehlhorn, K., Schmidt, E., Las Vegas is better than determinism in VLSI and distributed computing. *Proc. 14th Annual ACM Symp. on Theory of Computing*, 330-337, 1982.

28. Micali, S., Two-way deterministic finite automata are exponentially more succinct than sweeping automata. *Information Processing Letters* 12(2), 103-105, 1981.

29. Miltersen, P.B., Lower bounds for union-split-find related related problems on random access machines. *Proc. 26th Annual ACM Symp. on Theory of Computing*, 625-634, 1994.

30. Miltersen, P.B., Nisan, N., Safra, S., Wigderson, A., On data structures and asymmetric communication complexity. *Proc. 27th Annual ACM Symp. on Theory of Computing*, 103-111, 1995.

31. Nisan, N., The communication complexity of threshold gates. *Technical Report*, Dept. of Comp. Sci., Hebrew Univ., 1994.

32. Nisan, N., Wigderson, A.: On rank versus communication complexity. *Combinatorica* 15, 557-565, 1995.

33. Orlitsky, A., El Gamal, A., Communication Complexity, In: *Complexity in Information Theory* (Y. Abu-Mostafa, ed.), Springer-Verlag 1988.

34. Raz, R., Wigderson, A., Monotone circuits for matching require linear depth, *J. of the ACM* 39(3), 736-744, 1992.

35. Razborov A.A., On the distributed complexity of disjointness, *Theo. Comput. Sci.* 106(2), 385-390, 1992.

36. Roychowdhury, V.P., Orlitzky, Siu K.Y., Lower bounds on threshold and related circuits via communication complexity. *IEEE Transactions on Information Theory*, 1994.

37. Schönhage, A. A., Storage modification machines. *SIAM J. Computing 9*, 490-508, 1980.

38. Thompson, C.D., A complexity theory for VLSI, Doctoral dissertation. CMU-CS-80-140, Computer Science Department, Carnegie-Mellon University, Pittsburgh, August 1980.

39. Turan, G., On the complexity of planar Boolean circuits. *Computational Complexity*, 1995.

40. Ullman, J. D., *Computational Aspects of VLSI*. Computer Science Press, 1984.

41. Wigderson, A., Information theoretic reasons for computational difficulty or communication complexity for circuit complexity. *Proc. of the International Congress of Mathematicians*, 1537-1548, 1990.

42. Yannakakis, M., Expressing combinatorial optimization problems by linear programs, *J. of Comput. Syst. Sci.* 43(3), 441-466, 1991.

43. Yao, A. C., Some complexity questions related to distributive computing. *Proc. 11th Annual ACM Symp. on Theory of Computing*, 209-213, 1979.

44. Yao, A.C., Should tables be sorted? *J. of ACM* 28, 615-628, 1981.

Lower Bounds for a Proof System with an Exponential Speed-up over Constant-Depth Frege Systems and over Polynomial Calculus

Jan Krajíček*

Mathematical Institute, Oxford**

Abstract. We prove lower bounds for a proof system having exponential speed-up over both polynomial calculus and constant-depth Frege systems in DeMorgan language.

Introduction

An interesting open problem is to prove a lower bound for constant-depth subsystems of $F(MOD_p)$ (cf. [5, Definition 12.6.1] or [3]), a system that combines boolean reasoning (DeMorgan language) with algebraic reasoning (counting modulo p). In this note we show that the lower bound for the degree of Nullstellensatz proofs for the onto pigeonhole principle PHP_n^{n+m} from [2], as well as the bound for the polynomial calculus proofs of the counting principles $Count_q^n$ from [8], imply lower bounds for a weak subsystem of $F(MOD_p)$ that has nevertheless an exponential speed-up over constant-depth Frege systems in DeMorgan language and over polynomial calculus.

Namely, the proofs in the system, call it $F_d^c(MOD_p)$ here, are $F(MOD_p)$ - proofs that may use only formulas that can be obtained by substituting DeMorgan formulas of depth at most d into formulas in the language

$$\{0, 1, \neg, \vee, \wedge, MOD_{p,0}, \ldots, MOD_{p,p-1}\}$$

$\vee$ and $\wedge$ binary and $MOD_{p,i}$ of arbitrary arity, of *logical depth* at most c, where d and c are independent constants.

The speed-ups are witnessed by $Count_p^n$ (see [5, Thms. 12.5.4 and 12.6.2]) and $PHP_n^{n^2}$ (see [13, Thm 3.1] and [5, Thm. 12.1.2]) respectively, and they are achieved already by tree-like $F_d^c(MOD_p)$-proofs.

This note complements my lecture at this meeting. Most of the material covered there can be found in literature (a good starting point is either [5], or one of [6, 7, 11, 12], or introductions to [1, 3, 13]) except the theorem and its proof given here.

---

* Partially supported by cooperative research grant INT-9600919/ME-103 from the NSF (USA) and the MŠMT (Czech republic) and by the grant #A1019602 of the Academy of Sciences of the Czech Republic.

** On leave from the Mathematical Institute of the Academy of Sciences at Prague.

The paper is self-contained in giving all details but presupposes familiarity with the forcing method in propositional proof complexity as formalized by the notion of k-evaluations, in particular with the material covered in [2] or [5, Secs.12.4-7.]. Using these known results and terminology the proof is rather concise.

1 The lower bound

The formulas PHP_n^{n+m} and $Count_q^n$ are defined in [5, Def. 12.5.1]. In particular, PHP_n^{n+m} is the onto-version

$$\bigvee_i \bigwedge_j \neg x_{ij} \lor \bigvee_j \bigwedge_i \neg x_{ij} \lor$$

$$\bigvee_{i_1 \neq i_2} \bigvee_j (x_{i_1 j} \land x_{i_2 j}) \lor \bigvee_i \bigvee_{j_1 \neq j_2} (x_{ij_1} \land x_{ij_2})$$

with i, i_1, i_2 ranging over $n + m$ and j, j_1, j_2 ranging over n.

Theorem 1. *Let c, d and a prime p be fixed, and let q be a number not divisible by p. Then*

1. *There is $\delta > 0$ such that for all n large enough it holds: There is $m \leq n$ such that in every tree - like $F_d^c(MOD_p)$ - proof of PHP_n^{n+m} at least $\exp(n^\delta)$ different formulas must occur.*
2. *For every k the following holds: For all n large enough and not divisible by q, in every $F_d^c(MOD_p)$ - proof of $Count_q^n$ at least n^k different formulas must occur.*

Proof. We shall use without further explanation the notation and the results about k-evaluations as given in [5, Secs. 12.4.-12.7] (they are presented uniformly for PHP and $Count_q$) and in [2]. In particular, matching decision trees of height at most k of [2, 10] are particular k-complete systems of [9, 5], and sets denoted S_ω, H_ω in [5] are $Br(T_\omega)$, $Br_1(T_\omega)$ of [2, Defs. 3.1 and 4.1].

We prove the first statement only. The proof of the second statement is completely analogous replacing Nullstellensatz provability from $\neg PHP_{n'}^{n'+m}$ by provability in polynomial calculus from the (N, q)-system (see [1, 3]) and the bound from [2, Thm. 12] with [8, Cor. 3.3]. Assume for the simplicity that modus ponens is the only inference rule of the proof system. We consider Nullstellensatz proofs over $\mathbf{F}_p$.

Let $\pi = \sigma_1, \ldots, \sigma_\ell$ be a tree-like $F_d^c(MOD_p)$ - proof of PHP_n^{n+m}, where n is large enough and $m = p^\ell$ for suitable ℓ (to be chosen later). We are going to show that if the number of different formulas in π is less than 2^{n^δ}, $\delta > 0$ a rational depending only on d, then for some n' there is Nullstellensatz refutation of

$\neg PHP_{n'}^{n'+m}$ of degree less than $2^\ell - 1$. By $\neg PHP_{n'}^{n'+m}$ we denote the polynomial system denoted $onto - \mathcal{PHP}_{n'}^{n'+m}$ in [2], i.e.,

$$1 - \sum_j x_{ij} = 0 \text{ for each } i \tag{1}$$

$$1 - \sum_i x_{ij} = 0 \text{ for each } j \tag{2}$$

$$x_{i_1 j} x_{i_2 j} = 0 \text{ for each } i_1 \neq i_2, j \tag{3}$$

$$x_{i j_2} x_{i j_2} = 0 \text{ for each } i, j_1 \neq j_2 \tag{4}$$

with i, i_1, i_2 ranging over $n' + m$ and j, j_1, j_2 ranging over n'. That will contradict the bound [2, Thm. 12].

Note that it is more convenient to work over the ring $R_{n'} := \mathbf{F}_p[\bar{x}]/I_{n'}$, where $I_{n'}$ is the ideal generated by left-hand sides of equations (3) and (4) together with all $x_{ij}^2 - x_{ij}$. The degree of Nullstellensatz refutation of $\neg PHP_{n'}^{n'+m}$ in $\mathbf{F}_p[\bar{x}]$ is the same as of refutations of equations (1) and (2) over $R_{n'}$ (cf. [3, L. 3.1] holding identically for PHP). Monomials in $R_{n'}$ can be naturally identified with partial one-to-one maps from $n' + m$ to n'.

Pick $\frac{1}{5} > \epsilon > 0$ and $\epsilon^{d+1} \frac{\log_p 2}{2} > \delta > 0$ small enough that the hypotheses of [2, L. 6] (corresponds to [5, Thm. 12.4.3]) are fulfilled for $n_i := n^{\epsilon^i}$, $\ell := \lfloor \epsilon^{d+1} \log_p(n) \rfloor$ and $m := p^\ell$. By that lemma there is a k - evaluation of the set of all DeMorgan formulas in π, with $k = O(n^\delta)$, restricted by a partial one-to-one map ρ of size $n - n'$, for $n' = \Omega(n^{\epsilon^d})$.

Assume that $\sigma_i = \phi_i(\psi_{i,1}, \ldots, \psi_{i,j_i})$ where $\psi_{i,j}$ are DeMorgan of depth at most d and ϕ_i are of logical depth at most c.

We give first an intuitive argument motivating the proof ($m = 1$ here). Let n be a non - standard number in some countable model M of true arithmetic. By [5, L. 12.7.3] for any generic set G, (I, f_G) satisfies $\psi_{i,j}^\rho$ iff it satisfies the disjunction $\bigvee H_{i,j}$, where $H_{i,j}$ is the set assigned to $\psi_{i,j}^\rho$ by the k - evaluation. As all partial maps in $H_{i,j}$ are incompatible, the disjunction holds in (I, f_G) iff the equality $h_{i,j} = 1$ holds there, where $h_{i,j} := \sum H_{i,j}$, the sum of monomials corresponding to maps in $H_{i,j}$. Hence, in (I, f_G), $(\phi_i(\bar{\psi}))^\rho$ holds iff $g_i = 1$ where $g_i := \phi_i^*(h_{i,1}, \ldots, h_{i,j_i})$. Here ϕ_i^* is a polynomial of degree at most $(\max(2, p-1))^c \leq p^c$ straightforwardly translating the formula ϕ_i into a polynomial: $0^* := 0$, $1^* := 1$, $(f \wedge g)^* := f^* \cdot g^*$, $(\neg f)^* := 1 - f^*$, $(f \vee g)^* := f + g - fg$ and $(MOD_{p,i}(f_1, \ldots, f_k))^* := (\sum_{i \leq k} f_i)^{p-1}$. As both k -evaluation and the $*$ - translation are sound w.r.t generic extension (i.e., modulo provability from $\neg PHP_{n'}^{n'+1}$) the sequence $g_1, \ldots, g_\ell$ acts as a tree-like semantic refutation of $\neg PHP_{n'}^{n'+1}$ as $g_\ell = 0$ by [5, L. 12.5.2]. This is then turned into a low degree Nullstellensatz refutation using [3, Thms. 2.6 and 5.4].

The proof shall follow these lines replacing validity in generic extensions by provability, and the statements about $PHP_{n'}^{n'+1}$ from [5] by corresponding statements from [2] about $PHP_{n'}^{n'+m}$.

Claim 1. *Every $1 - g_i$ has a Nullstellensatz proof from $\neg PHP_{n'}^{n'+m}$ of degree at most $O(p^c k^2)$.*

Note that as $O(p^c k^2) << 2^\ell - 1$ we get the required contradiction.

We need to show that any axiom translates into a polynomial g such that $1 - g$ has a Nullstellensatz proof and that modus ponens can be simulated by Nullstellensatz such that the degrees of these simulations are small, i.e. proportional to the degrees of polynomials involved. Then we get tree-like polynomial calculus refutation of $\neg PHP_{n'}^{n'+m}$ of size at most 2^k and degree $O(p^c k)$ and applying [3, Thm. 5.4] yields the required Nullstellensatz refutation.

If only the translation $*$ were involved then this is trivial, see also [3]. A problem can arise when the $*$ translation is combined with the translation via k-evaluations. The following claim clarifies this situation. Note that it also explains why it is irrelevant how exactly are σ_i represented in the form $\phi_i(\psi_{i,1}, \ldots, \psi_{i,j_i})$, provided that the total degree is small.

Claim 2. *Let $\alpha = \beta(\gamma_1, \ldots, \gamma_t)$ be a DeMorgan formula and let $h_\alpha := \sum H_\alpha$, $h_{\gamma_i} := \sum H_{\gamma_i}$ be the polynomials assigned to α, γ_i via the k-evaluation. Then $h_\alpha - \beta^*(h_{\gamma_1}, \ldots, h_{\gamma_t})$ has a Nullstellensatz proof from $\neg PHP_{n'}^{n'+m}$ of degree at most $O(k)$.*

Let h_ω denotes $\sum H_\omega$, whenever defined by the k-evaluation. By [2, L. 11] $1 - \sum S_\omega$ has a Nullstellensatz proof from $\neg PHP_{n'}^{n'+m}$ of degree bounded by the height of S_ω, i.e. by $\max(deg(h_\omega, deg(h_{\neg\omega}))$ (in case of $Count_p$ this is replaced by [1, L. 4.3] or [5, L. 12.6.3]). By [2, Def. 4.1] (or [5, Def.12.4.1]) $h_\omega + h_{\neg\omega} = \sum S_\omega$. Thus $h_{\neg\omega} - (1 - h_\omega)$ has degree $\max(deg(h_\omega), deg(h_{\neg\omega}))$ Nullstellensatz proof. As $1 - h_\omega$ is by definition also $(\neg h_\omega)^*$ this shows that the translations behave equally w.r.t. the negation.

It remains to show the same for the disjunction (the conjunction is in both translations defined by DeMorgan rules from the negation and the disjunction). Let $\eta = \omega_1 \vee \omega_2$ We want to show that $h_\eta - (h_{\omega_1} \vee h_{\omega_2})^*$, i.e.

$$h_\eta - (h_{\omega_1} + h_{\omega_2} - h_{\omega_1} h_{\omega_2}) \tag{5}$$

has Nullstellensatz proof of degree at most $O(k)$. Note that monomials in $h_{\omega_1} h_{\omega_2}$ that do not cancel by $I_{n'}$ have degree at most the height of S_η, i.e., $O(k)$. This is because such monomials are of the form $\mu_1 \cup \mu_2$, $\mu_i \in H_{\omega_i}$ compatible, and thus contained in some $\xi \in S_\eta$.

To prove this we need to exploit the definition of k-evaluations. Recall that we identify monomials with partial one-to-one maps.

Claim 3. *Let μ be a monomial and let S be a k-complete system such that for all $\xi \in S$, $\xi \supseteq \mu$ or $\xi \perp \mu$. Then*

$$\mu - \sum_{\xi \in S, \xi \supseteq \mu} \xi$$

has a Nullstellensatz proof of degree bounded by the height of S, i.e., by $O(k)$.

To prove Claim 3 note that $S^\mu = \{\xi \setminus \mu \mid \xi \in S \wedge \xi \supseteq \mu\}$. By [2, L. 1] (or [5, L. 12.3.9]) S^μ is k-complete too (on the universe without the elements of μ) and thus by [2, L. 11] $1 - \sum_{\xi \setminus \mu \in S^\mu} \xi \setminus \mu$ has a Nullstellensatz proof of degree less then the height of S^μ (in case of $Count_p$ one uses here [1, L. 4.3]). Multiplying the proof by monomial μ gives a proof required in Claim 3.

Let us return to (5). Take $H_{\omega_1} \cup H_{\omega_2}$, monomials occurring in one of $h_{\omega_1}, h_{\omega_2}$. By the definition of k-evaluation ([5, Def. 12.4.1]), S_η refines $H_{\omega_1} \cup H_{\omega_2}$ and hence any $\xi \in S_\eta$ either contains some $\mu \in H_{\omega_1} \cup H_{\omega_2}$ or is incompatible to all such μ. If it contains some $\mu_i \in H_{\omega_i}$ ($i = 1, 2$) it contains no other $\mu \in H_{\omega_i}$ (as $\mu, \mu_i \in S_{\omega_i}$ and thus are incompatible). Thus for any $\xi \in H_\eta$ one of the two cases occurs:

1. ξ contains one $\mu_i \in H_{\omega_i}$ and no $\mu \in H_{\omega_{2-i}}$, some $i = 1, 2$
2. ξ contains one $\mu_i \in H_{\omega_i}$ for both $i = 1, 2$

In the former case ξ contains no monomial of $h_{\omega_1} h_{\omega_2}$ (as it would also contain a monomial from $H_{\omega_{2-i}}$). In the later case ξ contains exactly one monomial of $h_{\omega_1} h_{\omega_2}$, namely $\mu_1 \cup \mu_2$. Hence applying Claim 3 for S_η and each monomial in $h_{\omega_1}, h_{\omega_2}$ and $h_{\omega_1} h_{\omega_2}$ yields the Nullstellensatz proof required for (5).

This completes the proof of Claim 2 and of the theorem.

$\square$

Note that exponential lower bounds for sequence-like $F_d^c(MOD_p)$ - proofs of either formula would follow from $n^{\Omega(1)}$ lower bounds for the degree of polynomial calculus refutations of the onto pigeonhole principle and of the counting principles. Unfortunately, neither [8] nor [13] seem to yield such bounds ([13] gives $\frac{n}{2}$ lower bound for the pigeonhole principle but not for the onto-version needed for the transformation from k - evaluations to Nullstellensatz provability as above).

Acknowledgements: It is stated in [4, Sec.8] and again in [13, Sec.1] that T.Pitassi suggested that a lower bound for polynomial calculus proofs of the pigeonhole principle or counting principles would yield a lower bound for constant - depth proofs operating with formulas that are DeMorgan except the top connective that might be the parity. That remark inspired somewhat stronger Theorem 1.

References

1. Beame, P., Impagliazzo, R., Krajíček, J., Pitassi, T., Pudlák, P.: Lower bounds on Hilbert's Nullstellensatz and propositional proofs. Proceedings of the London Mathematical Society **73(3)** (1996) 1–26
2. Beame, P., Riis, S.: More on the relative strength of counting principles. (submitted)
3. Buss, S., Impagliazzo, R., Krajíček, J., Pudlák, P., Razborov, A. A., Sgall, J.: Proof complexity in algebraic systems and bounded depth Frege systems with modular counting. Computational Complexity (to appear)

4. Clegg, M, Edmonds, J., Impagliazzo, R.: Using the Groebner basis algorithm to find proofs of unsatisfiability. In: *Proceedings of the 28th ACM Symposium on Theory of Computing*, ACM (1996) 174–183

5. Krajíček, J.: *Bounded arithmetic, propositional logic, and complexity theory.* Encyclopedia of Mathematics and Its Applications, Vol. **60** Cambridge University Press (1995)

6. Krajíček, J.: A fundamental problem of mathematical logic. Annals of the Kurt Gödel Society, Springer-Verlag, Collegium Logicum, **2** (1995) 56–64

7. Krajíček, J.: On methods for proving lower bounds in propositional logic. In: *Logic and Scientific Methods* Eds. M. L. Dalla Chiara et al., (Vol. 1 of Proc. of the Tenth International Congress of Logic, Methodology and Philosophy of Science, Florence (August 19-25, 1995)), Synthese Library, **259** Kluwer Academic Publ., Dordrecht (1997) 69–83

8. Krajíček, J.: On the degree of ideal membership proofs from uniform families of polynomials over a finite field. (in preparation)

9. Krajíček, J.,Pudlák, P., Woods, A.: Exponential lower bound to the size of bounded depth Frege proofs of the pigeonhole principle. Random Structures and Algorithms **7(1)** (1995) 15–39

10. Pitassi, T., Beame, P., and Impagliazzo, R.: Exponential lower bounds for the pigeonhole principle. Computational Complexity **3(2)** (1993) 97–140

11. Pudlák, P.: The lengths of proofs. In: *Handbook of Proof Theory*, Ed. S. Buss, (to appear)

12. Razborov, A. A.: Lower bounds for propositional proofs and independence results in bounded arithmetic. In: *Proc. of the 23rd ICALP*, F.Meyer auf der Heide and B. Monien eds., LN in Computer Science, **1099**, Springer-Verlag, (1996) 48–62

13. Razborov, A. A.: Lower bounds for the polynomial calculus. (submitted)

Current address:

Mathematical Institute
University of Oxford
24-29 St.Giles'
Oxford, OX1 3LB, U.K.
E.mail: krajicek@maths.ox.ac.uk

Computational Limitations of Stochastic Turing Machines and Arthur-Merlin Games with Small Space Bounds

Maciej Liśkiewicz

Uniwersytet Wrocławski
Instytut Informatyki
51-151 Wrocław, Poland
liskiewi@tcs.uni.wroc.pl

Rüdiger Reischuk

Med. Universität zu Lübeck
Institut für Theoretische Informatik
D-23560 Lübeck, Germany
reischuk@informatik.mu-luebeck.de

Abstract. A *Stochastic Turing machine* (STM) is a Turing machine that can perform nondeterministic and probabilistic moves and alternate between both types. Such devices are also called games against nature, Arthur-Merlin games, or interactive proof systems with public coins. We give an overview on complexity classes defined by STMs with space resources between constant and logarithmic size and constant or sublinear bounds on the number of alternations. New lower space bounds are shown for a specific family of languages by exploiting combinatorial properties. These results imply an infinite hierarchy with respect to the number of alternations of STMs, and nonclosure properties of certain classes.

1 The Computational Model

A stochastic Turing machine (STM) is a nondeterministic machine with the additional ability to perform random moves, also called games against nature by Papadimitriou [Pa85]. Alternative characterizations can be given by so called Arthur-Merlin-games [Ba85, Co89] and interactive proof systems with public coins [GoSi86]. A stochastic Turing machine M models a 2-person game where one of the players determines his moves randomly. In a probabilistic move M chooses among the successor configurations with equal probability. A computation of M on an input X can be described by a computation tree. To define acceptance of X, for each nondeterministic – also called existential – configuration one chooses a successor that maximizes the probability of reaching an accepting leaf. The acceptance probability of X is then given by the acceptance probability of the starting configuration in this truncated tree. In this paper, we will concentrate on machines with bounded error probability. M accepts a language L in space S if for every X it never uses more than $S(|X|)$ space and strings in L are accepted with probability more than $3/4$, while for strings not in L this probability is less than $1/4$. Later, we will also consider machines with space bounds below the twice iterated logarithm. In this case the space bounds can only be maintained with high probability since those functions are no longer space-constructible.

Definition. Let $MA_kSpace(S)$ (resp. $AM_kSpace(S)$) denote the set of languages that can be accepted by an S-space-bounded STM that makes at most $k-1$ alternations between nondeterministic and probabilistic configurations and starts with a nondeterministic (resp. probabilistic) move. For such a machine we also say that it works in space S and k rounds. The corresponding time classes will be denoted by $MA_kTime(T)$ and $AM_kTime(T)$. If there is no bound on the number of alternations we simply write $AMSpace(S)$ and $AMTime(T)$. ☐

The power of polynomial time bounded stochastic machines has been investigated quite extensively using the characterization by Arthur-Merlin-games and interactive proof systems. The complexity classes defined by stochastic machines can equivalently be described with the help of existential and probabilistic quantifiers. For these classes strong connections to standard classes have been obtained. Machines that arbitrarily often may alternate between existential and probabilistic configurations have exactly the same power as alternating Turing machines. Alternating machines work in existential and universal configurations, and one can directly interpret them as games, in which both players try to optimize their moves (existential and universal quantifiers). The corresponding complexity classes are denoted by $ATime(T)$. Thus it holds

$$AM\ Time(\text{POL}) = A\ Time(\text{POL}) = \mathcal{PSPACE},$$

where **POL** denotes the set of polynomial time bounds.

One could therefore say that given polynomial time and an unbounded number of quantifiers, universal quantifiers can be "interchanged" with probabilistic ones. If one however bounds the number of quantifiers the situation is different. For the combination of existential/universal quantifiers it is still unknown whether an interchange of quantifiers or the addition of a quantifier changes the complexity classes. Since it is generally believed that the Polynomial Hierarchy, which can be obtained this way, is strict it has been surprising that on the contrary for existential/probabilistic quantifier expressions any number of quantifiers can be replaced by just two [Ba85], more precisely using **CON** for the union of all constant bounds

$$MA_2Time(\text{POL}) \subseteq AM_2Time(\text{POL}) = AM_{\text{CON}}Time(\text{POL}).$$

2 Sublogarithmic Space Classes: What is Known?

Switching from time bounds to space bounds changes the situation completely. Now, due to Immerman's and Szelépcsenyi's simulation result alternating machines with a constant number of alternations have no more power than nondeterministic machines – at least for space bounds above the logarithm. In other words, existential and universal quantifiers can be interchanged and then merged to a single one:

$$\Sigma_kSpace(S) = \Pi_kSpace(S) = \Sigma_1Space(S) = NSpace(S)$$

for every $k \geq 1$ and $S \geq \log$, where $\Sigma_k Space(S)$ (resp. $\Pi_k Space(S)$) denotes the alternating S -space bounded complexity class, where the machine starts in an existential (resp. universal) state and alternates $k-1$. However, for space bounds below the logarithm this is no longer true.

Definition. Let $\log^{(i)}$ denote the i -times iterated logarithmic function $n \mapsto \log_2 \ldots \log_2 n$. For the twice iterated logarithm instead of $\log^{(2)}$ we will also use the shorter notation **llog** .

$$\text{SUBLOG} \ := \ \Omega(\text{llog}) \cap o(\log)$$

denotes the set of all nontrivial (with resp. to DTM and NTM) sublogarithmic space bounds. □

Several authors, Geffert, von Braunmühl/Gengler/Rettinger, and we have shown that the space hierarchy defined by such alternating machines is infinite [BGR93, Ge94, LiRe93, LiRe96a]:

$$\Sigma_1 Space(\text{SUBLOG}) \ \subset \ \Sigma_2 Space(\text{SUBLOG}) \ \subset \ \Sigma_3 Space(\text{SUBLOG}) \ \subset \ \ldots$$

Furthermore the situation becomes trivial for space bounds below the twice iterated logarithm, since then even alternating machines accept only regular languages [Iw93]. On the other hand, it is easy to see that all these classes lie inside $\mathcal{P}$ since, by definition, they are contained in $\mathcal{AL} := ASpace(\log)$, which has shown to be equal to $\mathcal{P}$ [CKS81]. For a recent overview on alternating sublogarithmic space classes see [LiRe97].

Space-bounded stochastic machines seem to be quite different. This even holds for the subclass of probabilistic machines without any nondeterminism. Freivalds has shown that probabilistic machines with constant space (probabilistic finite automata) are quite powerful already. They can accept nonregular languages [Fr81]. An example is the language

$$\text{COUNT} \ := \ \{1^n 01^m | n = m\}$$

which can be accepted by a probabilistic TM in constant space with an arbitrarily small error probability, that means

$$\text{COUNT} \ \in \ AM_1 Space(\text{CON}) \ = \ BPSpace(\text{CON}) ,$$

where BP stands for bounded error probabilistic machines (Monte Carlo).

Furthermore, probabilistic space classes cannot be separated by standard diagonalization techniques. Only recently, the first nontrivial separation results for such classes have been established, and these apply only to sublogarithmic space classes [DwSt92], [FrKa94].

Dwork and Stockmeyer have investigated stochastic Turing machines and interactive proof systems with small space bounds first. Extending an impossibility result for 2-way probabilistic finite automata they have shown that the language

$$\text{CENTER} \ := \ \{w0x \mid w, x \in \{0,1\}^* \quad \text{and} \quad |w| = |x|\}$$

cannot be recognized by a probabilistic TM even when sublogarithmic space is available [DwSt92]. However, a stochastic automata alternating between probabilistic and nondeterministic states can recognize this language easily, that is

$$\text{CENTER} \in AMSpace(\text{CON}) \setminus AM_1Space(\text{SUBLOG}) .$$

It has been conjectured to be crucial that constant space-bounded STMs use (expected) exponential time for nonregular languages. This has been known for probabilistic machines, and has been shown for stochastic machines with a 1-way input restriction in [CHPW94] only recently, but in general is still open.

For sublogarithmic space bounds it makes a dramatic difference whether the random moves are known, as it is the case for stochastic machines and Arthur-Merlin games, or whether they are hidden as in standard interactive proof systems. Let use denote the corresponding complexity classes for the secret randomness by $IPSpace(S)$. The language PALINDROME of all strings that are palindromes, is an example that can easily be recognized by an interactive proof system with a constant space-bounded verifier and hidden random moves, but requires logarithmic space in case of public coins [DwSt92]:

$$\text{PALINDROME} \in IPSpace(\text{CON}) \setminus AMSpace(\text{SUBLOG}) .$$

Note that for the corresponding time classes public or private coins do not make an essential difference [GoSi86].

By results of Condon/Lipton [CoLi89] and [DwSt92] (for an overview see [Co93]) we know that interactive proof systems even with only constant space – that means the verifier is a probabilistic automata – are already extremely powerful. A precise characterization by standard complexity classes, however, is still open. Let EXL , resp. EEXL denote the class of linear exponential ($2^{O(n)}$), resp. linear double exponential bounds. The best known upper and lower bounds seem to be

$$DTime(\text{EXL}) \subseteq IPSpace(\text{CON}) \subseteq ATime(\text{EEXL}) .$$

Fortunately, in case of stochastic machines with public random moves the gap is not that huge. Condon has shown that such machines with logarithmic space still stay inside $\mathcal{P}$, more precisely [Co89]

$$AM\,Space(\log) = \mathcal{P} = A\,Space(\log) .$$

This means that for logarithmic space – as we have seen above for polynomial time – unbounded sequences of alternating quantifiers, either existential/probabilistic or existential/universal, define the same complexity class. For quantifier sequences of fixed length the situation is not clear.

We will consider this question for sublogarithmic space bounds, where nothing has been known so far. Motivated by the separation results for alternating Turing machines, we will investigate stochastic space complexity classes without any restrictions on the time complexity, but with bounds on the number of alternations, or equivalently interactions between the prover and the verifier.

A sublogarithmic STM may run for exponential expected time. It has been shown in [DwSt92] that some languages cannot be recognized faster – CENTER is one of such example. On the other hand, the power of sublogarithmic space-bounded STMs restricted to polynomial expected time is still an open problem.

Our approach will be to define a sequence of simple languages parameterized by an index k, and to show that in order for a stochastic machine to accept them a certain number of alternations are necessary and sufficient, where the bound depends on k. As a consequence, almost tight separations for the corresponding complexity classes and an infinite hierarchy are obtained. These results and proof methods will be discussed in the following chapters.

3   Lower Bounds for STMs

The results for the languages CENTER and PALINDROME cited above, for any space bound $S \in o(\log)$, yield the separations

$$BPSpace(S) = AM_1Space(S) \subset AMSpace(S) \subset AMSpace(\log) = \mathcal{P} \; .$$

By limiting the number of alternations this chain will be refined.

Definition. Let Σ be an alphabet, and $bin(i)$ denote the binary representation of integer i of length $\max\{1, \lceil \log_2(i+1) \rceil\}$ and

$$BIN(m) \; := \; bin(0)\# bin(1)\# bin(2)\# \ldots \# bin(m) \; .$$

For a natural number k and strings $U, u \in \Sigma^+$ we say that u *occurs k times in* U if there exist $W_0, W_1, \ldots, W_k \in \Sigma^*$ such that $U = W_0 u W_1 u W_2 \ldots W_{k-1} u W_k$, and define the following languages

$$PATTERN_k \; := \; \{ \, U \# u \# BIN(2^d) \mid U, u \in \{0,1,2\}^+, \; d \in \mathbb{N}, \; |u| = d,$$
$$u \text{ occurs } k \text{ times in } U \, \}. \qquad \Box$$

As a first step towards separating the round hierarchy for space-bounded stochastic machines we have shown in [LiRe96b]

Theorem 1. *For $S \in o(\log)$, no S–space-bounded STM starting with a probabilistic move can recognize* PATTERN$_1$ *in 2 rounds, that is*

$$PATTERN_1 \; \notin \; AM_2Space(o(\log)) \; .$$

This result has been extended in [Li96] by showing that for every $k > 1$ the language PATTERN$_{k^{k-1}}$ requires more than k rounds on sublogarithmic STMs. Developing a simpler and more general proof technique we are now able to improve this separation substantially [LiRe96c].

Theorem 2. *For $S \in o(\log)$ and for any integer $k \geq 2$, no S–space-bounded STM can recognize* PATTERN$_k$ *in $\lceil 2k/3 \rceil$ rounds, that is*

$$PATTERN_k \; \notin \; AM_{\lceil 2k/3 \rceil}Space(o(\log)) \; \cup \; MA_{\lceil 2k/3 \rceil}Space(o(\log)) \; .$$

These impossibility results for space bounds in SUBLOG can be translated to arbitrary small bounds (remember functions in SUBLOG grow at least as the twice iterated logarithm). Such a technique will not work for deterministic or nondeterministic machines. Similarly to [FrKa94] for Monte Carlo machines, this can be done by a translation technique based on a suitable encoding.

Definition. For a string $W = w_0 w_1 \ldots w_n$ over Σ and a symbol $a \notin \Sigma$ define

$$E(W, a) := w_0 a^{2^0} w_1 a^{2^1} w_2 a^{2^2} \ldots w_i a^{2^i} \ldots w_n a^{2^n} .$$

Let $\#_1, \#_2, \ldots$ be symbols not in Σ. Then

$$E_0(W) := W , \quad \text{and for } i \geq 1 \quad E_i(W) := E(E_{i-1}(W), \#_i) .$$

For integers k and i a padded version of PATTERN is constructed by

$$\text{PATTERN}_k(i) := \{ U \# u \# E_i(\text{BIN}(2^d)) \mid U, u \in \{0, 1, 2\}^+, \, d \in \mathbb{N}, \, |u| = d,$$
$$u \text{ occurs at least } k \text{ times in } U \}. \qquad \Box$$

Note that $\text{PATTERN}_k(0)$ is just the language PATTERN_k. Combining the lower bound argument and this translation technique one can show

Theorem 3. *For $k \geq 1$ and $i \geq 0$, it holds*

$$\text{PATTERN}_{2k}(i) \notin AM_k Space(o(\log^{(i+1)})) \cup MA_k Space(o(\log^{(i+1)})) .$$

4 An Infinite Hierarchy and Nonclosure under Complement

With a larger number of alternations the languages PATTERN_k can be accepted with little space [Li96].

Theorem 4. *For any integer $k \geq 1$, and for arbitrary small $\epsilon > 0$, there exists a* llog *-space-bounded STM for* PATTERN$_k$ *with error probability ϵ that works in $2k$ rounds starting with a nondeterministic move, that is*

$$\text{PATTERN}_k \in MA_{2k} Space(\text{llog}) .$$

Moreover, the machine works in polynomial time (actually almost linear time).

Therefore, the following separations are obtained:

Theorem 5. *For any integer $k \geq 1$ it holds*

$$MA_{3k+2} SpaceTime(\text{llog}, \text{POL}) \not\subseteq AM_k Space(o(\log)) \cup MA_k Space(o(\log)) .$$

In addition, it holds $\quad MA_2 SpaceTime(\text{llog}, \text{POL}) \not\subseteq AM_2 Space(o(\log)) .$

This implies that the round/alternation hierarchy for sublogarithmic space-bounded AM_k machines is infinite – similar to alternating TMs.

Corollary 6. *For any function* $S \in$ SUBLOG *and any integer* $k \geq 1$, *it holds*

$$AM_k Space(S) \cup MA_k Space(S) \subset MA_{3k+2} Space(S) .$$

Thus we get an infinite hierarchy with a factor 3 distance between different levels. Note that such a round hierarchy does not hold for the corresponding polynomial time classes $AM_k Time$(POL). In [Ba85] Babai has shown that for polynomial time any constant number k of rounds can be reduced to two rounds, that is

$$AM_2 Time(\text{POL}) = \bigcup_k AM_k Time(\text{POL}) .$$

In particular, in contrast to sublogarithmic space the AM_2-class contains the MA_2-class.

One can also obtain strong lower bounds on the space and the number of rounds for recognizing the complement of the language PATTERN$_1$. Since PATTERN$_1$ itself lies in $MA_2 Space$(llog), this shows that for some languages that can be accepted in few rounds and little space even a drastic increase of the number of alternations does not help to recognize their complement with small space.

Theorem 7. *For any function* $R \in O(\sqrt{n})$ *it holds*
$$\overline{\text{PATTERN}}_1 \notin AM_R Space(o(\log)) .$$

Obviously, this result implies the following

Theorem 8. *For any* $S \in$ SUBLOG, *and for any function* R *with* $2 \leq R \leq O(\sqrt{n})$ *the complexity classes* $AM_R Space(S)$ *and* $MA_R Space(S)$ *are not closed under complement.*

For the padded languages one can show a suitable upper bound by combining the recognition algorithm for the unpadded languages with Freivalds' probabilistic finite automaton for the language

$$\{01^{2^0} 01^{2^1} 01^{2^2} \ldots 01^{2^i} 01^{2^{i+1}} \ldots 01^{2^n} \mid n \in \mathbb{N}\} .$$

This yields for any integer $k \geq 1$ and any $i \geq 1$

$$\text{PATTERN}_k(i) \in AM_{2k+1} Space(\log^{(i+2)}) .$$

Thus, not only for space bounds in SUBLOG, stochastic Turing machines possess an infinite round/quantifier hierarchy, but also for arbitrary small nonconstant bounds $\log \ldots \log$ with an exponential gap from $\log^{(i)}$ to $o(\log^{(i+1)})$.

Theorem 9. *For any nonconstant space bound* S *growing suitably smoothly, that means for some* i *it holds* $\Omega(\log^{i+1}) \leq S \leq o(\log^i)$, *the stochastic complexity classes form an infinite hierarchy by*

$$AM_k Space(S) \cup MA_k Space(S) \subset MA_{4k+1} Space(S) .$$

In the remainder of this paper we will present the basic new ideas to obtain these separation results. Finally, we will discuss some open questions concerning space-bounded stochastic Turing machines.

5 Technical Formalism to Analyse STMs

Let us start with some basic notion concerning stochastic Turing machines. A stochastic machine looks like an ordinary Turing machine except that its states are divided into the 4 types probabilistic (or random), nondeterministic (or existential), accepting, and rejecting.

A **memory state** of a STM is an ordered triple $\alpha = (q, u, i)$, where q is a state of the machine, u a string over the work tape alphabet, and i a position in u, the location of the work tape head. By $|\alpha|$ we denote the length of the work tape contents u in α, that is the space complexity of α.

Given an input X, a **configuration** of a machine M is a pair $C = (\alpha, j)$ consisting of a memory state α and a position $j \in [0..|X|+1]$ of the input head. $j = 0$ (resp. $j = |X|+1$) means that the input head scans the left (resp. the right) end-marker of the input tape. Let $\text{head}(C) := j$ denote the input head position in this configuration.

To each configuration $((q, u, i), j)$ we associate a type, which is the type of its state q. Let us assume that a machine starts with its input head on the left end-marker of the input tape and that every accepting or rejecting configuration is terminal.

The computation of a stochastic machine M on an input X can be described by a tree $T_M(X)$ similar to what is done for alternating machines. The root is the initial configuration $C_0(X)$ of M on X, the sons of a node are its successor configurations, and leaves are the terminal configurations. Nodes are labeled as being probabilistic, nondeterministic, etc. according to the type of the configuration. In the following we will consider a fixed STM M.

A **phase** of a computation of M is a linear sequence of consecutive steps. We call a phase Φ a **probabilistic (resp. nondeterministic) round** if M starts the phase in a probabilistic (resp. nondeterministic) configuration and makes only probabilistic (resp. nondeterministic) moves during Φ (note that the last configuration in Φ may be of a different type). For configurations C, D, where C is probabilistic, let $\mathbf{Rmove}_X[C, D]$ denote the probability that with X on its input tape and starting in C the machine reaches the configuration D in a probabilistic round. We are particular interested in configurations D that are maximal with respect to reachability from C in probabilistic rounds. This means that D is not a probabilistic configuration (hence nondeterministic, accepting or rejecting), but the predecessors of D are. For such a D let $\mathbf{Rmove}_X^*[C, D] = \text{Rmove}_X[C, D]$, whereas for non-maximal D $\text{Rmove}_X^*[C, D] := 0$.

For a nondeterministic configuration C let $\mathbf{Nmove}_X(C, D)$ denote the predicate that is true iff starting in C with X the machine can reach D in a nondeterministic round. For the predicate $\mathbf{Nmove}_X^*(C, D)$ to be true we require in addition that D is maximal, that means D is a probabilistic or terminal configuration.

In order to simplify the analysis we will compact a computation tree to its essential parts, which are the steps where the machine alternates between nondeterministic and probabilistic mode.

A "subtree" S of $T_M(X)$ is called a **strategy** of M on X if the following holds:

- the root of S equals the root of $T_M(X)$,
- for a probabilistic node C every son D fulfills $\text{Rmove}_X^*[C, D] > 0$,
- a nondeterministic node C has at most one son D and for it the predicate $\text{Nmove}_X^*(C, D)$ is true.

A strategy S is a **full strategy** if it is a maximal subtree of $T_M(X)$ with respect to these conditions.

Note that S is not a subtree in the strict graph-theoretic sense. A path of the computation tree may be collapsed to an edge, thus a son of a node C in S in general is a descendent of C in the computation tree. Obviously, the depth of S is bounded by k if M works in k rounds.

Let **Reach**$_X[C, S]$ denote the probability that the node C is reached by a path starting at the root. Formally, $\text{Reach}_X[C_0(X), S] := 1$. For any other node D with father C,

$$\text{Reach}_X[D, S] := \text{Reach}_X[C, S] \cdot \text{Rmove}_X^*[C, D] \quad \text{if } C \text{ is probabilistic,}$$
$$\text{Reach}_X[D, S] := \text{Reach}_X[C, S] \quad \text{if } C \text{ is nondeterministic.}$$

Finally, define **Accept**$_X[S]$ as the probability that the strategy S for input X is accepting, i.e.

$$\text{Accept}_X[S] := \sum_{D \text{ accepting}} \text{Reach}_X[D, S].$$

We say that M accepts X with probability

$$p[X] := \max_S \{\text{Accept}_X[S] \mid S \text{ is a full strategy on } X\}.$$

The following terminology is basically taken from [GrWe86, DwSt92].

Let Z be a string. A **probabilistic starting event** is a pair $\langle \alpha, h \rangle$, where α is a probabilistic memory state and $h \in \{1, |Z|\}$. The interpretation is that M starts in memory state α according to the value of h on the leftmost, resp. on the rightmost symbol of Z. A **stopping event** is either

1. a pair $\langle \alpha, h \rangle$ as above meaning that the input head crosses the left (resp. right) boundary of Z with M being in memory state α, or
2. "Accept" meaning that M halts in the accepting state, or
3. "Reject" meaning that M halts in the rejecting state, or
4. "Alter" meaning that within Z the machine alternates from a probabilistic to a nondeterministic round, or
5. "R-Loop" meaning that a probabilistic round loops forever within Z.

For each starting event σ and each stopping event τ, let $p(Z, \sigma, \tau)$ be the probability that M started on Z in σ reaches the event τ within a probabilistic round, in which the input head does not leave Z (except possibly in τ itself). These numbers are called **word probabilities** of Z.

Let s be a positive integer and consider all pairs (σ, τ) of starting and stopping events that are generated by memory states α of space complexity at most s in some fixed arbitrary ordering. Let $\mathtt{WP}_s(Z)$ be the vector of all word probabilities of these pairs.

An s-space-bounded computation within a probabilistic round of M can be modeled by a Markov chain with a finite state space, say $\sigma_1, \ldots, \sigma_m$ for some number $m \le \exp O(s)$. The Markov chain is completely defined by the transition matrix $R = \{r_{ij}\}_{1 \le i,j \le m}$, where r_{ij} denotes the probability that a single step moves from state σ_i to state σ_j. Let σ_1 be the designated starting state. There is also a set T_R of **trapping states** σ_t with the property $r_{tt} = 1$. For $\sigma_t \in T_R$, let $p^*[t, R]$ denote the probability that the Markov chain starting in σ_1 is trapped in state σ_t.

With respect to nondeterministic rounds a word Z is characterized by **word transitions**. A nondeterministic starting event is a pair $\langle \alpha, h \rangle$ similar to above, but now α has to be a nondeterministic memory state. A stopping event is an event from above of type 1., 2. or 3. For a starting event σ and a stopping event τ, define the predicate $t(Z, \sigma, \tau)$ to be true iff M starting on Z in σ can reach τ in a nondeterministic round without leaving Z.

Let $\mathtt{WT}_s(Z)$ denote the vector of all word transitions involving memory states of space complexity at most s, similar to the vector of word probabilities.

6 Exploiting the Limitations of Stochastic Machines with Small Memory

In this section we will present the main line of reasoning to prove the lower bounds. For this purpose, methods developed in [GrWe86, DwSt92] and [LiRe96a] will be applied and extended to sublogarithmic space-bounded stochastic machines. For a real number $\beta \ge 1$ two numbers r and r' are called β-**close** if either

    (i) $r = r' = 0$, or
    (ii) $r > 0$, $r' > 0$, and $\beta^{-1} \le r/r' \le \beta$.

Two word probabilities $\mathtt{WP}_s(w)$ and $\mathtt{WP}_s(\bar{w})$ are β-close if the vectors are componentwise β-close. Similarly, two Markov chains $R = \{r_{ij}\}_{1 \le i,j \le m}$ and $R' = \{r'_{ij}\}_{1 \le i,j \le m}$ with identical state spaces are β-close if r_{ij} and r'_{ij} are β-close for all pairs i, j. Dwork/Stockmeyer have shown that close transition probabilities guarantee that the trapping probabilities are also somehow close.

Lemma 10 [GrWe86, DwSt92]. *Let R and R' be two m-state Markov chains which are β-close, and let σ_t be a trapping state of both R and R'. Then $p^*[t, R]$ and $p^*[t, R']$ are β^{2m}-close.*

Suppose M is a STM of space complexity S, where for simplicity we may assume that S is nondecreasing. Let $\mathrm{Vol}(\ell)$ denote the number of possible $S(\ell)$-space bounded memory states of M on inputs of length ℓ. It holds $\mathrm{Vol}(\ell) \le$

$\exp O(S(\ell))$. We restrict all word probabilities, word transitions, and strategies considered through this section to this fixed machine.

Below we will consider strings w and $\bar{w}$ of some bounded length that have the same word transitions and almost the same word probabilities, and investigate k-round strategies of M on inputs X of the following periodic structure

$$X \quad := \quad w \; u_1 \; w \; u_2 \; w \ldots u_{b-1} \; w \; u_b \; ,$$

where k and b are positive integers and $u_1, \ldots, u_b$ are arbitrary strings. The key argument will show that for some position i in X one can replace the i-th substring w by $\bar{w}$ such that the probability for a specific k-round strategy to accept the distorted input does not decrease drastically. We can show that the loss is small enough if b exceeds $\frac{3}{2}k$, but further improvements of this bound seem possible. In other words, if the verifier asks too few questions to the prover there is a good chance that the prover can cheat successfully.

Proposition 11. *Let ℓ, k, and b be positive integers with $k \geq b$, and define $s := S(\ell)$. Assume that w and $\bar{w}$ are words with identical word transitions, that is $\mathtt{WT}_s(w) = \mathtt{WT}_s(\bar{w})$, and β-close word probabilities $\mathtt{WP}_s(w)$ and $\mathtt{WP}_s(\bar{w})$ for some $\beta \geq 1$. Furthermore, let $X := w \; u_1 \; w \; u_2 \ldots w \; u_b$ be of length at most ℓ, and assume that this bound on the length also holds if in X the occurrence of w is replaced by $\bar{w}$.*

Then for any k-round strategy S on X there exists an integer $i \in [1..b]$ and a k-round strategy S' such that for the input $X_i := w \; u_1 \; w \ldots u_{i-1} \; \bar{w} \; u_i \ldots w \; u_b$ and for $z := 10 \cdot \mathrm{Vol}(\ell) + 20$ it holds

$$Accept_{X_i}[\, S' \,] \quad \geq \quad \frac{b-k}{b} \cdot \beta^{-2z\lceil k/2\rceil} \cdot Accept_X[\, S \,] \; .$$

This proposition provides the main technical lower bound tool to obtain a sufficient condition for a language not to be recognizable by STMs of low space and round complexity.

Theorem 12. *Let $L \subseteq \Sigma^*$ be a language, and let S and R be nondecreasing integer functions. Suppose there exists a sequence of integers $\ell_1 < \ell_2 < \ldots$ such that the following holds*

- *$m \geq (\log R(\ell_m))^3$ for allmost all m and $m \geq \exp cS(\ell_m)$ for any constant c and all m exceeding a constant m_c that depends on c, and*

- *for a constant c_L and infinitely many m one can find a set $W_m = \{w_1, w_2, \ldots, w_{2^m}\}$ of 2^m strings each of length at most m^{c_L} with the property: for every pair $w \neq \bar{w} \in W_m$ there exists $b \geq \frac{3}{2}R(\ell_m) + 1$ words $u_1, \ldots, u_b \in \Sigma^*$ such that $X := w \; u_1 \; w \ldots w \; u_b \in L$ and for all $i \in [1..b]$ $X_i := w \; u_1 \ldots u_{i-1} \; \bar{w} \; u_i \ldots w \; u_b \notin L$ and all words X, X_i are of length at most ℓ_m.*

Then $L \notin AM_R Space(S) \cup MA_R Space(S)$.

Sketch of proof. Let L and m fulfill the requirements of the theorem. Suppose M is a S-space-bounded STM accepting L in R rounds. A counting argument, similar to the one used in [DwSt92], shows that for $s := S(\ell_m)$ there exist two words w and $\bar{w}$ in W_m such that $\mathtt{WT}_s(w) = \mathtt{WT}_s(\bar{w})$ and that the word probabilities $\mathtt{WP}_s(w)$ and $\mathtt{WP}_s(\bar{w})$ are β-close, where β can be chosen as small as $2^{2^{-\sqrt{m}}}$. For $b \geq \frac{3}{2}R(\ell_m) + 1$ select strings $u_1, \dots, u_b$ such that $X = w\, u_1\, w \dots w\, u_b \in L$ and $X_i = w\, u_1 \dots u_{i-1}\, \bar{w}\, u_i \dots w\, u_b \notin L$ for every i. Now apply Proposition 11 with $\ell := \ell_m$, $k := R(\ell_m)$, and b.

Since $X \in L$ the machine M accepts X with probability at least $3/4$, in other words $\mathrm{Accept}_X[\mathcal{S}] \geq 3/4$ for some full k-round strategy $\mathcal{S}$ on X. From Proposition 11, however, we conclude that there exists an integer $i \in [1..b]$ and a k-round strategy $\mathcal{S}'$ such that

$$\mathrm{Accept}_{X_i}[\mathcal{S}'] \geq \frac{b-k}{b} \cdot \beta^{-2z\lceil k/2 \rceil} \cdot \frac{3}{4} \geq \left(\frac{1}{3} + \frac{2}{3R(\ell_m)} \right) \cdot 2^{-2^{-\sqrt{m}} \cdot O(m \cdot R(\ell_m))} \cdot \frac{3}{4} > \frac{1}{4}$$

for m large enough. But, by assumption, the input X_i does not belong to L – a contradiction. ∎

7 The Upper Bound for the Pattern Languages

It is easy to check that for any nondecreasing function $S \in o(\log)$ and every constant $k \geq 1$, the languages $\mathtt{PATTERN}_k$ fulfill the assumptions of theorem 12 if we choose $R(\ell) := \lceil 2k/3 \rceil$. Hence theorem 2 holds. It is also easy to verify the assumtions of theorem 12 for $S \in o(\log)$, $R \in O(\sqrt{\ })$ and the complement of the language $\mathtt{PATTERN}_1$. Therefore, theorem 7 holds, too.

On the other hand, to accept $\mathtt{PATTERN}_k$ according to theorem 4, one can apply the following stochastic algorithm:
First, check deterministically whether the input is of the form $U\#u\#\mathtt{BIN}(2^d)$, for some words $U, u \in \{0, 1, 2\}^+$, and whether $|u| = d$.
Reject and stop if this condition does not hold.
Otherwise, move the input head to the left end-marker, set $i := 1$ and execute the following loop.

1. Moving the input head to the right, nondeterministically choose a starting position for a substring u_i of length d in U and interpret u_i and u as natural numbers between 0 and $2^d - 1$.
2. Randomly choose a prime $q_i \in [2..d^2]$, compute $r_i := u_i \bmod q_i$ and store both q_i and r_i.
3. If $i = k$ or if the whole string U has been scanned, go to step 4. Otherwise go to the right end of u_i, increase i by 1 and go to step 1.
4. If $i = k$ and for all $j \in [1..k]$ the equality $r_j = u \bmod q_j$ holds then accept, otherwise reject.

The correctness of the procedure can be seen easily. If for every i the strings w_i and u are equal then of course $w_i = u$ mod q_i for any modulus q_i, and the machine accepts as desired. If $w_i \neq u$ for some i then it still could happen that $n_{w_i} = n_u$ mod q_i and M accepts incorrectly in step 3. However, this happens with small probability. Since $|n_{w_i} - n_u| \leq 2^d$ the integer $n_{w_i} - n_u$ has at most d different prime divisors, while M selects q_i uniformly among $\Theta(d^2/\log d)$ many primes. Thus, the probability for choosing a bad prime q_i tends to 0. Obviously, M uses $O(\text{llog})$ space and works in $2k$ rounds.

8 Conclusions and Open Problems

Considering the infinite hierarchy for the sublogarithmic $AM_k Space$ classes, an interesting open problem is whether the distance given by the factor 3 can be reduced. One is tempted to conjecture that a maximal separation

$$AM_k Space(S) \overset{?}{\subset} AM_{k+1} Space(S)$$

holds for any integer k and any sublogarithmic space bound S. Unfortunately, standard translation techniques do not work for these classes.

However, the lower and upper bounds for the languages $\textbf{PATTERN}_k$ can be improved. If we insist on a quite small error probability then the lower bound for the number of rounds can be sharpened to $k - O(1)$. On the other hand, for a simple structured sublanguage of $\textbf{PATTERN}_k$, for which the lower bound still holds, an upper bound of $5/3\, k$ rounds can be obtained, and relaxing the error probability to $1/2 - \epsilon$ this can further be reduced to something like $(1 + \delta)k$ for some small δ that depends on ϵ. These methods, however, do not seem to lead to the ultimate goal k versus $k + 1$.

How looks the round/alternation hierarchy for at least logarithmic space bounds? Using a simple simulation of space-bounded NTMs by one-sided-error probabilistic TMs (see e.g. [Gil77] or the survey paper [Ma95]) one can easily show that

$$\mathcal{NL} = M A_1 Space(\log) \subseteq AM_2 Space(\log) = AM_1 Space(\log) = \mathcal{BPL} \ ,$$

where $\mathcal{BPL}$ denotes the probabilistic bounded error logarithmic space class. This means that the AM_2-class is quite weak in case of space bounds – just opposite to the time bounded case. Is it also true that

$$AM_2 Space(\text{SUBLOG}) \overset{?}{=} AM_1 Space(\text{SUBLOG}) \ ,$$

or at least $\qquad M A_1 Space(\text{SUBLOG}) \overset{?}{\subseteq} AM_1 Space(\text{SUBLOG}) \ ,$

which in the standard notation reads as

$$N Space(\text{SUBLOG}) \overset{?}{\subseteq} BP Space(\text{SUBLOG}) \ .$$

What is the situation for constant space bounds? It has been well known that $MA_1Space(\text{CON}) = NSpace(\text{CON})$ coincides with the class of regular languages. As we have already discussed at the beginning this result does not extend to the probabilistic classes because of Freivalds' algorithm for the language COUNT. However, it requires exponential expected time and this cannot be avoided. Is there a language that for some $k > 1$ separates probabilistic from k-round stochastic automata, that means

$$AM_1Space(\text{CON}) \overset{?}{\subset} AM_kSpace(\text{CON}) .$$

From [DwSt92] we know that CENTER does not belong to $AM_1Space(\text{CON})$ and that there exists a constant space Arthur-Merlin game for this language. It, however, requires exponential expected time and a similar number of alternations. Since it is also shown that the time complexity of an stochastic automata recognizing CENTER cannot be polynomial it seems unlikely that the number of rounds required for this language can be improved to a bounded number.

References

[Ba85] L. Babai, *Trading Group Theory for Randomness*, Proc. 17. ACM Symp. on Theory of Computing, 1985, 421-429.

[BGR93] B. von Braunmühl, R. Gengler, R. Rettinger *The Alternation Hierarchy for Machines with Sublogarithmic Space is Infinite*, Comp. Compl. 3, 1993, 207-230.

[CKS81] A. Chandra, D. Kozen, L. Stockmeyer, *Alternation*, J. ACM 28, 1981, 114-133.

[Co89] A. Condon, *Computational Model of Games*, MIT Press, 1989.

[Co93] A. Condon, *The Complexity of Space Bounded Interactive Proof Systems*, in *Complexity Theory: Current Research*, S. Homer, U. Schöning, K. Ambos-Spies (Eds.), Cambridge Univ. Press, 1993, 147-190.

[CoLi89] A. Condon, R. Lipton, *On the Complexity of Space Bounded Interactive Proofs*, Proc. 30. IEEE Symp. on Found. of Comp. Science, 1989, 462-467.

[CHPW94] A. Condon, L. Hellerstein, S. Pottle, A. Wigderson, *On the Power of Finite Automata with both Nondeterministic and Probabilistic States*, Proc. 26. ACM Symp. on Theory of Computing, 1994, 676-685.

[DwSt92] S. Dwork, L. Stockmeyer, *Finite State Verifiers I: the Power of Interaction*, J. ACM 39, 1992, 800-828.

[Fr79] R. Freivalds, *Fast Probabilistic Algorithms*, Proc. 8. Int. Symp. on Math. Found. of Comp. Science, 1979, LNCS, 57-69.

[Fr81] R. Freivalds, *Probabilistic 2-way Machines*, Proc. 10. Int. Symp. on Math. Found. of Comp. Science, 1981, LNCS, 33-45.

[FrKa94] R. Freivalds, M. Karpinski, *Lower Space Bounds for Randomized Computation*, Proc. 21. EATCS Int. Colloq. on Automata, Languages, and Programming, 1994, LNCS, 580-592.

[Ge94] V. Geffert, *A Hierarchy that Does not Collaps: Alternation in Low Level Space*, Theo. Information and Applications 28, 1994, 465-512.

[GrWe86] A. Greenberg, A. Weiss, *A Lower Bound for Probabilistic Algorithms for Finite State Machines*, J. Comput. Syst. Sci. 33, 1986, 88-105.

[Gil77] J. Gill, *Computational Complexity of Probabilistic Turing Machines*, SIAM J. Computing 7, 1977, 675-695.

[GMR89] S. Goldwasser, S. Micali, C. Rackoff, *The Knowledge Complexity of Interactive Proof Systems*, SIAM J. Computing 18, 1989, 186-208.

[GoSi86] S. Goldwasser, M. Sipser, *Private Coins versus Public Coins in Interactive Proof Systems*, Proc. 18. ACM Symp. on Theory of Computing, 1986, 59-68.

[Iw93] K. Iwama, *A Space ($o(\log \log n)$) is Regular,* SIAM J. Computing 22, 1993, 136-146.

[Li96] M. Liśkiewicz, *Interactive Proof Systems with Public Coins: Lower Space Bounds and Hierarchies of Complexity Classes,* ICSI Technical Report 1996, also Proc. 14. GI-AFCET Symp. on Theo. Aspects of Comp. Science, 1997, LNCS 1200,129-140.

[LiRe93] M. Liśkiewicz, R. Reischuk, *Separating the Lower Levels of the Sublogarithmic Space Hierarchy,* Proc. 10. GI-AFCET Symp. on Theo. Aspects of Comp. Science, 1993, LNCS, 16-27.

[LiRe96a] M. Liśkiewicz, R. Reischuk, *The Sublogarithmic Alternating Space World,* SIAM J. Computing 24, 1996, 828-861.

[LiRe96b] M. Liśkiewicz and R. Reischuk, *Space Bounds for Interactive Proof Systems with Public Coins and Bounded Number of Rounds,* ICSI Technical Report No. TR-96-025, Berkeley, July 1996.

[LiRe96c] M. Liśkiewicz and R. Reischuk, *Separating Small Space Complexity Classes of Stochastic Turing Machines,* Technical Report Informatik/Mathematik A-96-17, Med. Universität zu Lübeck, November 1997.

[LiRe97] M. Liśkiewicz and R. Reischuk, *Computing with Sublogarithmic Space,* in *Complexity Theory Retrospective II*, A. Selman, L. Hemaspaandra (Eds), Springer Verlag, 1997.

[Ma95] I. Macarie, *Space-bounded Probabilistic Computation: Old and New Stories,* SIGACT News 26, 1995, 2-12.

[Pa85] C. Papadimitriou, *Games against Nature,* J. CSS 31, 1985, 288-301.

Appendix

Sketch of the Proof of Proposition 11:

Let V be the set of all configurations in S. Partition V into subsets V_j for $j \in [0..b]$ according to the input head position as follows. Let V_0 contain all configurations in which the machine reads a substring u_j or one of the end markers. For $j > 0$ define

$$V_j := \{C \in V \mid |w\, u_1 \ldots u_{j-1}| < \mathbf{head}(C) \leq |w\, u_1 \ldots u_{j-1}\, w| \} \,.$$

Since M starts reading the left end-marker the root of S belongs to V_0.

Let V_{acc} be the set of all accepting configurations in S. Then for any $C \in V_{acc}$ let $P(C)$ be the set of nodes that lie on the path from the root to C, including C, but excluding the root. Define $J(C)$ as the subset of $[0..b]$ containing all indices j such that an element from $P(C)$ is in V_j, i.e.

$$J(C) := \{j \in [0..b] \mid P(C) \cap V_j \neq \emptyset\} \,.$$

Since the length of each path in S is bounded by k, we get $1 \leq |J(v)| \leq k$. For $J \subseteq [0..b]$ of size at most k, and for $d \in [1..b]$ define

$$A_J := \sum_{C \in V_{acc},\, J(C)=J} \mathrm{Reach}_X[C, S] \quad \text{and} \quad B_d := \sum_{\substack{J \subseteq [0..b] \\ 1 \leq |J| \leq k \text{ and } d \in J}} A_J \,.$$

Obviously, it holds
$$\mathrm{Accept}_X[S] = \sum_{J \subseteq [0..b],\, 1 \leq |J| \leq k} A_J \,.$$

We claim that there is an index $i \in [1..b]$ such that

$$B_i \leq \frac{k}{b} \cdot \mathrm{Accept}_X[S] \,.$$

To see this consider the sum $B_1 + B_2 + \ldots + B_b$. According to the definition of B_d, and because each subset J has at most k numbers, we conclude that each element A_J occurs in the sum above at most k times. Therefore

$$B_1 + B_2 + \ldots + B_b \leq \sum_{J \subseteq [0..b],\, 1 \leq |J| \leq k} k \cdot A_J = k \cdot \mathrm{Accept}_X[S] \,.$$

Hence, by the Pigeon-Hole-Principle the required i exists. Remove from S all nodes $v \in N_i$ together with all their descendants. Denote the strategy obtained this way by $\mathcal{S}'$. Since only nodes C with $i \in J(C)$ have been removed from V_{acc} it holds

$$\text{Accept}_X[\mathcal{S}'] = \text{Accept}_X[\mathcal{S}] - B_i \geq \frac{b-k}{b} \cdot \text{Accept}_X[\mathcal{S}] .$$

Now consider the input $X_i = w\, u_1\, w \ldots u_{i-1}\, \bar{w}\, u_i \ldots w\, u_b$. Remember that w and $\bar{w}$ were assumed to have the same word transitions and that their word probabilities are β-close. Using these properties one can prove the following

Lemma 13. Let C_1, C_2 be two configurations in $\mathcal{S}'$ such that C_2 is a son of C_1. Then it holds:

1. if C_1 is nondeterministic then $\quad Nmove_X^*(C_1, C_2) \iff Nmove_{X_i}^*(C_1, C_2)$,

2. if C_1 is probabilistic then the values $Rmove_X^*[C_1, C_2]$ and $Rmove_{X_i}^*[C_1, C_2]$ are β^{2z}-close.

Sketch of the Proof. Claim (1) follows directy from the assumption that $\text{WT}_s(w) = \text{WT}_s(\bar{w})$. To prove (2), we construct two z-state Markov chains R and $\bar{R}$, with $z = 10 \cdot \text{Vol}(\ell) + 20$, which model a probabilistic round of M on inputs X (resp. X_i) when the machine starts in C_1. R and $\bar{R}$ are constructed in such a way that they are β-close and that there is a trapping state σ_t of both R and $\bar{R}$ which corresponds to the configuration C_2. Then using Lemma 10 we obtain that $p^*[t, R]$ and $p^*[t, \bar{R}]$ are β^{2z}-close which proves (2). ∎

As a corollary we obtain that $\mathcal{S}'$ is a strategy of M on the input X_i, too. To estimate $\text{Accept}_{X_i}[\mathcal{S}']$ first note that for any node C in $\mathcal{S}'$ it holds

$$\text{Reach}_{X_i}[C, \mathcal{S}'] \geq \beta^{-2z\lceil k/2 \rceil} \text{Reach}_X[C, \mathcal{S}'] .$$

This follows from lemma 13 since for the subsequence $C_1, \ldots, C_{r-1}$ of probabilistic configurations on the path from the root C_0 to configuration $C = C_r$, where $r \leq \lceil k/2 \rceil$, it holds:

$$\text{Reach}_{X_i}[C, \mathcal{S}'] = \prod_{0 \leq t < r} Rmove_{X_i}^*[C_t, C_{t+1}] \geq \prod_{0 \leq t < r} \beta^{-2z} \cdot Rmove_X^*[C_t, C_{t+1}]$$

$$\geq \beta^{-2z\lceil k/2 \rceil} \prod_{0 \leq t < r} Rmove_X^*[C_t, C_{t+1}] = \beta^{-2z\lceil k/2 \rceil} \cdot \text{Reach}_X[C, \mathcal{S}'] .$$

Adding up proves the proposition:

$$\text{Accept}_{X_i}[\mathcal{S}'] = \sum_{C \in \mathcal{S}' \text{ accepting}} \text{Reach}_{X_i}[C, \mathcal{S}'] \geq \sum \beta^{-2z\lceil k/2 \rceil} \text{Reach}_X[C, \mathcal{S}']$$

$$= \beta^{-2z\lceil k/2 \rceil} \cdot \text{Accept}_X[\mathcal{S}'] \geq \frac{b-k}{b} \cdot \beta^{-2z\lceil k/2 \rceil} \cdot \text{Accept}_X[\mathcal{S}] .$$

Learning to Perform Knowledge-Intensive Inferences

Dan Roth

[1] Dept. of Appl. Math. & CS, Weizmann Institute of Science, Rehovot 76100, Israel
[2] Department of Computer Science, University of Illinois, Urbana, IL 61801, U.S.A
danr@wisdom.weizmann.ac.il
http://www.wisdom.weizmann.ac.il/~danr

Consider, for example, the task of language understanding, which humans perform effortlessly and effectively. It depends upon our ability to disambiguate word meanings, recognize speaker's plans, perform predictions and generate explanations. In these and other "high level" cognitive tasks, such as high level vision, inference heavily depends on "knowledge" about the language and the world. We call these tasks *Knowledge-Intensive Inferences*.

While the central role of learning in cognition is acknowledged by many, most lines of research nevertheless study reasoning phenomena separately from learning phenomena. We present a new framework for the study of knowledge-intensive inferences — an integrated theory of learning, knowledge representation and reasoning. In this framework for the study of reasoning, the learning component is given a principle role. T:q he result is shown to be more powerful computationally than traditional approaches.

The presentation consists of three parts: brief analysis of the past, overview of the theoretical approach and a hint on applications.

The generally accepted framework for the study of reasoning in intelligent systems is the knowledge-based system approach [McC58, Nil91]. It is assumed that the knowledge is given to the system, stored in some *representation language* with a well defined meaning assigned to its sentences. The sentences are stored in a Knowledge Base (KB) which is combined with a reasoning mechanism, used to determine what can be inferred from the sentences in the KB. The question of how this knowledge might be acquired and whether this should influence how the performance of the reasoning system is measured is normally not considered [Kir91]. However, experience in AI research over the past few decades, as well as many recent computational studies, show that it is unlikely that hand programming or any form of knowledge engineering will generate robust, non-brittle and tractable reasoning system in complex domains.

The *Learning to Reason* (L2R) approach, in contrast to earlier approaches to reasoning, views learning as an integral part of the process, and suggests to study the entire process of *learning* some knowledge representation and *reasoning* with it. The theoretical work on this framework started by Khardon and Roth in [KR94]. For a survey of works in this framework, see [Rot96].

In the Learning to Reason framework it is not assumed that the knowledge representation describing the "world" is *given* to the agent. Instead, the agent constructs the knowledge representation while interacting with the world.

In this way the reasoning task is no longer a "stand alone" process, and the agent does not reason from a previously defined "general purpose" knowledge representation. Rather, it can choose a knowledge representation that facilitates the reasoning task at hand. Moreover, we take the view that a reasoner need not answer efficiently *all* possible queries, but only those that are "relevant", or "common", in a well defined sense. This relaxation can be used by the agent in selecting its knowledge representation. In addition, the interaction with the environment (which is formally defined), allows the performance of the agent to be measured relative to the environment it interacts with. We prove the usefulness of the Learning to Reason approach by showing that through interaction with the world, the agent truly gains additional reasoning power, over what is possible in the traditional setting. Several results are presented to substantiate this claim, exhibiting cases where learning to reason about the world is feasible but either (1) reasoning from a given representation of the world or (2) learning representations of the world do not have efficient solutions.

Perhaps the major difference between the traditional approach to intelligent inference and the L2R approach is that the latter approach suggests that for large scale reasoning to work in practice, reasoning systems need to be trained over a large number of examples and various kinds of interactions. To respond to this challenge we present an architecture and an algorithmic framework which is used in a large scale experimental study of knowledge intensive inferences in the language understanding domain. Rather than viewing a knowledge base as a starting point on which inference procedures uniformly act, we now interact with various kinds of knowledge bases (such as dictionaries and raw text) and incorporate relevant knowledge as we learn new representations. We show how these ideas are used in a study of context sensitive spelling correction [GR96].

References

[GR96]   A. R. Golding and D. Roth. Applying winnow to context-sensitive spelling correction. In *Proc. of the International Conference on Machine Learning*, 1996.

[Kir91]   D. Kirsh. Foundations of AI: the big issues. Foundations of AI: the big issues. *Artificial Intelligence*, 47:3–30, 1991.

[KR94]   R. Khardon and D. Roth. Learning to reason. In *Proc. of the National Conference on Artificial Intelligence*, pages 682–687, 1994. To appear in Journal of the ACM.

[McC58]   J. McCarthy. Programs with common sense. Reprinted in R. Brachman and H. Levesque, editors, *Readings in Knowledge Representation, 1985.*

[Nil91]   N. J. Nilsson. Logic and artificial intelligence. *Artificial Intelligence*, 47:31–56, 1991.

[Rot96]   D. Roth. Learning in order to reason: The approach. In K. G. Jeffery, J. Kral, and M. Bartosek, editors, *SOFSEM 96: Theory and Practice of Informatics*, Springer-verlag, 1996. Lecture notes in Artificial Intelligence, vol. 1175.

Resolution Proofs, Exponential Bounds, and Kolmogorov Complexity

Uwe Schöning

Abteilung Theoretische Informatik,
Universität Ulm,
89069 Ulm, Germany
schoenin@informatik.uni-ulm.de

Abstract. We prove an exponential lower bound for the length of any resolution proof for the same set of clauses as the one used by Urquhart [13]. Our contribution is a significant simplification in the proof and strengthening of the bound, as compared to [13]. We use on the one hand a simplification similar to the one suggested by Beame and Pitassi in [1] for the case of the pidgeon hole clauses. Additionally, we base our construction on a simpler version of expander graphs than the ones used in [13]. These expander graphs are located in the core of the construction. We show the existence of our expanders by a Kolmogorov complexity argument which has not been used before in this context and might be of independent interest since the applicability of this method is quite general.

1 Introduction

Resolution is a particularly simple calculus for propositional logic (and predicate logic), and has been used in automatic theorem proving and logic programming systems. It took more than a decade (after the paper [12] appeared) to succeed to prove an exponential lower bound on the length of resolution proofs, for the clauses expressing the pidgeon hole principle [5]. Later, other sets of clauses have been defined which can be shown to require exponentially long proofs [13, 4]. Haken's original construction and counting argument [5] is quite complicated and difficult to apply to other sets of clauses. Therefore, several iterations in simplifying or modifying the lower bound proof have been made [13, 4, 2, 1]. This is a worth-while project since we are still far away from characterizing the type of clause sets which require long proofs. This paper is a further step in this project. Particularly interesting might be the use of Kolmogorov complexity.

Recall the basic definitions regarding clauses and resolution. A *literal* is a variable or negated variable. A *clause* is a set of literals (to be understood as being a logical *disjunction* of the literals). If a set of clauses is given, it is understood that the clauses are connected *conjunctively*. Given two clauses C_1 and C_2 such that for a literal l, $l \in C_1$ and $\bar{l} \in C_2$, the new clause $C_3 = (C_1 - \{l\}) \cup (C_2 - \{\bar{l}\})$ is called a *resolvent* of C_1, C_2. Formally we denote such a resolution step by $C_1, C_2 \vdash C_3$. It is known that resolution is *refutation complete*, i.e. for every unsatisfiable set of clauses there is a sequence of resolution steps, starting with the given set of clauses and ending up in the empty clause, denoted by $\square$.

A well-known technical construction to be needed later is the following. If a resolution refutation R for an unsatisfiable clause set $\mathcal{F}$ is given, and if l is some literal occuring in $\mathcal{F}$, by assigning the truth value, say 1, to l, by trivial modifications, a new resolution refutation R' is obtained. All clauses in R (including the initial clause set $\mathcal{F}$) which contained the literal l have disappeared in R'.

2   The Graph-Based Formulas

Let $G = (V, E)$ be any undirected graph on $n = |V|$ vertices. Let $m : V \to \{0, 1\}$ be a "marking" of the vertices of G such that $\Sigma_{x \in V} \, m(x) \equiv 1 \pmod 2$. Given any such graph G and a marking m, one can define a unsatisfiable formula (set of clauses) $\mathcal{F}(G, m)$ as follows. The formula contains as variables the names of the edges in E. Therefore, any assignment to these variables can be interpreted as selecting a subgraph of G. The formula $\mathcal{F}$ is a conjunction of formulas F_x, $x \in V$, where $F_x = e_1(x) \oplus \ldots \oplus e_d(x)$ if $m(x) = 1$ and $F_x = \overline{e_1(x) \oplus \ldots \oplus e_d(x)}$ if $m(x) = 0$, otherwise. Here $e_1(x), \ldots, e_d(x)$ are all the edges (=variables) which are incident with the vertex x. Each such formula F_x can be converted into conjunctive normal form. This normal form for F_x has exactly 2^{d-1} many clauses. If every vertex in G has degree at most d (a constant), then the conjunctive normal form of $\mathcal{F}(G, m) = \bigwedge_{x \in V} F_x$ has at most $n 2^{d-1}$ many clauses, each containing at most d many literals. Therefore, if the degree d is constant, the size of the clause set $\mathcal{F}(G, m)$ is $O(n)$. These graph-based formulas have been originally introduced by Tseitin [12] and later been used by Urquhart [13] to prove an exponential lower bound on the length of any resolution proof for $\mathcal{F}(G, m)$ where G is an especially constructed graph.

It is easy to see that $\mathcal{F} = \mathcal{F}(G, m)$ is unsatisfiable. Suppose to the contrary that φ is an assignment with $\varphi(F_x) = 1$ for all $x \in V$. The assignment φ specifies a subgraph G_φ. Since $\varphi(F_x) = 1$ for all $x \in V$, each vertex in G_φ has an odd (even) number of neighbors iff $m(x) = 1$ (resp. $m(x) = 0$). Summing up over the degrees of each vertex in G_φ must result in an odd number since $\Sigma_{x \in V} \, m(x)$ is odd. On the other hand, we have counted every edge twice. Therefore the sum should be even. Contradiction!

It is an interesting aspect to be used later that for every edge $\{x, y\}$ in G we can obtain another unsatisfiable formula $\mathcal{F}(G', m')$, either by simply deleting this edge from G and letting $m' = m$, or by deleting the edge from G and toggling the value of $m(x)$ and of $m(y)$ in m'. These two operations correspond to assigning the literal $\{x, y\}$ the truth value 0 or 1, respectively.

Our construction will be based on graphs G with n vertices which have a certain expansion property. For a constant $\gamma \in (0, \frac{1}{2})$ to be determined later it holds true that for every subset $S \subseteq V$, $\gamma n \leq |S| \leq 2\gamma n$, there are more than n edges $\{x, y\}$ such that $x \in S$ and $y \notin S$.

For these graphs it will be shown that every resolution proof requires more than $2^{\beta n}$ many steps where $\beta = 1/28$ and $d = 10$.

3   Large Clauses and Capture

Let R be a resolution proof for $\mathcal{F}(G, m)$. Such R exists since $\mathcal{F}(G, m)$ is unsatisfiable and the resolution calculus is refutation complete.

A clause in R is called *large* if it consists of at least $n/2$ many literals. Suppose there are no more than $2^{\beta n}$ many large clauses in R, for a constant β to be determined later. We will show below that this assumption (for a small enough β) leads to a contraction. Therefore, we will be able to conclude that R has more than $2^{\beta n}$ many large clauses, and therefore has more than $2^{\beta n}$ many clauses altogether.

The clauses in $\mathcal{F}(G, m)$ are built up from $dn/2$ many variables, therefore there are dn many different potential literals. Hence, by definition, a large clause contains a fraction of at least $\frac{n/2}{dn} = 1/2d$ of all literals. This implies that there exists a literal, say l, that appears in a fraction of $1/2d$ of all large clauses. By setting this literal to 1 we obtain a new (unsatisfiable) clause set $\mathcal{F}(G', m')$ and a corresponding resolution proof R'. All large clauses in R that contain l are not any more present in R'. Therefore, the number of large clauses in R' goes down by a factor of at least $1/2d$. We repeat this process a suitable number of times t until no large clauses appear any more, and we end up with a clause set $\mathcal{F}(G^*, m^*)$ and a corresponding resolution proof R^*. An inequality to determine the number t is

$$2^{\beta n}(1 - 1/2d)^t < 1$$

This gives us

$$t > \frac{\beta}{\log_2\left(\frac{1}{1 - 1/2d}\right)} \cdot n$$

Using $\ln\left(\frac{1}{1-x}\right) > x$ for $x \in (0, 1)$, it can be seen that it suffices to choose $t \geq (2 \ln 2)\beta dn$.

Every clause C in the resolution proof R^* captures in some sense some aspects of (some subset) of the initial clauses in $\mathcal{F}(G^*, m^*)$. The following definition serves as a measure of how much is captured by a clause C.

$$capture(C) = |\{x \in V \mid \exists \varphi : \varphi(C) = 0, \varphi(F_x) = 0, \text{ and } \varphi(F_y) = 1 \text{ for all } y \neq x\}|$$

We observe that for every $x \in V$ there exists an assignment φ with $\varphi(F_x) = 0$ and $\varphi(F_y) = 1$ for all $y \in V$, $y \neq x$ (a so-called *critical* assignment). Such an assignment can be obtained as follows. Start with an arbitrary assignment such that the value of F_x is 0. For an odd number of vertices z (including x), the value of F_z is 0. Select any two vertices u, v among them, except x. These nodes are connected by some path p. Toggling the truth value on all the edges (=variables) along p, produces an assignment under which F_u and F_v become true, while the value of all other F_z remain the same. Repeating this process yields the desired assignment. This argument shows that for the last clause in the resolution proof we have $capture(\Box) = n$. It is easy to see that for the initial clauses, the clauses C in $\mathcal{F}(G^*, m^*)$, we have $capture(C) = 1$. Furthermore, if $C_1, C_2 \mid\!\!- C_3$ is a resolution step in R^*, we have

$$capture(C_3) \leq capture(C_1) + capture(C_2)$$

Therefore, for every constant $\gamma \in (0, \frac{1}{2})$ there exists a clause $C \in R^*$ with $capture(C) \in [\gamma n, 2\gamma n]$. (Just choose the *first* clause C in the resolution proof R^* with $capture(C) \geq \gamma n$.)

Compared to G, in the graph G^* there have been t edges eliminated. Therefore the expander property of G^* is somewhat weaker. For every subset $S \subseteq V^*$, $|S| \in [\gamma n, 2\gamma n]$, there exist $> n - t$ edges $\{x, y\}$ with $x \in S$ and $y \notin S$. Now choose S from the definition of $capture(C)$, i.e.

$$S = \{x \in V \mid \exists \varphi : \varphi(C) = 0, \ \varphi(F_x) = 0, \ \text{and} \ \varphi(F_y) = 1 \ \text{for all} \ y \neq x\}$$

For each edge $l = \{x, y\}$, $x \in S$, $y \notin S$, given by the expander property of G^*, by toggling the truth value of the literal l in the assignment φ with $\varphi(F_x) = 0$ and $\varphi(F_y) = 1$ for all $y \neq x$ (given by the definition of $capture(C)$) we obtain a new assignment φ' with $\varphi'(F_y) = 0$ and $\varphi'(F_z) = 1$ for all $z \neq y$. By the fact that $y \notin S$ it follows $\varphi'(C) = 1$. That is, toggling the truth value of one literal l changes the truth value of C. This means, C must contain the literal l or $\bar{l}$. Since we can repeat this argument for $> n - t$ many different edges $\{x, y\}$, this clause C must have more than $n - t$ literals.

Now, the contradiction is coming up, if $n - t \geq n/2$, this means that C is a large clause, but by the above elimination process, no large clause in the resolution proof can exist. Therefore, we examine the condition $n - t \geq n/2$ or $n/2 \geq t$ which leads to the contradiction. Using $t \geq (2 \ln 2)\beta dn$ we obtain that choosing $\beta \leq \frac{1}{(4 \ln 2)d}$ yields the contradiction. As discussed above, the resolution proof must have more than $2^{\beta n}$ many clauses.

4 Construction of Expanders

It remains to show that such expander graphs as needed in our construction exist. These are graphs $G = (V, E)$ that should have $n = |V|$ vertices, each vertex should have degree $\leq d$, and furthermore they should satisfy the following expansion property (for a constant γ to be determined later):

$$\forall S \subseteq V, |S| \in [\gamma n, 2\gamma n] \ \exists > n \ \text{edges} \ \{x, y\} \in E : x \in S, y \notin S$$

We "construct" these graphs (rather, show their existence) by a Kolmogorov complexity argument. Our method of using Kolmogorov complexity to construct expanders seems to be new, and might be of independent interest. The approach works for other versions of expanders studied in the literature (e.g. [9, 8, 7, 11]) as well, see [10]. The reader is refered to [6] for the basic properties of Kolmogorov complexity, and to [3] for a survey on expander graphs.

Consider a bipartite graph H with n left vertices and $dn/2$ right vertices. Each vertex on the left has degree d, and each vertex on the right has degree 2. It is clear that any such a graph H can be interpreted as a graph G as it is needed in the construction above. Namely, the n left vertices of H correspond to the vertices of G, and the $dn/2$ right vertices of H correspond to the edges of G. See the following figure.

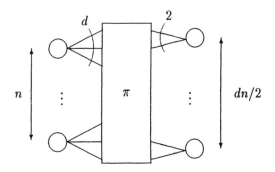

In the graph H we have dn many "connections" on the left and the same number of connections on the right. We choose a Kolmogorov-incompressible permutation $\pi : [dn] \to [dn]$ and use it to connect the left and right connections. The result is the graph $H = H_\pi$. More precisely, choose π such that

$$C(\pi \mid d, n) \geq \log_2 ((dn)!)$$

Here C denotes the (conditional) Kolmogorov complexity. From the theory of Kolmogorov complexity it is known that there exist such permutations. (In fact, *almost all* permutations achieve almost this Kolmogorov complexity.) We will show that the graph G that corresponds to the so constructed graph H_π satisfies the expansion property.

Suppose to the contrary that G does not have the expansion property. In this case we will be able to split π into two permutations $\pi_1 : A \to A'$, $\pi_2 : B \to B'$ where $A \cup B = A' \cup B' = [dn]$. The point is that A, A', B, B' are compactly describable, and that such permutations π_1, π_2 (given A, A', B, B') are more compactly describable than π. This leads to a contradiction to the choice of π.

The following lemma about the Kolmogorov complexity of permutations might be interesting in its own right.

Lemma. Let $\pi : [m] \to [m]$ be a permutation and let $k \in \{1, \ldots, m\}$. Suppose π can be represented by two permutations $\pi_1 : [k] \to [k]$ and $\pi_2 : [m-k] \to [m-k]$ in the following sense:

$$\pi(i) = \begin{cases} \pi_1(i), & i \leq k, \\ \pi_2(i - k) + k, & i > k. \end{cases}$$

Then

$$C(\pi \mid m, k) \leq \log(m!) - h\left(\frac{k}{m}\right) m + O(\log m)$$

where $h(\alpha) = -\alpha \log \alpha - (1 - \alpha) \log(1 - \alpha)$ is the binary entropy function.

Proof. Using Stirling's formula $n! \sim \sqrt{2\pi n}(n/e)^n$, we obtain

$C(\pi \mid m, k)$
$= \log(k!(m-k)!) + O(1)$
$= k \log k - k \log e + (m-k) \log(m-k) - (m-k) \log e + O(\log m)$
$= k \log m - k \log \frac{k}{m} - m \log e + (m-k) \log m - (m-k) \log \frac{m-k}{m} + O(\log m)$
$= m \log m - m \log e - k \log \frac{k}{m} - (m-k) \log \frac{m-k}{m} + O(\log m)$
$= \log(m!) - h(\frac{k}{m})m + O(\log m)$

$\square$

Loosely speaking, this means that if we know that π can be split into two permutations as described in the lemma, then we loose about $h(\frac{k}{n})n$ many bits of entropy.

If the constructed graph G does not have the expansion property, then there is a set $S \subseteq V$, $|S| \in [\gamma n, 2\gamma n]$, which has at most n edges $\{x, y\}$ such that $x \in S$, $y \notin S$.

Now we can describe the permutation π by the following information.

- A description of the subset $S \subseteq V$. This requires at most $h(\alpha)n$ bits. (Here and in the following we are using the inequality $\sum_{i=0}^{\lambda n} \binom{n}{i} \le 2^{h(\lambda)n}$, for $\lambda \le \frac{1}{2}$, cf. [14]. This also implies $\log \binom{n}{\alpha n} \le h(\alpha)n$.)
- A description of those n connections leaving S in the graph H_π which go to "outer edges" (i.e. vertices on the right which have exactly one connection to S). We identify these n connections among the $d\alpha n$ connection which leave S. This requires $\log \binom{d\alpha n}{n} \le h(\frac{1}{d\alpha})d\alpha n$ bits.
- A description of the set T on the right side to which the remaining $d\alpha n - n$ connections from S go. This set T has size $(d\alpha - 1)n/2$. This requires $\log \binom{dn/2}{(d\alpha-1)n/2} \le h(\alpha - \frac{1}{d})\frac{d}{2}n$ many bits.
- A description of the two permutations π_1, π_2 into which π can be split where π_1 connects the $d\alpha n - n$ many links on the left (the above specified subset of S's links which go to the to the links of T). The permutation π_2 connects the remaining links. By the lemma this requires $\log((dn)!) - h(\alpha - \frac{1}{d})dn + O(\log dn)$ many bits.
- An additional $O(\log n)$ term is needed to make all these descriptions self-delimiting.

It is clear that π can be reconstructed from the above given information. Altogether we obtain

$C(\pi \mid n, d)$
$\le h(\alpha)n + h\left(\frac{1}{d\alpha}\right)d\alpha n + h\left(\alpha - \frac{1}{d}\right)\frac{dn}{2} + \log((dn)!) - h\left(\alpha - \frac{1}{d}\right)dn + O(\log dn)$
$= \log((dn)!) + \left[h(\alpha) + h\left(\frac{1}{d\alpha}\right)d\alpha - h\left(\alpha - \frac{1}{d}\right)\frac{d}{2}\right] \cdot n + O(\log dn)$

A contradiction occurs if this expression is strictly less than $\log((dn)!)$. It suffices to satisfy that

$$f(d, \alpha) := h(\alpha) + h\left(\frac{1}{d\alpha}\right)d\alpha - h\left(\alpha - \frac{1}{d}\right)\frac{d}{2}$$

is strictly less than 0. Substituting $d = 10$, we see that $f(10, \alpha)$ for $\alpha \in [0.32, 0.64]$ is strictly less than 0. (Hence, choose $\gamma = 0.32$.) This shows that the desired expanders, using degree 10, exist.

Remark: The above calculation has been designed to get d as small as possible. Actually, using trivial estimations, like $\binom{n}{\alpha n} \leq 2^n$, etc., the above Kolmogorov complexity-theoretic approach is still successful, and leads to simpler formulas, but results in a larger value of d.

References

1. P. Beame, T. Pitassi. Simplified and improved resolution lower bounds. *Proc. 37th IEEE Sympos. Foundations of Computer Science 1996.*
2. S. Cook, T. Pitassi. A feasible constructive lower bound proof for resolution proofs. *Information Processing Letters* **34** (1990) 81–85.
3. F.R.K. Chung. Constructing random-like graphs. In: B. Bollobs (ed.): *Probabilistic Combinatorics and Applications*, Proceedings of Symposia in Applied Mathematics, Vol. 44, pp. 21–55. American Math. Society, 1991.
4. V. Chvátal, E. Szemerédi. Many hard examples for resolution. *Journal of the Assoc. for Comput. Mach.* **35** (1988) 759–768.
5. A. Haken. The intractability of resolution. *Theoretical Computer Science* **39** (1985) 297–308.
6. M. Li, P. Vitanyi. *An Introduction to Kolmogorov Complexity and its Applications.* Springer-Verlag, 1993.
7. M. Paterson. Improved sorting networks with O(log n) depth. *Algorithmica*, to appear.
8. M. Pinsker. On the complexity of a concentrator. *7th Int. Teletraffic Conference.* 1973, 318/1–318/4.
9. N. Pippenger. Superconcentrators. *SIAM J. Comput.* **6** (1977) 298–304.
10. U. Schöning. Better expanders and superconcentrators by Kolmogorov complexity. *4th International Colloq. on Structural Information and Communication Complexity*, 1997.
11. M. Sipser, D.A. Spielman. Expander codes. *Proc. 35th IEEE Sympos. Foundations of Computer Science*, 566–576, 1994.
12. G.S. Tseitin. On the complexity of derivation in propositional calculus. In *Studies in Constructive Mathematics and Mathematical Logic*, Part 2, 1968. Reprinted in Siekmann, Wrightson: *Automization of Reasoning*, Springer-Verlag, 1983.
13. A. Urquhart. Hard examples for resolution. *Journal of the Assoc. Comput. Mach.* **34** (1987) 209–219.
14. D. Welsh. *Codes and Cryptography.* Oxford University Press, 1988.

CONTRIBUTED PAPERS

The Expressiveness of Datalog Cicuits (DAC)

Foto AFRATI [1], Irène GUESSARIAN [2] and Michel de ROUGEMONT [3]

ABSTRACT : We define a new logic query language, called DAC, which is an extension of Datalog. We exhibit queries which are not Datalog expressible but are DAC expressible. We also prove non-expressiveness results for DAC and we infer various strict hierarchies obtained by allowing more rapidly growing functions on the bound parameters.

I Introduction

A Datalog program is an existential inductive definition on a class of finite relational structures which defines a monotone query computable in polynomial time. It has been shown in [3] that there are monotone PTIME-computable queries that are not Datalog expressible, for example the chain query that defines on the class of graphs the existence of a path of length q^2 for some q. We observed that if we fix n, there is a *Datalog* program P_n that defines this query for the class of graphs of size n, and there is a PTIME-computable function g such that for each n, $g(n)$ describes P_n as a string. We will say that such a query is expressible by *DAtalog Circuits* and is in the class *DAC*. The class *DAC* extends *Datalog* by allowing (i) non-constant (with respect to the size of the input structure) size of programs and (ii) an infinite sequence of programs $P_1, \ldots, P_n, \ldots$, each P_n defining the query on structures (or databases) of size n. Namely, program P_n will give the answers whenever the size of the data domain is n. The parameters of the program P_n, now, are functions of n. We describe P_n in terms of the number of rules –we call it height and denote it $h(n)$–, the maximum number of atoms per rule –we call it width and denote it $w(n)$–, and the recursion depth –we call it bound of recursion and denote it $b(n)$. $DAC(w(n), h(n))(b(n))$ is the class of queries that are accepted by such families of programs $\mathcal{P} = \{P_n, n \in \mathbb{N}\}$.

Of particular interest is the class $smDAC(1, poly)(\log n)$ where the programs have constant width, polynomial height and $\log n$ recursion depth and are strongly monotone. This class subsumes (i) both the parallelism and logarithmic depth of the class mAC^1 (monotone AC^1 in [5]) of queries accepted by positive circuits of polynomial size and $\log n$ recursion depth, and (ii) the strong monotonicity of *Datalog*. We first study the case of bounded *DAC*, i.e. the case when the recursion depth is bounded and then the case when the recursion depth is $\log n$. Our main results are:

- $smDAC(1, h(n))(1)$ coincides with Bounded Datalog (Theorem II.1).
- The perfect p-th power query separates the classes $smDAC(1, n^{12})(\log n)$ and $smDAC\big(1, (\log n)^{1-\varepsilon}\big)(\log n)$ for $\varepsilon > 0$ (Theorem III.3).

[1] N.T.U.A. 157 73 Zographou, Athens, Greece. Part of this work was done while visiting Université Paris VI and Université Paris XIII. E-mail: afrati@softlab.ece.ntua.gr
[2] L.I.A.F.A. Université Paris VI, 4 Place Jussieu, 75252 Paris Cedex 05, France. Work supported by Esprit Gentzen project no 7232. E-mail: ig@litp.ibp.fr.
[3] L.R.I. Université Paris XI, 91405 Orsay Cedex, France. E-mail: mdr@lri.fr

- Let, for $i > 0$, $n_i = \overbrace{\log\log\ldots\log n}^{i\ \text{times}}$; the classes $DAC(1, n_i)(n_i)$ form a strictly decreasing hierarchy (Theorem IV.4).

It is well known [3] that *Datalog* expressible queries are computed by polynomial size monotone circuits. The fragment of DAC that we investigate also expresses queries which can be computed by polynomial size monotone circuits. DAC programs that belong to this fragment are of interest, because they enhance largely the expressive power of *Datalog*, while being at most as expressive as polynomial size monotone circuits. DAC programs allow for trade-offs between the width, the height and the depth of recursion, as illustrated in Example I.1 below:

EXAMPLE I.1 On a graph with a binary edge relation E, the transitive and reflective closure of E, denoted by TC, is

(i) $DAC(1, \log n)(\log n)$. The following program $\mathcal{P} = \{P_n, n \in \mathbb{N}\}$ which has constant width 2 and height $2\lfloor \log n \rfloor + 2$, hence is in $DAC(1, \log n)$, defines TC in $\log n$ parallel computation steps.

$$P_n \begin{cases} TC(X, Y) \longleftarrow TC_{\lfloor \log n \rfloor - 1}(X, Z)\, TC_{\lfloor \log n \rfloor - 1}(Z, Y) \\ TC(X, Y) \longleftarrow TC_{\lfloor \log n \rfloor - 1}(X, Y) \\ RC(X, Y) \longleftarrow E(X, Y) \\ RC(X, X) \longleftarrow \\ TC_1(X, Y) \longleftarrow RC(X, Z)\, RC(Z, Y) \\ TC_1(X, Y) \longleftarrow RC(X, Y) \\ \qquad\qquad \vdots \\ TC_{\lfloor \log n \rfloor - 1}(X, Y) \longleftarrow TC_{\lfloor \log n \rfloor - 2}(X, Z)\, TC_{\lfloor \log n \rfloor - 2}(Z, Y) \\ TC_{\lfloor \log n \rfloor - 1}(X, Y) \longleftarrow TC_{\lfloor \log n \rfloor - 2}(X, Y) \end{cases}$$

(ii) $DAC(n, 1)(1)$. The following program $\mathcal{P}' = \{P'_n, n \in \mathbb{N}\}$ which has width n and height 3, hence is in $DAC(n, 1)$, defines TC in recursion depth 2.

$$P'_n \begin{cases} TC(X, Y) \longleftarrow RC(X, X_1)\, RC(X_1, X_2) \ldots RC(X_{n-1}, Y) \\ RC(X, Y) \longleftarrow E(X, Y) \\ RC(X, X) \longleftarrow \end{cases}$$

(iii) $DAC(\log n / \log \log n, 1)(\log n / \log \log n)$. For any n, let $k(n)$ be the least natural number such that: $k(n)^{k(n)} < n \leq (k(n) + 1)^{k(n)+1}$. The following program $\mathcal{P}'' = \{P''_n, n \in \mathbb{N}\}$ which has width $k(n) + 1$, two IDB predicates and height 4 defines TC in recursion depth $k(n) + 3$. The result follows because $k(n)$ is in $O(\log n / \log \log n)$.

$$P''_n \begin{cases} TC(X, Y) \longleftarrow RC(X, Y) \\ TC(X, Y) \longleftarrow TC(X, Z_1)\, TC(Z_1, Z_2) \ldots TC(Z_{k(n)-1}, Z_{k(n)})\, TC(Z_{k(n)}, Y) \\ RC(X, X) \longleftarrow \\ RC(X, Y) \longleftarrow E(X, Y) \end{cases}$$

(iv) Transitive closure is **not** $DAC(\log n, 1)(1)$. Assume $TC \in DAC(\log n, 1)(1)$ and let $\mathcal{P} = \{P_n, n \in \mathbb{N}\} \in DAC(\log n, 1)(1)$ be a DAC program defining TC. Consider now the fringe (see Definition II.2) of a proof tree corresponding to a pair (X, Y) in

$TC_{P_n}(\mathcal{D}_n)$: it consists of a set of atoms $E(Z_i, Z_j)$ which form a path going from X to Y. Because proof trees of programs in $DAC(\log n, 1)(1)$ have depth at most k', and branching width at most $c \log n$, their fringes will correspond to simple paths of length at most $(c \log n)^{k'} < n$; hence P_n cannot capture e.g. the transitive closure on a graph consisting of a single simple path of length n.

(v) Note that TC is a strongly monotone query (see Section II.3), and all our programs are strongly monotone; hence we have the more precise results that TC is in $smDAC(1, \log n)(\log n)$, in $smDAC(n, 1)(1)$ and also in $smDAC(\log n/ \log \log n, 1)$ $(\log n/ \log \log n)$.

The paper is organized as follows. In the second section, we define DAC programs, we study bounded and monotone DAC programs and state Theorem II.1. In the third section, we state the pumping lemma for DAC programs and Theorem III.3. In the fourth section, we show our hierarchy theorems and Theorem IV.4. The technical proofs are in the full paper, which can be obtained from the authors.

II The definition of DAC

II.1 Datalog

Let D be a finite set; a **database** $\mathcal{D}$ over the domain D is a finite relational structure $\mathcal{D} = (D, R_1, \ldots, R_k)$, the sequence $a = (a_1, \ldots, a_k)$ of arities of the R_is is the type of the database. $R_1, \ldots, R_k$ are called extensional predicates (or EDBs). In the sequel, $\mathcal{D}_n$, or just $\mathcal{D}$, will denote a database having n elements (or entities): n is said to be the **size** of $\mathcal{D}_n$; the size here refers to the number of entities in the database, and **not** to the number of ground facts, or tuples in the database. $\mathbb{D}_n$ will denote the set of databases $\mathcal{D}_n$ having n elements. A **query** Q (of arity p) is a function associating with a database $\mathcal{D}$ a relation $Q(\mathcal{D})$ of arity p over the domain D of $\mathcal{D}$.

Let $T_1, \ldots, T_l$ be new predicate symbols, called intensional predicates (or IDBs). A *Datalog* **program** P is a finite set of function-free Horn clauses, called rules, of the form:

$$Q(X_1, \ldots, X_n) \longleftarrow Q_1(Y_{1,1}, \ldots, Y_{1,n_1}), \ldots, Q_p(Y_{p,1}, \ldots, Y_{p,n_p})$$

where the X_is and the $Y_{i,j}$s are either variables or constants, Q is an IDB, the Q_is are either IDBs or EDBs. For each database $\mathcal{D}$:

- $F(\mathcal{D}) = \{R_j(d_1, \ldots, d_n) \mid d_i \in D, \text{ for } i = 1, \ldots, n \text{ and } (d_1, \ldots, d_n) \in R_j\}$ is the set of ground atoms which hold in $\mathcal{D}$, also called facts;
- for each *Datalog* program P, the Herbrand basis is the set of ground atoms $B(\mathcal{D}) = \{Q(d_1, \ldots, d_n) \mid d_i \in D, \text{ for } i = 1, \ldots, n \text{ and } Q \in \{R_1, \ldots, R_k\} \cup \{T_1, \ldots, T_l\}\}$.

With $P, \mathcal{D}$ we associate the program $P(\mathcal{D}) = P \cup \{A \longleftarrow \mid \text{ for } A \in F(\mathcal{D})\}$. For any subset I of $B(\mathcal{D})$, let $T_P(I)$ be the set of immediate consequences of I using $P(\mathcal{D})$, i.e. the set of atoms which can be deduced from I using only one application of one of the rules in $P(\mathcal{D})$. Define $T_P^0(\mathcal{D}) = F(\mathcal{D})$, $T_P^i(\mathcal{D}) = T_P(T_P^{i-1}(\mathcal{D}))$, and $T_P^\infty(\mathcal{D}) = \bigcup_{i \geq 0} T_P^i(\mathcal{D})$. Let $Q_P^i(\mathcal{D})$ be the consequences about Q in $T_P^i(\mathcal{D})$, i.e. the atoms of the form $Q(d_1, \ldots, d_n)$ in $T_P^i(\mathcal{D})$. The depth of recursion is said to be *bounded by* $b(n)$ if and only if $T_P^\infty(\mathcal{D}) = \bigcup_{0 \leq i \leq b(n)} T_P^i(\mathcal{D})$, for all $\mathcal{D} \in \mathbb{D}_n$. The depth of recursion is said to be *bounded* if and only if it is bounded by a constant c for all n. A pair (P, Q) consisting of a *Datalog* program P together with one of its IDB predicates Q called **principal predicate** defines a query $Q_P: \mathcal{D} \mapsto Q_P(\mathcal{D})$. If no ambiguity can arise the query Q_P will simply be denoted by Q.

We will adopt a vector notation, and write $Q(\vec{x})$ instead of $Q(x_1, \ldots, x_k)$ whenever this clarifies the notations. A *proof tree* for the IDB atom $Q(\vec{x})$ serves as a witness of a computation proving $Q(\vec{x})$ from the rules of P.

DEFINITION II.2 *A proof tree for the IDB atom $Q(x_1, \ldots, x_n)$ is a tree whose root is labeled by $Q(x_1, \ldots, x_n)$, whose internal nodes are labeled by IDB atoms, whose leaves are labeled by EDB atoms, and which is such that, if node v is labeled by $S(\vec{x})$, then the children of node v are labeled by $Q_{i_1}(\vec{x}, \vec{y}_1)$, $\ldots$, $Q_{i_r}(\vec{x}, \vec{y}_r)$ where $S(\vec{x}) \longleftarrow Q_{i_1}(\vec{x}, \vec{y}_1), \ldots, Q_{i_r}(\vec{x}, \vec{y}_r)$ is one of the rules of P defining $S(\vec{x})$. The fringe of a proof tree consists of all its leaves.*

EXAMPLE II.3 For instance, letting P define the binary query TC_1 by the rules

$$P_1 \quad \begin{cases} TC_1(X, Y) \longleftarrow TC_0(X, Z) \, TC_0(Z, Y) \\ TC_0(X, Y) \longleftarrow E(X, Y) \end{cases}$$

A proof tree for $TC_1(X, Y)$ whose fringe is $E(X, Z) \, E(Z, Y)$ is depicted in Figure 1.

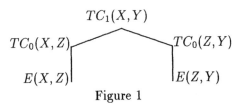

Figure 1

II.2 *DAC* Programs

DEFINITION II.4 *The class $DAC(w(n), h(n))$ is the set of queries Q, each query Q being expressible as follows:*
1. *there exists a family $\mathcal{P}$ of Datalog programs $\mathcal{P} = \{P_n, n \in \mathbb{N}\}$, each P_n having a principal predicate Q_n of fixed arity, and such that*
(i) *Q_n defines Q on databases $\mathcal{D}_n$ of size n, i.e. $Q_n(\mathcal{D}_n) = Q(\mathcal{D}_n)$,*
(ii) *P_n has width bounded by a function in $O(w(n))$, i.e. all rule bodies of the programs P_n have at most $O(w(n))$ atoms,*
(iii) *P_n has height bounded by a function in $O(h(n))$, i.e. there are at most $O(h(n))$ rules in each P_n.*
2. *There exists a PTIME-computable function f which, for any n given in unary notation, produces the program P_n.*

Note that if we drop the second condition (uniformity condition) in the above definition, then all monotone queries are defined with polynomial width, exponential height.

If we assume that each IDB is defined by a finite number of rules bounded by a constant k, then we may replace condition (iii) by the following weaker condition: there are at most $O(h(n))$ IDB predicates in each program P_n.

Let $DAC(w(n), h(n))(b(n))$ be the set of $DAC(w(n), h(n))$ programs having depth of recursion bounded by a function in $O(b(n))$.

$\mathcal{P} = \{P_n, n \in \mathbb{N}\}$ in $DAC(w(n), h(n))$ ($DAC(w(n), h(n))(b(n))$) will be called a DAC-program (with depth of recursion $O(b(n))$); in other words, for $\mathcal{P}$ in $DAC(w(n), h(n))(b(n))$, the maximum depth of a proof tree of P_n is $b(n)$, i.e. $T_{P_n}^{\infty}(\mathcal{D}) = \bigcup_{0 \le i \le b(n)} T_{P_n}^i(\mathcal{D})$ for all $\mathcal{D} \in \mathbb{D}_n$. $b(n)$ is also called the stage function [2]. Note that the bound $b(n)$ is related to the depth parameter in circuits.

PROPOSITION II.5 *Any query defined by a DAC-program is in* $L_{\infty \omega}(\exists, \ne, \forall)$.

Proof. We refer the reader to [4] or [6] for the definition of $L_{\infty \omega}$. Every *Datalog* program is in $L_{\infty \omega}^{\omega}(\exists)$. Let $\mathcal{P} = \{P_n, n \in \mathbb{N}\}$ be a DAC-program and let $\phi_n = Q(\vec{X})$ be a query on the principal IDB predicate of P_n; the query corresponding to $\mathcal{P}$ is defined by $\phi = \bigvee_{n \in \mathbb{N}}(card(D) = n \wedge \phi_n)$ which is a formula of $L_{\infty \omega}(\exists, \ne, \forall)$. $\square$

II.3 Monotonicity

Recall that [3] a query is said to be **strongly monotone** if it is preserved by adding elements to the database domain, adding tuples to the database EDB relations (i.e. it is **monotone**), and moreover, it is also preserved by possibly identifying elements of the database domain. *Datalog* programs define strongly monotone queries [3]. We shall restrict our attention, here, in DAC queries which are strongly monotone.

We recall first some useful notation: Database $\mathcal{D}' = (D', R'_1, \ldots, R'_k)$ is an **extension** of database $\mathcal{D} = (D, R_1, \ldots, R_k)$, (denoted by $\mathcal{D} \subseteq \mathcal{D}'$), if $D \subseteq D'$ and for $i = 1, \ldots, k$, $R_i \subseteq R'_i$. If query Q is **monotone**, then: $\mathcal{D} \subseteq \mathcal{D}' \implies Q(\mathcal{D}) \subseteq Q(\mathcal{D}')$.

DEFINITION II.6 *1. A DAC-program $\mathcal{P} = \{P_n, n \in \mathbb{N}\}$ defining the principal predicate Q is said to be* **monotone** *if for all $n \le m$, for all $\mathcal{D}$: $Q_{P_n}(\mathcal{D}) \subseteq Q_{P_m}(\mathcal{D})$. The monotone DAC-programs will be denoted by mDAC.*
2. A DAC-program $\mathcal{P} = \{P_n, n \in \mathbb{N}\}$ defining the principal predicate Q is said to be **strongly monotone**, *denoted by $\mathcal{P} \in smDAC$, if and only if it is monotone, and moreover $\forall n, m \in \mathbb{N}$ with $n \le m$: $Q_{P_m}(\mathcal{D}_n) \subseteq Q_{P_n}(\mathcal{D}_n)$. The strongly monotone DAC-programs will be denoted by smDAC.*

Intuitively, *monotonicity* means that the P_ns are preserved under extensions of the domain and the database, in other words if a 'small' program P_n can prove the query A, then a 'large' program P_m can also prove A. *Strong monotonicity* means that the P_ns are preserved under extensions of the domain and the database, and moreover, if for $n \le m$, a query A can be proven on domain D_n using the program P_m, then it can also be proven using the program P_n. In other words, a strongly monotone program is monotone, and moreover, if a 'large' program P_m can prove the query A on a 'small' database, then a 'small' program P_n can also prove A. Note that the (strong) monotonicity of a DAC-program with principal predicate Q is a priori a notion different from the (strong) monotonicity of query Q; our terminology is however consistent because (see Proposition II.10) a DAC-program with principal predicate Q is strongly monotone if and only if the corresponding query Q is strongly monotone.

PROPOSITION II.7 *If the DAC-program* $\mathcal{P} = \{P_n, n \in \mathbb{N}\}$ *defining the principal predicate* Q *is monotone, then for all* $n \leq m$, *for all* $\mathcal{D}, \mathcal{D}'$: $\qquad \mathcal{D} \subseteq \mathcal{D}' \implies Q_{P_n}(\mathcal{D}) \subseteq Q_{P_m}(\mathcal{D}')$.

Let $DAC^{\neq}$ (resp. *Datalog*$^{\neq}$) be the language obtained by adjoining a symbol $\neq$ to our basic language DAC (resp. *Datalog*). The following examples offer some more intuition on DAC queries:

EXAMPLE II.8 1. Consider a graph as a database with a single binary relation, and the query $Q(x, y) = \{(x, y) \mid$ there exists a path of length $\lfloor \log n \rfloor$ between x and $y\}$. This query is $DAC(\log n, 1)(1)$ but it is not *Datalog* definable because it is not strongly monotone: indeed, let $n = 2^k$ and let $\mathcal{D}_n$ be a database of size n, then there may be a path of length $\lfloor \log n \rfloor = k$ between x and y, but if we collapse 2 points of $\mathcal{D}_n$, we obtain a database $\mathcal{D}_{n-1}$ of size $n - 1$ and there may no longer be a path of length $\lfloor \log(n-1) \rfloor \leq k - 1$ between x and y. Because query Q coincides with the query Q_1^0 of Theorem IV.5, query Q is also $DAC(1, \log \log n)(\log \log n)$.

2. On the previous class of graphs with a single binary relation,
- parity, i.e. the query which is true iff $\mathcal{D}$ has an even number of elements, is a non monotone $DAC(1, 1)(1)$ query: take the trivial program that always accepts or always rejects according to the parity of n,
- connectivity, i.e. the query which is true iff for any d, d' in $\mathcal{D}$ there is a path from d to d', is not a DAC query: the argument applied to show that it is not *Datalog* (see [1]) can also be applied to show that it is not DAC.
3. The query Q^k defined by

$$Q^k(\mathcal{D}) = \begin{cases} true & \text{if } \text{card}(\mathcal{D}) \geq k \\ false & \text{if } \text{card}(\mathcal{D}) < k \end{cases}$$

is a monotone $DAC(1, 1)$ query, defined by the DAC-program $\mathcal{P}^k = \{P_n^k, n \in \mathbb{N}\}$ with

$$P_n^k = \begin{cases} Q^k \longleftarrow & \text{for } n \geq k \\ \square & \text{for } n < k \end{cases}$$

$\mathcal{P}^k$ is monotone $DAC(1, 1)$ but not strongly monotone, since for $n < k \leq m$, $Q_{P_m^k}(\mathcal{D}_n) = true$, but $Q_{P_n^k}(\mathcal{D}_n) = false$. Note that query Q^k is expressible in *Datalog*$^{\neq}$, but it is not expressible in *Datalog*, because it is not strongly monotone.

II.4 General properties of DAC-programs

PROPOSITION II.9 $smDAC(1, 1) = Datalog$.
It can be shown that strongly monotone DAC-programs correspond to strongly monotone queries.

PROPOSITION II.10 *A strongly monotone DAC-program* $\mathcal{P} = \{P_n, n \in \mathbb{N}\}$ *defines a strongly monotone query. Conversely, if a DAC-program* $\mathcal{P}$ *defines a strongly monotone query, then* $\mathcal{P}$ *is strongly monotone.*

The following proposition shows a trade-off between the parameters of a DAC-program.

PROPOSITION II.11

$$DAC(w(n), h(n))(b(n)) \subseteq DAC\big(1, h(n)(w(n) - 1)\big)\big(b(n) \log w(n)\big)$$

The idea is to use a 'divide-and-conquer' strategy to simulate one rule of width w by $\log w$ rules of width 2. In other words, DAC allows for trade-offs between width, height and recursion depth:

- Proposition II.11 shows that a rule of width $w(n)$ can be simulated by $w(n)-1$ rules defining new IDBs of width 2 at the cost of an explosion in the number of arguments of the predicates;
- similarly, recursion depth could be decreased at the cost of increasing the number of IDB predicates together with the number of arguments of the predicates.

COROLLARY II.12 1. $DAC(w(n),1)(1) \subseteq DAC(1,w(n))(\log w(n))$ and
2. $DAC(1,n)(n) \not\subseteq DAC(n,1)(1)$.

The above corollary asserts that width can be simulated by height (and number of IDB predicates), but that the converse is false.

II.5 Bounded and strongly monotone DAC

Let n be the number of elements (entities) in the database, and N be the number of ground facts in the database; assuming every database has a fixed number p of EDB relations of maximum arity k, then $N \leq pn^k$ and that implies $O(\log n) = O(\log N)$. Hence, whenever we are interested in computations of at most logarithmic complexity, the complexity can be evaluated with respect to n or N indifferently.

A DAC-program is said to be **bounded** if its depth of recursion is bounded by a finite constant b, i.e. all computations of the programs P_n stop after at most b recursive calls, for all databases, and all n.

PROPOSITION II.13 *The classes Datalog and $mDAC(1,1)(1)$ are incomparable.*

Proof. 1. Note that (bounded) $mDAC(1,1)(1) \not\subseteq Datalog$: this is shown by the query Q^k: card$(\mathcal{D}) \geq k$ (of Example II.8 3)), which is monotone and bounded DAC, but is not $Datalog$, because it is not strongly monotone.
2. Conversely, $Datalog \not\subseteq mDAC(1,1)(1)$, this follows from Example I.1 (iv). □

THEOREM II.1 *For any function h, $smDAC(1,h(n))(1)$ coincides with bounded Datalog.*

III The expressiveness of DAC-programs on chain queries

Queries defining paths on databases consisting of graphs are quite useful in separating various classes of DAC-programs.

Let $\mathcal{D}_n = (D_n, R)$ be a database of type 2 on a domain D_n consisting of n vertices, i.e. our databases $\mathcal{D}_n$ will have a single edge relation R. On such a database, we will define some queries. We first give some useful definitions.

- *Paths* – On a database consisting of a graph given by an edge relation R, let $p = u_1 \cdots u_n$ be a path from x to y, i.e. a sequence such that for $k = 1, \ldots, n$, $u_k = (x_k, x_{k+1}) \in R$ and $x_1 = x$, $x_{n+1} = y$.
- *Paths accepted by queries or programs* – Let p be a path from x to y. Let Q be a binary query: p is said to be *accepted by the query Q* if and only if $(x, y) \in Q(\mathcal{D}_n)$.

DEFINITION III.14 *Let $f(x)$ be a strictly increasing function on $\mathbb{N}$. We define the chain query Q_f by:*

$$Q_f = \{(x,y)| \text{ there is a path of length } f(q) \text{ from } x \text{ to } y \text{ for some } q \in \mathbb{N}\}. \quad (1)$$

Such chain queries are completely described by languages $L \subset \{R\}^*$, or subsets of $\mathbb{N}$: to the query Q_f corresponds the language $L_f = \{R^{f(q)} \mid q \in \mathbb{N}\}$, or the subset $\{f(q) \mid q \in \mathbb{N}\}$ of $\mathbb{N}$. Note that chain queries are strongly monotone: they are clearly monotone, and if some elements of the database are collapsed, the edges connecting them are not deleted, but possibly give rise to cycles, hence the lengths of the paths are preserved.

III.1 Pumping Lemma

Let $\mathcal{D} = (D, R_1, \ldots, R_k)$ be a database of type $a = (2, \ldots, 2)$ on a domain D consisting of n vertices, i.e. all the R_is are binary relations. Let P be a *Datalog* program with principal predicate Q. Consider the fringe e of a proof tree for $Q(X, Y)$ (see Definition II.2). A fringe e which is a path of the form $R_{i_1}(x, x_1) \ldots R_{i_p}(x_{p-1}, y)$ is called a *word fringe* and $\omega = R_{i_1} \ldots R_{i_p}$ is said to be the corresponding word over the EDB symbols; we say that P *accepts* word ω via fringe e; if moreover ω can be pumped in the sense of the context–free language pumping lemma, keeping the size l of the pumped part less than N, i.e. $0 < l < N$, ω is said to *have a pumping string of length less than N*. In [3], the following technical result is proven:

LEMMA III.15 *Let P be a Datalog program, and assume that P accepts word ω via a fringe. There exists an integer N, depending only on P, such that, if ω is longer than N, then ω has a pumping string of length $< N$.*

Let P be a Datalog program with the following parameters: ν is the number of IDB predicate symbols, w is the maximum width (=number of arguments) of IDB predicate symbols, d is the maximum number of occurrences of IDB predicate symbols per rule, μ is the maximum number of EDB predicate symbols per rule. Let $N = \mu d^{\nu w^{4(w!3^w)^2}}$ and let ω be a word accepted by P via a fringe. If ω is longer than N, then ω has a pumping string of length $< N$.

Moreover, if we assume that all other parameters except ν are bounded from above by c_0 and $c' = c_0^{4(c_0!3^{c_0})^2}$, then we can take $N = c_0^{1+\nu c'}$. Consider, now, a strongly monotone $DAC(1, h(n))(b(n))$-program $\mathcal{P}$. For each P_n of $\mathcal{P}$ the above lemma holds if we can guarantee that ω_n (i.e., now the word has to be of length n) is longer than $N = c_0^{1+h(n)c'} = c2^{ch(n)}$, where c_0, c', c are constants of $\mathcal{P}$ (i.e., not depending on n). This is the case if $h(n) \in O(\log n)$. In addition, we need strong monotonicity to guarantee that P_n gives the right answer also if applied to databases smaller than n. Hence the following pumping lemma:

LEMMA III.16 *Let $h(n)$ be a function on $\mathbb{N}$; assume that $h(n) \in O(\log n)$. Let $\mathcal{P}$ be a strongly monotone $smDAC(1, h(n))$-program, and let $l(i)$ be an increasing function on $\mathbb{N}$. If $\mathcal{P}$ accepts the query Q_l, then there exist natural numbers i_0 and c such that, for any natural number $i > i_0$, $l(i+1) - l(i) \leq c2^{ch(l(i+1))}$, where c is a function of the constant parameters of $\mathcal{P}$.*

III.2 Applications

We use Lemma III.16 to separate DAC families by showing that 'too small' programs cannot capture chain queries computing paths whose length grow 'too fast'.

EXAMPLE III.17 The perfect square query Q_f defined by (1) (see Definition III.14) where $f(q) = q^2$, is strongly monotone and PTIME; it is not $Datalog$ ([3]), but we will see that it is $smDAC(\log n, n)(\log n)$ (cf. Corollary III.18).

THEOREM III.2 *Let Q_f be defined as in Definition III.14 with f PTIME-computable. Assume that, for any database $\mathcal{D}_n$ on a domain D_n of size n, if there exists a path of length $f(q)$ between two elements x and y of D_n, then there exists a path of length $f(q') \leq 2^{g(n)}$. Then Q_f is in the class $DAC(g(n), 2^{g(n)})(g(n))$.*

Proof. A program using a 'divide-and-conquer' strategy can accept the paths of length 2^i for $i \leq g(n)$ in recursion depth $g(n)$ (see Example I.1 (i)). Then we can combine paths of length 2^i for each bit i of the binary representation of $f(n)$. □

So this means for instance that each Q_f corresponding to an $f(n)$ with $|f(n)| < n$ is expressible in $smDAC(\log n, n)(\log n)$.

We first study the perfect p-th power query Q_p defined by (1) (see Definition III.14) where $f(q) = q^p$.

COROLLARY III.18 *The perfect p-th power query is in $smDAC(\log n, n)(\log n)$.*

THEOREM III.3 *The perfect p-th power query Q_p (i) is expressible in $smDAC(1, n^{12})(\log n)$, (ii) is not expressible in $smDAC(1, (\log n)^{1-\varepsilon})(\log n)$ for $\varepsilon > 0$, and (iii) is not expressible in Datalog.*

Proof. (i) It has been shown in [3] that, if a directed graph with n vertices has a 'long' path from x to y whose length is a perfect p-th power, then it also has a 'short' path of length l' a perfect p-th power with $l' \leq n^6$; hence it suffices to find all paths of length l'^p, with $l' \leq n^6$ to answer the query. We then apply (with some care) the above Theorem III.2 and Proposition II.11. Non-expressibility follows from Lemma III.16 for (ii) and from [3] for (iii). □

We now study the $Q_{2^\bullet}$ query, namely the query Q_f as defined in Definition III.14, where $f(x) = 2^x$.

PROPOSITION III.19 *Assume $\mathcal{D}_n$ is a size n database containing m loops of lengths $i_1, \ldots, i_k$ with $k \leq n$, k fixed. The perfect power-of-two query defined by $Q_{2^\bullet}(x, y)$ if and only if there exist an integer q and a path of length 2^q from x to y is in $DAC(1, poly)(\log n)$, and is not in $DAC(1, (\log n)^{1-\varepsilon})(\log n)$.*

IV A hierarchy of DAC-programs

Let $\mathcal{D}_n = (D_n, R)$ be a database consisting of a domain D_n of n vertices with a single edge relation R. On $\mathcal{D}_n$, consider the query $Q_0 = \{(x, y) |$ there is a path of length n from x to $y\}$. Using this query, we show that the following exponential hierarchy is strict.

THEOREM IV.4 *We have, for any $i > 0$, the strict inclusion:*

$$DAC(1, \underbrace{\log\log\ldots\log n}_{i \text{ times}})(\underbrace{\log\log\ldots\log n}_{i \text{ times}}) \subset DAC(1, \underbrace{\log\log\ldots\log n}_{i-1 \text{ times}})(\underbrace{\log\log\ldots\log n}_{i-1 \text{ times}})$$

In between any two levels of the exponential hierarchy of Theorem IV.4, we can squeeze a strict polynomial subhierarchy, namely we have strict sublogarithmic polylog hierarchies as stated in the following

THEOREM IV.5 *Let $n_0 = n$, and, for $i > 0$, $n_i = \underbrace{\log\log\ldots\log n}_{i \text{ times}}$; we have, for any $i \geq 2$ and $k \geq 1$:*

$$DAC(1, n_i)(n_i) \subset \cdots \subset DAC(1, n_i^k)(n_i^k)$$
$$\subset DAC(1, n_i^{k+1})(n_i^{k+1}) \subset \cdots \subset DAC(1, n_{i-1})(n_{i-1}).$$

V Conclusion

We introduced the class DAC of queries definable by a class of *Datalog* programs with various restrictions on the width, height and depth. We first showed that bounded $smDAC(1, h(n))$, for any h, coincides with bounded *Datalog* and we studied the class $DAC(1, poly)(\log n)$: it contains many natural queries which are not *Datalog* expressible.

We then proved non-expressiveness results for the class $DAC(1, \log\log n)(\log n)$ with the pumping technique and we inferred various strict hierarchies. DAC-definability of queries may turn out to be interesting in practice: we can compute transitive closure by a DAC-program of recursion depth $\log n/\log\log n$ as opposed to the usual $\log n$ (see Example I.1 (iii)) and other gains may be possible for other queries.

Acknowledgments: We thank Patrick Cegielski, Yuri Gurevich, Miklos Santha and Dan Suciu for interesting discussions related to this paper.

VI References

[1] S. Abiteboul, R. Hull, V. Vianu, *Foundations of databases*, Addison-Wesley (1995).

[2] S. Abiteboul, P. Kanellakis, Deciding bounded recursion in database logic programs, *SIGACT News* 20 (1989), 17–23.

[3] F. Afrati, S. Cosmadakis, M. Yannakakis, On *Datalog* vs. polynomial time, *J. Comput. Syst. Sc.* **51** (1995), 177–196.

[4] J. Barwise, The Syntax and Semantics of Infinitary Languages, Springer-Verlag (1968).

[5] R. Boppana, M. Sipser, The complexity of finite functions, *Handbook of Theoretical Computer Science* **14** (1990), 757–804.

[6] H. Ebbinghaus, J. Flum, *Finite Model Theory*, Springer-Verlag (1995).

The Complexity of Policy Evaluation for Finite-Horizon Partially-Observable Markov Decision Processes

Martin Mundhenk[1]* and Judy Goldsmith[2]** and Eric Allender[3]***

[1] Universität Trier, FB IV - Informatik, D-54286 Trier, Germany
[2] Dept. of Computer Science, University of Kentucky, Lexington KY 40506-0046
[3] Dept. of Computer Science, Rutgers University, Piscataway NJ 08855-1179

Abstract. A partially-observable Markov decision process (POMDP) is a generalization of a Markov decision process that allows for incomplete information regarding the state of the system. We consider several flavors of finite-horizon POMDPs. Our results concern the complexity of the policy evaluation and policy existence problems, which are characterized in terms of completeness for complexity classes.

We prove a new upper bound for the policy evaluation problem for POMDPs, showing it is complete for Probabilistic Logspace. From this, we prove policy existence problems for several variants of unobservable, succinctly represented MDPs to be complete for NP^{PP}, a class for which not many natural problems are known to be complete.

1 Introduction

Partially observable Markov Decision Processes (POMDPs) model controllable stochastic processes: there is a set of *states;* a controller chooses among some number of *actions,* each of which has an associated probability matrix of state transitions; associated with each state and action pair is a *reward.* Unlike Markov Decision Processes (MDPs), the controller may not get complete information about the system's current state, i.e. the state space is partitioned into *observations.* The basic goal, given such a model, is to find a strategy or *policy* for choosing actions, that maximizes the total expected reward over some fixed time horizon. Policies may consider the current observation only (stationary), or the observation and time (time-dependent). Although a large body of literature in mathematics, operations research, and engineering deals with optimization and

---

* Supported in part by the Office of the Vice Chancellor for Research and Graduate Studies at the University of Kentucky, and by the Deutsche Forschungsgemeinschaft (DFG), grant Mu 1226/2-1. Part of the work was done at University of Kentucky.
** Supported in part by NSF grant CCR-9315354.
*** Supported in part by NSF grant 9509603. Portions of the work were performed while at the Institute of Mathematical Sciences, Chennai (Madras), India, and at the Wilhelm-Schickard Institut für Informatik, Universität Tübingen (supported by DFG grant TU 7/117-1).

approximation strategies for POMDPs (see [18] for an overview), there has been little work aimed at characterizing the complexity of these problems and proving lower bounds.

We consider the complexity of the following decision problems. The *policy evaluation problem* asks whether a given POMDP has positive expected reward under a given policy. The *policy existence problem* asks whether – given a POMDP and type of policy – there exists a policy of the given type under which the POMDP has positive expected reward. All of the problems considered here are for POMDPs that run for a fixed, finite time horizon, equal to the number of states of the system or the size of the representation of the system.

For a given POMDP and policy, the expected reward is known to be computable in polynomial time. We show that this Dynamic Programming approach for POMDPs can be performed in a parallel manner. Therefore we obtain an algorithm for the policy evaluation problem in probabilistic logarithmic space (PL) and prove completeness for PL (Section 3).

We then apply our technique to succinctly represented POMDPs. This representation allows compact encoding of e.g. graphs with regularities, which arise in many practical and theoretical areas of computer science (see [3, 12, 24]). Our interest in succinct representations is motivated by work in AI/planning, including e.g. [7, 8, 17]. We show that the policy evaluation problem for succinctly encoded POMDPs is complete for polynomial space (PSPACE) (Section 4). We even show completeness for probabilistic polynomial time (PP), if the POMDPs have "concise" transition probabilities (Section 5). As consequences, we get completeness results for the policy existence problems for unobservable MDPs, i.e. POMDPs where the controller does not receive any information about the system's state.

Our results are summarized in tables 1 and 2. Each line stands for the completeness of some type(s) of MDP problems. The number of states of the MDP is denoted by n. There are many factors that contribute to the complexity of the problems. In the "policy" column, "s" stands for stationary, "t" for time-dependent. The "succinct" column says whether the encoding of the MDP is not succinct (−), or whether the encoding is succinct and the rewards and transition probabilities have $\log n$ bits (i.e. "regular" transitions) or n bits. The "rewards" column says whether we consider MDPs with nonnegative rewards, or with unrestricted (i.e. rational) rewards. The "horizon" and the number of "actions" may be restricted accordingly.

Our work presents a uniform approach to a wide variety of planning problems, and differs from previous works as follows. Papadimitriou and Tsitsiklis [21] considered POMDPs with non-positive rewards, and asked whether there is a policy with expected value of 0. Their maximization problem can be stated as a minimization problem for POMDPs with nonnegative rewards. As with many other optimization problems, the complexity of the minimization and maximization versions differs strongly (see Section 6). For time-dependent policies they showed the policy existence problem for (fully-observable) MDPs to be P-complete (for finite or infinite horizon), that for unobservable MDPs to be NP-complete (compare with Theorem 6), and for history-dependent policies they proved the policy

policy	succinct	reward	horizon	completeness
s/t	–	+	n	NL
s/t	–	–/+	n	PL
s/t	n	+	$\log n$	NP
s/t	$\log n$	–/+	$\log n$	PP
s/t	n	+	n	PSPACE
s/t	n	–/+	n	PSPACE

Table 1. Completeness results of policy evaluation problems for POMDPs

policy	succinct	reward	actions	horizon	completeness
s/t	–	+	n	n	NL
s	–	–/+	n	n	PL
t	–	–/+	n	n	NP
s/t	n	+	n	$\log n$	NP
s	$\log n$	–/+	$\log n$	$\log n$	PP
t	$\log n$	–/+	$\log n$	$\log n$	NP^{PP}
s/t	$\log n$	–/+	n	$\log n$	NP^{PP}
s/t	n	+	n	n	PSPACE
s	n	–/+	n	n	PSPACE
t	n	–/+	n	n	NEXP

Table 2. Completeness results of policy existence problems for unobservable MDPs

existence problem for POMDPs to be PSPACE-complete [21]. They also showed an NP-completeness result for the stationary, finite-horizon case [20]. Beauquier et al. [5] considered different optimality criteria. Most of the related AI literature, especially [8, 9, 10], consider computationally simpler problems without probabilistic transitions. Thus, the complexity of their problems is generally lower than of ours (except [13, 17] and one theorem in [8]).

2 Definitions and Preliminaries

For definitions of standard complexity classes, reductions, and results from complexity theory we refer to [19]. We give a short review of definitions of counting classes we use. The class #L was defined in [2]. It is the class of functions f such that, for some nondeterministic logarithmic space-bounded machine N, the number of accepting paths of N on x equals $f(x)$. The class GapL (defined in [23]) is the class of functions f such that, for some nondeterministic logarithmic space-bounded machine N, the number of accepting paths of N on x minus the number of rejecting paths of N on x equals $f(x)$. An equivalent characterization of GapL is as the class of functions representable as difference of #L functions: GapL = #L − #L (see [1]). Note that every logspace computable function is in #L, and #L is contained in GapL. Whereas all #L functions map to nonnegative integers, GapL functions map to all integers. All GapL functions are computable in polynomial time.

Probabilistic logspace, PL, is defined as the class of sets A for which there exists a nondeterministic logarithmic space-bounded machine N, such that $x \in A$ if and only if the number of accepting paths of N on x is greater than the number of rejecting paths. (Thus PL is the class of sets A, for which there is a GapL function f such that for every x, $x \in A$ iff $f(x) > 0$.) In apparent contrast to P-complete sets, sets in PL can be decided using very fast parallel computations (see [6]).

Counting classes for polynomial time or space are defined analogously. #P, GapP, and PP are defined accordingly with respect to polynomial time [11].

For polynomial space-bounded computations, the respective definitions can be shown to yield the same class, so PSPACE = PPSPACE, and #PSPACE is the same as the class of polynomial space computable functions (see [16]).

2.1 Markov Decision Processes

A partially-observable Markov decision process (POMDP) describes a controlled stochastic system by its states and the consequences of actions on the system. It is denoted as a tuple $M = (\mathcal{S}, s_0, \mathcal{A}, \mathcal{O}, t, o, r)$, where

- $\mathcal{S}$, $\mathcal{A}$ and $\mathcal{O}$ are finite sets of *states*, *actions* and *observations*,
- $s_0 \in \mathcal{S}$ is the *initial state*,
- $t : \mathcal{S} \times \mathcal{A} \times \mathcal{S} \to [0, 1]$ is the *state transition function*, where $t(s, a, s')$ is the probability to reach state s' from state s on action a (where $\Sigma_{s' \in \mathcal{S}} t(s, a, s') \in \{0, 1\}$ for $s \in \mathcal{S}, a \in \mathcal{A}$),
- $o : \mathcal{S} \to \mathcal{O}$ is the *observation function*, where $o(s)$ is the observation made in state s,
- $r : \mathcal{S} \times \mathcal{A} \to \mathbf{Z}$ is the *reward function*, where $r(s, a)$ is the reward gained by taking action a in state s.

We distinguish two special types of observability of a POMDP. If states and observations are identical, i.e. $\mathcal{O} = \mathcal{S}$ and o is the identity function (or a bijection), then the POMDP is called a *fully-observable* MDP. It is called *unobservable*, if the set of observations contains only one element or the observation function is constant, i.e. in every state the same observation is made.[4] If the reward function maps to nonnegative integers, we say it has *nonnegative rewards*.

2.2 Policies and Performances

A policy describes how to act depending on observations. We distinguish two types of policies. A *stationary policy* π_s (for M) is a function $\pi_s : \mathcal{O} \to \mathcal{A}$,

---

[4] Making observations probabilistically does not add any power to POMDPs. Any probabilistically observable MDP can be turned into one with deterministic observations with only a polynomial increase in its size.

mapping each observation to an action. A *time-dependent policy* π_t is a function $\pi_t : \mathcal{O} \times \mathbf{N} \to \mathcal{A}$, mapping each pair ⟨observation, time⟩ to an action.

Let $M = (\mathcal{S}, s_0, \mathcal{A}, \mathcal{O}, t, o, r)$ be a POMDP. A *trajectory* θ for M is a finite sequence of states $\theta = \sigma_1, \sigma_2, \ldots, \sigma_m$ $(m \geq 0, \sigma_i \in \mathcal{S})$. The probability $\mathrm{prob}_\pi(M, \theta)$ of a trajectory $\theta = \sigma_1, \sigma_2, \ldots, \sigma_m$ under policy π is

 – $\mathrm{prob}_\pi(M, \theta) = \Pi_{i=1}^{m-1} t(\sigma_i, \pi(o(\sigma_i)), \sigma_{i+1})$, if π is a stationary policy, and
 – $\mathrm{prob}_\pi(M, \theta) = \Pi_{i=1}^{m-1} t(\sigma_i, \pi(o(\sigma_i), i), \sigma_{i+1})$, if π is a time-dependent policy.

The *reward* $\mathrm{rew}_\pi(M, \theta)$ of trajectory θ under π is the sum of its rewards, i.e. dependent on the type of policy π, $\mathrm{rew}_\pi(M, \theta) = \Sigma_{i=1}^{m-1} r(\sigma_i, \pi(\cdot))$. The *performance of a policy* π for finite-horizon k with initial state σ is the expected sum of rewards received on the next k steps by following the policy π, i.e. $\mathrm{perf}(M, \sigma, k, \pi) = \sum_{\theta \in \Theta(\sigma, k)} \mathrm{prob}_\pi(M, \theta) \cdot \mathrm{rew}_\pi(M, \theta)$, where $\Theta(\sigma, k)$ is the set of all length k trajectories beginning with state σ. The α-value $\mathrm{val}_\alpha(M, k)$ of M for horizon k is M's maximal performance under a policy π of type α for horizon k when started in its initial state, i.e. $\mathrm{val}_\alpha(M, k) = \max_{\pi \in \Pi_\alpha} \mathrm{perf}(M, s_0, k, \pi)$, where Π_α is the set of all α policies.

For simplicity and w.l.o.g. we assume that the size of a POMDP is determined by the size $|\mathcal{S}|$ of its state space, that there are no more actions than states, and that each state transition probability is given as binary fraction with $|\mathcal{S}|$ bits and each reward is an integer of $|\mathcal{S}|$ bits.

3   "Flat" Represented POMDPs

There are different methods how a POMDP can be represented. In this section, we consider the complexity of policy evaluation and existence problems for problem instances being represented in a straightforward way. Namely a POMDP with n states is represented by a set of $n \times n$ tables for the transition function (one table for each action) and similar tables for the reward and observation functions. In the same way, stationary policies can be encoded as lists with n entries, and time-dependent policies for horizon n as $n \times n$ tables.

The α *policy evaluation problem for partially-observable MDPs* is the set of all pairs (M, π) consisting of a POMDP $M = (\mathcal{S}, \mathcal{O}, s_0, \mathcal{A}, o, t, r)$ and an α policy π with $\mathrm{perf}(M, s_0, |\mathcal{S}|, \pi) > 0$.

The α *policy existence problem for unobservable MDPs* is the set of all unobservable MDPs $M = (\mathcal{S}, \mathcal{O}, s_0, \mathcal{A}, o, t, r)$ with $\mathrm{val}_\alpha(M, |\mathcal{S}|) > 0$.

The standard polynomial time algorithm for evaluating a given policy for a given POMDP uses dynamic programming [21, 22]. We show that for POMDPs this evaluation can be performed quickly in parallel. Eventually this yields the policy evaluation problem being complete for PL. We begin with a technical lemma about the complexity of matrix powering.

Lemma 1. *(cf. [23]) Let T be an $n \times n$ matrix of nonnegative binary integers, each of length n, and let $1 \leq i, j \leq n$, $0 \leq m \leq n$. The function mapping (T, m, i, j) to $(T^m)_{(i,j)}$ is in #L.*

Lemma 2. *The stationary policy evaluation problem for partially-observable MDPs is in* PL.

Proof. Let $\hat{M} = (\mathcal{S}, s_0, \mathcal{A}, \mathcal{O}, \hat{t}, o, \hat{r})$ be a POMDP, and let π be a stationary policy. We show that $\text{perf}(\hat{M}, s_0, |\mathcal{S}|, \pi) > 0$ can be decided in PL.

We transform $\hat{M}$ to an unobservable MDP M with the same states and rewards as $\hat{M}$, having the same performance as $\hat{M}$ under policy π. This can be achieved by "hard-wiring" the actions chosen by π into the MDP. Then $M = (\mathcal{S}, s_0, \{a\}, \mathcal{O}, t, o, r)$ with $t(s, a, s') = \hat{t}(s, \pi(o(s)), s')$ and $r(s, a) = \hat{r}(s, \pi(o(s)))$. Since M has only one action, the only policy to consider is the constant function mapping each observation to that action a. It is clear that $\hat{M}$ under policy π has the same performance as M under (constant) policy a. This performance can be calculated using a recursive definition of perf, namely $\text{perf}(M, i, m, a) = r(i, a) + \sum_{j \in \mathcal{S}} t(i, a, j) \cdot \text{perf}(M, j, m-1, a)$ and $\text{perf}(M, i, 0, a) = 0$.

The state transition probabilities are given as binary fractions of length $h = |\mathcal{S}|$. Let T be the matrix obtained from the transition matrix of M by multiplying all entries by 2^h. The recursive definition of perf can be resolved to

$$\text{perf}(M, i, m) = 2^{-hm} \cdot \sum_{k=1}^{m} \sum_{j \in \mathcal{S}} (T^{k-1})_{(i,j)} \cdot r(j, a) \cdot 2^{(m-k+1) \cdot h} \quad .$$

In order to complete the proof, we have to show that the sum on the right hand side of the above equation is in GapL.

Each $T_{(i,j)}$ is logspace computable from the input $\hat{M}$. From Lemma 1 we get that $(T^{k-1})_{(i,j)} \in \#L$. The reward function is part of the input too, thus r is in GapL (note that rewards may be negative integers). Because GapL is closed under multiplication and polynomial summation (see [1]), the proof is complete.

Lemma 3. *The stationary policy evaluation problem for partially-observable MDPs is* PL*-hard.*

Proof. Consider $A \in$ PL. Then there exists a probabilistic logspace machine N accepting A, and a polynomial p such that each computation of N on x uses at most $p(|x|)$ steps [15]. For any input x, we construct an (unobservable) MDP $M(x)$ which has only one action and models the behavior of N on x. Each of the polynomially many states (c, i) of $M(x)$ consists of a configuration c of N on x and an integer i used as a counter for the number of random decisions made to reach this configuration. The state transition function of $M(x)$ is defined according to the state transition function of N on x, so that each halting computation of N on x corresponds to a trajectory of $M(x)$ and vice versa. The probability of a trajectory equals 2^{-i} depending on the number i of random decisions made by the corresponding computation. Therefore, chosing the reward function such that $r(c, i)$ equals 2^i if c is an accepting configuration, $-(2^i)$ if c is rejecting, and 0 otherwise, yields that $\text{rew}(\theta) \cdot \text{prob}(\theta)$ equals 1 for trajectories θ corresponding to accepting computations, or -1 for rejecting computations, or 0 otherwise. It then follows that $x \in A$ if and only if $\text{perf}(M(x), s_0, |\mathcal{S}|, a) > 0$.

The same technique applies for time-dependent policy evaluation problems, and for unobservable MDPs.

Theorem 4. *The stationary and time-dependent policy evaluation problems for unobservable and for partially-observable MDPs are* PL-*complete.*

For unobservable MDPs, there are only a few stationary policies. Hence the evaluation and the existence problem are of the same complexity.

Theorem 5. *The stationary policy existence problem for unobservable MDPs is* PL-*complete.*

The time-dependent policy existence problem for unobservable MDPs turns out to be as complex as the one for partially-observable MDPs (see [14]).

Theorem 6. *The time-dependent policy existence problem for unobservable MDPs is* NP-*complete.*

Papadimitriou and Tsitsiklis proved a result very similar to Theorem 6 [21].

4 Succinctly Represented POMDPs

Succinct representations of MDPs arise when the system being modeled has sufficient structure, e.g., when a state is modeled by the states of n variables, and for each action, the state of each variable depends on only a few of the variables. This is true, for instance, in systems modeled as a Bayes belief net, or when actions are given in the STRIPS model [8, 17, 13].

Changing the way in which POMDPs (and policies) are represented may change the complexities of the considered problems too. We focus on the concept of *succinct representations*, which was introduced independently in [12, 24]. For a POMDP, it can be seen as a Boolean circuit C such that $C(s, a, s', l)$ outputs the l-th bit of $t(s, a, s')$. These circuits are no larger than the straightforward matrix encodings. But for POMDPs with sufficient structure, they may be much smaller, namely size order of the logarithm of the state space. In going from straightforward to succinct representations, the complexity of the corresponding decision problems increases at most exponentially. More importantly (and less obviously), there are many problems for which this exponential increase in complexity is inherent (see [4] for conditions).

The policy evaluation and existence problems for partially-observable succinctly represented MDPs (sMDPs) are analogous to flat POMDPs. The horizon now may be exponential in the size of the representation of the POMDP.

Put succinctly, all results from Section 3 translate to succinctly represented POMDPs.

Theorem 7. *1. The following problems are complete for* PSPACE:
- *The stationary and time-dependent policy evaluation problems for unobservable and for partially-observable sMDPs, and*
- *the stationary policy existence problem for unobservable sMDPs.*
2. The time-dependent policy existence problem for unobservable sMDPs is NEXP-*complete.*

5 Succinctly Represented Limited POMDPs

A special case of succinctly represented POMDPs are those where the circuits representing the transition tables have many output gates from which on input s, a, s' the transition probability $t(s, a, s')$ can be read. Because the number of output gates is bounded by the size of the circuit, for a POMDP with n states, which is represented by such circuits of size order of $\log n$, the transition probabilities must be represented with only $\log n$ bits. Our completeness proof techniques from the last section do not apply for these limited POMDPs. But we show that for succinctly represented limited POMDPs we get completeness results intermediate between those of flat and succinct POMDPs, if we restrict the horizon to be logarithmic in the size of the state space. It turns out that the number of bits needed to represent the actions and rewards contributes to the complexity too. Since there are so many factors now, the problem descriptions are more involved.

The α *policy evaluation problem for partially-observable* $s_{\log}$MDP$[f(n)]$ *with* $g(n)$ *horizon* is the set of all pairs (M, π) consisting of a partially-observable succinctly represented MDP $M = (\mathcal{S}, \mathcal{O}, s_0, \mathcal{A}, o, t, r)$ where each transition probability and reward takes $\log |\mathcal{S}|$ many bits and $|\mathcal{A}| \leq f(|\mathcal{S}|)$, and an α policy π, such that $\mathrm{perf}(M, s_0, g(|\mathcal{S}|), \pi) > 0$. The α *policy existence problem for unobservable* $s_{\log}$MDP$[f(n)]$ *with* $g(n)$ *horizon* is defined accordingly.

If f or g is the identity function, we omit it in the problem description.

The complexity of the policy evaluation problems can be shown to be between NP and PSPACE.

Theorem 8. *The stationary and time-dependent policy evaluation problems for unobservable and for partially-observable* $s_{\log}$MDP$[\log n]$ *and* $\log n$ *horizon are PP-complete.*

For the complexity of policy existence problems, the number of actions of the MDP is important.

Theorem 9. *The stationary policy existence problem for unobservable* $s_{\log}$MDP$[\log n]$ *with* $\log n$ *horizon is PP-complete.*

To show containment in PP, we use arguments similar to those in the proof of Lemma 2. (The respective matrix powering problem can be shown to be in #P.)

Omitting the restriction on the number of actions, the complexity of the problem rises to NP^{PP}.

Theorem 10. *The stationary policy existence problem for unobservable* $s_{\log}$MDP *with* $\log n$ *horizon is* NP^{PP}*-complete.*

If the horizon increases to the size of the state space (which may be exponential in the size of the MDP's representation), we get completeness for PSPACE. Note that the same problem for unobservable MDPs with nonnegative rewards is PSPACE-complete too (see Theorem 13).

Theorem 11. *The stationary policy existence problem for unobservable* $s_{\log}$MDP *is PSPACE-complete.*

As a consequence we obtain that the policy existence problem for flat MDPs with exponential horizon is in PSPACE.

The complexity gap between the stationary and the time-dependent policy existence problems for $s_{\log}$MDP[$\log n$]'s is as big as that for flat MDPs, but the difference no longer depends on the number of actions.

Theorem 12. *The time-dependent policy existence problems for unobservable* $s_{\log}$MDP[$\log n$] *with* $\log n$ *horizon and for unobservable* $s_{\log}$MDP *with* $\log n$ *horizon are* NP^{PP}-*complete.*

6 POMDPs with Nonnegative Rewards

For POMDPs with nonnegative rewards, the policy evaluation and existence problems reduce in a straightforward way to graph accessibility problems. In order to determine whether an MDP has performance greater than 0 under a given policy, it suffices to find a reachable state with positive transition probability and a corresponding positive reward. Moreover, the transition probabilities and rewards do not need to be calculated exactly.

Theorem 13. *1. The stationary and time-dependent policy evaluation problems for unobservable and for partially-observable MDPs (resp. sMDPs) with nonnegative rewards are complete for NL (resp. PSPACE).*
2. The stationary and time-dependent policy existence problems for unobservable MDPs (resp. sMDPs) with nonnegative rewards are complete for NL (resp. PSPACE).
3. The stationary and time-dependent policy evaluation problems for unobservable and for partially-observable sMDPs with nonnegative rewards for horizon $\log n$ *are complete for NP.*
4. The stationary and time-dependent policy existence problems for unobservable sMDPs with nonnegative rewards for horizon $\log n$ *are complete for NP.*

Acknowledgements

We would like to thank Anne Condon, Chris Lusena, Matthew Levy, and especially Michael Littman for discussions and suggestions on this material.

References

1. E. Allender and M. Ogihara. Relationships among PL, #L, and the determinant. *RAIRO Theoretical Informatics and Applications*, 30(1):1–21, 1996.
2. C. Àlvarez and B. Jenner. A very hard log-space counting class. *Theoretical Computer Science*, 107:3–30, 1993.

3. J.L. Balcázar. The complexity of searching implicit graphs. *Artificial Intelligence*, 86:171–188, 1996.

4. J.L. Balcázar, A. Lozano, and J. Torán. The complexity of algorithmic problems on succinct instances. In R. Baeza-Yates and U. Manber, editors, *Computer Science*, pages 351–377. Plenum Press, 1992.

5. D. Beauquier, D. Burago, and A. Slissenko. On the complexity of finite memory policies for Markov decision processes. In *Math. Foundations of Computer Science*, pages 191–200. Lecture Notes in Computer Science #969, Springer-Verlag, 1995.

6. A. Borodin, S. Cook, and N. Pippenger. Parallel computation for well-endowed rings and space-bounded probabilistic machines. *Information and Control*, 58(1–3):113–136, 1983.

7. C. Boutilier and D. Poole. Computing optimal policies for partially observable decision processes using compact representations. In *Proc. 13th National Conference on Artificial Intelligence*, pages 1168–1175. AAAI Press / MIT Press, 1996.

8. T. Bylander. The computational complexity of propositional STRIPS planning. *Artificial Intelligence*, 69:165–204, 1994.

9. K. Erol, J. Hendler, and D. Nau. Complexity results for hierarchical task-network planning. *Annals of Mathematics and Artificial Intelligence*, 1996.

10. K. Erol, D. Nau, and V. S. Subrahmanian. Complexity, decidability and undecidability results for domain-independent planning. *Artificial Intelligence*, 76:75–88, 1995.

11. S. Fenner, L. Fortnow, and S. Kurtz. Gap-definable counting classes. *Journal of Computer and System Sciences*, 48(1):116–148, 1994.

12. H. Galperin and A. Wigderson. Succinct representation of graphs. *Information and Control*, 56:183–198, 1983.

13. J. Goldsmith, M. Littman, and M. Mundhenk. The complexity of plan existence and evaluation in probabilistic domains. In *Proc. 13th Conf. on Uncertainty in AI*. Morgan Kaufmann Publishers, 1997.

14. J. Goldsmith, C. Lusena, and M. Mundhenk. The complexity of deterministically observable finite-horizon Markov decision processes. Technical Report 269-96, University of Kentucky Department of Computer Science, 1996.

15. H. Jung. On probabilistic time and space. In *Proceedings 12th ICALP*, pages 281–291. Lecture Notes in Computer Science #194, Springer-Verlag, 1985.

16. R. Ladner. Polynomial space counting problems. *SIAM Journal on Computing*, 18:1087–1097, 1989.

17. M.L. Littman. Probabilistic propositional planning: Representations and complexity. In *Proc. 14th National Conference on AI*. AAAI Press / MIT Press, 1997.

18. W.S. Lovejoy. A survey of algorithmic methods for partially observed Markov decision processes. *Annals of Operations Research*, 28:47–66, 1991.

19. C.H. Papadimitriou. *Computational Complexity*. Addison-Wesley, 1994.

20. C.H. Papadimitriou and J.N. Tsitsiklis. Intractable problems in control theory. *SIAM Journal of Control and Optimization*, pages 639–654, 1986.

21. C.H. Papadimitriou and J.N. Tsitsiklis. The complexity of Markov decision processes. *Mathematics of Operations Research*, 12(3):441–450, 1987.

22. M.L. Puterman. *Markov decision processes*. John Wiley & Sons, New York, 1994.

23. V. Vinay. Counting auxiliary pushdown automata and semi-unbounded arithmetic circuits. In *Proc. 6th Structure in Complexity Theory Conference*, pages 270–284. IEEE, 1991.

24. K. W. Wagner. The complexity of combinatorial problems with succinct input representation. *Acta Informatica*, 23:325–356, 1986.

A Category of Transition Systems and Its Relations with Orthomodular Posets

Luca Bernardinello and Lucia Pomello *

Dipartimento di Scienze dell'Informazione, Università di Milano
Via Comelico 39 - 20135 Milano (Italy)
e-mail: bernardi@socrate.usr.dsi.unimi.it, pomello@dsi.unimi.it

Abstract. Two categories are defined and their relationships are studied. The objects of the first category, PCOS, are prime coherent orthomodular posets, which have been mainly studied in connection with quantum logic. Morphisms in PCOS are homomorphisms in the usual sense, preserving order and a binary relation, named compatibility.
The second category, denoted by CETS, comprises the class of labelled transition systems that can be generated, up to isomorphism, by case graphs of CE systems. Two contravariant functors linking the two categories are defined. The functor from CETS to PCOS is given by the calculus of regions, according to Ehrenfeucht and Rozenberg. The functor from PCOS to CETS defines a procedure which builds a labelled transition system, given an abstract set of regions with their order relation. We show that the two functors form an adjunction.

1 Introduction

In this paper, the relationships between a class of transition systems and a class of partially ordered sets, whose elements can be interpreted as propositions denoting properties of a system, are studied in the categorical framework.

We consider the subclass of transition systems that can be generated, up to isomorphism, as case graphs of Condition Event net systems (CE systems) [Pet73], [Pet96]. A CE system is a model of a distributed system, characterized by local states and transitions, as opposed to global states and transitions in transition systems. The theory of such net systems, initiated by C.A. Petri in the early '60s, is founded on the relation of concurrency between net elements and on a small set of axioms. From our perspective, the net model of a system consists in a set of propositions (conditions), which, at a given instant, may be true or false, and by a set of events, which modify the truth value of conditions. The axiom of extensionality of events states that an event is fully characterized by two sets of conditions: its pre-conditions, which must be true before the event takes place and become false after its occurrence, and its post-conditions, which must be false before the event takes place and become true after its occurrence. The behaviour of such a net system can be described by its case graph, that

---

* This work was partially supported by the MURST

is a transition system whose global states are sets of conditions, and whose transitions are occurrences of events.

In 1990, Ehrenfeucht and Rozenberg [ER90] introduced 2-structures, which can be seen as transition systems. They introduced the notion of region; a region is a subset of states verifying a uniformity condition with respect to event occurrences. Regions correspond very strictly to conditions in CE systems. They are at the heart of the solution to the following problem [ER90]: given a labelled transition system A, decide whether there exists a CE system whose case graph is isomorphic to A, and, if the answer is positive, exhibit such a net system.

Later, Nielsen, Rozenberg, and Thiagarajan [NRT92] showed that the construction of the case graph of a CE system, and the regional construction of a CE system from a labelled transition system are functorial, with respect to morphisms intuitively corresponding to behavioural simulation. They also proved that the two functors form an adjunction.

These results fostered a series of studies on the problem of synthesis, see [BD96]. Further studies have shown that the set of regions of a CE transition system, ordered by set inclusion, is a prime coherent orthomodular poset (denoted PCO poset) [BF95]. Orthomodular posets are partially ordered sets with a least and a greatest element, and a complement operation. They have been studied mainly in connection with the algebraic approach to quantum logic [PP91], [Hug89]. It was then natural to ask the question whether any PCO poset is isomorphic to the set of regions of a CE transition system. Even if the answer is negative, it was shown that, given an arbitrary PCO poset, it is possible to build a CE transition system such that the original poset embeds into its set of regions (see [BF95], [Ber96], [Fer96]). The key mathematical tool in the construction is the set of *prime filters* of a PCO poset. If we regard the elements of a PCO poset as propositions, and the order relation as implication, then a prime filter is a maximal set of consistent propositions. Therefore, we regard prime filters as states of the transition system to be built. According to the axiom of extensionality for events, which was mentioned above, a transition between two states is labelled by the pair of set differences between prime filters corresponding to the two states. They are precisely the set of propositions which cease to hold and the set of propositions which begin to hold as a consequence of the event occurrence.

In this paper, we show that the two constructions, of a PCO poset from a CE transition system, via the calculus of regions, and of a CE transition system from a PCO poset, via the calculus of prime filters, are functorial, the two functors being contravariant. The main result of the paper is the proof that they form an adjunction.

The paper is organized as follows. Section 2 collects basic definitions and results on orthomodular posets and CE transition systems, and the definition of the two considered categories PCOS and CETS. Section 3 contains the definition of a contravariant functor from CETS to PCOS. In Section 4, the synthesis of a CE transition system from a PCO poset is recalled, and it is shown that this construction can be lifted to a contravariant functor from PCOS to CETS.

Section 5 contains the main results of this paper, concerning the relationships between the two functors. Finally, in the last section, some open problems and expected future developments are briefly discussed.

2 Preliminaries

In this section we introduce the two categories which will be compared in the subsequent sections.

2.1 Orthomodular Posets

Orthomodular posets can be considered as a generalization of Boolean algebras, where meet and join are partial operations. Orthomodular posets are equipped with a unary operator corresponding to the complement in set theory.

Definition 1. An *orthomodular poset* $P = <P, \leq, 0, 1, '>$ is a partially ordered set $< P, \leq>$, equipped with a minimum and a maximum element, respectively denoted by 0 and 1, and with a map $(.)' : P \to P$, such that the following conditions are verified, where $\vee$ and $\wedge$ denote, respectively, the least upper bound and the greatest lower bound with respect to $\leq$, when they exist: $\forall x, y \in P$

1. $(x')' = x$;
2. $x \leq y \Rightarrow (y' \leq x' \text{ and } y = x \vee (y \wedge x'))$;
3. $x \leq y' \Rightarrow x \vee y \in P$;
4. $x \wedge x' = 0$.

Definition 2. Let $P = <P, \leq, 0, 1, '>$ be an orthomodular poset and $x, y \in P$.

- x and y are *orthogonal*, denoted $x \perp y$, iff $x \leq y'$;
- x and y are *compatible*, denoted $x \$ y$, iff $\exists x_0, y_0, z \in P$ such that $x_0 \perp y_0 \perp z \perp x_0$ and $x = x_0 \vee z$ and $y = y_0 \vee z$.

From the previous definitions it follows that: $x \perp x'$; $x \$ x'$; $x \$ y \Rightarrow (x \vee y \in P$ and $x \wedge y \in P$ and $x' \wedge y' = (x \vee y)'$ and $x' \vee y' = (x \wedge y)')$; $x \leq y \Rightarrow x \$ y$; $x \perp y \Rightarrow x \$ y$.

Definition 3. An orthomodular poset $P = <P, \leq, 0, 1, '>$ is *coherent* iff $\forall x, y, z \in P$ such that $x \$ y \$ z \$ x$ it holds: $(x \vee y) \$ z$.

Definition 4. Let P_1 and P_2 be orthomodular posets. A *morphism* from P_1 to P_2 is a map $\beta : P_1 \to P_2$ such that:

1. $\beta(0) = 0$;
2. $\beta(x') = (\beta(x))'$;
3. $\forall x, y \in P : x \perp y \Rightarrow \beta(x \vee y) = \beta(x) \vee \beta(y)$.

Morphisms defined as above preserve order, compatibility and orthogonality. Prime filters in orthomodular posets generalize ultrafilters in Boolean algebras.

Definition 5. Let $P = <P, \leq, 0, 1, '>$ be an orthomodular poset and $f \subseteq P$. Then f is a *prime filter* iff

1. $f \neq \emptyset$;
2. $(x \leq y$ and $x \in f) \Rightarrow y \in f$;
3. $(x \$ y$ and $x, y \in f) \Rightarrow x \wedge y \in f$;
4. $x \in f \Leftrightarrow x' \notin f$.

Conditions 1. and 2. of Definition 5 imply $1 \in f$; moreover, condition 4. and $1 \in f$ implies $0 \notin f$.

The set of all prime filters of P will be denoted by $\text{FP}(P)$, while the set of all prime filters containing an $x \in P$ will be denoted by $F_x = \{f \in \text{FP}(P) | x \in f\}$.

Definition 6. An orthomodular poset $P = <P, \leq, 0, 1, '>$ is *prime* iff $\forall x, y \in P : x \neq y \Rightarrow (\exists f \in \text{FP}(P)$ such that $x \in f \Leftrightarrow y \notin f)$.

Lemma 7. *[Fer96] Let P_1 and P_2 be two prime and coherent orthomodular posets and $\beta : P_1 \rightarrow P_2$ be a morphism. Let $f \in \text{FP}(P_2)$, then $\beta^{-1}(f) \in \text{FP}(P_1)$.*

In the following we will consider only the class of finite prime and coherent orthomodular posets. This class, together with morphisms as defined above will be denoted by PCOS; PCOS is a category [BF95].

2.2 Transition Systems

In this section we present CE transition systems, which describe the sequential behaviour of CE net systems [Pet73]. CE transition systems are strictly related to elementary transition systems [NRT92]. The properties of elementary transition systems can be easily extended to CE transition systems.

Definition 8. A *transition system* is a structure $A = (S, E, T)$, where S is a set of *states*, E is a set of *events*, $T \subseteq S \times E \times S$ is a set of *transitions*, and

1. the underlying graph of the transition system is simply connected;
2. $\forall (s, e, s') \in T \ s \neq s'$;
3. $\forall (s, e_1, s_1), (s, e_2, s_2) \in T : s_1 = s_2 \Rightarrow e_1 = e_2$;
4. $\forall e \in E \ \exists (s, e, s') \in T$.

A region is a set of states such that all occurrences of a given event have the same crossing relation (entering, leaving or non crossing) with respect to the region itself, and this property holds for all events [ER90].

Definition 9. Let $A = (S, E, T)$ be a transition system. A set of states $r \subseteq S$ is said to be a *region* iff $\forall e \in E, \forall (s_1, e, s_1), (s_2, e, s_2) \in T$ we have $(s_1 \in r$ and $s_1' \notin r) \Rightarrow (s_2 \in r$ and $s_2' \notin r)$ and $(s_1 \notin r$ and $s_1' \in r) \Rightarrow (s_2 \notin r$ and $s_2' \in r)$.

It is easy to verify that, for each transition system, both the empty set and the set of all states are regions. The set of all regions of A will be denoted by R_A. For each $s \in S$, R_s will denote the set of regions containing s.

Definition 10. Let $A = (S, E, T)$ be a transition system. Let $r \in R_A$. Then the *pre-set* of r, denoted by $\bullet r$, and the *post-set* of r, denoted by $r\bullet$, are defined by: $\bullet r = \{e \in E | \; \exists (s, e, s') \in T : s \notin r \text{ and } s' \in r\} \; r\bullet = \{e \in E | \; \exists (s, e, s') \in T : s \in r \text{ and } s' \notin r\}$. Let $e \in E$. Then the *pre-set* and the *post-set* of e, denoted by, respectively, $\bullet e$ and $e\bullet$, are defined by: $\bullet e = \{r | \; r \in R_A \text{ and } e \in r\bullet\}$, $e\bullet = \{r | \; r \in R_A \text{ and } e \in \bullet r\}$

Definition 11. A finite transition system $A = (S, E, T)$ is a *Condition Event Transition System* (CE transition system) iff it satisfies the following axioms:

A1. $\forall s, s' \in S : R_s = R_s \Rightarrow s = s'$
A2. $\forall s \in S \; \forall e \in E : \bullet e \subseteq R_s \Rightarrow \exists s' \in S \; (s, e, s') \in T$
A3. $\forall s \in S \; \forall e \in E : e\bullet \subseteq R_s \Rightarrow \exists s' \in S \; (s', e, s) \in T$

Definition 12. Let $A_1 = (S_1, E_1, T_1)$ and $A_2 = (S_2, E_2, T_2)$ be CE transition systems. A *morphism* from A_1 to A_2 is a total function $\alpha : S_1 \to S_2$ such that: $\forall (s_1, e_1, s_2), (s_3, e_1, s_4) \in T_1 \; (\alpha(s_1) = \alpha(s_2) \text{ and } \alpha(s_3) = \alpha(s_4))$ or $\exists e_2 \in E_2 : (\alpha(s_1), e_2, \alpha(s_2)), (\alpha(s_3), e_2, \alpha(s_4)) \in T_2$.

The proof of the following lemma can be easily derived from [NRT92].

Lemma 13. *Let A_1 and A_2 be CE transition systems and α be a morphism from A_1 to A_2. Let $r \in R_{A_2}$, then $\alpha^{-1}(r) \in R_{A_1}$.*

CETS denotes CE transition systems together with their morphisms. It is easy to prove that CETS is a category, [NRT92].

3 From CETS to PCOS

In this section we define a contravariant functor from the category CETS to the category PCOS. An example of its application is given in the next section. The first result of this section states that the set of regions of a CE transition system is a PCO poset. We shall prove that this may be lifted to a functor. The functor will be denoted by **H**.

Lemma 14. *[Ber96] Let $A = (S, E, T)$ be an element of CETS. Define $\mathbf{H}(A) = \; <R_A, \subseteq, \emptyset, S,'>$. Then $\mathbf{H}(A)$ is an element of PCOS.*

Orthogonality and compatibility can be interpreted as follows in the ortho-modular poset of regions $< R_A, \subseteq, \emptyset, S,' >$. For each $x, y \in R_A$

$- \; x \perp y \Leftrightarrow x \cap y = \emptyset; \; x \; \$ \; y \Leftrightarrow x \cup y \in R_A \Leftrightarrow x \cap y \in R_A;$
$- \;$ if $x \cap y \in R_A$ then $x \cap y = x \wedge y;$ if $x \cup y \in R_A$ then $x \cup y = x \vee y.$

The following figure shows a CE transition system A and the PCO poset associated to it by the functor **H**. We have, for example, $u = \{1, 2\}, w = \{5\} \in R_A; R_2 = \{u, x, v', w', y', 1\} \in \mathrm{FP}(\mathbf{H}(A)).$

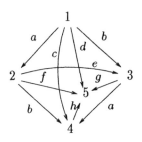 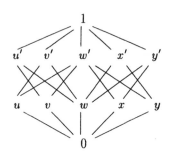

The following lemma justifies the definition of **H** on morphisms.

Lemma 15. *Let $A_i = (S_i, E_i, T_i)$ be an element of CETS for $i = 1, 2$, and $\alpha : S_1 \to S_2$ be a morphism in CETS. Then α^{-1} restricted to R_{A_2} is a morphism in PCOS from $\mathbf{H}(A_2)$ to $\mathbf{H}(A_1)$.*

Proof. 1. $\alpha^{-1}(\emptyset) = \emptyset$.
2. $\alpha^{-1}(r') = \alpha^{-1}(S_2 - r) = \alpha^{-1}(S_2) - \alpha^{-1}(r) = S_1 - \alpha^{-1}(r) = (\alpha^{-1}(r))'$.
3. Let r_1, r_2 be disjoint regions of A_2; then $(r_1 \cup r_2)$ is a region of A_2, and $\alpha^{-1}(r_1 \cup r_2) = \alpha^{-1}(r_1) \cup \alpha^{-1}(r_2)$.
Therefore the thesis is proved (see Definition 4). □

Definition 16. Let $\alpha : A_1 \to A_2$ be a morphism in CETS. Define $\mathbf{H}(\alpha) : \mathbf{H}(A_2) \to \mathbf{H}(A_1)$ by $\forall r \in R_{A_2} : \mathbf{H}(\alpha)(r) = \alpha^{-1}(r)$.

In order to assert that **H** is actually a functor, we have to prove that **H** preserves identities and compositions of morphism. This is accomplished in the following two lemmas, whose proofs are easily derived.

Lemma 17. *Let $i : A \to A$ be an identity morphism in CETS. Then $\mathbf{H}(i)$ is the identity morphism for $\mathbf{H}(A)$ in PCOS.*

Lemma 18. *Let $\alpha_1 : A_1 \to A_2$ and $\alpha_2 : A_2 \to A_3$ be morphisms in CETS, and $\alpha : A_1 \to A_3$ be their composition. Then $\mathbf{H}(\alpha)$ is the composition of $\mathbf{H}(\alpha_2)$ and $\mathbf{H}(\alpha_1)$ in PCOS.*

4 From PCOS to CETS

In this section, we define a contravariant functor from the category PCOS to the category CETS. The action of the functor on objects can be seen as a synthesis procedure, which builds a transition system from a given logical structure, expressed as an orthomodular poset. If we interpret the elements of an orthomodular poset as propositions, then we would like to define the transition system so that each proposition corresponds to a region.

Intuitively, prime filters of an orthomodular poset are maximal sets of mutually consistent propositions; hence they can be considered as states. The transition from a state to another is characterized by the propositions that cease to hold and the propositions that begin to hold.

Let $P = <P, \leq, 0, 1, '>$ be a prime coherent orthomodular poset. Define $E_P = \{< f_1 - f_2, f_2 - f_1 > | f_1, f_2 \in \text{FP}(P), f_1 \neq f_2\}$ and $T_P = \{(f_1, e, f_2) | f_1, f_2 \in \text{FP}(P), f_1 \neq f_2$ and $e = < f_1 - f_2, f_2 - f_1 >\}$.

Lemma 19. *[Ber96] Let $P = <P, \leq, 0, 1, '>$ be a prime coherent orthomodular poset. Define $\mathbf{J}(P) = (\text{FP}(P), E_P, T_P)$. Then $\mathbf{J}(P)$ is an element of CETS.*

The transition system associated to P contains a transition for each pair of states. The labelling of events reflects the principle of extensionality: any event is characterized by its pre-set and post-set, given by a symmetric set difference.

With respect to the example given in Section 3, $\mathbf{J}\mathbf{H}(A)$ can be obtained by symmetric and transitive closure of the underlying graph of A. In general $\mathbf{J}\mathbf{H}(A)$ may have not only more transitions, but also more states than A.

Lemma 20. *[Ber96] Let $y \in P$; then F_y is a region of $\mathbf{J}(P)$.*

The proof of Lemma 20 is a simple verification of the conditions in Def. 9. The following definition is well-founded by Lemma 7.

Definition 21. Let $\beta : P_1 \rightarrow P_2$ be a morphism in PCOS. Define $\mathbf{J}(\beta) : \mathbf{J}(P_2) \rightarrow \mathbf{J}(P_1)$ by $\forall f \in \text{FP}(P_2) : \mathbf{J}(\beta)(f) = \beta^{-1}(f)$.

Lemma 22. *Let $\beta : P_1 \rightarrow P_2$ be a morphism in PCOS. Then $\mathbf{J}(\beta)$ is a morphism from $\mathbf{J}(P_2)$ to $\mathbf{J}(P_1)$ in CETS.*

Proof. Let $(f_1, e, f_2), (f_3, e, f_4) \in T_{P_2}$. If $\beta^{-1}(f_1) \leq \beta^{-1}(f_2)$, then, by construction of $\mathbf{J}(P_1)$, there is $x \in E_{P_1}$ such that $(\beta^{-1}(f_1), x, \beta^{-1}(f_2)) \in T_{P_1}$. We have $f_1 - f_2 = f_3 - f_4$, and $f_2 - f_1 = f_4 - f_3$. Since β^{-1} preserves set difference, $\beta^{-1}(f_1 - f_2) = \beta^{-1}(f_1) - \beta^{-1}(f_2) = \beta^{-1}(f_3) - \beta^{-1}(f_4)$, hence $(\beta^{-1}(f_3), x, \beta^{-1}(f_4)) \in T_{P_1}$. $\square$

The following immediate lemmas complete the proof that $\mathbf{J}$ is a functor.

Lemma 23. *Let $i : P \rightarrow P$ be an identity morphism in PCOS. Then $\mathbf{J}(i)$ is the identity morphism of $\mathbf{J}(P)$ in CETS.*

Lemma 24. *Let $\beta_1 : P_1 \rightarrow P_2$ and $\beta_2 : P_2 \rightarrow P_3$ be morphisms in PCOS, and $\beta : P_1 \rightarrow P_3$ their composition. Then $\mathbf{J}(\beta) = \mathbf{J}(\beta_1)\mathbf{J}(\beta_2)$.*

5 Adjunctions

In this section, we study the relations between $\mathbf{H}$ and $\mathbf{J}$.

We start by recalling some constructions from [Mac71] which will be used later on. To a category C we associate the opposite category C^{op}; the objects of C^{op} are the objects of C, the arrows of C^{op} are of the form f^{op}, where f is an arrow of C. If $f : a \rightarrow b$, then $f^{\text{op}} : b \rightarrow a$.

To a contravariant functor $F : C \rightarrow D$ we associate two covariant functors, $F^{\text{op}} : C \rightarrow D^{\text{op}}$ and $\hat{F} : C^{\text{op}} \rightarrow D$, defined as follows. For each object a of C,

$F^{\mathrm{op}}(a) = \hat{F}(a) = F(a)$; for each arrow f in C, $F^{\mathrm{op}}(f) = (F(f))^{\mathrm{op}}$; for each arrow f^{op} in C^{op}, $\hat{F}(f^{\mathrm{op}}) = F(f)$.

Since $\mathbf{H}$ and $\mathbf{J}$ are contravariant, following [Mac71], in this section we will actually prove that $\mathbf{H}^{\mathrm{op}}$ is a left adjoint for $\hat{\mathbf{J}}$, and $\mathbf{J}^{\mathrm{op}}$ is a left adjoint for $\hat{\mathbf{H}}$. In the proofs we apply the previous equalities without explicit mention. The two following lemmas, [Ber96], define the unity of each adjunction.

Lemma 25. *Let* $A = (S, E, T)$ *be an object of CETS. Define* $u_A : S \to \mathrm{FP}(\mathbf{H}(A))$ *by* $\forall s \in S : u_A(s) = R_s$. *Then* u_A *is an injective morphism in CETS from* A *to* $\mathbf{J}\,\mathbf{H}(A)$.

Lemma 26. *Let* P *be an object of PCOS. Define* $\eta_P : P \to \mathcal{P}(\mathrm{FP}(P))$ *by* $\forall x \in P : \eta_P(x) = F_x$. *Then* η_P *is an injective morphism in PCOS from* P *to* $\mathbf{H}\,\mathbf{J}(P)$. *(Where* $\mathcal{P}(X)$ *denotes the powerset of* X.*)*

Theorem 27. $\mathbf{H}^{\mathrm{op}}$ *is a left adjoint for* $\hat{\mathbf{J}}$.

Proof. Following [Mac71], proving the thesis amounts to show that, given an object A in CETS, an object P in PCOS, and a morphism $\alpha : A \to \mathbf{J}(P)$, there is exactly one $\beta : P \to \mathbf{H}(A)$ in PCOS such that $\mathbf{J}(\beta)u_A = \alpha$

First we show that such a β exists. Let $y \in P$. By Lemma 20, we know that F_y is a region of $\mathbf{J}(P)$. By Lemma 13, $\alpha^{-1}(F_y)$ is a region of A, hence an element of $\mathbf{H}(A)$. Define $\beta(y) = \alpha^{-1}(F_y)$, then β is a morphism in PCOS:

1. $\beta(0) = \alpha^{-1}(F_0) = \alpha^{-1}(\emptyset) = \emptyset$;
2. $\beta(y') = \alpha^{-1}(F_{y'}) = \alpha^{-1}(\mathrm{FP}(P) - F_y) = \alpha^{-1}(\mathrm{FP}(P)) - \alpha^{-1}(F_y) = S - \beta(y) = (\beta(y))'$;
3. let $y_1, y_2 \in P$ such that $y_1 \perp y_2$; then $\beta(y_1 \vee y_2) = \alpha^{-1}(F_{y_1 \vee y_2}) = \alpha^{-1}(F_{y_1} \cup F_{y_2}) = \alpha^{-1}(F_{y_1}) \cup \alpha^{-1}(F_{y_2}) = \beta(y_1) \cup \beta(y_2)$.

We will now prove that $\mathbf{J}(\beta)u_A = \alpha$. Let $s \in S$; by Lemma 25, $\mathbf{J}(\beta)(u_A(s)) = \mathbf{J}(\beta)(R_s) = \mathbf{J}(\beta)(\{r \in R_A | s \in r\}) = \{x \in P | s \in \beta(x)\} = \{x \in P | s \in \alpha^{-1}(Fx)\} = \{x \in P | s \in \{q \in S | x \in \alpha(q)\}\}$.

We have now to prove that this last set $M = \{x \in P | s \in \{q \in S | x \in \alpha(q)\}\}$ is equal to $\alpha(s)$. Let $z \in M$, then $s \in \{q \in S | z \in \alpha(q)\}$ and therefore $z \in \alpha(s)$, that implies $M \subseteq \alpha(s)$; let $z \in \alpha(s)$; then $s \in \{q \in S | z \in \alpha(q)\}$ and therefore $z \in M$, that implies $\alpha(s) \subseteq M$.

We now prove that such a β is unique. Let $\gamma : P \to \mathbf{H}(A)$ be a morphism in PCOS such that $\mathbf{J}(\gamma)u_A = \alpha$, i.e.: such that for each $s \in S : \gamma^{-1}(u_A(s)) = \alpha(s)$, i.e.: for each $s \in S : \gamma^{-1}(R_s) = \alpha(s)$.

We shall prove that, under this hypothesis, for each $y \in P : \gamma(y) = \alpha^{-1}(F_y)$, thus showing that $\gamma = \beta$. Let $s \in \alpha^{-1}(F_y)$ then $\alpha(s) \in F_y$. We have for the previous hypothesis that $\alpha(s) = \gamma^{-1}(R_s)$, therefore, being $y \in \alpha(s)$, $\gamma(y) \in R_s$ and $s \in \gamma(y)$, and then we have proved $\alpha^{-1}(F_y) \subseteq \gamma(y)$.

Let now $s \in \gamma(y)$; then $\gamma(y) \in R_s$. By hypothesis $\gamma^{-1}(R_s) = \alpha(s)$; then, since $y \in \gamma^{-1}(R_s)$, it follows $y \in \alpha(s)$, then $\alpha(s) \in F_y$ and $s \in \alpha^{-1}(F_y)$, and then we have proved also $\gamma(y) \subseteq \alpha^{-1}(F_y)$, and then $\gamma(y) = \alpha^{-1}(F_y)$.

$\square$

Theorem 28. $\mathbf{J}^{\mathrm{op}}$ *is a left adjoint for* $\hat{\mathbf{H}}$.

Proof. Along the same line of Theorem 27, we shall prove that, given an object P in PCOS, an object $A = (S, E, T)$ in CETS, and a morphism $\beta : P \to \mathbf{H}(A)$ in PCOS, there is a unique morphism $\alpha : A \to \mathbf{J}(P)$ in CETS such that $\beta = \mathbf{H}(\alpha)\eta_P$ (η_P is the same as in Lemma 26).

First we define α. Let $s \in S$; then, by Lemmas 25 and 7, $R_s \in \mathrm{FP}(\mathbf{H}(A))$ and $\beta^{-1}(R_s) \in \mathrm{FP}(P)$; therefore $\beta^{-1}(R_s)$ is a state of $\mathbf{J}(P)$. Define $\alpha(s) = \beta^{-1}(R_s)$.

We prove that α is a morphism in CETS. Take $(s_1, e, s_2), (s_3, e, s_4) \in T$. We have (see [NRT92]) $R_{s_1} - R_{s_2} = R_{s_3} - R_{s_4}$ and $R_{s_2} - R_{s_1} = R_{s_4} - R_{s_3}$. Since β^{-1} preserves set difference, $\alpha(s_1) - \alpha(s_2) = \alpha(s_3) - \alpha(s_4)$ and $\alpha(s_2) - \alpha(s_1) = \alpha(s_4) - \alpha(s_3)$, whence $(\alpha(s_1), a, \alpha(s_2)), (\alpha(s_3), a, \alpha(s_4)) \in T_P$, for $a \in E_P$.

Commutativity: Let $x \in P$. $\mathbf{H}(\alpha)\eta_P(x) = \mathbf{H}(\alpha)(F_x) = \alpha^{-1}(F_x) = \alpha^{-1}(\{f \in \mathrm{FP}(P) | x \in f\}) = \{s \in S | x \in \alpha(s)\} = \{s \in S | x \in \beta^{-1}(R_s)\}$. Let $z \in \{s \in S | x \in \beta^{-1}(R_s)\}$. Then $x \in \beta^{-1}(R_z)$, that is $\beta(x) \in R_z$, so $z \in \beta(x)$. This proves that $\mathbf{H}(\alpha)\eta_P(x) \subseteq \beta(x)$. Now let $z \in \beta(x)$; then $\beta(x) \in R_z$, hence $x \in \beta^{-1}(R_z)$, whence $z \in \{s \in S | x \in \beta^{-1}(R_s)\}$. So $\beta(x) \subseteq \mathbf{H}(\alpha)\eta_P(x)$. We can conclude that $\beta(x) = \mathbf{H}(\alpha)\eta_P(x)$. Uniqueness: Let $\gamma : A \to \mathbf{J}(P)$ be a morphism in CETS (hence γ is a map of S into $\mathrm{FP}(P)$), such that for each $x \in P \gamma^{-1}(\eta_P(x)) = \beta(x)$. Then, we prove that, for each $s \in S, \gamma(s) = \beta^{-1}(R_s)$, thus showing that $\gamma = \alpha$. From the hypothesis, we get $\beta(x) = \gamma^{-1}(F_x) = \{s \in S | \gamma(s) \in F_x\} = \{s \in S | x \in \gamma(s)\}$. Let $z \in S$ (hence $\gamma(z) \in \mathrm{FP}(P)$); take $w \in \gamma(z)$. Then $z \in \gamma^{-1}(F_w) = \beta(w)$, so $w \in \beta^{-1}(R_z)$. This shows that $\gamma(z) \subseteq \beta^{-1}(R_z)$. On the other hand, let $w \in \beta^{-1}(R_z)$; then $z \in \beta(w) = \gamma^{-1}(F_w)$, whence $w \in \gamma(z)$. Therefore $\beta^{-1}(R_z) \subseteq \gamma(z)$. Finally, we have $\gamma(z) = \beta^{-1}(R_z)$, which completes the proof that $\gamma = \alpha$. $\qquad\square$

6   Conclusion

In this paper, we have introduced two categories. The objects of the first category, PCOS, are prime coherent orthomodular posets. Morphisms in PCOS are homomorphisms in the usual sense, preserving order and a binary relation, named compatibility. The second category, denoted by CETS, consists in the class of labelled transition systems that can be generated, up to isomorphism, by case graphs of CE systems. Morphisms in this category correspond to the notion of partial simulation. Two contravariant functors linking the two categories have been defined. The functor from CETS to PCOS is given by the calculus of regions, according to Ehrenfeucht and Rozenberg. The functor from PCOS to CETS defines a procedure which builds a labelled transition system given an abstract set of regions with their order relation. We have shown that the two functors form an adjunction.

The result raises some open problems. First of all, we would like to characterize the images of the functors, which we know do not coincide with the whole categories. More important, from our point of view, is the question whether composing the two functors, in both orders, yields idempotent functors. This would allow us to interpret those compositions as a sort of saturation operations.

Apart from the open problems, future development should investigate Boolean subalgebras of prime coherent orthomodular posets, which are connected with so-called state-machine subnets in Petri net theory. Moreover, we plan to study some categorical aspects, in particular the existence of limits in the two categories and their interpretation. We are also interested in comparing this approach with the one proposed in [Pra94] and based on Chu Spaces, a very general mathematical structure formalizing the duality between states and properties of a system.

References

[BD96] E. Badouel, P. Darondeau, A survey on net synthesis, In Borne et al., Symposium on Discrete Events and Manufacturing Systems, CESA '96 IMACS Multiconference, Lille, France, July 1996, pp. 309-316.

[Ber96] L. Bernardinello, Propriétés algébriques et combinatoires des régions dans les graphes, et leurs application à la synthèse de réseaux, PhD thesis, Université de Rennes I, to be discussed, 1996.

[BF95] L. Bernardinello, C. Ferigato, Automata and Quantum Logics, DSI Int.Rep., n. 38-95, abstract in 10th Int.Congress of Logic, Methodology and Philosophy of Science, Florence, 1995.

[Fer96] C. Ferigato, Note su alcune Proprietà Algebriche, Logiche, Topologiche della Concorrenza, Tesi di Dottorato di Ricerca in Informatica, Università degli Studi di Milano, 1996.

[ER90] A. Ehrenfeucht, G. Rozenberg, Partial (set) 2-Structures I & II, Acta Informatica, 27, 4, pp. 315-368, 1990.

[Hug89] R.I.G. Hughes, The structure and Interpretation of Quantum Mechanics, Harvard University Press, 1989.

[Mac71] S. MacLane, Categories for the working mathematician, Graduate Text in Mathematics 5, Springer-Verlag, 1971.

[NRT92] M. Nielsen, G. Rozenberg, P.S. Thiagarajan, Elementary Transition Systems, in Theoretical Computer Science 96 (1), pp. 3-33, 1992.

[Pet73] C.A. Petri, Concepts in net theory, in MFCS'73, Mathematical Institute of Slovak Academy of Sciences, pp. 137-146, 1973.

[Pet96] C.A. Petri, Nets, time and space, in TCS 153 (1-2), pp.3-48, 1996.

[Pra94] V.R. Pratt, Chu spaces: Automata with quantum aspects, Proc. PhysComp'94, IEEE 1994.

[PP91] P. Pták, S. Pulmannová, Orthomodular structures as Quantum Logics, Kluwer Academic Publishers, 1991.

Accepting Zeno Words
Without Making Time Stand Still

Béatrice Bérard and Claudine Picaronny

LSV, CNRS URA 2236, ENS de Cachan, 61 av. du Prés. Wilson,
F-94235 Cachan Cedex, France, Fax: +33 (0)1 47 40 24 64
E-mails: Beatrice.Berard@lsv.ens-cachan.fr,
Claudine.Picaronny@lsv.ens-cachan.fr

Abstract. Timed models were introduced to describe the behaviors of real-time systems and, up to now, they were usually required to produce only executions with divergent sequences of times. However, some physical phenomena, as the movements of a damped oscillator, can be represented by convergent executions, producing Zeno words in a natural way. Moreover, time can progress if such an execution can be followed by other ones. We extend the definition of timed automata, allowing to generate sequences of infinite convergent executions, while keeping good properties for the verification of systems: emptiness is still decidable. We introduce a new notion of refinement for timed systems, in which actions are replaced by recognizable Zeno languages and we prove that the corresponding class of languages is the closure of the usual one under refinement.

1   Introduction

Timed models have been intensively studied for the specification and verification of real time systems. Contrary to classical (untimed) transition systems, these models contain an explicit notion of time and allow to describe time requirements. For instance, timed Petri nets [19], timed transition systems [18,15], timed automata [1,2] or timed I/O automata [17] have been discussed and successfully used for the verification of real-time systems [11,16].

In these systems, executions can be observed by finite or infinite sequences of timed actions $(a_1, t_1)(a_2, t_2) \cdots$, where t_i is the date at which action a_i takes place. Surprisingly, executions for which the time sequence $t_1 t_2 \cdots$ is convergent appear in these models. The corresponding sequences, called Zeno words in reference to Zeno's paradox, are usually forbidden because they seem to prevent the course of time.

Nevertheless, Zeno words could be needed to describe some physical phenomena that produce convergent time sequences: for instance when a continuous action is represented by an infinite discrete sequence or, from another point of view, when an infinite number of actions takes place within a finite interval of time. Of course, time does not stop, because other phenomena can be observed later on. This idea was already expressed in [13] where systems, in which the state can change infinitely often in a finite time, are investigated within the framework of the duration calculus. Different examples (like the fall of a satellite towards a planet) are proposed there and the Car-Bee problem is studied in details. As another illustration, consider an elastic ball which is dropped on an horizontal plane, bounces while losing amplitude, stops after some time and can then be dropped again or trigger the beginning of another observation. If action a represents the movement of the ball between two contacts, we obtain an infinite sequence $(a, t_1)(a, t_2) \cdots$ with convergent time, followed by other sequences, thus producing what is called a transfinite timed word. In fact, in the usual timed models, a Zeno word makes time stop only because executions are reduced to one infinite sequence.

On the other hand, automata producing transfinite untimed words have already been studied ([9,10,21,14]). Even though they do not take explicitly into account the notion of time, these automata suppose implicitly that an infinite number of actions may occur in a finite interval of time, in order to be followed by other actions. From this point of view, adding an explicit notion of time in such models is interesting in itself.

In this paper, our purpose is to define a timed version of the automata of Choueka [10], which can generate transfinite timed words, and to study the class of timed languages accepted by these automata. We prove that they have convenient closure properties, particularly under the concatenation and the derived star and ω-power operations, which become possible with Zeno words. Moreover, we show that emptiness remains decidable for these languages, which is an important point regarding the verification of systems.

The rest of the paper is devoted to the important notion of refinement. A refinement consists in the replacement of a single action (intended to represent some high level abstract action) by a language. Although many results have been obtained in the classical framework of transition systems, only a few cases have been investigated for timed models [7]. We introduce a special class of timed automata, which can be used to describe timed refinement and include the cases in [7]. Finally, the main result of the paper is the following: the class of languages accepted by timed transfinite automata is exactly the closure under refinement of the class of languages accepted by classical timed automata. All detailed proofs are in [8].

2 Preliminaries

In this section, we define timed words (with time domain $\mathbb{R}_+$) of ordinal length and we extend the basic operations for languages of timed words. Ordinals are used mainly in order to number sequences and all facts concerning ordinals in this paper can be found in [20].

Recall that the finite ordinals are the natural numbers and the first non-finite ordinal is denoted by ω. In general, an ordinal α represents a well-ordered set (up to isomorphism) and two ordinals α, β may be summed ($\alpha + \beta$) by considering (the isomorphism class of) the union $\alpha \cup \beta$, where $\alpha < \beta$. The ordinal $\alpha + 1$

is called the *successor* of α. This addition defines a product (as for the natural numbers). For instance, $\omega.2 = \omega + \omega$ and $\omega^2 = \omega + \omega + ...$, ω times.

In this work, we consider only ordinals smaller than ω^ω: such an ordinal has a decomposition $\alpha = \sum_{k=p}^{0} \omega^k.n_k$, where $p, n_0, n_1 \cdots$ are natural numbers and the type of α is the least integer k such that $n_k \neq 0$. If the type of α is positive, i.e. α does not have a greatest element, then it is called a *limit* ordinal. Thus, a limit ordinal of type k can be written $\alpha = \beta + \omega^k$, where $\beta = 0$ or β is an ordinal of type $\geq k$.

Let Σ be a finite alphabet of actions and let α be an ordinal. A *timed α-word* (or simply *timed word*) over Σ is an α-sequence $w = ((a_i, t_i))_{i<\alpha}$, where a_i is in Σ for each $i < \alpha$ and $(t_i)_{i<\alpha}$ is a non decreasing sequence of non negative real numbers. The value t_i represents the time at which the action a_i ends, and $t_\alpha = sup(t_i, i < \alpha)$ is called the *ending time* of w and denoted by $\tau(w)$. A *Zeno (timed) word* is a timed α-word for which the ending time is finite.

As time progresses, such a word can be followed by other timed actions, so that we extend the usual operation of concatenation after Zeno words: if $w = (a_i, t_i)_{i<\alpha}$ and $w' = (a'_i, t'_i)_{i<\alpha'}$ are timed words such that the ending time of w is finite, then $ww' = ((a''_i, t''_i))_{i<\alpha+\alpha'}$ with $(a''_i, t''_i) = (a_i, t_i), \forall i < \alpha$ and $(a''_{i+\alpha'}, t''_{i+\alpha'}) = (a'_i, \tau(w) + t'_i), \forall i < \alpha'$.

A *timed language* is a set of timed words. For two timed languages L and L', the concatenation (with the associated star and ω-product operations) becomes possible after Zeno words:

- concatenation: $LL' = \{ww'/w \in L$ such that $\tau(w)$ is finite and $w' \in L'\}$,
- star operation: $L^* = \bigcup_{i \geq 0} L^i$, where $L^0 = \{\varepsilon\}$ and $L^{i+1} = L^i L$,
- ω-power: $L^\omega = \{\prod_{i<\omega} w_i/w_i \in L\}$.

3 Timed n-automata

Different models of automata accepting transfinite (untimed) words have been studied ([10,21,14,4,3]), since their first introduction by Büchi ([9]) in order to prove the decidability of monadic second order logics. Restricted to ordinals less than ω^ω, these models have the same expressive power: they accept tranfinite languages having rational expressions. We chose to use Choueka's model [10] as a basis, because it can be described as a natural generalization of classical Muller automata (for infinite words), and provides an intuitive view of the accepting mechanism.

If n is a natural number, a Choueka n-automaton is a finite automaton with a global set of states S split into $n+1$ layers. A state in the kth layer is called a state of *type k* and, for $k > 0$, such a state is in fact a set of states of type $k-1$. An execution begins as in a usual automaton in the first layer of states. After an infinite (ω) sequence of actions, the set of infinitely repeated states of the path is considered as a state of the second layer. A new action may then be performed to get down to the first layer and the execution can go on. A state of the third layer will be reached after ω^2 actions, and so on.

More precisely, for a finite set Q, we define inductively $[Q]^0 = Q$ and, for any natural number k, $[Q]^{k+1}$ is the powerset of $[Q]^k$, without the empty set, and we write $[Q]_0^k = Q \cup [Q]^1 \cup \cdots \cup [Q]^k$.

A *(Choueka) n-automaton* is a tuple $\mathcal{A} = (\Sigma, Q, I, F, \Delta)$, where

Σ is a finite alphabet of actions,
Q is a finite set of states (of type 0),
$I \subseteq Q$ is a subset of initial states,
$F \subseteq [Q]_0^n$ is a set of final states,
$\Delta \subseteq [Q]_0^{n-1} \times \Sigma \times Q$ is the transition relation.

Note that the total set of states of $\mathcal{A}$ is $S = [Q]_0^n$.

A *continuous run* of $\mathcal{A}$ on a α-word $w = (a_i)_{i<\alpha}$ is an α-sequence $(q_i)_{0 \leq i \leq \alpha}$ of states such that:

- $q_0 \in [Q]_0^{n-1}$
- if i is the successor of some ordinal $i-1$, there is a transition $(q_{i-1}, a_i, q_i) \in \Delta$,
- if i is a limit ordinal of type $k > 0$, recall that $i = j + \omega^k$, where $j = 0$ or j is an ordinal of type greater than or equal to k. In this case, q_i is a state of type k defined by: $q_i = inf\{q_{j+\omega^{k-1}.p}/p \in \mathbb{N}\}$, where $inf\{z_1, z_2, \cdots\}$ is the set of elements z_i which appear infinitely often in the sequence.

The run is *accepting* if $q_0 \in I$ and $q_\alpha \in F$ and in this case, the word w is *accepted* by $\mathcal{A}$. The language $L(\mathcal{A})$ is the set of words accepted by $\mathcal{A}$. As $F \subset [Q]_0^n$, the words in $L(\mathcal{A})$ have a length smaller than ω^n.

Remark. The third point above explains why such a run is called continuous: the state corresponding to a limit ordinal is itself a limit state (of the same type), which is reached in an implicit way, when an execution has gone infinitely often through some set of states of the type just below. In particular, a 1-automaton is a Muller automaton where a usual Muller acceptance condition is just a final state of type 1.

Example. The 2-automaton in Figure 1, with initial state 1 and final state 3, accepts the word $a^\omega b^\omega c$ with length $\omega.2 + 1$:

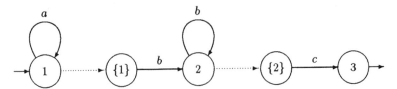

Fig. 1. A 2-automaton for $a^\omega b^\omega c$

Timed models are usually obtained by a now classical method: starting from a class of untimed transition systems, time constraints and reset about a finite set of variables called clocks are added in the transition labels.

A timed n-automaton is obtained from an untimed one, using a finite set X

$x \# c$, for $x \in X$, c some constant in $\mathbb{Q}$, and $\# \in \{<, =, >\}$. For the global time and the time values of clocks, we use non-negative real numbers. For a clock $x \in X$ and a time $t \in \mathbb{R}_+$, we denote by $x(t) \in \mathbb{R}_+$ the clock value of x, and by $X(t)$ the tuple $(x(t))_{x \in X}$ of all clock values of X. As usual, for a real number d, $X(t) + d$ is the tuple $(x(t) + d)_{x \in X}$. If the number of clocks in X is p, we may identify a constraint A with the subset of $\mathbb{R}_+^p$ of all values of $X(t)$ satisfying A.

Extending Choueka's automata and classical timed automata introduced in [1], we now define the notion of timed n-automata. A *timed n-automaton* (over $\mathbb{R}_+$) is a tuple $\mathcal{A} = (\Sigma, Q, I, F, \Delta, X)$, where Σ, Q, I and F are as in a n-automaton, X is a finite set of clocks and the transition relation Δ contains elements of the form (q, A, a, ρ, q'), also written $q \xrightarrow{A, a, \rho} q'$, where $q \in [Q]_0^{n-1}$, A is a constraint, $a \in \Sigma$, $\rho \subseteq X$, and $q' \in Q$.

Let us explain the execution of a transition $q \xrightarrow{A, a, \rho} q'$, when q' is a state of type 0. Assume the automaton has entered state q at time t with clock values $X(t)$, the pair $(q, X(t))$ is called an *extended state* of $\mathcal{A}$. Then, the automaton may execute the transition at time $t' \geq t$, if the constraint A is satisfied by the clock values $x(t) + (t' - t)$ for all clocks $x \in X$. The automaton switches to state q' and enters this state at time t'. Moreover, the clocks in ρ are reset, so that the new clock values are $x(t') = 0$ for all $x \in \rho$ and $x(t') = x(t) + (t' - t)$ otherwise, and the automaton has reached the extended state $(q', X(t'))$.

As for timed automata, we wish to define a (continuous) run of the timed n-automaton $\mathcal{A}$ as an α-sequence of extended states:

$$(q_0, X(t_0)) \xrightarrow[t_1]{A_1, a_1, \rho_1} (q_1, X(t_1)) \xrightarrow[t_2]{A_2, a_2, \rho_2} (q_2, X(t_2)) \rightarrow \quad \cdots \quad \rightarrow (q_\alpha, X(t_\alpha))$$

However, in order to do this, it must be possible to determine the clock values obtained when reaching a limit state after an infinite convergent run. The following lemma (for which the proof is rather natural) gives the answer. Note that, as in classical timed automata, it is not necessary to define limits of clock values for divergent sequences of times.

Lemma 1. *Let i be a limit ordinal of type $k > 0$, with $i = j + \omega^k$, where $j = 0$ or j is an ordinal of type greater than or equal to k. Assume that t_i is finite. Let x be a clock in X. According to the possible positions where the clock x has been reset, one of these two cases is verified:*

- *Case 1: There exists some integer N such that, for each ordinal h, $\omega^{k-1}.N < h < \omega^k$, $x(t_{j+h}) \neq 0$. Then, the sequence $(x(t_{j+\omega^{k-1}.n}))_{n \in \mathbb{N}}$ is non decreasing for $n \geq N$, bounded by t_i, so that it is convergent.*
- *Case 2: For each integer N, there exists an ordinal h, $\omega^{k-1}.N < h < \omega^k$, such that $x(t_{j+h}) = 0$. In this case, the sequence $(x(t_{j+\omega^{k-1}.n}))_{n \in \mathbb{N}}$ is the term of a convergent series, so that its limit is zero.*

Finally, a run of a timed n-automaton on a α-timed word $w = ((a_i, t_i))_{i < \alpha}$ is an α-sequence $(q_i, X(t_i))_{0 \leq i \leq \alpha}$, such that:

- the first state is $q_0 \in [Q]_0^{n-1}$, at time $t_0 \in \mathbb{R}_+$, with values $X(t_0)$,

- if $i \geq 1$ is the successor of some ordinal $i-1$, there is a transition $q_{i-1} \xrightarrow{A_i, a_i, \rho_i} q_i$, executed at time t_i, with $t_{i-1} \leq t_i$. The clock values $X(t_{i-1}) + t_i - t_{i-1}$ satisfy the constraint A_i and $x(t_i) = 0$ if $x \in \rho_i$, $x(t_i) = x(t_{i-1}) + t_i - t_{i-1}$ otherwise.

- if $i = j + \omega^k$ is a limit ordinal of type $k > 0$, with $j = 0$ or j is of type $\geq k$, q_i is defined as in the untimed case. Moreover, thanks to lemma 1, we have: $x(t_i) = lim_p x(t_{j+\omega^{k-1}p})$, for all $x \in X$, if t_i is finite, and $X(t_i)$ is undefined otherwise.

The run is *convergent* if t_α is finite. It is *accepting* if $q_0 \in I$, $t_0 = 0$, the initial valuation is $X(0)$, with $x(0) = 0$ for each clock $x \in X$ and $q_\alpha \in F$. The word w is then *accepted* and the timed language $L(\mathcal{A})$ is the set of accepted words.

Examples.

1. In the complete paper, we give a timed 2-automaton which accepts the set L of timed words $(a_i, t_i)_{i<\omega^2}$ such that,
   - $(a_i)_{i<\omega^2}$ belongs to $(a^\omega b^\omega c)^\omega$, and
   - for the jth factor of the form $a^\omega b^\omega c$, the time of c is $2j$ and the time difference between the first a and the first b is less than 1.

2. The timed 2-automaton in Figure 2 can be used to describe the successive steps in testing the resisting power of a spring: after it has been extended, the spring oscillates an infinite number of times (states 1 and 2) until it stops in the state $\{1, 2\}$. If the oscillation time is too short (compared with some time unit), the spring is faulty (state $F1$). Else, the operation is repeated until the spring breaks. If it breaks too early, it is again faulty (state $F2$), else the test is successful (state S).

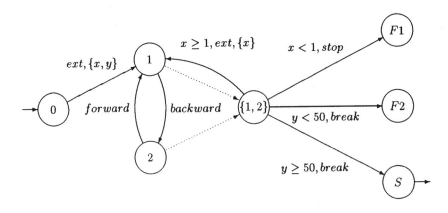

Fig. 2. A timed 2-automaton for testing a spring

In the sequel, we denote by $TA(n)$ the class of (Choueka) timed n-automata. The family of timed languages accepted by automata from $TA(n)$ is $TL(n)$ and $TL = \cup_{n \geq 1} TL(n)$.

As for usual timed automata, the class TL is closed under union and intersection. The interesting new fact about using Zeno words is the closure under concatenation, as well as under star and ω-power operations:

Proposition 2. *The family TL is closed under union, intersection, concatenation, star and ω-power.*

In the evaluation of a model, the test for emptiness is an important question. Indeed, a positive answer allows the design of verification algorithms. Thus, Theorem 3 shows that our extension of timed automata retains the property of the original class. The proof follows the standard technique introduced in [1] and extended in [5]: from a timed n-automaton accepting L, we build an untimed n-automaton $\mathcal{A}'$, called a region automaton, accepting the language $Untime(L)$ obtained from L by removing the dates of actions. The result follows from the property that $Untime(L)$ is empty if and only if L is empty and the decidability of emptiness for the corresponding class of untimed languages [10].

Theorem 3. *Emptiness is decidable in the class TL.*

4 Generalized transitions and refinement in the class TL

In the framework of untimed automata, the refinement of some action a corresponds to the replacement of each transition labeled by a, by some finite automaton describing the details of the operations performed by a. When dealing with timed executions, the refinement of an action may also contain time requirements, so that a transition must be replaced by a timed automaton, say $\mathcal{A}_a$. Moreover, like a transition, the automaton $\mathcal{A}_a$ is accessed at some time t, with (possibly non zero) values $X(t)$ of the clocks. Starting from these values, a convergent run of $\mathcal{A}_a$ assigns a new value to each clock. Each execution of such an automaton is thus used as a function on clock values and must therefore behave like a transition. For this reason, we define a subclass, denoted by GT, of timed automata which we shall use in the refinement operations.

A *generalized transition* is a timed n-automaton $\mathcal{A} = (\Sigma, Q, I, F, \Delta, X)$, such that:

1. the empty word ε does not belong to $L(\mathcal{A})$, and
2. for each convergent run of $\mathcal{A}$, starting at some time t_0 in a state $(q_0, X(t_0))$, with $q_0 \in I$, and ending in a state $(q_\alpha, X(t_\alpha))$, with $q_\alpha \in F$, for each clock x in X, if x was reset at least once along the run, then its final value is zero.

Remarks.

- Point 1 is required because we do not allow ε-transitions in timed n-automata. Timed automata, where the empty word ε may label a transition, have been studied recently and proved to be strictly more expressive than the usual model ([6], [12]).

- Point 2 expresses the fact that a run in the timed automaton simulates a transition if the final clock values are either 0 if there was a reset, or $x(t_\alpha) = x(t_0) + t_\alpha - t_0$ otherwise.
- Given a timed n-automaton $\mathcal{A}$, it can be decided if $\mathcal{A}$ is a generalized transition, for instance by looking at the region automaton (see e.g. [2] and Theorem 3).

Examples.

1. A reset-free automaton (transitions do not reset clocks) is a generalized transition.
2. In [7], two particular cases of generalized transitions were considered. The first one corresponds to the refinement of an instantaneous action into a finite sequence of actions occurring at the same time. The second case corresponds to the refinement of an action a, assumed to have some duration d, into some sequence of actions having the same global duration.

A generalized transition associates with each tuple of initial clock values $X(t)$ the language $L(\mathcal{A})_{X(t)}$, obtained by starting the executions of $\mathcal{A}$ at time t, with clock values $(x(t))_{x \in X}$. Note that, again from lemma 1, each run over a Zeno word $w = ((a_i, t_i))_{i<\alpha}$ in the language $L(\mathcal{A})_{X(t)}$ yields final clock values defined, for each $x \in X$, by $x(\tau(w)) = x(t) + \tau(w) - t$ if x has not been reset and $x(\tau(w)) = 0$ otherwise (recall that $\tau(w) = t_\alpha$ is the ending time of w). For a Zeno word $w \in L(\mathcal{A})_{X(t)}$, we denote by $\mathcal{X}(w, X(t))$ the finite set of clock values obtained by all runs of $\mathcal{A}$ over w starting from $X(t)$.

Let L be a language in TL over an alphabet Σ and $\mathcal{A}$ a timed n-automaton, with a set X of clocks, accepting L. A *refinement* of L is given by a family $(\mathcal{A}_a)_{a \in \Sigma}$ of generalized transitions over some other alphabet Σ', with X as set of clocks, represented by a mapping $\sigma : \Sigma \longrightarrow GT$ such that $\sigma(a) = \mathcal{A}_a$. The refinement of a timed action (a, t), denoted by $\sigma(a, t)$, is then defined for given clock values $X(t_0)$, $t \geq t_0$, by: $\sigma(a, t)_{X(t_0)} = \{w \in (L(\mathcal{A}_a))_{X(t_0)} / \tau(w) = t\}$.
It should be noticed that, if we want to be able to refine the last timed action (a, t) of a timed word, by a non Zeno timed word, then we must allow its time to be infinite.

In order to define the refinement of a timed word, we introduce an operation, similar to a concatenation between words, abusively called *composition* by: if $w = (a_i, t_i)_{i<\alpha}$ and $w' = (a'_i, t'_i)_{i<\alpha'}$, with $\tau(w) \leq t'_1$, then $w \circ w' = ((a''_i, t''_i))_{i<\alpha+\alpha'}$ with $(a''_i, t''_i) = (a_i, t_i), \forall i < \alpha$ and $(a''_{i+\alpha'}, t''_{i+\alpha'}) = (a'_i, t'_i), \forall i < \alpha'$. Note that the execution of two consecutive transitions over (a, t) and (b, t') respectively, accepts $(a, t) \circ (b, t')$, if $t' \geq t$. If these transitions are replaced respectively by $\mathcal{A}_a$ and $\mathcal{A}_b$, the corresponding executions, starting from clock values $X(t_0)$, accept what we call the refinement of $(a, t) \circ (b, t')$, defined by:

$$\sigma((a, t) \circ (b, t'))_{X(t_0)} = \{w \circ w' / w \in (L(\mathcal{A}_a))_{X(t_0)}, \tau(w) = t, w' \in (L(\mathcal{A}_b))_{X(t)}$$
$$\text{for some } X(t) \in \mathcal{X}(w, X(t_0)) \text{ and } \tau(w') = t'\}$$

The operation $\circ$ composes, in fact, runs of the automata $\mathcal{A}_a$ and $\mathcal{A}_b$.

We extend this definition to a timed word $w \in L$, inductively on its length. Finally, the refinement of L is: $\sigma(L) = \cup_{w \in L} \sigma(w)_{X(0)}$. Note that, for a language, the initial time is chosen as $t_0 = 0$ and the initial values of the clocks are all equal to zero: $x(0) = 0$, for each clock $x \in X$.

If L and L' are two languages of transfinite words, we say that L' is a refinement of L if there exists a timed refinement σ such that $L' = \sigma(L)$. As already remarked above, this definition of refinement generalizes those introduced in [7].

A good notion of refinement must preserve the property of being recognizable. As in [7], the question we are mainly interested in, is the closure of the class TL under refinement. We can now give a positive answer: starting from a suitable Muller timed automaton and replacing recursively some transitions by Muller timed automata in GT, we obtain a Choueka timed automaton. Therefore the refinement, when defined, preserves the recognizability of timed languages. Moreover, each class $TL(n)$ is closed under refinements using only finite timed automata in GT, so the notion of refinement proposed here is interesting in the usual class of timed languages $TL(1)$. One of the problems arising in the construction concerns the fact that, for a given interval I of $\mathbb{R}_+$, the language $\{w/w$ is finite and $\tau(w) \in I\}$ is recognizable while this is not always the case for $L_I^{inf} = \{w/w$ is infinite and $\tau(w) \in I\}$. It is proved in [12] that L_I^{inf} cannot be accepted by a timed automaton, when $I = [0, c_2[$ and, in fact, we have the following precise result:

Lemma 4. *The language $L_I^{inf} = \{w/w$ is infinite and $\tau(w) \in I\}$ is accepted by a timed automaton if and only if the interval I is of the form $]c_1, c_2]$, $[0, c_2]$ or $]c_1, \infty[$.*

This lemma explains why condition $2(b)$ in Theorem 5 is necessary to build a refinement, if condition $2(a)$ does not hold.

Theorem 5. *Let L be a language in $TL(n)$ over an alphabet Σ, accepted by a timed n-automaton $\mathcal{A}$ with a set X of clocks, and let σ be a timed refinement. For each $a \in \Sigma$, we consider the generalized transition $\sigma(a) = \mathcal{A}_a$ and the language $L_a = L(\mathcal{A}_a)$ and the subset α_a of clocks of X reset in $\mathcal{A}_a$. If, for each $a \in \Sigma$,*

1. *for each transition $\delta = (A, a, \alpha)$ in $\mathcal{A}$ labelled by a, we have $\alpha_a \subset \alpha$ and,*
2. *(a) either the language L_a contains only finite words,*
   (b) or for each transition δ in $\mathcal{A}$ labelled by a, for each clock $x \in X$, the clock constraint associated with δ is a disjunction of subformulas of the form $x \in I$, where $I =]c_1, c_2]$ or $I = [0, c_2]$ or $I =]c_1, \infty[$,

then $\sigma(L)$ belongs to TL.

The last result indicates a property of decomposition: any language L in $TL(n + 1)$ is obtained from a language accepted by a Muller timed automaton, using successive refinements.

Theorem 6. *Let n be an integer, $n \geq 1$. Any timed language in $TL(n + 1)$ is the refinement of some language in $TL(n)$.*

Note that the conditions of Theorem 5 hold for the refinement constructed in Theorem 6.

References

1. R. Alur and D.L. Dill. Automata for modeling real-time systems. In *Proceedings of ICALP'90*, number 443 in Lecture Notes in Computer Science, pages 322–335. Springer Verlag, 1990.
2. R. Alur and D.L. Dill. A theory of timed automata. *Theoretical Computer Science*, 126:183–235, 1994.
3. N. Bedon. Automata, semi-groups and recognizability of words on ordinals. *International Journal of Algebra and Computation*. to appear.
4. N. Bedon. Finite automata and ordinals. *Theoretical Computer Science*, 156:119–144, 1996.
5. B. Bérard. Untiming timed languages. *Information Processing Letters*, 55:129–135, 1995.
6. B. Bérard, P. Gastin, and A. Petit. On the power of non observable actions in timed automata. In *Proceedings of STACS'96*, number 1046 in Lecture Notes in Computer Science, pages 257–268. Springer Verlag, 1996.
7. B. Bérard, P. Gastin, and A. Petit. Refinement and abstraction for timed languages. Technical report, LSV, CNRS URA 2236, ENS de Cachan, 1997.
8. B. Bérard and C. Picaronny. Accepting zeno words without stopping time. Technical report, LSV, CNRS URA 2236, ENS de Cachan, 1997.
9. R. Büchi. On a decision method in restricted second order arithmetic. In *Proceedings of the International Congress on Logic, Methodology and Philosophy*, pages 1–11. Standford University Press, 1962.
10. Y. Choueka. Finite automata, definable sets and regular expressions over ω^n-tapes. *Journal of Computer and System Sciences*, 17:81–97, 1978.
11. C. Courcoubetis and M. Yannakakis. Minimum and maximum delay problems in real-time systems. In *Proceedings of CAV'91*, number 575 in Lecture Notes in Computer Science, pages 399–409. Springer Verlag, 1991.
12. V. Diekert, P. Gastin, and A. Petit. Removing ϵ-transitions in timed automata. In *Proceedings of the 14th Annual Symposium on Theoretical Aspects of Computer Science (STACS'97)*, number 1200 in Lecture Notes in Computer Science, pages 583–594. Springer Verlag, 1997.
13. M.R. Hansen, P.K. Pandya, and C. Zhou. Finite divergence. *Theoretical Computer Science*, 138:113–139, 1995.
14. J. C. Hemmer and P. Wolper. Ordinal finite automata and languages. Technical report, University of Liege, 1992.
15. T.A. Henzinger, Z. Manna, and A. Pnueli. Temporal proofs methodologies for real-time systems. In *Proceedings of POPL'91*, pages 353–366, 1991.
16. T.A. Henzinger, X. Nicollin, J. Sifakis, and S. Yovine. Symbolic model checking for real-time systems. *Information and Computation*, 111(2):193–244, 1994.
17. N. Lynch and H. Attiya. Using mappings to prove timing properties. In *Proceedings of PODC'90*, pages 265–280, 1990.
18. J. Ostroff. *Temporal Logic of Real-time Systems*. Research Studies Press, 1990.
19. C. Ramchandani. Analysis of asynchronous concurrent systems by Petri nets. Technical report, Massachusetts Institute of Technology, 1974.
20. J.G. Rosenstein. *Linear orderings*. Academic Press, New York, 1982.
21. J. Wojciechowski. Finite automata on transfinite sequences and regular expressions. *Fundamenta Informaticae*, 8.3-4:379–396, 1985.

Complexity Theoretical Results on Partitioned (Nondeterministic) Binary Decision Diagrams

Beate Bollig[*] and Ingo Wegener[*]

bollig/wegener@ls2.informatik.uni-dortmund.de

FB Informatik, LS2, Univ. Dortmund, 44221 Dortmund, Germany

Abstract. Ordered binary decision diagrams (OBDDs) have found a lot of applications in the verification of combinational and sequential circuits, protocols, and the synthesis and analysis of systems. The applications are limited, since the expressive power of polynomial-size OBDDs is too restricted. Therefore, several more general BDD models are used. Partitioned OBDDs are an OBDD model allowing a restricted use of nondeterminism and different variable orderings. They are restricted enough such that the essential operations can be performed efficiently and they allow polynomial-size representations for many more functions than OBDDs. Here the expressive power of polynomial-size partitioned OBDDs is investigated. A tight hierarchy with respect to the number of parts in the partition is proved and partitioned OBDDs are compared with other BDD models.

1   INTRODUCTION

If one would like to compute Boolean functions, one is interested in representations of small size allowing a fast evaluation. In this scenario circuits and their variants like depth-restricted circuits are the best model. In other situations, e.g., the formal verification where we like to prove that a design meets the specification, we are interested in representations fulfilling more requirements. Obviously small size for many Boolean functions is still an essential issue. In order to transform a formal specification (perhaps a circuit) into the representation it is necessary to have efficient algorithms to combine Boolean functions in the given representation by Boolean operators (like conjunction etc.). Moreover, one needs efficient algorithms to check whether two representations describe the same function and to check whether the described function is satisfiable.

Since the last problems are coNP-hard resp. NP-hard for circuits and for general branching programs, a lot of restricted types of branching programs (or binary decision diagrams which only is another notion for branching programs) have been investigated. In order to describe our results and the organization of our paper we first define the different models of binary decision diagrams which we discuss in this paper.

[*] Supported in part by DFG grant We 1066/7.

Definition 1. A binary decision diagram (BDD) or branching program is a directed acyclic graph with one source. Each sink is labeled by a Boolean constant and each other node by a Boolean variable from $\{x_1, \ldots, x_n\}$. These nodes have two outgoing edges one labeled by 0 and the other by 1. The BDD represents the Boolean function $f : \{0,1\}^n \to \{0,1\}$ defined in the following way. An input $a \in \{0,1\}^n$ activates all edges consistent with a, i.e., the edges labeled by a_i which leave nodes labeled by x_i. The value of the sink reached by the unique path which starts at the source and is activated by a is defined as $f(a)$.

Definition 2. (i) A free BDD (FBDD or read-once branching program) is a BDD where each path contains for each variable x_i at most one node labeled by x_i.

(ii) An ordered BDD (OBDD) is an FBDD where the orderings of the variables on all paths are consistent with one ordering.

(iii) A k-OBDD consists of k layers of OBDDs respecting the same ordering.

(iv) A k-IBDD consists of k layers of OBDDs respecting perhaps different orderings.

The size of a BDD G is the number of its nodes and is denoted by $|G|$.

OBDDs introduced by Bryant (1986) are the most popular representation for formal circuit verification. Gergov and Meinel (1994) and Sieling and Wegener (1995) independently have shown how FBDDs can be used for verification. In k-OBDDs and k-IBDDs the condition that variables are tested only once is relaxed. For k-OBDDs and constant k the test whether two decision diagrams represent the same function can be performed in polynomial time. This problem is coNP-complete already for 2-IBDDs. On the other hand k-IBDDs allow a more compact representation of many functions but the equality test only is based on heuristic algorithms. Bollig, Sauerhoff, Sieling, and Wegener (1997) present results on the representational power of these types of BDDs and Jain, Abadir, Bitner, Fussell, and Abraham (1992) have performed and described experiments.

Even more recently it has been shown that BDD models allowing restricted nondeterminism can be handled efficiently.

Definition 3. A $\oplus$-OBDD is a generalized OBDD where the number of edges leaving a node is not restricted. The output value for an input a is defined as 1 iff the number of paths between the source and the 1-sink activated by a is odd.

Here nondeterminism (inputs can activate many paths) is combined with the $\oplus$-acceptance mode. Nevertheless, Waack (1997) has shown that all important operations can be performed in polynomial time on $\oplus$-OBDDs. We have to admit that the number of edges of a $\oplus$-OBDD can be quadratic with respect to the number of nodes and it is still open whether the algorithms on $\oplus$-OBDDs are practically efficient.

The main drawback of OBDDs, k-OBDDs, and $\oplus$-OBDDs is that they only can use one variable ordering. Different variable orderings can be used explicitly in k-IBDDs and implicitly in FBDDs. These models have other drawbacks. For k-IBDDs the equality test is difficult and for FBDDs it is difficult to find a good oracle graph (Sieling and Wegener (1995)) or a good type (Gergov and Meinel (1994)). Partitioned OBDDs introduced by Jain (1993) and Jain, Bitner, Fussell, and Abraham (1992) and more intensively investigated by Narayan, Jain, Fujita, and Sangiovanni-Vincentelli (1996) try to overcome these difficulties.

Definition 4. A k-PBDD (partitioned BDD with k parts where k may depend on the number of variables) consists of k OBDDs whose variable orderings may be different. The output value for an input a is defined as 1 iff at least one of the OBDDs computes 1 on a. A PBDD is a k-PBDD for some k. The size of a k-PBDD is the sum of the sizes of the k OBDDs.

The model of k-PBDDs allows a restricted form of nondeterminism. In the case of OBDDs it is a restriction that nondeterminism only can be used at the source. Such a form of nondeterminism can be useful even if all OBDDs use the same variable ordering, see, e.g., Hu, York, and Dill (1994). The model of k-PBDDs in its general form has several practical drawbacks. Even if the k variable orderings are fixed, it is difficult to check whether two k-PBDDs G' and G'' represent the same function. An input a may lead in the i-th part of G' to 1 but only in the j-th part of G'' where $j \neq i$. For practical purposes it is necessary to work with so-called window functions $w_1, \ldots, w_k$ whose disjunction is the constant 1. The i-th part G_i of a k-PBDD G for f is responsible for the part $w_i^{-1}(1)$, i.e., G_i has to represent the function $f_i := f \wedge w_i$. This allows an efficient equality check.

In certain applications like combinatorial problems it is necessary to count the number of inputs satisfying a function f. This is efficiently possible for OBDDs but not for k-PBDDs, since many parts may compute 1 for an input a. The window functions are called disjoint, if $w_i \wedge w_j = 0$ for all $i \neq j$. Then the number of satisfying inputs is the sum of the number of satisfying inputs for the parts. Only in this case the notion "partitioned" is really justified. This kind of nondeterminism is known in complexity theory as unique nondeterminism. We use the notions k-PBDD^w and k-PBDD^d, if we work with fixed window functions resp. fixed disjoint window functions. If nothing else is explicitly mentioned, we prove our upper bounds for the weakest model of k-PBDD^{d}s and our lower bounds for the strongest model of k-PBDDs. Moreover, the upper bounds hold for window functions with small OBDD representations.

In the following we denote for a BDD model M by $P(M)$ the class of all (sequences of) Boolean functions representable by polynomial-size BDDs of type M.

In Section 2 we shortly compare the different types of PBDDs. In Section 3 we prove our main result that $P((k-1)$-PBDD) is a proper subclass of $P(k$-PBDD), i.e., the classes $P(k$-PBDD) build a tight hierarchy. The motivation of this result is twofold. First, the hierachy theorem itself as a typical complexity theoretical result. Second, we present lower bound techniques which are subtle enough to obtain large lower bounds for $k - 1$ parts even if the function can be easily and efficiently represented with k parts. In Section 4 we compare the expressive power of partitioned OBDDs with the expressive power of the other mentioned BDD-models. In Section 5 we state some structural results.

2   THE DIFFERENT MODELS OF PARTITIONED BDDs

We present an example which shows that we loose expressive power if we ask for efficiently representable window functions. The function PERM is defined on quadratic Boolean matrices X and outputs 1 iff X is a permutation matrix. By NPERM we denote the negation of PERM.

Theorem 5. *(i)* $NPERM \in P(2\text{-}PBDD)$.
(ii) $NPERM \notin P(PBDD^w)$ *for window functions representable by polynomial-size OBBDs for the corresponding variable orderings.*

Proof. By definition $NPERM(X) = 1$ iff there exists a row containing not exactly one 1 or there exists a column containing not exactly one 1. This leads directly to a 2-PBDD of linear size, where in the first part the first condition is tested by a rowwise variable ordering and in the second part the second condition is tested by a columnwise variable ordering.

For the lower bound we apply a result of Krause, Meinel, and Waack (1991) that PERM has no polynomial-size nondeterministic read-once branching programs. This implies that $PERM \notin P(\text{PBDD})$. Hence, it is sufficient to prove that for functions f in $P(\text{PBDD}^w)$ with efficiently representable window function the negative functions $\overline{f}$ are also contained in $P(\text{PBDD}^w)$. Let G_i represent $w_i \wedge f$. It is easy to obtain an OBDD $\overline{G_i}$ for $\overline{w_i \wedge f} = \overline{w_i} \vee \overline{f}$ of the same size as G_i. By assumption there exists a polynomial-size OBDD G_i^w for w_i with the same variable ordering as G_i and $\overline{G_i}$. Already Bryant (1986) has proved that we can obtain in time $O(|G_i||G_i^w|)$ an OBDD for $w_i \wedge (\overline{w_i} \vee \overline{f}) = w_i \wedge \overline{f}$.   □

3   THE HIERARCHY THEOREM

With respect to the results in Section 4 we try to define a function in $P(k\text{-}PBDD^d) - P((k-1)\text{-}PBDD)$ which is also easy for many other BDD-models. The idea is to start with a function in $P(\text{OBDD})$ which has small size OBDD representations only if most of a certain set of variables are tested at the beginning of the variable ordering. Then k variants of this function where for each variant another set of variables has to be tested first are combined to one function.

The function describes a graph problem on graphs with $k - 1 + (k + 1)n$ vertices $u_1, \ldots, u_{k-1}$ and $v_{i,j}, 1 \leq i \leq k+1, 0 \leq j \leq n - 1$. Let U denote the set of the u-vertices and V_i the set of the vertices $v_{i,j}, 0 \leq j \leq n - 1$. The variables $z_1, \ldots, z_{k-1}$ and $x_{i,j}, 1 \leq i \leq k, 0 \leq j \leq n - 1$, may take values in $\{0, \ldots, n-1\}$ and describe the pointers starting at the vertices $u_1, \ldots, u_{k-1}$ and $v_{i,j}$ resp. If $z_i = j$, the pointer from u_i leads to $x_{1,j}$. If $x_{i,j} = m$, the pointer from $v_{i,j}$ leads to $v_{i+1,m}$. Moreover, there are n Boolean so-called color variables $c_0, \ldots, c_{n-1}$ describing the colors of the nodes in V_{k+1}, i.e., the color of $v_{k+1,j}$ is black if $c_j = 1$ and white otherwise. Each input describes a graph with $k - 1$ paths starting in U and ending in V_{k+1}. The output of the function $f_{k,n}^1$ is 1 if the number of paths reaching a black vertex is odd. In order to obtain a Boolean function we have to replace each z- and x-variable by $\log n$ Boolean variables (w.l.o.g. we assume that n is a power of 2).

Lemma 6. *The function $f_{k,n}^1$ can be represented by OBDDs of size $O(2^k k^2 n^k)$.*

Proof. We use the natural variable ordering where the pointers are tested with respect to the vertex sets $U, V_1, \ldots, V_k$ and where finally the color variables are tested. The z-variables are tested in a complete binary tree with n^{k-1} leaves. During the tests of the $x_{i,\cdot}$-variables we store about each path the information which vertex in V_i or, if this is already known, in V_{i+1} is reached. These are $(2n)^{k-1}$ possible informations. It is only necessary to test the pointers leaving vertices in V_i which are reached by at least one path. This leads to an additional factor of $O(kn)$. The situation for the color variables is even simper. We get the last factor $O(k)$ for the different vertex sets $V_1, \ldots, V_{k+1}$. □

The function $f_{k,n}^i$ is defined similar to $f_{k,n}^1$ with another meaning of the x-variables. The pointers leaving V_1 are now described by the $x_{i,\cdot}$-variables, more generally the pointers leaving V_j are described by the $x_{i+j-1,\cdot}$-variables, if $i+j-1 \leq k$, and by the $x_{i+j-1-k,\cdot}$-variables else. The x-variables are rotated blockwise during the definition of $f_{k,n}^1, \ldots, f_{k,n}^k$. Obviously, Lemma 6 holds also for $f_{k,n}^i$.

For the definition of the path function $P_{k,n}$ we add $\lceil \log k \rceil$ additional selection variables $s_0, \ldots, s_{\lceil \log k \rceil - 1}$. If the value $||s||$ of the s-variables interpreted as binary number is equal to $i \leq k - 1$, we have to evaluate $f_{k,n}^{i+1}$. In the other case we have to output 0.

Lemma 7. *The function $P_{k,n}$ can be represented by k-PBDD^{d}s of size $O(2^k k^3 n^k)$.*

Proof. The window function w_i for $1 \leq i \leq k - 1$ tests whether $||s|| = i - 1$ and w_k tests whether $||s|| \geq k - 1$. These disjoint window functions have for all variable orderings an OBDD size of $O(\log k)$. The i-th part of the PBDD can realize $w_i^* \wedge f_{k,n}^i$, where w_i^* tests whether $||s|| = i - 1$, with an appropriate variable ordering by Lemma 6 in size $O(2^k k^2 n^k)$. □

The proof of the lower bound for $(k-1)$PBDDs representing $P_{k,n}$ is technically involved. We present the proof for the function $P_{k,n}^*$ where the pointer variables may take values in $\{0, \ldots, n-1\}$ and are not divided into $\log n$ Boolean variables. This implies that a pointer is tested as a whole. In the Boolean case the $\log n$ variables of a pointer may be scattered over the variable ordering. Later we discuss how our proof works in the Boolean case. In the proof decision diagrams are used where the nodes for the pointer variables have outdegree n. This only is done for the ease of description. Each of these decision nodes with its edges easily can be replaced by a complete binary tree of depth $\log n$.

Theorem 8. *Each* $(k-1)$*-PBDD for* $P_{k,n}^*$ *has size* $2^{\Omega(n/k^4)}$.

Proof. Let G be a $(k-1)$-PBDD for $P_{k,n}^*$ whose parts $G_1, \ldots, G_{k-1}$ respect the variable orderings $\pi_1, \ldots, \pi_{k-1}$. W.l.o.g. $k^4 \le n$. Otherwise the bound is trivial.

The number of x-variables equals nk. Let A_j be the set of those n x-variables which are tested at first with respect to π_j. Hence, we can choose at least n x-variables not contained in any A_j. Moreover, we choose an index i such that at least n/k $x_{i,\cdot}$-variables are not tested among the first n x-variables with respect to any of the $k-1$ variable orderings. We fix the s-variables such that $\|s\| = i-1$ and the z-variables such that the pointers from $u_1, \ldots, u_{k-1}$ point for $f_{k,n}^i$ to $k-1$ different vertices in V_i whose pointers belong to the chosen $x_{i,\cdot}$-variables. W.l.o.g. $i = 1$ and the u-pointers point to $v_{1,0}, \ldots, v_{1,k-2}$. In the following we consider the resulting $(k-1)$-PBDD for the obtained subfunction $f_{k,n}^*$ of $f_{k,n}^1$. We still have $k-1$ parts to represent $f_{k,n}^*$ but we know that all variable orderings are intuitively bad.

With respect to π_j there are n/k $x_{1,\cdot}$-variables not tested among the first n x-variables. Hence, at least n/k of the first n x-variables with respect to π_j do not belong to pointers leaving V_1. We choose a set B_j of at least n/k^2 pointers which leave the same set $V_{j'}$, $j' \ne 1$, and which are tested before the pointers leaving $V_1^* = \{v_{1,0}, \ldots, v_{1,k-2}\}$. Let C_j be a subset of B_j of at least n/k^3 pointers such that the sets $C_1, \ldots, C_{k-1}$ are pairwise disjoint. We associate with C_j the index set I_j of all indices m such that the pointer leaving $v_{j',m}$ belongs to C_j. We may construct the sets C_j even in such a way that the sets $I_1, \ldots, I_{k-1}$ are pairwise disjoint. Let $K := \lceil n/k^3 \rceil$.

Now we further restrict our function. The pointer from $v_{1,j}$ may point only to a vertex $v_{2,m}$ where $m \in I_j$. The pointer from $v_{i,m}$, where $2 \le i \le k-1$, $m \in I_j$, and $i \ne j'$, leads to $v_{i+1,m}$. The pointer from $v_{j',m}$, where $m \in I_j$, may point only to a vertex $v_{j'+1,l}$ where $l \in I_j$. Now the paths have been made independent. Finally, we fix the color variables in such a way that half of the vertices $v_{k+1,m}$, $m \in I_j$, are white and the other half is black. All paths have length k, since the start vertices belong to V_1. For the j-th path the first pointer is not fixed and may lead to one of K vertices. Then the path does not change the index of the vertex until $V_{j'}$ is reached. There we again have K possibilities and afterwards the path does not change the index of the vertex. Hence, we may interpret the path as a path of length 2.

For the j-th part G_j perhaps the color of several paths (more exactly their last vertices) can easily be determined but for the j-th path all variables describing the second edge of the path are tested before the first edge of the path. By our restrictions we have made the paths completely independent. In order to know whether the number of black paths is odd we have to know the color of all paths.

Our restricted OBDDs $G_1, \ldots, G_{k-1}$ all have to output 0 if the function $f_{k,n}^*$ outputs 0 and at least one of them has to output 1 if the function $f_{k,n}^*$ does so. We simplify the task for the OBDD G_j by giving the information about the color of all paths but the j-th one for free. Then G_j is left with one path starting at a definite vertex whose pointer may lead to K vertices which may lead to K vertices half of which are white and black resp. But G_j has to test all possible second edges before the first edge. The OBDD G_j has to output one color, w.l.o.g. white, correctly and for the other color it may make mistakes. Since for each input there has to be one OBDD G_j giving the correct answer, there has to exist one G_j which answers correctly for at least a fraction of $1/(k-1)$ of the inputs belonging to a black path. W.l.o.g. $j = 1$.

In G_1 we investigate the cut where all possible second edges have been tested. Here we may distinguish 2^K input sets describing which of the K vertices lead to a black vertex. If the PBDD size is bounded by S these 2^K information patterns are collected in at most S disjoint nodes where the first edge is tested. By assumption G_1 computes 1 for at least $K2^K/(2(k-1))$ of the $K2^K/2$ "inputs" consisting of the first edge and the set of vertices leading to a black vertex. Let w be a node of the cut of G_1. If less than $K/(4(k-1))$ of the edges leaving w lead to the 1-sink, the node w is called "poor". By definition the other nodes called "rich" nodes have to ensure that G_1 computes for at least $K2^K/(4(k-1))$ of the inputs the output 1. Let w be a rich node. Then we can assume w.l.o.g. that the first $N' \geq N := \lceil K/(4(k-1)) \rceil$ edges leaving w lead to the 1-sink. Hence, we have to be sure that the N' vertices which are reached if the first edge takes one of the values $0, \ldots, N'-1$ lead to a black vertex. This implies that such a node is passed by at most $N'2^{K-N'} \leq N2^{K-N}$ inputs. Hence,

$$ S \geq \frac{K2^K}{4(k-1)N2^{K-N}} = \frac{K2^N}{4(k-1)N} \geq 2^{N-1} \geq 2^{\Omega(n/k^4)}. \qquad \square $$

From Lemma 7 and Theorem 8 we obtain a polynomial upper bound for k-PBDD^{d}s and an exponential lower bound for $(k-1)$-PBDDs only for constant k. For a separation result between $P(k\text{-PBDD}^d)$ and $P((k-1)\text{-PBDD})$ a non polynomial lower bound is sufficient. We apply the well-known padding technique. E.g., we can work with $r = \log^6 n$ "real" vertices in each V_j and $n - r$ dummy vertices. Then the upper bound of Lemma 7 is polynomial for $k = O((\log n)/\log\log n)$. The lower bound of Theorem 8 is larger than each polynomial.

It is more difficult to obtain a lower bound on the size of $(k-1)$-PBDDs for $P_{k,n}$. Then we only can ensure that a few, namely $(\log n)/k$, bit variables of the pointers leaving w.l.o.g. $v_{1,0}, \ldots, v_{1,k-2}$ for $f^1_{k,n}$ are tested later than $(n\log n)/k$ x-variables which do not belong to V_1-pointers for each of the orderings $\pi_1, \ldots, \pi_{k-1}$. In order to obtain a situation of $k-1$ independent paths like in the proof of Theorem 8 we have to replace more variables by constants. Instead of $K = \lceil n/k^3 \rceil$ interior vertices on the paths of length 2 our proof for the Boolean case only can ensure the existence of $\lceil n^{1/k}(\log^{-1} n)k^{-4}/2 \rceil$ interior vertices on the independent paths.

Theorem 8*. *Each $(k-1)$-PBDD for $P_{k,n}$ has size at least 2^B where $B = \Omega(n^{1/k}k^{-5}\log^{-1} n)$.*

For a hierarchy theorem we restrict the vertex sets to $r = 2^{(\log^{1/2} n)\alpha(n)}$ "real" vertices where $\alpha(n) = \omega((\log\log n)^{1/2})$. For $k = O((\log^{1/2} n)\alpha^{-1}(n))$ the upper bound of Lemma 7 becomes polynomial while the lower bound of Theorem 8* grows faster than each polynomial. This leads to the following hierarchy result.

Theorem 9. *There are Boolean functions contained in $P(k\text{-PBDD}^d)$ but not in $P((k-1)\text{-PBDD})$, if $k = o(((\log n)/\log\log n)^{1/2})$.*

4   A COMPARISON WITH OTHER BDD MODELS

The function $P_{k,n}$ is not only easy for k-PBDDs but also for 2-OBDDs, FBDDs, and $\oplus$-OBDDs.

Theorem 10. *The function $P_{k,n}$ can be represented by k-PBDDs, 2-OBDDs, and FBDDs of size $O(2^k k^3 n^k)$, and by $\oplus$-OBDDs of size $O(2^{3k} k^9 n^{3k})$.*

Proof. The claim for k-PBDDs is proved in Lemma 7. The FBDD and the 2-OBDD start with a complete binary tree of the s-variables. If $||s|| = i \le k - 1$, $f_{k,n}^{i+1}$ is represented by the OBDD constructed in the proof of Lemma 6. This obviously leads to an FBDD. Using the variable ordering of the OBDD for $f_{k,n}^1$ we obtain 2-OBDDs for all $f_{k,n}^i$. In the representation for $f_{k,n}^i$ we test all variables $x_{m,\cdot}$ where $m \ge i$ in the right order in the first layer and all other x-variables in the right order in the second layer. Now the result for $\oplus$-OBDDs follows from the proof of Theorem 12, part viii).   □

Now we look for a function which is easy for 2-PBDD^{d}s and difficult for many other models.

Theorem 11. *There exists a function f_n on n^2 Boolean variables which can be represented in 2-PBDD^{d}s of size $O(n^2)$ and which needs size $2^{\Omega(n^{1/2})}$ for FBDDs, size $2^{\Omega(n/k)}$ for k-OBDDs, and size $2^{\Omega(n)}$ for $\oplus$-OBDDs.*

Sketch of proof. The function f_n is defined on $n \times n$ Boolean matrices X and outputs 1 iff the matrix X contains an odd number of ones and a row consisting of ones only or an even number of ones and a column consisting of ones only.

For the 2-PBDD^d we use the disjoint window functions testing whether the number of ones in X is odd and even resp. They have linear OBDD size for all variable orderings. The upper bound for 2-PBDD^{d}s is obvious.

The standard lower bound technique for FBDDs shows that the corresponding tree looks like a tree up to a certain depth d. This technique cannot be applied here. Instead of that we consider $\binom{n-1}{n^{1/2}}$ paths with special properties and prove that at most $(n^{1/2})!$ of them can be merged in one FBDD node. This implies the lower bound.

For the other lower bounds we apply a lower bound technique due to Krause (1991). For a given variable ordering we consider the cut where for the first time $n - 2$ rows or columns have a tested variable. W.l.o.g. these are the first $n - 2$ rows. We consider the following assignments of the variables tested before the cut. Exactly half of the variables tested first in the first $n - 2$ rows are replaced by ones and the other by zeros. All other tested variables are replaced by ones. For each of these assignments we define a corresponding assignment of the remaining variables in the following way. The rows consisting up to now only of ones are filled with zeros and the rows with a zero are filled with ones. The last but one row is filled with zeros ensuring that no column contains only ones and the last row is filled always in a way ensuring that the total number of ones is odd. Each assignment leads to the output 0. But if we combine one assignment of the variables before the cut with a non-corresponding assignment of the other variables this leads to the output 1. Hence, we obtain a fooling set of size $\binom{n-2}{(n-2)/2} = 2^{\Omega(n)}$. The lower bound for k-OBDDs follows from the lower bound of Krause (1991). The lower bound for $\oplus$-OBDDs follows from a new result of Dietzfelbinger and Savický (1997) that such fooling sets also lead to lower bounds for $\oplus$-OBDDs. $\qquad\square$

Theorem 12. *(i)* $P(2\text{-PBDD}^d) \not\subseteq P(\text{FBDD})$.
(ii) $P(\text{FBDD}) \not\subseteq P(k\text{-PBDD})$, *if* $k = o(((\log n)/\log\log n)^{1/2})$.
(iii) $P(2\text{-PBDD}^d) \not\subseteq P(\oplus\text{-OBDD})$.
(iv) $P(\oplus\text{-OBDD}) \not\subseteq P(k\text{-OBDD})$ *if* $k = o(((\log n)/\log\log n)^{1/2})$.
(v) $P(2\text{-PBDD}^d) \not\subseteq P(k\text{-OBDD})$ *if* $k = o(n/\log n)$.
(vi) $P(2\text{-OBDD}) \not\subseteq P(k\text{-PBDD})$ *if* $k = o(((\log n)/\log\log n)^{1/2})$.
(vii) $P(k\text{-OBDD}) \subseteq P(\text{PBDD}^d)$ *if* $k = O(1)$.
(viii) $P(k\text{-OBDD}) \subseteq P(\oplus\text{-OBDD})$ *if* $k = O(1)$.
(ix) $P(k\text{-PBDD}) \subseteq P(k\text{-IBDD})$ *for all* k.

Sketch of proof. Part i), iii), and v) follow from Theorem 11. Part ii), iv), and vi) follow from Theorem 8* and Theorem 10. For part vii) consider the at most $s = |G|^{k-1}$ possibilities to switch between the layers of a k-OBDD G. The i-th window function, $1 \le i \le s$, equals 1 for the inputs choosing the i-th possibility. Each part of the considered function can be represented by an OBDD of size $|G|^k$. All these OBDDs use the variable ordering of G. Therefore, it is easy to obtain also a $\oplus$-OBDD of size $|G|^{2k-1}$. Part ix) is easy. $\qquad\square$

5   READ-ONCE PROJECTIONS

Definition 13. The function $f = (f_n)$ is a read-once projection of $g = (g_n)$, $f \le_{rop} g$, if $f_n(x_1, \ldots, x_n) = g_{p(n)}(y_1, \ldots, y_{p(n)})$ for some polynomial p, $y_j \in \{x_1, \overline{x_1}, \ldots, x_n, \overline{x_n}, 0, 1\}$, and for all i the restriction that $y_j \in \{x_i, \overline{x_i}\}$ for at most one j is fulfilled.

Bollig and Wegener (1996) have shown that read-once projections are an adequate reduction concept for BDD models. If $f \le_{rop} g$ and $g \in P(k\text{-PBDD})$, then also $f \in P(k\text{-PBDD})$. The same holds for the classes $P(k\text{-PBDD}^w)$, $P(k\text{-PBDD}^d)$, and $P(\oplus\text{-PBDD})$.

Theorem 14. *The complexity classes $P(k\text{-PBDD})$ and $P(\oplus\text{-OBDD})$ have complete problems with respect to read-once projections.*

Sketch of proof. The proof for $P(\oplus\text{-OBDD})$ follows the lines of the proof of the completeness result for $P(\text{OBDD})$ (Bollig and Wegener (1996)) and the proof for $P(k\text{-PBDD})$ is an extension of the proof of the completeness result for $P(k\text{-IBDD})$ (Bollig and Wegener (1996)). $\qquad\qquad\qquad\qquad\qquad\qquad$ □

References

Bollig, B., Sauerhoff, M., Sieling, D., and Wegener, I.: Hierarchy theorems for kOBDDs and kIBDDs. Accepted for publication in Theoretical Computer Science (1997). (Preliminary version: On the power of different types of restricted branching programs. Electronic Colloquium on Computational Complexity. TR 94-026).

Bollig, B. and Wegener, I.: Read-once projections and formal circuit verification with binary decision diagrams. STACS'96, Lecture Notes in Computer Science **1046** (1996), 491 – 502.

Bryant, R.E.: Graph-based algorithms for Boolean function manipulation. IEEE Trans. on Computers **35** (1986), 677 – 691.

Dietzfelbinger, M. and Savický, P.: Parity OBDDs cannot represent the multiplication succinctly. Manuscript, Univ. Dortmund (1997).

Gergov, J. and Meinel, C.: Efficient Boolean manipulation with OBDDs can be extended to FBDDs. IEEE Trans. on Computers **43** (1994), 1197 – 1209.

Hu, A.J., York, G., and Dill, D.L.: New techniques for efficient verification with implicitly conjoined BDDs. Proc. of 31st ACM/IEEE Design Automation Conference (1994), 276 – 282.

Jain, J.: On Analysis of Boolean Functions. Ph.D. Thesis, Dept. of Electrical and Computer Engineering, University of Texas at Austin (1993).

Jain, J., Abadir, M., Bitner, J., Fussell, D.S., and Abraham, J.A.: IBDDs: an efficient functional representation for digital circuits. Proc. European Design Automation Conference (1992), 440 – 446.

Jain, J., Bitner, J., Fussell, D.S., and Abraham, J.A.: Functional partitioning for verification and related problems. Brown/MIT VLSI Conf. (1992), 210 – 226.

Krause, M.: Lower bounds for depth-restricted branching programs. Information and Computation **91** (1991), 1 – 14.

Krause, M., Meinel, C., and Waack, S.: Separating the eraser Turing machine classes L_e, NL_e, co-NL_e and P_e. Theoretical Computer Science **86** (1991), 267 –275.

Narayan, A., Jain, J., Fujita, M., and Sangiovanni-Vincentelli, A.: Partitioned ROBDDs - a compact, canonical and efficiently manipulable representation for Boolean functions. Proc. of IEEE/ACM Int. Conf. on Computer Aided Design ICCAD'96 (1996), 547 – 554.

Sieling, D. and Wegener, I.: Graph driven BDDs - a new data structure for Boolean functions. Theoretical Computer Science **141** (1995), 283 – 310.

Waack, S.: On the descriptive and algorithmic power of parity ordered binary decision diagrams. Proc. of STACS'97, LNCS 1200 (1997), 201 – 212.

Specifying Computations
Using Hyper Transition Systems

Marcello M. Bonsangue and Joost N. Kok

Department of Computer Science, Rijks Universiteit Leiden
P.O. Box 9512, 2300 RA Leiden, The Netherlands
marcello@cs.leidenuniv.nl and joost@cs.leidenuniv.nl

Abstract. We study hyper transition systems as a formalism to give
semantics to specification languages which support both unbounded angelic and unbounded demonic non-determinism as well as recursion. Hyper transition are a generalization of transition systems and are suited
for the specification of computations by means of properties that atomic
steps in a computation have to satisfy. As an application we use a hyper transition system to give an operational semantics to the language
of Back's refinement calculus. This operational semantics abstracts from
the internal configurations and we prove it to be equivalent to the standard weakest precondition semantics. Finally, we propose a refinement
relation that preserves the atomic step of a computation and generalizes
the simulation relation on ordinary transition systems. This can be used
to augment specification languages with a form of concurrency.

1   Introduction

In this paper we study operational semantics for specification languages based on
hyper transition systems. With specification languages we mean languages that
have specification constructs like angelic choice or unbounded non-determinism.
Such specification languages are useful as a starting point for program refinement [BW90]: in a typical program development by refinement, one starts from a
program in a specification language and develops in correctness-preserving steps
a program that is efficiently executable on a target architecture. One of the well-
known specification languages is the language of the refinement calculus [BW97].
For this language several semantic models based on input/output relations have
been proposed, including a weakest precondition semantic and game-theoretic
semantics based on a transition system.

We use hyper transition systems because they provide a natural way to define operational semantics with both angelic and demonic non-determinism. As
an application we present a hyper transition system and associated operational
semantics for the specification language of the refinement calculus. From this operational semantics we derive a weakest precondition semantics which coincides
with the ordinary semantics of the refinement calculus [Bac80]. Working with a
hyper transition system has the advantage that we can follow the computation
step-by-step. Hence it is possible to introduce interleaving operators in the spec-

ification languages, and define a notion of simulation needed for a compositional semantical treatment of the interleaving operator.

Our operational interpretation differs in the following aspects from the game semantics of Back and Von Wright [BW97] and the game semantics of Hesselink [Hes94]. Back and Von Wright define a game interpretation of the commands of the refinement calculus using a standard transition system. A transition step corresponds to a move in the game. A configuration is said to be angelic if only the angel can make a move and is said to be demonic otherwise. This suggests a close relation to our hyper transition system model. However, every sequence of transitions in the game interpretation of Back and Von Wright is finite (in fact infinite plays are not possible), and we allow also infinite sequences. The game semantics for the refinement calculus given by Hesselink uses hyper transition systems which allow for infinite games. However, both the hyper transition system induced by the refinement calculus and the way of collecting the information from it are different from our operational approach. Hyper transition systems can also be interpreted as games. We refer the interested reader to Chapter 4 of [Bon96] where also most of the proofs of the statements in this paper can be found.

2 Transition systems and hyper transition systems

Transition systems are a useful mathematical structure to describe the atomic steps of a computation of a program [Plo81].

Definition 1. A *transition system* with deadlock is a tuple $\langle X, \delta, \longrightarrow \rangle$ where X is the class of all proper configurations for a program, $\delta \notin X$ denotes a deadlock configuration, and $\longrightarrow \subseteq (X \times X) \cup (X \times \{\delta\})$ is a transition relation.

The idea is that configurations represent states of a computation, whereas a transition $x \longrightarrow y$ (read 'x goes to y') indicates a possible atomic step which a computation can do, changing the configuration x into the configuration y. If $x \longrightarrow \delta$ then the computation in the configuration x may deadlock. If there is no $y \in X \cup \{\delta\}$ such that $x \longrightarrow y$ then the computation is undefined in the configuration x. Essentially, a computation can be undefined in a configuration x either because it terminates successfully or because it aborts.

Let us now be more precise about what we mean by 'computation'. Let $T = \langle X, \delta, \longrightarrow \rangle$ be a transition system, $F \subseteq X$ be a set of final configuration, and $x \in X$. Define a *finite computation* of T starting at x to be a finite sequence $(x_n)_{n \leq k}$ in $X \cup \{\delta\}$ such that
   (i) $x = x_0$,
   (ii) $x_n \longrightarrow x_{n+1}$ for all $n < k$, and
   (iii) for all $y \in X \cup \{\delta\}$ there is no transition $x_k \longrightarrow y$ in T.
A finite computation $(x_n)_{n \leq k}$ of T starting at x_0 is terminating in the configuration x_k if either $x_k \in F$ or $x_k = \delta$, otherwise the computation is undefined. Not every computation of a program need to be finite. An *infinite computation* of T starting at x is a countable sequence $(x_n)_{n \in \mathbb{N}}$ in X such that

(i) $x = x_0$, and

(ii) $x_n \longrightarrow x_{n+1}$ for all $n \in \mathbb{N}$.

In general, a *computation* of a transition system T is a finite or infinite computation of T. In other words, a computation of T is a transition sequence of T that cannot be extended.

To specify set of computations by means of the properties that their atomic steps have to satisfy, hyper transition system can be used. Hyper transition systems occur under the name of AND/OR graphs or hyper-graphs in logic programming and artificial intelligence [Nil82].

Definition 2. A *hyper transition system* is a pair $H = \langle X, \longrightarrow_\in \rangle$ where X is the class of all possible configurations in which a computation is allowed to work, and $\longrightarrow_\in \subseteq X \times \mathcal{P}(X)$ is a transition relation which specifies the atomic steps of a computation.

A hyper transition system specifies a set of computations by specifying their atomic steps. The idea is that a computation specified by a hyper transition system $H = \langle X, \longrightarrow_\in \rangle$ can change a configuration x into a configuration y if the configuration y satisfies all and at least one of the predicates $W \subseteq X$ such that $x \longrightarrow_\in W$ (read 'x goes into W'). More formally, the set of all computations specified by a hyper transition system H can be modeled by the following transition system $TS(H)$.

Definition 3. For a hyper transition system $H = \langle X, \longrightarrow_\in \rangle$ define the induced transition system $TS(H) = \langle X, \delta, \longrightarrow \rangle$ as follows. For all $x, y \in X$,

$$x \longrightarrow \delta \Leftrightarrow \bigcap \{ W \mid x \longrightarrow_\in W \} = \emptyset,$$

$$x \longrightarrow y \Leftrightarrow (\exists W : x \longrightarrow_\in W) \ \& \ y \in \bigcap \{ W \mid x \longrightarrow_\in W \}.$$

A computation of $TS(H)$ (or, equivalently, a computation that satisfies the specification of the hyper transition system H) in a configuration x has four possibilities with respect to a set $F \subseteq X$ of final configurations:

(i) it terminates in a deadlock configuration because there is no configuration $y \in X$ satisfying all $W \subseteq X$ such that $x \longrightarrow_\in W$;

(ii) it terminates because $x \in F$ and there is no $W \subseteq X$ such that $x \longrightarrow_\in W$;

(iii) it is undefined because $x \notin F$ and there is no $W \subseteq X$ such that $x \longrightarrow_\in W$;

(iv) it goes to a configuration y satisfying all $W \subseteq X$ such that $x \longrightarrow_\in W$.

Observe that, by definition, exactly one of the above four possibilities is possible. Indeed, for every $x \in X$, if $x \longrightarrow_\in W$ then either $x \longrightarrow \delta$ or there exists $y \in W$ such that $x \longrightarrow y$. Conversely, there exists $W \subseteq X$ such that $x \longrightarrow_\in W$ only if either $x \longrightarrow \delta$ or $x \longrightarrow y$ (and in this case $y \in W$). It follows that a computation specified by a hyper transition system H either terminates or is undefined in a configuration x if and only if there is no $W \subseteq X$ such that $x \longrightarrow_\in W$.

Under the above interpretation of hyper transition systems it is natural to consider hyper transition systems such that the transition relation $\longrightarrow_\in$ is *upper closed on the right hand side*, that is,

$$x \longrightarrow_\in V \ \& \ V \subseteq W \text{ implies } x \longrightarrow_\in W. \tag{1}$$

Essentially, the above closure property is due to the fact that $V \subseteq W$ if and only if $V = V \cap W$. No extra information is added by upper closing on the right hand side of the transition relation of a hyper transition system.

Hyper transition systems specify computations at the level of the properties that an atomic step has to satisfy, whereas transition systems specify computations at the level of the configurations that an atomic step may reach. Because of this difference a hyper transition system $H = \langle X, \longrightarrow_\epsilon \rangle$ can model two different kinds of non-determinism: one at the level of the computations specified and one at the level of the specification. The non-determinism of the computations specified by H in a configuration x depends on all the sets $W \subseteq X$ such that $x \longrightarrow_\epsilon W$: the bigger these sets, the more computations are specified. The non-determinism of the specification depends on the number of transitions starting from the same configuration: the more a specification is non-deterministic, the less is the number of computations that it specifies.

In the next section we will see that the non-determinism of the specification is related to the angelic non-determinism, and the non-determinism of the computations is related to the demonic non-determinism. Moreover, the possibility of describing two different kinds of non-determinism in a single framework will allow for a compositional specification of a computation in terms of the properties that the atomic steps of the computation have to satisfy.

First we compare hyper transition systems to transition systems. We have already seen that a hyper transition system H induces a transition system $TS(H)$ representing all the computations specified by H. Clearly, different hyper transition systems can specify the same sets of computations. Conversely, every transition system T induces a canonical hyper transition system $HTS(T)$ which specifies exactly all computations of T.

Definition 4. For a transition system $T = \langle X, \delta, \longrightarrow \rangle$ define the hyper transition system $HTS(T) = \langle X, \longrightarrow_\epsilon \rangle$ by putting $x \longrightarrow_\epsilon W$ if and only if

$$x \longrightarrow \delta \text{ or } (\exists y \in X : x \longrightarrow y) \& (\forall y \in X : x \longrightarrow y \Rightarrow y \in W)$$

for every $x \in X$ and $W \subseteq X$.

The computations specified by $HTS(T)$ coincide with the computations of T. This is a consequence of the following lemma.

Lemma 5. Let $T = \langle X, \delta, \longrightarrow \rangle$ be a transition system with deadlock. Then $TS(HTS(T)) = T$.

In order to characterize the class of hyper transition systems equivalent to transition systems, notice that, for every transition system $T = \langle X, \delta, \longrightarrow \rangle$ the transition relation $\longrightarrow_\epsilon$ of the hyper transition system $HTS(T)$ is upper closed on the right hand side and it satisfies the following property for every $x \in X$:

$$\exists W \subseteq X : x \longrightarrow_\epsilon W \Rightarrow x \longrightarrow_\epsilon \bigcap \{ V \subseteq X \mid x \longrightarrow_\epsilon V \}. \tag{2}$$

Lemma 6. Let $H = \langle X, \longrightarrow_\epsilon \rangle$ be a hyper transition system satisfying (2) and with $\longrightarrow_\epsilon$ upper closed on the right hand side. Then $HTS(TS(H)) = H$.

3 Operational semantics for a specification language

We now consider a language $\mathcal{L}$ with the specification constructs of the refinement calculus [Bac80]. The language has two conditionals '$B\to$' and '$\{B\}$', a state transformer command '$\langle f\rangle$', sequential composition ';', angelic choice '$\bigvee_I$', demonic choice '$\bigwedge_I$', and recursion through procedure variables. The main difference with the language of the refinement calculus is that we have procedure variables in the language.

Definition 7. Let St be a set of states and let $PVar$ be a set of procedure variables. The class $(S \in) Stat$ of *statements* is given by

$$S ::= B\to \mid \{B\} \mid \langle f\rangle \mid x \mid \bigvee_I S_i \mid \bigwedge_I S_i \mid S\,;\,S,$$

where $B \subseteq St$, $f : St \to St$, $x \in PVar$, and I is an arbitrary set. A *command* in the language $\mathcal{L}$ is a pair $\langle d, S\rangle$, where d is a declaration in $Decl = PVar \to Stat$ and S a statement in $Stat$.

Notice that the language $\mathcal{L}$ is a proper class since the index I in the $\bigvee$ and $\bigwedge$ constructs can be any set. We give the semantics of the language $\mathcal{L}$ by means of a hyper transition system, from which, by the results of the previous section, a semantics based on ordinary transition system can be derived. We consider configurations to be either states in St, representing the final outcomes of the computations, or pairs $\langle S, s\rangle$ where $s \in St$ is a possible initial or intermediate state of a computation and $S \in Stat$ is the specification of the remainder of the computation to be executed. A finite computation which terminates in a configuration $\langle S, s\rangle \in Stat \times St$ is undefined.

Definition 8. Let $(c \in) Conf = (Stat \times St) \cup St$ be the class of configurations and define, for every declaration $d : PVar \to Stat$ the hyper transition system $\langle Conf, \longrightarrow_d\rangle$ by taking $\longrightarrow_d$ to be the least relation between configurations in Conf and subsets of configurations of Conf satisfying the following axioms and rules:

$$\langle B\to, s\rangle \longrightarrow_d W \qquad \text{if } s \in B \text{ implies } s \in W$$

$$\langle \{B\}, s\rangle \longrightarrow_d W \qquad \text{if } s \in B \cap W$$

$$\langle \langle f\rangle, s\rangle \longrightarrow_d W \qquad \text{if } f(s) \in W$$

$$\langle x, s\rangle \longrightarrow_d W \qquad \text{if } \langle d(x), s\rangle \in W, \text{ for } x \in PVar$$

$$\frac{\langle S_i, s\rangle \longrightarrow_d W}{\langle \bigvee_I S_i, s\rangle \longrightarrow_d W} \qquad \text{if } i \in I$$

$$\frac{\{\langle S_i, s\rangle \longrightarrow_d W_i \mid i \in I\}}{\langle \bigwedge_I S_i, s\rangle \longrightarrow_d \bigcup\{W_i \mid i \in I\}}$$

$$\frac{\langle S_1, s\rangle \longrightarrow_d W}{\langle S_1\,;\,S_2, s\rangle \longrightarrow_d \{\langle S_2, t\rangle \mid t \in W \cap St\} \cup \{\langle S_1'\,;\,S_2, t\rangle \mid \langle S_1', t\rangle \in W\}}.$$

An explanation is in order here. According to our interpretation of hyper transition systems, the command $\langle d, B \rightarrow \rangle$ specifies a computation that when started at input $s \in B$ terminates in one step with the state s as the only outcome because $\langle B \rightarrow, s \rangle \longrightarrow_{\epsilon d} \{s\}$. However, if the computation is started at input $s \notin B$ then it must deadlock because $\langle B \rightarrow, s \rangle \longrightarrow_{\epsilon} \emptyset$.

The command $\langle d, \{B\} \rangle$ is similar except that the computations specified by $\langle d, \{B\} \rangle$ are undefined at input $s \notin B$ because no transition is possible from the configuration $\langle \{B\}, s \rangle$.

The command $\langle d, \langle f \rangle \rangle$ specifies a computation that at input s terminates in one step, with as only output the state $f(s)$ (because $\langle \langle f \rangle, s \rangle \longrightarrow_{\epsilon d} \{f(s)\}$).

The command $\langle d, x \rangle$ specifies a computation that at input s goes to the configuration $\langle d(x), s \rangle$ (because $\langle x, s \rangle \longrightarrow_{\epsilon d} \{\langle d(x), s \rangle\}$).

The command $\langle d, \bigvee_I S_i \rangle$ specifies those computations which are specified by all $\langle d, S_i \rangle$ for $i \in I$. It increases the non-determinism of the specification and hence restricts the non-determinism of the computations. The computations specified by $\langle d, \bigvee_I S_i \rangle$ are undefined at input s only if the computations specified by all $\langle d, S_i \rangle$ for $i \in I$ are undefined at input s. The computations specified by $\langle d, \bigvee_I S_i \rangle$ must deadlock at input s if there is one $\langle d, S_k \rangle$ for $k \in I$ which specifies a computation which must deadlock.

The command $\langle d, \bigwedge_I S_i \rangle$ increases the non-determinism at the level of the specified computations. It specifies computations which behave as any of the computations specified by $\langle d, S_i \rangle$ for $i \in I$. Dual to the command $\langle d, \bigvee_I S_i \rangle$, the computations specified by $\langle d, \bigwedge_I S_i \rangle$ are undefined at input s if there is one $\langle d, S_k \rangle$ for $k \in I$ which specifies a computation undefined at input s. Also, the computations specified by $\langle d, \bigwedge_I S_i \rangle$ must deadlock at input s only if the computations specified by all $\langle d, S_i \rangle$ for $i \in I$ must deadlock at input s.

Finally, the command $\langle d, S_1 ; S_2 \rangle$ specifies computations that at input s may either deadlock, or go to a configuration $\langle S_2, s' \rangle$ if S_1 specifies a computation which at input s terminates in a state s', or goes to a configuration $\langle S ; S_2, s' \rangle$ if S_1 specifies a computation which at input s may go in a state s' with $\langle d, S \rangle$ the command specifying the remainder of the computation to be executed.

Notice that because of the angelic choice, $\langle Conf, \longrightarrow_{\epsilon d} \rangle$ does not satisfy (2). In general, the transition relation $\longrightarrow_{\epsilon d}$ is upper closed on the right hand side. As a consequence, $\langle \bigwedge_I S_i, s \rangle \longrightarrow_{\epsilon d} W$ if and only if $\langle S_i, s \rangle \longrightarrow_{\epsilon d} W$ for all $i \in I$. Dually, by Definition 8, $\langle \bigvee_I S_i, s \rangle \longrightarrow_{\epsilon d} W$ if and only if there exists $k \in I$ such that $\langle S_k, s \rangle \longrightarrow_{\epsilon d} W$.

If we want to capture the input-output behaviour of the language $\mathcal{L}$ we need to abstract from the intermediate configurations recorded by the transition relation of the hyper transition system $(Conf, \longrightarrow_{\epsilon d})$.

Definition 9. Let $\langle X, \longrightarrow_{\epsilon} \rangle$ be a hyper transition system. For all ordinal $\lambda \geq 0$ define the relation $\xrightarrow{\lambda}_{\epsilon}$ on $X \times \mathcal{P}(X)$, for $x \in X$ and $W \subseteq X$, inductively by

$$x \xrightarrow{0}_{\epsilon} W \equiv x \in W,$$
$$x \xrightarrow{\lambda+1}_{\epsilon} W \equiv \exists V \subseteq X : x \longrightarrow_{\epsilon} V \; \& \; \forall y \in V \exists \alpha \leq \lambda : y \xrightarrow{\alpha}_{\epsilon} W,$$
$$x \xrightarrow{\lambda}_{\epsilon} W \equiv \exists \alpha < \lambda : x \xrightarrow{\alpha}_{\epsilon} W \qquad \text{where } \lambda \text{ is a limit ordinal.}$$

By induction on λ it is easy to see that, for every ordinal $\lambda \geq 0$, the relation $\xrightarrow{\lambda}_\in$ is upper closed on the right hand side if the relation $\longrightarrow_\in$ is upper closed on the right hand side.

The ordinal used to label the transition relation $x \xrightarrow{\lambda}_\in W$ is not equal to the number of atomic steps which a computation specified by a hyper transition system starting in a configuration x need to execute in order to satisfy the predicate W. Rather, the label takes in account both the length of the computation specified which starts in a configuration x and the non-determinism of the computations. Since we allow for unbounded demonic nondeterminism, this label need not to be a finite ordinal.

We can now define a semantics $Op[\![\cdot]\!]$ for the language $\mathcal{L}$ in terms of the hyper transition system $(Conf, \longrightarrow_{\in d})$.

Definition 10. Put $Sem = Decl \times Conf \to \mathcal{P}(\mathcal{P}(St))$ and define $Op \in Sem$, for $d \in Decl$ and $c \in Conf$, by

$$Op(d, c) = \{P \subseteq St \mid \exists\lambda: c \xrightarrow{\lambda}_{\in d} P\}.$$

The operational semantics $Op[\![\cdot]\!] : \mathcal{L} \to (St \to \mathcal{P}(\mathcal{P}(St)))$ is given by

$$Op[\![\langle d, S\rangle]\!](s) = Op(d, \langle S, s\rangle).$$

The idea behind the above operational semantics is that of total correctness (considering programs which deadlock as terminating and satisfying every postcondition).

Theorem 11. Let $T = \langle Conf, \delta, \longrightarrow_{\in d}\rangle$ be the transition system induced by the hyper transition system associated to $\mathcal{L}$ according to Definition 3. For all $\langle d, S\rangle \in \mathcal{L}$, $P \subseteq St$ and $s \in St$ if $P \in Op[\![\langle d, S\rangle]\!](s)$ then every computation of T starting at $\langle S, s\rangle$ is finite and terminates either in the configuration δ or in a state $t \in P$.

We conjecture that also the converse of the above theorem holds.

Next we give some properties of our operational semantics $Op[\![\cdot]\!]$. The semantics $Op[\![\langle d, S\rangle]\!](s)$ of a command $\langle d, S\rangle$ in $\mathcal{L}$ at input $s \in St$ abstracts from the intermediate configurations reached by a transition sequence starting from $\langle d, S\rangle$ and collects only the final outcomes. This can be seen by characterizing the function $Op(\cdot)$ as the least solution of a fixed point equation.

Theorem 12. The function $Op(\cdot)$ is the least function in Sem such that, for $d \in Decl$, $s \in St$, and $S \in Stat$,

$$Op(d, s) \quad = \{P \subseteq St \mid s \in P\},$$
$$Op(d, \langle S, s\rangle) = \bigcup\{\bigcap\{Op(d, c') \mid c' \in W\} \mid \langle S, s\rangle \longrightarrow_{\in d} W\}.$$

To prove that the semantics $Op[\![\cdot]\!]$ is compositional we first show that every semantics for $\mathcal{L}$ which satisfies the above fixed point characterization is a compositional fixed point semantics with respect to both the angelic and demonic choice operators. In a second step we show the compositionality of $Op[\![\cdot]\!]$ with respect to the sequential composition operator.

Lemma 13. *Let* $F : Decl \times Conf \to \mathcal{P}(\mathcal{P}(\text{St}))$ *be a function such that, for* $d \in$ *Decl,* $s \in$ St, *and* $S \in$ *Stat,*

$$F(d, s) \qquad = \{P \subseteq \text{St} \mid s \in P\}$$
$$F(d, \langle S, s \rangle) = \bigcup \{\bigcap \{F(d, c') \mid c' \in W\} \mid \langle S, s \rangle \longrightarrow_{\epsilon d} W\}.$$

Then, for every $d \in$ *Decl and* $s \in$ St,

(i) $F(d, \langle x, s \rangle) = F(d, \langle d(x), s \rangle),$
(ii) $F(d, \langle \bigvee_I S_i, s \rangle) = \bigcup \{F(d, \langle S_i, s \rangle) \mid i \in I\},$
(iii) $F(d, \langle \bigwedge_I S_i, s \rangle) = \bigcap \{F(d, \langle S_i, s \rangle) \mid i \in I\}.$

Compositionality of $Op[\![\cdot]\!]$ with respect to the sequential composition is proved by induction on the ordinal λ in the transition $\langle S_1 ; S_2, s \rangle \xrightarrow{\lambda}_{\epsilon d} P$ for $P \in$ $Op[\![\langle d, S_1 ; S_2 \rangle]\!](s)$.

Lemma 14. *For* $d \in$ *Decl,* $s \in$ St, *and* $S_1, S_2 \in$ *Stat,*
$$Op[\![\langle d, S_1 ; S_2 \rangle]\!](s) = \bigcup \{\bigcap \{Op[\![\langle d, S_2 \rangle]\!](t) \mid t \in Q\} \mid Q \in Op[\![\langle d, S_1 \rangle]\!](s)\}.$$

From the above lemma and Lemma 13 applied to the function $Op(\cdot)$ it follows that $Op[\![\cdot]\!]$ is a compositional fixed point semantics for $\mathcal{L}$.

We conclude this section by deriving a weakest precondition semantics for the language $\mathcal{L}$ from its operational semantics $Op[\![\cdot]\!]$

Definition 15. Define the weakest precondition of a command $\langle d, S \rangle \in \mathcal{L}$ for a postcondition $P \subseteq$ St by

$$Wp[\![\langle d, S \rangle]\!](P) = \{s \in \text{St} \mid P \in Op[\![\langle d, S \rangle]\!](s)\}.$$

The above definition is justified by Theorem 11: the predicate $Wp[\![\langle d, S \rangle]\!](P)$ holds for those input sates $s \in St$ for which each computation of the program specified by the command $\langle d, S \rangle$ terminates in a final state satisfying the postcondition P.

The semantics $Wp[\![\cdot]\!]$ can be characterized uniquely by means of axioms given in the theorem below. These axioms show that our weakest precondition semantics of $\mathcal{L}$ induced by the operational semantics $Op[\![\cdot]\!]$ coincides with the ordinary lattice theoretical interpretation of the refinement calculus [Bac80].

Theorem 16. *The semantics* $Wp[\![\cdot]\!]$ *is the least function from* $\mathcal{L} \to (\mathcal{P}(\text{St}) \to$ $\mathcal{P}(\text{St}))$ *such that, for all* $\langle d, S \rangle \in \mathcal{L}$ *and* $P \subseteq$ St,

$$
\begin{aligned}
Wp[\![\langle d, B \to \rangle]\!](P) \quad &= \{s \in \text{St} \mid s \in B \Rightarrow s \in P\}, \\
Wp[\![\langle d, \{B\} \rangle]\!](P) \quad &= \{s \in \text{St} \mid s \in B \cap P\}, \\
Wp[\![\langle d, \langle f \rangle \rangle]\!](P) \quad &= \{s \in \text{St} \mid f(s) \in P\}, \\
Wp[\![\langle d, x \rangle]\!](P) \quad &= Wp[\![\langle d, d(x) \rangle]\!](P), \\
Wp[\![\langle d, \bigvee_I S_i \rangle]\!](P) \quad &= \bigcup \{Wp[\![\langle d, S_i \rangle]\!](P) \mid i \in I\}, \\
Wp[\![\langle d, \bigwedge_I S_i \rangle]\!](P) \quad &= \bigcap \{Wp[\![\langle d, S_i \rangle]\!](P) \mid i \in I\}, \\
Wp[\![\langle d, S_1 ; S_2 \rangle]\!](P) &= Wp[\![\langle d, S_1 \rangle]\!](Wp[\![\langle d, S_2 \rangle]\!](P)).
\end{aligned}
$$

4 Simulation for hyper transition systems

In Definition 9 we abstracted from the intermediate configurations recorded by the transition relation of a hyper transition system in order to capture the input-output behaviour of the computations specified by it. In this section we propose a relation between hyper transition systems which preserves the specification of the atomic steps of a computation. This relation is a generalization of a simulation relation between ordinary transition systems which take in account also deadlock configurations and undefined transitions. By considering a symmetric version of it one obtains the ordinary notion of bisimulation [Mil80, Par81] which can be similarly generalized to hyper transition systems.

For an ordinary transition system $\langle X, \delta, \longrightarrow \rangle$, a relation $R \subseteq X \times X$ is called a *simulation* whenever, if $\langle x, y \rangle \in R$ then

(i) $x \longrightarrow \delta \Rightarrow \exists y' \in X \cup \{\delta\}: y \longrightarrow y'$,

(ii) $x \not\longrightarrow \Rightarrow y \not\longrightarrow$,

(iii) $x \longrightarrow x' \Rightarrow \exists y' \in X: y \longrightarrow y' \ \& \ \langle x', y' \rangle \in R$,

where $x \not\longrightarrow$ means that there is no $x' \in X \cup \{\delta\}$ such that $x \longrightarrow x'$.

If there exists a simulation relation containing the pair $\langle x, y \rangle$ then the computations of T starting in the configuration x refine those starting in y. According to the usual law of the refinement calculus, a computation which deadlocks (in one step) can refine every other computation, while an undefined or terminating computation cannot refine one which does not terminate. In general, if x refines y then computations starting in a configuration x are less non-deterministic than similar computations starting in y.

The same idea of refinement holds also for a hyper transition system: the non-determinism of of the computations specified by a hyper transition system H starting in a configuration x must be less than the non-determinism of the computations starting in a *hyper similar* configuration y.

Definition 17. Given a hyper transition system $\langle X, \text{---}\epsilon \rangle$, a binary relation R on X is called a *hyper simulation* whenever, if $\langle x, y \rangle \in R$ then

(i) $x \text{---}\epsilon \ W \Rightarrow \exists V \subseteq X: y \text{---}\epsilon \ V$,

(ii) $x \not{\text{---}}\epsilon \Rightarrow y \not{\text{---}}\epsilon$

(iii) $x' \in \bigcap \{W \subseteq X \mid x \text{---}\epsilon \ W\} \Rightarrow \exists y' \in \bigcap \{V \subseteq X \mid y \text{---}\epsilon \ V\}: \langle x', y' \rangle \in R$,

where $x \not{\text{---}}\epsilon$ means that there is no $W \subseteq X$ such that $x \text{---}\epsilon \ W$.

The above definition of hyper simulation is conservative with respect to the notion of simulation: if two configurations x and y of a hyper transition system $H = \langle X, \text{---}\epsilon \rangle$ are hyper similar then they are similar in the induced transition system $TS(H) = \langle X, \delta, \longrightarrow \rangle$.

Conversely, two configurations x and y of a transition system $T = \langle X, \delta, \longrightarrow \rangle$ are similar if and only if they are hyper similar in the induced hyper transition system $HTS(T) = \langle X, \text{---}\epsilon \rangle$. Hence, for a hyper transition system $H = \langle X, \text{---}\epsilon \rangle$ satisfying (2) and with a transition relation upper closed on the right hand side, two configurations are hyper similar if and only if they are similar in $TS(H)$.

The identity relation between configurations is a hyper simulation between a hyper transition system H and $HTS(TS(H))$. Since $HTS(TS(H))$ is isomorphic to the transition system $TS(H)$ (Lemma 5 and 6), the above means that hyper transition systems are not more expressive than ordinary transition systems with respect to simulation. Rather, they describe computations in a different way: in terms of all the properties that the atomic steps of computations have to satisfy.

For the hyper transition system $\langle Conf, \longrightarrow_{\epsilon_d} \rangle$ describing the computations of the language $\mathcal{L}$, hyper simulation is a congruence with respect to the angelic and demonic choices as well as sequential composition. Moreover if we augment the specification for $\langle Conf, \longrightarrow_{\epsilon_d} \rangle$ with the rule below describing a simple interleaving operator, then hyper simulation is a congruence also for this operator.

$$\frac{\langle S_1, s \rangle \longrightarrow_{\epsilon_d} W}{\begin{array}{l} \langle S_1 \parallel S_2, s \rangle \longrightarrow_{\epsilon_d} \{\langle S_2, t \rangle \mid t \in W \cap \mathtt{St}\} \cup \{\langle S_1' \parallel S_2, t \rangle \mid \langle S_1', t \rangle \in W\} \\ \langle S_2 \parallel S_1, s \rangle \longrightarrow_{\epsilon_d} \{\langle S_2, t \rangle \mid t \in W \cap \mathtt{St}\} \cup \{\langle S_2 \parallel S_1', t \rangle \mid \langle S_1', t \rangle \in W\}. \end{array}}$$

Hence it is possible to introduce an interleaving operator in the specification language $\mathcal{L}$ while maintaining a compositional semantics for it.

Acknowledgments: The authors are grateful to the members of the Amsterdam Concurrency Group, headed by Jaco de Bakker, and to Ralph Back, Henk Goeman, Michael Mislove, Wim Hesselink, and the anonymous referees for comments and discussions on the subject of this paper.

References

[Bac80] Back, R.-J.R.: *Correctness Preserving Program Refinements: Proof Theory and Applications*, vol. 131 of *Mathematical Centre Tracts*, Amsterdam, 1980.

[Bon96] Bonsangue, M.M.: *Topological Dualities in Semantics*. PhD thesis, Vrije Universiteit Amsterdam, 1996.

[BW90] Back, R.-J.R., von Wright, J.: Dualities in specification languages: a lattice theoretical approach. *Acta Informatica*, 27:583–625, 1990.

[BW97] Back, R.-J.R., von Wright, J.: *Refinement Calculus: a Systematic Introduction*. Preliminary version of a book submitted for publication, 1997.

[Hes94] Hesselink, W.H.: Nondeterminacy and recursion via stacks and games. *Theoretical Computer Science*, 124(2):273–295, 1994.

[Mil80] Milner, R.: *A Calculus of Communicating Systems*, vol. 92 of *LNCS*. Springer-Verlag, 1980.

[Nil82] Nilsson, N.J.: *Principles of Artificial Intelligence*. Springer-Verlag, 1982.

[Par81] Park, D.M.: Concurrency and automata on infinite sequences. In P. Deussen, ed, *5th GI Conference*, vol. 104 of *LNCS*, pp. 167–183, Springer-Verlag, 1981.

[Plo81] Plotkin, G.D.: A structural approach to operational semantics. Technical Report DAIMI FN-19, Aarhus University, 1981.

A Shift-Invariant Metric on S^{zz} Inducing a Non-trivial Topology

G. Cattaneo[1], E. Formenti[2], L. Margara[3], and J. Mazoyer[2]

[1] Dipartimento di Scienze dell'Informazione, Laboratorio di Fondamenti Fisici,
39 Via Comelico, 20135 Milano, Italy
[2] Ecole Normale Supériéure de Lyon, Laboratoire de l'Informatique du Parallélisme,
46 Allée d'Italie, 69364 Lyon, FRANCE
[3] Dipartimento di Scienze dell'Informazione, Universitá di Bologna, Italy

Abstract. In this paper we discuss the meaning of sensitivity and its implications in CA behavior. A new shift-invariant metric is given. The metric topology induced by this metric is perfect but not compact. Moreover we prove that the new space is "suitable" for the study of the dynamical behavior of CA. In this context sensitivity assumes a stronger meaning than before (usually S^{ZZ} is given the product topology). Now cellular automata are sensitive if they are not only capable of "transporting" the information but if they are also able to create new information. We also provide an experimental evidence of the fact that (in the new topology) sensitivity is linked to the fractal dimension of the space-time pattern generated by cellular automata evolutions.

1 Introduction

One dimensional cellular automata (CA) are dynamical systems in which space and time are discrete. They consist of cells arranged in a bi-infinite array. The state of cells is chosen from a finite set S and is updated synchronously by a local interaction rule. A configuration is a snapshot of cells state i.e. a mapping from Z to S. Despite of their apparent simplicity CA may exhibit very complex behavior. In particular, CA provide easy to understand examples of chaotic behavior which capture the attention of many scientists. One of the main properties used to detect chaotic behavior is sensitivity to initial conditions [4,9,1]. In this paper we discuss the meaning of sensitivity and its implications in CA behavior.

Let us recall the definition of sensitivity to initial conditions for a discrete time dynamical system (DTDS) $\langle \mathcal{X}, F \rangle$. Here, we assume that $\mathcal{X}$ is equipped with a metric d and F is continuous according to the metric topology induced on $\mathcal{X}$ by d.

Definition 1. A discrete time dynamical system $\langle \mathcal{X}, F \rangle$ is *sensitive to initial conditions* (or, simply, sensitive) if there exists $\epsilon > 0$ such that for any $x \in \mathcal{X}$ and for any open ball of center x, $\mathcal{B}_\delta(x)$, there exists a point $y \in \mathcal{B}_\delta(x)$ and $n \in \mathbb{N}$ such that $d(F^n(x), F^n(y)) \geq \epsilon$. ϵ is called *sensitivity constant*.

Intuitively a map is sensitive to initial conditions if there exist points arbitrary close to x which eventually separate of at least ϵ under iteration of F. We also recall that if we equip S with the discrete metric then the metric

$$d^{P}(x, y) = \sum_{i=-\infty}^{+\infty} \frac{\delta(x_i, y_i)}{2^{|i|}} \tag{1}$$

induces the product topology on $S^{\mathbb{Z}}$ (δ is the discrete metric on S). Consider a sensitive CA $\langle S^{\mathbb{Z}}, G_f \rangle$ (in the product topology) with sensitivity constant $\epsilon > 0$. For all $x \in S^{\mathbb{Z}}$ and for all $\mathcal{B}_\delta(x)$, there exist $y \in N(x)$ and an integer $n > 0$ such that $d^{P}(G_f^n(x), G_f^n(y)) \geq \epsilon$. Let $i \in \mathbb{Z}$ be such that $x_i \neq y_i$. The value at the cell i is called *damage*. We remark that we can choose y such that it has only one damage with respect to x. In this context, the meaning of sensitivity is that damages move towards the cell of index 0. Because of this fact we say that in the product topology a rule is sensitive if it can "move damages". A paradigmatic example of this behavior is the shift map σ. There are sensitive CA that not only move damages but also they create new damages. We think that these systems capture the essence of sensitivity also from an intuitive point of view. An example of this behavior is the elementary rule 90. In Figure 1 we compare the two way of "interpreting" sensitivity, the one of the shift map and the one of rule 90. As a matter of fact, in the product topology, the two ways of interpreting the sensitivity may present chaotic behavior. The problem is how to enucleate the "real sensitivity" (i.e. the one as rule 90). We can follow two strategies:

1. adding stronger properties to the existing definition;
2. maintaining the old definition and changing completely the topology of the phase space in such a way that sensitivity matches our requirements.

In this paper we follow the latter approach and we prove the following results:

1. the key property for "real" sensitivity is the shift-invariance of the metric.
2. we provide a shift-invariant pseudo-metric for which the induced topology is not trivial, i.e. not discrete. We also prove that the new space is perfect but not compact.
3. if we take the quotient space with respect to the equivalence relation "being at null distance" then the quotient space has the quotient topology.
4. CA are uniformly continuous deterministic mappings of the quotient space into itself;
5. studying the behavior of CA on the quotient space is not meaningless since most of the interesting topological properties are preserved when passing from the quotient space to $S^{\mathbb{Z}}$.

Moreover we prove that some linear rules are sensitive. We also empirically obtain that sensitive rules (in the new topology) are the ones in which space-time patterns have fractal dimension. This fact, if proved mathematically, will be of great interest for future research.

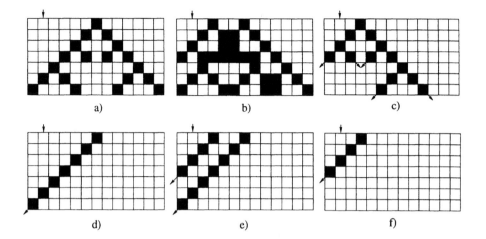

Fig. 1. Different ways of "interpreting" sensitivity: the elementary rule 90, diagrams a),b),c); the shift map σ, diagrams d),e),f). In diagram a) (resp. d)) the evolution starts from a single 1 centered in the cell of index 0; in b) (resp. e)) the evolution starts from a "damaged" configuration; c) is the site per site xor of a) and b) (resp. d) and e)). Arrows show the "direction" of the motion of damages. An arrow on the top of the diagrams represents the cell of index zero, filled boxes are cells in state 1, and empty boxes are cells in state 0. Time grows downward.

The paper is structured as follows. Section 2 gives basic definitions about cellular automata. Section 3 introduces the new pseudo-metric topology on $S^{\mathbb{Z}}$. In Section 4 the properties of the new topology are used to study the dynamical behavior of CA. In Section 5 we draw our conclusions and give some hints for future research.

We remark that were not explicitly indicated the proofs of theorems may be found in [3]. Here we give only the most significant ones.

2 Cellular automata

A one dimensional CA is a triple $\langle S, N, f \rangle$, where $S = \{0, 1, \ldots, p-1\}$ is the set of *states*, $N = \{-r, \ldots, 0, \ldots, r\}$ is the *neighborhood* structure with *rule radius* r, and $f: S^{2r+1} \to S$ is the *local rule*. The *global function* of a given CA with local rule f is the mapping $G_f: S^{\mathbb{Z}} \mapsto S^{\mathbb{Z}}$ associating to any bi-infinite S-valued sequence $c : \mathbb{Z} \mapsto S$, $i \to c_i$ the bi-infinite S-valued sequence $G_f(c) : \mathbb{Z} \mapsto S$, $i \to G_f(c)_i$ defined by $G_f(c)_i = f(c_{i-r}, \ldots, c_i, \ldots, c_{i+r})$. As an example, the global function of the *shift* CA, denoted by σ, with radius $r \geq 1$ and S as its set of states is defined $\forall c \in S^{\mathbb{Z}}, \forall i \in \mathbb{Z}$ by $\sigma(c)_i = c_{i+1}$. A configuration c is said to be *spatial periodic* of period p if it is a periodic point of period p of σ. In this case the set of cells $c_0, c_1, \ldots, c_{p-1}$ is the *basic pattern* of c.

The collection $S^{\mathbb{Z}}$ of all bi-infinite S-valued sequences is the *phase space* of the discrete time dynamics generated by iterations of the global function G_f.

Hedlund ([7],Thm. 3.4) proved that the following two statements are equivalent when $S^{\mathbb{Z}}$ is given the product topology.

1. $F : S^{\mathbb{Z}} \mapsto S^{\mathbb{Z}}$ is continuous and commutes with the shift map (i.e. $F \circ \sigma = \sigma \circ F$).
2. $F : S^{\mathbb{Z}} \mapsto S^{\mathbb{Z}}$ is a global function of a suitable local rule f, i.e., $F = G_f$.

For any $s \in S$ a configuration $c \in S^{\mathbb{Z}}$ is said to be s-quiescent if it has only finite a number of sites that are not in state s. A cellular automaton is *elementary* if $S = \{0,1\}$ and $r = 1$. Elementary CA can be conveniently represented by a unique decimal integer (for the representation algorithm see [11]).

A CA $\langle S, N, f \rangle$ is *linear* if $f(x_{-r}, \ldots, x_r) = \sum_{i=-r}^{r} \lambda_i \cdot x_i \mod |S|$, where $\lambda_i \in S$. The class of additive CA has been extensively studied [5,6,2] and, in our case, they provide a set of easy to understand examples.

3 A new metric on $S^{\mathbb{Z}}$

The first shift invariant metric over $S^{\mathbb{Z}}$ that comes into one's mind is the discrete metric. By the way, a discrete topology is not desirable as phase space of a dynamical system because of its many pathologies. For example, in discrete phase spaces, sensitivity looses its meaning completely, there are no more transitive systems and many others. Therefore another requirement for the new topology is non triviality, i.e. non discreteness.

Definition 2. For all configurations $x, y \in S^{\mathbb{Z}}$ and $\forall k \in \mathbb{N}$ let

$$\#_{-k,k}^{\neq}(x, y) = |\{i \in \{-k, -k+1, \ldots, 0, 1, \ldots, k\} \text{ s.t. } x_i \neq y_i\}| \ .$$

Roughly speaking, the quantity $\#_{-k,k}^{\neq}(x, y)$ is the Hamming distance between the two segments of length $2k + 1$ (and centered in zero) of the configurations x and y.

Let us consider the mapping $d \colon S^{\mathbb{Z}} \times S^{\mathbb{Z}} \to \mathbb{R}^+$ defined as follows

$$\forall x, y \in S^{\mathbb{Z}}, \, d(x, y) = \limsup_{k \to +\infty} \left\{ \frac{\#_{-k,k}^{\neq}(x, y)}{2k + 1} \right\} \ .$$

Proposition 3. *The mapping d is a shift-invariant pseudo-metric on $S^{\mathbb{Z}}$.*

Proof. First of all note that $\forall x, y \in S^{\mathbb{Z}}, 0 \leq d(x, y) < +\infty$ since $\forall k \in \mathbb{N}, 0 \leq \#_{-k,k}^{\neq}(x, y) \leq 2k + 1$. Moreover it is straightforward from the definition that $d(x, y) = 0$ if $x = y$ and $d(x, y) = d(y, x)$. We have to prove only the triangle inequality. For all $x, y, z \in S^{\mathbb{Z}}$ we have that

$$d(x,z) + d(z,y) = \limsup_{k \to \infty} \left\{ \frac{\#_{-k,k}^{\neq}(x,z)}{2k+1} \right\} + \limsup_{k \to \infty} \left\{ \frac{\#_{-k,k}^{\neq}(z,y)}{2k+1} \right\} \geq$$

$$\geq \limsup_{k \to \infty} \left\{ \frac{\#_{-k,k}^{\neq}(x,z) + \#_{-k,k}^{\neq}(z,y)}{2k+1} \right\} \geq$$

$$\geq \limsup_{k \to \infty} \left\{ \frac{\#_{-k,k}^{\neq}(x,y)}{2k+1} \right\} = d(x,y) \ .$$

Before proving that d is σ-invariant let us remark that for all $k \in \mathbb{N}$ and for all $x, y \in S^{\mathbb{Z}}$ we have $\#_{-k,k}^{\neq}(x,y) - 1 \leq \#_{-k,k}^{\neq}(\sigma(x), \sigma(y)) \leq \#_{-k,k}^{\neq}(x,y) + 1$. By the above remarks we have for all $x, y \in \mathcal{X}$

$$d(\sigma(x), \sigma(y)) = \limsup_{k \to \infty} \left\{ \frac{\#_{-k,k}^{\neq}(\sigma(x), \sigma(y))}{2k+1} \right\} =$$

$$= \limsup_{k \to \infty} \left\{ \frac{\#_{-k,k}^{\neq}(x,y) + c_{-k,k}(x,y)}{2k+1} \right\} = d(x,y)$$

where $\forall x, y \in \mathcal{X}$, $c_{-k,k}(x,y) \in \{-1, 0, +1\}$. $\qquad \Box$

Let us define the following relation:

$$\forall x, y \in S^{\mathbb{Z}}, \ x \doteq y \Leftrightarrow \limsup_{k \to +\infty} \left\{ \frac{\#_{-k,k}^{\neq}(x,y)}{2k+1} \right\} = 0 \ .$$

Proposition 4. *The relation $\doteq$ is an equivalence relation.*

The space $\langle S^{\mathbb{Z}}, d \rangle$ can be given the pseudo-metric topology induced by d. Now let $\mathcal{D}$ be the family of all the sets of the form $cl(x_{\dot{=}})$, where $cl(.)$ is the topological closure operator (in the pseudo-metric space $\langle S^{\mathbb{Z}}, d \rangle$) and $x_{\dot{=}} \in S^{\mathbb{Z}}/ \doteq$. For any two sets $cl(x_{\dot{=}})$ and $cl(y_{\dot{=}})$, let

$$D(cl(x_{\dot{=}}), cl(y_{\dot{=}})) = \inf \left\{ d(a,b), a \in cl(x_{\dot{=}}) \text{ and } b \in cl(y_{\dot{=}}) \right\} \ .$$

Then $\langle \mathcal{D}, D \rangle$ is a metric space whose topology is the quotient topology for $\mathcal{D}$, and the projection Π of $S^{\mathbb{Z}}$ onto $\mathcal{D}$ is an isometry. Moreover Π is both open and closed. (For more details about pseudo-metric topologies see [8]). Substantially we have three levels:

$$\langle S^{\mathbb{Z}}, d^{\mathcal{P}} \rangle \xrightarrow{I} \langle S^{\mathbb{Z}}, d \rangle \xrightarrow{\Pi} \langle S^{\mathbb{Z}}/ \doteq, d \rangle$$

where I is the identity and Π the canonical projection mapping defined as above. Let us denote $\langle S^{\mathbb{Z}}, d^{\mathcal{P}} \rangle$ simply by $S^{\mathbb{Z}}$, $\mathcal{X} = \langle S^{\mathbb{Z}}, d \rangle$ and $\tilde{\mathcal{X}} = \langle \mathcal{D}, D \rangle$. If S is a group (w.r.t. addition) then $S^{\mathbb{Z}}$ is a compact abelian topological group. As a consequence also $\mathcal{X}$ (and therefore $\tilde{\mathcal{X}}$) is a topological group. Neither $\mathcal{X}$ nor $\tilde{\mathcal{X}}$ are necessarily compact since I is not continuous. In effect we will see in Theorem 10 that $\tilde{\mathcal{X}}$ is not compact. When no confusion is likely we will denote the metric D with d and $x_{\dot{=}} \in \tilde{\mathcal{X}}$ by x.

Remark 5. The equivalence relation $\doteq$ has very broad equivalence classes. All quiescent configurations are in the same equivalence class. This very big gathering may result dangerous when considering the evolution of dynamical systems on $S^{\mathbb{Z}}$. For example, the behavior of CA on the set of quiescent configurations may vary significantly from the behavior on $\underline{0}_{\doteq}$. We will prove in Theorem 16 that most of the interesting dynamical properties are preserved when passing from $\tilde{\mathcal{X}}$ to $\mathcal{X}$.

Remark 6. From Remark 5 one can immediately deduce that all the configurations that differ in a finite number of sites are in the same equivalence class of $\doteq$. But things go even further. In fact, two configurations can be in the same equivalence class even if they differ in an infinite set of sites. For example, consider the quiescent configuration $\underline{0}$ and a configuration $\overline{t}$ $(t > 1)$ that differs from the quiescent configuration only in the cells of index $\pm t, \pm t^2, \pm t^3, \ldots, \pm t^n$, and so on. It is easy to see that $\underline{0} \doteq \overline{t}$, in fact

$$\limsup_{k \to \infty} \left\{ \frac{2 \cdot \left\lfloor \frac{\log_2 k}{\log_2 t} \right\rfloor}{2k+1} \right\} \leq \limsup_{k \to +\infty} \left\{ \frac{\#^{\neq}_{-k,k}(\underline{0}, \overline{t})}{2k+1} \right\} \leq \limsup_{k \to \infty} \left\{ \frac{2 \cdot \left\lfloor \frac{\log_2 k}{\log_2 t} \right\rfloor + 2}{2k+1} \right\} .$$

Proposition 7. *Let $x, y \in \tilde{\mathcal{X}}$ be two distinct spatial periodic configurations. Then $x \neq y$. Moreover the space $\tilde{\mathcal{X}}$ does not coincide with the class of spatial periodic configurations, and the class of spatial periodic configurations is not dense in $\tilde{\mathcal{X}}$.*

Proposition 8. *The space $\langle \tilde{\mathcal{X}}, d \rangle$ is not discrete.*

Proof. The idea of the proof is to show that $\underline{0}$ is an accumulation point for $\tilde{\mathcal{X}}$. This task is accomplished by proving that for every $x \in \tilde{\mathcal{X}}$ such that $d(\underline{0}, x) = \epsilon$ $(\epsilon > 0)$ there exists a configuration $y \in \tilde{\mathcal{X}}$ such that $d(\underline{0}, y) < \epsilon$. Let $\overline{n} \in \mathbb{N}$ be such that $\frac{1}{\overline{n}} < \epsilon$. Let x be the spatial periodic configuration with the basic pattern made by $\overline{n} - 1$ zeroes and a symbol s, with $s \neq 0$ (here 0 is considered as the quiescent state). Then we have, $c_{-k,k}(x) \in \{0, 1, 2, \ldots, \overline{n} - 1\}$,

$$d(\underline{0}, x) \leq \limsup_{k \to \infty} \frac{\left\lfloor \frac{k}{\overline{n}} \right\rfloor + c_{-k,k}(x)}{2k+1} = \frac{1}{2\overline{n}} < \epsilon .$$

$\square$

In the proof of Proposition 8 we have seen that the quiescent configuration is an accumulation point for $\tilde{\mathcal{X}}$. This property holds for all the points in $\tilde{\mathcal{X}}$ as it is proved in Theorem 9.

Theorem 9. *The metric space $\langle \tilde{\mathcal{X}}, d \rangle$ is perfect i.e. every point is an accumulation point.*

Proof. The idea of the proof is similar to the one of Proposition 8. For any $x \in \tilde{\mathcal{X}}$ let x_k be a configuration such that

$$\forall i \in \mathbb{Z}, \; x_i^k = \begin{cases} s \; (s \neq x_i), & \text{if } |i| \equiv 0 \bmod k, \\ x_i, & \text{otherwise .} \end{cases}$$

Then it is easy to prove that $\forall k \in \mathbb{N}, \; d(x, x^k) = \frac{1}{k}$ and $\lim_{k \to \infty} d(x, x^k) = 0$. $\qquad\square$

Theorem 10. *The metric space $\tilde{\mathcal{X}}$ is not compact.*

Proof. Let us consider the following sequence of spatial periodic configurations. Let $C^0 = \underline{0}$. Let $t \in \mathbb{N}$. For all $i \in \mathbb{N}$, C^i is the spatial periodic configuration with its basic pattern made of $t \cdot 2^{i-1}$ zeroes followed by $t \cdot 2^{i-1}$ ones. We are going to prove that for all $i, n \in \mathbb{N}$ $(i < n)$ the distance $d(C^i, C^n)$ is $\frac{1}{2}$. The first $t \cdot 2^{n-1}$ sites in state 0 of the basic pattern of C^n can be divided into 2^{n-i} blocks of length $t \cdot 2^{i-1}$. Now if we superpose 2^{n-i} times the basic pattern of C^i we have that there are 2^{n-i-1} blocks of different sites with $t \cdot 2^{i-1}$ cells per block, the total amount of different sites is $t \cdot 2^{n-2}$. In the same way we may reason for the second part of the basic pattern of C^n. Therefore we have

$$\limsup_{k \to \infty} \left\{ \frac{t \cdot 2^{n-1} \cdot \lfloor \frac{2k+1}{t \cdot 2^n} \rfloor}{2k+1} \right\} \leq d(C^i, C^n) \leq \limsup_{k \to \infty} \left\{ \frac{t \cdot 2^{n-1} \cdot \lfloor \frac{2k+1}{t \cdot 2^n} \rfloor + c_k}{2k+1} \right\}$$

where $\forall k \in \mathbb{N}, 0 \leq c_k \leq t \cdot 2^n$. Therefore $d(C^i, C^n) = \frac{t \cdot 2^{n-1}}{t \cdot 2^n} = \frac{1}{2}$. We may conclude that neither is $\{C^n\}_{n \in \mathbb{N}}$ converging nor it has converging subsequences.

$\square$

3.1 On the cardinality of $\tilde{\mathcal{X}}$

In this section we establish that the cardinality of the new "workspace" is not too small for our purposes. In fact in Proposition 11 we prove that the cardinality of $\tilde{\mathcal{X}}$ is at least countable. The result is furtherly stengthened by Theorem 13 which proves that $\tilde{\mathcal{X}}$ has the cardinality of $\mathbb{R}$, the same as $S^{\mathbb{Z}}$.

Proposition 11. *The index of $\doteq$ is infinite (at least countable).*

Proof. By proposition 7, two distinct spatial periodic configurations are in distinct classes of $\doteq$. Since the family of spatial periodic configurations is countable then $\doteq$ has at least a countable index.

For the sake of simplicity let us consider $S = \{0, 1\}$. The set $\mathcal{X}$ is a topological group (as we have already remarked). One can easily prove that (in this case) $d(x, y) = d(\underline{0}, x - y)$ i.e. d is left-invariant w.r.t. the translations on the group $\mathcal{X}$. Let $h: \tilde{\mathcal{X}} \to [0, 1]$ be define by

$$\forall x \in \tilde{\mathcal{X}}, \; h(x) = \limsup_{k \to \infty} \left\{ \frac{\sum_{-k}^{k} x_k}{2k+1} \right\} .$$

Lemma 12. *The mapping h is continuous and surjective.*

As an immediate consequence of Lemma 12 we have the following theorem.

Theorem 13. *The metric space $\mathcal{X}$ has the cardinality of $\mathbb{R}$.*

Moreover since Π is a homomorphism we have that all the equivalence classes of $\doteq$ have the same cardinality. In particular the class $0_{\doteq}$ is the class of the configurations with upper density of ones equal to zero and (Theorem of E. Borel) has null Lebesgue measure and cardinality continuum.

4 CA and the new topology on $S^{\mathbb{Z}}$

This section correlates the properties of the new topology on $\tilde{\mathcal{X}}$ with cellular automata and their dynamics. The key idea is to study the behavior of CA in the quotient space and then give theorems proving which properties are preserved when passing from the "rough" $\tilde{\mathcal{X}}$ to the "coarse" $\mathcal{X}$. The first thing to do is to guarantee that the computation of CA is "deterministic", i.e. points in the same equivalence class are mapped to the same equivalence class.

Proposition 14. *Let G_f be the global rule of a CA with radius r. For all $x, y \in \mathcal{X}$, if $x \doteq y$ then $G_f(x) \doteq G_f(y)$.*

Proof. If $x \doteq y$ then $\limsup_{k \to \infty} \left\{ \frac{\#^{\neq}_{-k,k}(x,y)}{2k+1} \right\} = 0$. For all $k \in \mathbb{N}$, the cardinality of the sites that can be changed by G_f is bounded by $(2r+1)\#^{\neq}_{-k,k}(x,y)$. Therefore we have

$$0 \leq \limsup_{k \to \infty} \left\{ \frac{\#^{\neq}_{-k,k}(G_f(x), G_f(y))}{2k+1} \right\} \leq \limsup_{k \to \infty} \left\{ \frac{(2r+1)\#^{\neq}_{-k,k}(x,y)}{2k+1} \right\} = 0 \ .$$

$\square$

Proposition 15. *CA are uniformly continuous in $\langle \tilde{\mathcal{X}}, d \rangle$.*

Proof. Let $x \in \tilde{\mathcal{X}}$. We are going to prove that a generic CA of global rule G_f and radius r is uniformly continuous in x. For all $\epsilon > 0$ let $\delta = \frac{\epsilon}{2r+1}$. Then for all $y \in \tilde{\mathcal{X}}$ such that $d(x, y) < \delta$ we have

$$d(G_f(x), G_f(y)) = \limsup_{k \to \infty} \left\{ \frac{\#^{\neq}_{-k,k}(G_f(x),G_f(y))}{2k+1} \right\} \leq$$
$$\leq \limsup_{k \to \infty} \left\{ \frac{(2r+1)\#^{\neq}_{-k,k}(x,y)}{2k+1} \right\} =$$
$$= (2r+1)d(x,y) < (2r+1)\delta = \epsilon \ .$$

$\square$

Let G_f be the global rule of a CA, then the global function of the same CA on the quotient space is $\tilde{G}_f = \Pi \circ G_f \circ \Pi^{-1}$ and its n-th iterate is $\tilde{G}_f{}^n = \Pi \circ G_f^n \circ \Pi^{-1}$. Given a one dimensional CA $\mathcal{A} = \langle S, r, G_f \rangle$, the CA $\tilde{\mathcal{A}} = \langle S, r, \tilde{G}_f \rangle$ is called *projection* or *quotient* CA of $\mathcal{A}$. Theorem 16 says that proving dynamical properties of CA on the quotient set $\tilde{\mathcal{X}}$ it is not meaningless since most part of the interesting topological properties are preserved when passing from $\tilde{\mathcal{X}}$ to $\mathcal{X}$. Before proving the theorem let us recall some more definitions. A DTDS $\mathcal{A} = \langle \mathcal{U}, F \rangle$ is *transitive* (resp. *strongly transitive*) if for all non-void open sets A we have $cl(\bigcup_{n \in \mathbb{N}} F^n(A)) = \mathcal{U}$ (resp. $\bigcup_{n \in \mathbb{N}} F^n(A) = \mathcal{U}$). We say that $\mathcal{A}$ is *equicontinuous at* $x \in \mathcal{U}$ if $\forall \epsilon > 0 \forall y \in \mathcal{B}_\epsilon(x) \forall t \in \mathbb{N}, d(F^t(x), F^t(y)) < \epsilon$. $\mathcal{A}$ is *equicontinuous* if it is equicontinuous at all points of $\mathcal{U}$.

Theorem 16. *Let G_f be a global function of a CA and $\tilde{G}_f$ the global function of its projection. If $\tilde{G}_f$ has the property $\mathcal{P}$ then also G_f has $\mathcal{P}$, where $\mathcal{P}$ is one of the following: equicontinuity, sensitivity to initial conditions, transitivity, strongly transitivity.*

A DTDS $\langle \mathcal{U}, F \rangle$ is *expansive* if there exists $\epsilon > 0$ s.t. for all $x, y \in \mathcal{U}$, there exists $t \in \mathbb{N}$ such that $d(F^t(x), F^t(y)) \geq \epsilon$. Theorem 16 does not mention the property of expansivity since there are no expansive CA in $\langle \mathcal{X}, d \rangle$! In fact any point has an infinity of points which are at null distance from itself.

Lemma 17 ([10]). *All periodic points of non-trivial linear CA are spatial periodic.*

We recall that a DTDS is *regular* if its periodic points set is dense in the phase space. The following proposition is an immediate consequence of Proposition 7 and Lemma 17.

Proposition 18. *No linear CA is regular.*

Example 19. Let us consider the elementary rule 90, i.e. a CA with the following local rule $\forall x, y, z \in \{0, 1\}$, $f(x, y, z) = x \oplus z$, where $\oplus$ is addition modulo 2. Let $\epsilon = \frac{1}{2}$. For any $x \in \mathcal{X}$ and any $\delta > 0$ let $y \in \mathcal{B}_\delta(x)$ be such that

$$\forall i \in \mathbb{Z}, \; y_i = \begin{cases} x_i & \text{if } i \not\equiv 0 \bmod 2^t \\ 1 \oplus x_i & \text{otherwise} \end{cases}$$

where t is such that $\frac{1}{2^t} < \delta$ and $\frac{1}{2^{t-1}} \geq \delta$. It is not difficult to see that from each damage after $2^{t-1} - 1$ steps there is an offspring of 2^{t-1} new damages. Therefore

$$d(G_{90}^t(x), G_{90}^t(y)) \geq \limsup_{k \to \infty} \left\{ \frac{2^{t-1} \cdot \left\lfloor \frac{2k+1}{2^t} \right\rfloor}{2k+1} \right\} = \frac{1}{2} \; .$$

We may conclude that rule 90 is sensitive with sensitivity constant equal to $\frac{1}{2}$.

We stress that with a technique similar to the one used in Example 19 we can prove the sensitivity of many linear CA. We noticed that this technique always works if the space-time pattern generated from an initial configuration consisting of a single non-quiescent symbol has fractal dimension. This fact is one of the subjects of our future research.

5   Conclusions

We have seen that requiring shift-invariance may strongly change the topological properties of the space. The dynamics of CA may be greatly perturbed as well. In this paper we have proved the existence of a shift-invariant metric which induces a non trivial topology. This topology turns out to be a suitable playground for the study of the dynamical behavior of CA.

Although the new metric topology provides a stronger and more intuitive interpretation of sensitivity to initial conditions than the usual product topology, there are still many open question that give possibilities for future research. Here we address some of them. A problem that appears to be difficult is how to prove the existence of transitive CA. And along this line of thought, is there any invariant measure on $\mathcal{X}$? If such a measure exists which are the measure preserving functions? Are there ergodic CA? The answers to these questions may shed new light on the chaotic behavior of CA.

Acknowledgements. The authors want to thank an anonymous referee for many helpful suggestions and for the ideas which have inspired Lemma 12 and Theorem 13.

References

1. Banks, J., Brooks, J., Cairns, G., Davis, G., Stacey, P.: On Devaney's definition of chaos. American Mathematical Montly **99** (1992) 332–334
2. Cattaneo, G., Formenti, E., Manzini, G., Margara, L.: On ergodic linear cellular automata over Z_m. Proceeding of STACS '97.
3. Cattaneo, G., Formenti, E., Margara, L., Mazoyer, J.: A shift-invariant metric on $S^{\mathbb{Z}}$. Ecole normal Superiéure de Lyon, Laboratoire de l'Informatique du Paralélisme rep. ?? (1997)
4. Devaney, R. L.: *Introduction to chaotic dynamical systems*. Addison-Wesley, second edition (1989)
5. v. Haeseler, H., Peitgen, H.-O., Skordev, G.: Cellular automata, matrix substitutions and fractals. Annales Mathematical Artificial Intelligence **8** (1993) 345–362
6. v. Haeseler, H., Peitgen, H.-O., Skordev, G.: Global analysis of self-similarity features of cellular automata: some selected examples. Physica D **86** (1995) 64–80
7. Hedlund, G. A.: Endomorphism and automorphism of the shift dynamical system. Mathematical System Theory, **3** (1969) 320–375
8. Kelley, J. L.: *General topology*, volume **27** of Graduate texts in Mathematics. Springer-Verlag (1975)
9. Knudsen, C.: Chaos without nonperiodicity. American Mathematical Montly **101** (1994) 563–565
10. Margara, L.: Cellular automata and non periodic orbits. Complex Systems (to appear)
11. Wolfram, S.: *Theory and Applications of Cellular Automata*. World Scientific, Singapore (1986)

Subtyping Calculus of Construction (Extended Abstract)

Gang Chen *

Ecole Normale Supérieure,Laboratoire d'Informatique, 45, rue d'Ulm, 75230 Paris, France, email: gang@dmi.ens.fr

Abstract. *We present in this paper a subtyping extension of Calculus of Construction. We prove that this system has good meta-theoretical properties: transitivity elimination, subject reduction, strong normalization and decidability of subtyping. This work provides a theoretical foundation for adding subtyping to proof checkers like Coq, LEGO etc.*

1 Introduction

In recent ten years, significant progress has been made in subtyping extensions to systems in λ-cube. These subtyping systems can be divided into two classes: those with bounded quantification and those without bounded quantification. For the former, there are second order systems, e.g., $F_\le$[CG92] and higher order systems, $F_\le^\omega$ [PS96], $F_\wedge^\omega$[Com94]. For the second group, there are second order subtyping system, e.g., $CO^\vdash$ [LMS95] and first order dependent type system with subtyping, e.g., $\lambda P_\le$[AC96]. Subtyping systems with bounded quantification have been extensively studied in the context of object oriented programming. Those without bounded quantification are more related to logical reasoning.

In this paper, we study a subtyping extension to Calculus of Construction(CC) without bounded quantification. CC can be viewed as a combination of F^ω and λP. Both F^ω and λP are dependent type systems. F^ω has types depending on types and λP has types depending on terms. In CC, types can depend on both types and terms. Dependent type is also called type family. The distinguished aspects in subtyping dependent types are the subtyping relation between type families (or simply subfamily relation) as well as subtyping induced by type conversion. These features complicate the meta-theoretical study. The main obstacle is the conflict between type conversion and the transitivity elimination.

Transitivity elimination is not only an important property of its own but also an intermediate step in the proofs of many meta-theoretical properties, including subject reduction, decidability of subtyping etc. The existence of type conversion makes it difficult to prove this property. The second order λ-calculus does not have type conversion, transitivity rule in both of its subtyping extensions

* This work is supported by Programme PRA, Association Franco-Chinoise pour la Recherche Scientifique et Technique, and Bourse du Ministère des Affaires Etrangères du Gouvernement Français.

$(F_\le, CO^\vdash)$ can be eliminated. There are type conversions in F^ω and λP, the subtyping systems $F^\omega_\le$ (as well as $F^\omega_\wedge$) and $\lambda P_\le$ do not have transitivity elimination property. In order to establish the meta-theoretical properties like subject reduction and decidability of subtyping, intermediate algorithmic systems based on normalized types have been adopted. Transitivity elimination holds in those algorithmic systems.

In the case of CC, it is difficult to adapt this technique since there are too many reduction relations involved in the type conversion.

F^ω has three reduction relations, but only one of them involved in type conversion. Although λP has only two reductions, both of them are involved in type conversion. The study of $\lambda P_\le$ is more difficult than $F^\omega_\le$. Since algorithmic system uses normalization in their rules, proofs of strong normalization, subject reduction and transitivity elimination depend on each other. To break the circle, in [AC96] reductions are splitted and treated one after the other. In CC, all four reductions are mixed together in type conversion, it would be hard to construct an algorithmic system based on normalized terms in order to get out from the circle.

Another difficulty for subtyping CC is the proof of termination of subtyping algorithm. The proofs in $F^\omega_\le$ and $\lambda P_\le$ are closely related to specific reductions. The proof in [PS96] uses a transformation from $\beta\Gamma$ reduction in $F^\omega_\le$ to β-reduction in F_ω. The delicate part is to alternate β and Γ reductions in such a way that all Γ reductions are moved to the front of the reduction sequence. In [AC96], the termination proof is achieved by introducing a new type operator *plus*. It is not clear how to combine these different techniques to construct a termination proof in the general case like subtyping extension of CC.

In [Che96], we have constructed a system $\lambda\Pi_\le$ which is equivalent to $\lambda P_\le$ in typing and subtyping, but its subtyping system has type level transitivity elimination property(in contrast, $F^\omega_\le$ and $\lambda P_\le$ have transitivity elimination at the level of normalized types). Besides, it gives a simple and general proof of $\beta\Gamma$ strong normalization which turns out to be a uniform method to prove the termination of subtyping algorithm. Like [PS96], we transform $\beta\Gamma$ reduction in $\lambda\Pi_\le$ to $\lambda\Pi$, but our mapping does not need to alternate β and Γ reduction. The proof is simpler than that of [PS96] and it has less dependence on specific features of reductions. So it can be applied to general case.

In this paper, we apply these techniques from [Che96] to realize a subtyping extension of Calculus of Construction, which is named $\lambda C_\le$. We prove that this system has good meta-theoretical properties: transitivity elimination, subject reduction, strong normalization and decidability of subtyping.

The main purpose of this system is for the proof assistants. Recently, there are lots of interests of adding subtyping to proof environments,like Coq, LEGO, Elf, Alf, Nuprl. Such interests come from the need of developing mathematics in large scale. We noticed that in each of these groups, there are studies of using various forms of subtyping. As pointed out by Luo[Luo96], the intention is two fold: 1. definition inheritance, such as defining group as an extension of monoid; 2. theorem inheritance, such as stating that any monoid theorem is a group theorem. From the existing investigations on using subsorting, implicit coercion and inheritance in proof checkers[Pfe93, Bar95, Bai96, Luo96], especially the work of Saibi [Sai97], we recognize the importance of subfamily relation and the decidability of subtyping checking. So we decide not to include bounded quantification and some delicate rules in $CO^{\vdash}$, on the other hand, we have considered type conversions and subfamily relations.

Another possible application area of this research is programming language design. Dependent types can represent certain data structures, such as the array type in PASCAL, e.g., Array[1..10]of Integer, which depends on both term and type.

The remained part of this paper is organized as follows. In Section 2, the definition of $\lambda C_{\leq}$ is presented. In Section 3, we give an example to show an application of this system in formalizing algebra. In Section 4, we show the meta-theoretical properties of this system. These include Church-Rosser property, well-formedness of subtyping, admissibility of transitivity, subject reduction, strong normalization and decidability of subtyping. Related works and concluding remarks are in Section 5 and Section 6 respectively.

2 Definition of $\lambda C_{\leq}$

Definition 2.1 (Types and terms of $\lambda C_{\leq}$) *1. The set of* pre-expressions *of $\lambda C_{\leq}$, notation, $\mathcal{T}$ is defined as follows*

$$\mathcal{T} = \mathcal{V} \mid \mathcal{S} \mid \mathcal{T}\mathcal{T} \mid \lambda \mathcal{V}{:}\mathcal{T}.\mathcal{T} \mid \Pi \mathcal{V}{:}\mathcal{T}.\mathcal{T}$$

where $\mathcal{V}$ is the collection of variables, $\mathcal{S}$ is that of sorts. No distinction between type- and term-variables is made. 2. There are exactly two elements: $\star$ and $\square$ in $\mathcal{S}$.

Intuitively, $\star$ and $\square$ represent the set of types and kinds respectively.

Definition 2.2 (Context for $\lambda C_{\leq}$) *A typing statement is of the form $A : B$ with $A, B \in \mathcal{T}$. A subtyping statement is of the form $A \leq B$ with $A, B \in \mathcal{T}$. A declaration is either of the form $x : A$ with $A \in \mathcal{T}$ and x a variable, or of the form $x{\leq}A{:}K$ with $A, K \in \mathcal{T}$ and x a variable. A pre-context is a finite linearly ordered declarations. The empty context is denoted by $<>$. Γ is a well-formed context (or legal context) if $\Gamma \vdash A : B$ is derivable.*

Definition 2.3 (Judgments) *There are two forms of judgments in this system: 1. Typing judgment: $\Gamma \vdash A : B$; 2. Subtyping judgment $\Gamma \vdash A \leq B$*

We use $A, B, C, a, b, \dots$ for arbitrary pre-expressions (or pre-terms), $\Gamma, \Delta \dots$ for precontexts, $x, y, z, \dots$ for arbitrary variables, sometimes, we will use α for the variable used in the subtyping declaration in the form $\alpha \leq A : K$, such α will not be used as quantification variable. Letters $s, s_1, s_2, ..$ are sorts ranging over $\{\star, \square\}$.

Conventions and abbreviations are listed in the following.

Notation 2.4 *Suppose that Γ is a pre-context.* 1) $x : A \in \Gamma$ *if* $\Gamma \equiv \Gamma_1, x : A, \Gamma_2$; 2) $\alpha \leq B \in \Gamma$ *if* $\Gamma \equiv \Gamma_1, \alpha \leq B : K, \Gamma_2$; 3) $Dom(\Gamma) = \{x \mid x : A \in \Gamma \vee x \leq A : K \in \Gamma\}$; 4) $Fv(A) \equiv$ *the set of free variables in* A; 5) $x \in \Gamma \equiv x \in Dom(\Gamma)$; 6) $\Gamma \vdash A, B : C \equiv \Gamma \vdash A : C \wedge \Gamma \vdash B : C$; 7) $\Gamma \vdash A : B : C \equiv \Gamma \vdash A : B \wedge \Gamma \vdash B : C$

Definition 2.5 (Reduction for $\lambda C_\leq$) *On $\mathcal{T}$ the notions of β-conversion and β-reduction are defined by the following contraction rule:* $C[(\lambda x{:}A.B)C] \rightarrow_\beta C[B[x := C]]$ *(C indicates a pre-expression with a hole in it).* $\overset{*}{\rightarrow}$ *and* $=_\beta$ *are reflexive and transitive closure of* $\rightarrow_\beta$ *and* $\leftarrow_\beta \cup \rightarrow_\beta$ *respectively.*

The typing rules of $\lambda C_\leq$ is formed by adding $start_\leq$, $weakening_\leq$ to CC for introducing subtyping declarations. The conversion rule in CC has been changed to subsumption rule.

Definition 2.6 (Rules for $\lambda C_\leq$) *Typing judgments are defined by the following rules. Letters s, s_1, s_2 range over $\{\star, \Box\}$.*

axiom $\dfrac{}{\langle\rangle \vdash \star : \Box}$

start $\dfrac{\Gamma \vdash A : s \quad x \notin \Gamma}{\Gamma, x : A \vdash x : A}$
$\qquad$
start$_\leq$ $\dfrac{\Gamma \vdash A : K : \Box \quad \alpha \notin \Gamma}{\Gamma, \alpha \leq A : K \vdash \alpha : K}$

weakening $\dfrac{\Gamma \vdash A : B \quad \Gamma \vdash C : s \quad x \notin \Gamma}{\Gamma, x : C \vdash A : B}$
$\qquad$
weakening$_\leq$ $\dfrac{\Gamma \vdash A : B \quad \Gamma \vdash C : K : \Box \quad \alpha \notin \Gamma}{\Gamma, \alpha \leq C : K \vdash A : B}$

application $\dfrac{\Gamma \vdash F : \Pi x{:}A.B \quad \Gamma \vdash a : A}{\Gamma \vdash Fa : B[x := a]}$
$\qquad$
abstraction $\dfrac{\Gamma, x : A \vdash b : B \quad \Gamma \vdash (\Pi x{:}A.B) : s}{\Gamma \vdash \lambda x{:}A.b : \Pi x{:}A.B}$

s_1, s_2 rule $\dfrac{\Gamma \vdash A : s_1 \quad \Gamma, x : A \vdash B : s_2}{\Gamma \vdash \Pi x{:}A.B : s_2}$
$\qquad$
subsumption $\dfrac{\Gamma \vdash a : A \quad \Gamma \vdash A \leq B \quad \Gamma \vdash A, B : s}{\Gamma \vdash a : B}$

Subtyping is introduced to the context in the form $\alpha \leq A : K$ where A is a type family and K is the kind of A. One may think of A as a function with n parameters: $A : \Pi x_1{:}A_1..\Pi x_n{:}A_n.\star$, $n \geq 0$. Its arguments may be either terms or types.

There are two equivalent presentations of subtyping relation. The more intuitive set of rules is:

S-Π $\dfrac{\Gamma \vdash A' \leq A \quad \Gamma, x : A' \vdash B \leq B'}{\Gamma \vdash \Pi x{:}A.B \leq \Pi x{:}A'.B'}$
$\qquad$
S-VAR $\dfrac{(\alpha \leq A : \Pi x_1{:}A_1..\Pi x_1{:}A_n.\star) \in \Gamma}{\Gamma \vdash \alpha M_1..M_n \leq A M_1..M_n}$

S-CONV $\dfrac{\Gamma \vdash A, B : s \quad A =_\beta B}{\Gamma \vdash A \leq B}$
$\qquad$
S-TRANS $\dfrac{\Gamma \vdash A \leq B \quad \Gamma \vdash B \leq C \quad \Gamma \vdash A, B, C : s}{\Gamma \vdash A \leq C}$

Another set of rules, which are less intuitive but more suitble for meta-theoretical study is the following.

Definition 2.7 (Subtyping rules)

S-Π $\dfrac{\Gamma \vdash A' \leq A \quad \Gamma, x : A' \vdash B \leq B'}{\Gamma \vdash \Pi x{:}A.B \leq \Pi x{:}A'.B'}$
$\qquad$
S-Γ $\dfrac{\Gamma \vdash P M_1..M_n \leq A \quad \alpha \leq P \in \Gamma}{\Gamma \vdash \alpha M_1..M_n \leq A}$

S-id $\dfrac{}{\Gamma \vdash A \leq A}$
$\qquad$
S-β $\dfrac{A =_\beta B \quad \Gamma \vdash B \leq C \quad C =_\beta D \quad \Gamma \vdash A, B, C, D : s}{\Gamma \vdash A \leq D}$

where s ranges over $\{\star, \Box\}$.

Since the main objective of this paper is to prove the meta-theoretical properties, we take this set of rules as our subtyping rules. If $\Gamma \vdash A, B : s$, then $\Gamma \vdash A \leq B$ is derivable in the first system iff it is derivable in the second. It is easy to verify that the second set of rules are derivable from the first set. For the other direction, the admissibilities of S-VAR, S-CONV are immediate, that of transitivity will be proved in the next section.

The rule S-Π is actually four rules corresponding to possible sortings of $\Pi x{:}A.B$, that is, A, B can be of sort $\star, \Box$. The rule S-Γ allows us use subfamily declaration. Note that a subtyping declaration in the context is expected to be of the form $\alpha \leq P : \Pi x_1{:}A_1..\Pi x_n{:}A_n.\star$, in which α, P can be viewed as two predicates with n arguments. From the subtyping as logic implication point of view[LMS95], we may consider such declaration as the logical assertion $\forall x_1{:}A_1..\forall x_n{:}(A_n.\alpha(x_1,..,x_n) \Rightarrow P(x_1,..,x_n))$ Therefore, the S-Γ rule actually states that if $PM_1..M_n$ implies A, then $\alpha M_1..M_n$ implies A. The S-β rule is a key contribution of this work, it allows us to eliminate transitivity.

In these subtyping rules we have kept sortings as less as possible. In Section 4, we will prove that a subtyping derivation will be well-formed if its last judgment is well-formed. That is, each subtyping judgment $\Gamma \vdash A \leq B$ is expected to be used when they are of the same sort.

We do not include the bounded quantification in our system for two reasons. One is the difficulty of meta-theoretical study. Another is that we expect our system will be used in proof checkers like Coq and LEGO. For this purpose, we believe that the subtyping system should be as close to the original system as possible.

Our system has included four features of subtyping : 1. type conversion (S-β, S-id, S-CONV); 2. transitivity (S-TRANS); 3. Π-type subtyping; 4. application of subfamily relation. In next section we use an example to illustrate application of subtyping in proof checking.

3   Example

At present, the major application of CC is to implement mathematical theories. To see how proof checkers can be benefited from using subtyping, let's consider the following example, which is adapted from [Sai97]. Suppose a monoid : $\mathcal{M} =< S, \cdot, e, Assoc, IdR, IdL >$, where S is a set, $\cdot$ is the binary operation over S, $e \in S$ is the unit of $\mathcal{M}$, $Assoc, IdR, IdL$ are monoid axioms: $a \cdot (b \cdot c) = (a \cdot b) \cdot c, x = e \cdot x, x = x \cdot e$ respectively.

Group can be considered as a subtype of monoid. This view has two aspects: 1. Group could be defined by adding certain components to monoid; 2. Any theorem in monoid is a theorem in group.

The class of monoids can be represented as:

$$\mathcal{M} : \star, \ S : \mathcal{M} \to \star, \ OP : \Pi M{:}\mathcal{M}.\ S(M) \to S(M) \to S(M),$$
$$E : \Pi M{:}\mathcal{M}.\ S(M), \ IdR : \Pi M{:}\mathcal{M}.\Pi x{:}M.\ OP(M)(x, E(M)) = x, \ ...$$

where $\mathcal{M}$ is the type of monoid. Given a monoid $M : \mathcal{M}$, $S(M)$ denotes the carrier set of M, $OP(M)$ and $E(M)$ are the binary operation and the unit respectively. The declaration $IdR : \Pi M{:}\mathcal{M}.\Pi x{:}M.\ OP(M)(x, E(M)) = x$ encodes the axiom $x \cdot e = x$. Encodings for other axioms are similar.

The group is defined by adding the following declarations: $\mathcal{G} < \mathcal{M}$, $Inv :$ $\Pi G{:}\mathcal{G}.\ S \to S$, ...

All declarations of monoid are inherited to group. For example, it is easy to verify that $G : \star, OP : \Pi G{:}\mathcal{G}.\ S(G) \to S(G) \to S(G)$ are all derivable. Furthermore, any monoid theorem is a group theorem, this assertion itself can be represented as a meta-theorem provable in the system:

$$\Gamma \vdash \lambda P{:}\mathcal{M}{\to}Prop.\lambda y{:}\Pi X{:}\mathcal{M}.P.y : \Pi P{:}\mathcal{M}{\to}Prop.(\Pi X{:}\mathcal{M}.P \to \Pi X{:}\mathcal{G}.P)$$

Next, we illustrate the use of subfamily relation by studying the inheritance of monoid morphisms to group morphism. The monoid morphism can be represented as a type family having two arguments: $\mathcal{MM} : \mathcal{M} \to \mathcal{M} \to \star$ such that the set of monoid morphisms between two monoids A, B is the type $\mathcal{MM}(A, B)$, along with the morphism axioms. Therefore, the group morphism could be declared as the subfamily of monoid morphism: $\mathcal{GM} \leq \mathcal{MM} : \mathcal{G} \to \mathcal{G} \to \star$.

4   Meta-theoretical Properties

First, the Church-Rosser property holds for pre-expressions.

Proposition 4.1 (Church-Rosser for $\overset{*}{\to}$) *Suppose that A, B are pre-expressions, then 1. $A \overset{*}{\to} B \wedge A \overset{*}{\to} C \Rightarrow \exists D$ s.t. $B \overset{*}{\to} D \wedge C \overset{*}{\to} D$; 2. $A =_\beta B \Rightarrow \exists C$ s.t. $A \overset{*}{\to} C \wedge B \overset{*}{\to} C$.*

Since subtyping system is defined over preterms, types in a derived subtyping judgement may not be well-formed, but, if the last judgement of a subtyping derivation is well formed, then so does each judgement in the derivation.

Proposition 4.2 (Well-formedness of subtyping) *Suppose that $\mathcal{D}$ is a derivation for $\Gamma \vdash A \leq B$ and $\Gamma \vdash A, B : s$, then in $\mathcal{D}$, 1. each judgement is well-formed; 2. sortings in the assumptions and the conclusion of each inference step agree.*

The admissibility of transitivity is one of the main contribution of this work. The result is due to our version of type conversion rule, the S-β rule. To prove this property, we add the transitivity rule S-TRANS to the subtyping system, then show that the new system has the transitivity elimination property. The proof is essentially a process of transformation from the derivation ended by transitivity application to a transitivity-free one.

Theorem 4.3 (Admissibility of transitivity) $\Gamma \vdash A, B, C : s \wedge \Gamma \vdash A \leq B \wedge \Gamma \vdash B \leq C \Rightarrow \Gamma \vdash A \leq C$.

The proof uses a simple induction measure to show its termination, which is the lexicographical order $< depth(\Gamma \vdash A \leq B) + depth(\Gamma \vdash B \leq C), depth(\Gamma \vdash A \leq B) >$.

To prove the subject reduction, we need the substitution property.

Lemma 4.4 (substitution) *Suppose* $\Gamma \vdash M : A$, *then* 1. $\Gamma_1, x : A, \Gamma_2 \vdash B \leq C \Rightarrow \Gamma_1, \Gamma_2[x := M] \vdash B[x := M] \leq C[x := M]$; 2. $\Gamma_1, x : A, \Gamma_2 \vdash B : C \Rightarrow \Gamma_1, \Gamma_2[x := M] \vdash B[x := M] : C[x := M]$.

The subject reduction can be proved by means of the admissibility of transitivity and the substitution property.

Theorem 4.5 (Subject reduction) $\Gamma \vdash B : C \wedge B \twoheadrightarrow_\beta^* B' \Rightarrow \Gamma \vdash B' : C$.

Well-formed terms have strong normalization property. The basic idea of the proof is to transform terms to those in λC and show that a reduction step in $\lambda C_\leq$ corresponds to one or more steps in λC. Note that λC can be obtained by removing from $\lambda C_\leq$ the subtyping rules, $start_\leq$, $weakening_\leq$ rules and changing the subtyping judgment in the subsumption rule to β-equivalence. Let Γ reduction be defined as

$$C(\alpha) \rightarrow_\Gamma C(\Gamma(\alpha))$$

and $FE_\Gamma(x)$ be defined as the Γ-strong normal form of its argument x. For well-formed context Γ, the function FE_Γ is well defined on terms. This function can be extended to context by 1. $FE(<>) = <>$; 2. $FE(\Gamma, x : A) = FE(\Gamma), x : FE_\Gamma(A)$; 3. $FE(\Gamma, \alpha \leq A : K) = FE(\Gamma)$. FE_Γ and FE have nice properties:

Lemma 4.6

1. $\Gamma, x : A \vdash N : B \wedge \Gamma \vdash M : A$
   $\Rightarrow FE_\Gamma(N[x := M]) = FE_{\Gamma, x:A}(N)[x := FE_\Gamma(M)]$;
2. $\Gamma \vdash U, V : A \wedge U =_\beta V \Rightarrow FE_\Gamma(U) =_\beta FE_\Gamma(V)$;
3. $\Gamma \vdash A \leq B \wedge \Gamma \vdash A, B : s \Rightarrow FE_\Gamma(A) =_\beta FE_\Gamma(B)$;
4. $\Gamma \vdash A : B \Rightarrow FE(\Gamma) \vdash_{\lambda C} FE_\Gamma(A) : FE_\Gamma(B)$

where $\vdash_{\lambda C}$ *denotes judgments in* λC.

Instead of proving β strong normalization directly, we prove a more general result: the strong normalization of $\beta\Gamma$ reduction. This result will also be used to prove the termination of the subtyping checking algorithm.

The $\beta\Gamma$-reduction is the union of β and Γ reduction: $\rightarrow_{\beta\Gamma} \equiv \rightarrow_\beta \cup \rightarrow_\Gamma$. We write $\twoheadrightarrow_R^*$ for the reflexive and transitive closure of relation R. The normal form of A with respect to the reduction relation R is denoted by A^R.

Proposition 4.7 ($\beta\Gamma$ subject reduction) $\Gamma \vdash A : B \wedge A \twoheadrightarrow_{\beta\Gamma}^* A' \Rightarrow \Gamma \vdash A' : B$.

The proof of strong normalization is achieved by using the correspondence of reductions between $\lambda C_\leq$ and λC , which is characterized by the following lemma.

Lemma 4.8 (Reduction correspondence with λC) 1. $\Gamma \vdash M : B \wedge M \rightarrow_\beta M' \Rightarrow FE_\Gamma(M) \rightarrow_\beta FE_\Gamma(M')$; 2. $\Gamma \vdash M : B \wedge M \rightarrow_\Gamma M' \Rightarrow FE_\Gamma(M) = FE_\Gamma(M')$.

Theorem 4.9 (Strong normalization) *If $\Gamma \vdash A : B$, then 1. A is Γ strongly normalizing; 2. A is $\beta\Gamma$ strongly normalizing; 3. A is β strongly normalizing.*

The proofs follows the sequence of these three results. Note that the 1. and 3. are implied by 2.

Algorithmic subtyping rules are formed by replacing the S-β by the following rule[2]

$$S\text{-}\beta' \quad \frac{A \xrightarrow{*}_\beta B \quad \Gamma \vdash B \leq C \quad C \xleftarrow{*}_\beta D}{\Gamma \vdash A \leq D}$$

It can be shown that a subtyping derivation is derivable iff it is derivable by the algorithmic system.

First, define a pair of rules S-β^+ and S-β^- :

$$S\text{-}\beta^+ \quad \frac{\Gamma \vdash A, B, C, D : s \quad A \xrightarrow{*} B \quad \Gamma \vdash B \leq C \quad C \xleftarrow{} D}{\Gamma \vdash A \leq D} \qquad S\text{-}\beta^- \quad \frac{\Gamma \vdash A, B, C, D : s \quad A \xleftarrow{} B \quad \Gamma \vdash B \leq C \quad C \xrightarrow{*} D}{\Gamma \vdash A \leq D}$$

Since these two rules are special cases of S-β, so they are admissible in the system. Conversely, it is easy to show that an application of S-β can be replaced by consecutive applications of S-β^- and S-β^+. Then, we can show that S-β^- and S-β^+ can be eliminated. Let $\vdash_\mathcal{A}$ denote the judgment in algorithmic system, and $\vdash_+$ the judgment in the system S-id, S-Π, S-Γ and S-β_+.

Lemma 4.10 (S-β^-, S-β^+ elimination) *1. If $\Gamma \vdash A \leq B$ is derived by the system S-id, S-Π, S-Γ,S-β^+ and S-β^-, then it can be derived by the system S-id, S-Π, S-Γ,S-β^+; 2. If $\Gamma \vdash A, B : s$, then $\Gamma \vdash_+ A \leq B \Leftrightarrow \Gamma \vdash_\mathcal{A} A \leq B$.*

Finally, we have the equivalence between algorithmic system and the original:

Theorem 4.11 (Algorithmic subtyping system) *If $\Gamma \vdash A, B : s$, then $\Gamma \vdash A \leq B \Leftrightarrow \Gamma \vdash_\mathcal{A} A \leq B$.*

Using $\beta\Gamma$ strong normalization, we can show the termination of the algorithmic system and get the decidability of subtyping.

Theorem 4.12 (Decidability of subtyping) *If $\Gamma \vdash A, B : s$, then $\Gamma \vdash A \leq B$ is decidable.*

---

[2] The rule S-β' does not have subformula property. When A, D are well-formed, there are finite number of possible pairs of B, C with which the rule is applicable. The set of algorithmic rules are not syntax oriented. But it is not hard to see that it can be transformed into a deterministic algorithm for checking subtyping.

5 Related Works

Our work is based on careful analysis of several existing subtyping systems.

The $F^\omega_\leq$ of Steffen and Pierce is a bounded subtyping extension of F^ω which has type family taking types as parameters. Its technique of using $\beta\Gamma$ strong normalization to prove the decidability of subtyping has been adopted here.

$\lambda P_\leq$ of Aspinall and Compagnogni[AC96] is a subtyping extension to first order dependent types. It has type family taking terms as parameters. This system has strong influence to $\lambda\Pi_\leq$[Che96], which directly leads to the present work. The main contribution of $\lambda\overline{\Pi}_\leq$ with respect to $\lambda P_\leq$ is the type-level transitivity elimination. Another important contribution of $\lambda\overline{\Pi}_\leq$ is the proof of $\beta\Gamma$ strong normalization, which is the key step to establish the termination of subtyping algorithm. The proof there is simple and general. This technique is also adopted here. $\lambda C_\leq$ can be viewed as an extension of $\lambda P_\leq$ or $\lambda\Pi_\leq$.

Both $\lambda P_\leq$ and $F^\omega_\leq$ achieve the decidability of subtyping by means of some intermediate subtyping systems defined on normalized types. The transitivity elimination property hold in normalized types. $\lambda C_\leq$ has type families taking both terms and types as parameters. Due to our choice of the S-β rule, $\lambda C_\leq$ has type level transitivity elimination property, which greatly simplifies the meta-theoretical study.

The pioneering work of $F_\leq$[CG92] has a long influence to many subtyping systems, including ours. Its technique, as well as that from $CO^\vdash$, for achieving the transitivity elimination property influenced our design of S-β.

Several researchers [Bar95, Bai96, Luo96, Sai97])have proposed some desirable subtyping features in CC and its extensions in the context of proof checking. These include: transitivity, type conversion, Π-type subtyping, subfamily, coercion, multiple inheritance, subtyping parameterized inductive types, subtyping between type universes, etc. Bailey[Bai96] and Saibi[Sai97] have implemented subtyping algorithms in LEGO and Coq respectively, they have also given many interesting examples of proof development by means of coercive subtyping. Luo[Luo96] has studied coercive subtyping in a meta-language which is capable of encoding CC, ECC and inductive types. But these works have not dealt with the meta-theoretical properties of their subtyping systems. Barthe[Bar95] has proved some meta-theoretical properties, but his system is rather restricted (e.g. Π-type subtyping and subfamily relation are not included). Our system has included the features of transitivity, type conversion, Π-type and use of subfamily declaration. We have proved that this system has nice properties.

6 Conclusion and Future Works

Since the works on $F^\omega_\leq$ and $\lambda P_\leq$, the type normalization method seems to be standard in adding subtyping to systems containing reduction relations on types. But in the case of Calculus of Construction, there are four different kinds of β-reductions, the existing technique become difficult to work.

In the study of $\lambda \Pi_{\leq}$ [Che96], we have developed a subtyping system with type-level transitivity elimination property. This becomes a new approach to establish meta-theoretical properties for systems with type conversions. In this paper, we have adapted this technique to establish a subtyping extension of Calculus of Construction.

We have proved that this system has good meta-theoretical properties: transitivity elimination, subject reduction, strong normalization and decidability of subtyping. The system is an extension of $\lambda P_{\leq}$ [AC96] or $\lambda \Pi_{\leq}$ [Che96].

To extend this work, we would like to construct a coercive version of this system and prove the decidability of type checking via such version. Coercive subtyping seems particularly useful in the proof checkers [Luo96, Bar95, Bai96, Sai97]. On the other hand, in the logical interpretation of subtyping[LMS95], a subtyping judgment corresponds to logical implication and coercion corresponds to its proof. We hope to implement the coercion version in proof checkers.

Acknowledgement Thanks to D. Aspinall, A. Compagnoni for explaining to me $\lambda P_{\leq}$ and useful discussions. Thanks to J. Courant, Z. Luo, C. Paulin-Mohring, A. Saibi and B. Werner for helping me understand the need of subtyping in proof checkers. I am very grateful to my supervisor G. Longo and G. Castagna for their support in this work. Many thanks to anonymous refrees who have given valuable comments and suggestions.

References

[AC96] D. Aspinall and A. Compagnoni. Subtyping dependent types. In LICS'96.

[Bai96] A. Bailey. Lego with implicit coercions, 1996. draft.

[Bar92] H. Barendregt. Lambda calculi with types. In *Handbook of Logic in Computer Science*. Oxford University Press, 1992.

[Bar95] G. Barthe. Inheritance in type theory, TYPES'95, 1995.

[Car87] L. Cardelli. Typechecking dependent types and subtypes, December 1987.

[CG92] P. L. Curien and G. Ghelli. Coherence of subsumption, minimum typing and the type checking in $F_{\leq}$. MSCS, 2(1), 1992.

[Che96] G. Chen. Dependent type system with subtyping. To appear in KIT Summer School and Workshop on Formal Models of Programming and their Applications, Beijing, 1997.

[Com94] A. Compagnoni. Subtyping in $F_{\wedge}^{\omega}$ is decidable. CSL'94.

[LMS95] G. Longo, K. Milsted, and S. Soloviev. A Logic of Subtyping. In LICS'95.

[Luo96] Z. Luo. Coercive subtyping in type theory. In CSL'96.

[Mit88] J.C. Mitchell. Polymorphic type inference and containment. *Information and Computation*, 76:211–249, 1988.

[Pfe93] F. Pfenning. Refinement types for logical frameworks. In *Informal Proceedings of the 1993 Workshop on Types for Proofs and Programs*, May 1993.

[Sai97] A. Saibi. Typing algorithm in type theory with inheritance, 1996. in the 24th Annual SIGPLAN-SIGACT Symposium on principles of Programming Languages, Paris, France, January 15-17, 1997.

[PS96] B. Pierce and M. Steffen. Higher-order subtyping. To appear in: *Theoretical Computer Science*, 1996.

Distances Between Languages
and Reflexivity of Relations

Christian Choffrut[1] and Giovanni Pighizzini[2]

[1] LIAFA, Université Paris 7
Tour 55–56, 2 Place Jussieu, 75251 Paris Cedex 05 (`cc@litp.ibp.fr`)
[2] Dipartimento di Scienze dell'Informazione, Università degli Studi di Milano
via Comelico, 39 – 20135 Milano – Italy (`pighizzi@dsi.unimi.it`)

Abstract. In this paper, the notions of k–reflexivity and almost reflexivity of binary relations over the free monoid A^* are introduced, which are natural extensions of the usual reflexivity. They can be defined relative to arbitrary distances over A^*. The problems of deciding whether or not a relation is almost reflexive and whether or not it is k–reflexive for a given integer k, are studied. It is shown that both problems are unsolvable in the case of deterministic rational relations. Moreover, the latter problem remains undecidable even when an oracle asserting that the relation is almost reflexive is provided. On the other hand, for the subclass of recognizable relations, both problems are shown to be solvable: as a consequence, the distance between two rational languages can be effectively computed. Further decidability results concerning the intermediate class of synchronized rational relations are proved.

1   Introduction

Comparing strings has been intensively studied during the past two decades, for example in connection with bibliographic search, text editing, alignments of biologic sequences *etc.* The problem usually consists in associating with two strings a non negative integer which reports how close they are from each other. This usually defines a distance between words, the most popular of which is certainly the one based on the maximal common subsequence, the so-called Levenshtein distance.

The basic question concerning the subsequence distance and its extension the edit distance [Kru83] asks whether or not we can design a sequential algorithm that computes the distance in time less than $O(n^2)$, where n is the maximal length of the two words.

Here we tackle the problem of computing the distance between two subsets of words. We assume these subsets to be effectively given. Actually we restrict the effectiveness of the subsets to some types of finite memory device. This is a reasonable assumption since there is no hope of actually computing the distance outside this class as we easily show in this paper.

Let A be a finite alphabet and let A^* be the free monoid it generates. The passage from the distance between strings to the distance between subsets $X, Y \subseteq A^*$ of strings is done in the standard way by resorting to the Hausdorff distance. In particular, whenever X and Y are rational subsets of A^* and whenever we endow the free monoid with the Hamming distance, the problem can be reduced to Hashiguchi's result on distance-automata (caveat: the term distance here refers to a different concept, weight would be more appropriate).

It is worthwhile observing that some fundamental problems can be stated in terms of distances. For example, asking whether or not the distance, whatever the distance, between two subsets is equal to 0 amounts to asking whether or not they are equal. This question is undecidable for context-free languages, so this gives a limit to our expectations and shows that we are quickly facing difficult problems.

Intuitively, asserting that two subsets X and Y have finite distance is a claim that the binary relation $X \times Y$ is "almost" reflexive with respect to the distance under consideration. Say a relation $R \subseteq A^* \times A^*$ is k–reflexive, for some integer $k \leq \infty$, if every element x of the domain is at distance at most k from some element y of the range, with $(x, y) \in R$, and vice versa. It is almost reflexive if $k < \infty$. We are led to consider the more general question of asking whether or not a given relation over A^* is almost reflexive. J. Johnson [Joh86] proved that reflexivity is undecidable for rational relations. In this work, we strengthen this result by showing that the problem remains undecidable even when restricted to *deterministic* rational relations. This result is a new contribution to the theory of rational relations and has some important consequences concerning the problems we are interested in in this paper. First of all, we cannot decide whether or not a deterministic rational relation is k–reflexive, for a given integer k, with respect to any distance. Moreover, in the case of the distances studied in this work, namely the Hamming, subword and edit distances, this problem remains undecidable even in the presence of an oracle asserting that the relation is almost reflexive. In addition, we prove that it is undecidable whether or not a deterministic rational relation is almost reflexive, i.e., whether or not $k < \infty$.

Subsequently, in order to identify the largest classes of relations for which the previously mentioned properties are decidable, we restrict our attention to proper subfamilies of deterministic rational relations, in particular that of synchronized rational relations [FS93] and that of recognizable relations. Recall that the class of recognizable relations is a proper subclass of that of synchronized rational relations which is itself a proper subclass of that of deterministic rational relations (assuming A and B are non empty).

We show that Hashiguchi's above mentioned problem is very close to the problem of deciding whether or not a synchronized rational relation is almost reflexive with respect to the Hamming distance. In fact, each problem can be immediately encoded into the other. As a consequence, we prove that the smallest $k \leq \infty$ such that a synchronized rational relation is k–reflexive with respect to the Hamming distance can be effectively computed.

In the case of the subword and of the edit distances, the largest class for which we show that this computation can be done effectively is that of recognizable relations. This result does not seem to be extendable to synchronized rational relations. However, we are able to prove that given an oracle asserting that a synchronized rational relation is almost reflexive, then it is decidable whether or not it is k–reflexive, for a given integer k, with respect to the subword or edit distance. These results are summarized in the following table

	Rec	Syn	DRat
Is R k–reflexive?	S	S	U
Is R almost reflexive?	S	S/?	U
Find the smallest k s.t. R is k–reflexive	S	S/?	U
Given R almost reflexive, find the smallest k s.t. R is k–reflexive	S	S	U

S=Solvable, U=Unsolvable
S/?=Solvable for the Hamming distance, unknown for the edit and the subword distance
Rec=Recognizable relations   Syn=Synchronized rational relations
DRat=Deterministic rational relations

We conclude this introduction, by mentioning that in the unary case, namely in the case of relations defined over a one letter alphabet, rational relations are equivalent to Presburger formulas [GS66]. Using this fact, we have shown that in the unary case all the considered problems are solvable. For lack of space, this result is omitted in this version of the paper. For the same reason, some of the proofs have been just outlined.

2 Preliminary definitions and results

Given a finite alphabet A, let A^* be the free monoid of strings (or words) over A, and let ϵ be the empty word. Given a string $x \in A^*$, we denote by $|x|$ its length and by x_i its i-th symbol (or letter), $i = 1, \ldots, |x|$. We recall that a string $z \in A^*$ is a *factor* or *subword* of x if there exist two strings $u, v \in A^*$ such that $x = uzv$.

A *distance* over A^* is a function $d : A^* \times A^* \to \mathbf{N} \cup \{\infty\}$ which is reflexive, symmetric, and which satisfies the triangular inequality: for all x, y, z, $d(x, y) \leq d(x, z) + d(z, y)$ holds.

In this paper, we consider three distances over A^*. Given two strings $x, y \in A^*$, the *Hamming distance* reports the number of positions where the symbols of x differ from those of y; the *subword distance* is based on the longest common subword of x and y; the *edit distance* is defined as the minimum number of *edit operations* (substitution, insertion, deletion of a symbol) which transform x into y. Formally

Definition 1. Given $x, y \in A^*$, the *Hamming*, *subword*, and *edit distance* between x and y, denoted by $d_H(x, y)$, $d_s(x, y)$, and $d_e(x, y)$, respectively, are defined as

- $d_H(x, y) = \begin{cases} \#\{i \mid x_i \neq y_i\} & \text{if } |x| = |y| \\ \infty & \text{otherwise;} \end{cases}$
- $d_s(x, y) = |x| + |y| - 2 \max \{|z| \mid z \text{ is a subword of both } x \text{ and } y\}$;
- $d_e(x, y) = \min \{k \mid x \vdash^{\underline{k}} y\}$, where for $w, z \in A^*$, $w \vdash z$ if and only if there exist $u, v \in A^*$, $a, b \in A$, with $a \neq b$, such that either $w = uav$, $z = ubv$ (substitution), or $w = uav$, $z = uv$ (deletion), or $w = uv$, $z = ubv$ (insertion), and, as usual, $\vdash^{\underline{k}}$ denotes the k-th power of $\vdash$.

Every distance over A^* can be extended to a distance between a string and a language or between two languages. The distance between a string x and a language L is defined as the minimum of the distances between x and the strings belonging to L; the distance between two languages can be defined in a standard way, by resorting to the Hausdorff distance. More precisely

Definition 2. Let A be a finite alphabet and $d : A^* \times A^* \to \mathbf{N} \cup \{\infty\}$ a distance. Given a string $x \in A^*$ and a language $L \subseteq A^*$, the *distance* $d(x, L)$ between x and L is defined as

$$d(x, L) = \begin{cases} \min \{d(x, y) \mid y \in L\} & \text{if } L \neq \emptyset \\ \infty & \text{otherwise.} \end{cases}$$

Given two languages $L', L'' \subseteq A^*$, the *distance* $d(L', L'')$ between L' and L'' is defined as

$$d(L', L'') = \max \{\sup \{d(x, L'') \mid x \in L'\}, \sup \{d(x, L') \mid x \in L''\}\}.$$

Recalling that the equivalence problem is undecidable for context–free languages, it is immediate that the distance between context–free languages is not computable. So, it is natural to restrict our attention to subclasses of context–free languages. On the other hand, the study of the distance between two languages can be generalized to the study of reflexivity properties of relations. To this aim, we now recall some basic definitions concerning binary relations over strings, in order to introduce the notion of k–reflexive relation.

Given a binary relation $R \subseteq A^* \times B^*$, let $Dom(R)$ and $Ran(R)$ be the domain and the range of R, respectively, namely, $Dom(R) = \{x \mid \exists y \text{ s.t. } (x, y) \in R\}$, and $Ran(R) = \{y \mid \exists x \text{ s.t. } (x, y) \in R\}$, and let R^{-1} be the relation $\{(y, x) \mid (x, y) \in R\}$.

Given a language $L \subseteq A^*$, let $i_L = \{(x, x) \mid x \in L\}$ be the identity relation restricted to L. A relation $R \subseteq A^* \times A^*$ is *reflexive* if and only if, for any $x \in Dom(R) \cup Ran(R)$, the pair (x, x) belongs to R, namely, $i_{Dom(R) \cup Ran(R)} \subseteq R$. We now extend the notion of reflexivity, by introducing the k–*reflexivity*

Definition 3. Given a distance d over A^*, a relation $R \subseteq A^* \times A^*$, and an integer $k \in \mathbf{N} \cup \{\infty\}$, the relation R is k–*reflexive* with respect to d, if and only if the following two statements are true
- $\forall x \in Dom(R) \; \exists y \in A^*$ s.t. $(x, y) \in R$ and $d(x, y) \leq k$;
- $\forall y \in Ran(R) \; \exists x \in A^*$ s.t. $(x, y) \in R$ and $d(x, y) \leq k$.

The relation R is *almost* reflexive (with respect to d) if and only if it is k–reflexive for some $k < \infty$.

Note that, for any distance d, a relation is reflexive if and only if it is 0–reflexive. Furthermore, it is not difficult to observe that the distance between two languages L' and L'' is bounded by k if and only if the relation $L' \times L''$ is k–reflexive. Hence, L' and L'' have finite distance if and only if the relation $L' \times L''$ is almost reflexive.

We now briefly recall the definition of the main classes of relations we are interested in. For more details the reader is referred to [Eil74, Joh86, FS93].

As usual, given a monoid M, we denote by $Rec(M)$ the family of *recognizable* subsets of M (i.e., all subsets of M that are the union of some equivalence classes of a congruence of finite index) and by $Rat(M)$ the family of *rational* subsets of M (i.e., all the subsets that are obtained from singletons by a finite application of the operations of set union, concatenation and Kleene star). The family $Rat(A^* \times B^*)$ of *rational relations* over $A^* \times B^*$ can be characterized in terms of rational transducers, i.e., of two-tape automata. By considering deterministic and synchronized transducers, we define, respectively, the classes $DRat(A^* \times B^*)$ and $Syn(A^* \times B^*)$ of *deterministic* and *synchronized rational relations* over $A^* \times B^*$ [FS93].

The following proper inclusions are well known [RS59, Eil74, Joh86, FS93] (we assume $\#A \times \#B \geq 2$ where $\#A$ denotes the cardinality of A)

$$Rec(A^* \times B^*) \subset Syn(A^* \times B^*) \subset DRat(A^* \times B^*) \subset Rat(A^* \times B^*).$$

We conclude this section by recalling that a *distance* automaton $\mathcal{A}$ (actually a finite automaton with a *weight* function) [Has82, Has90] is an ordinary nondeterministic automaton (A, Q, δ, I, F), along with a function $\omega : Q \times A \times Q \to \mathbf{N} \cup \{\infty\}$ such that, for $q, p \in Q$, $a \in A$, $\omega(q, a, p) = \infty$ if and only if $p \notin \delta(q, a)$.

The weight function ω can be extended to strings, by setting, for $q, p \in Q$, $x, w \in A^*$, $a \in A$

$$- \omega(q, \epsilon, p) = \begin{cases} 0 & \text{if } q = p \\ \infty & \text{otherwise}; \end{cases}$$
$$- \omega(q, wa, p) = \min \{\omega(q, w, r) + \omega(r, a, p) \mid r \in Q\};$$
$$- \omega(x) = \min \{\omega(q, x, p) \mid q \in I, p \in F\}.$$

The weight of $\mathcal{A}$, denoted by $\omega(\mathcal{A})$, is the maximum of the weights of the strings accepted by the underlying automaton, namely $\omega(\mathcal{A}) = \sup \{\omega(x) \mid x \in L(\mathcal{A})\}$, with the convention that if $\mathcal{A}$ accepts the empty set, then $\omega(\mathcal{A}) = \infty$. The following result was proved by Hashiguchi in [Has82, Has90].

Theorem 4. *Given an automaton $\mathcal{A}$ with a weight function ω, the weight $\omega(\mathcal{A})$ is computable.*

3 Undecidability results

In this section, we begin the study by showing that k–reflexivity is undecidable, for deterministic rational relations. The results we state are mainly based on the following theorem, which extends to deterministic rational relations a result proved in [Joh86] for rational relations

Theorem 5. *It is undecidable whether or not a deterministic rational relation $R \subseteq A^* \times A^*$ is reflexive, i.e., $i_{Dom(R) \cup Ran(R)} \subseteq R$.*

Proof. Given a deterministic rational relation $R \subseteq A^* \times A^*$, let $X = A^* - Dom(R)$ and $Y = A^* - Ran(R)$. Then $R \cup (X \times Y)$ is deterministic too. Moreover, R is reflexive if and only if the diagonal i_{A^*} is contained in $R \cup (X \times Y)$. Now, we show that the problem of deciding whether or not a deterministic rational relation contains the diagonal is undecidable.

Given an instance $u_1, \ldots, u_p, v_1, \ldots, v_p \in A^*$ of Post Correspondence Problem (PCP) such that $\{u_1, \ldots, u_p\}$ and $\{v_1, \ldots, v_p\}$ are prefix codes, consider the relation $R = \{(u_i, v_i) \mid i = 1, \ldots, n\}^+$. It is not difficult to see that $R \in DRat(A^* \times A^*)$. Moreover, $DRat(A^* \times A^*)$ is closed under complementation. Hence $R^c \in DRat(A^* \times A^*)$. Note that R^c contains the diagonal i_{A^*} if and only if the given instance of PCP does not have a solution. Since PCP is undecidable even for prefix codes [Ruo85], we conclude that the reflexivity is undecidable. $\square$

As an immediate consequence of Theorem 5, we can state the following undecidability result

Corollary 6. *Let d be a fixed distance over A^*. It is undecidable, given a deterministic rational relation $R \in DRat(A^* \times A^*)$ and an integer k, whether or not R is k-reflexive, relative to d.*

Now, in the case of the Hamming, subword and edit distance, we strengthen the result obtained in Corollary 6 in two directions. On the one hand, we show that it is undecidable whether or not a deterministic rational relation is almost reflexive (namely, we just require the existence of a finite bound k, we do not require its value). On the other hand, we show that the problem of deciding whether or not a deterministic rational relation is k-reflexive remains undecidable even when we know it is almost reflexive.

Theorem 7. *It is undecidable whether or not a deterministic rational relation is almost reflexive, relative to the Hamming, the subword, and the edit distance, respectively.*

Proof. (outline) Given $R \in DRat(A^* \times A^*)$, let $R_\$ = \{(\$u\$, \$v\$) \mid (u, v) \in R\}$, where $\$ \notin A$. Of course, $R_\$$ and its Kleene closure $R_\$^*$ are deterministic.

We prove that R is reflexive if and only if $R_\$^*$ is almost reflexive, and then, by Theorem 5, we conclude that this property is undecidable. The "only if" part is trivial. To prove the converse implication, suppose that $R_\$^*$ is k-reflexive, for an integer $k < \infty$. Thus, given $u \in Dom(R)$ and $x = (\$u\$)^{k+1}$, there exists $y \in (A \cup \{\$\})^*$ such that $(x, y) \in R_\$^*$ and $d(x, y) \le k$.

It is not difficult to see that in the case of the Hamming and the edit distances, the string y can be decomposed as the product of $k+1$ factors $y_0, y_1, \ldots, y_k$ such that $d(x, y) = d(\$u\$, y_0) + d(\$u\$, y_1) + \ldots + d(\$u\$, y_k)$. This implies that $y_i = \$u\$$, for some i, $0 \le i \le k$. Thus, $(\$u\$, \$u\$) \in R_\$$ and $(u, u) \in R$.

In the case of the subword distance, for $k = 0$ it is immediate to observe that $(u, u) \in R$. For $k \ge 1$, we have $d_s(x, y) = |x| + |y| - 2l_s(x, y) \le k$, where $l_s(x, y)$ is the length of the longest common subword of x and y. So $2l_s(x, y) \ge |x| + |y| - k \ge (|u| + 2)(k + 1) + l_s(x, y) - k$, and then $l_s(x, y) \ge 2|u| + 3$. Thus, $\$u\$$ should be a subword of both x and y, and, by construction, $(\$u\$, \$u\$) \in R_\$$, $(u, u) \in R$.

Dually, if $R_\$^*$ is k-reflexive, then $v \in Ran(R)$ implies $(v, v) \in R$. $\square$

Theorem 8. *It is undecidable whether or not a k-reflexive deterministic rational relation $R \in DRat(A^* \times A^*)$ is $(k-1)$-reflexive, relative to the Hamming, the subword, and the edit distance, respectively.*

Proof. We prove the result by showing that given a 1–reflexive deterministic rational relation, it is undecidable whether it is reflexive too.

First, we observe that the problem of deciding whether a deterministic rational relation R is reflexive is undecidable even if $\epsilon \notin Dom(R) \cup Ran(R)$. Let R be a deterministic rational relation satisfying such a restriction, and let $\sharp$ be a symbol not belonging to A. Consider the relation $R^\sharp = R \cup R_0 \cup R_1 \cup R_2$, where $R_0 = \{(u, u^\sharp) \mid u \in Dom(R)\}$, $R_1 = \{(u^\sharp, u) \mid u \in Ran(R)\}$, $R_2 = \{(u^\sharp, u^\sharp) \mid u \in Dom(R) \cup Ran(R)\}$, and, for any $u \in A^*$, $i = 1, \ldots, k$, $u^\sharp$ denotes the word $\sharp u_2 \ldots u_{|u|}$.

It is not difficult to observe that $R^\sharp$ is a 1–reflexive deterministic rational relation, with respect to the considered distances. Furthermore, $R^\sharp$ is reflexive if and only if R is reflexive, but, by Theorem 5, this is undecidable. $\square$

Since the relation $R^\sharp$ used to prove Theorem 8 is almost reflexive, actually 1–reflexive, we immediately obtain

Corollary 9. *Given an almost reflexive deterministic rational relation R and an integer $k \in \mathbb{N}$, it is undecidable whether or not R is k–reflexive, with respect to the Hamming, the subword and the edit distance, respectively.*

4 k–reflexivity relative to the Hamming distance

In Section 3, we have shown that the problem of deciding whether a deterministic rational relation is k–reflexive with respect to the Hamming distance is unsolvable. Now, we prove that this problem turns out to be solvable when we restrict our attention to the class of synchronized rational relations.

Theorem 10. *For any synchronized rational relation R, it is possible to compute the smallest $k \leq \infty$, such that R is k–reflexive relative to the Hamming distance.*

Proof. (outline) Given $R \in Syn(A^* \times A^*)$, let $\mathcal{A} = (A, A, Q, I, F, E)$ be a synchronized transducer accepting R. Thus, for any transition $(q, x, y, p) \in E$ of $\mathcal{A}$, $(x, y) \in (A \cup \{\epsilon\}) \times (A \cup \{\epsilon\})$, and $(x, y) \neq (\epsilon, \epsilon)$ holds [FS93].

We consider an automaton $\mathcal{A}'$ obtained by "forgetting" the third component of $\mathcal{A}$-transitions, namely for any $(q, a, b, p) \in E$, with $a, b \in A$, the automaton $\mathcal{A}'$ has a transition from q to p on a. If $a = b$ then the weight of this transition is 0, otherwise it is 1. This automaton recognizes $Dom(R)$. Dually, we can define an automaton $\mathcal{A}''$, which recognizes $Ran(R)$, with a suitable weight function ω''.

It is not difficult to verify that R is k–reflexive if and only if $\omega'(\mathcal{A}') \leq k$ and $\omega''(\mathcal{A}'') \leq k$. $\square$

Actually, the problem of deciding whether or not a synchronized rational relation is k–reflexive with respect to the Hamming distance and the problem of computing weights of automata are closely related. Indeed, Theorem 10 shows a reduction in one direction. The converse also holds.

Theorem 11. *Given an automaton $\mathcal{A} = (A, Q, \delta, I, F)$, with a weight function ω, there exists a synchronized rational relation R such that $\omega(\mathcal{A})$ is finite if and only if R is almost reflexive.*

5 k–reflexivity relative to the edit and to the subword distance

In this section we study the problem of deciding whether or not a relation is k–reflexive with respect to the edit and to the subword distance, respectively. In particular, we show that this problem is solvable for recognizable relations. As in the case of the Hamming distance and of synchronized relations, these results are obtained by means of reductions to the problem of Hashiguchi.

We start by showing the following lemma

Lemma 12. *For any recognizable relation $R \subseteq A^* \times A^*$ there exists an automaton $\mathcal{A}$ with a weight function ω, such that $\mathcal{A}$ recognizes $Dom(R)$, and for any $x \in Dom(R)$, $x \neq \epsilon$, $\omega(x) = \min \{d_e(x, y) \mid (x, y) \in R\}$.*

Proof. (outline) We sketch the proof for $R = L' \times L''$ with L' and L'' rational languages and $L'' \neq \emptyset$. The proof can be easily extended to the general case (all recognizable relations are finite unions of relations of this type).

Let $\mathcal{A}' = (A, Q', \delta', q_0', F')$ and $\mathcal{A}'' = (A, Q'', \delta'', q_0'', F'')$ be two finite deterministic automata accepting the languages L' and L'', respectively. We consider the product automaton $\mathcal{A} = (A, Q, \delta, I, F)$, with $Q = Q' \times Q''$, $I = \{(q_0', q_0'')\}$, $F = F' \times F''$, and $\delta((q', q''), a) = \{(\delta'(q', a), \delta''(q'', y)) \mid y \in A^*\}$ for $a \in A$, $q' \in Q'$, $q'' \in Q''$. Intuitively, each $\mathcal{A}'$ transition is "padded" with $\mathcal{A}''$-paths. The language accepted by $\mathcal{A}$ coincides with the language L' accepted by $\mathcal{A}'$. Moreover, each $\mathcal{A}$-accepting path corresponds to some $\mathcal{A}''$-accepting path.

We define the weight function ω of $\mathcal{A}$, in such a way that the edit distance between a string $x \in L'$ and the language L'' can be obtained as the minimum of the weights of accepting paths on x in $\mathcal{A}$. As in the dynamic programming, we obtain the minimum from "local" minima. To this aim, we define ω for $q', p' \in Q'$, $q'', p'' \in Q''$, $a \in A$, as follows

$$
\omega((q', q''), a, (p', p'')) = \begin{cases} \infty & \text{if } \delta'(q', a) \neq p' \text{ or } \{y \mid \delta''(q'', y) = p''\} = \emptyset \\ 1 & \text{if } \delta'(q', a) = p', q'' = p'', \text{ and } \delta''(q'', a) \neq q'' \\ n & \text{otherwise,} \end{cases}
$$

where $n = \min(\{|y| \mid y \in (A - \{a\})^* \text{ and } \delta''(q'', y) = p''\} \cup \{|y| - 1 \mid y \in A^* a A^* \text{ and } \delta''(q'', y) = p''\})$.

Some comments are now in order. In the first case, the transition from (q', q'') to (p', p'') on a in $\mathcal{A}$ is forbidden either because there is no transition from q' to p' on a in $\mathcal{A}'$, or because p'' is not reachable from q'' in $\mathcal{A}''$.

In the second and in the third cases, the symbol a is compared with each string y such that, in the automaton $\mathcal{A}''$, p'' is reachable from q'' on y. Let us denote by $L''_{q''p''}$ the set of such strings. If q'' and p'' coincide, then $L''_{q''p''}$ contains the empty string, so the distance between a and $L''_{q''p''}$ is at most 1. More precisely, if $a \notin L''_{q''p''}$, i.e., $q'' \neq \delta''(q'', a)$, then this distance is exactly 1 (second case: the symbol a is deleted); otherwise, it is 0 (third case, with $y = a$).

If states q'' and p'' do not coincide, then each string y belonging to $L''_{q''p''}$ contains at least one symbol. If no symbol of y coincides with a, then $|y|$ edit operations are needed to obtain y from a: one operation to replace a by a symbol of y, the other $|y| - 1$ operations to insert the remaining symbols of y. On the other hand, if a symbol of y coincides with a, then the substitution is not needed; thus, the total number of edit operations is $|y| - 1$.

It is possible to prove that, for $q', p' \in Q'$, $q'', p'' \in Q''$, $x \in A^+$, the following equality holds

$$\omega((q', q''), x, (p', p'')) = \begin{cases} \min \{d_e(x, y) \mid \delta''(q'', y) = p''\} \\ \quad \text{if } \{y \mid \delta''(q'', y) = p''\} \neq \emptyset \text{ and } \delta'(q', x) = p' \\ \infty \quad \text{otherwise.} \end{cases}$$

Thus, for $x \in L'$, $x \neq \epsilon$, the weight of x in $\mathcal{A}$ can be computed as

$$\omega(x) = \min \{\omega((q_0', q_0''), x, (p', p'')) \mid (p', p'') \in F\} = \min \{d_e(x, y) \mid y \in L''\}. \quad \square$$

Now, we are able to prove the main result of this section.

Theorem 13. *For any recognizable relation R, it is possible to compute the smallest integer k such that R is k-reflexive with respect to the edit distance.*

Proof. (outline) Given $R \in Rec(A^* \times A^*)$, we can define two automata $\mathcal{A}$, $\mathcal{A}'$, with weight functions ω, ω', respectively, as in Lemma 12. If $\epsilon \notin Dom(R) \cup Ran(R)$, then we can show that, for $k \in \mathbf{N}$, R is k-reflexive if and only if $\omega(\mathcal{A}) \leq k$ and $\omega'(\mathcal{A}') \leq k$. This allows to compute the smallest k such that R is k-reflexive. The proof can be easily adapted to the case of recognizable relations which contain the empty word in the domain or in the range. $\quad \square$

Now, it is natural to investigate whether or not Theorem 13 can be extended to the class of synchronized rational relations. On the one hand, in the case of the Hamming distance we have to compare symbols in corresponding positions, so the fact of having a synchronized transducer is very useful (see the proof of Theorem 10), while in the case of the edit distance we have to compare symbols in different positions: hence, dealing with synchronized relations is of no special help. On the other hand, we are able to prove that given a constant k, the problem of deciding whether or not a synchronized rational relation is k-reflexive relative to the edit distance is solvable. As a consequence, if we know that a synchronized rational relation R is almost reflexive, then we are able to obtain the minimum integer k such that R is k-reflexive (recall that, by Theorem 8, this problem is unsolvable for deteterministic rational relations).

Theorem 14. *For any synchronized rational relation $R \subseteq A^* \times A^*$ and any integer $k \geq 0$, it is possible to decide whether or not R is k-reflexive, with respect to the edit distance.*

Proof. For $j \geq 0$, let $I_j = \{(u, v) \in A^* \times A^* \mid d_e(u, v) \leq j\}$. Observe that $I_0 = i_{A^*}$, $I_1 = i_{A^*} \cdot ((A \cup \{\epsilon\}) \times (A \cup \{\epsilon\})) \cdot i_{A^*}$, and $I_j = I_{j-1} \circ I_1$, where $\circ$ denotes the composition of relations.

So, the relation I_1 is rational and has length difference bounded by 1, namely $||u| - |v|| \leq 1$, for $(u, v) \in I_1$. By Corollary 4.2 in [FS93], this implies that I_1 is a synchronized rational relation. Furthermore, the class $Syn(A^* \times A^*)$ is closed under composition and intersection. So, $R \cap I_k = \{(x,y) \in R \mid d_e(x,y) \leq k\} \in Syn(A^* \times A^*)$, for any $k \geq 0$. Hence, R is k–reflexive if and only if the domain and the range of $R \cap I_k$ coincide with the domain and the range of R, respectively. Since domains and ranges of rational relations are rational languages, this property is decidable. $\qquad\Box$

The decidability results relative to the edit distance can be extended even to the subword distance. In particular, by considering a different reduction to the problem of Hashiguchi, it can be shown that the smallest integer k such that a recognizable relation R is k–reflexive, with respect to the subword distance, can be effectively computed. Furthermore, the proof of Theorem 14 can be easily adapted to the subword distance.

Finally, we point out that using the results of [Web94], it is possible to evaluate the complexity of the decision procedures presented in this paper.

References

[Eil74] S. Eilenberg. *Automata, languages and machines*, volume A. Academic Press, 1974.

[FS93] C. Frougny and J. Sakarovitch. Synchronized rational relations of finite and infinite words. *Theoretical Computer Science*, 108:45–82, 1993.

[GS66] S. Ginsburg and E. Spanier. Semigroups, Presburger formulas, and languages. *Pacific Journal of Mathematics*, 16:285–296, 1966.

[Has82] K. Hashiguchi. Limitedness theorem on finite automata with distance functions. *Journal of Computer and System Sciences*, 24:233–244, 1982.

[Has90] K. Hashiguchi. Improved limitedness theorems on finite automata with distance functions. *Theoretical Computer Science*, 72:27–38, 1990.

[Joh86] J. Johnson. Rational equivalence relations. *Theoretical Computer Science*, 47:39–60, 1986.

[Kru83] J. Kruskal. An overview of sequence comparison. In D. Sankoff and J. Kruskal, editors, *Time warps, string edits, and macromolecules: the theory and practice of sequence comparison*, pages 1–44. Addison–Wesley, 1983.

[RS59] M.O. Rabin and D. Scott. Finite automata and their decision problems. *IBM J. Res. Develop*, 3:125–144, 1959. Also in: E.F. Moore. *Sequential machines*. Addison–Wesley, 1964.

[Ruo85] K. Ruohonen. Reversible machines and Post's correspondence problem for biprefix morphisms. *J. Inform. Process. Cybernet. EIK*, 21:579–595, 1985.

[Web94] A. Weber. Finite–valued distance automata. *Theoretical Computer Science*, 134:225–251, 1994.

Partial Characterization
of Synchronization Languages[*]

Isabelle Ryl, Yves Roos, and Mireille Clerbout

C.N.R.S. U.A. 369, L.I.F.L. Université de Lille I
Bât. M3, Cité Scientifique
59655 Villeneuve d'Ascq Cedex, FRANCE
e-mail : {ryl,yroos,clerbout}@lifl.fr

Abstract. Synchronization languages are associated with synchronization expressions, a high-level construct which allows a programmer to express synchronization constraints in a distributed context. We give a negative answer to a conjecture enunciated by L. Guo, K. Salomaa and S. Yu which aims at characterizing synchronization languages in terms of regular languages closed under a rewriting system. Then we propose an extension of the system which gives a positive answer for a class of regular languages.

Synchronization expressions, introduced in [GGYW91] within the framework of the *ParC* project, are a high-level construct which allows a programmer to express minimal synchronization constraints of a program in a distributed context. The study and the implementation of these expressions are based on their associated synchronization languages.

Synchronization languages have been introduced in [GSY94] and [GSY96]. They give a way to implement synchronization expressions and specify their semantics. These languages represent distributed systems whose behaviour respects synchronization constraints expressed by the programmer with synchronization expressions. So these languages describe all the correct executions of a program. In [GSY94] and [GSY96] L. Guo et al. propose a characterization of synchronization languages based on a rewriting system named R which generalizes partial commutations. This system gives a way to rewrite a word representing a parallel execution into a word with a lower or equal degree of parallelism. L. Guo et al. show that every synchronization language is closed under the system R and they conjecture that it is sufficient for a regular language to be closed under R to be a synchronization language.

First, we show that the conjecture is true in the particular case of languages expressing the synchronization between two distinct actions but not in the general case. Then we propose an extension of the closure function which keeps good properties of the old one. Moreover, we prove that this function satisfies classical properties of commutation systems. We also show that in the general case when a smallest subset of a language L generating L belongs to the family of well-formed languages defined in [GSY96] the conjecture becomes true.

---

[*] This work is partially supported by the group MOSYDIS of the PRC-GDR AMI

1 Preliminaries

In the following text Σ is the used alphabet. We shall denote by $|w|$ the length of the word w and by $alph(w)$ its alphabet. A word u is a factor of a word v if there exist two words v_1 and v_2 such that $v = v_1 u v_2$. A language M is a factor of a language L if there exists two languages L_1 and L_2 such that $L = L_1.M.L_2$.

We denote $if(L)$ the set of iterating factors of L, that is:
$$if(L) = \{u \in \Sigma^* \mid \exists v, w \in \Sigma^* \text{ such that } vu^*w \subseteq L\}$$

We denote by $\Pi_Y(w)$ the *projection* of the word w onto the *subalphabet* Y, i.e. the image of w by the homomorphism Π_Y which is defined by: for each $x \in \Sigma$, $\Pi_Y(x) = x$ if $x \in Y$ and $\Pi_Y(x) = \varepsilon$ otherwise.

We denote by $u \shuffle v$ the *shuffle* of the two words u and v that is $u \shuffle v = \{u_1 v_1 u_2 v_2 ... u_n v_n \mid u_i \in \Sigma^*, v_i \in \Sigma^*, u = u_1 u_2 ... u_n, v = v_1 v_2 ... v_n\}$.

Let us consider a rewriting system R. We shall write $u \xrightarrow{R} v$ if there is a rule $\alpha \longrightarrow \beta$ in R and two words w and w' such that $u = w\alpha w'$ and $v = w\beta w'$.

We say that there is a derivation from u to v denoted by $u \xrightarrow{*}{R} v$ if there are words $w_0, w_1, ..., w_n (n \geq 0)$ such that $w_0 = u, w_n = v$, and for each $i < n, w_i \xrightarrow{R} w_{i+1}$. The integer n is named the *derivation length*. When we have $u \xrightarrow{*}{R} v$ with a known derivation of length n, we shall also write: $u \xrightarrow{n}{R} v$. We denote $f_R(u) = \{v \in \Sigma^* \mid u \xrightarrow{*}{R} v\}$ and $f_R(L) = \bigcup_{u \in L} f_R(u)$.

2 Synchronization expressions and languages

2.1 Synchronization expressions description

Synchronization constructs developed in the *ParC* parallel programming language relieve programmer of the implementation of synchronization constraints, he just has to specify necessary constraints using expressions over statement tags named synchronization expressions [GGYW91]. The choice of tags as base of synchronization expressions is interesting because the size of the grain can vary, there is also no need to modify the program to express synchronization constraints, and constraints can be easily modified.

Then synchronization expressions allow to describe the behaviour of the application: a statement can be executed immediately if its execution satisfies constraints described by the expression, if it does not the execution is delayed.

A synchronization expression may be:

1. a statement tag or ε for no action,
2. if e_1 and e_2 are synchronization expressions:
   - $(e_1 \rightarrow e_2)$ which imposes that the execution of e_2 starts only after the end of the execution of e_1,
   - $(e_1 \parallel e_2)$ which allows the executions of e_1 and e_2 to overlap. Because of the definition of $\parallel$, a same statement tag can not appear in both operands of $\parallel$,

- $(e_1 \mid e_2)$ which specifies that either e_1 or e_2 can be executed but not both,
- $(e_1 \& e_2)$ which imposes that the execution satisfies both expressions e_1 and e_2,
- (e_1^*) which allows the execution of e_1 to be repeated an arbitrary number of times.

Example 1. Let us consider the following constraints: a statement b can be executed only after the end of a statement a and a statement d can be executed only after the end of the statements a and c, this leads to the expression $((a \to b) \mid\mid (c \to d)) \& ((a \to d) \mid\mid b \mid\mid c)$.

Formally, the set of expressions over an alphabet Σ, denoted by $SE(\Sigma)$, is inductively defined by:

Definition 2.
$$SE(\Sigma) \subseteq (\Sigma \cup \{\to, \&, \mid, \mid\mid, *, (,)\})^*$$

is the smallest set such that:

- $\Sigma \cup \{\varepsilon\} \subseteq SE(\Sigma)$,
- $\forall e_1, e_2 \in SE(\Sigma), (e_1 \to e_2), (e_1 \mid e_2), (e_1 \& e_2), (e_1^*) \in SE(\Sigma)$,
- $\forall e_1, e_2 \in SE(\Sigma)$ such that $alph(e_1) \cap alph(e_2) = \emptyset$ (with $alph(e)$ the set of tags used to write e) , $(e_1 \mid\mid e_2) \in SE(\Sigma)$.

In $ParC$, the execution of a program is controlled by an automaton which recognizes a regular language named synchronization language corresponding with the expression.

2.2 Synchronization languages

The language associated with an expression describes all the executions which respect the synchronization constraints given by the expression, it corresponds with the set of execution traces. From an expression $e \in SE(\Sigma)$, we construct a language $L(e) \in (\Sigma_s \cup \Sigma_t)^*$ with Σ_s and Σ_t defined by:

Definition 3. Let Σ be a finite alphabet. The alphabets Σ_s and Σ_t are defined by the relation:
$$a \in \Sigma \Leftrightarrow a_s \in \Sigma_s \Leftrightarrow a_t \in \Sigma_t$$

With each action labelled by a letter a, corresponds in a synchronization language the letters a_s (s meaning *start*) and a_t (t meaning *termination*) which indicate the start and the termination of the action a. Words of a synchronization language show the duration of actions as well as their real concurrency.

Definition 4. Let Σ be the alphabet of tags. We define the synchronization language $L(e) \subseteq (\Sigma_s \cup \Sigma_t)^*$ associated with a synchronization expression $e \in SE(\Sigma)$ by:

- $L(\varepsilon) = \varepsilon$,
- $\forall a \in \Sigma$, $L(a) = a_s a_t$,
- if $e = e_1 \rightarrow e_2$ then $L(e) = L(e_1).L(e_2)$,
- if $e = e_1 \mid e_2$ then $L(e) = L(e_1) \cup L(e_2)$,
- if $e = e_1 \& e_2$ then $L(e) = L(e_1) \cap L(e_2)$,
- if $e = e_1 \parallel e_2$ then $L(e) = L(e_1) \amalg L(e_2)$,
- if $e = e_1^*$ then $L(e) = (L(e_1))^*$.

Clearly, by construction, synchronization languages are regular languages, moreover, they are st-languages:

Definition 5. A word $u \subseteq (\Sigma_s \cup \Sigma_t)^*$ is an st-word if and only if for each $x \in \Sigma$, $\Pi_{\{x_s, x_t\}}(u) \subseteq (x_s x_t)^*$. A language $L \subseteq (\Sigma_s \cup \Sigma_t)^*$ is an st-language if and only if for each $x \in \Sigma$, $\Pi_{\{x_s, x_t\}}(L) \subseteq (x_s x_t)^*$.

Example 6. For $e = a \parallel b$ we obtain the language $L(e) = \{a_s a_t b_s b_t, a_s b_s a_t b_t, a_s b_s b_t a_t, b_s a_s a_t b_t, b_s a_s b_t a_t, b_s b_t a_s a_t\}$.

2.3 Synchronization languages characterization

In [GSY94], L. Guo et al. try to characterize synchronization languages in terms of languages closed under a rewriting system. They define the following rewriting system:

Definition 7. The rewriting system R over the alphabet Σ is defined by $R = R_1 \cup R_4$ with:

- $R_1 \subseteq (\Sigma_s \cup \Sigma_t)^2 \times (\Sigma_s \cup \Sigma_t)^2$ contains all the rules r_1, r_2, r_3 for every $a, b \in \Sigma, a \neq b$:

  $r_1 : a_s b_s \longrightarrow b_s a_s$

  $r_2 : a_t b_t \longrightarrow b_t a_t$

  $r_3 : a_s b_t \longrightarrow b_t a_s$

- $R_4 \subseteq (\Sigma_s \cup \Sigma_t)^* \times (\Sigma_s \cup \Sigma_t)^*$ is defined in the following way:
  for each sequence $a_1, \ldots, a_m, b_1, \ldots, b_n$ of pairwise distinct elements of Σ with $m, n \geq 1$, R_4 contains the rule $a_{1t} \ldots a_{mt} a_{1s} \ldots a_{ms} b_{1t} \ldots b_{nt} b_{1s} \ldots b_{ns}$ $\longrightarrow b_{1t} \ldots b_{nt} b_{1s} \ldots b_{ns} a_{1t} \ldots a_{mt} a_{1s} \ldots a_{ms}$

L. Guo et al. give also the following proposition and conjecture:

Proposition 8. *[GSY96] An arbitrary synchronization language defined over Σ is closed under the rewriting system R.*

Conjecture 9. *[GSY96] An arbitrary regular st-language defined over Σ and closed under R is a synchronization language.*

The family of synchronization languages is interesting for itself, so the proof of Conjecture 9 would give us a new characterization of this family and would allow us to decide if a given language is a synchronization language. On the other hand, synchronization languages have been used to implement control of execution in parallel programming languages. A method to construct a synchronization expression corresponding with a given language should allow us to propose a debugging tool based on synchronization languages.

3 Languages closed under R

We first show that Conjecture 9 is satisfied in the case of languages defined over alphabets of size four corresponding with two actions. We can suppose that $\Sigma = \{a, b\}$.

Lemma 10. *Let* $L \subseteq (a_s + a_t + b_s + b_t)^*$ *be a regular st-language closed under R. There exists a regular language* $K \subseteq \mathbf{M} = (a_s a_t + b_s b_t + a_s b_s (a_t a_s)^* (b_t b_s)^* a_t b_t)^*$ *such that* $L = f_R(K)$.

In order to prove this lemma we show that each word $u \in L$ can be factorized in $u = u_1 u_2 \ldots u_n$ such that for each $1 \leq i \leq n$, u_i is an st-word with no proper left st-factor and there exists $v_i \in \mathbf{M}$ such that $v_1 \xrightarrow[R]{*} u_i$.

Then we show that we can factorize a language included in $\mathbf{M}$ into union, product and star of elementary languages:

Lemma 11. *Let* $K \subseteq \mathbf{M}$ *be a regular language. Then* K *can be expressed as product, union and star of regular languages* K_i *with for each i,* $K_i = a_s a_t$ *or* $K_i = b_s b_t$ *or* $K_i \subseteq a_s b_s (a_t a_s)^* (b_t b_s)^* a_t b_t$.

In the factorization of a language included in $\mathbf{M}$ into elementary languages, $a_s a_t$ and $b_s b_t$ are clearly synchronization languages corresponding with the expressions a and b. We also show that the closure of a regular language included in $a_s b_s (a_t a_s)^* (b_t b_s)^* a_t b_t$ corresponds with an expression over the alphabet $\{a\}$ in parallel with an expression over the alphabet $\{b\}$ which can effectively be constructed.

It remains to prove that the closure of an union (resp. product or star) of such elementary languages corresponds with the union (resp. product or star) of their closures:

Lemma 12. *Let* K_1, K_2 *be regular languages. If* $K_i = a_s a_t$, $K_i = b_s b_t$ *or* $K_i \subseteq a_s b_s (a_t a_s)^* (b_t b_s)^* a_t b_t$ *for* $i \in \{1, 2\}$, *then:*

- $f_R(K_1 + K_2) = f_R(K_1) + f_R(K_2)$
- $f_R(K_1.K_2) = f_R(K_1).f_R(K_2)$
- $f_R((K_1)^*) = (f_R(K_1))^*$

We have shown that if a language L is an st-language, regular and closed under R then it is the closure of a regular language K included in $\mathbf{M}$ (Lemma 10). Moreover K can be factorized into elementary languages (Lemma 11) and each part of this factorization can be closed separately to obtain elementary synchronization languages (Lemma 12). The language L can be expressed as product, union and star of a finite number of synchronization languages, therefore it is a synchronization language. At last, using this factorization, we can construct a synchronization expression substituting synchronization expression for each language, exclusion operator for union operator and sequence operator for concatenation operator. We obtain:

Proposition 13. *A regular, st-language defined over an alphabet of two actions and closed under R is a synchronization language.*

We have given a positive answer to the Conjecture 9 in the case of languages defined over alphabets of two actions, but the conjecture does not hold in the general case. We use two examples to show two different problems.

Counter-example 14. *The language $f_R(b_s(a_s a_t)^* c_s b_t(a_s a_t)^* c_t)$ is an example of regular st-language closed under R which is not a synchronization language.*

For the proof it is easy to see that in this case $f_R(b_s(a_s a_t)^* c_s b_t(a_s a_t)^* c_t) = f_{R_1}(b_s(a_s a_t)^* c_s b_t(a_s a_t)^* c_t)$. So using properties of semi-commutations [CLR95] we show that this language is regular. Then we use several technical lemmas to show that a synchronization language containing $b_s(a_s a_t)^* c_s b_t(a_s a_t)^* c_t$ contains also $b_s(a_s a_t)^* c_s a_s a_t b_t(a_s a_t)^* c_t$ and this is not the case in our example (see [CRR]) .

In this example, we cannot differentiate the two groups of actions a, if the expression allows us to execute as much a as we want during the b, as much a as we want during the c and the b during the c, then we can execute an a during the b and the c. The problem would be solved if we could differentiate the two groups of a.

Secondly, projections onto subalphabets of two actions do not satisfy any projection property as shown in the following example:

Example 15. Let $u, v \in (\Sigma_s \cup \Sigma_t)^*$ such that $u = a_s b_s c_s a_t b_t b_s b_t b_s a_s c_t c_s a_t b_t c_t$ and $v = a_s b_s c_s c_t c_s a_t b_t b_s b_t b_s a_s a_t b_t c_t$. We have:

$$a_s b_s a_t b_t b_s b_t b_s a_s a_t b_t \xrightarrow[R]{0} a_s b_s a_t b_t b_s b_t b_s a_s a_t b_t$$

$$a_s c_s a_t a_s c_t c_s a_t c_t \xrightarrow[R]{1} a_s c_s c_t c_s a_t a_s a_t c_t$$

$$b_s c_s b_t b_s b_t b_s c_t c_s b_t c_t \xrightarrow[R]{2} b_s c_s c_t c_s b_t b_s b_t b_s b_t c_t$$

but u cannot be rewritten into v using the rewriting system R.

Nevertheless, the semantics of synchronization expressions makes us think such properties exist.

So we propose to extend the function f_R to solve this problem. We want to define a function $f_{R'}$ satisfying projection properties and we propose to use a morphism to solve the first problem.

4 Extension of the function f_R

4.1 Definitions

We define a new alphabet which associates bounded memories, represented by booleans, with each letter. We use the following alphabets:

$$\Sigma = \{a_1, \ldots, a_n\},$$
$$\Sigma' = \{(x, p_1, \ldots, p_n) \mid x \in \Sigma_s \cup \Sigma_t \text{ and } \forall i \in \{1, \ldots, n\}, p_i \in \{0, 1\}\},$$
$$\Sigma_0 = \{(x, p_1, \ldots, p_n) \mid x \in \Sigma_s \cup \Sigma_t \text{ and } \forall i \in \{1, \ldots, n\}, p_i = 0\}.$$

For example, the letter $(a_{is}, p_1, \ldots, p_n)$ corresponds with the letter a_{is} and each p_j, with $j \in \{1, \ldots, n\}$, allows to remember the last commutation between this letter and a letter of the subalphabet $\{a_{js}, a_{jt}\}$ (note that p_i will always be equal to 0).

In order to go from an alphabet to the other one, we use two morphisms. The morphism h allows to remove booleans, it is defined from Σ'^* to $(\Sigma_s \cup \Sigma_t)^*$ by $h((x, p_1, \ldots, p_n)) = x$. The morphism g adds null booleans to "normal" letters in order to initialize them, it is defined from $(\Sigma_s \cup \Sigma_t)^*$ to Σ_0^* by $g(x) = (x, 0, \ldots, 0)$.

We also extend the definition of st-words to the words defined over Σ', a word u of Σ'^* is an st-word if and only if $h(u)$ is an st-word.

For each $j \in \{1, \ldots, n\}$, we denote $i(a_j) = i(a_{js}) = i(a_{jt}) = j$ and for each $x \in \Sigma'$, $x = (y, p_1, \ldots, p_n)$, $i(x) = i(y)$.

Definition 16. The symmetrical rewriting system R_e defined over Σ' contains each rule $(x, p_1, \ldots, p_n)(y, q_1, \ldots, q_n) \to (y, q'_1, \ldots, q'_n)(x, p'_1, \ldots, p'_n)$ with $x, y \in \Sigma_s \cup \Sigma_t$, $i(x) \neq i(y)$, for each $k \neq i(x)$, $q_k = q'_k$, for each $k \neq i(y)$, $p_k = p'_k$ and

$$\begin{cases} \text{if } x \in \Sigma_s, \ y \in \Sigma_s, \ q_{i(x)} = 0 \text{ then } p'_{i(y)} = 0, q'_{i(x)} = p_{i(y)} \\ \text{if } x \in \Sigma_t, \ y \in \Sigma_t, \ p_{i(y)} = 0 \text{ then } p'_{i(y)} = q_{i(x)}, q'_{i(x)} = 0 \\ \text{if } x \in \Sigma_s, \ y \in \Sigma_t, \ p_{i(y)} = q_{i(x)} = 0 \text{ then } p'_{i(y)} = q'_{i(x)} = 1 \\ \text{if } x \in \Sigma_t, \ y \in \Sigma_s, \ p_{i(y)} = q_{i(x)} = 1 \text{ then } p'_{i(y)} = q'_{i(x)} = 0 \end{cases}$$

At last the new function $f_{R'}$, extension of f_R, is defined by:

Definition 17. For each $u \in (\Sigma_s \cup \Sigma_t)^*$, $(v \in f_{R'}(u)) \Leftrightarrow (\exists w \in h^{-1}(v) \mid g(u) \xrightarrow[R_e]{*} w)$ and we denote $u \xmapsto[R']{} v$.

The projection onto a subalphabet of two actions of letters of Σ' is defined by:

Definition 18. Let $x = (a, p_1, \ldots, p_n)$ be in Σ'

$$\Pi_{ij}(x) = \begin{cases} (a, 0, \ldots, 0, p_j, 0, \ldots, 0) \text{ if } a \in \{a_{is}, a_{it}\} \\ (a, 0, \ldots, 0, p_i, 0, \ldots, 0) \text{ if } a \in \{a_{js}, a_{jt}\} \\ \varepsilon \text{ otherwise} \end{cases}$$

4.2 Properties of R_e

The transformation $f_{R'}$ is based on the rewriting system R_e which satisfies classical properties of commutation systems. In particular, we have a distance lemma. The distance between two words is the minimal number of permutations of two letters which allow to go from one to the other. We extend the definition of classical distance (see [CLR95]) to the words of Σ'^* by $d(u,v) = d(h(u), h(v))$. And we have the lemma:

Lemma 19. *(Distance lemma) Let* u,v *be in* Σ'^* *such that* $u \xrightarrow{*}{R_e} v$. *If* $d(u,v) = k$ *then* $u \xrightarrow{k}{R_e} v$.

We also show an equivalent of the Levi lemma :

Lemma 20. *(Levi lemma) Let* u_1, u_2, v_1, v_2 *be in* Σ'^* *such that* $u_1 u_2 \xrightarrow{*}{R_e} v_1 v_2$

then $\begin{cases} u_1 \xrightarrow{*}{R_e} \alpha_1\alpha_2 \\ u_2 \xrightarrow{*}{R_e} \beta_1\beta_2 \end{cases}$ and $\begin{cases} \alpha_2\beta_1 \xrightarrow{*}{R_e} \beta'_1\alpha'_2 \\ \alpha_1\beta'_1 \xrightarrow{*}{R_e} v_1 \\ \alpha'_2\beta_2 \xrightarrow{*}{R_e} v_2 \end{cases}$ with $\begin{cases} h(\alpha_1), h(\alpha_2) \text{ subwords of } h(u_1) \\ h(\beta_1), h(\beta_2) \text{ subwords of } h(u_2) \\ h(\alpha_2) = h(\alpha'_2), h(\beta_1) = h(\beta'_1) \\ \forall x \in alph(\alpha_2), \forall y \in alph(\beta'_1), \\ i(x) \neq i(y) \end{cases}$

At last, our new system satisfies a projection lemma :

Lemma 21. *(Projection lemma) Let* u,v *be in* Σ'^*

$$(u \xrightarrow{*}{R_e} v) \Leftrightarrow (\forall i,j \in \{1,\ldots,n\}, \ \Pi_{ij}(u) \xrightarrow{*}{R_e} \Pi_{ij}(v))$$

We prove the tree above-mentioned lemmas by induction on the length of words and using the following lemma:

Lemma 22. *(Basic Lemma) Let* xu *be a word of* Σ'^* *with* $x \in \Sigma'$. *If* $xu \xrightarrow{*}{R_e} u_1 x' u_2$ *with* $h(x) = h(x')$ *and for each* $y \in alph(u_1)$, $i(y) \neq i(x)$, *then* $u \xrightarrow{*}{R_e} u'_1 u_2$ *and* $xu'_1 \xrightarrow{|u_1|}{R_e} u_1 x'$ *with* $h(u_1) = h(u'_1)$.

We prove this lemma by induction on the length of the derivation $xu \xrightarrow{*}{R_e} u_1 x' u_2$. We just have to consider all the possible last steps.

4.3 Results with the new function

We have shown that the old system gives a good characterization for synchronization languages defined over alphabets of two actions, we show that our function has the same behaviour in this case :

Proposition 23. *Let* u,v *be st-words of* $(\Sigma_s \cup \Sigma_t)^*$ *with* $|\Sigma| = 2$.

$$(u \xmapsto{}{R'} v) \Leftrightarrow (u \xrightarrow{*}{R} v)$$

We also have the equivalent of Proposition 8 :

Proposition 24. *An arbitrary synchronization language defined over $\Sigma_s \cup \Sigma_t$ is closed under the function $f_{R'}$.*

Now we are interested in the converse of this proposition. In order to build a synchronization expression for a regular closed language L, we use a generator of L which exactly contains one word of each $f_{R'}(u)$ in L.

Definition 25. Let u be in $f_{R_e}(\Sigma_0^*)$. We denote $lex(u)$ the smallest word of $f_{R_e}(u) \cap \Sigma_0^*$ with respect to lexicographical order induced by an arbitrary order over Σ.

If L is an st-language closed under R_e, we denote $lex(L) = \bigcup_{u \in L} lex(u)$.

Definition 26. Let L be a regular st-language defined over $(\Sigma_s \cup \Sigma_t)^*$ such that $L = f_{R'}(L)$.

We define

$$K = f_{R_e}(g(L))$$

$$M = K \setminus [(K \cap \Sigma_0^*) \cap g(h(K \cap \Sigma'^*(\Sigma' \setminus \Sigma_0)\Sigma'^*))]$$

$$Max_L = h(lex(M \cap \Sigma_0^*))$$

The language M contains the words of $K \cap \Sigma_0^*$ whose image under h cannot be obtained using a word which is not a word of Σ_0^*. We can show that Max_L allows to generate L and that it is minimal with respect to inclusion:

Lemma 27. *Let $L \subseteq (\Sigma_s \cup \Sigma_t)^*$ be a regular st-language closed under $f_{R'}$.*

$$f_{R'}(Max_L) = L \text{ and } \forall u, v \in Max_L, \ (u \xrightarrow{R'} v) \Rightarrow (u = v)$$

We have not yet solved the converse of Proposition 24 in the general case but we can give an answer for the family of well-formed languages defined by L. Guo et al. in [GSY96]:

Definition 28. A rational expression α is said to be well-formed if for each subexpression $(\beta)^*$ of α, β is the rational expression of an st-language. A language is said to be well-formed if it can be defined by a well-formed rational expression.

They conjecture that:

Conjecture 29. *[GSY96] The closure under R of an arbitrary well-formed st-language is a synchronization language. An arbitrary synchronization language is the closure under R of some well-formed language.*

We have shown in [CRR] that this conjecture does not hold. Nevertheless we can show that:

Proposition 30. *The image by $f_{R'}$ of a regular well-formed st-language is regular.*

The proof uses ideas of proofs of Y. Métivier [Mét88] and R. Cori and D. Perrin [CP85] in the case of partial commutations.

At last we show the following proposition:

Proposition 31. *Let $L = f_{R'}(L)$ be a regular st-language such that Max_L is well-formed. Then L is the image by a strictly alphabetical morphism of a synchronization language. Moreover we can effectively construct an expression for this language.*

The proof is based on an induction on the construction of a well-formed regular expression of Max_L. We restrict us to the case of well-formed languages first because we know that the images of these languages are regular and secondly the fact that the factors of the language in the form M^* are st-languages simplifies the construction.

Conclusion

We have now a sufficient condition to decide whether a language is a synchronization language. Precisely, we have given a positive answer to the conjecture of L. Guo, K. Salomaa and S. Yu in the case of languages defined over alphabets of two actions and a negative answer in the general case. We define an extension of their transformation which keeps the good properties of the initial system and which gives us a positive answer in the case of languages whose Max is well-formed. Moreover, the rewriting system R_e has several classical properties like projection lemma or Levi lemma. So we have a complete tool and our aim is now to extend the results to the general case using the same definition of Max_L and the partial result obtained when Max_L is well-formed.

References

[CLR95]　M. Clerbout, M. Latteux, and Y. Roos. Semi-commutations. In V. Diekert and G. Rozenberg, editors, *The Book of Traces*, chapter 12, pages 487–551. World Scientific, 1995.

[CP85]　R. Cori and D. Perrin. Automates et commutations partielles. *R.A.I.R.O. — Theoretical Informatics and Applications*, 19:21–32, 1985.

[CRR]　M. Clerbout, Y. Roos, and I. Ryl. Synchronization languages. *Theoretical Computer Science*. to appear.

[GGYW91]　R. Govindarajan, L. Guo, S. Yu, and P. Wang. ParC project : Practical constructs for parallel programming languages. In *Proc. IEEE of the 15th Annual Internationnal Computer Software & Applications Conference*, pages 183–189, 1991.

[GSY94]　L. Guo, K. Salomaa, and S. Yu. Synchronization expressions and languages. In *Proc. of the 6th Symposium on Parallel and Distributed Processing*, pages 257–264, 1994.

[GSY96]　L. Guo, K. Salomaa, and S. Yu. On synchronization languages. *Fundamenta Informaticae*, 25:423–436, 1996.

[Mét88]　Y. Métivier. *Contribution à l'étude des monoïdes de commutations*. PhD thesis, Université de Bordeaux I, 1988.

Integrating the Specification Techniques of Graph Transformation and Temporal Logic [*]

Reiko Heckel[1], Hartmut Ehrig[1], Uwe Wolter[1], Andrea Corradini[2][**]

[1] TU Berlin, FR 6-1, Franklinstrasse 28/29, 10587 Berlin, Germany
{reiko, ehrig, wolter}@cs.tu-berlin.de
[2] CWI, Kruislaan 413, 1098 SJ Amsterdam, The Netherlands
andrea@cwi.nl

Abstract. The aim of this paper is an integration of graph grammars with different kinds of behavioural constraints, in particular with temporal logic constraints. Since the usual algebraic semantics of graph transformation systems is not able to express constrained behaviour we introduce - in analogy to other approaches - a coalgebraic semantics which associates with each system a category of models (deterministic transition systems). Such category has a final object, which includes all finite and infinite transition sequences. The coalgebraic framework makes it possible to introduce a general notion of 'logic of behavioural constraints'. Instances include, for example, graphical consistency constraints and temporal logic constraints. We show that the considered semantics can be restricted to a final coalgebra semantics for systems with behavioural constraints. This result can be instantiated in order to obtain a final coalgebra semantics for graph grammars with temporal logic constraints.

1 Introduction

The *theory of graph transformation* basically studies a variety of formalisms which extend the theory of formal languages in order to deal with structures more general than strings, like graphs and maps. A graph transformation system allows one to describe finitely a (possibly infinite) collection of graphs, i.e., those which can be obtained from a start graph through repeated applications of graph productions. Each production can be applied to a graph by replacing an occurrence of its left-hand side with its right-hand side. The form of graph productions and the mechanisms stating how a production can be applied to a graph and what the resulting graph is, depend on the specific formalism. In this paper we use the "algebraic, Double Pushout (DPO) approach" [1,2].

The classical theory of the DPO approach is mainly concerned with structural properties and the sequential and parallel composition of transformation

---

[*] Research partially supported by the German Research Council (DFG) and the TMR network GETRATS

[**] A. Corradini is on leave from Dipartimento di Informatica, Pisa. He is also supported by the EC Fixed Contribution Contract n. EBRFMBICT960840

steps. More recently, also distributed graph transformations and means of synchronisation have been investigated, see [2] for a recent survey. It has been shown that graph transformations are well-suited for modelling the state transformation aspect of software systems. In a purely rule-based framework, however, it is difficult to control the order and frequency of rule applications. Such control aspects are most important for the specification of software systems, in particular of concurrent and distributed ones. For this reason we study behavioural constraints for graph grammars.

Instead of inventing new techniques for specifying the behavioural aspects of a system we shall provide an interface for the integration of graph transformation with existing specification techniques. In this paper we shall elaborate in some detail a temporal logic for graph grammars, extending the approach of [3]. The advantage of this approach is two-fold. First, it combines graphical (visual) and logical means which makes it much easier to work with than purely textual expressions (at least for non-experts). Second, all graph-specific aspects are formulated categorically, which makes the approach applicable also to other kinds of structures.

The integration of graph grammars and behavioural constraints is done both syntactically and semantically. On the syntactical level, a notion of behavioural constraint is assumed together with a satisfaction relation for derivation sequences. On the semantical level, a coalgebraic semantics of graph grammars provides a model for the restricted behaviour of systems. This coalgebraic semantics is defined in analogy to a recently developed coalgebraic semantics for graph grammars in [4], where *graph transitions* are used as a kind of "loose transformation step". Since, the coalgebraic idea is orthogonal to the kind of transformation one considers, we may reuse the constructions of [4] without additional effort. The proofs for the results of this paper can also be found there.

The coalgebraic loose semantics allows us to handle in a natural way a large class of constraints imposed on the behaviour of a graph transformation system: It is shown that for each set of behavioural constraints, the restriction to the behaviours that satisfy these constraints is a cofree construction. This implies the existence of a final (maximal) model for each graph grammar with behavioural constraints.

2   Basic Notions

This section reviews basic notions and definitions of the algebraic double-pushout (DPO) approach to the transformation of typed graphs [2,5].

Let **Graph** be the category of graphs and graph morphisms. Given a graph $TG \in |\textbf{Graph}|$, the category $\textbf{Graph}_{TG}$ of *typed graphs* over TG and *typed graph morphisms* is the comma category ($\textbf{Graph} \downarrow TG$). If not stated otherwise, graphs and graph morphisms will be assumed to be typed over TG in the following.

A *graph production* $p : s$ is composed of a *production name* p and of a span of injective graph morphisms $s = (L \xleftarrow{l} K \xrightarrow{r} R)$, called *production span*. A *(typed) graph transformation system* $\mathbf{G} = \langle TG, P, \pi \rangle$ consists of a *type graph* TG, a set of production names P, and a mapping π associating with each production name p a production span $\pi(p)$. A *(typed) graph grammar* $\mathbf{GG} = \langle \mathbf{G}, G_0 \rangle$ is a typed graph transformation system $\mathbf{G} = \langle TG, P, \pi \rangle$ together with a start graph $G_0 \in |\mathbf{Graph}_{TG}|$.

Derivation steps in the DPO approach are defined as double pushout constructions: A *double-pushout* d is a diagram like below, where top and bottom are production spans and (1) and (2) are pushouts. If $p : (L \xleftarrow{l} K \xrightarrow{r} R)$ is a production, a *derivation step* from G to H is denoted by $G \xRightarrow{p/d} H$. We also write p/d if G and H are understood, and denote by In, Out, and pn the projections $In(p/d) = G, Out(p/d) = H$, and $pn(p/d) = p$. The set of all derivation steps in $\mathbf{G}$ is $\mathbf{G}^{\Rightarrow}$.

$$
\begin{array}{ccccc}
L & \xleftarrow{\;\;l\;\;} & K & \xrightarrow{\;\;r\;\;} & R \\
\downarrow{\scriptstyle d_L} & (1) & \downarrow{\scriptstyle d_K} & (2) & \downarrow{\scriptstyle d_R} \\
G & \xleftarrow{\;\;l^*\;\;} & D & \xrightarrow{\;\;r^*\;\;} & H
\end{array}
$$

A *derivation* in a graph transformation system $\mathbf{G}$ is a finite or infinite sequence of derivation steps $G_0 \xRightarrow{p_1/d_1} G_1 \xRightarrow{p_2/d_2} G_2 \ldots$ where $p_1, p_2, \ldots$ are production names of $\mathbf{G}$. A *derivation in a grammar* $\mathbf{GG} = \langle \mathbf{G}, G_0 \rangle$ is a derivation in $\mathbf{G}$ that starts in G_0.

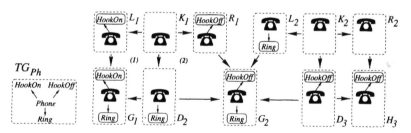

Fig. 1. Telephone example.

Our sample graph transformation system $\mathbf{G}_{Ph} = \langle TG_{Ph}, \{HookOff, Mute\}, \pi_{Ph} \rangle$ models (part of) the user's interaction with a telephone (see [6] for the full case study). Type graph TG_{Ph} and productions $HookOff : (L_1 \leftarrow K_1 \rightarrow R_1)$ and $Mute : (L_2 \leftarrow K_2 \rightarrow R_2)$ are shown in Figure 1. The typing is indicated by the inscription of vertices and the phone icon. Using production $HookOff$, the user may change the hook status of the phone, while $Mute$ models the phone's reaction of turning the bell off. A derivation step using $HookOff$ is given by the pushout diagrams (1) and (2).

3   Coalgebraic Semantics of Graph Grammars

In analogy to [4] we present now a coalgebraic loose semantics for graph grammars. A model for a graph grammar $\mathbf{GG}$ is a deterministic transition system with terminal states that at each step outputs a derivation step. Such models are defined as coalgebras over a suitable functor, and they form a category having a final object, the *full transition system* over **G**, that includes all finite and infinite derivation sequences.

Let T be an endofunctor on a category **C**. A *T-coalgebra* (see e.g., [7]) is a pair $\langle C, \alpha \rangle$ consisting of an object C and a morphism $\alpha : C \to T(C)$ in **C**. A morphism $f : \langle C, \alpha \rangle \to \langle C', \alpha' \rangle$ of coalgebras is a morphism $f : C \to C'$ in **C** such that $T(f) \circ \alpha = f \circ \alpha'$. This defines category $\mathbf{C}^T$, called the category of T-coalgebras.

Definition 1 (category of graph transition systems). Let $T_\mathbf{G} : \mathbf{Set} \to \mathbf{Set}$ be the endofunctor defined for each graph transformation system **G** by $T_\mathbf{G}(S) = (\mathbf{G}^\Rightarrow \times S) + 1$ on objects and $T_\mathbf{G}(f) = (id_{\mathbf{G}^\Rightarrow} \times f) + id_1$ on arrows (recall that $\mathbf{G}^\Rightarrow$ is the set of all derivation steps in **G**, while 1 denotes the final object $\{\perp\}$ in **Set**). For each $T_\mathbf{G}$-coalgebra $\mathbf{T} = \langle S, step : S \to (\mathbf{G}^\Rightarrow \times S) + 1 \rangle$, the partial functions $first$ and $next$ are defined for all states $s \in S$ where $step(s) = \langle t, s' \rangle$, and in this case $next(s) = s'$ and $first(s) = t$. The ith iteration $next^i$ of $next$ is defined by $next^0 = id_S$ and $next^{i+1} = next \circ next^i$.

A *graph transition system* over **G** is a $T_\mathbf{G}$-coalgebra such that for all $s \in S$, $Out(first(s)) = In(first(next(s)))$, i.e., the output graph of the first step equals the input graph of the second, etc. If $\mathbf{GG} = \langle \mathbf{G}, G_0 \rangle$ is graph grammar, a *graph transition system over* **GG** is a transition system over **G** where for all $s \in S$ there is $n \in \mathbb{N}$ (including zero) and $s_0 \in S$ such that $In(first(s_0)) = G_0$ and $next^n(s_0) = s$, i.e., all sequences are reachable from the start graph G_0.

The category $\mathbf{GraTS(GG)}$ of graph transition systems over **GG** is the full subcategory of the category $\mathbf{Set}^{T_\mathbf{G}}$ of $T_\mathbf{G}$-coalgebras having all graph transition systems over **GG** as objects. △

Intuitively, a graph transition system $\mathbf{T} = \langle S, step \rangle$ is a deterministic automaton with final states $\langle S, next : S \to S, first : S \to \mathbf{G}^\Rightarrow \rangle$ where $step(s) = \perp$ represents the case where both $next(s)$ and $first(s)$ are undefined, i.e., the termination of the automaton [7]. A state transition from s to s' in S requires no input, but produces a derivation step in $\mathbf{G}^\Rightarrow$ as output (observation). Due to the absence of input, the future behaviour of the system is fully determined by its current state.

Theorem 2 (final coalgebra semantics). *For each graph grammar* **GG**, *the category* $\mathbf{GraTS(GG)}$ *has a final object, the full transition system* $\mathcal{TS}(\mathbf{GG})$. *The unique morphism for some* $\mathbf{T} \in |\mathbf{GraTS(GG)}|$ *is denoted by* $\phi_\mathbf{T} : \mathbf{T} \to \mathcal{TS}(\mathbf{GG})$.

The category $\mathbf{GraTS(GG)}$ of all graph transition systems over the grammar **GG** is regarded as the category of its "models". In contrast to this loose semantics, the full transition system $\mathcal{TS}(\mathbf{GG})$ is the canonical model of **GG** representing all finite and infinite derivation sequences in **GG**.

For any transition system **T**, the morphism induced by the finality of $\mathcal{TS}(\mathbf{GG})$ is denoted by $\phi_\mathbf{T} : \mathbf{T} \to \mathcal{TS}(\mathbf{GG})$. If $s' \in S'$ is a state of **T**, then

$\phi_{\mathbf{T}}(s')$ is a derivation sequence representing the *behaviour* of s'. Two states with the same behaviour are bisimilar [8]. For the full transition system $\mathcal{TS}(\mathbf{GG})$ this means that if two derivation sequences are bisimilar, then they are already equal, i.e., $\mathcal{TS}(\mathbf{GG})$ provides a minimal representation of all possible behaviours of $\mathbf{GG}$.

It is worth summarising here some advantages of having defined the category of models $\mathbf{GraTS}(\mathbf{GG})$ via coalgebraic instead of algebraic techniques, obtaining for example an initial model by a free construction, as in [9,10]. First of all, the free construction in the mentioned papers only generates finite sequences, while the full transition system contains both finite and infinite sequences. But more importantly, in the algebraic approach all models have to include a homomorphic image of all the computations of the initial model, thus there are no models corresponding to some kind of restriction of behaviour. On the contrary, it is well-known (see e.g., [7,8]) that the coalgebraic framework allows for the definition of various expressive techniques for considering models that realize only part of the behaviours of the final model. Such techniques, based on various kinds of constraints, will be described in more detail in the next section.

4 Behavioural Constraints for Graph Grammars

In this section we introduce logics of behavioural constraints as syntactical interfaces for the integration of graph grammars with behavioural specification techniques. The main result of this section (see Theorem 4) shows that the full transition system can be restricted to those derivation sequences satisfying the constraints, such that we obtain a final coalgebra semantics with behavioural constraints. Particular instances of behavioural constraints, like graphical consistency constraints and temporal logic constraints are studied in the next section.

Definition 3 (logic of behavioural constraints). A *logic of behavioural constraints for graph grammars* $LOBC = \langle Constr, \models \rangle$ is given by a class $Constr(\mathbf{GG})$ of *behavioural constraints* and a *satisfaction relation* $\models_{\mathbf{GG}} \subseteq$ $|\mathbf{GraTS}(\mathbf{GG})| \times Constr(\mathbf{GG})$ for each graph grammar $\mathbf{GG}$, such that for each $c \in Constr(\mathbf{GG})$ the empty graph transition system satisfies c, and satisfaction is closed under homomorphic images and union of subcoalgebras.[1]

The satisfaction relation is extended to sets of behavioural constraints $C \subseteq$ $Constr(\mathbf{GG})$ in the obvious way. $\triangle$

A *graph grammar with behavioural constraints* $\mathbf{GC} = \langle \mathbf{GG}, C \rangle$ consists of a graph grammar $\mathbf{GG}$ together with a set of behavioural constraints

---

[1] A transition system $\mathbf{T}' = \langle S', step' \rangle$ is a subcoalgebra of $\mathbf{T} = \langle S, step \rangle$, written $\mathbf{T}' \subseteq \mathbf{T}$, if $S' \subseteq S$ and $step' = step|_{S'}$ is the restriction of $step$ to S. The homomorphic image of a coalgebra $\mathbf{T}$ under a morphism $f : \mathbf{T} \to \mathbf{T}'$ is the subcoalgebra $f(\mathbf{T}) \subseteq \mathbf{T}'$ determined by $f(S)$, the set-theoretical image of S. Coalgebras are closed under set-theoretical union just like algebras are closed under set-theoretical intersection. It has been shown in [4] that this also holds for transition systems over grammars.

$C \subseteq Constr(\mathbf{GG})$. The category of *(constrained) graph transition systems* $\mathbf{GraTS(GC)}$ over $\mathbf{GC}$ is the full subcategory of $\mathbf{GraTS(GG)}$ where for each $\mathbf{T} \in |\mathbf{GraTS(GC)}|$ we have that $\mathbf{T} \models_{\mathbf{GG}} C$.

The final coalgebra semantics can be lifted to graph grammars with constraints:

Theorem 4 (final coalgebra semantics with beh. constraints).
For each graph grammar with constraints $\mathbf{GC} = \langle \mathbf{GG}, C \rangle$, *the inclusion functor* $\mathcal{E}_{\mathbf{GC}} : \mathbf{GraTS(GC)} \rightarrow \mathbf{GraTS(GG)}$ *has a right adjoint* $\_|_C : \mathbf{GraTS(GG)} \rightarrow \mathbf{GraTS(GC)}$. *Consequently,* $\mathbf{GraTS(GC)}$ *has a final object, the constrained transition system* $\mathcal{TS}(\mathbf{GG})|_C$. $\triangle$

Roughly speaking, $\mathcal{TS}(\mathbf{GG})|_C$ is obtained as the largest subcoalgebra of $\mathcal{TS}(\mathbf{GG})$ where all derivation sequences satisfy the constraints C.

In the next section, examples of behavioural constraints are defined following a generic scheme: Given a graph grammar $\mathbf{GG} = \langle \mathbf{G}, G_0 \rangle$, constraints and their satisfaction are defined for derivation sequences in $\mathbf{G}$ first. Then, a logic of behavioural constraints is derived from this in one of three possible ways, by universal quantification either (I) over *all sequences*, (II) over *runs* (i.e., maximal sequences), or (III) over *initial runs* (i.e., runs that start in the start graph of a grammar). Hence, the same constraint may become more and more "permissive" when the logic is defined according to II or III instead of I.

Let for each graph transformation system $\sqsubseteq \subseteq S \times S$ denote the usual prefix order on derivation sequences in $\mathbf{G}$. If $\mathbf{T} = \langle S_{\mathbf{T}}, step_{\mathbf{T}} \rangle$ is a transition system over $\mathbf{G}$, some $s \in S_{\mathbf{T}}$ is called *run* if for all $s' \in S_{\mathbf{T}}$, $\phi_{\mathbf{T}}(s) \sqsubseteq \phi_{\mathbf{T}}(s')$ implies that s and s' are bisimilar, i.e., $\phi_{\mathbf{T}}(s) = \phi_{\mathbf{T}}(s')$. Let $\mathbf{GG} = \langle \mathbf{G}, G_0 \rangle$ be a graph grammar and $\mathbf{T} \in |\mathbf{GraTS(GG)}|$. Then, a run s in $\mathbf{T}$ is *initial* if $In(first(s)) = G_0$.

Proposition 5 (constraints for derivation sequences). *Assume for each graph transformation system* $\mathbf{G}$ *a class of behavioural constraints* $Constr(\mathbf{G})$ *together with a satisfaction relation* $\models_{\mathbf{G}}^{seq} \subseteq S \times Constr(\mathbf{G})$ *for derivation sequences (i.e.,* S *is the set of all derivation sequences in* $\mathbf{G}$). *Then, there is a logic of behavioural constraints for graph grammars* $LOBC^I = \langle Constr^I, \models^I \rangle$ *where* $Constr^I(\mathbf{GG}) = Constr(\mathbf{G})$ *and* $\models_{\mathbf{GG}}^I \subseteq |\mathbf{GraTS(GG)}| \times Constr^I(\mathbf{GG})$ *is defined by* $\mathbf{T} \models_{\mathbf{GG}}^I c \iff$ *for all* $s \in S_{\mathbf{T}} : \phi_{\mathbf{T}}(s) \models_{\mathbf{G}}^{seq} c$.

Moreover, $LOBC^{II} = \langle Constr^{II}, \models^{II} \rangle$ *forms a logic of behavioural constraints where* $Constr^{II}(\mathbf{GG}) = Constr(\mathbf{G})$ *and* $\models_{\mathbf{GG}}^{II} \subseteq |\mathbf{GraTS(GG)}| \times Constr(\mathbf{GG})$ *is given by* $\mathbf{T} \models_{\mathbf{GG}}^{II} c \iff$ *for all runs* r *in* $\mathbf{T}$: $\phi_{\mathbf{T}}(r) \models_{\mathbf{G}}^{seq} c$.

Finally, a logic of behavioural constraints $LOBC^{III} = \langle Constr^{III}, \models^{III} \rangle$ *is defined by* $Constr^{III}(\mathbf{GG}) = Constr(\mathbf{G})$ *and* $\models_{\mathbf{GG}}^{III} \subseteq |\mathbf{GraTS(GG)}| \times Constr(\mathbf{GG})$ *where* $\mathbf{T} \models_{\mathbf{GG}}^{III} c \iff$ *for all initial runs* r *in* $\mathbf{T}$: $\phi_{\mathbf{T}}(r) \models_{\mathbf{G}}^{seq} c$. $\triangle$

In order to define a particular logic it is now sufficient to provide a notion of constraints and satisfaction for derivation sequences and to specify if the logic shall be of type I, II, or III.

5 Temporal Logic Constraints and Other Examples

In this section several logics of behavioural constraints, especially temporal logic constraints, are defined and illustrated by the telephone example.

Definition 6 (graphical consistency constraints). Assume a graph transformation system $\mathbf{G} = \langle TG, P, \pi \rangle$ and denote by S the set of all derivation sequences in $\mathbf{G}$. Let $Constr(\mathbf{G})$ be the set of all *graphical consistency constraints* [11], i.e., injective morphisms $c : X \to Y \in \mathbf{Graph}_{TG}$. An *assignment* for X in a graph G is an injective morphism $a : X \to G$. It is a *solution* for c if there is an injective $b : Y \to G$ such that $b \circ c = a$. Now, define $\models^{seq}_{\mathbf{G}} \subseteq S \times Constr(\mathbf{G})$ by $s \models^{seq}_{\mathbf{G}} c$ iff each assignment $a : X \to In(first(s))$ is a solution for c. Then, the *logic of consistency constraints* is defined by $CC = \langle Constr^I, \models^I \rangle$ according to Proposition 5. $\triangle$

For the condition c on the left of Figure 2, a graph satisfies c if for each occurrence of X there is also an occurrence of Y, i.e., if a phone is ringing then the hook must be on. Hence, c is satisfied by the graph G_1 of Figure 1 but not by the derived graph G_2.

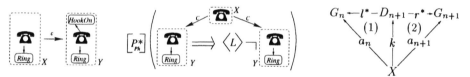

Fig. 2. Graphical consistency constraint, temporal logic constraint with $L = HookOff; Mute$, and derivation step (bottom span) with assignments.

Another logic of behavioural constraints of type I could be defined by *positive and negative application conditions* for graph productions in the sense of [12], where all steps in a derivation sequence would have to satisfy the application conditions of their corresponding productions.

In the rest of this section we define temporal logic [13] as a logic of behavioural constraints using case II and III of Proposition 5. Therefore, let $\mathbf{G} = \langle TG, P, \pi \rangle$ be a graph transformation system and $\mathbf{GG} = \langle \mathbf{G}, G_0 \rangle$ be a graph grammar.

If Q is a set of atomic propositions, the language of propositional temporal formulas over a graph transformation system $\mathbf{G}$ has the form

$$\Phi ::= Q \mid \neg \Phi \mid \Phi_1 \wedge \Phi_2 \mid [L]\Phi \mid \langle L \rangle \Phi$$

where $Q \in \mathcal{Q}$, and $L \subseteq P^*$ is a set of sequences over production names. As usual, we abbreviate $\neg \Phi \vee \Psi$ as $\Phi \Rightarrow \Psi$. The intended meaning of $[L]\Phi$ is, that after every application of a sequence of productions $w \in L$, Φ holds. Dually, $\langle L \rangle \Phi$ means that there exists an application of a production sequence $w \in L$ such that Φ holds.

This propositional language is, of course, not able to express any property of the graphs in a derivation sequences. Hence, we have to combine it with a calculus like graphical consistency constraints. The resulting language shall be a graphical temporal logic where consistency constraints for graphs are connected by the propositional temporal operators defined above.

Assume an interpretation I of all atomic propositions $Q \in \mathcal{Q}$ into graphical consistency constraints $I(Q)$ over $\mathbf{G}$ (cf. Definition 6). Then, for each graph $X \in |\mathbf{Graph}_{TG}|$ the class $Constr_X(\mathbf{G})$ of *interpreted temporal formulas over* X is given by all those formulas Φ of the propositional temporal language where for all atomic proposition Q in Φ the interpretation of Q is a constraint $I(Q)$: $X \to Y$ with domain X.

A sample temporal logic constraint is shown in Figure 2 in the middle. It says that every state reachable from the initial state by productions in P_{Ph} has to satisfy the following condition: If the phone is ringing, then *HookOff* is applied before *Mute* and the phone will stop ringing. In this way we are able to specify, in a loose, axiomatic way, the effect of the application of a production and the order in which productions shall be applied.

Temporal formulas are evaluated over transition systems. For each graph transformation system $\mathbf{G} = \langle TG, P, \pi \rangle$ and each graph $X \in |\mathbf{Graph}_{TG}|$, the *labelled transition system* $\mathcal{T}_X(\mathbf{G}) = (\mathcal{A}_X, (\xrightarrow{w})_{w \in P^*})$ of $\mathbf{G}$ over X consists of

- the set $\mathcal{A}_X$ of all pairs $\langle s, a_0 \rangle$ of a derivation sequence s of $\mathbf{G}$ and an assignment (i.e., an injective morphisms) $a_0 : X \to G_0$ into the first graph of the sequence.
- for each sequence of production names $w \in P^*$ a transition relation $\xrightarrow{w} \subseteq \mathcal{A}_X \times \mathcal{A}_X$, which is defined inductively by $\langle s, a_0 \rangle \xrightarrow{\lambda} \langle s, a_0 \rangle$ (where λ denotes the empty sequence) for all $\langle s, a_0 \rangle \in \mathcal{A}_X$, and $\langle s, a_0 \rangle \xrightarrow{wp_{n+1}} \langle s', a_{n+1} \rangle$ for $w = p_0 \ldots p_n \in P^*$ iff $\langle s, a_0 \rangle \xrightarrow{w} \langle s'', a_n \rangle$ and for step $n+1$ of the sequence there exists $k : X \to D_{n+1}$ such that (1) and (2) in Figure 2 on the right commute.

Notice that the assignment a_{n+1}, if it exists, is uniquely determined by a_0 and the commutativity of (1) and (2).

Now we are able to define the satisfaction of interpreted temporal formulas by derivation sequences in $\mathbf{G}$:

Definition 7 (logic of temporal constraints). Let $\mathcal{T}_X(\mathbf{G}) = (\mathcal{A}_X, (\xrightarrow{w})_{w \in P^*})$ be the labelled transition system of a graph transformation system $\mathbf{G}$ over $X \in |\mathbf{Graph}_{TG}|$ and let I be an interpretation of the atomic propositions $Q \in \mathcal{Q}$. We define the satisfaction of X-interpreted temporal formulas $\Phi \in Constr_X(\mathbf{G})$ by $\langle s, a_0 \rangle \in \mathcal{A}_X$ inductively as

- $\langle s, a_0 \rangle \models_{\mathbf{G}, I} Q$ iff a_0 is a solution for $I(Q)$ (cf. Definition 6),
- $\langle s, a_0 \rangle \models_{\mathbf{G}, I} \neg \Phi$ iff $\langle s, a_0 \rangle \not\models_{\mathbf{G}, I} \Phi$,
- $\langle s, a_0 \rangle \models_{\mathbf{G}, I} \Phi_1 \wedge \Phi_2$ iff $\langle s, a_0 \rangle \models_{\mathbf{G}, I} \Phi_1$ and $\langle s, a_0 \rangle \models_{\mathbf{G}, I} \Phi_2$,
- $\langle s, a_0 \rangle \models_{\mathbf{G}, I} [L]\Phi$ iff for all $\langle s', a_n \rangle \in \mathcal{A}_X$, $\langle s, a_0 \rangle \xrightarrow{w} \langle s', a_n \rangle$ for some $w \in L$ implies that $\langle s', a_n \rangle \models_{\mathbf{G}, I} \Phi$,
- $\langle s, a_0 \rangle \models_{\mathbf{G}, I} \langle L \rangle \Phi$ iff there exists $\langle s', a_n \rangle \in \mathcal{A}_X$ with $\langle s, a_0 \rangle \xrightarrow{w} \langle s', a_n \rangle$ for some $w \in L$ and $\langle s', a_n \rangle \models_{\mathbf{G}, I} \Phi$.

Now, a derivation sequence s satisfies an X-interpreted temporal formula Φ, written $s \models^{seq}_{\mathbf{G},I} \Phi$, iff $\langle s, a_0 \rangle \models_{\mathbf{G},I} \Phi$ for all assignments $a : X \rightarrow G_0$ into the first graph of s.

According to Proposition 5, this satisfaction relation defines two logics of behavioural constraints $Temp^{II} = \langle Constr^{II}, \models^{II} \rangle$ and $Temp^{III} = \langle Constr^{III}, \models^{III} \rangle$. $\triangle$

By Definition 7 and Theorem 4 this implies that for each graph grammar with temporal logic constraints $\mathbf{GC} = \langle \mathbf{GG}, C \rangle$, the restriction $\mathcal{TS}(\mathbf{GG})|_C$ of the full transition system $\mathcal{TS}(\mathbf{GG})$ is a final object in $\mathbf{GraTS(GC)}$.

In $Temp^{II}$ the constraints are evaluated w.r.t. behaviours of maximal length only. It can be shown that this property is closed under reachability, i.e., if s is maximal in a transition system $\mathbf{T}$, then also $next(s)$ is maximal in $\mathbf{T}$. This means that all formulas are implicitly quantified over all subsequent states, i.e., there is the sound inference rule that $\mathbf{T} \models^{II} \Phi$ implies $[P^*]\mathbf{T} \models^{II} \Phi$. Contrastingly for logic $Temp^{III}$, the evaluation of formulas is restricted to those maximal sequences that begin in the start graph G_0 of a grammar, i.e., the constraints are less restrictive than in $Temp^{II}$.

6 Conclusion

In this paper graph grammars have been integrated with behavioural constraints. A final coalgebraic semantics for graph transformation systems with constraints in general has been developed and specialised to temporal logic constraints. It shall be stressed that these notions and results can be generalised to arbitrary HLR categories [14] in order to transfer them to the transformation other kinds of objects like hypergraphs, attributed graphs, and relational structures.

Behavioural constraints, as an abstract framework, are comparable to control conditions in GRACE, a graph and rule based specification language that is being developed independent of the particular graph transformation approach [15]. It could be an interesting future topic to extend our approach of coalgebraic semantics to this language.

Moreover, the coalgebraic semantics has still to be extended to graph grammar morphisms. There are various notions of such morphisms to contend with in literature ([9,16,10,6]). We only mention that in the category proposed in [6], pullbacks are used for describing a sort of *parallel composition with synchronisation* of systems. Since the final coalgebra semantics can be characterised via a cofree construction (that preserves pullbacks) this semantics would be compositional with respect to such parallel composition.

The authors are greatful to Fabio Gadducci, Martin Große-Rhode and Manuel Koch for their valuable remarks and contributions.

References

1. H. Ehrig, M. Pfender, and H. Schneider, "Graph grammars: an algebraic approach," in *14th Annual IEEE Symposium on Switching and Automata Theory*, pp. 167–180, IEEE, 1973.
2. A. Corradini, U. Montanari, F. Rossi, H. Ehrig, R. Heckel, and M. Löwe, "Algebraic approaches to graph transformation I: Basic concepts and double pushout approach," in *Handbook of Graph Grammars and Computing by Graph transformation, Volume 1: Foundations* (G. Rozenberg, ed.), World Scientific, 1997.
3. M. Koch, "Modellierung und Nachweis der Konsistenz von verteilten Transaktionsmodellen für Datenbanksysteme mit algebraischen Graphgrammatiken," Master's thesis, TU Berlin, 1996.
4. R. Heckel, H. Ehrig, U. Wolter, and A. Corradini, "Loose semantics and constraints for graph transformation systems," Tech. Rep. 97-07, TU Berlin, 1997. http://www.cs.tu-berlin.de/cs/ifb/TechnBerichteListe.html.
5. A. Corradini, U. Montanari, and F. Rossi, "Graph processes," *Special issue of Fundamenta Informaticae*, vol. 26, no. 3,4, pp. 241–266, 1996.
6. L. Ribeiro, *Parallel Composition and Unfolding Semantics of Graph Grammars*. PhD thesis, TU Berlin, 1996.
7. H. Reichel, "An approach to object semantics based on terminal co-algebras," *Math. Struc. in Comp. Science*, vol. 5, pp. 129–152, 1995.
8. J. Rutten, "Universal coalgebra: a theory of systems," Tech. Rep. CS-R9652, CWI, 1996.
9. A. Corradini, H. Ehrig, M. Löwe, U. Montanari, and J. Padberg, "The category of typed graph grammars and their adjunction with categories of derivations," in *5th Int. Workshop on Graph Grammars and their Application to Computer Science, Williamsburg '94, LNCS 1073*, 1996.
10. R. Heckel, A. Corradini, H. Ehrig, and M. Löwe, "Horizontal and vertical structuring of typed graph transformation systems," *Math. Struc. in Comp. Science*, vol. 6, no. 6, pp. 613–648, 1996.
11. R. Heckel and A. Wagner, "Ensuring consistency of conditional graph grammars – a constructive approach," *Proc. of SEGRAGRA'95 "Graph Rewriting and Computation", Electronic Notes of TCS*, vol. 2, 1995. http://www.elsevier.nl/locate/entcs/volume2.html.
12. A. Habel, R. Heckel, and G. Taentzer, "Graph grammars with negative application conditions," *Special issue of Fundamenta Informaticae*, vol. 26, no. 3,4, 1996.
13. Z. Manna and A. Pnueli, *The Temporal Logic of Reactive and Concurrent Systems, Specification*. Springer-Verlag, 1992.
14. H. Ehrig, A. Habel, H.-J. Kreowski, and F. Parisi-Presicce, "Parallelism and concurrency in High Level Replacement Systems," *Math. Struc. in Comp. Science*, vol. 1, pp. 361–404, 1991.
15. M. Andries, G. Engels, A. Habel, B. Hoffmann, H.-J. Kreowski, S. Kuske, D. Plump, A. Schürr, and G. Taentzer, "Graph transformation for specification and programming", Tech. Rep. 7/96, University of Bremen, 1996.
16. F. Parisi-Presicce, "Transformation of graph grammars," in *5th Int. Workshop on Graph Grammars and their Application to Computer Science, Williamsburg '94, LNCS 1073*, 1996.

On the Generation of Trees
by Hyperedge Replacement[*]

Frank Drewes

Department of Computer Science, University of Bremen,
P.O. Box 33 04 40, D–28334 Bremen, Germany
e-mail: drewes@informatik.uni-bremen.de

Abstract. A characterization of the sets of hypertrees generated by hyperedge-replacement graph grammars is established. It says that these sets are exactly those of the form $val(T)$, where T, a set of terms over hyperedge-replacement operations, is the output language of a finite-copying top-down tree transducer. Furthermore, the terms in T may be required to consist of hyperedge-replacement operations whose underlying hypergraphs are hypertrees. The result is closely related to a similar characterization that was obtained for the case of string graphs by Engelfriet and Heyker some years ago. In fact, the results of this paper also yield a new proof for the characterization by Engelfriet and Heyker.

1 Introduction

Hyperedge-replacement graph grammars (hr grammars), also called context-free hypergraph grammars, are well-studied devices for the generation of graph and hypergraph languages (see, e.g., [Hab92,Eng97,DHK97]). Their basic operation is the replacement of a non-terminal hyperedge by a hypergraph. This works by removing the replaced hyperedge e and adding the replacing hypergraph R in such a way that, if e has k incident nodes, these nodes are identified with k distinguished nodes of R called its *points*. So, the points are the only places where R is connected to the host hypergraph. This suggests immediately that the power of hr grammars depends on the maximum type of right-hand side hypergraphs (called the *order* of the grammar), where the *type* of a hypergraph is the number of points. In fact, even if only string graphs (i.e., edge-labelled, chain-like graphs of type 2) are generated, the power of hr grammars increases properly whenever the order is increased by 2 (see [EH91,Hab92]).

The string case is interesting because even the use of string graphs as right-hand sides suffices to generate the set of *all* string graphs over a fixed alphabet. However, making use of larger and larger types, one can specify *more and more sophisticated subsets* of this basic set. For example, the set of string graphs corresponding to $\{a_1^n \cdots a_{2k}^n \mid n \geq 1\}$ $(k \geq 1)$ can be generated, but it requires a grammar of order $2k$ (cf. [EH91,Hab92]).

---

[*] Partially supported by the EC TMR Network GETGRATS (General Theory of Graph Transformation Systems) through the University of Bremen.

For a closer look at this phenomenon it is useful to turn to a slightly different view of hr grammars, where the right-hand sides of productions are seen as operations on hypergraphs (cf. [Cou91]), so-called *hr operations*. If a right-hand side has n non-terminal hyperedges then it corresponds to an operation of arity n. Applying this operation to n hypergraphs means to replace the non-terminal hyperedges by the argument hypergraphs. Given a set Σ of such hr operations a set $T \subseteq \mathrm{T}_\Sigma$ of terms over Σ defines the hypergraph language $val(T) = \{val(t) \mid t \in T\}$, where val is the evaluation mapping on terms. The languages of the form $val(T)$ where T is a *regular* set of terms are exactly those which are generated by hr grammars (cf. [Cou91]), and are called *context-free*. (The set T corresponds closely to the set of derivation trees of the respective hr grammar.)

Now, the observations concerning string graphs can be formulated as follows. Let $\Sigma_{(k)}$ denote the set of all hr operations of type $\leq k$. Then $val(\mathrm{T}_{\Sigma_{(2)}})$ contains all string graphs, but not every context-free string-graph language L admits a regular subset of $\mathrm{T}_{\Sigma_{(2)}}$ defining L. In other words, if $L = val(T)$ for a regular set $T \subseteq \mathrm{T}_{\Sigma_{(k)}}$, one can obtain from T some $T' \subseteq \mathrm{T}_{\Sigma_{(2)}}$ with $L = val(T')$, but in general this construction cannot preserve regularity.

It is therefore natural to ask how complex the set T' must necessarily be. For the case of string graphs this question was answered in [EH91]: A string-graph language L is context-free iff $L = val(T')$ for a set $T' \subseteq \mathrm{T}_{\Sigma_{(2)}}$ which is the output language of a finite-copying top-down tree transducer (cf. [ERS80]). In this paper a similar result is obtained for the more general case of trees and hypertrees. Hypertrees are a straightforward generalization of trees, obtained by viewing a hyperedge of type $m \geq 1$ as a directed hyperedge with $m - 1$ sources and one target (i.e., pointing towards the root). Let $\mathbb{T}_m$ denote the set of all hypertrees of type $\leq m$, where the type of hyperedges is at most m, too. Furthermore, let $\Sigma_{\mathrm{ht}(m)}$ denote the signature consisting of all hr operations given by hypertrees in $\mathbb{T}_m$. The main result of this paper says that the following statements are equivalent for sets $L \subseteq \mathbb{T}_m$:

(i) L is context-free;
(ii) $L = val(T)$, where $T \subseteq \mathrm{T}_{\Sigma_{\mathrm{ht}(m)}}$ is the output language of a finite-copying top-down tree transducer;
(iii) $L = val(T)$, where T is the output language of a finite-copying top-down tree transducer (using arbitrary hyperedge-replacement operations).

Intuitively, the reason for these equivalences is that large hyperedges can be used to control several places in a hypergraph simultaneously during a derivation, which allows for a finite amount of synchronization between parts that otherwise would have to be derived independently. Top-down tree transducers with the finite-copying property can be used to achieve a similar effect by copying an input term (or a subterm thereof) and using the identical copies to synchronize sub-derivations.

This paper is a short version of [Dre97]. Due to the limitation of space, proofs are omitted or only sketched; details can be found in [Dre97].

2 Tree transductions

For $n \in \mathbb{N}$, $[n]$ denotes the set $\{1, \ldots, n\}$. The set of all finite sequences over a set S is denoted by S^*. For $l \in S^*$, $[l]$ denotes the smallest subset S_0 of S with $l \in S_0^*$. The length of l is denoted by $|l|$, and $l(i)$ denotes the ith element of l. If $f : A \to S^*$ is a mapping that yields sequences, $f(a, i)$ denotes $f(a)(i)$.

A (one-sorted) *signature* is a set Σ of symbols such that for every $f \in \Sigma$ a natural number, called its *rank*, is specified. To indicate that f has rank n the notation $f^{(n)}$ will be used. The set of *variables* is $X = \{x_1, x_2, \ldots\}$, and $X_n = \{x_1, \ldots, x_n\}$ for all $n \in \mathbb{N}$. A term consisting of an outermost symbol f and n subterms $t_1, \ldots, t_n$ is denoted by $f[t_1, \ldots, t_n]$. The set of all terms over Σ with variables in $Y \subseteq X$ (defined as usual) is denoted by $T_\Sigma(Y)$, and T_Σ denotes $T_\Sigma(\emptyset)$. For $t \in T_\Sigma(X_n)$ and $t_1, \ldots, t_n \in T_{\Sigma'}(Y)$ the substitution of t_i for x_i in t ($i \in [n]$) is defined as usual and is denoted by $t[\![t_1, \ldots, t_n]\!]$.

A *regular tree grammar*[1] is a system $g = (N, \Sigma, P, S)$, where N and Σ are finite signatures, all symbols in N (the *non-terminals*) have rank 0, $P \subseteq N \times T_{\Sigma \cup N}$ is a finite set of term rewrite rules (the *productions*), and $S \in N$ (the *start symbol*). The set of terms generated by g is $L(g) = \{t \in T_\Sigma \mid S \to_P^* t\}$, where $\to_P$ denotes the term rewrite relation determined by P and $\to_P^*$ its reflexive and transitive closure. The set $L(g)$ is called a *regular set of terms*.

A *top-down tree transducer with regular look-ahead* (tdr transducer, see [Eng77]) is a system $\mathsf{tdr} = (\Sigma, \Sigma', \Gamma, R, \gamma_0)$, where Σ and Σ' are finite signatures of *input* and *output* symbols, respectively, Γ with $\Gamma \cap (\Sigma \cup \Sigma') = \emptyset$ is a finite signature of *states* each of which has rank 1, $\gamma_0 \in \Gamma$ is the *initial state*, and R is a finite set of *rules* of the form

$$\gamma[f[g_1, \ldots, g_n]] \to t[\![\gamma_1[x_{i_1}], \ldots, \gamma_m[x_{i_m}]]\!].$$

Here, $f^{(n)} \in \Sigma$, $\gamma, \gamma_1, \ldots, \gamma_m \in \Gamma$, $t \in T_{\Sigma'}(X_m)$ for some m, $x_{i_1}, \ldots, x_{i_m} \in X_n$, and $g_1, \ldots, g_n$ are regular tree grammars. For terms s, s', $s \to_R s'$ if R contains a rule as above, such that $s = s_0[\![\gamma[f[t_1, \ldots, t_n]]]\!]$ with $t_1 \in L(g_1), \ldots, t_n \in L(g_n)$ and $s' = s_0[\![t[\![\gamma_1[t_{i_1}], \ldots, \gamma_m[t_{i_m}]]\!]]\!]$. Thus, the rule has the effect of the term rewrite rule $\gamma[f[x_1, \ldots, x_n]] \to t[\![\gamma_1[x_{i_1}], \ldots, \gamma_m[x_{i_m}]]\!]$, but it applies only if every subterm t_i matched by the variable x_i is an element of $L(g_i)$. The *tree transduction* computed by tdr is the binary relation $tdr \subseteq T_\Sigma \times T_{\Sigma'}$ given by $tdr = \{(t, t') \in T_\Sigma \times T_{\Sigma'} \mid \gamma_0[t] \to_R^* t'\}$.

The *domain* of tdr is $dom(tdr) = \{t \mid (t, t') \in tdr$ for some $t' \in T_{\Sigma'}\}$, and its *output language* is $out(tdr) = \{t' \mid (t, t') \in tdr$ for some $t \in T_\Sigma\}$. If C is a class of tree transductions and Σ is a signature $OUT_\Sigma(C)$ denotes the set $\{out(\tau) \subseteq T_\Sigma \mid \tau \in C\}$.

A tdr transducer is *deterministic* if for every pair of distinct rules in R with left-hand sides $\gamma[f[g_1, \ldots, g_n]]$ and $\gamma[f[g_1', \ldots, g_n']]$, respectively, there is some $i \in [n]$ such that $L(g_i) \cap L(g_i') = \emptyset$. Note that tdr is a partial function if tdr is

---

[1] For references concerning tree grammars and tree transducers see the survey paper [GS97].

deterministic. Tdr transducers with rules like, e.g., $\gamma[s[g_1]] \to f[\gamma[x_1], \gamma[x_1]]$, have the ability to copy subterms an exponential number of times during a derivation (in the indicated case the subterm 0 of $s^n[0]$ would be copied 2^n times). A derivation of tdr is said to be k-*copying* ($k \in \mathbb{N}$) if it does not copy any subterm of the input term more than k times (for a formal definition see [ERS80]). If every derivation of tdr is k-copying then tdr is said to be k-copying or, to express that some such k exists, *finite-copying*. A tree transduction is a (deterministic, k-copying, finite-copying) tdr transduction if it is computed by a (deterministic, k-copying, finite-copying) tdr transducer.

A tdr transducer $\mathsf{td} = (\Sigma, \Sigma', \Gamma, R, \gamma_0)$ is a *top-down tree transducer* (td transducer) if all right-hand sides of rules in R have the form $\gamma[f[g, \ldots, g]]$ with $L(g) = \mathrm{T}_\Sigma$ (i.e., no regular look-ahead is used). The class of all k-copying (finite-copying) td transductions is denoted by $\mathrm{TD}_{\mathrm{fc}(k)}$ ($\mathrm{TD}_{\mathrm{fc}}$, respectively).

3 Hyperedge replacement

A *hypergraph* is a tuple $H = (V, E, att, lab, point)$, where V and E are finite sets of *nodes* and *hyperedges*, respectively, $att \colon E \to V^*$ is the *attachment function*, $lab \colon E \to L$ is the *labelling function* (where L is an alphabet of labels), and $point \in V^*$ is a finite sequence of distinguished nodes called the *points* of H. Isomorphism of hypergraphs H and H' is defined as usual and is denoted by $H \cong H'$. For technical convenience it is required that *point* and, for all $e \in E$, $att(e)$ do not contain repetitions. The *type* of H, denoted by $type(H)$, is its number of points, and the type of $e \in E$, $type_H(e)$, is the number of attached nodes. Furthermore, $maxtype(G) = \max(\{type(G)\} \cup \{type_G(e) \mid e \in E_G\})$. An *unlabelled hypergraph* is a hypergraph all of whose hyperedges are labelled with a special symbol $\square$ which intuitively stands for "no label". The components of a hypergraph H are also denoted by V_H, E_H, att_H, lab_H, and $point_H$, respectively.

Let H be a hypergraph and let $\nu_1, \ldots, \nu_k \in E_H$ be pairwise distinct. Then $H_{\nu_1 \cdots \nu_k}$ denotes the k-ary partial operation on hypergraphs which is defined as follows (see also [Cou91]). For hypergraphs $H_1, \ldots, H_k$ such that $type(H_i) = type_H(\nu_i)$ for $i \in [k]$, $H_{\nu_1 \cdots \nu_k}(H_1, \ldots, H_k)$ is obtained from H by (1) removing $\nu_1, \ldots, \nu_k$, (2) adding $H_1, \ldots, H_k$ disjointly and (3) identifying $att_H(\nu_i, j)$ and $point_{H_i}(j)$ for all $i \in [k]$ and $j \in [type(H_i)]$. The points of $H_{\nu_1 \cdots \nu_k}(H_1, \ldots, H_k)$ are those of H.

$H_{\nu_1 \cdots \nu_k}$ is called a *hyperedge-replacement operation* (hr operation) with *underlying hypergraph* H and *virtual hyperedges* $\nu_1, \ldots, \nu_k$. Note that $H_{\nu_1 \cdots \nu_k}$ is an operation on abstract hypergraphs, and that $H_{\nu_1 \cdots \nu_k}(H_1, \ldots, H_k)$ is defined only if the types of argument hypergraphs coincide with the types of the corresponding virtual hyperedges.

A *hyperedge-replacement signature* (hr signature) is a signature Σ of hr operations, where ranks coincide with arities. For $t \in \mathrm{T}_\Sigma$, $val(t)$ denotes the value of t, i.e., $val(H_{\nu_1 \cdots \nu_k}[t_1, \ldots, t_k]) = H_{\nu_1 \cdots \nu_k}(val(t_1), \ldots, val(t_k))$. In order to prevent unnecessary technicalities, we shall only be interested in terms with a defined value. Thus, in the following, T_Σ always stands for this subset of the set of all terms over Σ, and only these terms are allowed to appear as input or output terms of tree transductions or regular tree grammars.

A set of hypergraphs is said to be Σ-*context-free* if it has the form $val(T)$ for a regular subset T of T_Σ (where $val(T) = \{val(t) \,|\, t \in T\}$), and is called *context-free* if it is Σ-context-free for some hr signature Σ. It follows from general results by Mezei and Wright ([MW67], see also Corollary 1.10 of [Cou91]) that a set of hypergraphs is context-free if and only if it is generated by a hyperedge-replacement graph grammar.

4 Hypertrees generated by hyperedge replacement

The main result of the paper is obtained by constructing tdr transducers that take as input terms defining hypertrees (see below) and yield as output equivalent terms that contain only a specific sort of hr operations. For this purpose, three sets $\mathbb{G}$, $\bar{\mathbb{G}}$, and $\mathbb{G}'$ of hypergraphs and a partial function $^-\colon \mathbb{G} \to \bar{\mathbb{G}}$ are considered. If these sets are related via $^-$ in a certain way then tdr transducers can transform terms with operations in $\mathbb{G}$ into equivalent terms where the outer-most operation is in $\bar{\mathbb{G}}$ and all others are in $\mathbb{G}'$. In the following, let $\bar{\mathbb{G}} \subseteq \mathbb{G}$ and assume that (T1)–(T5) below are satisfied.

(T1) For all $G, G' \in \mathbb{G}$ $(G, G' \in \mathbb{G}')$ with $\nu \in E_G$ and $type(G') = type_G(\nu)$ it holds that $G_\nu(G') \in \mathbb{G}$ $(G_\nu(G') \in \mathbb{G}'$, respectively).

(T2) For all $G \in \mathbb{G}$, if $\bar{G}$ is defined and $E_{\bar{G}} = \{\nu_1, \dots, \nu_k\}$ then there are $G_1, \dots, G_k \in \mathbb{G}'$ with $type(G_i) = type_{\bar{G}}(\nu_i)$ for all $i \in [k]$, such that $G \cong \bar{G}_{\nu_1 \cdots \nu_k}(G_1, \dots, G_k)$.

(T3) For all $G \in \bar{\mathbb{G}}$ with $E_G = \{\nu_1, \dots, \nu_k\}$ and $G_1, \dots, G_k, G_1', \dots, G_k' \in \mathbb{G}'$ with $type(G_i) = type_G(\nu_i) = type(G_i')$ for $i \in [k]$, $G_{\nu_1 \cdots \nu_k}(G_1, \dots, G_k) \cong G_{\nu_1 \cdots \nu_k}(G_1', \dots, G_k')$ implies $G_i \cong G_i'$ for all $i \in [k]$.

(T4) For all $G, G' \in \mathbb{G}$ with $\nu \in E_G$ and $type(G') = type_G(\nu)$, $\overline{G_\nu(G')}$ is defined if and only if $\overline{G_\nu(\bar{G}')}$ is defined, and in this case $\overline{G_\nu(\bar{G}')} = G_\nu(\bar{G}')$.

(T5) For every finite hr signature Σ whose operations are given by hypergraphs in $\mathbb{G}$, $\overline{val(T_\Sigma)} = \{\overline{val(t)} \,|\, t \in T_\Sigma, \overline{val(t)} \text{ defined}\}$ is finite.

The essential conditions are (T2)–(T4): For $G = H_{\mu_1 \cdots \mu_l}(H_1, \dots, H_l)$, if $\bar{G}$ is defined then G has a unique decomposition $\bar{G}_{\nu_1 \cdots \nu_k}(G_1, \dots, G_k)$ with $G_1, \dots, G_k \in \mathbb{G}'$, and $\bar{G} = H_{\mu_1 \cdots \mu_l}(\bar{H}_1, \dots, \bar{H}_l)$. The latter follows from (T4) since $H_{\mu_1 \cdots \mu_l}(H_1, \dots, H_l) = (H_{\mu_2 \cdots \mu_l}(H_2, \dots, H_l))_{\mu_1}(H_1)$. Below, the hypergraphs $G_1, \dots, G_k \in \mathbb{G}'$ in the decomposition $\bar{G}_{\nu_1 \cdots \nu_k}(G_1, \dots, G_k)$ of G are denoted by $G\langle\nu_1\rangle, \dots, G\langle\nu_k\rangle$. The following lemma can be shown.

Lemma 1. *Let $\mathbb{G}$, $\bar{\mathbb{G}}$, $\mathbb{G}'$, and $^-$ satisfy (T1)–(T5). Let Σ be a finite hr signature all of whose operations are given by hypergraphs in $\mathbb{G}$. Then there is a deterministic finite-copying tdr transduction $tdr\colon T_\Sigma \to T_{\Sigma'}$ (for some hr signature Σ') such that*

(i) $dom(tdr) = \{t \in T_\Sigma \mid \overline{val(t)}$ is defined$\}$,

(ii) $val(t) = val(tdr(t))$ for all $t \in dom(tdr)$, and

(iii) every term $t \in out(tdr)$ has the form $G_{\nu_1 \cdots \nu_l}[t_1, \ldots, t_l]$, where the underlying hypergraphs of operations in $t_1, \ldots, t_l$ are in $\mathbb{G}'$, $G = \overline{val(t)}$ (in particular, $G \in \bar{\mathbb{G}}$), and $E_G = \{\nu_1, \ldots, \nu_l\}$.

Moreover, for every term $t \in T_\Sigma$ the derivation of tdr on input t is m-copying, where $m = \max\{|E_G| \mid G = \overline{val(t')}$ for a subterm t' of $t\}$.

To consider the case of hypertrees, some more notions are required. From now on, only hypergraphs G are considered such that $type(G) \geq 1$ and $type_G(e) \geq 1$ for all $e \in E_G$. If G is such a hypergraph and $e \in E_G$ with $att_G(e) = v_1 \cdots v_k$, then the sequence $s_G(e)$ of *sources* of e is $v_1 \cdots v_{k-1}$ and its *target* is $t_G(e) = v_k$. Furthermore, if $point_G = v'_1 \cdots v'_l$ then $s(G) = v'_1 \cdots v'_{l-1}$ and $t(G) = v'_l$. A (directed) vv'-*path* in G (for $v, v' \in V_G$) is an alternating sequence $v_0 e_1 v_1 \cdots e_n v_n$ of nodes and hyperedges such that $v_0 = v$, $v_n = v'$, and $v_{i-1} \in [s_G(e_i)]$, $v_i = t_G(e_i)$ for $i \in [n]$. The path is said to *pass* $v_1, \ldots, v_{n-1}$ and $e_1, \ldots, e_n$. As usual, it is called *simple* if no node occurs twice, and *empty* if $n = 0$. G is called *acyclic* if the only vv-paths in G ($v \in V_G$) are the empty ones. The *in-degree* of $v \in V_G$ is the number of hyperedges $e \in E_G$ with $t_G(e) = v$ and the *out-degree* is the number of pairs (e, i) with $e \in E_G$ and $i \in |s_G(e)|$ such that $s_G(e, i) = v$.

Definition 2. (1) A hypergraph G is a *hyperforest* if

    (i) G is acyclic,

    (ii) for all nodes $v \in V_G$ the out-degree of v in G is at most 1, and

    (iii) for all nodes $v \in V_G$ there is a vv'-path in G for some $v' \in [point_G]$.

(2) A hypergraph G is a *hypertree* if

    (i) G is acyclic,

    (ii) for all nodes $v \in V_G$ the out-degree of v in G is at most 1,

    (iii) for all nodes $v \in V_G$ there is a $vt(G)$-path in G, and

    (iv) the in-degree of every node in $[s(G)]$ is 0.

$\mathbb{T}$ denotes the set of all hypertrees and $\mathbb{T}_m = \{G \in \mathbb{T} \mid maxtype(G) \leq m\}$ for $m \in \mathbb{N}$. An hr operation whose underlying hypergraph is a hypertree is called a *hypertree operation*. For all $m \in \mathbb{N}$, $\Sigma_{(m)}$ denotes the hr signature consisting of all hr operations whose underlying hypergraph G satisfies $maxtype(G) \leq m$, and $\Sigma_{\mathrm{ht}(m)}$ denotes the signature of hypertree operations in $\Sigma_{(m)}$. Furthermore, $\Sigma_{(*)} = \bigcup_{m \in \mathbb{N}} \Sigma_{(m)}$ and $\Sigma_{\mathrm{ht}(*)} = \bigcup_{m \in \mathbb{N}} \Sigma_{\mathrm{ht}(m)}$. By an appropriate choice of $\bar{\ }$, $\mathbb{G}$, $\bar{\mathbb{G}}$, and $\mathbb{G}'$ the first result can be shown.

Theorem 3. *For every finite hr signature Σ there is a deterministic finite-copying tdr transduction* $tdr: T_\Sigma \to T_{\Sigma_{\mathrm{ht}(*)}}$ *such that* $val(tdr(t)) = val(t)$ *for all* $t \in T_\Sigma$ *for which* $val(t) \in \mathbb{T}$ *(and $tdr(t)$ is undefined otherwise).*

Proof (sketch). The proof uses Lemma 1 with the following choices of $\bar{\ }$, $\mathbb{G}$, $\bar{\mathbb{G}}$, and $\mathbb{G}'$. $\mathbb{G}$ is simply the set of all hypergraphs. $\bar{\mathbb{G}}$ is the set of all unlabelled hyperforests G such that, if $type(G) = n$,

(i) $V_G = [point_G]$,

(ii) $E_G = \{\nu_1, \ldots, \nu_n\}$ with $t_G(\nu_i) = point_G(i)$ for all $i \in [n]$, and

(iii) $s_G(\nu_i) = point_G(i_1) \cdots point_G(i_k)$ implies $i_1 < \cdots < i_k$ for all $i \in [n]$.

$\mathbb{G}'$ equals $\mathbb{T}$ and $\bar{G} \in \bar{\mathbb{G}}$ is defined if and only if G is a hyperforest. In this case $\bar{G}$ is given as follows: $V_{\bar{G}} = [point_G]$, $point_{\bar{G}} = point_G$, and $[s_{\bar{G}}(\nu_i)]$ is the set of all $point_G(j)$ such that there is a nonempty $point_G(j)point_G(i)$-path in G that does not pass any point.

In order to understand the intuition behind these choices consider a hyperforest G and a point $v \in [point_G]$. From v one can descend down the hyperforest, passing hyperedges against their direction on a reverse path, until another point is reached (or until the path cannot be extended any further). The sub-hypergraph H' of G consisting of all nodes and hyperedges reachable from v in this manner must have the structure of a hypertree. In fact, if v is chosen as $t(H')$ and the points of G reached on the mentioned paths are taken as $s(H')$ this sub-hypergraph becomes a hypertree. Now, imagine that all hyperedges and all non-points of H' are deleted from G, while a hyperedge ν is added whose attachment is the sequence of points of H'. Then, the resulting hypergraph H is a hyperforest satisfying $H_\nu(H') = G$. If all k points of G are treated in this way then $\bar{G}$ and the corresponding components $G\langle\nu_1\rangle, \ldots, G\langle\nu_k\rangle$ are obtained.

It is not very hard to verify that (T1)-(T5) are satisfied. In particular, the decomposition $\bar{G}_{\nu_1 \cdots \nu_k}(G\langle\nu_1\rangle, \ldots, G\langle\nu_k\rangle)$ of a hyperforest is unique and, because of the rather obvious correspondence between the $point_G(i)point_G(j)$-paths in G' and $\bar{G}'$, (T4) is satisfied.

Let $tdr: T_\Sigma \to T_{\Sigma'}$ be as in Lemma 1. Then, $dom(tdr) = \{s \in T_\Sigma \mid val(s)$ is a hyperforest$\}$ and for all $s \in dom(tdr)$, $t = tdr(s)$ has the form $G_{\nu_1 \cdots \nu_k}[t_1, \ldots, t_k]$, where G is a hyperforest and $t_1, \ldots, t_k \in T_{\Sigma_{ht(*)}}$. Thus, $val(t) = val(s)$ is a hypertree if and only if $att_G(\nu_k) = point_G$ and, for all $i \in [k-1]$, $val(t_i)$ is the single-node hypertree without hyperedges. It is easy to see that the terms in $T_{\Sigma'}$ with the latter property form a regular set. Therefore, a very simple construction yields a linear (i.e., 1-copying) deterministic tdr transduction which is undefined if $val(t) \notin \mathbb{T}$, and yields $t_k \in T_{\Sigma_{ht(*)}}$ otherwise (note that $val(t_k) = val(t) = val(s)$). Since the class of deterministic finite-copying tdr transductions is closed under composition (see [ERS80, Theorem 5.4]) this proves the result. □

The types of operations in an output term of the tdr transducer tdr in Theorem 3 can be larger than the type of the defined hypergraph and the types of its hyperedges. The second result states that this can be avoided.

Theorem 4. *The tdr transducer* tdr *in Theorem 3 can be constructed in such a way that for all* $t \in out(tdr)$ *with* $val(t) = G$ *the type of operations in* t *is at most* $maxtype(G)$.

Proof (sketch). Since the class of deterministic finite-copying tdr transductions is effectively closed under composition, using Theorem 3 the proof need to take care only for input terms in $T_{\Sigma_{ht(*)}}$. Therefore, choose $\mathbb{G} = \mathbb{T}$. For the definition of $^-$, $\bar{\mathbb{G}}$, and $\mathbb{G}'$ call $v \in V_G$ (where G is a hypertree) a *fork node* if $v \in [point_G]$ or there are paths from distinct points to v which do not pass a common node. Furthermore, for $e \in E_G$, let $F_G(e)$ denote the sequence $v_1 \cdots v_n$ of pairwise distinct fork nodes such that there are $v_i t_G(e)$-paths which pass e but do not pass any fork node. The order on $v_1, \ldots, v_n$ is as follows. If the $v_i t_G(e)$-path passes $s_G(e, j_i)$ for every $i \in [n]$ then $j_1 < \cdots < j_n$. (Note that the j_i are pairwise distinct because the paths do not pass fork nodes.) Now, the definitions are as follows.

$\bar{\mathbb{G}}$ is the set of all hypertrees G such that every node is a fork node and for all $v \in V_G$ there is at most one hyperedge $e \in E_G$ of type 1 such that $t_G(e) = v$.

$\mathbb{G}'$ is the set of all hypertrees G such that $|E_G| \geq 1$ and if $type(G) > 1$ then there is a hyperedge e with $\{e\} = \{e \in E_G \mid t_G(e) = t(G)\}$ and $F_G(e) = s(G)$ (which in particular implies that the fork nodes of G are the points).

For all $G \in \mathbb{T}$ whose set of fork nodes is V, $\bar{G}$ is the unlabelled hypertree with $V_{\bar{G}} = V$ and $point_{\bar{G}} = point_G$, such that $E_{\bar{G}}$ contains for every node $v \in V$ and every sequence $v_1 \cdots v_n \in \{F_G(e) \mid e \in E_G, t_G(e) = v\}$ a hyperedge e with $att_{\bar{G}}(e) = v_1 \cdots v_n v$ (and, of course, $\bar{G}$ contains no further hyperedges).

Again, it can be shown that the decomposition $\bar{G}_{\nu_1 \cdots \nu_k}(G\langle \nu_1 \rangle, \ldots, G\langle \nu_k \rangle)$ of $G \in \mathbb{G}$ is uniquely determined, i.e., (T2) and (T3) are satisfied. Furthermore, the fork nodes and the structure on them induced by the paths in a hypertree G' remain the same in $\bar{G}'$. Since, intuitively, these are the only aspects on which the definition of $\overline{G_\nu(G')}$ depends one can also show (T4), i.e., that $\overline{G_\nu(G')} = \bar{G}_\nu(\bar{G}')$. (T1) and (T5) are easy to prove.

If G and e are as in the definition of $\mathbb{G}'$ then $|s(G)| = |F_G(e)| \leq |s_G(e)|$. Therefore, all $G \in \mathbb{G}'$ satisfy $type(G) \leq \max\{type_G(e) \mid e \in E_G\}$. Thus, if tdr' is the tdr transducer Lemma 1 yields, every proper subterm t' of $t \in out(tdr)$ satisfies $type(val(t')) \leq \max\{type_{val(t')}(e) \mid e \in E_{val(t')}\}$ (because $val(t') \in \mathbb{G}'$ by (T1)). Therefore, the output terms of tdr' satisfy the requirement formulated in the theorem and, hence, the composition of tdr and tdr' (where tdr is as in Theorem 3) yields the tree transduction sought. $\square$

Using Theorem 4 the main result of this paper can be proved.

Theorem 5. *For all $L \subseteq \mathbb{T}_m$ ($m \in \mathbb{N}$) the following are equivalent.*

(i) L is context-free
(ii) $L = val(T)$ for some $T \in OUT_{\Sigma_{ht(m)}}(\mathrm{TD}_{fc})$
(iii) $L = val(T)$ for some $T \in OUT_{\Sigma_{(*)}}(\mathrm{TD}_{fc})$

Proof (sketch). For (i)$\Rightarrow$(ii), if $L = val(T')$ for a regular set $T' \subseteq T_\Sigma$, then $T = tdr(T')$ (where tdr is a k-copying tdr transducer for some $k \in \mathbb{N}$, obtained by Theorem 4) yields $val(T) = L$. By [Eng77, Theorem 4.2] and [ERS80, Theorem 3.2.1], $T = tdr(T')$ implies that T is the output language of a k-copying td transducer.

The implication (ii)⇒(iii) is trivial.

For (iii)⇒(i) it suffices to show how to turn a k-copying td transducer td into a linear td transducer td' satisfying $val(out(td')) = val(out(td))$ and $out(td') \subseteq T_{\Sigma_{(mk)}}$, because $out(td')$ is regular for linear td transducers td'. The idea is to construct td' in such a way that, intuitively, the operations of $\Sigma_{(mk)}$ used in its rules are obtained by combining at most k operations of $\Sigma_{(m)}$ into a single operation. Every virtual hyperedge in the new operation comprises $\leq k$ hyperedges of the old operations, and thus has type $\leq mk$. □

5 The string case

A *string graph* is a hypertree G of type 2 all of whose hyperedges have type 2 and all of whose nodes lie on the $point_G(1)point_G(2)$-path in G. Let S_k $(k \geq 2)$ denote the set of $\Sigma_{(k)}$-context-free string-graph languages L. Furthermore, let Σ_{str} be the hr signature consisting of all hr operations given by string graphs, and denote by $yTD_{fc(k)}$ the set of all hypergraph languages of the form $val(T)$, where $T \in OUT_{\Sigma_{str}}(TD_{fc(k)})$.[2] In [EH91] Engelfriet and Heyker showed that $S_{2k} = yTD_{fc(k)} = S_{2k+1}$ for all $k \geq 1$ by proving the inclusions $yTD_{fc(k)} \subseteq S_{2k}$ and $S_{2k+1} \subseteq yTD_{fc(k)}$ (which is formulated in terms of the tree-walking transducers of [AU71] in [EH91], however). The same can be done using the results of the previous section.

Theorem 6 (Theorem 6.1 of [EH91]). $S_{2k} = yTD_{fc(k)} = S_{2k+1}$

Proof (sketch). By the proof of Theorem 5(iii)⇒(i) $yTD_{fc(k)} \subseteq S_{2k}$ because $m = 2$ for $\Sigma = \Sigma_{str}$. For the proof of $S_{2k+1} \subseteq yTD_{fc(k)}$ Lemma 1 can be used as follows. Let $L \in S_{2k+1}$ with $L = val(T)$ for some regular subset T of $\Sigma_{(2k+1)}$. Since every node of $val(t)$ $(t \in T)$ has degree[3] at most 2 it is easy to see that T can be modified so that for all subterms t' of a term in T the hypergraph $val(t)$ contains no path that passes a point. (This can be done by applying an appropriate linear tdr transduction to T—a class of tree transductions that perserve regularity—in order to turn such points into ordinary nodes.)

Now, assume that the terms in T have the mentioned form. Choose $\bar{\ }$, G, $\bar{G}$, G' as in the proof of Theorem 3, with the following changes: G, $\bar{G}$, and G' contain only hypergraphs all of whose hyperedges are passed by a path between points, all hypertrees in G' have type ≥ 2, and all hyperedges of type 1 are removed from $\bar{G}$. Then (T1)–(T5) can be verified as before, with only minor modifications to the proofs. Furthermore, every derivation of the

---

[2] $yTD_{fc(k)}$ is the set of yields of output languages of k-copying td transducers, represented as string graphs. The definition, which is chosen here because it allows for a concise presentation of the results, is easily seen to be equivalent to the usual one (cf. [ERS80]).

[3] The *degree* of a node is the sum of its in- and its out-degree.

resulting tdr transducer on an input term $t \in T$ is k-copying since, by the form of terms in T, every subterm t' of t satisfies $\left| E_{\overline{val(t')}} \right| \leq k$ (since the hypergraph $val(t')$ has type $2k + 1$ at most and no distinct $e, e' \in E_{\overline{val(t')}}$ share an attached node).

Therefore, as in Theorem 5(i)⇒(ii) one can obtain from tdr a k-copying td transducer td such that $out(td) = tdr(T)$ and thus $val(out(td)) = L$. Moreover, all operations in a term $t \in out(td)$ are hypertree operations of type ≥ 2, which in fact means that they are given by string graphs (as $val(t)$ is a string graph). In other words, $out(td) \subseteq T_{\Sigma_{str}}$, as required. □

Acknowledgement. I thank Joost Engelfriet for numerous valuable and interesting suggestions. Furthermore, I thank an anonymous referee for the improvements he/she proposed.

References

[AU71]   Alfred V. Aho and Jeffrey D. Ullman. Translations on a context free grammar. *Information and Control*, 19:439–475, 1971.

[Cou91] Bruno Courcelle. The monadic second-order logic of graphs V: on closing the gap between definability and recognizability. *Theoretical Computer Science*, 80:153–202, 1991.

[DHK97] Frank Drewes, Annegret Habel, and Hans-Jörg Kreowski. Hyperedge replacement graph grammars. In G. Rozenberg, editor, *Handbook of Graph Grammars and Computing by Graph Transformations. Vol. I: Foundations*, chapter 2, pages 95–162. World Scientific, 1997.

[Dre97] Frank Drewes. A characterization of the sets of hypertrees generated by hyperedge-replacement graph grammars. Report 3/97, Univ. Bremen, 1997.

[EH91]   Joost Engelfriet and Linda Heyker. The string generating power of context–free hypergraph grammars. *Journal of Computer and System Sciences*, 43:328–360, 1991.

[Eng77] Joost Engelfriet. Top-down tree transducers with regular look-ahead. *Mathematical Systems Theory*, 10:289–303, 1977.

[Eng97] Joost Engelfriet. Context-free graph grammars. In G. Rozenberg and A. Salomaa, editors, *Handbook of Formal Languages. Vol. III: Beyond Words*, chapter 3, pages 125–213. Springer, 1997.

[ERS80] Joost Engelfriet, Grzegorz Rozenberg, and Giora Slutzki. Tree transducers, L systems, and two-way machines. *Journal of Computer and System Sciences*, 20:150–202, 1980.

[GS97]   Ferenc Gécseg and Magnus Steinby. Tree languages. In G. Rozenberg and A. Salomaa, editors, *Handbook of Formal Languages. Vol. III: Beyond Words*, chapter 1, pages 1–68. Springer, 1997.

[Hab92] Annegret Habel. *Hyperedge Replacement: Grammars and Languages*, volume 643 of *Lecture Notes in Computer Science*. Springer-Verlag, 1992.

[MW67]   J. Mezei and Jesse B. Wright. Algebraic automata and context-free sets. *Information and Control*, 11:3–29, 1967.

Regulation by Valences

Henning Fernau[*1] and Ralf Stiebe[2]

[1] Wilhelm-Schickhard-Institut für Informatik, Universität Tübingen,
Sand 13, D-72076 Tübingen, Germany
email: fernau@informatik.uni-tuebingen.de
[2] Institut für Informatik, Martin-Luther-Universität Halle-Wittenberg,
Kurt-Mothes-Str. 1, D-06120 Halle, Germany
email: stiebe@informatik.uni-halle.de

Abstract. Valences are a very simple and yet powerful method of regulated rewriting. In this paper we give an overview on different aspects of this subject. We discuss closure properties of valence languages. It is shown that matrix grammars can be simulated by valence grammars over finite monoids. A Chomsky normal form theorem is proved for multiplicative valence grammars, thereby solving the open question of the existence of normal forms for unordered vector grammars. This also gives an alternative proof of the inclusion of context-free unordered vector languages in LOG(CFL). Moreover, we investigate valences in parallel systems, thereby solving part of open problems posted in [5, p. 267].

1 Introduction

Valences were introduced in 1980 by Păun [13] as a mechanism of regulated rewriting. The original idea was to assign to a context-free core rule an integer, the valence, and to compute for a derivation a value by adding the valences of the applied rule. A derivation is valid iff this sum evaluates to 0, reflecting the balance of positive and negative valences in chemical molecules or in directed graphs. This mechanism can be easily extended to monoids different from $(\mathbb{Z}, +, 0)$. Already in [13], multiplicative valence grammars (over $(\mathbb{Q}_+, \cdot, 1)$) have been discussed.

Valence languages have been in the focus of several papers [6, 10, 11, 14, 15, 22, 23]. We think for several reasons that valence grammars are worth a deeper investigation. First of all, valences are a very natural and simple mechanism. The context-free derivation process is not changed at all; the validity of a derivation is only checked at the end. Thus, many attractive properties of context-free grammars can immediately be transferred. Moreover, it is possible to describe several language families by valence grammars over different monoids, and one can hope to simplify investigations concerning these families by studying the respective valence grammars. For example, unordered vector languages can be characterized via valence languages over the monoid of positive rational numbers with multiplication, and Greibach's family *BLIND* of languages accepted by blind multicounter automata can be generated by regular valence grammars over $(\mathbb{Q}_+, \cdot, 1)$. Finally,

---

[*] Supported by Deutsche Forschungsgemeinschaft grant DFG La 618/3-1.

valences are very flexible and can be incorporated into parallel systems, grammars with other means of regulation and machine models (in fact, "finite valence automata" were discussed by several authors even before the introduction of valence grammars, e.g., in [7, 9] and in [16, Section 8]).

This paper discusses several of the mentioned aspects. After giving the necessary notations we discuss valence languages over arbitrary monoids. The known results regarding closure properties for specific monoids are extended and generalized, while the proofs are simplified. It is shown that valence grammars over finite monoids are equivalent to matrix grammars. We show how to construct normal forms for valence grammars over $(\mathbb{Q}_+, \cdot, 1)$, a result which also applies to the equivalent class of unordered vector grammars. Finally, we investigate several variants of valences in parallel systems. For reasons of space, we can only list further interesting connections: we showed that valence languages over the monoid of integers correspond to homomorphic images of vertex languages of edge grammars as introduced in [3]. Taking abelian monoids as valence regulators, we found algebraic characterizations of the corresponding language families in the spirit of [1, 2].

2   Preliminaries

Throughout the paper we assume the reader to be familiar with the theory of context-free languages, see e.g., [19].

Let $V = \{a_1, \ldots, a_n\}$, $n \geq 1$, be an alphabet. The set of all words over V is denoted by V^*, the empty word by λ, and $V^+ = V^* \setminus \{\lambda\}$. For $w \in V^*$, the length of w is denoted by $|w|$, the number of appearances of the letter $a \in V$ in w is denoted by $|w|_a$. The *Parikh mapping* associated with V is a map $\Psi : V^* \to \mathbb{N}^n$ such that $\Psi(w) = (|w|_{a_1}, \ldots, |w|_{a_n})$. For a language $L \subseteq V^*$, we define the *Parikh set of L* by $\Psi(L) = \{\Psi(w) : w \in L\}$. Two languages $L_1, L_2 \subseteq V^*$ are called *letter equivalent* iff their Parikh sets are equal. For a word w, let $\text{Perm}(w)$ denote the set of all words obtained by permuting the symbols of w. For a language L, we define $\text{Perm}(L) = \bigcup_{w \in L} \text{Perm}(w)$.

Let G be a grammar. For a derivation Δ applying the rules $p_1, p_2, \ldots, p_n$ (in this sequence), the control word of Δ is defined as $c(\Delta) = p_1 p_2 \cdots p_n$.

A *valence grammar* (over the monoid $\mathbf{M} = (M, \circ, e)$) is a construct $G = (N, T, P, S, \varphi, \mathbf{M})$, where $G' = (N, T, P, S)$ is a context-free grammar, and $\varphi : P \to M$ is the valence mapping (which can be extended to a monoid morphism from P^* to M). To avoid explicit reference to the monoid, we write $\text{Lab}(G)$ instead of $\varphi(P)$. P is called the set of core rules; a pair (p, m), $p = (A, \alpha)$, $m = \varphi(p)$ is called a valence rule and usually denoted by $(m : A \to \alpha)$. The yield relation $\Rightarrow$ over $(N \cup T)^* \times M$ is defined as: $(w, m) \Rightarrow (w', m')$ iff there is a rule $(n : A \to \alpha)$ such that $w = w_1 A w_2$ and $w' = w_1 \alpha w_2$ and $m' = m \circ n$. The language generated by G is $L(G) = \{w \in T^* : (S, e) \overset{*}{\Rightarrow} (w, e)\}$. A valence grammar is called regular if all its core rules are regular; a valence grammar is λ-free if it has no core rule of the form $A \to \lambda$. The families generated by regular, context-free λ-free, and context-free, respectively, valence grammars over $\mathbf{M}$, are denoted by $\mathcal{L}(\text{Val}, \text{REG}, \mathbf{M})$,

$\mathcal{L}(\text{Val}, \text{CF} - \lambda, \mathbf{M})$, $\mathcal{L}(\text{Val}, \text{CF}, \mathbf{M})$, respectively. For brevity, let $\mathbf{Z}^0$ denote the trivial monoid. Then, $\mathcal{L}(\text{Val}, X, \mathbf{Z}^0) = \mathcal{L}(X)$ for $X \in \{\text{REG}, \text{CF} - \lambda, \text{CF}\}$.

In terms of control words, a derivation in the underlying context-free grammar is valid in a valence grammar iff its control word is mapped by the valence morphism to the neutral element. Next, we define some other regulation mechanisms depending on control words, and hence related to valence grammars. Now, let $X \in \{\text{REG}, \text{CF} - \lambda, \text{CF}\}$. A *matrix grammar* is a quintuple $G = (N, T, P, S, M)$ where $G' = (N, T, P, S)$ is a type-X grammar, and $M \subset P^*$ is a finite set of so-called matrices. A terminal derivation Δ in G' is valid in G iff $c(\Delta) \in M^*$. $L(G)$ consists of all words obtained by valid derivations. An *unordered vector grammar* is defined like a matrix grammar, with the difference that a terminal derivation Δ in G' is valid in G iff $c(\Delta) \in \text{Perm}(M^*)$. The families of unordered vector languages of type X are denoted by $\mathcal{L}(\text{UV}, X)$. A *grammar with regular control* is a quintuple $G = (N, T, P, S, R)$, where (N, T, P, S) is a type-X grammar, and R is a regular subset of P^*. A terminal derivation Δ is valid iff $c(\Delta) \in R$. Grammars with regular control are equivalent to matrix grammars [5]. The corresponding language classes are denoted by $\mathcal{L}(\text{rC}, X)$.

In analogy to valence grammars, one can define valence transducers, where a valence is assigned to each element of the valence relation, and a transduction is valid iff the valence product evaluates to the neutral element.

We close this section with some conventions used in this paper. The inclusion relation is denoted by $\subseteq$, proper inclusion by $\subset$. If $\mathbf{M} = (M, \circ, e)$ is a monoid and $M' \subseteq M$, the monoid generated by M' is denoted by $\langle M' \rangle$. The monoids $(\mathbf{Z}^k, +, \mathbf{0})$ and $(\mathbb{Q}_+, \cdot, 1)$ are sometimes simply denoted by $\mathbf{Z}^k$ and $\mathbb{Q}_+$.

For a vector $\mathbf{r} = (r_1, r_2, \ldots, r_k) \in \mathbf{Z}^k$, we define the max-norm by $\|\mathbf{r}\|_{\max} = \max\{|r_i| : 1 \le i \le k\}$ and the 1-norm by $\|\mathbf{r}\|_1 = \sum_{i=1}^k |r_i|$. We define the modulo and integer division operations for vectors, $\text{mod}, \text{div} : \mathbf{Z}^k \times \mathbf{Z} \to \mathbf{Z}^k$, as the componentwise application of the integer operations $\text{mod}, \text{div} : \mathbf{Z} \times \mathbf{Z} \to \mathbf{Z}$, and denote them by $\mathbf{r} \bmod m$ and $\mathbf{r} \operatorname{div} m$, for $\mathbf{r} \in \mathbf{Z}^k, m \in \mathbf{Z}$.

3 Valence grammars over monoids

Up to now, valence grammars have been investigated only for a few specific groups. In this section, we are first going to prove some general closure properties of classes of valence languages over arbitrary monoids. Later we discuss valence grammars over particular monoids: the monoid $(\mathbb{Q}_+, \cdot, 1)$, abelian monoids and finite monoids.

In short, valence language classes form semi-AFLs, i.e., they are closed under union and finite transducer mappings, and this fact can be shown quite generally, not requiring a separate proof for each monoid, see [11, 23].

Theorem 1. *For each monoid* $\mathbf{M}$ *and each* $X \in \{\text{REG}, \text{CF} - \lambda, \text{CF}\}$, *the class* $\mathcal{L}(\text{Val}, X, \mathbf{M})$ *is a semi-AFL. Moreover,* $\mathcal{L}(\text{Val}, X, \mathbf{M})$ *is closed under mirror image and substitution by* $\mathcal{L}(X)$*-languages.*

Proof. The triple constructions for closure under transducer mappings of the standard families can be adapted for the families of valence languages. Thus

$\mathcal{L}(\text{Val}, X, \mathbf{M})$ is closed under λ-free tranductions for $X = \text{CF} - \lambda$, and closed under arbitrary transducer mappings for $X \in \{\text{REG}, \text{CF}\}$. As regards the other properties, the standard constructions known from the theory of context-free languages can easily be carried over. Analogous results for finite automata with valences can be found in [12]. □

The classes of valence languages over $\mathbb{Q}_+$ are, additionally, closed under catenation. Another interesting closure property is:

Theorem 2. *The families* $\mathcal{L}(\text{Val}, X, \mathbb{Q}_+)$, $X \in \{\text{REG}, \text{CF}\}$, *are closed under valence transductions over* $\mathbb{Q}_+$.

Proof. Note that, for a valence grammar (transducer) over $\mathbb{Q}_+$, there exists an equivalent valence grammar (transducer) over some $\mathbb{Z}^k$, and vice versa. Let G be a valence grammar over $(\mathbb{Z}^k, +, \mathbf{0})$ and $\mathcal{A}$ be a valence transducer over $(\mathbb{Z}^\ell, +, \mathbf{0})$. Again, the triple construction for the classical language families can be modified to obtain a valence grammar H over $(\mathbb{Z}^{k+\ell}, +, \mathbf{0})$ such that $L(H) = T_{\mathcal{A}}(L(G))$. Note that the commutativity of addition is essential for the construction. □

Corollary 3. *Each of the families* $\mathcal{L}(\text{Val}, X, \mathbb{Q}_+)$, $X \in \{\text{REG}, \text{CF}\}$, *is closed under permutation and under intersection by languages from* $\mathcal{L}(\text{Val}, \text{REG}, \mathbb{Q}_+)$.

Proof. It can easily be shown that the mentioned operations are λ-free valence transductions. This generalizes the older result that *BLIND* is closed under intersection [7]. □

Before discussing valences over specific monoids, we mention two simple but useful observations. Let $\mathbf{M} = (M, \circ, e)$ be a monoid. An element $m \in M$ is called *completable* iff there are $m_1, m_2 \in M$ such that $m_1 \cdot m \cdot m_2 = e$. Obviously, only rules with a completable valence can be applied in a valid derivation. Therefore, we can restrict to grammars G where $\text{Lab}(G)$ contains only completable elements. Moreover, any valence grammar G over $\mathbf{M}$ is a valence grammar over the finitely generated monoid $\langle \text{Lab}(G) \rangle$. Hence, $\mathcal{L}(\text{Val}, X, \mathbf{M}) = \bigcup_{\mathbf{M}' \in \mathcal{F}(\mathbf{M})} \mathcal{L}(\text{Val}, X, \mathbf{M}')$, where the set of finitely generated submonoids of $\mathbf{M}$ is denoted by $\mathcal{F}(\mathbf{M})$.

Next we are going to show that the generative power of valence grammars is not enlarged when admitting arbitrary abelian valence monoids instead of $(\mathbb{Q}_+, \cdot, 1)$.

Theorem 4. *Let* $\mathbf{M}$ *be an abelian monoid and* $X \in \{\text{REG}, \text{CF} - \lambda, \text{CF}\}$. *Then* $\mathcal{L}(\text{Val}, X, \mathbf{M})$ *equals either* $\mathcal{L}(\text{Val}, X, \mathbb{Q}_+)$ *or* $\mathcal{L}(\text{Val}, X, \mathbb{Z}^k)$, *for some* $k \geq 0$.

Proof. Let G be a valence grammar over the abelian monoid $\mathbf{M}$ such that all elements of $\text{Lab}(G)$ are completable. Note that in an abelian monoid an element is completable iff it has an inverse. Hence, $\langle \text{Lab}(G) \cup \{m' : m \circ m' = e, m \in \text{Lab}(G)\} \rangle$ is a finitely generated abelian group and G can be considered as a valence grammar over this group. By the well-known fundamental theorem for finitely generated abelian groups, see e.g., [17], $\langle \text{Lab}(G) \rangle$ is either finite or isomorphic to some $(\mathbb{Z}^k, +, \mathbf{0})$, $k \geq 1$. It is easy to show that valences over finite abelian groups do not increase the power of the basic model. □

As mentioned in the definitions, regulation by valences is very similar to regulation by matrices and unordered vectors. This is stressed by proving that valence grammars over finite monoids are as powerful as matrix grammars.

Theorem 5. *For the class of finite monoids* **FM** *and* $X \in \{\text{REG}, \text{CF} - \lambda, \text{CF}\}$, $\bigcup_{\mathbf{M} \in \mathbf{FM}} \mathcal{L}(\text{Val}, X, \mathbf{M}) = \mathcal{L}(\text{rC}, X)$. *So,* $\bigcup_{\mathbf{M} \in \mathbf{FM}} \mathcal{L}(\text{Val}, \text{REG}, \mathbf{M}) = \mathcal{L}(\text{REG})$.

Proof. The proof is based on the equivalence of matrix grammars and grammars with regular control [5]. Let $\mathbf{M} = (M, \circ, e)$ be some finite monoid, and let $G = (N, T, P, S, \varphi, \mathbf{M})$ be a valence grammar over $\mathbf{M}$. One can easily construct a finite automaton $\mathcal{A}_{\mathbf{M}}$ accepting $m_1 m_2 \ldots m_k \in M$ iff $m_1 \circ m_2 \circ \ldots \circ m_k = e$, and starting with $\mathcal{A}_{\mathbf{M}}$, a finite automaton $\mathcal{A}_G$ accepting $\pi \in P^*$ iff $\varphi(\pi) = e$. The equivalent grammar with regular control is $G' = (N, T, P, S, L(\mathcal{A}_G))$.

On the other hand, let $G = (N, T, M, S)$ be a matrix grammar in 2-normal form, cf. [5]. This means, especially, that terminal words can only be derived after a complete application of a matrix. So, there is an equivalent grammar $G' = (N, T, P, S, R)$ with regular control such that $\pi \in P^*$ is in R iff it is syntactically equivalent to λ (i.e., iff R is accepted by a deterministic finite automaton whose only accepting state is the initial state). Hence, there is a straightforward construction of an equivalent valence grammar over the syntactic monoid of R. It is known that regular matrix grammars only generate regular languages [5]. □

4 Normal forms for valence grammars

Since the introduction of unordered vector grammars, it has been an open question whether there are normal form representations. Sudborough [21] showed that, in analogy to the context-free case, the membership problem can be efficiently solved for unordered vector grammars without erasing and unit productions, more exactly, it was shown to be in LOG(CFL). Satta [20] could prove that complexity result for arbitrary unordered vector grammars, while leaving open the problem of the existence of normal forms. Our result yields an alternative proof for the relation $\mathcal{L}(\text{Val}, \text{CF}, \mathbb{Q}_+) \subseteq \text{LOG(CFL)}$. In what follows, we shall outline the construction of Chomsky normal forms for valence grammars over $\mathbb{Z}^k$, $k \geq 1$. The general idea is the same as in the classical case. i.e., first λ-rules, then unit core rules are eliminated. Due to valences, the construction gets more involved compared to the plain context-free case. In addition, the valence vectors are normalized.

Theorem 6. *For any valence grammar over* $\mathbb{Z}^k$, *there is an equivalent valence grammar over* $\mathbb{Z}^k$ *with core rules of the form* $A \rightarrow BC$ *or* $A \rightarrow a$, *and valence vectors with 1-norms not greater than 1.*

Proof (Sketch). Let $G_0 = (N_0, T, P_0, S_0, \varphi_0, \mathbb{Z}^k)$ be a valence grammar. We shall construct a sequence of equivalent valence grammars $G_i = (N_i, T, P_i, S_i, \varphi_i, \mathbb{Z}^k)$. In a first step, the core rules can be simplified such that an equivalent grammar G_1 with core rules of the forms $A \rightarrow BC$, $A \rightarrow B$, $A \rightarrow a$, $A \rightarrow \lambda$ is obtained. Moreover, N_1 can be partitioned into two disjoint sets N_1' and N_1'', where symbols from N'_1 generate only nonempty strings, while the only terminal string derivable from letters in N''_1 is λ. For all $B \in N''_1$, let $E(B) = \{\mathbf{r} \mid (B, 0) \overset{*}{\Rightarrow}_{G_1} (\lambda, \mathbf{r})\}$. One can find a *regular* valence grammar $G_B = (N_B, \emptyset, P_B, S_B, \varphi_i, \mathbb{Z}^k)$ such that $(B, 0) \overset{*}{\Rightarrow}_{G_B} (\lambda, \mathbf{r})$ iff $\mathbf{r} \in E(B)$. This construction is based on the existence of a letter equivalent regular language for every context-free language. One can choose N_B and $N_{B'}$ to be disjoint for $B \neq B'$.

Now in G_1 the rules envolving symbols from N''_1 are removed. For any rule $(\mathbf{r} : A \to BC)$ or $(\mathbf{r} : A \to CB)$, $A, C \in N'_1$, $B \in N''_1$, the rule $(\mathbf{r} : A \to (S_B, C))$ is introduced. For any rule $(\mathbf{r} : X_B \to Y_B)$, $(\mathbf{r} : X_B \to \lambda)$, respectively, in G_B, we introduce $(\mathbf{r} : (X_B, C) \to (Y_B, C))$, $(\mathbf{r} : (X_B, C) \to C)$, respectively. The thus obtained grammar G_2 has core rules of the forms $A \to BC$, $A \to B$, $A \to a$. Note that a left-derivation in G_1, $(A, 0) \Rightarrow (BC, \mathbf{q}) \overset{*}{\Rightarrow} (C, \mathbf{q}+\mathbf{r})$, $A, C \in N'$, $B \in N''$, $\mathbf{r} \in E(B)$, can be simulated in G_2 by the derivation $(A, 0) \Rightarrow ((S_B, C), \mathbf{q}) \overset{*}{\Rightarrow} (C, \mathbf{q}+\mathbf{r})$.

The next goal is to remove core rules of the form $A \to B$. For all $A \in N_2$, let $LOOP(A) = \{\mathbf{r} : (A, 0) \overset{*}{\Rightarrow} (A, \mathbf{r})\}$, for $M \subseteq N_2$, define $LOOP(M) = \{\sum_{A \in M} \mathbf{r}_A : \mathbf{r}_A \in LOOP(A)\}$. A left-derivation is loop-free if it has no subderivation $(\alpha, \mathbf{q}) \overset{*}{\Rightarrow} (\alpha, \mathbf{q} + \mathbf{r})$. It can be shown that, for any derivation $(S, 0) \overset{*}{\Rightarrow} (w, 0)$, there is a loop-free left-derivation $\Delta : (S, 0) \overset{*}{\Rightarrow} (w, -\mathbf{r})$ with $\mathbf{r} \in LOOP(M)$, where M is the set of nonterminals appearing in Δ.

For any $M \subseteq N_2$, one can construct a valence grammar G_M with core rules of the forms $A \to BC$, $A \to a$, such that a pair $(w, -\mathbf{r})$ is generated in G_M iff there is a loop-free derivation in G_2 in which all symbols of M appear. It can also be found a *regular* valence grammar H_M generating $(w, \mathbf{r})$ iff $w = \lambda$ and $\mathbf{r} \in LOOP(M)$. Moreover, it can be guaranteed that the derivation of $(\lambda, \mathbf{r})$ in H_M does not take more steps than the derivation of $(w, -\mathbf{r})$ in G_M.

By the last remark, it is possible to "incorporate" the rules of H_M into the rules of G_M. The core rules of the obtained grammar K_M have the forms $A \to BC$, $A \to a$. A pair $(w, 0)$ is derivable in K_M iff, for some $\mathbf{r}$, $(w, -\mathbf{r})$ and $(\lambda, \mathbf{r})$ are derivable in G_M and K_M. It is easy to construct, preserving the shapes of the rules, a valence grammar G_3 with $L(G_3) = \bigcup_{M \subseteq N_2} L(K_M)$, and by the above arguments, $L(G_3) = L(G_2)$.

It remains to normalize the valence vectors. In a first step, if $k \geq 2$, we construct from G_3 the valence grammar G_4 with core rules of the forms $(\mathbf{r} : A \to B_1 B_2 \ldots B_\ell)$, $\ell > k$ and $(0 : A \to a)$. Let $m = \max\{\|\mathbf{r}\|_{\max} : \mathbf{r} \in \text{Lab}(G_4)\}$.

The next valence grammar G_5 is obtained by setting $N_5 = \{0, \ldots, m-1\}^k \times N_4 \times \{0, \ldots, m-1\}^k$, $S_5 = (0, S_4, 0)$, each valence rule $(\mathbf{r} : A \to B_1 \ldots B_\ell)$, $\ell \geq k$, is replaced by the set of all rules $(\mathbf{s} : (\mathbf{r}_0, A, \mathbf{r}_{\ell+1}) \to (\mathbf{r}_1, B_1, \mathbf{r}_2) \ldots (\mathbf{r}_\ell, B_\ell, \mathbf{r}_{\ell+1}))$, where $\mathbf{r}_1 = (\mathbf{r}_0 + \mathbf{r}) \bmod m$ and $\mathbf{s} = (\mathbf{r}_0 + \mathbf{r}) \operatorname{div} m$, each valence rule $(0 : A \to a)$, is replaced by the set of all rules $(0 : (\mathbf{r}, A, \mathbf{r}) \to a)$. Finally, from G_5 we can easily construct a grammar with core rules of the forms $A \to a$, $A \to BC$ and unit valence vectors. $\qquad\square$

Note that the proof mainly depends on the commutativity of the monoid $(\mathbb{Z}^k, +, 0)$. Thus it cannot be modified for valence grammars over, e.g., finite monoids (and hence for matrix grammars). This remains a challenging open question.

In [5] it is shown that, for any valence grammar with unit or zero valence vectors, there is an equivalent unordered vector grammar with the same set of core rules. So, we have the following consequences via [21, 22]:

Corollary 7. *For any unordered vector grammar, there is an equivalent unordered vector grammar with core rules of the form $A \to BC$ or $A \to a$.* □

Corollary 8. *Let $k \geq 0$. Then, we have: $\mathcal{L}(\text{Val}, \text{CF} - \lambda, \mathbb{Z}^k) = \mathcal{L}(\text{Val}, \text{CF}, \mathbb{Z}^k) \subset \mathcal{L}(\text{Val}, \text{CF} - \lambda, \mathbb{Z}^{k+1}) = \mathcal{L}(\text{Val}, \text{CF}, \mathbb{Z}^{k+1}) \subset \mathcal{L}(\text{Val}, \text{CF} - \lambda, \mathbb{Q}_+) = \mathcal{L}(\text{Val}, \text{CF}, \mathbb{Q}_+) = \mathcal{L}(\text{UV}, \text{CF} - \lambda) = \mathcal{L}(\text{UV}, \text{CF}) \subset \text{LOG}(\text{CFL})$.* □

Furthermore, let us mention that it is possible to extend the construction in order to obtain a Greibach normal form result for these language classes. This immediately yields a simple machine characterization of the corresponding classes via pushdown machines endowed with a certain number of blind counters.

5 Valences in parallel systems

Valences can also be attached to systems with parallel derivations. There are several modes to be discussed. In the plain mode, the validity of a derivation is checked in the end, while in the test mode the valence product has to be the neutral element after each parallel step. Finally, in table systems, the valences are attached to the tables (instead to individual rules).

We assume that the reader is familiar with the basics of Lindenmayer systems, see e.g., [18]. We use the common notation for variants of Lindenmayer systems, such as 0L, E0L, D0L, T0L, P0L, ET0L, and so on. For a derivation in an ET0L system $G = (\Sigma, \Delta, \mathbf{P}, \omega)$, the control word over $\mathbf{P}$ is defined as the sequence of applied tables; the Szilard language is the set of all control words of terminal derivations.

A valence ET0L system (over $\mathbb{Q}_+$), $G = (\Sigma, \Delta, \mathbf{P}, \omega)$ is defined similar to an ET0L system with the difference that the tables $P \in \mathbf{P}$ are finite sets of valence rules, i.e., $P \subset \Sigma \times \Sigma^* \times \mathbb{Q}_+$. As usual a valence rule (a, w, r) is denoted by $(r : a \to w)$. We say that the pair $(w, r) \in \Sigma^* \times \mathbb{Q}_+$ directly derives (w', r') in *plain mode* and write $(w, r) \Rightarrow_P (w', r')$ iff there is some $P \in \mathbf{P}$ such that $w = a_1 a_2 \ldots a_n$, $w' = (w_1 w_2 \ldots w_n)$, $r' = r_1 \cdot r_2 \cdot \ldots \cdot r_n \cdot r$, and $(r_i : a_i \to w_i) \in P$, for $1 \leq i \leq n$. The language generated by G in plain mode is $L(G) = \{w \in \Delta^* : (\omega, 1) \overset{*}{\Rightarrow}_\pi (w, 1), \pi \in P^*\}$. We say that the word $w \in \Sigma^*$ directly derives $w' \in \Sigma^*$ in *test mode* and write $w \Rightarrow_t w'$ iff there is some $P \in \mathbf{P}$ such that $w = a_1 a_2 \ldots a_n$, $w' = w_1 w_2 \ldots w_n$, $r_1 \cdot r_2 \cdot \ldots \cdot r_n = 1$, and $(r_i : a_i \to w_i) \in P$, for $1 \leq i \leq n$. The language generated by G in test mode is $L_t(G) = \{w \in \Delta^* : \omega \overset{*}{\Rightarrow}_t w\}$.

Finally, an *ET0L system with table valences* is a quintuple $G = (\Sigma, \Delta, \mathbf{P}, \omega, \varphi)$, where $G' = (\Sigma, \Delta, \mathbf{P}, \omega)$ is an ET0L system and φ is a mapping from $\mathbf{P}$ into $\mathbb{Q}_+$. We say that the pair $(w, r) \in \Sigma^* \times \mathbb{Q}_+$ directly derives (w', r') and write $(w, r) \Rightarrow_P (w', r')$ iff there is some $P \in \mathbf{P}$ such that $w = a_1 a_2 \ldots a_n$, $w' = (w_1 w_2 \ldots w_n)$, $r' = \varphi(P) \cdot r$, and $a_i \to w_i \in P$, for $1 \leq i \leq n$. The language generated by G is $L(G) = \{w \in \Delta^* : (\omega, 1) \overset{*}{\Rightarrow} (w, 1)\}$.

The families of languages generated by valence ET0L systems in plain mode, in test mode, and by ET0L systems with table valences, respectively, are denoted by $\mathcal{L}(\text{Val}, \text{ET0L})$, $\mathcal{L}(\text{Val}_t, \text{ET0L})$, and $\mathcal{L}(\text{Tab-Val}, \text{ET0L})$, respectively.

Theorem 9.

$$\mathcal{L}(\text{ET0L}) \subset \mathcal{L}(\text{Tab-Val}, \text{ET0L}) \subseteq \mathcal{L}(\text{Val}, \text{ET0L}) \subseteq \mathcal{L}(\text{Val}_t, \text{ET0L}) = \mathcal{L}(\text{RE}).$$

We conjecture that the second inclusion is also proper, but have no proof for it.

Proof. The chain of inclusions $\mathcal{L}(\text{ET0L}) \subset \mathcal{L}(\text{Tab-Val}, \text{ET0L}) \subseteq \mathcal{L}(\text{Val}, \text{ET0L})$ is easily obtained. The language $\{(a^m b)^n : m \geq n \geq 1\}$ is a witness for the properness of the first inclusion. Namely, consider the ET0L system with table valences $G = (\Sigma, \Delta, P, \omega, \varphi)$ where $\Sigma = \{a, b, A, B, C_1, C_2, X\}$, $\Delta = \{a, b\}$, $P = \{P_1, P_1', P_2, P_3\}$ with $P_1 = \{B \rightarrow B, C_1 \rightarrow BC_1\} \cup \{Y \rightarrow X : Y \in \{a, b, A, C_2, X\}\}$, $P_1' = P_1$, $P_2 = \{B \rightarrow AB, A \rightarrow A, C_1 \rightarrow C_2, C_2 \rightarrow C_2\} \cup \{Y \rightarrow X : Y \in \{a, b, X\}\}$, $P_3 = \{A \rightarrow a, B \rightarrow b, C_1 \rightarrow \lambda, C_2 \rightarrow \lambda\} \cup \{Y \rightarrow X : Y \in \{a, b, X\}\}$, $\omega = C_1$, $\varphi(P_1) = 2$, $\varphi(P_1') = 1$, $\varphi(P_2) = 1/2$, $\varphi(P_3) = 1$. The inclusion $\mathcal{L}(\text{RE}) \subseteq \mathcal{L}(\text{Val}_t, \text{ET0L})$ is shown by simulating a programmed grammar with appearance checking by a valence ET0L system in test mode. □

As regards closure properties, $\mathcal{L}(\text{Val}_t, \text{ET0L})$ is by the above theorem a full AFL. Similar to the case of context-free valence grammars, the other variants of valence ET0L languages are semi-AFL's. (All proofs known from the theory of ET0L systems, see [18], can be adapted to the valence variants.)

Theorem 10. $\mathcal{L}(\text{Tab-Val}, \text{ET0L})$ *and* $\mathcal{L}(\text{Val}, \text{ET0L})$ *are catenation-closed semi-AFLs which are, in addition, closed under substitution by ET0L languages and mirror image. Moreover,* $\mathcal{L}(\text{Val}, \text{ET0L})$ *is closed under valence transductions, in particular under permutations and intersection by regular valence languages.* □

Note that the construction for closure under valence transductions works only if valences are attached to the productions, and therefore not for table valence systems.

(Valence) ET0L systems are the most general case of (valence) Lindenmayer systems. It is therefore natural to discuss the existence of normal forms (i.e., the comparision with propagating, nonextended, deterministic, ..., variants). We shall give here a few results.

Theorem 11. *The families* $\mathcal{L}(\text{Val}_t, \text{T0L})$ *and* $\mathcal{L}(\text{Val}, \text{T0L})$, *respectively, are strictly included in* $\mathcal{L}(\text{Val}_t, \text{ET0L})$ *and* $\mathcal{L}(\text{Val}, \text{ET0L})$, *respectively.*
For $L \in \mathcal{L}(\text{Val}, \text{ET0L})$ $L \subseteq \Delta^*, S \notin \Delta, L \cup \{S\} \in \mathcal{L}(\text{Val}, \text{T0L})$.

Proof. Let $G = (\Sigma, \Delta, \mathbf{P}, \omega)$ be a valence ET0L system. By introducing a new symbol S as axiom and a prime number q not appearing in the factorization of some number in $\text{Lab}(G)$, it is possible to keep track of the number of nonterminals in the derived sentential form by the power of q in the factorization of the valence product. Hence, the product 1 in a derivation in the T0L system (with alphabet

Σ) can only be computed if the sentential form contains only symbols from Δ. The language of the T0L system is $L(G) \cup \{S\}$.

The strict inclusions can be shown by the example $\{a^{2^n} : n \geq 0\} \cup \{a^{3^n} : n \geq 0\}$ which is in $\mathcal{L}(\text{ET0L})$ but neither in $\mathcal{L}(\text{Val}_t, \text{T0L})$ nor $\mathcal{L}(\text{Val}, \text{T0L})$. $\quad\square$

Theorem 12. *We have* $\mathcal{L}(\text{Val}_t, \text{EPT0L}) \subset \mathcal{L}(\text{Val}_t, \text{ET0L})$, *while on the other hand,* $\mathcal{L}(\text{Tab-Val}, \text{EPT0L}) = \mathcal{L}(\text{Tab-Val}, \text{ET0L})$.

Proof. The strictness of the inclusion in the valence test case follows from the equivalence of (propagating) valence test languages and (λ-free) matrix grammars with appearance checking. To show equivalence in the case of table valence families, one can simply adapt the proof for classical L systems. $\quad\square$

Theorem 13. *The membership, the emptiness, and the finiteness problems are decidable for table valence ET0L systems.*

Proof. By the previous theorem, we can restrict to EPT0L table valence systems. For such systems, the membership problem is immediately reduced to the membership problem for monotone valence grammars over $\mathbb{Q}_+$, which is decidable.

To show decidability of the emptiness and finiteness problems we can, similar to the case without valences [18], assign to an EPT0L table valence system G an EDPT0L table valence system H, the so called combinatorially complete version. It can be shown that $L(G)$ is empty (finite) iff $L(H)$ is empty (finite). The problems with respect to H can be reduced to the corresponding problems for the Szilard language of H, which is a regular valence language over $\mathbb{Q}_+$. $\quad\square$

6 Conclusions

We have given an overview of the potentials of valence grammars. Many problems remain open. It would, for instance, be very interesting to investigate valence grammars over other specific monoids than discussed here in order to describe different language classes. Regarding parallel systems, decision questions for valence systems in plain mode are completely open, as well as questions with respect to determinism. (Of course, one has to be aware of the close equivalences to *k-BLIND* and reversal-bounded multicounter machines and their deterministic variants, see [7, 8].) The latter question could be very interesting keeping in mind the correspondence of Chomsky normal form grammars and machine models, since we can hope for efficient LR-style parsing algorithms for valence grammars. Valences can such be used as attributes which can be easily handled.

It should be mentioned that similar results as given in this paper for ET0L systems with valences hold also for systems with limited parallelism. Moreover, there are close connections to regulated valence grammars, which have been also investigated by us, hence answering questions raised in [5, p. 267]. An interesting special case of valence grammars are grammars with valuations, where values are assigned to terminals. We have identified such languages as homomorphic pre-images of special groupoids.

Finally, we would like our colleagues, especially H. Bordihn, M. Holzer, and K. Reinhardt, for stimulating discussions.

References

1. D. A. M. Barrington and D. Thérien. Finite monoids and the fine structure of NC^1. *Journal of the Association for Computing Machinery*, 35(4):941–952, October 1988.

2. F. Bédard, F. Lémieux, and P. McKenzie. Extensions to Barrington's M-program model. *Theoretical Computer Science*, 107:31–61, 1993.

3. F. Berman. Edge Grammars and parallel computation. In: *Proceedings of the 1983 Allerton Conference*, pages 214–223, 1983.

4. J. Berstel. *Transductions and Context-Free Languages*. Stuttgart: Teubner, 1979.

5. J. Dassow and Gh. Păun. *Regulated Rewriting in Formal Language Theory*. Berlin: Springer, 1989.

6. M. Gheorge. Linear valence grammars. In: *International Meeting of Young Computer Scientists*, pages 281–285, 1986.

7. S. Greibach. Remarks on blind and partially blind one-way multicounter machines. *Theoretical Computer Science*, 7:311–324, 1978.

8. E. M. Gurari and O. H. Ibarra. The complexity of decision problems for finite-turn multicounter machines. *Journal of Computer and System Sciences*, 22:220–229, 1981.

9. O. H. Ibarra, S. K. Sahni, and C. E. Kim. Finite automata with multiplication. *Theoretical Computer Science*, 2:271–296, 1976.

10. M. Marcus and Gh. Păun. Valence gsm-mappings. *Bull. Math. Soc. Sci. Math. Roumanie*, 31(3):219–229, 1987.

11. V. Mitrana. Valence grammars on a free generated group. *EATCS Bulletin*, 47:174–179, 1992.

12. V. Mitrana and R. Stiebe. Extended finite automata over groups. Accepted for *First International Conference on Semigroups and Algebraic Engeneering*, Aizu, Japan: 1997.

13. Gh. Păun. A new generative device: valence grammars. *Rev. Roumaine Math. Pures Appl.*, XXV(6):911–924, 1980.

14. Gh. Păun. On a class of valence grammars. *Stud. cerc. mat.*, 42:255–268, 1990.

15. Gh. Păun. Valences: increasing the power of grammars, transducers, grammar systems. *EATCS Bulletin*, 48:143–156, 1992.

16. V. Red'ko and L. Lisovik. Regular events in semigroup. *Problems of Cybernetics*, 37:155–184, 1980. In Russian.

17. J. J. Rotman. *An Introduction to the Theory of Groups*. New York: Springer, 5th edition, 1995.

18. G. Rozenberg and A. K. Salomaa. *The Mathematical Theory of L Systems*. Academic Press, 1980.

19. A. K. Salomaa. *Formal Languages*. Academic Press, 1973.

20. G. Satta. The membership problem for unordered vector grammars. In J. Dassow, G. Rozenberg, and A. Salomaa, editors, *Developments in Language Theory II*, pages 267–275, 1996.

21. I. H. Sudborough. The complexity of the membership problem for some extensions of context-free languages. *International Journal of Computer Mathematics*, 6:191–215, 1977.

22. S. Vicolov. Hierarchies of valence languages. In J. Dassow and A. Kelemenova, editors, *Developments in Theoretical Computer Science*, pages 191–196. Basel: Gordon and Breach, 1994.

23. S. Vicolov-Dumitrescu. Grammars, grammar systems, and gsm mappings with valences. In Gh. Păun, editor, *Mathematical Aspects of Natural and Formal Languages*, pages 473–491. Singapore: World Scientific, 1994.

Simulation as a Correct Transformation of Rewrite Systems

Wan Fokkink

University of Wales Swansea
Department of Computer Science
Singleton Park, Swansea SA2 8PP, Wales
e-mail: w.j.fokkink@swan.ac.uk

Jaco van de Pol

Eindhoven University of Technology
Department of Computer Science
PO Box 513, 5600 MB Eindhoven, The Netherlands
e-mail: jaco@win.tue.nl

Abstract. Kamperman and Walters proposed the notion of a *simulation* of one rewrite system by another one, whereby each term of the simulating rewrite system is related to a term in the original rewrite system. In this paper it is shown that if such a simulation is *sound* and *complete* and *preserves termination*, then the transformation of the original into the simulating rewrite system constitutes a correct step in the compilation of the original rewrite system. That is, the normal forms of a term in the original rewrite system can then be obtained by computing the normal forms of a related term in the simulating rewrite system.

1 Introduction

Questions on the correctness of compilation of programming languages date back to McCarthy [12]. In this paper we present a technique to deduce the correctness of compilation steps for functional programming languages which stay inside the domain of rewrite systems.

Quite a number of papers deal with particular examples of transformations of rewrite systems, usually with the aim to obtain a rewrite system which satisfies some desirable property, e.g. [17, 10, 18, 16, 19, 22, 7]. In most of these papers, correctness of the transformation is stated, meaning that the original and the transformed rewrite system are in some sense 'equivalent'. This claim is based on the observation either that desirable properties such as confluence and termination are preserved by the transformation, or that the transformed rewrite system can somehow simulate the original rewrite system, so that the reduction graph of an original and a simulating term have the same structure.

Recently, Kamperman and Walters [7, 8, 6] proposed a notion of simulation of one rewrite system by another rewrite system. A simulation basically consists of a surjective mapping ϕ which relates each term in the simulating rewrite system to a term in the original rewrite system. A simulation should be *sound*, meaning that if a term t can be rewritten to a term t' in one step in the simulating rewrite system, then $\phi(t)$ can be rewritten to $\phi(t')$ in zero or more steps in the original rewrite system. Furthermore, a simulation should be *complete*, meaning that if a term $\phi(t)$ can be rewritten to a term s in one step in the original rewrite system, then t can be rewritten to t' with $\phi(t') = s$ in one or more steps in the simulating rewrite system. Finally, a simulation should *preserve termination*, meaning that if the original rewrite system is terminating for a term $\phi(t)$, then the simulating rewrite system should be terminating for t. (The other way around is guaranteed by completeness.)

Kamperman and Walters apply simulation to transform a left-linear rewrite system into a form which is more suitable for compilation, as a first step in the implementation of their equational programming language EPIC [20, 21]. Kamperman and Walters state, for example in the title of [8], that if a simulation is sound and complete and preserves termination, then it constitutes a correct transformation of rewrite systems. However, they do not yet provide a foundation for this claim. At first sight, the link between the original and the simulating rewrite system is unclear. For example, in general the syntax of the original and of the simulating rewrite system differ. Furthermore, the original rewrite system may be confluent, while the simulating one is not. Hence, the question arises what it means to state that such a transformation of rewrite systems is 'correct'.

Although preservation of reduction graphs underlies simulation, reduction graphs are usually not of interest in applications of rewrite systems. Especially if a rewrite system is used to implement a functional language, then one is solely interested in the input/output behaviour of the system, where the input is any term, and the output is (one of) its normal form(s). So if a rewrite system is transformed as part of a compilation project, then the main interest is that the transformation preserves normal forms. We propose the notion of a correct transformation of rewrite systems, based on ideas on compiler correctness by Burstall and Landin [2] and Morris [13]. We say that the transformation of one rewrite system into another is correct if no information on normal forms in the original rewrite system is lost. That is, it should be possible to provide mappings *parse* from original to transformed terms and *print* from transformed to original terms such that for each original term t its normal forms can be computed as follows: compute the normal forms of $parse(t)$, and apply the *print* function to them. In order to make sure that the simulating rewrite system returns an answer whenever the original rewrite system does, it is required that a correct transformation also preserves termination properties.

We show that the notion of a simulation as proposed in [7, 8, 6] constitutes a correct transformation, under the conditions that it is sound and complete and preserves termination. Hence, the notion of simulation constitutes a useful tool for proving correctness of compilation of rewrite systems. Namely, such a compilation may involve a chain of transformations of rewrite systems. In order to prove correctness of one such transformation, it is sufficient to find a simulation relation that is sound, complete and termination preserving. The intuitive link between the two rewrite systems before and after a transformation can often be materialized in an explicit simulation relation. Simulation and its properties soundness, completeness and termination preservation are all conserved under composition, so that it suffices to determine these properties for each consecutive step of a transformation chain.

We will generalize and simplify existing simulation definitions considerably. The proof of the correctness of simulation will use almost in full the criteria for soundness and completeness and preservation of termination. One could therefore argue that these criteria were designed to satisfy the requirements of a correct transformation implicitly.

In practical cases, a simulation is often not immediately sound, complete and termination preserving, due to the fact that the simulating rewrite system contains 'junk'. In such cases, a reachability restriction on the elements in the simulating rewrite system can help to make the simulation sound, complete and termination preserving. We will formalize this reachability notion.

Related Work. In [4, 14], a transformation of an equational specification of abstract data types is called a 'correct implementation' if the initial algebras of the original and the transformed specification are isomorphic. This notion is considerably stronger than our notion of a correct transformation.

Thatte [17, 18] defined a transformation of certain types of rewrite systems, over a signature Σ, into rewrite systems that are constructor based, over an extended signature Σ^*. The relation between the original and the transformed rewrite system, with the latter restricted to the part that is reachable from Σ, is given by a mapping $\phi : \Sigma^* \to \Sigma$, which is the identity on Σ. Since this mapping is surjective, it is a simulation. Thatte shows that this simulation is sound, and satisfies a weaker completeness notion: if a term s can be rewritten to a term s' in one step in the original rewrite system, then s can be rewritten to s' in zero or more steps in the simulating rewrite system. This weaker completeness notion (which was also used by Sekar et al. [16]) does not imply that a transformation is correct.

Luttik [11] proposed a series of stronger simulation notions, and shows that they preserve termination and confluence.

In the technical report version of this paper [5], more information is provided on so-called 'weak correctness' of transformations, which basically means that at least one normal form of each term in the original rewrite system is conserved by the transformation.

Acknowledgements. Jasper Kamperman, Bas Luttik, Karen Stephenson and Pum Walters are thanked for useful comments, and Jan Bergstra for his support. A considerable part of this research was carried out when both authors worked at the Philosophy Department of Utrecht University.

2 Abstract Reduction Systems

This section introduces some preliminaries from rewriting [3, 9].

Definition 1. An *abstract reduction system (ARS)* consists of a collection A of elements, together with a binary reduction relation R between elements in A.

R^+ denotes the transitive closure of a reduction relation R, and R^* the reflexive transitive closure of R. In the following definitions, we assume an ARS (A, R).

Definition 2. $a \in A$ is a *normal form* for R if there does not exist an $a' \in A$ with aRa'. $a \in A$ is a *normal form of* $a' \in A$ if $a'R^*a$ and a is a normal form.

$nf_R : A \to \mathcal{P}(A)$ maps each $a \in A$ to its collection of normal forms for R.

Definition 3. R is *terminating* for $a \in A$ if there does not exist an infinite reduction $aRa_1 Ra_2 R \cdots$. (This is also known as strong normalization.)

3 Correctness of Transformations

We formulate general conditions which ensure that a transformation of rewrite systems is correct. We adopt the point of view that such a transformation is correct if it constitutes a sensible step in a compilation procedure. This is the case if the input/output behaviour of the original rewrite system is maintained, where the input is any term, and the output is (one of) its normal form(s). Hence, for compilation of rewrite systems, the prime interest of a transformation is that it preserves normal forms.

We note that rewriting is mostly concerned with the computational aspect, that is, a rewrite system is characterized by the normal forms that it attaches to terms, together with its termination properties. Justifications of this claim abound in the literature:

- equational theorem proving is mostly concerned with terminating rewrite systems which yield unique normal forms [15];
- if rewriting is applied to implement abstract data types, then the meaning of a term is fixed by its normal forms [1];
- in [3] it is remarked that "rewrite systems defining at most one normal form for any input term can serve as functional programs".

As explained in the introduction, we propose a notion of correctness of transformations, which requires parse and print functions that allow the normal forms of the original rewrite system to be computed via the simulating rewrite system.

In the following definition, we assume that a mapping $f : V \to W$ extends to a mapping $f : \mathcal{P}(V) \to \mathcal{P}(W)$ as expected: $f(V_0) = \{f(v) \mid v \in V_0\}$.

Definition 4. An ARS (B, S) is a *correct transformation* of an ARS (A, R) if there exist mappings *parse* $: A \to B$ and *print* $: B \to A$ such that:

1. if R is terminating for $a \in A$, then S is terminating for *parse*(a);
2. $print(nf_S(parse(a))) = nf_R(a)$ for $a \in A$, that is, the diagram below commutes:

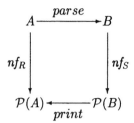

For this notion of a correct transformation, the definition of the *print* function outside of $nf_S(parse(A))$, and the S-relation outside the range of *parse*, are irrelevant.

4 Correctness of Simulation

4.1 Simulation

Kamperman and Walters [7, 8, 6] propose a notion of simulation for rewrite systems. We present simplified and more general versions of their definitions in the next sections. A simulation of an ARS (A, R) by an ARS (B, S) is characterized by a surjective mapping $\phi : B \to A$. The intuition for this mapping ϕ is that the reduction graph of $a \in A$ with respect to R is simulated by the reduction graphs of all $b \in \phi^{-1}(a)$ with respect to S.

Definition 5. A *simulation* of an ARS (A, R) by an ARS (B, S) is a surjective mapping $\phi : B \to A$.

Simulation is a transitive relation, simply because the composition of two surjective mappings is again surjective.

4.2 Soundness

In this and the following sections we assume as general notation that the ARS (A, R) is simulated by the ARS (B, S) by means of the surjective mapping $\phi : B \to A$. Soundness of a simulation ensures that each S-step of $b \in B$ is a simulation of some finite R-reduction of $\phi(b)$.

Definition 6. The simulation ϕ is *sound* if for each $b, b' \in B$ with bSb' we have $\phi(b)R^*\phi(b')$.

Soundness can be depicted as follows:

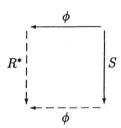

Let ϕ be a simulation of (A, R) by (B, S), and ϕ' a simulation of (B, S) by (C, T). If both ϕ and ϕ' are sound, then their composition $\phi \circ \phi'$ is also sound, which can be seen from the following graphical outline.

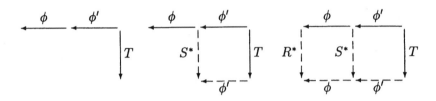

The first step is due to soundness of ϕ', and the second to soundness of ϕ, applied to each step in the S^*-reduction.

4.3 Completeness

As opposed to soundness, completeness ensures that each R-step from $\phi(b)$ is simulated by a finite S-reduction of b with length greater than zero. Note that, due to the surjectivity of ϕ, each $a \in A$ can be written as $\phi(b)$ for some b.

Definition 7. The simulation ϕ is *complete* if for each $a \in A$ and $b \in B$ with $\phi(b)Ra$ there is a $b' \in B$ with bS^+b' and $\phi(b') = a$.

Completeness can be depicted as follows:

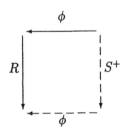

Let ϕ be a simulation of (A, R) by (B, S), and ϕ' a simulation of (B, S) by (C, T). If both ϕ and ϕ' are complete, then their composition $\phi \circ \phi'$ is also complete, which can be seen from the following graphical outline.

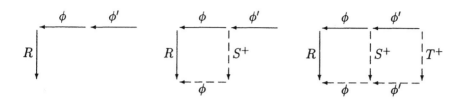

The first step is due to completeness of ϕ, and the second to completeness of ϕ', applied to each step in the S^+-reduction.

4.4 Termination Preservation

Termination preservation of a simulation ensures that termination properties for the original rewrite system are preserved by the simulating rewrite system.

Definition 8. A simulation *preserves termination* if for each $a \in A$ for which R is terminating, S is terminating for each $b \in \phi^{-1}(a)$.

Let ϕ be a simulation of (A, R) by (B, S), and ϕ' a simulation of (B, S) by (C, T). If both ϕ and ϕ' preserve termination, then their composition $\phi \circ \phi'$ also preserves termination. Namely, suppose that R is terminating for $a \in A$, and let $\phi \circ \phi'(c) = a$. Then termination preservation of ϕ yields that S is terminating for $\phi'(c)$, and termination preservation of ϕ' yields that T is terminating for c.

4.5 Correctness

Theorem 9. *If a simulation is sound and complete and preserves termination, then it is a correct transformation.*

Proof. Let ϕ be a simulation of (A, R) by (B, S) which is sound and complete and preserves termination. Choose *print* to be ϕ, and let *parse* be any inverse of *print* (i.e., $parse(a) \in \phi^{-1}(a)$ for each $a \in A$). We show that these mappings satisfy the requirements of a correct transformation.

First we prove $print(nf_S(parse(a))) \subseteq nf_R(a)$ for $a \in A$. Let $b \in nf_S(parse(a))$; we have to show that $print(b) \in nf_R(a)$. Since $parse(a)S^*b$, soundness yields $aR^*print(b)$. Since b is a normal form for S, completeness implies that $print(b)$ is a normal form for R. So, since $aR^*print(b)$, it follows that $print(b) \in nf_R(a)$.

Next we prove $nf_R(a) \subseteq print(nf_S(parse(a)))$ for $a \in A$. Let $a' \in nf_R(a)$; we show that $a' \in print(nf_S(parse(a)))$. Since aR^*a', completeness yields that there exists a $b \in B$ such that $parse(a)S^*b$ and $print(b) = a'$. Since a' is a normal form for R, preservation of termination yields that S is terminating for b. Hence, there exists a $b' \in B$ which is a normal form for S such that bS^*b'. Since $parse(a)S^*bS^*b'$, it follows that $b' \in nf_S(parse(a))$. Since bS^*b' and $print(b) = a'$, soundness yields $a'R^*print(b')$. The fact that a' is a normal form for R then implies $print(b') = a'$. Hence, $a' \in print(nf_S(parse(a)))$.

Finally, if R is terminating for $a \in A$, then preservation of termination ensures that S is terminating for $parse(a)$. $\square$

4.6 Reachability

Definition 10. Let (B, S) be an ARS, and $B_0 \subseteq B$. An element $b \in B$ is S-*reachable from* B_0 if $b_0 S^* b$ for some $b_0 \in B_0$.

This reachability notion also occurs in [18, 19].

In practical cases, a simulation $\phi : B \to A$ of an ARS (A, R) by an ARS (B, S) is not always immediately sound, complete and termination preserving, due to the fact that B contains 'junk'. In such cases, it may be possible to define a collection $B_0 \subseteq B$ that is closed under S-reductions, such that ϕ restricted to B_0 is still surjective, and a sound, complete, termination preserving simulation. A particular case arises when A itself is a subset of B. In that case, restricting ϕ to the S-reachable terms of A often does the job.

We note that such a restriction of a simulation never loses any desirable properties. That is, let $\phi : B \to A$ be a simulation of (A, R) by (B, S) that is sound or complete or termination preserving. If the restriction of ϕ to a collection $B_0 \subseteq B$, closed under S-reductions, is still surjective, then this restricted simulation is still sound or complete or termination preserving, respectively.

5 An Example

We present a toy example of a transformation of a term rewriting system (TRS) [3, 9], which will be shown to be a sound, complete and termination preserving simulation, if a reachability restriction is imposed on the ARS induced by the transformed TRS.

Assume two constants c and c', and two unary functions f and g, and let A be the set of closed terms over the signature $\{c, c', f, g\}$. Consider the TRS R that consists of the following rewrite rule:

$$f(g(c)) \longrightarrow c'$$

This rewrite rule induces an ARS on A, also denoted by R, in the standard way. That is, a relation aRa' holds if a' can be obtained from a by replacing a subterm $f(g(c))$ of a by c'.

In order to compile (A, R), the TRS R can be transformed into a so-called 'minimal' TRS, using a strategy from [7, 8, 6]. In minimal TRSs, rewrite rules are not allowed to contain more than three function symbols, and no more than two function symbols at each side of the rewrite rule. These restrictions are convenient for compilation of the TRS, because matchings with respect to a minimal rewrite system can easily be expressed in basic instructions of an abstract machine. Note that the rewrite rule in R is not minimal. The minimization strategy transforms it into the following minimal TRS M:

$$f(g(x)) \longrightarrow f_g(x)$$
$$f(x) \longrightarrow f^d(x)$$
$$f_g(c) \longrightarrow c'$$
$$f_g(x) \longrightarrow f^d(g(x))$$

These rewrite rules generate an ARS, also denoted by M, which applies to the collection B of closed terms over the signature $\{c, c', f, f^d, f_g, g\}$. A relation bMb' holds if b' can be obtained from b by replacing a subterm b_0 of b, which is an instance of the left-hand side of a rewrite rule r in M, by the corresponding instance of the right-hand side of r. The rewriting strategy is *innermost*, which requires that the proper subterms of b_0 cannot be replaced. Furthermore, rewrite rules are applied with respect to *specificity ordering*, which requires that the left-hand side of r is the most specific of all left-hand sides of rules in M that match with b_0. Note that the left-hand side of the first and third rewrite rule in M are more specific than the left-hand side of the second and fourth rule, respectively.

B contains two new unary function symbols, f^d and f_g. The intuition behind the transformation, and the two new function symbols, is as follows. Basically, the transformation splits possible matchings with respect to the left-hand side of the rule in R into elementary steps. A term $f_g(b)$ is an encoded representation of the term $f(g(b))$; this is captured by the first rewrite rule of M. Furthermore, a term $f^d(b)$ is an encoded representation of the term $f(b)$, and expresses that this term cannot be rewritten by the rule in R. Therefore, the second rule of M reduces $f(b)$ to $f^d(b)$, if b is not of the form $g(b')$ (due to the fact that the first rule has priority over the second rule). The third rule mimics the rule in R, because $f_g(c)$ represents $f(g(c))$. Finally, the fourth rule of M expresses that a term $f(g(b))$ with $b \neq c$ cannot be rewritten by R, so that in M a term $f_g(b)$ is reduced to $f^d(g(t))$ if $b \neq c$ (due to the fact that the third rule has priority over the fourth rule).

In spite of the intuition presented above, it remains a natural question to ask whether the transformation of R into M is correct. In order to give a positive answer to this question, we capture this intuition by a simulation relation as follows. Let B_0 the set of elements in B that are M-reachable from A. Note that $A \subseteq B_0$. We define a surjective mapping $\phi : B_0 \to A$ inductively as follows:

$$
\begin{aligned}
\phi(c) &= c & \phi(f(t)) &= f(\phi(t)) \\
\phi(c') &= c' & \phi(f^d(t)) &= f(\phi(t)) \\
\phi(g(t)) &= g(\phi(t)) & \phi(f_g(t)) &= f(g(\phi(t)))
\end{aligned}
$$

This simulation is sound and complete and preserves termination. Hence, according to Theorem 9, the transformation of (A, R) into (B_0, M) is correct.

Completeness depends on the reachability restriction that was imposed on B_0. For example, $f^d(g(c))$ is a normal form for M, while $\phi(f^d(g(c))) = f(g(c))$ is not a normal form for R. However, $f^d(g(c))$ is not in B_0, because it is not M-reachable from A.

References

1. J.A. Bergstra, J. Heering, and P. Klint, eds. *Algebraic Specification*. ACM Press in cooperation with Addison Wesley, 1989.
2. R.M. Burstall and P.J. Landin. Programs and their proofs: an algebraic approach. In *Machine Intelligence*, Volume 4, pp. 17–43. Edinburgh University Press, 1969.

3. N. Dershowitz and J.-P. Jouannaud. Rewrite systems. In J. van Leeuwen, ed., *Handbook of Theoretical Computer Science, Volume B*, pp. 243–320. Elsevier, 1990.

4. H. Ehrig, H.-J. Kreowski, and P. Padawitz. Stepwise specification and implementation of abstract data types. In *Proceedings ICALP'78*, LNCS 62, pp. 205–226. Springer, 1978.

5. W.J. Fokkink and J.C. van de Pol. Correct transformation of rewrite systems for implementation purposes. Logic Group Preprint Series 164, Utrecht University, 1996. Available at http://www.phil.ruu.nl.

6. J.F.Th. Kamperman. *Compilation of Term Rewriting Systems*. PhD thesis, University of Amsterdam, 1996.

7. J.F.Th. Kamperman and H.R. Walters. Minimal term rewriting systems. In *Proceedings WADT'95*, LNCS 1130, pp. 274–290. Springer, 1996.

8. J.F.Th. Kamperman and H.R. Walters. Simulating TRSs by minimal TRSs: a simple, efficient, and correct compilation technique. Report CS-R9605, CWI, 1996. Available at http://www.cwi.nl/epic.

9. J.W. Klop. Term rewriting systems. In *Handbook of Logic in Computer Science, Volume I*, pp. 1–116. Oxford University Press, 1992.

10. A. Laville. Lazy pattern matching in the ML language. In *Proceedings FSTTCS'87*, LNCS 287, pp. 400–419. Springer, 1987.

11. B. Luttik. Transformation of reduction systems: a view on proving correctness. Master's Thesis, University of Amsterdam, 1996.

12. J. McCarthy. Towards a mathematical science of computation. In *Proceedings Information Processing '62*, pp. 21–28. North-Holland, 1963.

13. F.L. Morris. Advice on structuring compilers and proving them correct. In *Proceedings POPL'73*, pp. 144–152. ACM Press, 1973.

14. C.F. Nourani. Abstract implementations and their correctness proofs. *Journal of the ACM*, 30:343–359, 1983.

15. M.J. O'Donnell. *Equational Logic as a Programming Language*. MIT Press, 1985.

16. R.C. Sekar, S. Pawagi, and I.V. Ramakrishnan. Transforming strongly sequential rewrite systems with constructors for efficient parallel execution. In *Proceedings RTA'89*, LNCS 355, pp. 404–418. Springer, 1989.

17. S.R. Thatte. On the correspondence between two classes of reduction systems. *Information Processing Letters*, 20(2):83–85, 1985.

18. S.R. Thatte. Implementing first-order rewriting with constructor systems. *Theoretical Computer Science*, 61(1):83–92, 1988.

19. R.M. Verma. Transformations and confluence for rewrite systems. *Theoretical Computer Science*, 152(2):269–283, 1995.

20. H.R. Walters and J.F.Th. Kamperman. EPIC: an equational language – abstract machine and supporting tools. In *Proceedings RTA'96*, LNCS 1103, pp. 424–427. Springer, 1996.

21. H.R. Walters and J.F.Th. Kamperman. EPIC 1.0 (unconditional), an equational programming language. Report CS-R9604, CWI, 1996. Available at http://www.cwi.nl/epic.

22. H. Zantema. Termination of term rewriting by semantic labelling. *Fundamenta Informaticae*, 24(1,2):89–105, 1995.

On the Dilation of Interval Routing

Cyril Gavoille

LaBRI, Université Bordeaux I, France. *gavoille@labri.u-bordeaux.fr*

Abstract. In this paper we deal with interval routing on n-node networks of diameter D. We show that for every fixed $D \geq 2$, there exists a network on which every interval routing scheme with $O(n/\log n)$ intervals per link has a routing path length at least $\lfloor 3D/2 \rfloor - 1$. It improves the lower bound on the routing path lengths for the range of very large number of intervals. No result was known about the path lengths whenever more than $\Theta(\sqrt{n})$ intervals per link was used. Best-known lower bounds for a small number of intervals are $2D - O(1)$ for 1 interval [11], and $3D/2 - O(1)$ up to $\Theta(\sqrt{n})$ intervals [5]. For $D = 2$, we show a network on which any interval routing scheme using less than $n/4 - o(n)$ intervals has a routing path of length at least 3. Moreover, we build a network of bounded degree on which every interval routing scheme with routing path lengths bounded by $3D/2 - o(D)$ requires $\Omega(n/\log^{2+\varepsilon} n)$ intervals per link, where ε is an arbitrary non-negative constant.

1 Introduction

The *dilation* of a routing scheme is the length of the longest routing path. Assuming that time cost of message delivery is function of the routing path length or of the number of routers crossed, dilation is a parameter of the worse-case time complexity. Concurrently, fast routers must be easily implemented with a small amount of hardware.

Interval Routing is a routing scheme implementing compact routing tables, and allowing fast routing decisions in each node [9, 12]. Interval routing consists in labeling nodes by an unique integer taken in $\{1, \ldots, n\}$, n the number of routers, and in assigning to each link at every router a set of intervals of destinations, such that any message can reach its destination from any source. Such a labeling scheme on a network G is so-called an *interval routing scheme* on G. Each router locally finds the next link to forward a message to its destination by choosing the link that contains the number of the destination in one of its intervals. At each node, the intervals must be pairwise disjoint, and cover the set of all the labels, maybe excepted the label of the node it-self. The local routing decision time is bounded by $O(d \log k \cdot \log n)$ bit operations[1] whereas the space complexity of the router is at most $O(kd \cdot \log n)$ bits, for a router of d links, and if at most k intervals per link are used. In particular, such a routing scheme is efficient, i.e., compact and fast, if the degree of the network and the number

---

[1] For each link the router can perform a binary search among the k sorted intervals, every operations being on integers of size $O(\log n)$ bits.

of intervals per link are both low[2] relatively to the number of routers of the network.

The interval routing scheme is used in the last generation of C104 routing chips used in the INMOS T9000 Transputer design [7]. Since the number of intervals is limited in each routing chip, we are interested in finding the minimum dilation for interval routing scheme using a fixed number of intervals. The dilation is expressed in term of the diameter of the networks which is a common lower bound of the dilation for all networks.

The following table summarizes the best lower bounds known, and our contribution about the dilation of interval routing schemes using at most k intervals per link on n-node networks of diameter D. Note that for every network, there is an interval routing scheme with one interval and of dilation $2D$, where D is the diameter of the network. Indeed, it is sufficient to route along a minimum spanning tree of the network which supports an interval routing scheme [9] with only 1 interval per link.

Number of intervals	Dilation (lower bound)	Reference
$k = 1$	$2D - 3$	[11]
$2 \le k \le \Theta(\sqrt{n})$	$3D/2 - 3$	[5]
$2 \le k \le \Omega(n/\log n)$	$\lfloor 3D/2 \rfloor - 1$	Theorem 1

The next section presents the notations, and previous works about the dilation. In Section 3, we prove the main theorem. Also, we extend the result to diameters which depend on n. We prove a trade-off of $\Omega(n/(D\log(n/D)))$ intervals required for every interval routing scheme of dilation less than $\lfloor 3D/2 \rfloor - 1$. In Section 4, we adapt our construction to bounded degree networks, and we improve the multiplicative constant of the result of [4] about the maximum compactness of n-node networks. Moreover the lower bound is proved for a network of diameter 2, and of maximum degree $n/2$.

2   Statement of the Problem

2.1   Notations

The model of networks is a undirected connected graph G, each node representing a router. The distance between any two nodes x and y is the minimum number of edges of paths connecting x and y, and is denoted $\text{dist}(x, y)$. In all the rest of the paper, n will denote the order of the graph, and D its diameter.

An *interval* means a set of consecutive integers taken in $\{1, \ldots, n\}$, n and 1 being considered as consecutive. For every arc[3] e, all the intervals associated to e form a set of integers, i.e., a set of labels of destinations, and is denoted by I_e. An interval routing scheme R on a graph G is denoted by a pair $(\mathcal{L}, \mathcal{I})$, where $\mathcal{L}$ is the labeling of vertices of G, and $\mathcal{I}$ is the set of all the I_e's.

---

[2] For instance k and d satisfying $kd/\log d = o(n/\log n)$ may provide a scheme more compact than the standard routing tables which need of $\Theta(n \log d)$ bits per router.

[3] For convenience, each edge of the graph is considered as a pair of two symmetric arcs.

2.2 A simple example

Let us consider the following example of interval routing on a graph G_0 of 7 vertices depicted on the left-hand side of Figure 1. Nodes are labeled by integers from 1 to 7, and intervals are assigned to each arc. If the node 5 sends a message to node 1, the message will successively be forwarded along the arc $(5,7)$, then along $(7,1)$, because $1 \in I_{(5,7)} = [7,2] = \{7,1,2\}$, and $1 \in I_{(7,1)} = [1] = \{1\}$. Each set of destinations I_e, e arc of G_0, is composed of at most two intervals of consecutive labels. One can check that every routing path on G_0 is of minimal length. Therefore, the dilation of this routing scheme is the diameter of G_0, here 2.

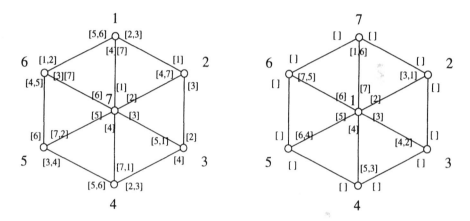

Fig. 1. Two interval routing schemes of dilation 2 on the same graph G_0. (We denote by [] the *empty interval*, corresponding to an empty set of destination.)

This interval routing scheme on G_0 qualifies as *shortest path* interval routing scheme because all the routing paths in the graph are of minimal length. A classical problem for interval routing is to compute the minimum number of intervals needed to guarantee a shortest path interval routing scheme on a given graph. A such number depends on the graph only, and is termed *compactness*. Compactness of a graph is at most $n/2$, because any set of destinations, and any labeling of these destinations, can not contain more than $n/2$ non consecutive integers. In [4], it has been built graphs of compactness at least $n/12$. Therefore $\Theta(n)$ is a tight lower bound of compactness of n-vertex graphs.

Actually, G_0 has no shortest path routing scheme with only one interval per link (see [2, page 793] for a proof). Hence the compactness of G_0 is 2. The right-hand side of Figure 1 shows another interval routing scheme on G_0, but with only one interval per arc. It has also a dilation 2, the diameter. The *dilation* problem is, given a graph G, and an integer k, k being less than the compactness of G, to determine an interval routing scheme on G using at most k intervals per link which minimizes the longest routing path. This general question is important in practice whenever a low number of intervals is forced by the hardware of the router, and whenever message delivery time must be as short as possible.

Fundamentally, the compactness problem consists in measuring the compression of the "routing information" whenever paths are of minimum length. Its dual problem, the dilation problem, consists in measuring the efficiency of the routing scheme when the compression rate is limited, which defines the size of the routing information in each node. Both the problems contribute to design some trade-off between time and space used by a router in a communication network.

2.3 Related works

Lower bound for the dilation problem for interval routing was first addressed by Ružička in [8]. He built a *globe-graph* on which every interval routing scheme with one interval has dilation at least $3D/2 - O(1)$. This result has been recently improved by Tse and Lau in [11], for one interval with a lower bound of $2D - 3$ (based on an extension of the globe-graph), and generalized in [10] to more than one interval with a lower bound still of $3D/2 - O(1)$ up to $\Theta(\log n)$ intervals, and of $5D/4 - O(1)$ up to $\Theta(\sqrt{n})$ intervals. Recently, in [5], Královič, Ružička, and Štefankovič improved the lower bound for the range of intervals from $\Theta(\log n)$ up to $\Theta(\sqrt{n})$, with $3D/2 - 3$, using a *multi-globe graph*. No result was known for a larger number of intervals. Intuitively, the more intervals are used, the lower the dilation. In [5], it is also proved that $O(\sqrt{n \log n})$ intervals suffice to guarantee a dilation of at most $\lceil 3D/2 \rceil$ on every network.

In this paper we extended the range of possible number of intervals up to $\Omega(n/\log n)$, and we prove a dilation of at least $\lfloor 3D/2 \rfloor - 1$. In [1], an independent and similar lower bound has been shown, however for a dilation $3D/2 - 2$, and for even D. Our result expresses the existence of a gap on the number of intervals required for a dilation close to $3/2$ the diameter. Indeed, the routing with a dilation δ might be done with only k intervals, whereas $k' = \omega(k)$ intervals might be required to route with a dilation $\delta - 1$.

3 Construction of the Worst-Case

The main Theorem of this section is the following:

Theorem 1. *For every integer $D \geq 2$, there exists an n-vertex graph of diameter D on which every interval routing scheme of dilation less than $\lfloor 3D/2 \rfloor - 1$ requires $\Omega(n/(D \log(n/D)))$ intervals.*

In Theorem 1, the number of intervals is expressed as a function of the diameter. It turns out more general results. For instance, Theorem 1 shows that for every constant k there exists an n-vertex graph of diameter $D = \Theta(n)$ on which every interval routing scheme using k intervals has a dilation $\lfloor 3D/2 \rfloor - 1$. Note that all the previous results were proved for an arbitrary but fixed diameter. The result of [5] is improved just by applying Theorem 1 with D an arbitrary constant.

Corollary 2. *For every constant integer $D \geq 2$, there exists an n-vertex graph of diameter D on which every interval routing scheme of dilation less than $\lfloor 3D/2 \rfloor - 1$ requires $\Omega(n/\log n)$ intervals.*

Sketch of the proof.

Basically we use a similar technique to establish lower bounds of the compactness, and to prove a lower bound of the dilation with a large number of intervals. Our construction is an adaptation of the graph defined in [4]. For the sketch we assume that D is an odd fixed constant ≥ 3. We build a graph which has the two following properties:

1) Some vertices require an interval routing scheme using $k = \Omega(n/\log n)$ intervals on some arcs to route along the shortest paths between vertices at distance $t = (D-1)/2$.
2) Any interval routing scheme which does not route along the shortest paths between these vertices has routing path lengths at least $3t$.

Any interval routing scheme of dilation $< 3t$ on this graph requires at least k intervals, or equivalently, any interval routing scheme that uses at most $k - 1$ intervals per link has dilation at least $3t$.

The graph construction.

Our construction is function of a boolean matrix M, and of an integer D. It is denoted by $G_{M,D}$. More precisely, for every $p \times q$ boolean matrix M and for every integer $D \geq 2$ we define the graph $G_{M,D}$ as follows: we associate with each row i of M, $i \in \{1, \ldots, p\}$, a vertex v_i in $G_{M,D}$. At each column j of M, $j \in \{1, \ldots, q\}$, we associate a pair of vertices $\{a_j, b_j\}$ which are connected by an edge. We set $t = 1$ if $D = 2$, and $t = \lfloor (D-1)/2 \rfloor$ for every $D \geq 3$. For every $i \in \{1, \ldots, p\}$ and $j \in \{1, \ldots, q\}$, we connect v_i to a_j by a path of length t if and only if $m_{i,j} = 0$. Similarly, we connect v_i to b_j by a path of length t if and only if $m_{i,j} = 1$. See Figure 2 for an example. Note that the graph obtained by contraction of the edges $\{a_j, b_j\}$, $j \in \{1, \ldots, q\}$, of the graph $G_{M,D}$ is a complete bipartite graph $K_{p,q}$. The construction is slightly different for D even. For every even $D \geq 4$, we subdivide each edge a_j, b_j with a new vertex c_j, and with the two new edges: $\{a_j, c_j\}$ and $\{c_j, b_j\}$. Also, we add p vertices of degree 1, w_i, for $i \in \{1, \ldots, p\}$. The vertex w_i is connected to v_i.

We define the *compactness* of a boolean matrix M as the smallest integer k such that there exists a matrix obtained by row permutation of M having at most k blocks of consecutive ones per column. The first and the last entry of a column are considered as consecutive. For example, the matrix M described on Figure 2 has its first and its third column composed of one blocks of consecutive ones, whereas its second column is composed of two blocks (each block being composed of only one 1-entry). The reader can check that whatever one permute the rows of M, there exists at least one column with two blocks of consecutive ones. Therefore, the compactness of M is 2. Intuitively, on this example, any interval routing scheme using only one interval on all the arcs of the form (a_j, b_j) cannot optimally reach all the v_i's, and should have a dilation of $3t$.

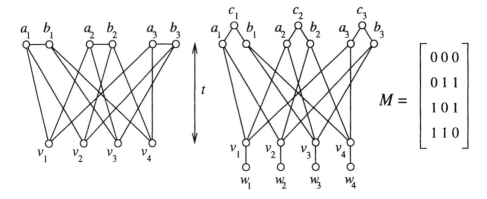

Fig. 2. The graph $G_{M,D}$: 1) of diameter $D \leq 2t + 1$, and 2) of diameter $D \leq 2t + 2$.

In all the following, columns of boolean matrices are seen as binary strings. log-functions are given for base two. We need of the following lemmas to prove Theorem 1 and Theorem 5:

Lemma 3. *Let p, q be two sufficiently large integers. Let $\mathcal{M}$ be the set of $p \times q$ boolean matrices having $\lfloor p/2 \rfloor$ 1-entries per column. Let $\mathcal{M}_1$ be the subset of matrices of $\mathcal{M}$ such that all the rows are pairwise non complemented. Let $\mathcal{M}_2$ be the subset of matrices of $\mathcal{M}$ such that for every pair of columns the $2 \times p$ matrix composed of the pair of columns contains the submatrix[4] $\left[\begin{smallmatrix} 0 & 0 & 1 & 1 \\ 0 & 1 & 0 & 1 \end{smallmatrix}\right]$ up to a column permutation. Then,*

- *if $p = o(2^{q/2})$, then $|\mathcal{M}_1| \sim |\mathcal{M}|$;*
- *if $q = o(2^{p/4})$, then $|\mathcal{M}_2| \sim |\mathcal{M}|$.*

This is a consequence of a result proved in [3]. In all the following of the paper, we set $\mathcal{M}_0 = \mathcal{M}_1 \cap \mathcal{M}_2$. The matrix M depicted on Figure 2 belongs to $\mathcal{M}_0$. The will see later that the graphs $G_{M,2}$ built from any $M \in \mathcal{M}_0$ have diameter 2 exactly. Futhermore, almost all matrices are in $\mathcal{M}_0$.

Lemma 4. *For every sufficiently large integer p, there exist a constant α, and a $p \times \lfloor \alpha \log p \rfloor$ matrix of $\mathcal{M}_0$ of compactness at least $p/5$.*

Proof. We use a counting argument which can be formalized by the Kolmogorov Complexity (see [6] for an introduction). Let $\mathcal{M}$ be the set of $p \times q$ boolean matrices with $\lfloor p/2 \rfloor$ 1-entries per columns. Let us begin to show that compactness of some matrices of $\mathcal{M}$ is linear in p for $q = \Theta(\log p)$.

For every $M \in \mathcal{M}$, we define cl(M) the subset of the matrices of $\mathcal{M}$ obtained by row permutation of M. There exists a matrix $M_0 \in \mathcal{M}$ such that all the matrices of cl(M_0) have a Kolmogorov Complexity at least $C = \log|\mathcal{M}| - \log(p!) -$

---

[4] *A is a submatrix of B if A can be obtained from B by removing some columns and rows in B.*

$3 \log p$. Indeed, by contradiction let $M_0' \in \text{cl}(M_0)$ be a matrix of Kolmogorov Complexity $C' < C$, for any $M_0 \in \mathcal{M}$. M_0 may be described by a pair (i_0, M_0'), where i_0 is the indexe of the row permutation of M_0' into M_0. Such an integer can be coded by $\log(p!) + 2 \log p + O(1)$ bits. $2\lceil \log p \rceil$ bits are sufficient to describe p, thus the length of any $i_0 \leq p!$, in a self-delimiting way. Hence the Kolmogorov Complexity of M_0 would be at most $C' + \log(p!) + 2 \log p + O(1) < \log |\mathcal{M}|$, that is impossible for any matrix $M_0 \in \mathcal{M}$.

$|\mathcal{M}| = \left(\binom{p}{\lfloor p/2 \rfloor}\right)^q = \Theta(2^p/\sqrt{p})^q$. By the Stirling's formula $\log |\mathcal{M}| = pq - O(q \log p)$, and $\log(p!) = p \log p - O(p)$. Hence,

$$C = \log |\mathcal{M}| - \log(p!) - 3 \log p = pq - p \log p - O(p + q \log p). \tag{1}$$

All the matrices of $\mathcal{M}$ have q columns, each one of Kolmogorov Complexity bounded by $p + O(1)$. Therefore there exists a matrix M_0 such that every matrix obtained by row permutation of M_0 has a column of Kolmogorov Complexity at least

$$C_0 = \frac{C}{q} - 2 \log p = p - \frac{p \log p}{q} - O\left(\frac{p}{q} + \log p\right). \tag{2}$$

The term $2 \log p$ codes the length of the description of such a column in a self-delimiting way. From [6, Theorem 2.15, page 131], every binary string of length p bits, and of Kolmogorov Complexity at least $p - \delta(p)$ contains at least

$$\frac{p}{4} - \sqrt{(\delta(p) + c)p\frac{3}{2} \ln 2} \tag{3}$$

occurrences of 01-sequences, for any *deficiency* function δ, and some constant c. Since each 01-sequence in a binary string starts necessarily a new block of consecutive ones, we get a lower bound on the number of blocks of consecutive ones for such strings.

By choosing $q = \lfloor \alpha \log p \rfloor$, Equation 2 provided

$$C_0 = p - \frac{p}{\alpha} - O\left(\frac{p}{\log p}\right) > p - \delta(p), \text{ with } \delta(p) = \frac{p}{\alpha} + o(p).$$

By Equation 3, it follows that M_0 has a compactness at least

$$\frac{p}{4} - \sqrt{(1 + o(1))\, p^2 \frac{3}{2\alpha} \ln 2} = p\left(\frac{1}{4} - \sqrt{\frac{3}{2\alpha} \ln 2}\right) - o(p) > \frac{p}{5}, \text{ for } \alpha = 416. \tag{4}$$

Let us show that the result holds also for the compactness of some matrices of $\mathcal{M}_0$. From Lemma 3, because $p = o(2^{q/2}) = o(2^{208 \log p}) = o(p^{208})$, we get that $|\mathcal{M}_1| \sim |\mathcal{M}|$. Similarly, $q = O(\log p) = o(2^{p/4})$, thus $|\mathcal{M}_2| \sim |\mathcal{M}|$, and therefore $|\mathcal{M}_0| \sim |\mathcal{M}|$. It implies that $\log |\mathcal{M}_0| = \log |\mathcal{M}| + o(1)$, and thus Equations 1, 2, 3, and 4 hold for $\mathcal{M}_0$ as well, that completes the proof. □

Proof of Theorem 1.

Let D be an integer ≥ 2, let p be an integer large enough, and let $M \in \mathcal{M}_0$ be a matrix satisfying Lemma 4. We consider the graph $G_{M,D}$. Let t be the length of the chains between the nodes $\{a_j, b_j\}$'s and the nodes v_i's. By the construction of $G_{M,D}$, $t = 1$ if $D = 2$, and $t = \lfloor (D-1)/2 \rfloor$ otherwise. Let us remark first that for every integer $D' > D$, and for every graph G of diameter D on which every interval routing scheme using k intervals has a dilation δ, there exists a graph G' of diameter D' on which every interval routing scheme using k intervals has a dilation at least δ. Indeed, to obtain G' it is sufficient to add a path of length $D' - D$ to any vertex of G which is of eccentricity D. Hence, to prove a lower bound for the dilation of graphs of diameter D it is sufficient to prove a lower bound for dilation in graphs of diameter at most D.

Fact 1. For odd $D \geq 3$, the diameter of $G_{M,D}$ is at most $2t + 1$.

By construction $G_{M,D}$ has no vertex w_i, and no vertex c_j. Let us show that, for every pair of vertices x and x', $\text{dist}(x, x') \leq 2t+1$. From the particular shape of the matrices of $\mathcal{M}_0$, no column is the string 0^p or the string 1^p. Thus, for $p \geq 2$, all the nodes have degree at least 2. Let us show that there is a cycle of length at most $4t + 2$ that cuts x and x'. x (respectively x') belongs to a chain of length t that directly connects a node, namely v_i (respectively $v_{i'}$), to a node $\alpha_j \in \{a_j, b_j\}$ (respectively $\alpha_{j'} \in \{a_{j'}, b_{j'}\}$). It follows that x and x' belongs to a cycle $(x, \alpha_j, \overline{\alpha_j}, v_{i'}, x', \alpha_{j'}, \overline{\alpha_{j'}}, v_i, x)$, where $\overline{\alpha_j}$ denotes the complement of α_j in $\{a_j, b_j\}$ (similarly for $\overline{\alpha_{j'}}$). Its length is clearly bounded by $4t + 2$. It follows that $\text{dist}(x, x') \leq (4t + 2)/2 = 2t + 1$. Therefore, for $D \geq 3$, $G_{M,D}$ has a diameter bounded by $D \geq 2t + 1$.

Fact 2. For any interval routing scheme $R = (\mathcal{L}, \mathcal{I})$ on $G_{M,D}$, for every arc (a_j, b_j), and for every vertex v_i, if $m_{i,j} = 1$ and $\mathcal{L}(v_i) \notin I_{(a_j, b_j)}$, or if $m_{i,j} = 0$ and $\mathcal{L}(v_i) \in I_{(a_j, b_j)}$, then the dilation of R is at least $3t$.

This results from the fact that there is no path shorter than $2t$ between two any vertices $v_i \neq v_{i'}$. Moreover any wrong decision for routing from a_j to v_i cuts a vertex $v_{i'} \neq v_i$ before to reach v_i.

Fact 3. Let k be the compactness of M, and $R = (\mathcal{L}, \mathcal{I})$ be any interval routing scheme on $G_{M,D}$. If R uses less than k intervals per arc, then its dilation is at least $3t$.

Assume R is fixed, and uses only $k - 1$ intervals. Let j_0 be a column of M composed of at least k blocks of consecutive ones. Such a column exists because the compactness of M is k. Let us consider the sequence $u = (u_1, \ldots, u_p)$ defined by: for every $i \in \{1, \ldots, p\}$, $u_i = 1$ if $\mathcal{L}(v_i) \in I_{(a_{j_0}, b_{j_0})}$, and $u_i = 0$ otherwise. Since $I_{(a_{j_0}, b_{j_0})}$ is composed of at most $k - 1$ intervals, u is composed of at most $k - 1$ blocks of consecutive ones. Thus the column j_0 and the sequence u differ in at least one place. Set i_0 the indexe such that $m_{i_0, j_0} \neq u_{i_0}$. If $u_{i_0} = 1$, then $\mathcal{L}(v_{i_0}) \in I_{(a_{j_0}, b_{j_0})}$, and $m_{i_0, j_0} = 0$. If $u_{i_0} = 0$, then $\mathcal{L}(v_{i_0}) \notin I_{(a_{j_0}, b_{j_0})}$, and $m_{i_0, j_0} = 1$. We conclude by applying Fact 2.

Fact 4. For every even $D \geq 4$, the diameter of $G_{M,D}$ is at most $2t + 2$. If R uses less than k intervals per arc, then its dilation is at least $3t + 2$.

Note that for $D = 2$, we do not need of Fact 4, because actually the statement of Theorem 1 holds for every graph of diameter 2.

The graph $G_{M,D}$ has it this case the vertices w_i of degree 1, and a path $\{a_j, c_j, b_j\}$ between a_j and b_j. Let x, x' be two non-w_i vertices. Similarly to Fact 1, there exists a cycle of length at most $4t + 4$ that cuts x and x'. Hence

$\operatorname{dist}(x, x') \leq 2t + 2$. Let us show that $\operatorname{dist}(v_i, v_{i'}) \leq 2t$. Indeed, $M \in \mathcal{M}_0$ has the property that the rows i and i' of M are not complemented. Thus there exists j such that $m_{i,j} = m_{i',j}$. It follows that either a_j either b_j is connected by a path of length t to v_i and $v_{i'}$. Hence $\operatorname{dist}(w_i, w_{i'}) \leq 2t + 2$. Let us show that $\operatorname{dist}(x, v_i) \leq 2t + 1$. Assume that x belongs to a path connecting $\alpha_j \in \{a_j, b_j\}$ to $v_{i'}$. Let $t_x \geq 0$ be the distance between x and $v_{i'}$. There is two paths between x and v_i: one trough $v_{i'}$ of length $t_x + 2t$, and another one trough α_j of length at most $t - t_x + 2 + t$. Thus $\operatorname{dist}(x, v_i) \leq \min\{t_x + 2t, 2t + 2 - t_x\} = 2t + \min\{t_x, 2 - t_x\} = 2t + 1$. It follows that $\operatorname{dist}(x, w_i) \leq 2t + 2$, and thus the diameter of $G_{M,D}$ is bounded by $2t + 2 = D$ for even $D \geq 4$. The dilation of R, using less than k intervals, is clearly two more longer than the case of odd diameter, because: 1) for a wrong routing path we start with a new edge of the form $\{c_j, a_j\}$ or $\{c_j, b_j\}$, and 2) the unique shortest path from c_j to any w_i cuts v_i.

The order of $G_{M,D}$ is $n \leq 2p + 2q + pq(t-1) + c$, with $q = \Theta \log p$, and where $c \in \{0, 1, 2\}$ is the number of nodes added to have a diameter D exactly. For $D \leq 4$, that is $t = 1$, $n = O(p + q)$, thus $p = \Omega(n)$. For $D \geq 5$, that is $t \geq 2$, $n = \Theta(Dp \log p)$, and thus $\log p = \Theta(\log(n/D))$. It follows that $p = \Theta(n/(D \log(n/D)))$. Therefore for every $D \geq 2$, $p = \Omega(n/(D \log(n/D)))$. By Lemma 4, the compactness of M is at least $p/5$. By Fact 3, for odd D, the dilation is $3t = 3 \lfloor (D-1)/2 \rfloor = 3D/2 - 3/2$, whereas by Fact 4, for even D, the dilation is $3t + 2 = 3 \lfloor (D-1)/2 \rfloor + 2 = 3(D/2 - 1) + 2 = 3D/2 - 1$. Therefore, in all the cases, the dilation is at least $\lfloor 3D/2 \rfloor - 1$, that completes the proof. □

Theorem 1 allows to establish a trade-off between the order, the diameter of the graph, and the number of intervals required for a dilation $\lfloor 3D/2 \rfloor - 1$. For instance, we saw that the matrix M of Figure 2 has a compactness 2. Fact 3 of Theorem 1 shows that the worst-case graph of Figure 2, which is of maximum degree 3 and of diameter $D \sim n/6$, has a dilation $\lfloor 3D/2 \rfloor - 1 \sim n/4$ for 1 interval. However in general, the graph $G_{M,D}$ has a maximum degree $\Theta(p + q)$ which is $\Theta(n/(D \log(n/D)))$, for $D \geq 5$.

4 Other Results

Let us recall that the compactness of a graph G is the smallest integer k such that G supports a shortest path interval routing scheme using at most k intervals per link.

Theorem 5. *For every sufficiently large integer n, there exists an n-vertex graph G of diameter 2, of maximum degree at most $n/2$, and of compactness at least k such that $k > n/4 - O(n^{3/4} \sqrt{\log n})$. Moreover, every interval routing scheme on G using $k - 1$ intervals per link has dilation at least 3.*

Proof. (outline) We use the structure of matrices of $\mathcal{M}_0$, and Equation 2 of the proof of Lemma 4 for $q = \lfloor \sqrt{n \log n} \rfloor$. □

Corollary 6. *For every sufficiently large integer n, and for every integer $D \geq 2$, $D = o(n)$, there exists an n-vertex graph of diameter D, of maximum degree at most $n/2$, and of compactness at least $n/4 - o(n)$.*

Proof. It is sufficient to add to any vertex of $G_{M,2}$ a path of length $D - 2$ to obtain a new graph G' of diameter D exactly. The number of nodes of G' is $n + D - 2 \sim n$, for $D = o(n)$. □

We saw that the degree of our construction is not bounded by a constant. We adapt our previous construction to show a lower bound for dilation in networks of degree bounded by 3, however with a little degradation of the number of intervals. A proof of the next theorem is given in the full version of the paper.

Theorem 7. *For every constant $\varepsilon > 0$, and for every sufficiently large integer D, there exists an n-vertex graph of diameter D, and of maximum degree 3 on which every interval routing scheme of dilation less than $3D/2 - O(D/\log^{\varepsilon} D)$ requires $\Omega(n/\log^{2+\varepsilon} n)$ intervals.*

References

1. M. FLAMMINI AND E. NARDELLI, *On the path length in interval routing schemes.* Manuscript, 1996.
2. P. FRAIGNIAUD AND C. GAVOILLE, *Optimal interval routing*, in Parallel Processing: CONPAR '94 - VAPP VI, B. Buchberger and J. Volkert, eds., vol. 854 of Lecture Notes in Computer Science, Springer-Verlag, Sept. 1994, pp. 785–796.
3. C. GAVOILLE AND M. GENGLER, *Space-efficiency of routing schemes of stretch factor three*, in 4th Colloquium on Structural Information & Communication Complexity (SIROCCO), July 1997.
4. C. GAVOILLE AND E. GUÉVREMONT, *Worst case bounds for shortest path interval routing*, Research Report 95-02, LIP, École Normale Supérieure de Lyon, 69364 Lyon Cedex 07, France, Jan. 1995.
5. R. KRÁĽOVIČ, P. RUŽIČKA, AND D. ŠTEFANKOVIČ, *The complexity of shortest path and dilation bounded interval routing*, in 3rd International Euro-Par Conference, Aug. 1997.
6. M. LI AND P. M. B. VITÁNYI, *An Introduction to Kolmogorov Complexity and its Applications*, Springer-Verlag, 1993.
7. D. MAY AND P. THOMPSON, *Transputers and routers: Components for concurrent machines*, INMOS Ltd., (1990).
8. P. RUŽIČKA, *On efficient of interval routing algorithms*, in Mathematical Foundations of Computer Science (MFCS), M. Chytil, L. Janiga, and V. Koubek, eds., vol. 324 of Lectures Notes in Computer Science, Springer-Verlag, 1988, pp. 492–500.
9. N. SANTORO AND R. KHATIB, *Labelling and implicit routing in networks*, The Computer Journal, 28 (1985), pp. 5–8.
10. S. S. H. TSE AND F. C. M. LAU, *Two lower bounds for multi-interval routing*, in Computing: The Australasian Theory Symposium (CATS), Sydney, Australia, Feb. 1996.
11. S. S. H. TSE AND F. C. M. LAU, *An optimal lower bound for interval routing in general networks*, in 4th Colloquium on Structural Information & Communication Complexity (SIROCCO), July 1997.
12. J. VAN LEEUWEN AND R. B. TAN, *Interval routing*, The Computer Journal, 30 (1987), pp. 298–307.

Relating Conflict-Free Stable Transition and Event Models
(Extended Abstract)

Zurab Khasidashvili[1] and John Glauert[2]

[1] NTT Basic Research Laboratories, Atsugi, Kanagawa, 243-01, Japan
zurab@theory.brl.ntt.co.jp
[2] School of Information Systems, UEA, Norwich NR4 7TJ England
jrwg@sys.uea.ac.uk ***

Abstract. We describe an event-style (or poset) semantics for conflict-free rewrite systems, such as the λ-calculus, and other stable transition systems with a residual relation. Our interpretation is based on considering redex-families as events. It treats permutation-equivalent reductions as representing the same concurrent computation. Due to erasure of redexes, Event Structures are inadequate for such an interpretation. We therefore extend the Prime Event Structure model by axiomatizing permutation-equivalence on finite configurations in two different ways, for the conflict-free case, and show that these extended models are equivalent to known stable transition models with axiomatized residual and family relations.

1 Introduction

The goal of this paper is to provide a *fully adequate* concurrent semantics for orthogonal rewrite systems. We chose Event Structures, as the most successful and commonly accepted model of concurrency [Win80, NPW81, Win86/89, WN95]. Conflict-free Prime Event Structures (PESs) are sufficient for our purposes. A (conflict-free) PES is simply a set of events partially ordered by a *causal dependency* relation where an event can only dominate a finite number of events (the *axiom of finite causes*). The idea is to view a process as performing events (some *atomic* tasks), and an event can only occur after all events on which it depends. Causally independent events are *concurrent* events and can be evaluated in parallel. Stages of a computation are thus represented by *configurations* – left-closed event sets w.r.t. the causal order, and the same configuration represents all sequential computations executing events in that configuration, ignoring the order in which the concurrent events occur.

In the theory of orthogonal rewrite systems, such as the λ-calculus, there is the well established concept of 'redex-family', due to Lévy [Lév80], which formalizes the idea of the 'same atomic task'. Therefore it is natural to base our interpretation on considering redex-families as events, and to interpret (complete) family-reductions (contracting entire families as multi-steps) as configurations of the corresponding PES. Indeed, reductions in optimal graph-implementations of orthogonal systems precisely correspond to family-reductions in the original system.

*** Part of this work was supported by the Engineering and Physical Sciences Research Council of Great Britain under grant GR/H 41300

Recent advances in the abstract study of orthogonal rewrite systems [Sta89, GLM92, GK96, KG96, KG97] allow us to address the problem in an *entirely* abstract setting. Stable Deterministic Residual Structures (SDRSs) and Deterministic Family Structures (DFSs) model computation in orthogonal rewrite systems. DRSs are Abstract Rewrite Systems with an axiomatized residual relation; DFSs are DRSs where in addition the concept of redex-family is axiomatized. Stable DRSs allow [GK96, KG96] proofs of analogs of the Normalization and Standardization Theorems [Bar84]. In DFSs one can further prove Lévy's Optimality Theorem [Lév80] and the Unique Families Lemma [GK96]. The latter states that any family can be contracted at most once in a family-reduction, and corresponds to the fact that any event may occur at most once in the course of a computation.

Since we want to restrict our attention to family-reductions, we define *implementation* of DFSs [KG97a]. We show that the implementation of any DFS is a non-duplicating, or *affine*, DFS, with *zig-zag* [Lév80] as the family relation (AZDFS). Family-reductions in a DFS precisely correspond to reductions in its implementation, allowing us to concentrate on construction of event semantics for AZDFSs only. We have shown in [KG97] that any affine SDRS (or ASDRS) is in fact an AZDFS.

To view a process as the domain of all its configurations, an *adequacy* requirement is desirable. Namely, that *runs* of a (finite) configuration (i.e., all its possible sequentializations) should be in one-to-one correspondence with all reductions in the equivalence class which the configuration is supposed to represent. Because of the huge importance of *permutation* or *Lévy* equivalence [Lév80] to the whole theory of (not only orthogonal) rewrite systems [HL91, Bou85], and to concurrency theory as well [Sta89], we want to treat permutation-equivalent reductions as representing the same concurrent computation, as indeed permutation-equivalent reductions result one from another by permuting concurrent consecutive steps. However, this is clearly impossible in the present PES framework [Win86/89], as a configuration in a PES has only a finite number of runs while, for example, a permutation-equivalence class of a (finite) reduction may have an infinite number of elements.

In order for an equivalence relation to satisfy the adequacy requirement, the corresponding equivalence classes of (finite) reductions must be partially ordered so as to form distributive lattices, since the domains of (finite) configurations of a PES ordered by inclusion are such lattices [Win86/89]. However, it is well known that Lévy's reduction space is not in general a lattice, but only an upper semi-lattice [Lév80]. This problem, caused by erasure of redexes, is well discussed in [Lan94], where it is argued that event semantics say for the λ-calculus cannot therefore be based on permutation-equivalence.

Several authors have proposed different ways to get round this problem. Laneve [Lan94] introduces *distributive* equivalence on β-reductions which only allows permutation of steps that cannot erase or duplicate one another, and consequently all equivalent sequences have the same length. Kennaway et al [KKSV93] work with permutation equivalence for orthogonal Term Graph Rewriting, but they only restrict themselves to needed events of a normalizable term to cope with problems with erasure. As needed redexes cannot be erased, all needed normalizing reductions are of the same length, and on needed reductions, the permutation and distributive equivalences coincide (since the system is affine). Similarly, Clark and Kennaway [CK96] work with permutation-equivalence and allow infinite reductions but restrict themselves to standard reductions only. Schied [Sch94] and Corradini et al [CELMR94] work with linear rewrite systems (no duplication, no erasure), mainly graph grammars, and do not therefore encounter the erasure problems.

In order to cope with the adequacy problem, we extend the PES model by axiomatizing permutation-equivalence on event configurations. We do it in two different ways. One by directly constraining the class of all equivalence relations on configurations, and another by axiomatizing the *inessentiality* or *erasure* relation expressing redundancy of an event for a configuration. The latter extends by further axioms an incomplete axiomatization of the inessentiality relation in [GK96]. In the resulting extended models, Deterministic *Permutation* and *Erasure* Event Structures, DPESs and DEESs, we can identify configurations with different numbers of events but still representing the same permutation-equivalence class of reductions. We show that permutation-equivalence classes of configurations in a DPES/DEES, appropriately ordered, are isomorphic to Lévy's reduction space of the corresponding ASDRSs, and therefore the former give a *fully adequate* semantics to ASDRSs.

2   Deterministic Residual and Family Structures

Let us first recall some basic theory for DRSs and DFSs developed in [GK96, KG96, KG97]. We start by introducing *Abstract Reduction Systems* [Hin69]. An ARS is a triple $A = (Ter, Red, \rightarrow)$ where Ter is a set of *terms*, ranged over by t, s, o, e; Red is a set of *redexes* (or *redex occurrences*), ranged over by u, v, w; and $\rightarrow: Red \mapsto (Ter \times Ter)$ is a function such that for any $t \in Ter$ there is only a finite set of $u \in Red$ such that $\rightarrow (u) = (t, s)$, written $t \overset{u}{\rightarrow} s$. This set will be known as the redexes of term t, where $u \in t$ denotes that u is a member of the redexes of t and $U \subseteq t$ denotes that U is a subset of the redexes. Note that $\rightarrow$ is a *total* function, so one can identify u with the triple $t \overset{u}{\rightarrow} s$. A *reduction* is a sequence $t \overset{u_1}{\rightarrow} t_2 \overset{u_2}{\rightarrow} \ldots$. Reductions are denoted by P, Q, N. We write $P : t \twoheadrightarrow s$ or $t \overset{P}{\twoheadrightarrow} s$ if P denotes a reduction (sequence) from t to s. $P + Q$ denotes the concatenation of P and Q. u also denotes the reduction that contracts u.

DRSs axiomatize exactly the properties of residuals in orthogonal rewrite systems which are used in [Lév80, HL91] to define permutation equivalence on reductions, and the Lévy-embedding relation on permutation-equivalence classes, i.e., to construct Lévy's reduction space. Hence DRSs are the minimal axiomatization of the residual concept enabling one to relate the operational and denotational semantics of orthogonal rewrite systems. The most closely related models are Stark's *Determinate Concurrent Transition Systems* [Sta89], where the residual relation is non-duplicating, and *Abstract Reduction Systems* of Gonthier et al. [GLM92], where in addition the nesting relation on redexes is axiomatized.

Definition 2.1 A *Deterministic Residual Structure* (DRS) is a pair $\mathcal{R} = (A, /)$, where A is an ARS and $/$ is a *residual* relation on redexes relating redexes in the source and target term of every reduction $t \overset{u}{\rightarrow} s \in A$, such that for $v \in t$, the set v/u of *residuals of v under u* is a set of redexes of s; a redex in s may be a residual of only one redex in t under u, and $u/u = \emptyset$. If v has more than one u-residual, then u duplicates v. If $v/u = \emptyset$, then u erases v. A redex of s which is not a residual of any $v \in t$ under u is said to be u-new or created by u. The set u/P of residuals of u under any reduction P is defined by transitivity.

A *development* of $U \subseteq t$ is a reduction $P : t \twoheadrightarrow$ that only contracts residuals of redexes from U; it is *complete* if $U/P = \cup_{u \in U} u/P = \emptyset$. Development of $\emptyset$ is identified with the empty reduction. U will also denote a complete development of $U \subseteq t$. The residual relation satisfies the following two axioms:

• [FD] (*Finite Developments* [GLM92]) All developments are terminating; all complete developments of $U \subseteq t$ end at the same term; and residuals of a redex $v \in t$ under all complete developments of U are the same.

• [weak acyclicity] ([Sta89]) Let $u, v \in t$, $u \neq v$, and $u/v = \emptyset$. Then $v/u \neq \emptyset$.

We call a DRS $\mathcal{R}$ *stable* (SDRS) if:

• [stability] If $u, v \in t$ are different redexes, $t \xrightarrow{u} e$, $t \xrightarrow{v} s$, and u creates a redex $w \in e$, then the redexes in $w/(v/u)$ are not u/v-residuals of redexes of s.

Non-duplicating stable DRSs will be called *affine* SDRSs, or ASDRSs.

In a DRS $\mathcal{R}$, the residual relation on redexes is extended to all co-initial reductions exactly as in syntactic orthogonal rewrite systems [HL91, Lév78, Lév80, Sta89]: $(P_1 + P_2)/Q = P_1/Q + P_2/(Q/P_1)$ and $P/(Q_1 + Q_2) = (P/Q_1)/Q_2$, and *Lévy-equivalence* or *permutation-equivalence* is defined as the smallest relation on co-initial reductions satisfying: $U + V/U \approx V + U/V$ and $Q \approx Q' \Rightarrow P + Q + N \approx P + Q' + N$, where U and V are complete developments of redex sets in the same term. Further, one defines $P \trianglelefteq Q$ iff $P/Q = \emptyset$, and can show that $P \approx Q$ iff $P \trianglelefteq Q$ and $Q \trianglelefteq P$; and $P \trianglelefteq Q$ iff $Q \approx P + N$ for some N. Finally, one shows that for any co-initial reductions P, Q, $P \sqcup Q \approx Q \sqcup P$, where $P \sqcup Q = P + Q/P$, and is the greatest lower bound of P and Q in the reduction space ordered by Lévy's embedding relation $\trianglelefteq$. The above relations can equivalently be defined also using Klop's method of commutative diagrams [Klo80, Bar84].

Definition 2.2 ([KG96]) • Let $P : t \twoheadrightarrow o$ and $u \in t$, in a DRS. We call u *erased* in P if $u/P = \emptyset$. We call P *external* to u if it does not contract its residuals. We say that P *discards* u if P is external to u and erases it.

• We call u *P-needed* if there is no $Q \approx P$ that is external to u, and call it *P-unneeded* otherwise. We call u *P-essential* if there is no $Q \approx P$ that discards u, and *P-inessential* otherwise.

• Let $Q : t \twoheadrightarrow o$, $P : t \xrightarrow{P'} s \twoheadrightarrow e$, and $u \in s$. We call u, or more precisely $P'u$, *Q-(un)needed* (resp. *Q-(in)essential*) if u is Q/P'-(un)needed (resp. (in)essential). We call P *Q-needed* if so is every redex contracted in P, and call *standard* if it is P-needed. The other concepts above are extended in the same way.

Definition 2.3 A *Deterministic Family Structure* (DFS) is a triple $\mathcal{F} = (\mathcal{R}, \simeq, \hookrightarrow)$, where $\mathcal{R}$ is a DRS; $\simeq$ is an equivalence relation on redexes with *histories*; and $\hookrightarrow$ is the *contribution* relation on co-initial families, defined as follows:

(1) For any co-initial reductions P and Q, a redex Qv in the final term of Q (read as v *with history* Q) is called a *copy* of a redex Pu if $P \trianglelefteq Q$, i.e., $P + Q/P \approx Q$, and v is a Q/P-residual of u; the zig-zag relation $\simeq_z$ is the symmetric and transitive closure of the copy relation. The *family* relation $\simeq$ is an equivalence relation among redexes with histories containing $\simeq_z$. A *family* is an equivalence class of the family relation; families are ranged over by $\phi, \psi, \ldots$. $Fam(\ )$ denotes the family of its argument.

(2) Further, $\simeq$ and $\hookrightarrow$ satisfy the following axioms:

• [initial] Let $u, v \in t$ and $u \neq v$, in $\mathcal{R}$. Then $Fam(\emptyset_t u) \neq Fam(\emptyset_t v)$, where $\emptyset_t$ is the empty reduction starting from t.

- [contribution] $\phi \hookrightarrow \phi'$ iff for any $Pu \in \phi'$, P contracts at least one redex in ϕ.
- [creation] if $e \xrightarrow{P} t \xrightarrow{u} s$ and u creates $v \in s$, then $Fam(Pu) \hookrightarrow Fam((P+u)v)$.
- [FFD] (*Finite Family Developments*) Any reduction that contracts redexes of a finite number of families is terminating.

We call a DFS $\mathcal{F}$ a *zig-zag* DFS, ZDFS, if its family relation is the zig-zag $\simeq_z$ (the corresponding contribution relation $\hookrightarrow_z$ is determined by [contribution]). We call a DFS *separable*, if, for any redex Pv, v cannot create two different redexes in the same family: if v creates w', w'' and $w' \neq w''$, then $Fam((P+v)w') \neq Fam((P+v)w'')$.

Convention In this paper we only consider *comma-DRS*s and *comma-DFS*s. That is, the term set of any DRS will be the *reduction graph* of a term, called the *initial term*, i.e., the set of terms to which the initial term is reducible. The initial term will often be denoted by $\emptyset$, by analogy with the initial state in an Event Structure, although $\emptyset$ will also be used to denote empty reductions. Further, in DFSs, families will always be considered relative to $\emptyset$, i.e., all histories start with $\emptyset$. By a *reduction* we mean a *finite* reduction. Reductions starting from $\emptyset$ will also be called *initial* reductions.

It is shown in [GK96] that every DFS is a stable DRS. DFSs model all known concepts of family relation [Lév78, Lév80, Klo80, KS89, Mar91, AL93, Oos96], and allow for an abstract proof [GK96] of the Optimality theorem [Lév80]. Further, we have shown in [KG97] that any ASDRS is in fact an affine ZDFS, AZDFS (i.e., for any ASDRS $\mathcal{R}$, $\mathcal{F}_\mathcal{R} = (\mathcal{R}, \simeq_z, \hookrightarrow_z)$ is an AZDFS), and that $\mathcal{F}_\mathcal{R}$ is the only separable DFS with R as the underlying SDRS. The proof of the above result is based on an abstract concept of *extraction* allowing to construct a *canonical* element Pu in every family ϕ: P is standard and $\forall P'u' \in \phi \exists P'' : P' \approx P + P'' \wedge u' = u/P''$ (for any finite Q in an ASDRS, there is a standard reduction $ST(Q) \approx Q$ [KG96]). Further, it enables us to associate an AZDFS $\mathcal{F}_I$ to a DFS $\mathcal{F}$, called the *implementation* of $\mathcal{F}$. The reduction steps in $\mathcal{F}_I$ correspond to (complete) family-reduction multi-steps of $\mathcal{F}$ (i.e., multi-steps contracting maximal sets of redexes of some family).

Definition 2.4 Let $\mathcal{F} = (\mathcal{R}, \simeq, \hookrightarrow)$ be a DFS. The *Lévy implementation* or simply *implementation* of $\mathcal{F}$ is the AZDFS $\mathcal{F}_I = (\mathcal{R}_I, \simeq_I, \hookrightarrow_I)$, where
- the branches of the reduction graph of the underlying ARS A of $\mathcal{R}_I = (A, /)$ are family-reductions starting from $\emptyset$, each edge (i.e., a reduction step) being a multi-step contracting a family of redexes.
- the residual relation $/$ is defined as follows: let U and V be complete sets of redexes in two families, in a term s, and let $U : s \to o$ be the multi-step contracting U. Then V/U is the multi-step $o \to e$ contracting all members of the set V/U.
- the family and contribution relations $\simeq_I, \hookrightarrow_I$ are those induced by $\simeq$ and $\hookrightarrow$.

Theorem 2.5 ([KG97a]) For any DFS $\mathcal{F}$, $\mathcal{F}_I$ is an AZDFS.

3 Deterministic Permutation and Erasure Event Structures

In this section, we introduce Deterministic Permutation and Erasure Event Structures, DPESs and DEESs, which extend conflict-free Prime Event Structures by an axiomatized permutation-equivalence and erasure relations, respectively, and establish an equivalence between them.

The purpose of this extension is to capture *directly* the phenomenon of *erasure* in event/poset models. The phenomenon is typical for languages based on rewrite systems, but is not confined to them. In higher-order process calculi (e.g., [San93]) processes (that can fire events) can be passed around as messages and therefore erased. Event Structures are linear in nature and consequently the configuration domains enjoy nice lattice properties; therefore Event Structures have been very successful in modelling *linear* process calculi (no actions can be erased or duplicated), such as CCS [Mil89]. However, any attempt to adequately interpret higher-order process languages in Event Structures would face the same erasure problems discussed in the introduction. As in the case of reductions, the aim of axiomatizing permutation equivalence on event configurations is to be able to equate configurations which differ only by irrelevant or *inessential* events.

A *Prime Event Structure* (PES) [Win86/89] is a triple $\mathcal{E} = (E, Con, \leq)$, where E is a set of *events*, ranged over by $e, e_1, \ldots$; the *consistency predicate* Con is a nonempty set of finite subsets of E, denoted by $X, Y, \ldots$; and the *causal dependency relation* $\leq$ is a partial order on E, such that $\{e\} \in Con$, $Y \subseteq X \in Con \Rightarrow Y \in Con$, $X \in Con \wedge \exists e' \in X.e \leq e' \Rightarrow X \cup \{e\} \in Con$, and $\{e' \mid e' \leq e\}$ is finite for any $e \in E$.

We only consider *conflict-free* or *deterministic* PESs, where no event can prevent others from occurring, and therefore the consistency predicate is the set of all finite subsets of E, $Fin(E)$, and will be omitted. Finite *configurations* of $\mathcal{E}$ are finite *left-closed* subsets $\alpha, \beta, \ldots$ of E, i.e., subsets $\mathcal{L}(E) = \{\alpha \subseteq Fin(E) \mid e \in \alpha \wedge e' < e \Rightarrow e' \in \alpha\}$. We only consider finite configurations, and will omit 'finite'.

Below, for better readability, we write $\alpha + e$ for $\alpha \cup \{e\}$, and write $\alpha - e$ for $\alpha \setminus \{e\}$. Further, we define $\lceil e^{\leq} \rceil = \{e' \mid e' \leq e\}$, $\lceil e^{<} \rceil = \{e' \mid e' < e\}$, and $\lceil e^{\geq} \rceil = \{e' \mid e \geq e'\}$.

Definition 3.1 A *Deterministic Permutation Event Structure* (DPES) is a triple $\mathcal{P} = (E, \leq, \approx)$, where $(E, \leq)$ is a conflict-free PES and $\approx \subseteq \mathcal{L}(E) \times \mathcal{L}(E)$ is an equivalence relation, called *Lévy* or *permutation-equivalence*, satisfying the following axioms, where $\alpha, \beta, \gamma \in \mathcal{L}(E)$:

- [P0] $\forall e \in E : \lceil e^{<} \rceil \not\approx \lceil e^{\leq} \rceil$;
- [P1] $\alpha \approx \beta \wedge \alpha \cap \beta \subseteq \gamma \subseteq \alpha \cup \beta \Rightarrow \alpha \approx \gamma$;
- [P2] $\alpha \approx \beta \wedge \alpha + e, \beta + e \in \mathcal{L}(E) \Rightarrow \alpha + e \approx \beta + e$;
- [P3] $\alpha \approx \alpha - e \wedge e < e' \wedge \alpha + e' \in \mathcal{L}(E) \Rightarrow \alpha \approx \alpha + e'$;

DEESs extend *Deterministic Computation Structures* (DCS) [GK96] by further erasure axioms to enable a fully adequate treatment of erasure.

Definition 3.2 A *Deterministic Erasure Event Structure* (DEES) is a triple $\mathcal{C} = (E, \leq, \rhd)$, where $(E, \leq)$ is a conflict-free PES and $\rhd \subseteq \mathcal{L}(E) \times E$ is *inessentiality* or *erasure* relation, satisfying the following axioms, where $\alpha, \beta \in \mathcal{L}(E)$:

- [E0] $\forall e \in E : \emptyset \not\rhd e$;
- [E1] $\alpha \rhd e \wedge \alpha \subseteq \beta \Rightarrow \beta \rhd e$;
- [E2] $\alpha \rhd e' \wedge \alpha \rhd e \wedge \alpha - e' \in \mathcal{L}(E) \Rightarrow \alpha - e' \rhd e$;
- [E3] $\alpha \rhd e \wedge e < e' \Rightarrow \alpha \rhd e'$;
- [E4] $\alpha \cup \lceil e^{<} \rceil \rhd e \Rightarrow \alpha \rhd e$.

We read $\alpha \rhd e$ as: 'e is α-inessential'.

The next definition gives a translation between DPESs and DEESs:

Definition 3.3 (1) For a DPES $\mathcal{P} = (E, \leq, \approx)$, we define an *erasure* relation $\rhd_\approx \subseteq$ $\mathcal{L}(E) \times E$ and the corresponding structure $\mathcal{C_P} = (E, \leq, \rhd_\approx)$ as follows:
- $\alpha \rhd_\approx e$ iff $\alpha \cup \lceil e^\leq \rceil \approx (\alpha \cup \lceil e^\leq \rceil) \setminus \lceil e^\geq \rceil$.

(2) For a DEES $\mathcal{C} = (E, \leq, \rhd)$, we define an equivalence relation $\approx_\rhd \subseteq \mathcal{L}(E) \times$ $\mathcal{L}(E)$ and the corresponding structure $\mathcal{P_C} = (E, \leq, \approx_\rhd)$ as follows:
- $\alpha \approx_\rhd \beta$ iff $ST_\rhd(\alpha) = ST_\rhd(\beta)$, where $ST_\rhd(\alpha) = \{e \in \alpha \mid \alpha \not\rhd e\}$ is the set of α-*essential* events of α. (The subscript $\rhd$ in $ST_\rhd(\alpha)$ will often be omitted.) $ST(\alpha)$ is called the *standard variant* of α.

The DPES axiom [P0] is a counterpart of the DEES axioms [E0] and [E4], and similarly for [P3] and [E3]. Further, the combination of [P1] and [P2] have the same effect as that of [E1] and [E2]. [E4] does not have a 'direct' counterpart among DPES axioms, but it corresponds to the fact that, according to Definition 2.2, a redex with history Pv is Q-inessential iff v is Q/P-inessential.

The following theorem, which is proved using the DEES and DPES axioms, implies the correctness of Definition 3.3.

Theorem 3.4 (1) For any DEES $\mathcal{C} = (E, \leq, \rhd)$, $\mathcal{P_C} = (E, \leq, \approx_\rhd)$ is a DPES.
(2) For any DPES $\mathcal{P} = (E, \leq, \approx)$, $\mathcal{C_P} = (E, \leq, \rhd_\approx)$ is a DEES.

Further, the translations between DEESs and DPESs, in Definition 3.3, commute, implying the equivalence of the two event models:

Theorem 3.5 (1) For any DEES $\mathcal{C} = (E, \leq, \rhd)$, $\mathcal{C_{P_C}} = \mathcal{C}$.
(2) For any DPES $\mathcal{P} = (E, \leq, \approx)$, $\mathcal{P_{C_P}} = \mathcal{P}$.

4 Relating Conflict-Free Transition and Event models

We now define translations of ASDRSs and AZDFSs into DPESs and DEESs, define the converse translations, and show that the translations commute, implying equivalence of the four computational models.

Definition 4.1 To an AZDFS $\mathcal{F} = (\mathcal{R}, \simeq_z, \hookrightarrow_z)$, we associate the DPES $\mathcal{P_F} = (E_\mathcal{F}, \leq, \approx)$, and the DEES $\mathcal{C_F} = (E_\mathcal{F}, \leq, \rhd)$, where
- $E_\mathcal{F} = FAM(\mathcal{F})$, the set of all families of $\mathcal{F}$;
- $\phi < \psi$ iff $\phi \hookrightarrow_z \psi$;
- $\alpha \approx \beta$ iff $\forall \phi \in \alpha \cup \beta \setminus \alpha \cap \beta, \exists P : \emptyset \twoheadrightarrow s : FAM(P) \subseteq \alpha \cap \beta$ and P discards a canonical element of ϕ, where $FAM(P)$ denotes the set of families whose member redexes are contracted in P.
- $\alpha \rhd \phi$ iff $\exists P : \emptyset \twoheadrightarrow s : FAM(P) \subseteq \alpha$ and P discards a canonical element of ϕ.

Theorem 4.2 For any AZDFS $\mathcal{F}$, $\mathcal{C_F}$ is a DEES, $\mathcal{P_F}$ is a DPES, $\mathcal{P_F} = \mathcal{P_{C_F}}$ and $\mathcal{C_F} = \mathcal{C_{P_F}}$.

The following translation of DPESs and DEESs into ASDRSs resembles to the representation of events in a PES $(E, \leq)$ as prime intervals of the configuration domain $\mathcal{L}(E)$ [Win86/89].

Definition 4.3 To a DPES $\mathcal{P} = (E, \leq, \approx)$ (resp. a DEES $\mathcal{C} = (E, \leq, \rhd)$) we associate an SDRS $\mathcal{R}_\mathcal{P}$ (resp. $\mathcal{R}_\mathcal{C}$) as follows:

• The terms of $\mathcal{R}_\mathcal{P}$ ($\mathcal{R}_\mathcal{C}$) are Lévy-equivalence classes $\langle\alpha\rangle_L, \langle\beta\rangle_L, \ldots$ of configurations of $\mathcal{P}$ ($\mathcal{C}$);

• The reduction relation of $\mathcal{R}_\mathcal{P}$ ($\mathcal{R}_\mathcal{C}$) consists of sets of pairs $u = (\langle\alpha\rangle_L, e)$, where $\alpha, \alpha + e \in \mathcal{L}(E)$ and $\alpha \not\approx \alpha + e$ (resp. $\alpha \not\rhd e$). Pairs $u = (\langle\alpha\rangle_L, e)$ such that $\alpha \approx \alpha + e$ ($\alpha \rhd e$) are identified with the empty redex $\emptyset = (\langle\alpha\rangle_L, \emptyset)$ in $\langle\alpha\rangle_L$.

• The residual relation is defined as follows: if $u = (\langle\alpha\rangle_L, e)$ and $v = (\langle\alpha\rangle_L, e')$, then $u/v = (\langle\alpha + e'\rangle_L, e)$.

We show that $\mathcal{R}_\mathcal{P}$ and $\mathcal{R}_\mathcal{C}$ are ASDRSs, $\mathcal{R}_\mathcal{P} = \mathcal{R}_{\mathcal{C}_\mathcal{P}}$, and $\mathcal{R}_\mathcal{C} = \mathcal{R}_{\mathcal{P}_\mathcal{C}}$.

Let us call a left-closed enumeration (i.e., respecting the causality relation) $e_1, \ldots, e_k$ of a configuration α in a DEES a *reduction*, written $[\alpha]$, if no event e_i in the sequence is *vacuous*, i.e., $\forall i : \{e_1, \ldots, e_{i-1}\} \not\rhd e_i$. Any other left-closed enumeration of α will be called a *quasi-reduction*. Quasi-reductions are defined similarly in any DRS, as 'reductions containing empty steps corresponding to erased redexes'.

Note that initial reductions in $\mathcal{R}_\mathcal{P}$ and $\mathcal{R}_\mathcal{C}$ can be represented as event sequences $e_1, \ldots, e_n$ respecting the causal order – e_i represents the (contraction of) *non-empty* redex $(\{e_1, \ldots, e_{i-1}\}, e_i)$, i.e., they correspond to reductions in $\mathcal{C}$. As we will see below, not all left-closed enumerations of a configuration represent a reduction – they represent quasi-reductions.

Definition 4.4 To a DPES $\mathcal{P} = (E, \leq, \approx)$, we associate an affine DFS $\mathcal{F}_\mathcal{P} = (\mathcal{R}_\mathcal{P}, \simeq_\mathcal{P}, \hookrightarrow_\mathcal{P})$ as follows (where $\mathcal{R}_\mathcal{P}$ is the DRSs defined in Definition 4.3):

• $\simeq_\mathcal{P}$-families are sets $\phi_e = \{([\alpha], e) \mid \alpha + e \in \mathcal{L}(\mathcal{P}) \wedge (\langle\alpha\rangle_L, e) \neq \emptyset\}$, for any $e \in E$.

• $\hookrightarrow_\mathcal{P}$ is defined by: $\phi_e \hookrightarrow_\mathcal{P} \phi_{e'}$ iff $e < e'$.

One easily shows that translations $\mathcal{P} \to \mathcal{F}_\mathcal{P}$ and $\mathcal{P} \to \mathcal{R}_\mathcal{P}$ commute with the zig-zag embedding $\mathcal{R} \to \mathcal{F}_\mathcal{R}$, i.e., $\mathcal{F}_{\mathcal{R}_\mathcal{P}} = (\mathcal{R}_\mathcal{P}, \simeq_z, \hookrightarrow_z) = (\mathcal{R}_\mathcal{P}, \simeq_\mathcal{P}, \hookrightarrow_\mathcal{P}) = \mathcal{F}_\mathcal{P}$, implying in particular the correctness of Definition 4.4: $\mathcal{F}_\mathcal{P}$ is an AZDFS.

For any of the four models, there is an obvious concept of *isomorphism*, $\rightleftharpoons$. The following theorem, combined with the results of previous sections, implies equivalence of the considered four transition and event models.

Theorem 4.5 (1) For any DPES $\mathcal{P}$, $\mathcal{P} \rightleftharpoons \mathcal{P}_{\mathcal{F}_\mathcal{P}}$.

(2) For any AZDFS $\mathcal{F}$, $\mathcal{F} \rightleftharpoons \mathcal{F}_{\mathcal{P}_\mathcal{F}}$.

5 Configuration Domains

Note that, unlike PESs, not every configuration α in a DEES has a reduction. For example, consider the DEES consisting of four events $\{e_1, e_2, e_1', e_2'\}$, ordered as follows: $e_1 < e_1'$ and $e_2 < e_2'$, and let $\rhd$ be given by: $\alpha \rhd e_1'$ iff $e_2 \in \alpha$, and $\alpha \rhd e_2'$ iff $e_1 \in \alpha$. Then $\{e_1, e_2, e_1', e_2'\}$ is a configuration, but any of its is left-closed enumerations, such as e_1, e_2, e_1', e_2', ends either with e_1' or with e_2', and we have $\{e_1, e_2, e_2'\} \rhd e_1'$ and $\{e_1, e_2, e_1'\} \rhd e_2'$. This is not surprising as the vacuous events in left-closed sequences correspond to empty reductions in the corresponding ASDRS, performed in the 'garbage' (erased redexes). However, it is easy to see that, for any $\alpha \in \mathcal{L}(\mathcal{C})$, in a DEES $\mathcal{C}$, any left-closed enumeration of $ST(\alpha)$ is a reduction.

Definition 5.1 (1) Let $\mathcal{P}$ be a DPES. Then $\mathcal{L}^{\approx}(\mathcal{P}) = \mathcal{L}(\mathcal{P})/_{\approx} = \{\langle\alpha\rangle_L \mid \alpha \in \mathcal{L}(\mathcal{P})\}$ and $\mathcal{L}^{\trianglelefteq}(\mathcal{P}) = (\mathcal{L}^{\approx}(\mathcal{P}), \trianglelefteq)$, where $\trianglelefteq$ is a partial order defined by: $\langle\alpha\rangle_L \trianglelefteq \langle\beta\rangle_L$ iff $\exists\alpha' \in \langle\alpha\rangle_L, \exists\beta' \in \langle\beta\rangle_L : \alpha' \subseteq \beta'$.

(2) Let $\mathcal{R}$ be an ASDRS. Then $\mathcal{L}^{\approx}(\mathcal{R}) = \mathcal{L}(\mathcal{R})/_{\approx} = \{\langle P\rangle_L \mid P \in \mathcal{L}(\mathcal{R})\}$, where $\mathcal{L}(\mathcal{R})$ is the set of initial reductions in $\mathcal{R}$ and $\langle P\rangle_L$ is the Lévy-equivalence class of P, and $\mathcal{L}^{\trianglelefteq}(\mathcal{R}) = (\mathcal{L}^{\approx}(\mathcal{R}), \trianglelefteq)$, where $\trianglelefteq$ is Lévy's embedding relation.

It has been shown in [KG97] that $P \trianglelefteq Q$ iff $FAM(P) \subseteq FAM(Q')$ for some $Q' \approx Q$, implying that $\langle P\rangle_L \trianglelefteq \langle Q\rangle_L$ in an ASDRS $\mathcal{R}$ iff $\langle FAM(P)\rangle_L \trianglelefteq \langle FAM(Q)\rangle_L$ in $\mathcal{P}_{\mathcal{F}_\mathcal{R}}$. Therefore

Theorem 5.2 Let $\mathcal{R}$ be an ASDRS and let $\mathcal{P} = \mathcal{P}_{\mathcal{F}_\mathcal{R}}$ be its corresponding DPES. Then $f : \langle P\rangle_L \rightarrow \langle FAM(P)\rangle_L$ is an isomorphism between $\mathcal{L}^{\trianglelefteq}(\mathcal{R})$ and $\mathcal{L}^{\trianglelefteq}(\mathcal{P})$. Moreover, (quasi-)reductions in $\langle P\rangle_L$ are in one-to-one correspondence with these in $\langle FAM(P)\rangle_L$.

Note that $\mathcal{L}^{\trianglelefteq}(\mathcal{P})$ is not a lattice since $\mathcal{L}^{\trianglelefteq}(\mathcal{R})$ is not. However, depending on the set of 'results' we are interested in, one can construct a (distributive) lattice on configurations in $\mathcal{P}$ reflecting growth of information relative to that goal. Construction of such 'relativized' information domains will be addressed in a separate paper.

6 Conclusions

We have established an equivalence between deterministic stable operational (ASDRS) and domain-theoretic (DPES/DEES) models of computation, and based on this, constructed a fully adequate event style concurrent semantics for orthogonal rewrite systems. The correspondence between Event Structures and other models of concurrency is well studied [Win86/89, WN95], and we hope that our results contribute to better understanding of the relationship between operational and denotational semantics of sequential and concurrent languages. Further, we think that our axiomatization of permutation equivalence is interesting on its own right from the poset/lattice theoretic point of view.

We expect that the theory developed here for conflict-free systems can be extended to the general case, so that higher order process calculi in which processes can be passed and erased can be modelled as well.

Acknowledgments
We thank Richard Kennaway and Fer-Jan de Vries for useful comments.

References

[AL93] Asperti A., Laneve C. Interaction Systems I: The theory of optimal reductions. MSCS 11:1-48, 1993.

[Bar84] Barendregt H. P. The Lambda Calculus, its Syntax and Semantics. 1984.

[Bou85] Boudol G. Computational semantics of term rewriting systems. In: Algebraic methods in semantics. Nivat M., Reynolds J.C., eds. Camb. Univ. Press, 1985, p. 169-236.

[CK96] Clark D., Kennaway R. Event structures and non-orthogonal term graph rewriting. MSCS 6:545-578, 1996.

[CELMR94] Corradini A., Ehrig H., Löwe M., Montanari U., Rossi F. An event structure semantics for safe graph grammars. PROCOMET'94, IFIP Transactions A-56, 1994.

[GK96] Glauert J.R.W., Khasidashvili Z. Relative normalization in deterministic residual structures. CAAP'96, Springer LNCS, vol. 1059, H. Kirchner, ed. 1996, p. 180-195.

[GLM92] Gonthier G., Lévy J.-J., Melliès P.-A. An abstract Standardisation theorem. In: Proc. of LICS'92, Santa Cruz, California, 1992, p. 72-81.

[Hin69] Hindley R.J. An abstract form of the Church-Rosser theorem. JSL 34(4):545-560, 1969.

[HL91] Huet G., Lévy J.-J. Computations in Orthogonal Rewriting Systems. In: Computational Logic, Essays in Honor of Alan Robinson, J.-L. Lassez and G. Plotkin, eds. MIT Press, p. 394-443, 1991.

[KS89] Kennaway J. R., Sleep M. R. Neededness is hypernormalizing in regular combinatory reduction systems. Report. University of East Anglia, 1989.

[KKSV93] Kennaway J. R., Klop J. W., Sleep M. R, de Vries F.-J. Event structures and orthogonal term graph rewriting. In Sleep M. R., Plasmeijer M. J., van Eekelen M. C. J. D., eds. Term Graph Rewriting: Theory and Practice. John Wiley, p. 141-156, 1993.

[KG96] Khasidashvili Z., Glauert J. R. W. Discrete normalization and Standardization in Deterministic Residual Structures. ALP'96, Springer LNCS, vol. 1139, M. Hanus, M. Rodríguez-Artalejo, eds. 1996, p.135-149.

[KG97] Khasidashvili Z., Glauert J.R.W. Zig-zag, extraction and separable families in non-duplicating stable Deterministic Residual Structures. Technical Report IR-420, Free University, Amsterdam, February 1997.

[KG97a] Khasidashvili Z., Glauert J.R.W. An abstract concept of optimal implementation. Report SYS-C97-??, UEA, Norwich, 1997.

[Klo80] Klop J. W. Combinatory Reduction Systems. Mathematical Centre Tracts n. 127, CWI, Amsterdam, 1980.

[Lan94] Laneve C. Distributive evaluations of λ-calculus, Fundamenta Informaticae, 20(4):333 – 352, 1994.

[Lév78] Lévy J.-J. Réductions correctes et optimales dans le lambda-calcul, Thèse de l'Université de Paris VII, 1978.

[Lév80] Lévy J.-J. Optimal reductions in the Lambda-calculus. In: To H. B. Curry: Essays on Combinatory Logic, Lambda-calculus and Formalism, Hindley J. R., Seldin J. P. eds, Academic Press, 1980, p. 159-192.

[Mar91] Maranget L. Optimal derivations in weak λ-calculi and in orthogonal Term Rewriting Systems. In: Proc. of POPL'91, p. 255-269.

[Mil89] Milner R. Communication and concurrency. Prentice Hall, 1989.

[NPW81] Nielsen M., Plotkin G., Winskel G. Petri nets, event structures and domains. Part 1. TCS 13:85-108, 1981.

[Oos96] Van Oostrom V. Higher order families. In: Proc. of RTA'96, Springer LNCS, vol. 1103, Ganzinger, H., ed., 1996, p. 392–407.

[San93] Sangiorgi D. Expressing mobility in process algebras: first-order and higher-order paradigms. Ph.D. Thesis, Edinburgh University, 1993.

[Sch94] Schied G. On relating rewrite systems and graph grammars to event structures. Springer LNCS, vol. 776, Schneider H.-J. Ehrig H., eds, 1994, p. 326-340.

[Sta89] Stark E. W. Concurrent transition systems. TCS 64(3):221-270, 1989.

[Win80] Winskel G. Events in Computation. Ph.D. Thesis, University of Edinburgh, 1980.

[Win86/89] Winskel G. An introduction to Event Structures. Springer LNCS, vol. 354, 1989, p. 364-397. Full version: Cambr. Univ. Comp. Lab. Report n. 95, 1986.

[WN95] Winskel G., Nielsen M. Models for concurrency. In: S. Abramsky, D. Gabbay, T. Maibaum eds. Handbook of Logic in Computer Science, vol. 4, Oxford Univ. Press, p. 1-148, 1995.

The Giant Component Threshold for Random Regular Graphs with Edge Faults

Andreas Goerdt

TU Chemnitz, Theoretische Informatik
D-09107 Chemnitz, Germany

Abstract. Let G be a given graph (modelling a communication network) which we assume suffers from static edge faults: That is we let each edge of G be present independently with probability p (or absent with *fault probability* $f = 1 - p$). In particular we are interested in robustness results for the case that the graph G itself is a random member of the class of all regular graphs with given degree.

Our result is: If the degree d is fixed then $p = 1/(d-1)$ is a threshold probability for the existence of a linear-sized component in a faulty version of almost all random regular graphs. We show: If each edge of an *arbitrary* graph G with maximum degree bounded above by d is present with probability $p = \lambda/(d-1)$ where $\lambda < 1$ is fixed then the faulted version of G has only components whose size is at most logarithmic in the number of nodes with high probability. If on the other hand G is a random regular graph with degree d and $p = \lambda/(d-1)$ where $\lambda > 1$ then for almost all G the faulted version of G has a linear-sized component with high probability. Note that these results imply some kind of optimality of random regular graphs among the class of graphs with the same degree bound.

The theme is: Use the known expansion properties of almost all random regular graphs to obtain strong robustness results. This has not been done systematically before.

Introduction

Modern multiprocessor architectures and communication networks compute over structured interconnection graphs like meshes. Here several applications share the same network while executing concurrently. This may of course lead to unavailability of links and nodes in certain cases and we may assume to compute over a subnetwork being randomly assigned by the operating system. Moreover, this subnetwork may suffer from edge or node faults. Our work addresses robustness properties in case the subnetwork is a random regular graph suffering from edge faults. Random regular graphs make an (at least theoretically) popular choice because they combine low degree with high expansion almost always (see [AS 92] for an introduction to expansion and [Bo 85] to random regular graphs). Our study continues the work begun in [Ni et al. 94] [NiSp 95] which are the only papers known to us which investigate random regular graphs with edge faults. In particular we build on the first of the abovementioned papers, where the authors show: Almost all faulty versions (i. e. with probability tending to 1 when the number of nodes goes to infinity) of random regular graphs with degree d where $64 < d < (1/2)\sqrt{\log n}$ and $p \geq 32/d$ have a linear-sized component. The second of the abovementioned papers deals with expansion properties of such a linear-sized component. For the case that $d \geq 3$ is fixed, we strengthen the result of [Ni et al. 94] as far as possible. Precise formulations of our results are theorems 1 and 2.

Our result adds to the list of theorems giving threshold probabilities for the existence of a linear-sized component: If G is the fully connected graph we are in the realm of the classical theory of random graphs. Threshold for a linear size component is $p = 1/n$, n being the number of nodes. In case of the $d \times d$ 2−dimensional grid we have $p = 1/2$. (At least in [KaNeTa 94] it is assured that results from [Ke 80] can be looked at this way). If G is the hypercube in d dimensions, i. e. we have $n = 2^d$ nodes, the threshold for a linear-sized component is $1/d$ [AjKoSz 82]. Without giving the actual value, the existence of a threshold probability roughly between one third and one half in case of the butterfly network is shown in [KaNeTa 94].

Note that in case of the fully connected G and the hypercube the threshold probability is something like the inverse of the degree. Our threshold probability for linear-sized components in faulty random regular graphs fits in nicely.

In spite of the fact that the existence of a linear-sized component is not the only measure of the robustness of a network it is a *necessary* requirement for efficiently simulating computations of the fault free graph in the faulted version: Without a linear-sized component computation on the faulted version must be slower by more than a constant factor than computation on the unfaulted version. In case of the 2−dimensional grid the threshold probability of a linear-sized component coincides with the threshold probability of other properties related more closely to the efficiency of computation, such as routing or embedding the fault-free version into the faulty one [Ra 89] [Ma 92].

1   The threshold proof

Let $H = (V, F)$ be an undirected graph of maximum degree $\leq d$ where d is a fixed constant. According to our fault model we consider the probability space of faulty versions $G = (V, E)$ of H where $E \subseteq F$. The probability of G is defined by: Each edge of F is present, i. e. in E, independently with probability $p < 1/(d-1)$, where p is a fixed constant.

Chernoff bound considerations similar to those on p. 154 of [AS 92] analyzing a breadth-first search algorithm starting from a given node imply:

Theorem 1. The probability of those faulty versions $G = (V, E)$ of H which have a component with $O(\ln n)$ nodes is $O(1/n)$. That is *almost all* faulty versions G are such that *all* components have $O(\ln n)$ nodes, the constant factor in the O-notation depending on the probability p.

For the rest of the section we fix the following parameters.

- The degree d of our graphs. We assume $d \geq 3$.
- The probability that a given edge is *not* faulty:

$$p = \frac{\lambda}{d-1} > \frac{1}{d-1}, \text{ hence } \lambda > 1$$

The probability space of faulty random regular graphs $G_{n,p}^d$ is defined as in [Ni et al. 94]. It is important to understand that the probability of 2 faulty graphs with the same number of edges need not be the same . It is helpful to think of the space of faulty graphs as consisting of pairs (G, H) where H is a subgraph of G. Then 2 pairs where the H's have the same number of edges are equally likely.

In the sequel we prove the following theorem

Theorem 2. Let $d \geq 3$ be fixed. Almost all faulty random regular graphs have a connected component with at least $\varepsilon \cdot n$ nodes. The constant $\varepsilon = \varepsilon(\lambda)$ depends only on λ, not on d (and of course not on n).

We assume familiarity with the notion configuration. This can easily be obtained from the cited standard literature on random regular graphs.

Definition 3. Let $W_1, \ldots W_n$ be as above n disjoint d-element classes. A random configuration with edge fault probability $f = 1 - p$ is obtained by the following stochastic experiment: First, choose a configuration F from the probability space of all configurations. Second, delete each edge of F independently with the fault probability $f = 1 - p$.

Note that 1 faulty configuration can be obtained from seneral F's. A faulty configuration obtained by the second step from a given F is called a faulty version of F. Moreover each faulty configuration with k edges has the same probability. Nevertheless it may sometimes be helpful to look at the space of faulty configurations as a space of pairs as in the case of faulty random regular graphs described above.

Each configuration with edge faults induces a multigraph with edge faults. In particular we get: A property that holds for almost every configuration with edge faults holds for every faulty random regular graph, too [Ni et al. 94]. Accordingly, to prove theorem 2 we prove the analogous result for random configurations with edge faults: For d and p as fixed almost every faulty configuration has a connected component consisting of at least εn classes.

Our proof makes use of the basic technique as introduced in [AjKoSz 82] for the faulty d-dimensional cube. This technique has been applied to the faulty butterfly network in [KaNeTa 94]. We need to fix 2 additional parameters:

- A probability p' with

$$p > p' = \frac{\lambda'}{d-1} > \frac{1}{d-1}, \text{ hence } \lambda > \lambda' > 0$$

- And

$$p'' = 1 - \frac{1-p}{1-p'} > 0. \text{ We let } \delta = \delta(\lambda) = \lambda - \lambda'.$$

Note that $(1-p')(1-p'') = 1 - p$ and $p'' = p''(\lambda, d) \geq \frac{\delta}{d}$.

We consider a random configuration with fault probability $f = 1 - p$ as being obtained by 2 independent stochastic experiments:

- Experiment 1: We choose a configuration F. Then we construct a configuration F' with fault probabilty $f' = 1 - p' > 1 - p$ by throwing each edge of F independently with probability p'.
- Experiment 2: We throw each edge of F with probability p'' into F' to get F''. In F'' each edge of F is absent with probability $(1 - p')(1 - p'') = 1 - p$, hence F'' is a random configuration with fault probability $f = 1 - p$.

Our proof follows these 2 experiments above:

- Step 1: With high probability the following holds in F': We have at least $\varepsilon' n$ classes which are distributed over at most $\varepsilon' n / \omega(n)$ connected components. (Note that this is weaker than saying $\varepsilon' n$ classes belong to components of size at least $\omega(n)$ each).We call these components *atoms*. The funtion $\omega(n)$ is fixed just before 7, $\varepsilon' = \varepsilon'(\lambda)$ is fixed only after lemma 11. At the moment we only note that $\omega(n)$ is a function independent of d and going (slowly) to infinity. The constant ε' is a constant depending on λ only.
- Step 2: With high probability the edges thrown in by the second experiment induce paths such that the following holds: The atoms which are connected by these paths yield a component having at least εn classes. Again $\varepsilon = \varepsilon(\lambda)$ is fixed further below.

We prove step 2 first. Not surprisingly the proof relies on expansion properties of almost all random configurations [Bo 88] when $d \geq 3$:

Fact 4. There is a constant (independent of d and n) $c > 0$ such that for a random configuration F the following holds with high probability:
(a) For all subsets X of classes we have:

$$| \partial X | \geq c \cdot d \cdot \text{Min}\{| X |, | \text{Cpl } X |\}$$

where ∂X is the set of edges of F connecting a class from the set of classes X with a class from Cpl X, the complement of X.
(b)For all X we have

$$| \mathcal{N}(X) | \geq c \cdot \text{Min}\{| X |, | \text{Cpl } X |\},$$

where $\mathcal{N}(X)$ is the set of classes adjacent to X in F but not belonging to X.

Conditioning on the set of those configurations F which satisfy the following lemma yields step 2:

Lemma 5. For almost every configuration F holds: Let F' be a faulty version of F such that $\varepsilon'n$ classes belong to $\varepsilon'n/\omega(n)$ atoms. In the probability space of faulty versions of F obtained by throwing each edge of F with probability p'' into F' the following holds with high probability: The newly thrown in edges induce paths which merge at least one third of all classes in atoms (i. e. $(1/3)\cdot\varepsilon'n$ classes) into 1 connected component.

Proof. The proof relies essentially on fact 4.

Hence step 2 of the proof of theorem 2 is achieved by letting $\varepsilon = \frac{1}{3}\varepsilon'$.

Outline of step 1 of the proof of theorem 2: The existence of the $\varepsilon' \cdot n$ classes and the corresponding atoms is shown by running sufficiently many breadth-first searches on disjoint parts of a faulty configuration. We restrict the number of search steps of each search to $\omega(n)$, a slowly growing function defined below. The choice of $\omega(n)$ implies that the following holds with high probability for *all* breadth-first searches run on 1 faulty configuration: Each search hits at most 1 class 2-times. This allows us to restrict attention to those parts of each search, where no overlapping occurs. These parts have a tree-like structure and therefore behave like branching processes.

The following fact 6 (a) is proved in [AS 92] [Fe 68], (b) can be found as claim 3 in [AjKoSz 81].

Fact 6. We consider a branching process with offspring distribution $\mathrm{Bin}(m,p)$, the binomial distribution where m is the number of trials and p the probability of success.

(a) If the expectation $m\cdot p$ of the offspring distribution is > 1, then the probability that the process dies after a finite number of generations is < 1.

(b) We even have the following uniform version of (a): If $\lambda_0 > 1$ is fixed, there exists a $\mu = \mu_0(\lambda_0) < 1$ such that *for all* m, p with $m \cdot p = \lambda_0$ the probability that the process dies is $\leq \mu_0(\lambda_0) < 1$

So the idea is to run disjoint breadth-first searches each performing at most $\omega(n)$ search steps until altogether a linear number of classes has been detected. Fact 6 (and some tail bounds) ensure that a constant fraction of these classes belong to sufficiently few components.

But here, an additional point arises, as can be seen from the following figure. It shows a typical substructure discovered by a breadth-first search starting at the class $S.($ We assume $d = 4$, $\omega(n) = 9)$. Faulty edges are drawn as dotted lines, class W_4 is hit 2-times:

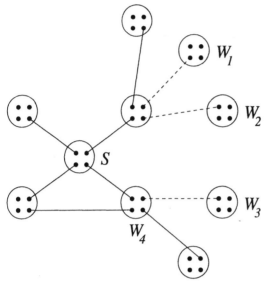

Interior classes are connected to S via non-faulty paths. Some technical problem is caused by the classes like W_1, W_2, W_3, which are detected by a faulty edge. The problem is that they do not belong to the connected component of S, nevertheless we do not want them to be considered by subsequent searches. Otherwise we would destroy the independence (and hence analyzability) of the various searches. Moreover, the number of classes found by a faulty edge can only be bounded (in the very worst-case) as being d-times the number of interior classes. If we would only use interior classes for our atoms, their number could be a fraction as low as $1/d$ of all classes looked at in our searches: The size of the component to be constructed would decrease in d.

Fortunately, we can allow each class to be hit by a constant fraction of d faulty edges, before it becomes a *bordering* class not to be considered any more by subsequent searches. This still ensures that with high probability a constant fraction of n classes are interior classes of sufficiently few components.

The preceding discussion refers to the probability space of configurations with fault probability $f' = 1 - p'$ where $p' = \frac{\lambda'}{d-1} > \frac{1}{d-1}$. We fix some more parameters:

- We let $\gamma = \lambda' - 1$, w.l.o.g. $\gamma < 1/2$ and

$$\beta = \beta(\lambda, d) = \lfloor \frac{\gamma}{2} \cdot (d-1) \rfloor.$$

  (A class becomes a bordering class only after being hit by $\beta + 1$ faulty edges and no non-faulty one.)
- The maximum number of bordering or interior classes discovered by our searches is bounded above by $\hat{\varepsilon}n$ where $\hat{\varepsilon}$ is specified by

$$\hat{\varepsilon} = \hat{\varepsilon}(\lambda) = \frac{1}{10}\frac{\gamma}{2}(1 - \gamma)$$

- The offspring distribution induced by expanding a class inside one of our branching processes turns out to be stochastically at least as large as

$$\text{Bin}(d - 1 - \beta \,, p' \cdot (1 - \frac{\hat{\varepsilon}}{1 - \hat{\varepsilon} - \frac{\gamma}{2}})).$$

The expectation of this random variable can be calculated to be $\geq \lambda_0(\lambda) > 1$. We let

$$\mu \geq \mu_0(\lambda) > 0$$

be the probability that a branching process with the offspring distribution as above does not go extinct (cf. fact 6). To simplify notation we let

$$\alpha = \alpha(\lambda) = 1 - \hat{\varepsilon} - \frac{\gamma}{2}.$$

- The number of interior or bordering classes identified by each search is bounded above by $\omega(n) = \log n$.

Algorithm 7. As in the proof of theorem 1 we formulate our breadth-first searches as breadth-first generation algorithms. We use the following global variables:

- E = the set of non-faulty edges generated
  $\neg E$ = the set of faulty edges generated.
- Free = the set of *vertices* which have not at all been looked at , that is which do not occur in edges from $E \cup \neg E$.
  $\neg$Free = the set of vertices which have already been looked at.
- Dis = the set of bordering or interior *classes* which have already been discovered.

The next 3 variables are local to each single breadth-first search:

- Atom = the set of classes discovered during the current breadth-first search. A class is discovered during a breadth-first search if it is hit by a non-faulty edge (interior class) or by the $\beta + 1$'th faulty edge (including the edges from previous searches) and has not yet been hit by a non-faulty edge (bordering class).
- Open = the set of classes in the current search front.
- Open' = the subset of classes already generated from the next search front.

A single breadth-first generation starting from the class S:

Procedure Bfg(S)

1. Open := $\{S\}$; Atom := $\{S\}$; Open' := $\emptyset$
2. while Open $\neq \emptyset$ do
3.   Pick a class T from Open according to some deterministic rule. (The class T will depend on the history of the computation.)

4.	while $T \cap$ Free $\neq \emptyset$ do
5.	Pick a vertex x from T deterministically.
6.	Choose the vertex y from Free$\setminus\{x\}$ according to the uniform distribution.
7.	Add the edge (x, y) either to E or to $\neg E$. To E with probability p', to $\neg E$ with probability $1 - p'$.
8.	If the class of y, we call it R, has been discovered before (i. e. $R \in$ Atom or $R \in$ Dis , see below) we do nothing with R. If however R has not been discovered before, we add R to the set Atom if $(x, y) \in E$. If however $(x, y) \notin E$ we add R to Atom only if the edge (x, y) is the $\beta + 1$'th faulty edge to hit R. We add R only to Open' if $(x, y) \in E$.
9.	if $\mid$ Atom $\mid = \omega(n)$ then exit (from the current call to Bfg(S)).
10.	if Open $= \emptyset$ then Open := Open'.

Note that Atom includes the bordering classes found. The procedure Gen (for generate) directs the iterated execution of Bfg(S):

Procedure Gen

1. for $i = 1$ to $\hat{\varepsilon} n / \omega(n)$ do
   (The maximal number of interior or bordering classes generated by 1 Bfg(S) is $\omega(n)$.)
2.   Pick a class S not from Dis deterministically.
3.   Bfg(S)
4.   Dis := Dis $\cup$ Atom.

Of course, initially Dis $= \emptyset$.

It is helpful, to visualize the possible computations of Gen as a probability tree: Each node corresponds to a probabilistic choice, the sons of a node represent the possible outcomes of a choice. A possible computation is represented by a path from the root to a leaf. Subsequent choices are independent and the probability of a computation is the product of the probabilities of all the choices made. Moreover the probability of the algorithm to generate a given set E is equal to the probability of the set of faulty configurations with edge set E.

For the subsequent discussion we collect the obvious:

- The number of classes in the set Dis $\cup$ Atom is always $\leq \hat{\varepsilon} \cdot n$. Each class not in Dis $\cup$ Atom can be hit by up to β faulty edges. Accordingly, the number of nodes x which have been touched by the algorithm (i. e. $(x, x') \in E \cup \neg E$ for any x') is always

$$\leq \hat{\varepsilon} \cdot n \cdot d + (1 - \hat{\varepsilon})n\beta \; < \; nd(\hat{\varepsilon} + \frac{\gamma}{2})$$

Accordingly always

$$| \text{ Free } | \geq nd - (\hat{\varepsilon} \cdot n \cdot d + (1 - \hat{\varepsilon})n\beta) > nd\alpha$$

- Whenever a choosing step in line 6 of Bfg(S) is executed, we have $| \text{ Atom } | < \omega(n)$. Hence, whenever a y is chosen in line 6, the number of free nodes belonging to classes from Atom is certainly $< \omega(n) \cdot d$.
- The number of choosing steps in 1 execution of Bfg(S) is certainly $\leq \omega(n)d$, as each class can be hit at most d-times.

Now conditional on any arbitrary previous history $\mathcal{H}$ the probability that a node belonging to a class from Atom is chosen in line 6 is equal to

$$\frac{\text{number of untouched nodes belonging to Atom}}{| \text{ Free } |} \leq \frac{\omega(n) \cdot d}{nd\alpha}$$

As each execution of Bfg(S) makes at most $\omega(n)d$ choosing steps the probability that in 1 given call of Bfg(S) 2 or more nodes belonging already to Atom are hit is at most

$$\binom{\omega(n)d}{2} \cdot (\frac{\omega(n)d}{nd\alpha})^2 = o(n^{-1})$$

As Bfg(S) is certainly called *not* more that $nd-$times in a single computation, we get that in almost all computations *no* call of Bfg(S) has 2 or more overlaps. In the following we restrict (i. e. condition) every probability to the set of these computations.

Lemma 8. The probability that a given call of Bfg(S) ends with $\omega(n)$ classes in Atom is at least $\mu_0(\lambda) \cdot \nu(\lambda) > 0$.

Proof. The claim follows because $\nu = \nu(\lambda) > 0$ is the probability that the search of Bfg(S) starts from S with ≥ 3 branches. Then 1 branch is overlap-free and fact 6 is applied.

Lemma 9. The number of classes discovered in a set Atom of size $\omega(n)$ is at least $\frac{1}{4}\mu_0 \cdot \nu \cdot \hat{\varepsilon} \cdot n$. with high probability.

Proof. From Chernoff bound considerations.

Due to the existence of the bordering classes, the different sets Atom are *not* necessarily connected subgraphs. We condition the remaining probabilities on the almost sure event that at least $s(n) = s(n, \lambda) = \frac{1}{4}\mu_0\nu\hat{\varepsilon}n$ classes belong to sets Atom with $\omega(n)$ classes. As each bordering class must be hit by a constant fraction of d faulty edges, we have not too many of those:

Lemma 10. There is a constant $\kappa = \kappa(\lambda)$ such that with high probability at least $(1 - \kappa) \cdot s(n)$ classes are distributed over at most $s(n)/\omega(n)$ connected subgraphs.

Proof. Outline of the argument: For any i with $1 \leq i \leq \hat{\varepsilon}n$ the following holds: Conditional on the set of those computations of the algorithm Gen for which the i'th choosing step (i. e. line 6 of $\text{Bfg}(S)$) takes place, the probability that in this step a bordering class is discovered is at most

$$(1 - \frac{\lambda}{d-1})^{\beta+1} \leq (\frac{1}{e})^{\gamma/2}$$

the upper bound being independent of d. The claim follows again from Chernoff bound considerations.

Finally setting

$$\varepsilon'n = (1 - \kappa) \cdot \frac{1}{4} \cdot \mu_0 \cdot \nu \cdot \hat{\varepsilon} \cdot n$$

yields step 1 of the proof of theorem 2.

References

[AS 92]   Noga Alon, Joel H. Spencer. *The Probabilistic Method.* Wiley, 1992.

[AjKoSz 81] M. Ajtai, J. Komlos, E. Szemeredi. The Longest Path in a Random Graph. *Combinatorica*, 1, 1981, 1-12.

[AjKoSz 82] M. Ajtai, J. Komlos, E. Szemeredi.Largest Random Component of a k-Cube. *Combinatorica*,2(1), 1982, 1-7.

[Bo 85]   Bela Bollobas. *Random Graphs.* Academic Press, 1985.

[Bo 88]   Bela Bollobas. The Isoperimetric Number of Random Regular Graphs. *European Journal of Combinatorics.* 9, 1988, 241-244.

[KaNeTa 94] A. Karlin, G. Nelson, H. Tamaki. On the Fault Tolerance of the Butterfly. In *Proceedings STOC 1994*, 125 -133.

[Ni et al. 94] S. Nikoletseas, K. Palem, P. Spirakis, M. Yung. Short Vertex Disjoint Paths and Multiconnectivity in Random Graphs: Reliable Network Computing. In *Proceedings ICALP 1994*. 508-519.

[NiSp 95]   S. Nikoletseas, P. Spirakis. Expander Properties in Random Regular Graphs with Edge Faults. In *Proceedings STACS 1995*. 422-423.

[Ke 80]   H. Kesten. The Critical Probability of Bond Percolation on the Square Lattice Equals 1/2. *Communication in Mathematical Physics.* 74, 1980, 41-59.

[Ma 92]   T. R. Mathies. Percolation Theory and Computing with Faulty Arrays of Processors. In *3rd SODA 1992.* 100-103.

[Ra 89]   P. Raghavan. Robust Algorithms for Packet Routing in a Mesh. In *Proceedings 1st SPAA 1989.* 344-350.

[Fe 68]   W. Feller. *An Introduction to Probability Theory an its Applications*, vol. 1. Wiley 1968.

A Topological Generalization of Propositional Linear Time Temporal Logic

Bernhard Heinemann

FernUniversität, D–58084 Hagen, Germany

Abstract. We study a logic of knowledge and time which is not only a (standard) combination of respective systems, but prescribes the interaction between the time operators and the knowledge operator in a way such that topological concepts appear. The underlying logical language could be suitable for the specification of linear time properties of programs involving knowledge. We present an axiomatization of the validities and prove its semantical completeness. Moreover, we show decidability of the logic and give a lower complexity bound for its satisfiability problem.

1 Introduction

In recent years, *temporal logic* has become an important tool in computer science and AI, respectively [5]. For example, on the one hand temporal logic represents a very appropriate formalism expressing properties of concurrent programs [11]; on the other hand, combined with logics of knowledge it is a useful language to specify *protocols* [4].

We present an extension of both the usual logic of knowledge of an agent and linear time temporal logic. Complementary to the standard systems of knowledge and time we start out from the following observation: In many situations the knowledge of an agent increases if computational *effort* is spent. As knowledge is represented by the set of alternative states, this assumption amounts to a successive *shrinking* of the set of states the agent considers possible in the course of time. In this way a *topological* component comes into play. — There is a corresponding axiom in the logical system which controls the above mentioned interaction between knowledge and effort in time. It says that an agent knows a formula at all future states whenever she knows that it is valid there. Axioms of this kind are typical of the systems formalizing topological reasoning.

An appropriate *topological modal logic* has been introduced in the papers [12] and [3] by Moss and Parikh. It comprises two modal operators K and $\Box$. The structures interpreting these operators are, however, quite different from those in ordinary (multi)modal logic. Underlying frames now consist of a non-empty set X and a distinguished set $\mathcal{O}$ of subsets of X, called *opens* (although they need not be open sets in the sense of topology). K then varies over the elements of an open set, whereas $\Box$ captures the shrinking of an open.

In the meantime, various systems of topological modal logic have been studied.

Sound and complete axiomatizations were proposed e.g. for the basic *subset–space logic* (where $\mathcal{O}$ may be an arbitrary set of subsets of X) [3], and for *topologic* (where $\mathcal{O}$ is in fact a topology on X) [3], [6]. Moreover, the topological modal theory of *treelike spaces* was investigated by GEORGATOS [7]. (A subset frame (as introduced above) is treelike by definition, iff $U \subseteq V$ or $V \subseteq U$ or $U \cap V = \emptyset$ holds for all opens U, V.) As to *finite–height trees* of opens and a "topological" variant of the modal system **G**, which is the basic **K** augmented by the "LÖB rule" $\square(\square\alpha \rightarrow \alpha) \rightarrow \square\alpha$, we refer the reader to [9].

In the present paper an extension of propositional linear time temporal logic with *nexttime* and $\square$ to the topological context is proposed. (We omit the *until–* operator only for the sake of simplicity.) The adequate semantical structures, however, are not linear in the sense that the opens form a chain w.r.t. set inclusion, as one could suspect at first glance. For the latter structures, only the *nexttime*–fragment of linear time temporal logic could be generalized correspondingly [10]. Instead, we have linearity only from the viewpoint of states, but a tree structure on the opens.

The paper requires some basic notions and techniques from modal and temporal logic like *canonical models* and *filtrations*. In any case, [8], §§1–5, or [1], §§1–4, provide sufficient background. In particular, we proceed as follows: In Section 2, the underlying logical language is defined. Then, giving a list of axioms and rules, a logical system **L** is introduced. The proof of the semantical completeness of **L** follows in Section 4. It uses a special filtration, allowing the subsequent model construction for a non–deducible formula. Afterwards, decidability of the logic is shown, and for its satisfiability problem a lower complexity bound is determined.

2 The Language

We introduce the syntax and semantics of a language, LK, which can express certain linear–time properties of knowledge. Let PV be a recursively enumerable set of strings, called *propositional variables* (denoted by upper case Roman letters). Based on PV, the set $\mathcal{F}$ of *LK–formulas* (denoted by lower case Greek letters) is defined by the following clauses:

- $PV \cup \{\top\} \subseteq \mathcal{F}$; $\alpha, \beta \in \mathcal{F} \Longrightarrow \neg\alpha, K\alpha, \bigcirc\alpha, \square\alpha, (\alpha\wedge\beta) \in \mathcal{F}$; no other strings belong to $\mathcal{F}$.

We use common conventions denoting formulas, and the following abbreviations (besides the usual ones from sentential logic): $L\alpha$ for $\neg K\neg\alpha$, $\boxtimes\alpha$ for $\neg\bigcirc\neg\alpha$, and $\diamondsuit\alpha$ for $\neg\square\neg\alpha$.

The semantical structures are triples (X, d, σ) specified by the subsequent definition.

Definition 1. Let X be a non–empty set, and let $\mathcal{P}(X)$ denote the powerset of X. Furthermore, let $d : X \times \mathbb{N} \longrightarrow \mathcal{P}(X)$ be a mapping such that for all $x, y \in X$ and $j \in \mathbb{N}$ the following conditions are satisfied:

$-\ x \in d(x,j),$
$-\ y \in d(x,j) \Longrightarrow d(x,j) = d(y,j),$ and
$-\ d(x,j) \supseteq d(x,j+1).$

Let a mapping $\sigma : PV \times X \longrightarrow \{0,1\}$ be given additionally. Then σ is called a *valuation*, and the triple $\mathcal{M} = (X,d,\sigma)$ is called a *pseudo–linear subset space* or a *pseudo–linear model*.

Note that in every pseudo–linear model the set of opens contained in $d(x,0)$ for a given $x \in X$ forms in fact an infinite tree, and no two opens on the same level intersect. (Thus pseudo–linear models are treelike in particular.)

Neighbourhood situations of pseudo–linear models are pairs x, U_j^x (designated without brackets), where $U_j^x := d(x,j)$. Using this notation we introduce the validity relation for LK–formulas (only the interesting cases are mentioned).

Definition 2 Semantics of LK. Let $\mathcal{M} = (X,d,\sigma)$ be a pseudo–linear model and x, U_j^x a neighbourhood situation. Then we define

$$
\begin{aligned}
x, U_j^x \models_{\mathcal{M}} A \quad &:\Longleftrightarrow \sigma(A,x) = 1\,, \\
x, U_j^x \models_{\mathcal{M}} K\alpha \quad &:\Longleftrightarrow y, U_j^y \models_{\mathcal{M}} \alpha \text{ for all } y \in U_j^x\,, \\
x, U_j^x \models_{\mathcal{M}} \bigcirc\alpha \quad &:\Longleftrightarrow x, U_{j+1}^x \models_{\mathcal{M}} \alpha\,, \\
x, U_j^x \models_{\mathcal{M}} \square\alpha \quad &:\Longleftrightarrow (\forall k > j)\ x, U_k^x \models_{\mathcal{M}} \alpha\,,
\end{aligned}
$$

for all $A \in PV$ and $\alpha \in \mathcal{F}$.

In case $x, U_i \models_{\mathcal{M}} \alpha$, we say that α *holds in $\mathcal{M}$ at the neighbourhood situation* x, U_i. The formula $\alpha \in \mathcal{F}$ *holds in $\mathcal{M}$* (denoted by $\models_{\mathcal{M}} \alpha$), iff it holds in $\mathcal{M}$ at every neighbourhood situation. If there is no ambiguity, we omit the index $\mathcal{M}$ subsequently. — Note that the semantics of the operator K is the intended one because of the second condition on d in Definition 1. The operator $\square$ quantifies over all future time points, i.e. the actual state is excluded; this simplifies proofs slightly.

3   The Logic

The following schemes of formulas are appropriate to axiomatize the set of LK–validities.

(1) All $\mathcal{F}$–instances of propositional tautologies
(2) $K(\alpha \rightarrow \beta) \rightarrow (K\alpha \rightarrow K\beta)$
(3) $K\alpha \rightarrow \alpha$
(4) $K\alpha \rightarrow KK\alpha$
(5) $L\alpha \rightarrow KL\alpha$
(6) $(A \rightarrow \bigcirc A) \wedge (\neg A \rightarrow \bigcirc \neg A)$
(7) $\bigcirc(\alpha \rightarrow \beta) \rightarrow (\bigcirc\alpha \rightarrow \bigcirc\beta)$
(8) $\bigcirc\alpha \leftrightarrow \boxtimes\alpha$
(9) $\bigcirc L\alpha \rightarrow L\bigcirc\alpha$

(10) $\Box(\alpha \to \beta) \to (\Box\alpha \to \Box\beta)$
(11) $\Box\alpha \to \bigcirc(\alpha \wedge \Box\alpha)$
(12) $\Box(\alpha \to \bigcirc\alpha) \to (\bigcirc\alpha \to \Box\alpha)$

for all $A \in PV$ and $\alpha, \beta \in \mathcal{F}$.

Adding rules, we get a logical system designated **L**. In fact, modus ponens as well as necessitation w.r.t. each modality are present:

$$(1) \quad \frac{\alpha, \alpha \to \beta}{\beta} \qquad (2) \quad \frac{\alpha}{K\alpha} \qquad (3) \quad \frac{\alpha}{\bigcirc\alpha} \qquad (4) \quad \frac{\alpha}{\Box\alpha},$$

for all $\alpha, \beta \in \mathcal{F}$. — For convenience, we comment on some of the axioms. The schemes (3), (4) and (5) represent the standard axioms of knowledge. They characterize reflexivity, transitivity, and the euclidean property, respectively, of the accessibility relation in standard modal logic. The scheme (8) corresponds in this sense with functionality. (9) relates the *nexttime*–operator to the dual of K, saying that $\bigcirc\alpha$ is possible at the actual neighbourhood situation whenever α is possible at the succeeding one. (The converse is not always true.) This axiom determines the interaction between knowledge and time; schemes of this typ are typical of the systems considered in topological modal logic; see also Lemma 4(c) below. Axioms (2), (7) and (10) first of all have a proof–theoretical meaning. Axioms (11) and (12) are suitable variants of the schemes *Mix* and *Ind*, respectively, from linear time temporal logic (compare with [8], §9). They express that the $\Box$–operator represents the transitive closure of the *next step* relation in a sense to be made precise later on. Finally, we should say a few words about the scheme (6). It has to be added because we defined the valuation independent of the time component of a neighbourhood situation. This simplifies the definition of the semantics, but it clearly implies that the system is not closed under substitution.

Soundness of the axioms w.r.t. the intended structures can easily be established.

Proposition 3. *Axioms (1) – (12) hold in every pseudo–linear model.*

To get to know the interplay between the operators $\Box, \bigcirc$ and K, respectively, we present some **L**–deducible formulas.

Lemma 4. *For all $\alpha \in \mathcal{F}$, the following formulas are derivable in the system* **L**:

(a) $\Box\alpha \to \Box\Box\alpha$
(b) $\bigcirc(\varphi \wedge \Box\varphi) \to \Box\varphi$
(c) $K\Box\alpha \to \Box K\alpha$.

Note that (a) expresses transitivity of the accessibility relation belonging to $\Box$ in usual modal logic. The scheme (b) is the reversal of axiom (10); it is applied in the completeness proof conclusively. Finally, (c) represents the *cross axiom* in topological modal logic (see [3]; (c) is precisely the axiom mentioned in the introduction).

4 Completeness

The completeness proof for the system **L** runs via a special filtration of a generated submodel C of the canonical model (as to the respective definitions see e.g. [8], §§3–5), which is essentially the MOSS–PARIKH filtration ([3], Section 2.3) suited to the presence of time. Let $\alpha \in \mathcal{F}$ be given. Then we define the following sets of formulas based on the set $sf(\alpha)$ of subformulas of α:

$$\Gamma_\bigcirc := \{\bigcirc\beta \mid \Box\beta \in sf(\alpha)\} \cup \{\bigcirc\Box\beta \mid \Box\beta \in sf(\alpha)\};$$
$$\Gamma := sf(\alpha) \cup \Gamma_\bigcirc;$$
$$\Gamma^\neg := \Gamma \cup \{\neg\beta \mid \beta \in \Gamma\};$$
$$\Gamma^\wedge := \Gamma^\neg \text{ joined with the set of all finite conjunctions of distinct}$$
$$\text{elements of } \Gamma^\neg;$$
$$\Gamma^L := \{L\beta \mid \beta \in \Gamma^\wedge\};$$
$$\widetilde{\Gamma} := \Gamma^\wedge \cup \Gamma^L.$$

Note that $\widetilde{\Gamma}$ is finite and closed under subformulas. — Now, for all $s, t \in C$, we let

$$s \sim t :\iff s \cap \Gamma = t \cap \Gamma.$$

Moreover, let $\bar{s}$ designate the $\sim$–equivalence class of s, and let $\bar{C} := \{\bar{s} \mid s \in C\}$. As Γ is finite, $\bar{C}$ is a finite set as well.

So far we have formed a filtration of C. Next we introduce filtrations of the accessibility relations. For convenience, we repeat the definition. Let $\triangle$ be a modal operator (i.e. $\triangle = K, \bigcirc$, or $\Box$) and ∇ its dual (note that $\bigcirc$ is self-dual because of axiom (8)). Then, designating the accessibility relation on the canonical model which belongs to the operator $\triangle$ by $\overset{\nabla}{\longrightarrow}$ as usual, a binary relation $\overset{\nabla}{\longmapsto}$ on $\bar{C}$ is called a $\widetilde{\Gamma}$-*filtration* of $\overset{\nabla}{\longrightarrow}$, iff the following two conditions are satisfied for all $s, t \in C$:

$$- \ s \overset{\nabla}{\longrightarrow} t \text{ implies } \bar{s} \overset{\nabla}{\longmapsto} \bar{t}; \ \bar{s} \overset{\nabla}{\longmapsto} \bar{t} \text{ implies } \{\beta \mid \triangle\beta \in s \cap \Gamma\} \subseteq t.$$

Now let the relation $\overset{\bigcirc}{\longmapsto}$ on $\bar{C}$ be a filtration of $\overset{\bigcirc}{\longrightarrow}$. Let $\overset{\bigcirc^+}{\longmapsto}$ designate the transitive closure of $\overset{\bigcirc}{\longmapsto}$. The essential advantage of passing from the canonical model to filtrations is captured by the following lemma.

Lemma 5. *The relation $\overset{\bigcirc^+}{\longmapsto}$ on $\bar{C}$ is a filtration of the relation $\overset{\Diamond}{\longrightarrow}$ on the set C.*

The lemma gives a precise meaning to our earlier remark that $\Box$ represents the transitive closure of $\bigcirc$. The special form of the above set Γ as well as the axioms (11) and (12) are used in its proof essentially.

The *minimal* filtration of the relation $\overset{\nabla}{\longrightarrow}$ plays a central role in our investigations. It is defined by

$$\bar{s} \overset{\nabla}{\longmapsto} \bar{t} :\iff (\exists s' \in \bar{s})(\exists t' \in \bar{t}) \ s \overset{\nabla}{\longrightarrow} t$$

for all $s, t \in C$, and it is in fact a filtration of $\overset{\triangledown}{\longmapsto}$ ($\triangledown = \bigcirc, L$). The so-called *Fun–lemma* ([8], 9.9) is valid for the minimal filtration of $\overset{\bigcirc}{\longmapsto}$. It is a consequence of the scheme (8).

Lemma 6. *Let $\overset{\bigcirc}{\longmapsto}$ be the minimal filtration of $\overset{\bigcirc}{\longrightarrow}$. Let $s \in C$ be given and assume that $\bigcirc\beta \in \Gamma$ holds. Then $\bigcirc\beta \in s$ iff $(\exists t \in C)\ \bar{s} \overset{\bigcirc}{\longmapsto} \bar{t}$ and $\beta \in t$.*

Passing to a filtration the functionality of $\overset{\bigcirc}{\longrightarrow}$ is lost generally. The above lemma represents its substitute.

Due to the definition of the set $\widetilde{\Gamma}$ the minimal filtration $\overset{L}{\longmapsto}$ of the relation $\overset{L}{\longrightarrow}$ satisfies the conditions stated in the subsequent proposition.

Proposition 7. *The relation $\overset{L}{\longmapsto}$ is an equivalence relation on $\bar{C}$. Moreover, for all $s, t, u \in C$ such that $\bar{s} \overset{\bigcirc}{\longmapsto} \bar{t} \overset{L}{\longmapsto} \bar{u}$ there exists $v \in C$ satisfying $\bar{s} \overset{L}{\longmapsto} \bar{v} \overset{\bigcirc}{\longmapsto} \bar{u}$.*

The second statement of the proposition asserts the so-called *cross property* to the accessibility relations $\overset{\bigcirc}{\longmapsto}$ and $\overset{L}{\longmapsto}$ of the filtration. This property is a consequence of axiom (9) and typical of topological modal logic.

The following considerations lead to a further property of the filtration $\overset{\bigcirc^+}{\longmapsto}$ of the relation $\overset{\diamond}{\longrightarrow}$, which will be applied later on. Since $\overset{\bigcirc^+}{\longmapsto}$ is a transitive relation, we may analyse its *clusters*. These are built as follows: The relation

$$\bar{s} \approx \bar{t} :\Longleftrightarrow \bar{s} = \bar{t} \text{ or } (\bar{s} \overset{\bigcirc^+}{\longmapsto} \bar{t} \text{ and } \bar{t} \overset{\bigcirc^+}{\longmapsto} \bar{s}) \quad (\forall s, t \in C)$$

is easily seen to be an equivalence relation on $\bar{C}$. The $\approx$–class of an element $\bar{s} \in \bar{C}$ is commonly called the cluster of $\bar{s}$, and it is designated $cl_{\bar{s}}$. Letting

$$cl_{\bar{s}} \leq cl_{\bar{t}} :\Longleftrightarrow \bar{s} \overset{\bigcirc^+}{\longmapsto} \bar{t}$$

gives a partial order $\leq$ on the set of all clusters. As usual, we write $cl_{\bar{s}} < cl_{\bar{t}}$ iff $cl_{\bar{s}} \leq cl_{\bar{t}}$ and $cl_{\bar{s}} \neq cl_{\bar{t}}$. The following *cluster lemma* can be proved with the aid of Lemma 3.2(b).

Lemma 8. *Let $s \in C$, and let $cl_{\bar{s}}$ be a cluster which is not a last element w.r.t. "$<$". Assume that for $\square\beta \in \Gamma$ it holds that $\square\beta \notin s$. Then for all $v \in C$ such that $cl_{\bar{v}}$ is an immediate $<$–successor of $cl_{\bar{s}}$ there exists $t \in C$ such that $cl_{\bar{v}} = cl_{\bar{t}}$ and $(\beta \notin t$ or $\square\beta \notin t)$. Moreover, if $cl_{\bar{v}}$ is a $<$–last cluster, then the first alternative is valid.*

We are going to show that every non–L–deducible formula can be falsified in a pseudo–linear subset space. So let $\gamma \in \mathcal{F}$ be given and assume $\nvdash_{\mathbf{L}} \gamma$. Then there exists an element s in the canonical model such that $\alpha := \neg\gamma \in s$. Form $\widetilde{\Gamma}$ dependent on α as above, and consider the filtration

$$(\bar{C}, \{\overset{L}{\longmapsto}, \overset{\bigcirc}{\longmapsto}, \overset{\bigcirc^+}{\longmapsto}\})$$

of $(C, \{\xrightarrow{L}, \xrightarrow{O}, \xrightarrow{\diamond}\})$, where the latter structure is the submodel of the canonical model generated by s. For every $\bar{v} \in \bar{C}$ choose an ordering of the set of $\xrightarrow{O}$–successors of $\bar{v}$ such that this set may be represented as a list $(t_0, \ldots, t_{\nu_{\bar{v}}})$. Then define for all $\bar{u}$ from the $\xrightarrow{L}$–equivalence class of $\bar{s}$ a finitely ramified tree $tr_{\bar{u}}$ of height ω by

- $root(tr_{\bar{u}}) := \bar{u}$
- if $\bar{v}$ is a node of $tr_{\bar{u}}$ on level n, then $(t_0, \ldots, t_{\nu_{\bar{v}}})$ are its sons (from left to right),

for all $n \in \mathbb{N}$. Let $\hat{C}$ be the set of all paths through these trees. Note that $\hat{C}$ can be identified with a set of functions $f : \mathbb{N} \longrightarrow \bar{C}$. The carrier set of the desired model will be a subset $\tilde{C}$ of $\hat{C}$, which is obtained in the following way:

- For each cluster cl, fix an element $v \in C$ such that $cl_{\bar{v}} = cl$; for each *first* cluster choose v such that $\bar{s} \xrightarrow{L} \bar{v}$ holds additionally (this is possible because we started with a generated submodel).

- For each cluster cl having more than one element forget the $\xrightarrow{O}$–transitions leading to the same point. Choose a shortest $\xrightarrow{O}$–circle ζ_{cl} containing $\bar{v}$, i.e. a shortest path *through* the cluster starting and ending in $\bar{v}$.

- Choose a node $\in \zeta_{cl}$ which has at least two $\xrightarrow{O}$–successors $\in \zeta_{cl}$ yet. Branch off the circle at the chosen node and get back to $\bar{v}$ along ζ_{cl} then; do this in all possible ways successively; put all such "circles" (which may have repetitions) one behind the other in any order and call the resulting $\xrightarrow{O}$–sequence a *cl–piece*.

- For each single–point cluster cl let its cl–piece consist of this point only.

- Now let $\tilde{C}$ be the set of all functions $f : \mathbb{N} \longrightarrow \bar{C}$ which pass faithfully through the cl–pieces of a sequence of clusters starting with a first and ending with a last $<$–element such that f eventually consists of a repetition of the cl–piece of the latter ad infinitum.

Note that two functions $f, g \in \tilde{C}$ belonging to the same sequence of clusters only differ in the connecting pieces between two successive clusters (which connect the chosen "base points"). — Define $d : \tilde{C} \times \mathbb{N} \longrightarrow \mathcal{P}(\tilde{C})$ by

$$g \in d(f, j) : \Longleftrightarrow (\forall k \leq j) \, g(k) \xrightarrow{L} f(k)$$

for all $f, g \in \tilde{C}$ and $j \in \mathbb{N}$. Furthermore, let a valuation σ be given which satisfies

$$\sigma(A, f) = 1 : \Longleftrightarrow A \in f(0),$$

for all $A \in PV \cap \Gamma$ and $f \in \tilde{C}$. Then we get the following theorem.

Theorem 9. *The structure* $\mathcal{M} := (\tilde{C}, d, \sigma)$ *is a pseudo–linear subset space. Moreover, for all* $\beta \in \tilde{\Gamma}$, $t \in C$, $f \in \tilde{C}$ *and* $j \in \mathbb{N}$ *it holds that*

$$f(j) = \bar{t} \Longrightarrow [\beta \in t \Longleftrightarrow f, U_j^f \models_{\mathcal{M}} \beta].$$

Applying the theorem on $\alpha \in s$ and an $f \in \tilde{C}$ satisfying $f(0) = \bar{s}$, completeness follows as an immediate corollary. Combining this result with Proposition 3 we get:

Corollary 10. *The system* **L** *is sound and complete w.r.t. the class of pseudo-linear subset spaces.*

5   Decidability and Complexity

The set of LK–formulas holding in all pseudo–linear subset spaces is decidable although the above completeness proof does not yield the finite model property for the system. Instead, we may proceed via suitable standard trimodal models, for which the axiomatization is complete as well. Maybe there are applications of the system w.r.t. this standard semantics, too.

Definition 11. Let $\mathcal{M} := (W, \{R, S, T\}, \sigma)$ be a trimodal model (i.e. W is a non–empty set, R, S and T are binary relations on W, and σ is a valuation). Then $\mathcal{M}$ is called an LK–*model, iff*

- R is an equivalence relation and S a function on W;
- T is the transitive closure of S;
- for all $s, t, u \in W$: if $(s, t) \in S$ and $(t, u) \in R$, then there exists a $v \in W$ such that $(s, v) \in R$ and $(v, u) \in S$ (this is the *cross property*);
- for all $s, t \in W$ such that $(s, t) \in T$ it holds that $\mathcal{M} \models A[s]$, iff $\mathcal{M} \models A[t]$ $(A \in PV)$.

Note that the relation R corresponds with the modality K; accordingly, S and $\bigcirc$ as well as T and $\square$ are related. — In fact, we have the following theorem.

Theorem 12. *A formula $\alpha \in \mathcal{F}$ is* **L**–*deducible, iff it holds in every LK–model.*

Apart from the functionality of the relation $\overset{\bigcirc}{\longmapsto}$ the finite structures obtained by filtration in the previous section are LK–models. We want to circumvent this deficiency left over next. For this purpose we mark every path $g \in \tilde{C}$ at the node where the cl–piece belonging to a $<$–terminal cluster is covered for the first time. Because of König's Lemma, a finite set of finite trees results. Now we "spread" these trees by introducing new points equivalent to the original ones (w.r.t. the valuation), such that the paths become "parallel lines", i.e. $\overset{\bigcirc}{\longmapsto}$ – sequences having no element in common. Afterwards, the last cl–piece is looped again. In this way we get a *finite* LK–model falsifying a given non–derivable formula. (In fact, the original model is a p–morphic image of the new one). — It should be mentioned that the proceeding here is similar to that in standard temporal logic, where the semantics has to be changed as well (letting *induction models* come into play; see [8], §9). Working out these ideas we get:

Theorem 13. *The set of LK–formulas holding in every pseudo–linear subset space is decidable.*

Unfortunately, the complexity of the satisfiability problem of the LK–logic is presumably high.

Theorem 14. *The set of all satisfiable LK–formulas is PSPACE–hard.*

The theorem is proved by reducing the satisfiability problem of propositional linear time temporal logic (see [2]) to that of the LK–logic.

6 Conclusion

We combined the modal system **S5**, which represents the logic of knowledge mostly used in computer science, with the well–known system of propositional linear time temporal logic such that the operators of time and knowledge, respectively, are connected by means of the *cross axiom*. We have got a sound and complete axiomatization w.r.t. the class of pseudo–linear subset spaces. Moreover, the set of theorems of the logic is decidable. However, the complexity of the satisfiability problem is high, as usual.

References

[1] Chellas, B. F. 1980. *Modal Logic: An Introduction.* Cambridge: Cambridge University Press.

[2] Clarke, E. M., and A. P. Sistla. 1985. The Complexity of Propositional Linear Temporal Logics. *Journal of the ACM* 32:733–749.

[3] Dabrowski, A., L. S. Moss, and R. Parikh. 1996. Topological Reasoning and The Logic of Knowledge. *Ann. Pure Appl. Logic* 78:73–110.

[4] Fagin, R., J. Y. Halpern, Y. Moses, and M. Y. Vardi. 1995. *Reasoning about Knowledge.* Cambridge(Mass.): MIT Press.

[5] Gabbay, D. M., I. Hodkinson, and M. Reynolds. 1994. *Temporal Logic – Mathematical Foundations and Computational Aspects.* Oxford: Clarendon Press.

[6] Georgatos, K. 1994. Knowledge Theoretic Properties of Topological Spaces. In *Knowledge Representation and Uncertainty,* ed. M. Masuch and L. Polos, 147–159. Springer. LNCS 808.

[7] Georgatos, K. 1994. Reasoning about Knowledge on Computation Trees. In *Proc. Logics in Artificial Intelligence (JELIA '94),* ed. C. MacNish, D. Pearce, and L. M. Pereira, 300–315. Springer. LNCS 838.

[8] Goldblatt, R. 1987. *Logics of Time and Computation.* CSLI Lecture Notes Number 7. Stanford: Center for the Study of Language and Information.

[9] Heinemann, B. 1996. 'Topological' Modal Logic of Subset Frames with Finite Descent. In *Proc. 4th Intern. Symp. on Artificial and Mathematics, AI/MATH-96,* 83–86. Fort Lauderdale.

[10] Heinemann, B. 1997. Topological Nexttime Logic. In *Proc. Advances in Modal Logic, AiML 96.* Kluwer series. to appear 1997.

[11] Manna, Z., and A. Pnueli. 1992. *The Temporal Logic of Reactive and Concurrent Systems.* New York: Springer.

[12] Moss, L. S., and R. Parikh. 1992. Topological Reasoning and The Logic of Knowledge. In *Proc. 4th Conf. on Theoretical Aspects of Reasoning about Knowledge (TARK 1992),* ed. Y. Moses, 95–105. Morgan Kaufmann.

Multi-Head Finite Automata: Data-Independent *Versus* Data-Dependent Computations

Markus Holzer

Wilhelm-Schickard Institut für Informatik, Universität Tübingen,
Sand 13, D-72076 Tübingen, Germany
email: holzer@informatik.uni-tuebingen.de

Abstract. We develop a framework on multi-head finite automata that allows us to study the relation of parallel logarithmic time and sequential logarithmic space in a uniform and nonuniform setting in more detail. In both settings it turns out that $\mathbf{NC}^1$ requires data-independent computations—the movement of the input-heads only depends on the length of the input—whereas logarithmic space is caught with data-dependent computations on multi-head finite state machines. This shed new light on the question whether these two classes coincide or not.

1 Introduction

Many important questions in the theory of computational complexity can be formulated as problems regarding the relationship between different classes of formal languages or automata. For example, in [12] it was shown that the equality of deterministic and nondeterministic logarithmic space is equivalent to the question whether every language accepted by some nondeterministic one-way two-head finite automaton (the corresponding language class is denoted by $1\mathbf{NFA}(2)$) is accepted by some deterministic two-way multi-head finite automaton (the corresponding language class is denoted by $\mathbf{DFA}(k)$). Using the introduced notation this result reads as follows:

$$1\mathbf{NFA}(2) \subseteq \bigcup_{k \in \mathbb{N}} \mathbf{DFA}(k) \quad \text{if and only if} \quad \mathbf{L} = \mathbf{NL}.$$

For the proof mainly the characterization of deterministic and nondeterministic logarithmic space in terms of multi-head finite automata was used.

Many other complexity classes as, e.g., $\mathrm{LOG}(\mathbf{DCFL})$, $\mathrm{LOG}(\mathbf{CFL})$, $\mathbf{P}$, or $\mathbf{NP}$, are also characterizable *via* multi-head machines, but up to now characterizations of $\mathbf{NC}^1$ and variants thereof in terms of multi-head devices (other than the constant-space sequential model [7] which mimics the evaluation of first-order-formulas with group quantifiers for solvable groups, an alternate characterization of $\mathbf{NC}^1$ first presented in [3]) are not known.

The key to a multi-head device characterization of $\mathbf{NC}^1$ in the uniform (here we stick to logspace uniformity) and nonuniform setting is *data-independence* or *obliviousness*, i.e., the movement of the input-heads only depends on the length of the input. This property is often required for Turing machine where it is easily achieved. To this end we introduce uniform and nonuniform multi-head finite automata—here a machine is nonuniform if the transition function varies in time—and study data-independent and data-dependent computations on these devices.

Although the obtained results are quite similar—data-independent multi-head finite automata characterize NC^1 whereas data-dependent machines catch logarithmic space—we also found some significant differences in the uniform and nonuniform world. On the one hand, the uniform hierarchies of data-independent and data-dependent finite automata with respect to the number of heads are strict. For nonuniform machines the data-independence head-hierarchy collapses to its first level and the one-headed machines are shown to be nothing else than Barrington's [2] model of a nonuniform finite state machine. In case of nonuniform data-dependent machines we do not know whether the head-hierarchy which links *nonuniform*-NC^1 with nonuniform logarithmic space is strict or not. Nevertheless, we show that even the lowest class of that hierarchy contains a logspace complete problem. Moreover, we give some translational results for that hierarchy.

The paper is organized as follows. In the next section we give the necessary notations. In Section 3 we present the results for the uniform world, and in Section 4 and 5 we turn our attention to nonuniform classes. In the former section we deal with nonuniform finite state machines, whereas in the latter section we make a short side-trip to Karp and Lipton's "universal" mechanism to define nonuniform classes from uniform ones. Both sections show that the results mainly rely on the "nonuniformization" mechanism one chooses. In the ultimate section we summarize and discuss our results.

2 Definitions

We assume the reader to be familiar with some basic notions of complexity theory, as contained in [1]. In particular we consider the following well-known sequence of containments: $NC^1 \subseteq L \subseteq NL$.

Here L (NL, respectively) is the set of problems accepted by deterministic (nondeterministic, respectively) logarithmic space bounded Turing machines, and NC^1 is the set of problems accepted by logspace uniform branching program families of constant width and polynomial depth (or equivalently by circuit families of logarithmic depth, polynomial size with AND- and OR-gates of bounded fan-in). Throughout this paper we assume that branching programs are *levelled*.

For a set A let $|A|$ be the cardinality of A, and the length of a word w is denoted by $|w|$.

Given a class $\mathbf{A}$ of languages over some finite alphabet Σ and a set F of advices $\alpha = (\alpha_n)$ of strings from Σ^*. According to Karp and Lipton [6] the class of sets of the form $L : \alpha = \{ w \mid \alpha_{|w|} w \in L \}$, with $L \in \mathbf{A}$ and $\alpha \in F$ is denoted $\mathbf{A}/F$. Mostly one considers the set of polynomial length bounded advices, which is denoted by *poly*.

A *two-way nonuniform nondeterministic finite automaton*[1] with k-heads, for $k \in \mathbb{N}$, is a quint-tuple $M = (Q, \Sigma, P, q_0, F)$, where Q is a finite nonempty set of states, Σ is a finite nonempty alphabet of input symbols such that Σ does not contain the symbols $\vdash$ and $\dashv$. These symbols are the left and right end-marker, respectively. The family $P = (P_n)$ contains programs, that are sequence of instructions, of polynomial length, i.e., there is a polynomial p such that $|P_n| \leq p(n)$ for all $n \geq 0$. An instruction is a mapping

$$\delta : Q \times \left(\Sigma \cup \{\vdash, \dashv\}\right)^k \to 2^{Q \times \{-1,0,1\}^k},$$

such that for $p, q \in Q$, $(q, d_1, \ldots, d_k) \in \delta(p, a_1, \ldots, a_k)$ implies $d_i \geq 0$ if $a_i = \vdash$ and $d_i \leq 0$ if $a_i = \dashv$. Further let $q_0 \in Q$ be the initial state, and $F \subseteq Q$ the set of final or accepting states.

Machine M accepts word w, if M started in its initial state and with all input-heads in their first position (that is the first letter of w) executes the entire program $P_{|w|}$ on input $\vdash w \dashv$ and eventually stops in an accepting state.

A machine M is *deterministic*, if all instructions that appear in a program satisfy: for each $(q, a_1, \ldots, a_k)$ in $Q \times (\Sigma \cup \{\vdash, \dashv\})^k$, $|\delta(q, a_1, \ldots, a_k)| \leq 1$, i.e., automaton M has at most one move defined at any moment of time. The machine is *uniform*, if all programs out of P are sequences of exactly one instruction δ only. In this case we simply write $M = (Q, \Sigma, \delta, q_0, F)$.

The class of languages accepted by deterministic (nondeterministic, respectively) nonuniform finite automata with k heads is denoted by **NuDFA**(k) (**NuNFA**(k), respectively). Skipping the prefix **Nu** denotes the corresponding uniform class.

Taking a machine M with multiple input-heads, we say that it works *data-independent* if and only if the position of the input-heads at any time depends on the length of the input only. The class of languages accepted by data-independent machines is referred as follows: In case of nonuniform machines we insert **Di** after the letters **Nu**, and for uniform machines we add a prefix **Di**.

It is easily seen that Barrington's version of a nonuniform finite state machine [2] is nothing else than a one-head nonuniform data-independent finite automaton. He showed that *nonuniform*-**NC**[1] (the nonuniform version of **NC**[1]) is identical with **NuDiDFA**(1) and **NuDiNFA**(1).

Throughout we only write down the results for the deterministic case. Unless otherwise stated, all presented results remain valid if nondeterminism is used instead of determinism.

3 Multi-Head Finite Automata

The power of multi-head finite automaton is well studied in the literature. From the complexity point of view, the two-way case is the most interesting one,

---

[1] Although the *terminus* nonuniform finite automaton (NuDFA) is common for Barrington's [2]'s automaton model, we deviate from this usual notation in literature and call his model *nonuniform data-independent finite automaton* (NuDiDFA). This is motivated by the mentioned fact at the end of this section.

because in [5, pages 338–339, Lemma] the following equality was shown:

Theorem 1. $\mathbf{L} = \bigcup_{k \in \mathbb{N}} \mathbf{DFA}(k)$.

This result is best possible, because $k + 1$ heads are better than k as shown in [9, page 69, Theorem 1]. Observe, that even tally languages may serve as a witness for $\mathbf{DFA}(k) \subset \mathbf{DFA}(k + 1)$.

Now let's turn our attention to data-independent machines. First we show that determinism and nondeterminism coincide.

Lemma 2. $\mathbf{DiDFA}(k) = \mathbf{DiNFA}(k)$ *for every* $k \in \mathbb{N}$.

Proof. Assume that the machine works data-independent, then one easily verifies by induction on the number of steps the machine has made, that there is only one possible trajectory of the heads for each input length. If the considered machine is nondeterministic, then the only nondeterminism which is left is the way the next state is chosen. Therefore, a power-set construction shows that even a deterministic machine can do already the job. $\square$

For the characterization of $\mathbf{NC}^1$ in terms of data-independent multi-head finite automata the following observation is very useful: Every multi-head deterministic finite automaton that works on a tally input is already data-independent by definition; hence $\mathrm{Tally}\mathbf{L} \subseteq \bigcup_{k \in \mathbb{N}} \mathbf{DiDFA}(k)$, where $\mathrm{Tally}\mathbf{A}$ is the set of all tally languages out of class $\mathbf{A}$.[2]

Theorem 3. $\mathbf{NC}^1 = \bigcup_{k \in \mathbb{N}} \mathbf{DiDFA}(k)$.

Sketch of Proof. We briefly sketch the construction of a multi-head finite automaton that accepts the same language as the $\mathbf{L}$-uniform family $B = (B_n)$. Here $\mathbf{L}$-uniformity means that the mapping $1^n \rightarrow \langle B_n \rangle$, where $\langle \cdot \rangle$ denotes the coding of branching programs, is computable by a deterministic logarithmic space bounded Turing machine M [11].

On input w of length n, the idea is to trace the computation path in the branching program B_n to one of the sink nodes. Since the width of the branching programs is constantly bounded, the actual location on a level can be stored in the finite control. To reach the next level we follow the appropriate edge using M to compute the necessary information. Finally, the input is accepted if the sink labelled with 1 is reached.

To implement this procedure on a multi-head finite automaton one has to be careful to achieve data-independency, because the traced path heavily relies on the input word. To avoid this, we do as follows: Assume that the actual position or node on the current level is stored in the finite control. Using M we cycle

---

[2] Observe that we do not known whether $\mathrm{Tally}\mathbf{NL} \subseteq \bigcup_{k \in \mathbb{N}} \mathbf{DiNFA}(k)$ holds. If so, it implies $\mathrm{Tally}\mathbf{NL} \subseteq \mathrm{Tally}\mathbf{L}$ by Theorem 2, which itself would lead by translational methods to a positive answer of the LBA problem. This is is the question whether every context-sensitive language can be accepted by a deterministic linear space bounded Turing machine.

through the nodes of the current level, compare the nodes computed by M with the actual node, query the desired input bits, and perform an update operation using the finite control only, if the computation path is traced one step further. With the observation given above, this ends up in a data-independent algorithm for multi-head machines. Thus, the desired claim for two-way machines follows.

Conversely, we argue as follows: multi-head finite automata can be regarded as branching programs, each state-changing instruction of the machine is simulated by one level of the branching program—nodes roughly correspond to states and edges model the next state relation or function. Obviously, the underlying graph structure is "highly" uniform. Hence it is sufficient to consider the computation of test-node labels, or in other words the position of the input-heads of the machine at some time t.

Thus, the uniformity machine uses beside some subroutines that generates the graph structure, another program part, that simulates the data-independent k-head finite automaton step-by-step on the given input 1^n in order to compute the positions of the input-heads at time t. Obviously, this can be done within deterministic logarithmic space. $\quad\square$

With TallyL $\subseteq \bigcup_{k \in \mathbb{N}}$ **DiDFA**(k) and the already mentioned fact that even tally languages may serve as witnesses for **DFA**$(k) \subset$ **DFA**$(k+1)$, we obtain:

Corollary 4. **DiDFA**$(k) \subset$ **DiDFA**$(k+1)$ *for every* $k \in \mathbb{N}$. $\quad\square$

Because 1**DFA**(2) contains, e.g., the deterministic logspace complete monotone graph accessibility problem for graphs with out-degree one, we immediately obtain from Theorem 3:

Corollary 5. $\mathbf{NC}^1 = \mathbf{L}$ *if and only if* 1**DFA**$(2) \subseteq \bigcup_{k \in \mathbb{N}}$ **DiDFA**(k). $\quad\square$

4 Nonuniform Multi-Head Finite Automata

In the work of Barrington [2] a new connection between the parallel complexity class $\mathbf{NC}^1$ and the theory of finite automata was discovered. He characterized *nonuniform-*$\mathbf{NC}^1$ as those languages recognized by a certain nonuniform version of a deterministic (nondeterministic) finite automaton, as mentioned in Section 2.

In most cases the use of more input-heads increase the computational power of devices, but this is definitely not the case for nonuniform data-independent finite automata. One may construct a program for a data-independent one-head automaton that simulates a k-headed machine of same type with a polynomially increase of program length. The main task in the simulation is to ask the desired and in advance known input-head positions in sequence and to perform afterwards the state-changing instruction of the multi-head device. This shows, that k heads are as good as one head. Moreover, even nondeterminism does not increase the computational power of the machines due to a power-set construction; this was also mentioned by Barrington for his one-head model.

Theorem 6. *nonuniform-*$\mathbf{NC}^1 = \mathbf{NuDiDFA}(k)$ *for every* $k \in \mathbb{N}$. □

Trivially, *nonuniform-*$\mathbf{NC}^1 \subseteq \mathbf{NuDFA}(1) \subseteq \cdots \subseteq \bigcup_{k\in\mathbb{N}} \mathbf{NuDFA}(k)$. In analogous to Theorem 1 we show that the latter class coincides with nonuniform deterministic logspace. To this end, we develop a description of k-head non-uniform finite automata in terms of $O(n^k)$ width bounded branching programs.

Theorem 7. *Language* L *is in* $\mathbf{NuDFA}(k)$ *if and only if* L *is accepted by a family* $B = (B_n)$ *of deterministic branching programs of polynomial depth and width* $O(n^k)$.

Sketch of Proof. Let $M = (Q, \Sigma, P, q_0, F)$ be a nonuniform finite automaton with k-heads. On inputs of length n let P_n be the sequence of $t(n)$ instructions, where the ith instruction is a mapping $\delta_i : Q \times (\{0, 1\} \cup \{\vdash, \dashv\})^k \to Q \times \{-1, 0, 1\}^k$.

For given $p \in Q$ and $\mathbf{i} = (i_1, \ldots, i_k) \in \{0, \ldots, n+1\}^k$ consider the deterministic branching program $B_{p,\mathbf{i}}$ which is a complete binary tree of $k+1$ levels, where on level j variable x_{i_j} is queried. Queries for x_0 and x_{n+1} are replaced by dummy queries. The 2^k sinks of $B_{p,\mathbf{i}}$ may be identified with words (induced by the corresponding queries) from $(\{0, 1\} \cup \{\vdash, \dashv\})^k$ in an obvious way.

For any level $r \in \{1, \ldots, t(n)\}$, $p \in Q$, and $\mathbf{i} \in \{0, \ldots, n+1\}^k$ let $[r; p, \mathbf{i}]$ be a source of a copy of $B_{p,\mathbf{i}}$ in B_n. Consider a particular copy of $B_{p,\mathbf{i}}$ with source $[r; p, \mathbf{i}]$. The sinks of $B_{p,\mathbf{i}}$ are hard-wired as follows: The sink that corresponds to word $a_1 \ldots a_k \in (\{0, 1\} \cup \{\vdash, \dashv\})^k$ is identified with $[r+1; q, \mathbf{j}]$ if $\delta_r(p, a_1, \ldots, a_k) = (q, d_1, \ldots, d_k)$ and $\mathbf{j} = \mathbf{i} + (d_1, \ldots, d_k)$. To complete the construction of B_n let $[1; q_0, \mathbf{1}]$ be the source node and label the nodes $[t(n) + 1; p, \mathbf{i}]$ for arbitrary $\mathbf{i}$ accepting if $p \in F$, and rejecting otherwise.

Inductively one verifies that the constructed branching program B_n simulates M on inputs of length n step-by-step. Further one observes that width and depth, respectively, of B_n is bounded by $O(2^k |Q| \cdot n^k)$ and $O(k \cdot t(n))$, respectively.

Due to lack of space we have to omit the other direction. The main part there is to simulate the branching program level-by-level, and to store the node that is reached. Since the branching program is $O(n^k)$ width bounded, the position of the node actually may be stored by the machine's input-head (positions). □

From Theorem 7 and the computational equivalence of $\mathbf{L}/poly$ and polynomial *size* deterministic branching programs [8, page 172, Theorem 5] (see also [10, page 16, Theorem 3.2]) it follows:

Theorem 8. $\mathbf{L}/poly = \bigcup_{k\in\mathbb{N}} \mathbf{NuDFA}(k)$. □

Compared to the uniform case, here we do not known whether the given characterization is best possible with respect to the number of heads. Unfortunately, for the head-hierarchies we can neither show a strictness nor a partial or total collapse result; a strict inclusion seems to be much harder, because this would separate $\mathbf{NC}^1$ and logspace in a nonuniform setting. Nevertheless, in the forthcoming we prove that even the lowest classes contain quite complicated problems. Then in the remainder we focus on translational results on the head-hierarchies for nonuniform finite automata.

Theorem 9. *The class* **NuDFA**(1) *contains a deterministic logarithmic space complete problem, w.r.t.* **DTime**(log n)*-reductions.*

Proof. The monotone graph accessibility problem for graphs with out-degree one and two were studied among other logspace complete problems in [8, page 178, Theorem 5] with respect to the source—he was mainly interested in results for the resource *size*—the problems need to be accepted by variants of branching programs. A careful analysis of the proof shows that for ordinary branching programs linear width and polynomial depth is sufficient to do the job. Then by Theorem 7 the result follows. □

As an immediately corollary we obtain:

Corollary 10. *nonuniform-***NC**^{1} = **L**/*poly iff* **NuDFA**(1) ⊆ **NuDiDFA**(1).

Proof. If *nonuniform-***NC**^{1} = **L**/*poly*, then by Theorem 6 and 8 we have **NuDiDFA**(1) = $\bigcup_{k \in \mathbb{N}}$ **NuDFA**(k). Thus, **NuDFA**(1) ⊆ **NuDiDFA**(1).

Conversely, if **NuDFA**(1) ⊆ **NuDiDFA**(1), then by Theorem 9 the class *nonuniform-***NC**^{1} contains a deterministic logspace complete problem with respect to **DTime**(log n)-reductions. Since *nonuniform-***NC**^{1} is closed under these reductions, *nonuniform-***NC**^{1} = **L**/*poly* follows. □

Most results on head-hierarchies build on translational techniques, e.g., see [9, page 69, Theorem 1]. They link "low" classes of the hierarchy with "higher" ones; in other words, a collapse of high classes implies a collapse of low classes. The padding we use is that of [9]: Let $T_k : \Sigma^* \to \Sigma^*$, for some alphabet Σ, be defined on words as $T_k(w) := w^{|w|^{k-1}}$, and on languages $L \subseteq \Sigma^*$ let $T_k(L)$ be $\{ T_k(w) \mid w \in L \}$.

Lemma 11. *For every* $k, \ell \in \mathbb{N}$, $L \in$ **NuDFA**($k \cdot \ell$) *if and only if* $T_k(L) \in$ **NuDFA**(ℓ).

Proof. If L is recognized by a $k \cdot \ell$-head nonuniform finite automaton, then by Theorem 7 there is a branching program family $C = (C_n)$ of width $O(n^{k \cdot \ell})$ and polynomial depth that accepts the same language. We describe how to construct a family $B = (B_n)$ that accepts $T_k(L)$.

For given $\mathbf{i} = (i_1, \ldots, i_{n^{k-1}}) \in \{1, \ldots, n^{k \cdot \ell}\}^{n^{k-1}}$ consider the deterministic branching program $E_{\mathbf{i}}$ which is a complete binary tree of $n^{k-1} + 1$ levels, where on level j variable x_{i_j} is queried. The $2^{n^{k-1}}$ sinks may be identified with words (induced by the corresponding queries) from $\{0, 1\}^{n^{k-1}}$. Finally let us identify the sinks $0^{n^{k-1}}$ and $1^{n^{k-1}}$ with an accepting node of $E_{\mathbf{i}}$, and the remaining nodes with a rejecting one. So for given $\mathbf{i} = (i_1, \ldots, i_{n^{k-1}})$ program $E_{\mathbf{i}}$ checks whether the assignment of the variables $x_{i_1}, \ldots, x_{i_{n^{k-1}}}$ are equal. Observe that E_i's construction can be modified such that it results in a branching program of constant width.

For every $1 \leq i \leq n$ define $\mathbf{d}_i := (i, n+i, 2 \cdot n+i, \ldots, (n^{k-1}-1) \cdot n+i)$. The source of B_n is the source of $E_{\mathbf{d}_1}$. Then for $1 \leq i \leq n-1$ identify the accepting sink of $E_{\mathbf{d}_i}$ with the source of $E_{\mathbf{d}_{i+1}}$, and finally identify the accepting sink of $E_{\mathbf{d}_n}$ with the source of C_n. To complete the construction all remaining rejecting sinks are identified. By the construction it is readily verified that $B = (B_n)$ accepts $T_k(L)$.

The constructed branching programs work on n^k variables. Hence, the width is $O(n^\ell)$ bounded and the depth remains polynomial. Thus by Theorem 7 language $T_k(L)$ is accepted by an ℓ-head nonuniform finite automaton.

Conversely, let $T_k(L)$ be recognized by an ℓ-head nonuniform finite automaton, and therefore also by a branching program family $C = (C_n)$ of width $O(n^\ell)$ and polynomial depth. To obtain the program family that accepts L, take C_{n^k} and rename every query $x_{i \cdot n+j}$, for $1 \leq i \leq n^{k-1}-1$ and $1 \leq j \leq n$, to x_j. Obviously, these programs have width $O(n^{k \cdot \ell})$ and polynomial depth. Again Theorem 7 gives the desired result. $\qquad\square$

Now we are ready to prove the mentioned downward separation result.

Theorem 12. *For every* $k, \ell_1, \ell_2 \in \mathbb{N}$, $\mathbf{NuDFA}(\ell_1) = \mathbf{NuDFA}(\ell_2)$ *implies* $\mathbf{NuDFA}(k \cdot \ell_1) = \mathbf{NuDFA}(k \cdot \ell_2)$.

Proof. Let $L \in \mathbf{NuDFA}(k \cdot \ell_2)$. Lemma 11 gives $T_k(L) \in \mathbf{NuDFA}(\ell_2)$, and by hypothesis $T_k(L) \in \mathbf{NuDFA}(\ell_1)$. Using Lemma 11 once again we obtain $L \in \mathbf{NuDFA}(k \cdot \ell_1)$. $\qquad\square$

An application of Theorem 12 is given next, which we state without proof.

Corollary 13. *For every* $k \in \mathbb{N}$, *(1)* $\mathbf{NuDFA}(k) = \mathbf{NuDFA}(2 \cdot k)$ *implies a collapse of the hierarchy to level* k, *and (2)* $\mathbf{NuDFA}(k) \subset \cdots \subset \mathbf{NuDFA}(2 \cdot k)$ *implies that the hierarchy up to level* k *is strict.* $\qquad\square$

Finally, observe that for instance it is not known whether any inclusion structure of the hierarchy that does not contradict Theorem 12 would have any consequences—besides either *nonuniform*-$\mathbf{NC}^1 \subset \mathbf{L}/poly$ or *nonuniform*-$\mathbf{NC}^1 = \mathbf{L}/poly$—on other complexity classes.

5 Multi-Head Finite Automata That Take Advice

One universal formalization of the idea of nonuniformity has been proposed and studied in depth by Karp and Lipton [6] in their seminal paper. First consider ordinary multi-head finite state machines relative to advices. It turns out that two-heads suffice to do all of nonuniform logspace.

Theorem 14. $\mathbf{L}/poly = \mathbf{DFA}(k)/poly$ *for every* $k \geq 2$.

Proof. It suffices to prove the inclusion $\mathbf{DFA}(k)/poly \subset \mathbf{DFA}(2)/poly$. Let $L :$ $\alpha \in \mathbf{DFA}(k)/poly$, for some $k \in \mathbb{N}$. Then there is a deterministic k-head finite automaton $M = (Q, \Sigma, \delta, q_0, F)$ that recognizes L. We construct a two-head finite automata M' of same type and an advice $\beta = (\beta_n)$ such that $L : \alpha = L(M') : \beta$.

The idea for the construction of the new advice is to code the old one and all possible head-positions into it. Assume $1 \in \Sigma$ and let \$, $\# \notin \Sigma$ be new input symbols. Define $pos : \mathbb{N}_0^k \to \{1, \$\}^*$ to be $pos(i_1, i_2, \ldots, i_k) := 1^{i_1}\$1^{i_2}\$\ldots\$1^{i_k}$. Advice β_n is set to $pos(1, \ldots, 1, 1)\#pos(1, \ldots, 1, 2)\#\ldots\#pos(m, \ldots, m, m)\#\alpha_n$, where $m = |\alpha_n| + n$. Observe, since \$, $\#$ are not in Σ, one has to code each β_n properly.

Now we briefly sketch the simulation of M by M'. In the starting configuration head two is positioned on the word $pos(1, \ldots, 1, 1)$. Then head one reads with the help of the second head the k bits of the coded positions (here $1, \ldots, 1$), and stores them in the finite control. Afterwards M' simulates the movement of M's input-heads according to the transition relation as follows: Head one scans the coded positions block by block and checks with the help of head two whether the currently read block codes the successor position of the previous one. After head one has found this position, head two is used to read the input bits in the next step. So both heads switch their roles after each simulated step. If M' eventually reaches an accepting state of M then it accepts, otherwise it rejects.

Clearly M' accepts $\beta_{|w|}w$ if and only if $\alpha_{|w|}w$ is accepted by M. Hence, $L : \alpha = L(M') : \beta \in \mathbf{DFA}(2)/poly$ for $\beta = (\beta_n) \in poly$. $\qquad\square$

The number of heads given in Theorem 14 is best possible and can not be amplified, because regular languages relative to polynomial length advices do not contain the set $\{0^n 1^n \mid n \geq 1\}$, as shown in [4].

Using Theorem 14 one may reformulate the question as to whether deterministic and nondeterministic logspace coincide in a nonuniform setting as follows:

Corollary 15. $\mathbf{L}/poly = \mathbf{NL}/poly$ *iff* $\mathbf{NFA}(2)/poly \subseteq \mathbf{DFA}(2)/poly$. $\qquad\square$

For data-independent machines it would be very gentle, if besides the trivial relation *nonuniform*-$\mathbf{NC}^1 = \bigcup_{k \in \mathbb{N}} \mathbf{DiDFA}(k)/poly$, an analogous result to Theorem 14 would hold, i.e., two heads can do all of *nonuniform*-$\mathbf{NC}^1$. The technique used in the previous proof does not carry over to data-independent finite state machines. Unfortunately we also do not know whether some other technique accomplishes the result wished. Hence, we can only state:

Corollary 16. *nonuniform*-$\mathbf{NC}^1 = \mathbf{L}/poly$ *if and only if* $\mathbf{DFA}(2)/poly \subseteq \bigcup_{k \in \mathbb{N}} \mathbf{DiDFA}(k)/poly$. $\qquad\square$

6 Conclusions

We have studied the computational power of various multi-head finite automata namely nonuniform machines, where the transition function is allowed to vary in time, data-independent machines, where the movement of the input-heads only depends on the length of the input, and combinations of both machine types. It turned out that within this framework the class $\mathbf{NC}^1$ can be characterized by data-independent multi-head finite state machines, whereas logarithmic space is caught with data-dependent computations on that model.

Based on the given results we feel it worth to study data-independent and data-dependent computations for other devices than multi-head finite state machines as, e.g., counter machines or various stack automata. In case of two-way pushdown machines (the pushdown-head may move freely) we found that data-independence is no restriction at all.

7 Acknowledgments

Thanks to C. Damm, H. Fernau, and K.-J. Lange for valuable hints and comments on that topic. Also thanks to P. McKenzie for a discussion on that topic.

References

1. J. L. Balcázar, J. Díaz, and J. Gabarró. *Structural Complexity I*, volume 11 of *EATCS Monographs on Theoretical Computer Science*. Springer, 1988.
2. D. A. Barrington. Bounded-width polynomial-size branching programs recognize exactly those languages in NC^1. *Journal of Computer and System Sciences*, 38:150–164, 1989.
3. D. A. M. Barrington, N. Immerman, and H. Straubing. On uniformity within NC^1. *Journal of Computer and System Sciences*, 41:274–306, 1990.
4. C. Damm and M. Holzer. Automata that take advice. In J. Wiedermann and P. Hájek, editors, *Proceedings of the 20th Conference on Mathematical Foundations of Computer Science*, number 969 in LNCS, pages 149–158, Prague, Czech Republic, August 1995. Springer.
5. J. Hartmanis. On non-determinancy in simple computing devices. *Acta Informatica*, 1:336–344, 1972.
6. R. M. Karp and R. J. Lipton. Turing machines that take advice. *L'Enseignement Mathématique*, 28:191–209, 1982.
7. S. Lindell. A constant-space sequential model of computation for first-order logic. Unpublished, 1995.
8. C. Meinel. Polynomial size ω-branching programs and their computational power. *Information and Computation*, 85(2):163–182, April 1990.
9. B. Monien. Two-way multihead automata over a one-letter alphabet. *R.A.I.R.O Informatique theòrique / Theoretical Informatics*, 14(1):67–82, 1980.
10. P. Pudlák and S. Žak. Space complexity of computations. Preprint, Charles University Prague, 1983.
11. W. Ruzzo. On uniform circuit complexity. *Journal of Computer and System Sciences*, 22:365–338, 1981.
12. I. H. Sudborough. On tape-bounded complexity classes and multihead finite automata. *Journal of Computer and System Sciences*, 10:62–76, 1975.

Complexity of Finding Short Resolution Proofs

Kazuo Iwama *

Kyushu University, Fukuoka 812, Japan

Abstract. This paper discusses the problem of finding a shortest Resolution proof for a CNF formula of n variables. It is shown that if there is a polynomial-time (superpolynomial-time or subexponential-time, respectively) approximation algorithm that finds a nearly shortest proof of length up to $S + O(n^d)$, where S is the length of the shortest proof and d may be any constant, then there is a polynomial-time (superpolynomial-time or subexponential-time, respectively) algorithm that solves the (conventional) satisfiability of CNF formulas. This immediately gives a positive answer to the open problem asking whether finding a shortest Resolution proof is NP-hard.

1 Introduction

Recently Clegg, Edmonds and Impagliazzo [CEI96] showed that there is a nontrivial deterministic algorithm of finding a Resolution proof for a CNF formula f with n variables. It runs in time $2^{O(\sqrt{n \log S} \cdot \log n)}$ if f has a Resolution proof of length up to S. Their algorithm takes advantage of Gröbner proofs. A direct algorithm of the same bound was also found by Beame and Pitassi [BP96]. Note that this bound is roughly $2^{\sqrt{n}}$ if S is a polynomial in n and is much better than the exponential (2^n) time-bound that is needed to solve this problem using the exhaustive method. It should be noted that whether or not natural NP-complete problems, such as CNF Satisfiability (SAT) and Maximum Clique, can be solved in time $2^{o(n)}$ (subexponential time) is a famous open problem [MS85, Jia86, Dub91, Rob86, Sch93]. (The current best bounds are $O(2^{0.465n})$ for 3SAT and $O(2^{0.426n})$ for 3-Coloring due to Beigel and Eppstein [BE95].) Thus this subexponential bound suggests that finding a Resolution proof is relatively easy if a short proof exists. In fact this "relative easiness" appears to be a fairly common perception of researchers who have been working on, for instance, automated theorem proving.

From a theoretical point of view, however, this subexponential-time algorithm should be understood with care: (1) The length of Resolution proofs can be as large as exponential for some formulas [Hak85], in which case the time upper bound also becomes exponential. (2) Even if the formula f has a short (say, of polynomial length) proof, the algorithm only guarantees finding *some* proof within the stated time-bound. It may not find that short proof but may find some other proof the length of which can be as large as the runtime of the algorithm, i.e., subexponential. Thus the algorithm should be regarded as "an approximation algorithm" for finding the shortest proof of a CNF formula. The approximation ratio currently achieved is not so good, but it may be improved in the future, which might make the algorithm much more attractive. A natural question is to what extent it can be improved with keeping the "relative easiness."

---

* Supported in part by Scientific Research Grant, Ministry of Education, Japan and Mitsubishi Foundation. Current address: Department of Information Science, Kyoto University, Kyoto 606, Japan, email: iwama@kuis.kyoto-u.ac.jp.

In this paper, we prove that there is a deterministic, polynomial-time algorithm that, for any given polynomial $g(n)$, translates a CNF formula f of n variables and t (bounded by some polynomial in n) clauses into another CNF formula F of $O(n \log n)$ variables that satisfies the following two conditions: (i) If f is satisfiable, then F has a Resolution proof of length at most S (bounded by some polynomial in n) the value of which can be calculated from n, t and $g(n)$ easily. (ii) Otherwise F does not have any Resolution proof of length up to $S + g(n)$. This result immediately implies: If there is a polynomial-time (superpolynomial-time or subexponential-time, respectively) approximation algorithm that finds a nearly shortest proof of length up to $S + O(n^d)$ for some d, then there is a polynomial-time (superpolynomial-time or subexponential-time, respectively) algorithm that solves the (conventional) satisfiability of CNF formulas. It should be noted that the reduction seldom increases the size of instances (from n to $n \log n$). If the increase would be, say, from n to n^2, then we could no longer claim preserving the subexponential time. Another corollary is that the problem of finding the shortest Resolution proof is NP-hard. The same problem is known to be NP-hard for a very restricted Resolution, called Read-Once Resolution [IM95], but was open for general Resolution.

It should be noted that $g(n)$ may be an arbitrary polynomial. That means that the difference between S ($=$ the length of the shortest proof) and $S + g(n)$ ($=$ the length of a nearly shortest proof that is nevertheless useful to decide if the original f is satisfiable) can be enlarged unlimitedly. However, if we enlarge $g(n)$ then S itself becomes also larger under the current translation and their ratio, $(S+g(n))/S$, is fairly close to 1. In order to exploit the approximation algorithms for finding short proofs as mentioned above, it is a key to enlarge the ratio. This enlargement, combined with the improved quality of the approximation algorithms, could result in, e.g., a subexponential-time satisfiability-checking algorithm; this might be interesting as a new attempt to attack NP-complete problems.

2 CNF Formulas and Resolution Proofs

A *literal* is a logic variable x or its negation $\bar{x}$. A *clause* is a sum of literals like $(x_1 + \overline{x_2} + \overline{x_3})$. The special clause that consists of 0 clauses is denoted by *nil*. A (CNF) formula is a product of clauses like $f_0 = (x_1 + x_2 + x_3)(x_1 + \overline{x_3})(\overline{x_1} + \overline{x_2} + \overline{x_3})(\overline{x_2} + x_3)(\overline{x_1} + x_2)$. A formula is said to be *satisfiable* if there is an *assignment* of 0 (*false*) and 1 (*true*) into each variable that makes all the clauses true and *unsatisfiable* otherwise.

Resolution is used to prove a given formula to be unsatisfiable. It is said that two clauses C_1 and C_2 *can be merged* if there is exactly one variable, say, x, such that C_1 contains x and C_2 contains $\bar{x}$. The clause, say, C_3, that contains all the literals of C_1 but x and all the literals of C_2 but $\bar{x}$ is called a *resolvent* of C_1 and C_2. Here it is said that the variable x *is deleted*. For example, $(x_1 + x_2 + x_3)$ and $(x_1 + \overline{x_3})$ can be merged and their resolvent is $(x_1 + x_2)$. A (Resolution) *proof* can be expressed by a directed acyclic graph (DAG). Its leaf nodes are *input nodes* corresponding to the clauses of the given formula and its root node is *nil*. Intermediate nodes can *be derived* from the input nodes. The *size* (or *length*) of a proof is the number of nodes, i.e., the sum of the number of original clauses used in the proof and the number of merges assuming that any two merges do not produce the same resolvent.

3 Main Result

We now prove our reduction result:

Lemma 1 (Main Lemma). Let $t(n)$ be an integer such that we can write as $t(n) = 2^T$ for some integer T and $t(n) = O(n^k)$ for some constant k, and also let $g(n)$ be a polynomial in n. Then there is a deterministic polynomial-time algorithm that translates a CNF formula f of n variables and $t(n)$ clauses into another CNF formula F of $O(n \log n)$ variables such that (i) if f is satisfiable, then F has a Resolution proof of length at most $S(n)$ the value of which can be calculated from n, $t(n)$ and $g(n)$ in polynomial time and is $O(\max(n^k \log n, ng(n)))$ and (ii) otherwise F does not have any Resolution proof of length up to $S(n) + g(n)$.

Remark. The restriction $(t(n) = O(n^k))$ on the number of clauses in f is not too strong. For example, kCNF formulas (each clause holds at most k literals) meets the condition. Many other problems, such as Clique, Hamiltonian and Coloring, of size w can be reduced into SAT of $w \log w$ variables and a polynomial number of clauses. The other condition $(t(n) = 2^T)$ does not cause any problem either. If the given formula has t' clauses for $2^T < t' < 2^{T+1}$, then we can add $2^{T+1} - t'$ trivial clauses without changing its satisfiability, by which the number of clauses increases at most twice and the number of variables only by a constant.

The following theorems are now immediate from this lemma.

Theorem 1. If there is a polynomial-time (superpolynomial-time or subexponential-time, respectively) approximation algorithm that finds a nearly shortest proof of length up to $S + O(n^d)$ for some constant d, where S is the length of the shortest proof, then there is a a polynomial-time (superpolynomial-time or subexponential-time, respectively) algorithm that solves the satisfiability of CNF formulas.

Theorem 2. For a given CNF formula f and an integer U bounded by some polynomial in n, the problem of deciding whether f has a Resolution proof of length up to U is NP-complete.

Remark. Although it is not described in this paper, we can also prove a similar lemma and similar theorems for Tree Resolution.

Proof (Sketch) of Lemma 1. For simplicity, we use the following example as the original formula f which uses five variables, α_1 through α_5, and eight $(= 2^3)$ clauses:

$$f = (\alpha_2 + \alpha_3 + \overline{\alpha_4})(\overline{\alpha_3} + \alpha_2 + \alpha_4)(\overline{\alpha_1} + \overline{\alpha_2} + \alpha_4 + \overline{\alpha_5})(\alpha_1 + \alpha_3 + \alpha_4)$$
$$(\overline{\alpha_1} + \overline{\alpha_3} + \alpha_4)(\overline{\alpha_1} + \overline{\alpha_4} + \alpha_5)(\overline{\alpha_2} + \overline{\alpha_3} + \alpha_4)(\alpha_3 + \overline{\alpha_4} + \overline{\alpha_5}).$$

In the following, we first describe what the target formula F looks like in Sec. 3.1. In Sec. 3.2, it is shown that if f is satisfiable then there is a (particular) Resolution proof for F. Its length, S, is also calculated. We then prove in Sec. 3.3 that if f is not satisfiable then any Resolution proof for F is of length at least $S + g(n) + 1$.

The idea is based on the following simple facts: Let us say that a clause C_1 is *necessary* if C_1 must appear in any proof for a formula f_1.

Fact 1. Suppose that an unsatisfied formula f_1 becomes satisfiable if a clause C_1 is removed from it. Then C_1 is necessary.

Fact 2. Suppose that h clauses C_1, C_2, $\cdots$, C_h are all necessary and that there are e different variables in them. Then all those e variables must be deleted (which takes at least e steps).

3.1 Translation from f to F

The target formula, F, consists of the following clauses (1)–(8), (9), (10)–(44), (52)–(61):

$$(\overline{x_1} + \overline{x_2} + \overline{x_3} + \overline{x_4} + \overline{x_5} + b_1 + b_2 + b_3 + a_1 + a_2 + a_3) \tag{1}$$

$$(\overline{x_1} + \overline{x_2} + \overline{x_3} + \overline{x_4} + \overline{x_5} + b_1 + b_2 + \overline{b_3} + a_1 + a_2 + \overline{a_3}) \tag{2}$$

$$(\overline{x_1} + \overline{x_2} + \overline{x_3} + \overline{x_4} + \overline{x_5} + b_1 + \overline{b_2} + b_3 + a_1 + \overline{a_2} + a_3) \tag{3}$$

$$(\overline{x_1} + \overline{x_2} + \overline{x_3} + \overline{x_4} + \overline{x_5} + b_1 + \overline{b_2} + \overline{b_3} + a_1 + \overline{a_2} + \overline{a_3}) \tag{4}$$

$$(\overline{x_1} + \overline{x_2} + \overline{x_3} + \overline{x_4} + \overline{x_5} + \overline{b_1} + b_2 + b_3 + \overline{a_1} + a_2 + a_3) \tag{5}$$

$$(\overline{x_1} + \overline{x_2} + \overline{x_3} + \overline{x_4} + \overline{x_5} + \overline{b_1} + b_2 + \overline{b_3} + \overline{a_1} + a_2 + \overline{a_3}) \tag{6}$$

$$(\overline{x_1} + \overline{x_2} + \overline{x_3} + \overline{x_4} + \overline{x_5} + \overline{b_1} + \overline{b_2} + b_3 + \overline{a_1} + \overline{a_2} + a_3) \tag{7}$$

$$(\overline{x_1} + \overline{x_2} + \overline{x_3} + \overline{x_4} + \overline{x_5} + \overline{b_1} + \overline{b_2} + \overline{b_3} + \overline{a_1} + \overline{a_2} + \overline{a_3}) \tag{8}$$

These eight clauses (the first group) are determined by only the number of variables and the number of clauses of the original formula f. The variables x_1 through x_5 are introduced since f has five variables. The number of the clauses in this group is the same as the number of the clauses of f, where $b_1 + b_2 + b_3$ (similarly for $a_1 + a_2 + a_3$) through $\overline{b_1} + \overline{b_2} + \overline{b_3}$ are distributed into these clauses in the obvious way.

The next group of clauses play an important role, where y_1 through y_{10}, z_1 through z_{10}, and σ are new variables. y_1 and y_2 are associated with α_1 and $\overline{\alpha_1}$, y_3 and y_4 with α_2 and $\overline{\alpha_2}$, and so on. Note that the labels α_1, $\overline{\alpha_1}$ and so on in (10)–(44) (α_1 is effective until right before $\overline{\alpha_1}$, i.e., for both (10) and (11), and similarly for others) are just for convenience; they are not necessary formally.

$$(x_1 + \overline{\sigma})(x_2 + \overline{\sigma})(x_3 + \overline{\sigma})(x_4 + \overline{\sigma})(x_5 + \overline{\sigma}) \tag{9}$$

$$\alpha_1 : (\overline{x_1} + A_{100,101} + y_1)(\overline{x_1} + A_{100,11} + y_1)(\overline{x_1} + A_{100,0} + y_1) \tag{10}$$

$$(\overline{z_1} + y_1) \ (V_1(1) + \overline{y_1}) \cdots (V_1(m) + \overline{y_1}) \tag{11}$$

$$\overline{\alpha_1} : (\overline{x_1} + A_{101,100} + y_2)(\overline{x_1} + A_{101,11} + y_2)(\overline{x_1} + A_{101,0} + y_2) \tag{12}$$

$$(\overline{x_1} + A_{011,010} + y_2)(\overline{x_1} + A_{011,00} + y_2)(\overline{x_1} + A_{011,1} + y_2) \tag{13}$$

$$(\overline{x_1} + A_{010,011} + y_2)(\overline{x_1} + A_{010,00} + y_2)(\overline{x_1} + A_{010,1} + y_2) \tag{14}$$

$$(\overline{z_2} + y_2) \ (V_2(1) + \overline{y_2}) \cdots (V_2(m) + \overline{y_2}) \tag{15}$$

$$\alpha_2 : (\overline{x_2} + A_{111,110} + y_3)(\overline{x_2} + A_{111,10} + y_3)(\overline{x_2} + A_{111,0} + y_3) \tag{16}$$

$$(\overline{x_2} + A_{110,111} + y_3)(\overline{x_2} + A_{110,10} + y_3)(\overline{x_2} + A_{110,0} + y_3) \tag{17}$$

$$(\overline{z_3} + y_3) \ (V_3(1) + \overline{y_3}) \cdots (V_3(m) + \overline{y_3}) \tag{18}$$

$$\overline{\alpha_2} : (\overline{x_2} + A_{101,100} + y_4)(\overline{x_2} + A_{101,11} + y_4)(\overline{x_2} + A_{101,0} + y_4) \tag{19}$$

$$(\overline{x_2} + A_{001,000} + y_4)(\overline{x_2} + A_{001,01} + y_4)(\overline{x_2} + A_{001,1} + y_4) \tag{20}$$

$$(\overline{z_4} + y_4) \ (V_4(1) + \overline{y_4}) \cdots (V_4(m) + \overline{y_4}) \tag{21}$$

$$\alpha_3 : (\overline{x_3} + A_{111,110} + y_5)(\overline{x_3} + A_{111,10} + y_5)(\overline{x_3} + A_{111,0} + y_5) \tag{22}$$

$$(\overline{x_3} + A_{100,101} + y_5)(\overline{x_3} + A_{100,11} + y_5)(\overline{x_3} + A_{100,0} + y_5) \tag{23}$$

$$(\overline{x_3} + A_{000,001} + y_5)(\overline{x_3} + A_{000,01} + y_5)(\overline{x_3} + A_{000,1} + y_5) \tag{24}$$

$$(\overline{z_5} + y_5) \ (V_5(1) + \overline{y_5}) \cdots (V_5(m) + \overline{y_5}) \tag{25}$$

$$\overline{\alpha_3} : (\overline{x_3} + A_{110,111} + y_6)(\overline{x_3} + A_{110,10} + y_6)(\overline{x_3} + A_{110,0} + y_6) \tag{26}$$

$$(\overline{x_3} + A_{011,010} + y_6)(\overline{x_3} + A_{011,00} + y_6)(\overline{x_3} + A_{011,1} + y_6) \tag{27}$$

$$(\overline{x_3} + A_{001,000} + y_6)(\overline{x_3} + A_{001,01} + y_6)(\overline{x_3} + A_{001,1} + y_6) \tag{28}$$

$$(\overline{z_6} + y_6) \ (V_6(1) + \overline{y_6}) \cdots (V_6(m) + \overline{y_6}) \tag{29}$$

$$\alpha_4 : (\overline{x_4} + A_{110,111} + y_7)(\overline{x_4} + A_{110,10} + y_7)(\overline{x_4} + A_{110,0} + y_7) \tag{30}$$

$$(\overline{x_4} + A_{101,100} + y_7)(\overline{x_4} + A_{101,11} + y_7)(\overline{x_4} + A_{101,0} + y_7) \tag{31}$$

$$(\overline{x_4} + A_{100,101} + y_7)(\overline{x_4} + A_{100,11} + y_7)(\overline{x_4} + A_{100,0} + y_7) \tag{32}$$

$$(\overline{x_4} + A_{011,010} + y_7)(\overline{x_4} + A_{011,00} + y_7)(\overline{x_4} + A_{011,1} + y_7) \tag{33}$$

$$(\overline{x_4} + A_{001,000} + y_7)(\overline{x_4} + A_{001,01} + y_7)(\overline{x_4} + A_{001,1} + y_7) \tag{34}$$

$$(\overline{z_7} + y_7) \ (V_7(1) + \overline{y_7}) \cdots (V_7(m) + \overline{y_7}) \tag{35}$$

$$\overline{\alpha_4} : (\overline{x_4} + A_{111,110} + y_8)(\overline{x_4} + A_{111,10} + y_8)(\overline{x_4} + A_{111,0} + y_8) \tag{36}$$

$$(\overline{x_4} + A_{010,011} + y_8)(\overline{x_4} + A_{010,00} + y_8)(\overline{x_4} + A_{010,1} + y_8) \tag{37}$$

$$(\overline{x_4} + A_{000,001} + y_8)(\overline{x_4} + A_{000,01} + y_8)(\overline{x_4} + A_{000,1} + y_8) \tag{38}$$

$$(\overline{z_8} + y_8) \ (V_8(1) + \overline{y_8}) \cdots (V_8(m) + \overline{y_8}) \tag{39}$$

$$\alpha_5 : (\overline{x_5} + A_{010,011} + y_9)(\overline{x_5} + A_{010,00} + y_9)(\overline{x_5} + A_{010,1} + y_9) \tag{40}$$

$$(\overline{z_9} + y_9) \ (V_9(1) + \overline{y_9}) \cdots (V_9(m) + \overline{y_9}) \tag{41}$$

$$\overline{\alpha_5} : (\overline{x_5} + A_{101,100} + y_{10})(\overline{x_5} + A_{101,11} + y_{10})(\overline{x_5} + A_{101,0} + y_{10}) \tag{42}$$

$$(\overline{x_5} + A_{000,001} + y_{10})(\overline{x_5} + A_{000,01} + y_{10})(\overline{x_5} + A_{000,1} + y_{10}) \tag{43}$$

$$(\overline{z_{10}} + y_{10}) \ (V_{10}(1) + \overline{y_{10}}) \cdots (V_{10}(m) + \overline{y_{10}}) \tag{44}$$

Here

$$A_{i_1 i_2 i_3, j_1 j_2 j_3} = b_{1,i_1} + b_{2,i_2} + b_{3,i_3} + a_{1,j_1} + a_{2,j_2} + a_{3,j_3} \tag{45}$$

$$A_{i_1 i_2 i_3, j_1 j_2} = b_{1,i_1} + b_{2,i_2} + b_{3,i_3} + a_{1,j_1} + a_{2,j_2} \tag{46}$$

$$A_{i_1 i_2 i_3, j_1} = b_{1,i_1} + b_{2,i_2} + b_{3,i_3} + a_{1,j_1} \tag{47}$$

where

$$b_{i,j} = b_i \text{ if } j = 1 \text{ and } b_{i,j} = \overline{b_i} \text{ if } j = 0 \tag{48}$$

and similarly for $a_{i,j}$ (a_i and b_j already appeared in (1)–(8)). For example, $A_{100,101} = b_1 + \overline{b_2} + \overline{b_3} + a_1 + \overline{a_2} + a_3$ and $A_{100,11} = b_1 + \overline{b_2} + \overline{b_3} + a_1 + a_2$. Their role is important and will be described later in this subsection. The value m must be written as 2^M for some integer M whose value will be given later. $V_i(1)$, $V_i(2)$, through $V_i(m)$ are defined as follows:

$$V_i(1) = c_{i,1} + c_{i,2} + \cdots + c_{i,M} \tag{49}$$

$$V_i(2) = c_{i,1} + c_{i,2} + \cdots + \overline{c_{i,M}} \tag{50}$$

$$\cdots$$

$$V_i(m) = \overline{c_{i,1}} + \overline{c_{i,2}} + \cdots + \overline{c_{i,M}} \tag{51}$$

where $c_{i,j}$ are also new variables. We need more clauses:

$\alpha_1 : (U(1) + \sigma + z_1)$ (52)		$\overline{\alpha_3} : (U(3) + \sigma + z_6)$ (57)
$\overline{\alpha_1} : (U(1) + \sigma + z_2)$ (53)		$\alpha_4 : (U(4) + \sigma + z_7)$ (58)
$\alpha_2 : (U(2) + \sigma + z_3)$ (54)		$\overline{\alpha_4} : (U(4) + \sigma + z_8)$ (59)
$\overline{\alpha_2} : (U(2) + \sigma + z_4)$ (55)		$\alpha_5 : (U(5) + \sigma + z_9)$ (60)
$\alpha_3 : (U(3) + \sigma + z_5)$ (56)		$\overline{\alpha_5} : (U(5) + \sigma + z_{10})$. (61)

Here, $U(1)$ through $U(5)$ are defined as follows (u_i are new variables), which concludes the description of F:

$$U(1) = u_1, \ U(2) = \overline{u_1} + u_2, \ U(3) = \overline{u_1} + \overline{u_2} + u_3, \tag{62}$$

$$U(4) = \overline{u_1} + \overline{u_2} + \overline{u_3} + u_4, \ U(5) = \overline{u_1} + \overline{u_2} + \overline{u_3} + \overline{u_4}, \tag{63}$$

Suppose that the original formula f uses n variables and consists of t clauses ($n = 5$ and $t = 8$ in this example). The construction is straightforward: Recall that (1)–(8), (9), (11), (15), (18), (21), (25), (29), (35), (39), (41), (44), (52)–(61) are determined only by n and t. The number of variables a_i (the same for b_i) is determined to be $\log t$. Since $t = O(n^k)$, $\log t$ is $O(\log n)$, which is denoted by N hereafter. The construction of the other clauses, (10), (12)–(14), $\cdots$, depends more deeply on the original f. For example, α_1 appears in the fourth clause of f. Associated with this, we introduce the three clauses in (10), or we will often call them a single *line* of clauses. Note that this line is labeled by α_1 and contains $A_{100,101}$, $A_{100,11}$ and $A_{100,0}$. The subscript 100 of those $A_{i,j}$ corresponds to $b_1 + \overline{b_2} + \overline{b_3}$ as described in (48) and this $b_1 + \overline{b_2} + \overline{b_3}$ appears in (4) that in turn corresponds to the fourth clause of f. As for $\overline{\alpha_1}$, we have three those lines, (12)–(14), because $\overline{\alpha_1}$ appears in the three different clauses in f. If α_1 does not appear at all (i.e., only $\overline{\alpha_1}$ appears), then we do not need the line (10) but we do need the line (11). As for the number of clauses in F, the number of clauses in (10)–(44) is dominant. That number is approximately $3\times$(the number of literals in f)$+2 \cdot n \cdot m$. As will be shown later, m is roughly $g(n)$ and hence the number of F's clauses is polynomial in n. Thus one can see that the reduction runs in polynomial time.

The basic idea of the construction is as follows: First of all, all (1)–(8) are necessary (by Fact 1). Then all a_i's and b_i's need to be deleted (by Fact 2), which can only be done by using $A_{i,j}$'s in (10)–(44). Then we need to delete y_i that appears with $A_{i,j}$. To do so, we have to use, for instance, $(\overline{z_1} + y_1)$ in (11) and then $\overline{y_1}$ as well to delete this y_1. The only way of getting y_1 is to merge $(V_1(1) + \overline{y_1}) \cdots (V_1(m) + \overline{y_1})$ in (11), which needs $m-1$ merges. A key observation is that we only need to do this for *either* (11) or (15), i.e., for *either* α_1 or $\overline{\alpha_1}$, for either α_2 or $\overline{\alpha_2}$ and so on if f is satisfiable. If f is not satisfiable, then we need to do this both for at least one pair of α_i *and* $\overline{\alpha_i}$. Thus extra $m - 1$ steps are needed to prove F.

3.2 Justification of the Translation (When f is Satisfiable)

To prove the correctness of this construction, we first show that if f is satisfiable then F can be proved by the following procedure: Since f is satisfiable, there must be a satisfying truth assignment. Let, for example, that assignment be

$$\alpha_1 = \alpha_2 = 1 \text{ and } \alpha_3 = \alpha_4 = \alpha_5 = 0. \tag{64}$$

Then we can obtain the following Resolution proof for F:

Step 1. As for clauses (10) to (44), we only use those labeled by α_1, α_2, $\overline{\alpha_3}$, $\overline{\alpha_4}$, and $\overline{\alpha_5}$, corresponding to the above satisfying truth assignment. Merge $(V_1(1) + \overline{y_1}) \cdots (V_1(m) + \overline{y_1})$ in (11) to get the single-literal clause $(\overline{y_1})$, for which $(m - 1)$ steps are obviously necessary and sufficient, and similarly for others to get $(\overline{y_3})$, $(\overline{y_6})$, $(\overline{y_8})$, and $(\overline{y_{10}})$. This step needs $(m-1) \cdot n$ merges. (Note that any two merges clearly do not produce the same clause. We will not mention this property below although it always holds.) Also we used mn original clauses.

Step 2. Delete a_1 through a_3 and b_1 through b_3 in (1)–(8). To do so, we can use $A_{i,j}$ in (10)–(44). For example, $a_1 + a_2 + a_3$ in (1) can be deleted using $(\overline{x_2} + A_{111,110} + y_3)$, $(\overline{x_2} + A_{111,10} + y_3)$ and $(\overline{x_2} + A_{111,0} + y_3)$ in (16), in three (or N in general) merges; namely merging $a_1 + a_2 + a_3$ with $A_{111,110}$ to get $a_1 + a_2$ and then merging it with $A_{111,10}$ to get a_1 and finally merging it with $A_{111,0}$. Note that y_3 is added to (1) in this process. Recall that $\alpha_2 = 1$ under the current assignment and this (16) is labeled by α_2 (not $\overline{\alpha_2}$). In general, since (64) is a satisfying truth assignment, there is a mapping, say δ, from $\{1, 2, \cdots, 8\}$ into

$\{\alpha_1, \alpha_2, \overline{\alpha_3}, \overline{\alpha_4}, \overline{\alpha_5}\}$ such that if $\delta(i) = \alpha'_j$ (α_j or $\overline{\alpha_j}$), then the ith clause of f includes α'_j. So, to delete a_j's in the clause (i) of F, we first look at the lines labeled by $\delta(i)$. Then it is guaranteed by the construction that we can find a line among those lines that includes $A_{i,j}$'s which can delete all a_j's of the clause (i) in the way described above. For example, the fifth clause of f (corresponding to 011 or (5)) includes $\overline{\alpha_3}$ (=1 under the current assignment) and the line (27) plays this role of deleting a_j's. Thus we can select another several lines like (16) from the groups labeled by α_1, α_2, $\overline{\alpha_3}$, $\overline{\alpha_4}$, $\overline{\alpha_5}$, to delete a_i's in (2)–(8). After deleting all a_j's, we can then delete all b_j's by merging (1) through (8) in the natural order. In general, we need $N \times t$ merges to delete all a_j's and another $t - 1$ merges to delete all b_j's, i.e.,

$$t(N + 1) - 1 \tag{65}$$

merges in total. Thus we have obtained a clause that includes $x_1, \cdots, x_5$ and some y_i's. The number of those remaining y_i's can change (for example, we can use (16) and (17) both of which contain y_3 to delete a_i's in (1) and (2)). However, its maximum value is obviously five (or n in general) since we only used the lines among (10)–(44) whose labels are α_1, α_2, $\overline{\alpha_3}$, $\overline{\alpha_4}$, and $\overline{\alpha_5}$, i.e., at most $y_1, y_3,$ $y_6, y_8,$ and y_{10} are added. Hence at most another n merges are needed to delete those y_i's using $(\overline{y_i})$ obtained in Step 1, and hence

$$t(N + 1) + n - 1 \tag{66}$$

merges are enough to derive $(\overline{x_1} + \overline{x_2} + \overline{x_3} + \overline{x_4} + \overline{x_5})$. Here we used at most $t + tN + n$ original clauses.

Step 3. Using five clauses in (9), we delete $\overline{x_i}$ in this $(\overline{x_1} + \overline{x_2} + \overline{x_3} + \overline{x_4} + \overline{x_5})$ and we get the clause including only $\overline{\sigma}$. This needs n merges and n original clauses

Step 4. Just as Step 1, we pick (52), (54), (57), (59) and (61) according to the satisfying truth assignment. Five z_i's are deleted using the corresponding $(\overline{z_i} + y_i)$ in (10), (18), (29), (39), and (44). Again y_i in those clauses should be deleted in advance using single-literal clauses, $(\overline{y_i})$, obtained in Step 1. We need $2n$ merges and n original clauses.

Step 5. Now we can merge the five clauses in Step 4 to get σ; we need $n - 1$ merges.

Step 6. Finally, we get nil from $\overline{\sigma}$ and σ, in one merge.

Thus, $(m-1)n + t(N+1) + n - 1 + n + 2n + n - 1 + 1 = (m+4)n + t(N+1) - 1$ merges are enough to prove F in total when the original formula f is satisfiable. The total number of original clauses used in this proof is at most $mn + t + tN + n + n + n = t(N + 1) + n(m + 3)$. Therefore, the total size of the proof is

$$[(m + 4)n + t(N + 1) - 1] + [t(N + 1) + n(m + 3)]$$
$$= n(2m + 7) + 2t(N + 1) - 1. \tag{67}$$

3.3 Justification of the Translation (When f is Not Satisfiable)

Now we shall show that if f is not satisfiable, then any Resolution proof needs more length. The idea is to show that there are many necessary clauses in F for which we can apply Facts 1 and 2.

Claim 1. If one clause in $(V_1(1) + \overline{y_1}) \cdots (V_1(m) + \overline{y_1})$ in (11) and one clause in $(V_2(1) + \overline{y_2}) \cdots (V_2(m) + \overline{y_2})$ in (15) are both missing, then the whole formula becomes satisfiable.

Proof. Suppose that $(V_1(i_1) + \overline{y_1})$ and $(V_2(i_2) + \overline{y_2})$ are both missing. Then there is an assignment to $c_{1,j}$'s and $c_{2,j}$'s such that all $V_1(j)$'s and $V_2(j)$'s for

$1 \leq j \leq m$ are true excepting $V_1(i_1)$ and $V_2(i_2)$. Now we can set $y_1 = y_2 = 1$. Then all the clauses in (10)–(15) are true, which means that we can set $z_1 = z_2 = 1$ without destroying that satisfaction. Then we can make all the clauses (52)–(61) true by the assignment that

$$\sigma = 0, \ z_1 = z_2 = 1, \ z_3 = \cdots = z_{10} = 0, \tag{68}$$

since there is an assignment to u_i's so that $U(1) = 0$ and $U(2) = \cdots = U(5) = 1$.

Now we set

$$x_1 = x_2 = x_3 = x_4 = x_5 = 0. \tag{69}$$

Then (1)–(8) are all true. Also all the clauses in (9) are already true by $\sigma = 0$. As for (16)–(44), the clauses including $\overline{x_i}$ or $\overline{z_i}$ are already true. So, by finally setting

$$y_3 = \cdots = y_{10} = 0 \tag{70}$$

we can make all the rest of the clauses true. □

This claim means we need at least either the whole of $(V_1(1)+\overline{y_1}) \cdots (V_1(m)+\overline{y_1})$ or the whole of $(V_2(1) + \overline{y_2}) \cdots (V_2(m) + \overline{y_2})$ to prove F. Suppose that we use the former, $(V_1(1) + \overline{y_1}) \cdots (V_1(m) + \overline{y_1})$, for the Resolution proof of F. Then we have to delete all the variables appearing in $V_1(1) \cdots V_1(m)$, namely, $c_{1,1}, \cdots, c_{1,M}$. Since those variables do not appear in any other portion, all we can do is to merge them with themselves, which clearly needs $m - 1$ merges (since, once again, if one of those are missing, then the rest can all be satisfied). This is the same for $(V_3(1) + \overline{y_3}) \cdots (V_3(m) + \overline{y_3})$ or $(V_4(1) + \overline{y_4}) \cdots (V_4(m) + \overline{y_4})$ and so on. That means we need

$$(m - 1) \cdot n \tag{71}$$

merges. Also we need mn original clauses.

Claim 2. If (52) and (53) are both missing then the whole formula becomes satisfiable.

Proof. Consider the following assignment:

$$u_1 = 0, \ \sigma = 0, \ z_1 = \cdots = z_{10} = 0, \tag{72}$$

$$y_1 = \cdots = y_{10} = 0, \ x_1 = \cdots = x_5 = 0. \tag{73}$$

It is not hard to verify that this assignment makes all the clauses true. □

A similar claim holds for (54) and (55) and so on. Namely, we need either (52) or (53), (54) or (55) and so on for the Resolution proof. Consequently, we have to delete five (n in general) z_i's, $(n - 1)$ u_i's and one σ, which requires at least $n + (n - 1) + 1$ merges. Furthermore, the only way of deleting z_i is to use $(\overline{z_i} + y_i)$ since we have no $\overline{z_i}$ elsewhere, which means we need another five $(\overline{z_i} + y_i)$'s and each y_i should also be deleted. Thus we need another n merges, i.e., we need

$$3n \tag{74}$$

merges in total. Also we need $2n$ original clauses here.

Claim 3. If (1) is missing, then the whole formula becomes satisfiable.

Proof. The following assignment makes the whole clauses (but (1)) true:

$$x_1 = \cdots = x_5 = 1, \ a_1 = a_2 = a_3 = b_1 = b_2 = b_3 = 0, \tag{75}$$

$$\sigma = 1, \ y_1 = \cdots = y_{10} = 0, \ z_1 = \cdots = z_{10} = 0 \tag{76}$$

Verification is easy but it should be noted that all $A_{i,j}$ in (10)–(44) become true by setting $a_1 = b_1$, $a_2 = b_2$ and $a_3 = b_3$. □

A similar claim holds also for (2)–(8). So, we need to delete x_1 through x_5, all a_i's and all b_i's. To delete x_1 through x_5, we have to use (9) and need at least n (= the number of different x_i's) merges. (Since each x_i is repeated several times, the total number of x_i's appearing there is much more than n. However, one cannot simply say that the number of necessary merges is larger than n since it may be possible to merge (1) to (8) first before deleting x_i's.) To delete a_i's and b_i's, we have to use $A_{i,j}$. Here is an important observation: Let $a_1 + a_2 + a_3 + b_1 + b_2 + b_3$ be denoted by $A_{111,111}$ and $a_1 + a_2 + \overline{a_3} + b_1 + b_2 + \overline{b_3}$ by $A_{110,110}$ and so on. Then we have eight different $A_{i,j}$'s in (1)–(8) and $3 \times 8 = 24$ different $A_{i,j}$'s in (10)–(44), which makes $(3 + 1) \times 8$, or $(N + 1) \cdot t$ in general, different $A_{i,j}$'s in total. One can now see that any two of those $(N+1) \cdot t$ $A_{i,j}$'s do not overlap, i.e., one contains some affirmative variable and the other its negation. That means the set of $A_{i,j}$'s becomes satisfiable even if only one of them is missing. Thus we can conclude that we need all of those $(N + 1) \cdot t$ $A_{i,j}$'s to delete all of a_1, a_2, a_3, b_1, b_2 and b_3 and therefore we need $(N + 1) \cdot t - 1$ merges. By adding the number of merges for deleting x_i's, we need

$$(N + 1) \cdot t + n - 1 \tag{77}$$

merges. Also we need $(N + 1)t$ original clauses.

In this procedure of deleting a_i's and b_i's, what we actually use is not $A_{i,j}$ itself but clauses in the form of $(\overline{x_{l_1}} + A_{i,j} + y_{l_2})$. Deleting $\overline{x_{l_1}}$ was already considered but how to deal with y_{l_2} is not yet. Now let Σ be the set of literals y_{l_2} that are included in the clauses $(\overline{x_{l_1}} + A_{i,j} + y_{l_2})$ which were actually used to delete a_i's and b_i's in (1)–(8). Then one can claim that Σ must contain both y_1 and y_2 or both y_3 and y_4, or $\cdots$, both y_9 and y_{10}, since otherwise the original formula f should be satisfiable by the following reason:

Since we need all the different 24 $A_{i,j}$'s, we of course need all of $A_{111,110}$, $A_{110,111}$, $\cdots$, $A_{000,001}$, which correspond to the first, second, $\cdots$, eighth clauses of f, respectively. Observe again that $A_{110,111}$, for example, appears in only the lines labeled by $\overline{\alpha_3}$, α_2 and α_4. Suppose, for example, that we used the clause placed in the line labeled by α_2 as the clause including $A_{110,111}$. Let us associate this with that 110 is mapped to α_2. Then we can obtain a mapping from $\{111, 110, \cdots, 000\}$ into $\{\alpha_1, \overline{\alpha_1}, \cdots, \alpha_5, \overline{\alpha_5}\}$. Now the key fact is that if Σ does not contain both y_{2i-1} and y_{2i} for any i, then the range of this mapping does not contain both α_i and $\overline{\alpha_i}$ for any i, which meets exactly the condition of the mapping δ described before (in Step 2) that guarantees the existence of a satisfying truth assignment. (End of the reason)

As a result, we must delete both y_{2i-1} and y_{2i} for at least one i. Recall that we already counted the merges for deleting $c_{i,j}$'s in (11) or (15), (18) or (21), and so on. Namely we can assume that we already have single-literal clauses $(\overline{y_1})$ or $(\overline{y_2})$, $(\overline{y_3})$ or $(\overline{y_4})$, and so on. But we do NOT have both $(\overline{y_1})$ and $(\overline{y_2})$ or $(\overline{y_3})$ and $(\overline{y_4})$, $\cdots$. Also recall that we already counted the merges to delete y_1 or y_2, y_3 or y_4 and so on. (Those are not in $(\overline{x_{l_1}} + A_{i,j} + y_{l_2})$ but in $(\overline{z_i} + y_i)$. Nevertheless this might have been enough to delete y_i in both clauses as mentioned before.) Again, however, we did ont count the merges to delete both y_{2i-1} and y_{2i}. So we need to delete at least one y_i that was not deleted anywhere before and to do so, we need a new single-clause $(\overline{y_i})$ by merging $(V_i(1) + \overline{y_1}) \cdots (V_i(m) + \overline{y_1})$, which adds another

$$m - 1 \tag{78}$$

merges and m original clauses.

Thus we have shown that we need at least

$$(m - 1)n + 3n + (N + 1)t + n - 1 + m - 1 = (m + 3)n + (N + 1)t + m - 2$$

merges that is the sum of (71), (74), (77) and (78). The number of original clauses which must be used is $mn + 2n + (N + 1)t + m$. So the size of the proof is at least

$$(m + 3)n + (N + 1)t + m - 2 + mn + 2n + (N + 1)t + m$$
$$= (2m + 5)n + 2t(N + 1) + 2m - 2 \qquad (79)$$

Thus the difference of (79) and (67) is $D = 2m - 2n - 1$. Consequently, if we choose the value of m so that it can be expressed as 2^M for some integer M and is larger than $g(n)/2 + 2n + 1$, then we can make the value of D larger than $g(n)$. That is what we wanted to prove.

3.4 The Number of Variables in F

Finally let us calculate the number of variables in F (recall that n is the number of variables in f, t is the number of clauses in f, $N = O(\log n)$ and m is bounded by a polynomial in n as described above):

(1)–(8): n x_i's, N a_i's, N b_i's, (9): one σ,
(10)–(44): $2n$ y_i's, $2n$ z_i's, $2n \times M$ $c_{i,j}$'s, (52)–(61): $(n-1)$ u_i's.

Since $M = O(\log n)$, the total number of variables is $O(n \log n)$. Also one can see that the number of proof–steps given by (67) is $O(\max(n^k \log n, ng(n)))$, which concludes the proof of the main lemma.   Q.E.D.

Acknowledgments. The author thanks Toni Pitassi for her valuable comments and suggestions at the beginning of this research.

References

[BE95] R. Beigel and D. Eppstein. 3-coloring in time $O(1.3446^n)$: a no-MIS algorithm. *Proc. 36th IEEE Symp. on Foundations of Computer Science*, pages 444–452, 1995.

[BP96] P. Beame and T. Pitassi. Simplified and improved resolution lower bounds. *Proc. 37th IEEE Symp. on Foundations of Computer Science*, 1996.

[CEI96] M. Clegg, J. Edmonds, and R. Impagliazzo. Using the Groebner basis algorithm to find proofs of unsatisfiability. *Proc. 28th ACM Symposium on Theory of Computing*, pages 174–183, 1996.

[Dub91] O. Dubois. Counting the number of solutions for instances of satisfiability. *Theor. Comput. Sci.*, 81:49–64, 1991.

[Hak85] A. Haken. The intractability of resolution. *Theor. Compout. Sci.*, pages 297–308, 1985.

[IM95] K. Iwama and E. Miyano. Intractability of read-once resolution. In *Proc. 10th IEEE Structure in Complexity Conference*, 1995.

[Jia86] T. Jian. An $O(2^{0.304n})$ algorithm for solving maximum independent set problem. *IEEE Trans. Comput.*, C-35:847–851, 1986.

[MS85] B. Monien and E. Speckenmeyer. Solving satisfiability in less than 2^n steps. *Discrete Appl. Math.*, 10:287–295, 1985.

[Rob86] J. Robson. Algorithms for maximum independent sets. *J. Algr.*, 7:425–440, 1986.

[Sch93] I. Schiermeyer. Solving 3-satisfiability in less than 1.579^n steps. *6th Worksh. Computer Science Logic*, pages 379–394, 1993.

On P versus NP∩co-NP for Decision Trees and Read-Once Branching Programs

S. Jukna* A. Razborov** P. Savický*** I. Wegener†

Abstract. It is known that if a Boolean function f in n variables has a DNF and a CNF of size $\leq N$ then f also has a (deterministic) decision tree of size $\exp\left(O(\log n \log^2 N)\right)$. We show that this simulation *cannot* be made polynomial: we exhibit explicit Boolean functions f that require deterministic trees of size $\exp\left(\Omega(\log^2 N)\right)$ where N is the total number of monomials in minimal DNFs for f and $\neg f$. Moreover, we exhibit new examples of explicit Boolean functions that require deterministic read-once branching programs of exponential size whereas both the functions and their negations have small nondeterministic read-once branching programs. One example results from the Bruen-Blokhuis bound on the size of nontrivial blocking sets in projective planes: it is remarkably simple and combinatorially clear. Whereas other examples have the additional property that f is in AC^0.

1 Introduction

The following general question has been widely studied for various computational models:

- *Suppose that both a computational problem and its complement possess an efficient nondeterministic computation in some model. Does this imply that the problem can also be computed efficiently (typically, with at most polynomial blow-up) and* **deterministically** *in the same model?*

We use for this question somewhat imprecise but very expressive the abbreviation *the P versus* NP∩*co*−NP *question*, and, since we study the models in Boolean (non-uniform) complexity, we assume in this notation polynomial size instead of polynomial time. More specifically, we study the P versus NP∩*co*−NP question for decision trees and read-once branching programs (see e.g. [18, 14] for definitions of these models).

In the context of decision trees Ehrenfeucht and Haussler [6] have proved that every Boolean function f in n variables has a deterministic decision tree of size

$$\exp(O(\log n \log^2 N)),$$

where N is the total number of monomials in the minimal DNFs for f and $\neg f$.[5] Since nondeterministic decision trees are essentially equivalent to DNFs, this upper bound states that for decision trees we have NP∩*co*−NP ⊆ P̃, where P̃ stands for *quasipolynomial* size. [6] asked whether their bound can be improved. We prove that it is optimal up to a factor of $\log n$ in the exponent. In particular, NP ∩ *co* − NP ⊈ P in the context of decision trees. For this purpose we show that the ITERATED MAJORITY function and the ITERATED NAND function considered for a related purpose already in [15] require deterministic decision trees of sizes $\exp\left(\Omega(\log^{1.58} N)\right)$ and $\exp\left(\Omega(\log^2 N)\right)$, respectively (Theorems 3, 4).

* Dept. of Computer Science, University of Trier, D-54286 Trier, Germany. Supported by DFG grant Me 1077/10–1. On leave from Institute of Mathematics, Vilnius, Lithuania. E-mail: jukna@ti.uni-trier.de

** Steklov Mathematical Institute, Gubkina 8, 117966, Moscow, Russia. Supported by RBRF grant #96-01-01222. E-mail: razborov@class.mi.ras.ru

*** Institute of Computer Science, Acad. of Sci. of Czech Republic, Pod vodárenskou věží 2, 182 07 Praha 8, Czech Republic. Supported by grant of GA the Czech Republic No. 201/95/0976. E-mail: savicky@uivt.cas.cz

† Dept. of Computer Science, University of Dortmund, D-44221 Dortmund, Germany. Supported by DFG grant We 1066/8-1. E-mail: wegener@ls2.informatik.uni-dortmund.de

[5] This result is not stated explicitly in [6], so we describe it more precisely in Theorem 1 below. For the analogous result about the *depth* of decision trees see [4, 9, 17].

In the rest of the paper, we deal with read-once branching programs. The first example of a function showing that in this context NP∩co−NP is not even in subexponential size was given in [11]. In the present paper, we provide new examples of this kind. One of them is remarkably simple and combinatorially clear, whereas the other two have the additional property that the separating function is in AC^0. More specifically:

1. We show that the characteristic function f of blocking sets of appropriately bounded cardinality in a finite projective plane requires 1-b.p. of exponential size whereas both f and $\neg f$ have polynomial size 1-n.b.p. of very transparent structure (Theorem 6).

2. We exhibit an explicit Boolean function f such that both f and $\neg f$ not only have 1-n.b.p. of polynomial size, but can also be computed by polynomial size depth-3 circuits. Whereas any 1-b.p. computing f still must have exponential size (Theorem 7).

3. We exhibit an explicit Boolean function f such that both f and $\neg f$ have extremely small 1-n.b.p., both can be computed by Σ_3^p-circuits whereas the minimal 1-b.p. computing f has quasipolynomial size (Theorem 8).

2 Decision Trees

In this section we establish the announced bounds for decision trees. Recall that a (deterministic) *decision tree* is a binary tree whose internal nodes have labels from $\{1, \ldots, n\}$ and whose leaves have labels from $\{0, 1\}$. If a node has label i then one of the outgoing edges is labelled by the test $x_i = 0$, and the other by $x_i = 1$. Every decision tree computes a Boolean function f in a natural way: for $a \in \{0, 1\}^n$ we simply follow the unique path p consistent with the input a (i.e. all edge labels along this path have the form $x_i = a(i)$), and output the label of the leaf finally reached by p. Let the size of a decision tree be the number of all its leaves, and let $dt(f)$ denote the minimum size of a decision tree for f.

Let $dnf(f)$ denote the minimum number of monomials in a DNF for f. As we already noted in the Introduction, this is essentially the minimum size of a *nondeterministic* decision tree for f, and we do not distinguish between these two complexity measures. The sum $\|f\| \rightleftharpoons dnf(f) + dnf(\neg f)$ will be called the *weight* of f.

The following result is due to Ehrenfeucht and Haussler [6]. Although not stated explicitly the result follows directly from Lemma 1 and Lemma 6 of that paper.

Theorem 1. *Let f be a nonconstant Boolean function in n variables, $s \rightleftharpoons dnf(f)$, $t \rightleftharpoons dnf(\neg f)$, and*

$$r(f) \rightleftharpoons \lfloor \log_2(s + t) \ln(st) \rfloor + 1.$$

Then f has a (deterministic) decision tree of a size bounded by

$$\sum_{i=0}^{r(f)} \binom{n}{i} = \exp\left(O(\log n \log^2 N)\right),$$

where $N \rightleftharpoons s + t$ is the weight of f.

In the following we will show that the simulation given by Theorem 1 cannot be made polynomial. We demonstrate this by proving lower bounds on the decision tree size for two explicit functions introduced in [15].

Our argument is based on spectral methods, so it will be convenient to switch to the $(-1, +1)$-notation, i.e., to consider Boolean functions as mappings from $\{-1, 1\}^n$ to $\{-1, 1\}$, where the correspondence *true* $= -1$ and *false* $= 1$ is assumed. These functions are treated as elements of a 2^n-dimensional real vector space with an inner product defined by $\langle f, g \rangle \rightleftharpoons 2^{-n} \sum_x f(x)g(x)$. The set of all monomials $X_S \rightleftharpoons \prod_{i \in S} x_i$ forms an orthonormal basis for this space. Hence, any function f on the cube $\{-1, 1\}^n$ can be uniquely expressed as $f = \sum_S \hat{f}(S) X_S$, where

$$\hat{f}(S) \rightleftharpoons \langle f, X_S \rangle = 2^{-n} \sum_x f(x) X_S(x)$$

is the Sth Fourier coefficient of f. The unique expression of f as a linear combination of X_S may be obtained from any real polynomial expressing f using identities $x_i^2 = 1$.

The following lemma combines [13, Lemma 4] and [12, Lemma 5.1].

Lemma 2. *For every Boolean function f in n variables and every $S \subseteq [n]$ we have the bound*

$$\mathrm{dt}(f) \geq 2^{|S|} \cdot \sum_{T \supseteq S} |\hat{f}(T)|. \tag{1}$$

Proof. Take a decision tree for f of size $\mathrm{dt}(f)$. For a leaf ℓ, let $\mathrm{val}(\ell) \in \{-1, 1\}$ be its label (recall that we are in $(-1, +1)$-notation), and let I_ℓ be the set of indices of those variables which are tested on the path to ℓ. Let $B_\ell \subseteq \{-1, 1\}^n$ be the set of all the inputs that reach leaf ℓ. We think of B_ℓ as of the corresponding $((0, 1)$-valued!) function; in particular, we let

$$\hat{B}_\ell(T) \rightleftharpoons 2^{-n} \sum_{x \in B_\ell} X_T(x).$$

Since each input reaches a unique leaf, the sets B_ℓ are mutually disjoint, and, hence, $\hat{f}(T) = \sum_\ell \mathrm{val}(\ell) \cdot \hat{B}_\ell(T)$ for every $T \subseteq [n]$. Now, if $T \not\subseteq I_\ell$ then $X_T(x) = +1$ for exactly half of the inputs $x \in B_\ell$, and, hence, $\hat{B}_\ell(T) = 0$. If $T \subseteq I_\ell$ then the value of X_T is fixed on B_ℓ to either $+1$ or -1, and hence, $|\hat{B}_\ell(T)| = 2^{-n} \cdot |B_\ell| = 2^{-|I_\ell|}$.

For any $S \subseteq [n]$ and any ℓ, there are at most $2^{|I_\ell| - |S|}$ sets T satisfying $S \subseteq T \subseteq I_\ell$. Hence,

$$\sum_{T \supseteq S} |\hat{f}(T)| \leq \sum_{T \supseteq S} \sum_\ell \left| \hat{B}_\ell(T) \right| = \sum_\ell \sum_{T : I_\ell \supseteq T \supseteq S} 2^{-|I_\ell|} = \sum_\ell 2^{-|S|} = 2^{-|S|} \cdot \mathrm{dt}(f),$$

and the desired bound (1) follows.

We are going to apply Lemma 2 for $S = [n]$ to the ITERATED MAJORITY function and for $S = \emptyset$ to the ITERATED NAND function.

Consider the monotone function in $n = 3^h$ variables which is computed by the balanced read-once formula of height h in which every gate is MAJ_3. the majority of 3 variables. Let us denote this function by F_h.

Theorem 3. $\mathrm{dt}(F_h) \geq \exp(\Omega(\log^\gamma N_h))$, *where* $N_h \rightleftharpoons \|F_h\|$ *and* $\gamma \rightleftharpoons \log_2 3 \geq 1.58$.

Proof. It is easy to see that $\mathrm{dnf}(F_h) = 3 \cdot \mathrm{dnf}(F_{h-1})^2$ and $\mathrm{dnf}(F_0) = 1$. Moreover, the function is self-dual and hence the DNF size of $\neg F_h$ coincides with that of F_h. Thus, $N_h = 2 \cdot 3^{2^h - 1}$ and $n = 3^h = \theta(\log^\gamma N_h)$. So, our goal is to prove the exponential bound

$$\mathrm{dt}(F_h) \geq \exp(\Omega(n)).$$

By Lemma 2 with $S = [n]$, it is sufficient to prove an appropriate lower bound on the leading Fourier coefficient $\hat{F}_h([n])$ of the polynomial $F_h(x_1, x_2, \ldots, x_n)$. We denote this coefficient by a_h and proceed by induction on h.

Clearly, $a_0 = 1$, since F_0 is a variable.

For the inductive step note that $MAJ_3(x_1, x_2, x_3) = (x_1 + x_2 + x_3 - x_1 x_2 x_3)/2$ in the $(-1, +1)$-representation. Since $F_h = MAJ_3(F_{h-1}^{(1)}, F_{h-1}^{(2)}, F_{h-1}^{(3)})$, where $F_{h-1}^{(\nu)}$ are three copies of F_{h-1} with disjoint sets of variables, we have

$$F_h = \frac{1}{2} \left(\sum_{\nu=1}^{3} F_{h-1}^{(\nu)} - \prod_{\nu=1}^{3} F_{h-1}^{(\nu)} \right).$$

Since F_{h-1} depends on less than $n = 3^h$ variables, the first summand does not contribute to the leading coefficient in F_h. Thus, we have the recursion

$$a_h = -\frac{1}{2}a_{h-1}^3.$$

This resolves to

$$a_h = (-1)^h \cdot \left(\frac{1}{2}\right)^{(3^h-1)/2} = (-1)^h \cdot \left(\frac{1}{2}\right)^{(n-1)/2}$$

and gives us (from Lemma 2) the bound $\mathrm{dt}(F_h) \geq 2^{(n+1)/2}$, as desired.

Consider the function in $n = 2^h$ variables which is computed by the balanced read-once formula of height h in which every gate is NAND, the negated AND operation $\mathrm{NAND}(x,y) = \neg x \vee \neg y$ (up to complementation of the inputs this is equivalent to a monotone read-once formula with alternating levels of AND and OR gates). Let us denote this function by G_h.

Theorem 4. $\mathrm{dt}(G_h) \geq \exp\left(\Omega(\log^2 N_h)\right)$, where $N_h \rightleftharpoons \|G_h\|$.

Proof. $\mathrm{dnf}(G_0) = \mathrm{dnf}(\neg G_0) = 1$ (since G_0 is a single variable), and it is easy to see that for every $h \geq 1$ we have $\mathrm{dnf}(G_h) \leq 2 \cdot \mathrm{dnf}(\neg G_{h-1})$ and $\mathrm{dnf}(\neg G_h) \leq \mathrm{dnf}(G_{h-1})^2$. By induction on h one obtains $\mathrm{dnf}(G_h) \leq 2^{2^{(h+1)/2}-1}$ and $\mathrm{dnf}(\neg G_h) \leq 2^{2^{(h/2)+1}-2}$. Hence, we have $N_h \leq 2^{2^{(h/2)+1}}$. Since $n = 2^h$, our statement boils down to showing

$$\mathrm{dt}(G_h) \geq \exp\left(\Omega(n)\right).$$

Let us say that a Fourier coefficient $\hat{G}_h(S)$ is *dense* if for every subtree of height 2, S contains the index of at least one of the four variables in that subtree. We are going to calculate exactly the sum of absolute values of dense coefficients. Denote this sum by c_h. Note that in the $(-1, +1)$-representation, we have $\mathrm{NAND}(x,y) = (xy - x - y - 1)/2$. Hence,

$$G_h = \frac{1}{2}\left(G_{h-1}^{(1)} \cdot G_{h-1}^{(2)} - G_{h-1}^{(1)} - G_{h-1}^{(2)} - 1\right), \tag{2}$$

where $G_{h-1}^{(1)}$, $G_{h-1}^{(2)}$ are two copies of G_{h-1} with disjoint sets of variables.

In order to compute c_2. we use the following transformation. Let $f_1 = G_1^{(1)} + 1/2$ and $f_2 = G_1^{(2)} + 1/2$. Then it follows from (2) that

$$G_2 = \frac{1}{2}f_1 f_2 - \frac{3}{4}f_1 - \frac{3}{4}f_2 + \frac{1}{8}.$$

Since each monomial in f_1 and f_2 contains at least one variable and the sets of variables of f_1 and f_2 are disjoint, there are no common monomials in the four terms in the above expression of G_2. Hence, it is easy to calculate the sum of the absolute values of the coefficients in the nonconstant monomials, which is $c_2 = 1/2 \cdot r_1 \cdot r_2 + 3/4 \cdot (r_1 + r_2) = 27/8 = 3.375$, where $r_1 = r_2 = 3/2$ is the sum of the absolute values of the coefficients in f_1 and f_2.

In order to compute c_h for $h > 2$, we use (2) directly. Only the first term $G_{h-1}^{(1)} \cdot G_{h-1}^{(2)}$ in this equation can contribute to dense coefficients, and its individual contributions do not cancel each other. Hence, we have the recursion

$$c_h = \frac{1}{2}c_{h-1}^2.$$

This resolves to $c_h = 2(c_2/2)^{2^{h-2}}$ which is $\exp\left(\Omega(n)\right)$ since $c_2 > 2$. The proof is now completed by applying Lemma 2 (this time with $S = \emptyset$).

3 Read-Once Branching Programs

In this section we investigate functions which separate P from NP ∩ co-NP for read-once branching programs. For simplicity, let us call such functions *separating*. The first example of a separating function was presented in [11, Example 6.14] and another separating function of a similar flavour was discovered in [16].

We first present a separating function of quite a different nature, which has a surprisingly compact combinatorial definition. It is the characteristic function of a certain system of point-sets in a projective plane $PG(2, q)$.

Then we present examples of separating functions in AC^0. Let Σ_d^p, Π_d^p be the classes of functions computable by polynomial size depth-d circuits over the de Morgan basis $\{AND, OR, NOT\}$ (negations are allowed only at the input variables and do not contribute to the depth) that have OR (respectively, AND) as output gate. Our first example is in Σ_3^p and demonstrates an exponential separation. The second example belongs to the smaller class $\Sigma_3^p \cap \Pi_3^p$, but the separation is only quasipolynomial.

All the lower bounds for 1-b.p. in this section are based on the following "folklore observation" (see [11], or [8] for a simple proof). Let us say that a Boolean function f in n variables is k-mixed if for any $I \subseteq [n]$ with $|I| = k$, and any two different assignments $a, b : I \to \{0, 1\}$, there is an assignment $c : [n] \setminus I \to \{0, 1\}$ for which $f(a, c) \neq f(b, c)$.

Lemma 5. *If f is k-mixed then any 1-b.p. computing f must have size at least $2^k - 1$.*

3.1 1-b.p. versus 1-n.b.p.

The size of DNF (or CNF) and the size of 1-b.p. are in general incomparable:

- the *parity* function has a small 1-b.p. but requires exponential size AC^0 circuits [7, 2, 19, 10], and
- the *exact-perfect-matching* function has a CNF of size $O(n^{3/2})$ but requires exponential size 1-b.p. [11].

Gál in [8] has proved that the characteristic function of all blocking sets in finite projective planes has even a CNF with a *linear* number of clauses, but still requires 1-b.p. of exponential size. We describe the function of Gál in more detail, since our first example is its modification.

Let $P \rightleftharpoons \{1, \ldots, n\}$ be the set of points of a projective plane $PG(2, q)$ of order q, and let $L_1, \ldots, L_n$ be the lines viewed as subsets of P; hence $n = q^2 + q + 1$. Recall that each line has exactly $q + 1$ points, every two lines intersect in exactly one point, and exactly $q + 1$ lines meet in one point. A *blocking set* is a set of points which intersects every line. The smallest blocking sets are just the lines. Gál [8] proved that the characteristic function

$$G(x_1, \ldots, x_n) \rightleftharpoons \bigwedge_{i=1}^{n} \bigvee_{j \in L_i} x_j$$

of all blocking sets has no 1-b.p. of size smaller than $2^{\sqrt{n}}$. It appears that, using known lower bounds on the size of non-trivial blocking sets due to Bruen [5] and Blokhuis [3], the argument of [8] can be easily modified to get another result, namely that $P \neq NP \cap co - NP$ in the context of read-once branching programs.

Blocking sets containing a line are called *trivial*. Bruen in [5] proved that any non-trivial blocking set in a projective plane of order q must have at least $q + \sqrt{q} + 1$ points, and this lower bound is known to be tight when q is a square. For the prime order q, Blokhuis [3] improved Bruen's bound to $3(q + 1)/2$ (which is also optimal). These results motivate the investigation of the following Boolean function:

$$B(x_1, \ldots, x_n) \rightleftharpoons \left(\bigwedge_{i=1}^{n} \bigvee_{j \in L_i} x_j \right) \wedge \neg T_{q+k+1}^n(x_1, \ldots, x_n),$$

where $k = (q + 1)/2$ if q is a prime, and $k = \lceil \sqrt{q} \rceil$ otherwise; $T_s^n(x_1, \ldots, x_n)$ is the usual threshold function which outputs 1 if and only if $x_1 + \cdots + x_n \geq s$. Thus, for any input $a : P \to \{0, 1\}$, $f(a) = 1$ if and only if the set $a^{-1}(1)$ is blocking and has at most $q + k$ points. This modified function has the required property:

Theorem 6. *Both B and $\neg B$ have 1-n.b.p. of size $O(n^{5/2})$ whereas any 1-b.p. computing B must have size at least $2^k - 1$.*

Proof. Upper bound. Associate with each of the n lines L_i ($i = 1, \ldots, n$) the following two Boolean functions: $\varphi_i \rightleftharpoons \bigwedge_{j \in L_i} x_j \wedge \neg T_k^{n-q-1}(\{x_j \mid j \notin L_i\})$ and $\psi_i \rightleftharpoons \bigwedge_{j \in L_i}(\neg x_j)$.

By the Bruen-Blokhuis bounds we have that $B(a) = 1$ iff a has at most $q + k$ ones and contains some line L, i.e., $a(i) = 1$ for all points $i \in L$. Thus, B is an OR of n functions $\varphi_1, \ldots, \varphi_n$, each of which has a 1-b.p. of size $O(n^{3/2})$. Hence, B has a 1-n.b.p. of size $O(n^{5/2})$. On the other hand, $\neg B(a) = 1$ if and only if either a has at least $q + k + 1$ ones or a avoids some line (or both). Thus, $\neg B$ is also an OR of the threshold function T_{q+k+1}^n and n functions $\psi_1, \ldots, \psi_n$. Each of the functions ψ_i has a 1-b.p. of size $q + 1$ and T_{q+k+1}^n has a 1-b.p. of size $O(nq)$. Hence, $\neg B$ has a 1-n.b.p. of size $O(n^{3/2})$.

Lower bound. To this end, we use an argument similar to that of Gál [8]. By Lemma 5, it is enough to verify that the function B is k-mixed. To show this, let $I \subseteq P$, $|I| = k$, and $a, b : I \to \{0, 1\}$ be two different assignments. Take an $i \in I$ where a and b differ. W.l.o.g. we may assume that $a(i) = 1$ and $b(i) = 0$. There are $q + 1$ lines containing the point i. Since $|I - \{i\}| = k - 1 \leq q - 1$ (the number of lines containing i, minus two) we can find among them two lines L_1 and L_2 such that $L_1 \cap I = L_2 \cap I = \{i\}$. Define the assignment $c : P \setminus I \to \{0, 1\}$ by letting $c(j) = 1$ if and only if $j \in L_1$. Then the inputs (a, c) and (b, c) both have at most $|I \cup L_1| \leq q + k$ ones. Thus, $f(a, c) = 1$, since (a, c) contains the line L_1, which in turn intersects all other lines. On the other hand, $f(b, c) = 0$ since (b, c) does not intersect the line L_2 in view of $L_2 \cap (I \cup L_1) = \{i\}$ and $b(i) = 0$.

3.2 1-b.p. versus 1-n.b.p. and AC⁰

In this subsection, we describe two separating functions in AC^0.

Theorem 7. *There is an explicit Boolean function in Σ_3^p such that both f and $\neg f$ have a 1-n.b.p. of polynomial size, whereas the 1-b.p. size of f is $\exp\left(\Omega(n/\log n)^{1/2}\right)$.*

Proof. Let $k = \lfloor \log_2 n \rfloor$ and choose arbitrarily n different subsets $\alpha(1), \alpha(2), \ldots, \alpha(n)$ of $\{1, 2, \ldots, 2k\}$ such that $|\alpha(i)| = k$ for all $i \in \{1, \ldots, n\}$. Split the n variables into $2k$ blocks $X_1, X_2, \ldots, X_{2k}$ of size $\lfloor n/(2k) \rfloor$. For every $j \in \{1, 2, \ldots, 2k\}$ we define a function $g_j(x)$ depending only on variables in X_j. For this purpose, we further split the variables in X_j into s subblocks of size s, where $s = \lfloor (n/2k)^{1/2} \rfloor$, and let g_j be the AND of ORs of variables in these subblocks. Finally, let

$$f(x) \rightleftharpoons \bigvee_{i=1}^n \left(x_i \wedge \bigwedge_{j \in \alpha(i)} g_j(x) \right).$$

Obviously, f is in Σ_3^p.

To compute f and $\neg f$ by polynomial size 1-n.b.p., we employ the fact that there are only logarithmically many functions $g_1, \ldots, g_{2k}$ and we can search exhaustively through all possible 2^{2k} outputs of this set. Formally, we re-write f in the form

$$f(x) = \bigvee_{\substack{\epsilon_1, \ldots, \epsilon_{2k} \\ \epsilon_j \in \{0,1\}}} \left(\bigvee_{i \in I(\epsilon)} x_i \wedge \bigwedge_{j=1}^{2k} (g_j(x) = \epsilon_j) \right), \tag{3}$$

where $I(\epsilon) \rightleftharpoons \{ i \in \{1, \ldots, n\} \mid \forall j \in \alpha(i)(\epsilon_j = 1) \}$. The expansion for $\neg f$ is obtained by replacing $\bigvee_{i \in I(\epsilon)} x_i$ with its negation. Now, $x_i \wedge \bigwedge_{j=1}^{2k} (g_j(x) = \epsilon_j)$ can be easily computed by a 1-b.p. of polynomial size, and taking OR of these programs over all $\epsilon \in \{0,1\}^{2k}$ and $i \in I(\epsilon)$ we get a poly-size 1-n.b.p. for f. In order to apply this argument to $\neg f$, we only need to compute $\bigwedge_{i \in I(\epsilon)} (\neg x_i) \wedge \bigwedge_{j=1}^{2k} (g_j(x) = \epsilon_j)$ by a poly-size 1-b.p. But this becomes obvious if we note that we can get rid of the double occurrences of x_i ($i \in I(\epsilon)$) by substituting in $\bigwedge_{j=1}^{2k} (g_j(x) = \epsilon_j)$ zeros for all these variables.

In order to prove the exponential lower bound on the 1-b.p. size of f we show that f is $(s-1)$-mixed. Assume that a and b are two different partial inputs setting some set I of $s - 1$ variables to constants. Let $l \in I$ be an index of a variable satisfying $a_l \neq b_l$. We need to prove that there is an assignment $c : [n] \setminus I \to \{0, 1\}$ such that $f(a, c) \neq f(b, c)$. To this end, we will construct in the next paragraph c in such a way that for any $j \in \{1, 2, \ldots, 2k\}$ c reduces g_j to the constant 1 if $j \in \alpha(l)$, and to the constant 0 if $j \notin \alpha(l)$. Then it is clear from (3) that $f(a, c) = a_l$, $f(b, c) = b_l$, and thus $f(a, c) \neq f(b, c)$.

Assignment c is constructed by considering each block X_j separately. If $j \in \alpha(l)$, select in each subblock of X_j one variable not in I and set this variable to 1. This guarantees $g_j|_c \equiv 1$. If $j \notin \alpha(l)$, find a subblock of X_j such that all its variables are not in I and set them to zero. This guarantees $g_j|_c \equiv 0$. By repeating this for all j and by setting all remaining variables in an arbitrary way, we obtain the required c.

Thus, the function f satisfies all requirements from the theorem.

Slightly modifying this construction, we get a separating function in $\Sigma_3^p \cap \Pi_3^p$.

Theorem 8. *There is an explicit Boolean function in $\Sigma_3^p \cap \Pi_3^p$ such that both f and $\neg f$ have a 1-n.b.p. of polylogarithmic (!) size whereas the 1-b.p. size of f is*

$$\exp\left(\Omega((\log n / \log\log n)^2)\right).$$

Proof. We will actually exhibit a Boolean function in m variables such that both f and $\neg f$ have polynomial (in m) size 1-n.b.p. and $\exp\left(O(m\log m)^{1/2}\right)$-sized Σ_3-circuits, whereas every 1-b.p. for f must have size $\exp\left(\Omega(m/\log m)\right)$. This function is transformed into the required form by the standard padding argument: simply introduce $2^{(m\log m)^{1/2}}$ extra dummy variables.

Let $k = \lceil \log_2 m \rceil$, and split the m variables into k blocks $X_1, \ldots, X_k$ of size $\lfloor m/k \rfloor$ each. As in the previous proof, we split the variables of X_j into s subblocks of size s, where $s = \lfloor (m/k)^{1/2} \rfloor$, and let g_j this time be the MAJORITY of MAJORITYs of variables in these subblocks.

f is defined by the following expansion that is analogous to (3):

$$f(x) \rightleftharpoons \bigvee_{\substack{\epsilon_1,\ldots,\epsilon_k \\ \epsilon_j \in \{0,1\}}} \left(x_{\overline{\epsilon_1 \ldots \epsilon_k}} \wedge \bigwedge_{j=1}^k (g_j(x) = \epsilon_j) \right), \tag{4}$$

where $\overline{\epsilon_1 \ldots \epsilon_k}$ is the integer with the binary representation $\epsilon_1 \ldots \epsilon_k$ (cf. [11, 16]).

The upper bound $O(m^3/\log m)$ on the size of 1-n.b.p. for f and $\neg f$ is clear (cf. the proof of the previous theorem).

For the lower bound on the size of 1-b.p., note that essentially the same argument as in the proof of Theorem 7 shows that f is $\left(\frac{s^2}{4} - 1\right)$-mixed. Indeed, if $|I| < s^2/4$ then for every particular j, there are at least $(s/2) + 1$ subblocks of X_j, each of them having at least $(s/2) + 1$ variables not in I. Setting all free variables in these subblocks by c to either 0 or 1 forces the value of g_j to the same constant.

Finally, MAJORITY of s variables has a trivial $\exp(O(s))$-sized Σ_2-circuit and Π_2-circuit. Merging these two, we get $\exp(O(s))$-sized Σ_3-circuit and Π_3-circuit for $g_j(x)$. Then we can get rid of the fan-in k AND gates in (4) at the cost of another factor k in the exponent. This results in Σ_3-circuits for both f and $\neg f$ of size $\exp(O(ks)) = \exp\left(O(m\log m)^{1/2}\right)$.

Remark. Note that in all our examples the nondeterministic 1-b.p. has a very special form: it is a disjunction of polynomially many small 1-b.p. Moreover, in the last example the nondeterminism is further restricted to so-called unique nondeterminism, i.e., at most one computation path may lead to a 1-leaf.

4 Open Problems

We conclude with several open questions. By Theorems 3 and 4, the functions F_h and G_h defined in Section 2 do not have a decision tree of size polynomial in $\|F_h\|$ and $\|G_h\|$. On the other hand, both have an oblivious 1-b.p. of size $O(n^2)$.

Question 1. Does there exist a Boolean function f which requires 1-b.p. of size super-polynomial in $\|f\|$?

In the terminology of Section 3, this is equivalent to the question if $\Sigma_2^p \cap \Pi_2^p$ contains a separating function. There are several further questions of the same flavour aimed at bridging the gap between our lower and upper bounds. Does Σ_2^p have a separating function? Does Σ_2^p or $\Sigma_3^p \cap \Pi_3^p$ contain a function separating $NP \cap co - NP$ from quasipolynomial time $\tilde{P}$ in the context of 1-b.p.? Finally, can one get rid of the peculiar $\log n$ factor in the exponent in Theorem 1?

Question 2. What can be said about the *probabilistic* complexity of the separating functions considered in Section 3? Optimistically, what do they separate: $NP \cap co\text{-}NP$ from BPP or $NP \cap co\text{-}NP \cap BPP(\cap AC^0)$ from P? A partial step in that direction was made by Ablayev in [1] where he proves that a Boolean function $f_\phi(\mathbf{x}) = x_{\phi(\mathbf{x})}$ with a particular "pointer" $\phi : \{0,1\}^n \to [n]$, requires *oblivious* two-sided small error randomized 1-b.p. of exponential size. The pointer ϕ was defined in [16], its graph has small 1-b.p., and hence, both f_ϕ and $\neg f_\phi$ have small 1-n.b.p. (cf. Section 3.2). Can this bound be extended to *non-oblivious* random 1-b.p.?

More modest but still (combinatorially) interesting is the following question.

Question 3. Does Gál's function (the characteristic function of all blocking sets in $PG(2, q)$) require exponential 1-n.b.p.? Note that any *minimal* 1-n.b.p. for this function (as well as for any monotone function) is *monotone*, i.e. edges labelled by $x_i = 0$ can lead only to reject sinks. Thus, the question is in fact about the combinatorial structure of blocking sets: take an acyclic directed graph and label its edges by points of $PG(2, q)$ so that no point appears in an s–t path twice. A graph is *blocking* if every s–t path corresponds to a blocking set and every such set has at least one s–t path. What is the minimal number of nodes in a blocking graph for $PG(2, q)$?

Acknowledgement

We want to thank Hans Ulrich Simon for giving us the reference to the paper of Ehrenfeucht and Haussler.

References

1. F. Ablayev (1996). Randomization and nondeterminism are incomparable for polynomial ordered binary decision diagrams. Manuscript.
2. M. Ajtai (1983). Σ_1^1-formulae on finite structures. *Annals of Pure and Applied Logic* **24** 1–48.
3. A. Blokhuis (1994). On the size of a blocking set in $PG(2, p)$. *Combinatorica* **14** 111–114.
4. M. Blum and R. Impagliazzo (1987). Generic oracles and oracle classes. In: *Proc. of 28th IEEE FOCS*, 118–126.
5. A. A. Bruen (1970). Baer subplanes and blocking sets. *Bull. Amer. Math. Soc.* **76** 342–344.
6. A. Ehrenfeucht and D. Haussler (1989). Learning decision trees from random examples. *Information and Computation* **82** 231–246.
7. M. Furst, J. Saxe and M. Sipser (1981). Parity, circuits and the polynomial time hierarchy. In: *Proc. of 22nd IEEE FOCS*, 260–270.
8. A. Gál (1997). A simple function that requires exponential size read-once branching programs. *Information Processing Letters* **62**, 13–16.
9. J. Hartmanis and L.A. Hemachandra (1987). One-way functions, robustness and non-isomorphism of NP-complete classes. Tech. Rep. DCS TR86-796, Cornell University.
10. J. Håstad (1989). Almost optimal lower bounds for small depth circuits. In: S. Micali, editor, *Randomness and Computation (Advances in Computing Research, Vol. 5)*, 143–170. JAI Press.
11. S. Jukna (1988). Entropy of contact circuits and lower bounds on their complexity. *Theor. Comput. Sci.* **57** 113–129.
12. E. Kushilevitz and Y. Mansour (1991). Learning decision trees using the Fourier spectrum. In: *Proc. of 23rd ACM STOC*, 455–464.
13. N. Linial, Y. Mansour and N. Nisan (1989). Constant depth circuits, Fourier transforms and learnability. In: *Proc. of 30th IEEE FOCS*, 574–579.
14. A. A. Razborov (1991). Lower bounds for deterministic and nondeterministic branching programs. In: *Proc. of FCT'91*, Lecture Notes in Computer Science **529** (Springer, Berlin), 47–60.
15. M. Saks and A. Wigderson (1986). Probabilistic Boolean decision trees and the complexity of evaluating games. In: *Proc. of 27th IEEE FOCS*, 29–38.
16. P. Savický and S. Žák (1996). A large lower bound for 1-branching programs, Electronic Colloquium on Computational Complexity (ECCC), Report # TR96-036, 1996.
17. G. Tardos (1989). Query complexity, or why is it difficult to separate $NP^A \cap co - NP^A$ from P^A by a random oracle A? *Combinatorica* **9** 385–392.
18. I. Wegener (1987). *The Complexity of Boolean Functions*. Wiley-Teubner.
19. A. Yao (1985). Separating the polynomial-time hierarchy by oracles. In: *Proc. of 26th IEEE FOCS*, 1–10.

A Characterization of Abstract Families of Algebraic Power Series

Georg Karner[1] and Werner Kuich[2]

[1] Alcatel Austria AG, Wien, Austria
[2] Abteilung für Theoretische Informatik
Institut für Algebra und Diskrete Mathematik
Technische Universität Wien

Abstract. Given a continuous semiring A and a collection $\mathfrak{H}$ of semiring morphisms mapping the elements of A into finite matrices with entries in A we define $\mathfrak{H}$-closed semirings. These are fully rationally closed semirings that are closed under the following operation: each morphism in $\mathfrak{H}$ maps an element of the $\mathfrak{H}$-closed semiring on a finite matrix whose entries are again in this $\mathfrak{H}$-closed semiring.

$\mathfrak{H}$-closed semirings coincide under certain conditions with abstract families of elements. If they contain only algebraic elements over some A', $A' \subseteq A$, then they are characterized by $\mathfrak{Rat}(A')$-algebraic systems of a specific form. The results are then applied to formal power series and formal languages.

1 Introduction

The purpose of this paper is to consider families of elements of a continuous semiring A (see Kuich [7, 8] and Karner, Kuich [5]) from an algebraic point of view and to characterize families of elements that are subsemirings of $\mathfrak{Alg}(A')$, where $A' \subseteq A$. This is achieved by dealing with $\mathfrak{H}$-closed semirings, where $\mathfrak{H}$ is a collection of semiring morphisms mapping the elements of A into finite matrices with entries in A. Here a semiring $\bar{A}$ is called $\mathfrak{H}$-closed if it is fully rationally closed and satisfies the following condition: For each $a \in \bar{A}$ and $h \in \mathfrak{H}$, the entries of $h(a)$ are again in $\bar{A}$.

In Section 2 we connect $\mathfrak{H}$-closed semirings with abstract families of elements and show that, under certain mild conditions, the concepts of "$\mathfrak{H}$-A'-abstract family of elements" and "$\mathfrak{H}$-closed semiring" coincide.

In Section 3 we characterize $\mathfrak{H}$-closed semirings that are subsemirings of $\mathfrak{Alg}(A')$ by families of $\mathfrak{Rat}(A')$-algebraic systems. This characterization is applied in Section 4 to the semirings of formal power series and formal languages and yields "normal forms" for abstract families of power series and abstract families of languages.

It is assumed that the reader is familiar with semiring theory. Notions and notations that are not defined are taken from Kuich [8]. In the sequel, A will always be a *continuous* semiring. This is a complete and naturally ordered semiring such that, for all index sets I and all families $(a_i \mid i \in I)$ the following condition is satisfied:

$$\sum_{i \in I} a_i = \sup\{\sum_{i \in E} a_i \mid E \subseteq I, E \text{ finite}\}.$$

Here "sup" denotes the least upper bound with respect to the natural order (see Goldstern [3], Sakarovitch [11], and Karner [4].)

A subset $\bar{A}$ of A is called *fully rationally closed* if for all $a \in \bar{A}$, we have $a^* := \sum_{i>0} a^i \in \bar{A}$. By definition, $\mathfrak{Rat}(A')$ is the smallest fully rationally closed subsemiring of A containing $A' \subseteq A$. Furthermore, the collection of the components of the least solutions of all A'-algebraic systems, where A' is a fixed subset of A, is denoted by $\mathfrak{Alg}(A')$. Here an A'-algebraic system is a system of formal equations $y_i = p_i$, $1 \le i \le n$, where $p_1, \ldots, p_n$ are polynomials in the polynomial semiring over A (with variables $y_1, \ldots, y_n$) with coefficients in A'. (See also Lausch, Nöbauer [10], Chapter 1, § 4.)

Given $A' \subseteq A$, we define $[A'] \subseteq A$ to be the least complete subsemiring of A that contains A'. The semiring $[A']$ is called the *complete semiring generated by A'*.

From now on, A' will be a fixed subset of A with $0, 1 \in A'$. Moreover, we assume that $[A'] = A$. Furthermore we make the notational convention that all sets Q, possibly provided with indices, are finite and nonempty index sets, and are subsets of some fixed countably infinite set Q_∞ with the following property: if $q_1, q_2 \in Q_\infty$ then $(q_1, q_2) \in Q_\infty$.

In the sequel, $\mathfrak{H}$ is a nonempty subfamily of the family of all semiring morphisms $h : A \to A^{Q \times Q}$, $Q \subset Q_\infty$, Q finite. We define

$$\mathcal{K}(\mathfrak{R}) = \{h(a)_{q_1, q_2} \mid a \in \mathfrak{R}, \ h : A \to A^{Q \times Q} \in \mathfrak{H}, \ q_1, q_2 \in Q\}$$

for $\mathfrak{R} \subseteq A$. Observe that the definition of $\mathcal{K}(\mathfrak{R})$ depends on $\mathfrak{H}$. A subsemiring $\mathfrak{L}$ of A is called $\mathfrak{H}$-*closed* if

(i) $\mathcal{K}(\mathfrak{L}) \subseteq \mathfrak{L}$,
(ii) $\mathfrak{Rat}(\mathfrak{L}) = \mathfrak{L}$.

The notion of an $\mathfrak{H}$-closed semiring is central for our paper. In the sequel we will characterize $\mathfrak{H}$-closed semirings for certain families $\mathfrak{H}$.

We define the subset $[\mathfrak{H}]$ of A as follows. For $h \in \mathfrak{H}$, $h : A \to A^{Q \times Q}$, let $B_h = [\{h(a)_{q_1, q_2} \mid a \in A, q_1, q_2 \in Q\}]$. Then $[\mathfrak{H}] = \bigcup_{h \in \mathfrak{H}} B_h$. A family of morphisms $\mathfrak{H}$ is called *closed under matricial composition* if the following conditions are satisfied for arbitrary morphisms $h : A \to A^{Q \times Q}$ and $h' : A \to A^{Q' \times Q'}$ in $\mathfrak{H}$:

(i) $A' \subseteq [\mathfrak{H}]$.
(ii)   For each $a \in [\mathfrak{H}]$ there is an $h_a \in \mathfrak{H}$ with $h_a(a) = a$.
(iii) If $\bar{Q} \subset Q_\infty$ and there exists a bijection $\pi : \bar{Q} \to Q$, then $\bar{h} : A \to A^{\bar{Q} \times \bar{Q}}$, defined by $\bar{h}(a)_{q_1, q_2} = h(a)_{\pi(q_1), \pi(q_2)}$ for all $a \in A$, $q_1, q_2 \in \bar{Q}$, is in $\mathfrak{H}$.
(iv)   The functional composition $h \circ h' : A \to A^{(Q' \times Q) \times (Q' \times Q)}$ is again in $\mathfrak{H}$.

(v) If $Q \cap Q' = \emptyset$ then the mapping $h + h' : A \to A^{(Q \cup Q') \times (Q \cup Q')}$ defined by

$$(h + h')(a) = \begin{pmatrix} h(a) & 0 \\ 0 & h'(a) \end{pmatrix}, \quad a \in A,$$

where the blocks are indexed by Q and Q', is again in $\mathfrak{H}$.

Observe that condition (ii) makes sure that $\mathfrak{R} \subseteq \mathcal{K}(\mathfrak{R})$. Condition (iii) will be needed when dealing with a finite number of morphisms $h_j : A \to A^{Q_j \times Q_j} \in \mathfrak{H}$, $1 \leq j \leq n$. Without loss of generality we may then assume that $Q_1, \ldots, Q_n$ are mutually disjoint index sets.

From now on, we assume that $\mathfrak{H}$ is a non-empty family of *complete A'-rational* semiring morphisms that is *closed under matricial composition*. Observe that $\mathfrak{Rat}(A')$ is $\mathfrak{H}$-closed by Theorem 7.3 of Kuich [8].

2 Connections with abstract families of elements

The following definitions are motivated by connections with automata theory (see Kuich [8] and Karner, Kuich [5]). An $\mathfrak{H}$-A'-*rational transducer* is a tuple $\mathfrak{T} = (Q, h, S, P)$ where $h : A \to A^{Q \times Q}$ is in $\mathfrak{H}$, $S \in \mathfrak{Rat}(A')^{1 \times Q}$, $P \in \mathfrak{Rat}(A')^{Q \times 1}$. The transducer defines a mapping $\|\mathfrak{T}\| : A \to A$ by

$$\|\mathfrak{T}\|(a) = Sh(a)P, \quad a \in A.$$

A mapping $\tau : A \to A$ is called an $\mathfrak{H}$-A'-*rational transduction* if there exists an $\mathfrak{H}$-A'-rational transducer $\mathfrak{T}$ such that $\tau(a) = \|\mathfrak{T}\|(a)$ for all $a \in A$.

Let $\mathfrak{L} \subseteq A$. We define

$$\mathcal{M}(\mathfrak{L}) = \{\tau(a) \mid a \in \mathfrak{L}, \tau : A \to A \text{ is an } \mathfrak{H}\text{-}A'\text{-rational transduction}\}.$$

Note that we have $\mathfrak{L} \subseteq \mathcal{M}(\mathfrak{L})$. The set $\mathfrak{L}$ is said to be *closed under $\mathfrak{H}$-A'-rational transductions* if $\mathcal{M}(\mathfrak{L}) \subseteq \mathfrak{L}$. Observe that the definition of $\mathcal{M}(\mathfrak{L})$ depends on $\mathfrak{H}$ and A'.

An $\mathfrak{H}$-A'-*family of elements* is just a subset of $[\mathfrak{H}]$. An $\mathfrak{H}$-A'-family of elements $\mathfrak{L}$ is called $\mathfrak{H}$-A'-*abstract family of elements* (briefly $\mathfrak{H}$-A'-AFE) if

(i) $\mathcal{M}(\mathfrak{L}) \subseteq \mathfrak{L}$,
(ii) $\mathfrak{Rat}(\mathfrak{L}) = \mathfrak{L}$.

Assume that $\mathfrak{L}$ is an $\mathfrak{H}$-A'-abstract family of elements. Then $\mathfrak{Rat}(A') \subseteq \mathfrak{L}$.

Let $\mathfrak{R} \subseteq A$. The notation $\mathcal{F}(\mathfrak{R})$ is used for the smallest $\mathfrak{H}$-A'-AFE containing $\mathfrak{R}$. For an $\mathfrak{H}$-A'-family of elements $\mathfrak{R}$, $\mathcal{M}(\mathfrak{R})$ and $\mathcal{F}(\mathfrak{R})$ are again $\mathfrak{H}$-A'-families of elements by Theorems 2 and 6 of Karner, Kuich [5].

Corollary 2 will state a connection between $\mathcal{M}(\mathfrak{R})$ and $\mathcal{F}(\mathfrak{R})$, where $\mathfrak{R}$ is an $\mathfrak{H}$-A'-family of elements. It is a generalization of Corollary 1 of Theorem 3.4.1 of Ginsburg [2], of Theorem VI.4.10 of Berstel [1] and of Theorem 11.30 of Kuich, Salomaa [9]. A theorem is needed before this result.

Theorem 1 *Let* $\mathfrak{R} \subseteq A$. *Then* $\mathcal{M}(\mathfrak{Rat}(\mathcal{M}(\mathfrak{R} \cup \{1\}))) \subseteq \mathfrak{Rat}(\mathcal{M}(\mathfrak{R} \cup \{1\}))$.

Proof. Since $\mathcal{M}(\mathfrak{R} \cup \{1\})$ contains 0 and 1 we can use the generalization of Kleene's Theorem, Theorem 4.10 of Kuich [8]. Hence, for a given $a \in \mathfrak{Rat}(\mathcal{M}(\mathfrak{R} \cup \{1\}))$ there exists a finite $\mathcal{M}(\mathfrak{R} \cup \{1\})$-automaton $\mathfrak{A} = (Q, M, S, P)$ such that $\|\mathfrak{A}\| = a$. In particular, we have $M_{q_1,q_2}, S_q, P_q \in \mathcal{M}(\mathfrak{R} \cup \{1\})$ for all $q_1, q_2, q \in Q$. Given any complete A'-rational semiring morphism $h : A \to A^{Q' \times Q'}$ in $\mathfrak{H}$, we infer by Theorem 7.9 of Kuich [8] that $h(M)_{(q_1,q_1'),(q_2,q_2')}$, $h(S)_{q_1',(q,q_2')}$, $h(P)_{(q,q_1'),q_2'} \in \mathcal{M}(\mathfrak{R} \cup \{1\})$. Consider now a $\mathfrak{H}$-A'-rational transducer $\mathfrak{T} = (Q', h, S', P')$. Then $\mathfrak{A}' = (Q \times Q', h(M), S'h(S), h(P)P')$ is a finite $\mathfrak{Rat}(\mathcal{M}(\mathfrak{R} \cup \{1\}))$-automaton. Since

$$\|\mathfrak{A}'\| = S'h(S)h(M)^*h(P)P' = S'h(SM^*P)P' = \|\mathfrak{T}\|(\|\mathfrak{A}\|) = \|\mathfrak{T}\|(a),$$

Corollary 4.9 of Kuich [8] proves that $\|\mathfrak{T}\|(a)$ is in $\mathfrak{Rat}(\mathcal{M}(\mathfrak{R} \cup \{1\}))$. $\square$

Corollary 2 *Let* $\mathfrak{R}$ *be an* $\mathfrak{H}$-A'-family of elements. Then

$$\mathcal{F}(\mathfrak{R}) = \mathfrak{Rat}(\mathcal{M}(\mathfrak{R} \cup \{1\})).$$

Theorem 3 *Let* $\mathfrak{L}$ *be an* $\mathfrak{H}$-A'-family of elements. Then $\mathfrak{L}$ is an $\mathfrak{H}$-A'-AFE iff $\mathfrak{Rat}(\mathcal{M}(\mathfrak{L})) = \mathfrak{L}$.

Proof. (i) Assume that $\mathfrak{L}$ is an $\mathfrak{H}$-A'-AFE. Then $\mathcal{M}(\mathfrak{L}) \subseteq \mathfrak{L}$ and $\mathfrak{Rat}(\mathfrak{L}) = \mathfrak{L}$ imply $\mathfrak{Rat}(\mathcal{M}(\mathfrak{L})) \subseteq \mathfrak{Rat}(\mathfrak{L}) = \mathfrak{L} \subseteq \mathcal{M}(\mathfrak{L}) \subseteq \mathfrak{Rat}(\mathcal{M}(\mathfrak{L}))$.

(ii) Conversely, $\mathcal{M}(\mathfrak{L}) \subseteq \mathfrak{Rat}(\mathcal{M}(\mathfrak{L})) = \mathfrak{L}$ and $\mathfrak{Rat}(\mathfrak{L}) = \mathfrak{Rat}(\mathfrak{Rat}(\mathcal{M}(\mathfrak{L}))) = \mathfrak{Rat}(\mathcal{M}(\mathfrak{L})) = \mathfrak{L}$ by Theorem 4.12 of Kuich [8]. $\square$

Lemma 4 *Let* $\mathfrak{R} \subseteq A$ *with* $A' \subseteq \mathfrak{R}$. *Then* $\mathfrak{Rat}(\mathcal{K}(\mathfrak{R})) = \mathfrak{Rat}(\mathcal{M}(\mathfrak{R}))$.

Proof. (i) The inequality $\mathcal{K}(\mathfrak{R}) \subseteq \mathcal{M}(\mathfrak{R})$ implies $\mathfrak{Rat}(\mathcal{K}(\mathfrak{R})) \subseteq \mathfrak{Rat}(\mathcal{M}(\mathfrak{R}))$.

(ii) We now show the converse inequality. Assume that $a \in \mathcal{M}(\mathfrak{R})$. Then there exists a $b \in \mathfrak{R}$ and an $\mathfrak{H}$-A'-rational transducer $\mathfrak{T} = (Q, h, S, P)$ such that

$$a = \|\mathfrak{T}\|(b) = \sum_{q_1,q_2 \in Q} S_{q_1} h(b)_{q_1,q_2} P_{q_2}.$$

Since $S_{q_1}, P_{q_2} \in \mathfrak{Rat}(A')$ for $q_1, q_2 \in Q$, we infer that $S_{q_1}, P_{q_2} \in \mathfrak{Rat}(\mathcal{K}(\mathfrak{R}))$. Hence, $\|\mathfrak{T}\|(b) \in \mathfrak{Rat}(\mathcal{K}(\mathfrak{R}))$ and we have shown that $\mathfrak{Rat}(\mathcal{M}(\mathfrak{R})) \subseteq \mathfrak{Rat}(\mathcal{K}(\mathfrak{R}))$. $\square$

Lemma 5 *Let* $\mathfrak{R} \subseteq A$ *with* $A' \subseteq \mathfrak{R}$. *Then* $\mathcal{K}(\mathfrak{Rat}(\mathcal{K}(\mathfrak{R}))) \subseteq \mathfrak{Rat}(\mathcal{K}(\mathfrak{R}))$.

Proof. $\mathcal{K}(\mathfrak{Rat}(\mathcal{K}(\mathfrak{R}))) \subseteq \mathcal{M}(\mathfrak{Rat}(\mathcal{K}(\mathfrak{R}))) = \mathcal{M}(\mathfrak{Rat}(\mathcal{M}(\mathfrak{R}))) \subseteq \mathfrak{Rat}(\mathcal{M}(\mathfrak{R})) = \mathfrak{Rat}(\mathcal{K}(\mathfrak{R}))$. $\square$

Theorem 6 *Let* $\mathfrak{L} \subseteq A$ *with* $A' \subseteq \mathfrak{L}$. *Then* $\mathfrak{L}$ *is an* $\mathfrak{H}$-closed semiring iff $\mathfrak{Rat}(\mathcal{K}(\mathfrak{L})) = \mathfrak{L}$.

Proof. (i) Assume that $\mathfrak{L}$ is $\mathfrak{H}$-closed. Then $\mathcal{K}(\mathfrak{L}) \subseteq \mathfrak{L}$ and $\mathfrak{Rat}(\mathfrak{L}) = \mathfrak{L}$ imply $\mathfrak{Rat}(\mathcal{K}(\mathfrak{L})) \subseteq \mathfrak{Rat}(\mathfrak{L}) = \mathfrak{L} \subseteq \mathcal{K}(\mathfrak{L}) \subseteq \mathfrak{Rat}(\mathcal{K}(\mathfrak{L}))$.

(ii) Conversely, $\mathcal{K}(\mathfrak{L}) = \mathcal{K}(\mathfrak{Rat}(\mathcal{K}(\mathfrak{L}))) \subseteq \mathfrak{Rat}(\mathcal{K}(\mathfrak{L})) = \mathfrak{L}$ by Lemma 5 and $\mathfrak{Rat}(\mathfrak{L}) = \mathfrak{Rat}(\mathfrak{Rat}(\mathcal{K}(\mathfrak{L}))) = \mathfrak{Rat}(\mathcal{K}(\mathfrak{L})) = \mathfrak{L}$ by Theorem 4.12 of Kuich [8]. □

The next corollary shows that $\mathfrak{H}$-A'-abstract families of elements and $\mathfrak{H}$-closed semirings coincide in certain situations.

Corollary 7 *Let $\mathfrak{L}$ be an $\mathfrak{H}$-A'-family of elements where $A' \subseteq \mathfrak{L}$. Then $\mathfrak{L}$ is an $\mathfrak{H}$-A'-AFE iff $\mathfrak{L}$ is an $\mathfrak{H}$-closed semiring.*

Proof. By Theorem 3, Lemma 4 and Theorem 6. □

Karner, Kuich [5], Corollary 8, characterized $\mathfrak{H}$-A'-AFEs by infinite automata of a certain type. By Corollary 7, this characterization applies also to $\mathfrak{H}$-closed semirings.

Corollary 8 *Let $\mathfrak{L} \subseteq A$, where $A' \subseteq \mathfrak{L} \subseteq [\mathfrak{H}]$. Then $\mathfrak{L}$ is an $\mathfrak{H}$-closed semiring iff there exists a restart type T such that*

$$\mathfrak{L} = \mathfrak{Rat}_T(A').$$

3 The characterization of semirings of algebraic elements

We will now characterize the $\mathfrak{H}$-closed semirings that are subsemirings of $\mathfrak{Alg}(A')$ and contain A'. So we consider $\mathfrak{H}$-closed semirings $\mathfrak{L}$ that are generated by some $\mathfrak{R} \subseteq \mathfrak{Alg}(A')$ in the sense that $\mathfrak{L} = \mathfrak{Rat}(\mathcal{K}(\mathfrak{R}))$. By Theorem 6, for each $\mathfrak{H}$-closed semiring $\mathfrak{L}$, where $\mathfrak{Rat}(A') \subseteq \mathfrak{L} \subseteq \mathfrak{Alg}(A')$, such an $\mathfrak{R}$ exists. Given an $a \in \mathfrak{L}$, we will construct a $\mathfrak{Rat}(A')$-algebraic system of a particular form such that a is the first component of its least solution.

Let now a be a fixed element in $\mathfrak{L}$. Then there exist, for some $k \geq 1$, elements $a_1, \ldots, a_k \in \mathfrak{R}$ (not necessarily distinct), morphisms $h_m : A \to A^{Q_m \times Q_m}$ in $\mathfrak{H}$ and $p_m, r_m \in Q_m$, $1 \leq m \leq k$, such that $a \in \mathfrak{Rat}(\{h_m(a_m)_{p_m,r_m} \mid 1 \leq m \leq k\})$. We assume without loss of generality that $Q_1, \ldots, Q_k$ are mutually disjoint. By Theorem 4.10 of Kuich [8], there exists a finite $\{h_m(a_m)_{p_m,r_m} \mid 1 \leq m \leq k\} \cup \{0, 1\}$-automaton $\mathfrak{A} = (Q, M, S, P)$ such that $\|\mathfrak{A}\| = a$.

Let $\mathbb{N}_A^\infty = \{\sum_{i<n} 1 \mid n \in \mathbb{N}^\infty\} \subseteq A$. By Satz 5.3 of Goldstern [3], $\mathbb{N}_A^\infty$ is a complete subsemiring of A. Observe that $\mathbb{N}_A^\infty \subseteq \mathfrak{Rat}(A')$. Furthermore, we have $na = an$ for all $n \in \mathbb{N}_A^\infty$ and $a \in A$.

Let $Z = \{z_m \mid 1 \leq m \leq k\}$ be an alphabet of variables and let $h : Z^* \to \mathfrak{Rat}(\{h_m(a_m)_{p_m,r_m} \mid 1 \leq m \leq k\})$ be the multiplicative monoid morphism defined by $h(z_m) = h_m(a_m)_{p_m,r_m}$, $1 \leq m \leq k$. Extend this monoid morphism in the usual manner to the complete semiring morphism $h : \mathbb{N}_A^\infty\langle\langle Z^*\rangle\rangle \to A$, i. e., $h(r) = \sum_{v \in Z^*}(r, v)h(v)$ for $r = \sum_{v \in Z^*}(r, v)v \in \mathbb{N}_A^\infty\langle\langle Z^*\rangle\rangle$.

By Theorems 4.5 and 4.13 of Kuich [8] there exists now a finite $\mathbb{N}_A^\infty\langle Z \cup \varepsilon\rangle$-automaton $\mathfrak{A}' = (Q', M', S', P')$, where $M'_{q_1,q_2} \in \mathbb{N}_A^\infty\langle Z\rangle$, $P'_{q_1}, S'_{q_2} \in \mathbb{N}_A^\infty\langle\varepsilon\rangle$,

$q_1, q_2 \in Q'$, with a single initial state such that $h(\|\mathfrak{A}'\|) = \|\mathfrak{A}\| = a$. Denote the initial state by q_0, i. e., $S'_{q_0} = \varepsilon$ and $S_q = 0$ for $q \neq q_0$.

We construct a finite $\mathfrak{Alg}(A')$-automaton $\hat{\mathfrak{A}} = (\hat{Q}, \hat{M}, \hat{S}, \hat{P})$ such that $\|\hat{\mathfrak{A}}\| = h(\|\mathfrak{A}'\|) = a$. Its specific form reflects the construction of our desired algebraic system for a.

(i) $\hat{Q} = Q' \cup (Q_1 \cup \ldots \cup Q_k) \times Q' \cup (Q_1 \cup \ldots \cup Q_k) \times \bar{Q}'$, where $\bar{Q}'$ is a copy of Q'.

(ii) The non-null blocks of $\hat{M}$ are

$$\hat{M}(Q', (Q_1 \cup \ldots \cup Q_k) \times Q') = \bar{S},$$
$$\hat{M}((Q_1 \cup \ldots \cup Q_k) \times Q', (Q_1 \cup \ldots \cup Q_k) \times \bar{Q}') = \bar{M},$$
$$\hat{M}((Q_1 \cup \ldots \cup Q_k) \times \bar{Q}', Q') = \bar{P},$$

where by definition,

$$\bar{S}(q_1, (Q_m, q_2)) = (M'_{q_1, q_2}, z_m) \otimes S_m,$$
$$\bar{M}((Q_{m_1}, q_1), (Q_{m_2}, \bar{q}_2)) = \delta_{m_1, m_2} \cdot \delta_{q_1, q_2} \otimes h_{m_1}(a_{m_1}),$$
$$\bar{P}((Q_m, \bar{q}_1), q_2) = \delta_{q_1, q_2} \otimes P_m,$$
$$1 \leq m, m_1, m_2 \leq k, \ q_1, q_2 \in Q'.$$

Here S_m denotes the p_m-th unit row vector of dimension Q_m, P_m denotes the r_m-th unit column vector of dimension Q_m, $\otimes$ denotes the Kronecker product and $\delta_{m_1, m_2}, \delta_{q_1, q_2}$ denote the Kronecker symbol. Furthermore, for $q \in Q'$, $\bar{q}$ denotes the corresponding state in the copy $\bar{Q}'$ of Q'.

(iii) The non-null blocks of $\hat{S}$ and $\hat{P}$ are $\hat{S}(Q') = h(S')$ and $\hat{P}(Q') = h(P')$, respectively. Note that the only non-null entry of $\hat{S}$ is $\hat{S}_{q_0} = 1$.

We obtain $\bar{S}\bar{M}\bar{P} = h(M')$ and $\hat{M}^*(Q', Q') = (\bar{S}\bar{M}\bar{P})^* = h(M')^*$.

This implies $\|\hat{\mathfrak{A}}\| = \hat{S}\hat{M}^*\hat{P} = \hat{S}(Q')\hat{M}^*(Q', Q')\hat{P}(Q') = h(S')h(M')^*h(P') = h(S'M'^*P') = h(\|\mathfrak{A}'\|) = a$.

We now proceed with the construction of a $\mathfrak{Rat}(A')$-algebraic system whose least solution has the first component $a \in \mathfrak{L}$. Since $a_m \in \mathfrak{Alg}(A')$, there exists, for each $1 \leq m \leq k$, an A'-algebraic system $y_i^m = p_i^m$, $1 \leq i \leq n_m$, such that a_m is the first component, i. e., the y_1^m-component of its least solution.

Consider now, for $1 \leq m \leq k$, the $\mathfrak{Rat}(A')$-algebraic system (written in matrix notation) $z_i^m = h_m(p_i^m)$, $1 \leq i \leq n_m$. Here z_i^m is, for $1 \leq m \leq k$, $1 \leq i \leq n_m$, a $Q_m \times Q_m$-matrix of variables, whose (q_1, q_2)-entry, $q_1, q_2 \in Q_m$, is denoted by $(z_i^m)_{q_1, q_2}$. By Theorem 7.4 of Kuich [8], the z_1^m-component of the least solution of $z_i^m = h_m(p_i^m)$, $1 \leq i \leq n_m$, is equal to $h_m(a_m) \in \mathfrak{Alg}(A')^{Q_m \times Q_m}$, $1 \leq m \leq k$. Hence, the $(z_1^m)_{q_1, q_2}$-entry of this least solution is equal to $h_m(a_m)_{q_1, q_2}$, $q_1, q_2 \in Q_m$, $1 \leq m \leq k$.

Let $Z_1 = \{(z_1^m)_{q_1, q_2} \mid q_1, q_2 \in Q_m, 1 \leq m \leq k\}$. We now define the matrix $\hat{M}' \in (\mathbb{N}_A^\infty \langle Z_1 \cup \varepsilon \rangle)^{\hat{Q} \times \hat{Q}}$. The non-null blocks of $\hat{M}'$ are

$$\hat{M}'(Q', (Q_1 \cup \ldots \cup Q_k) \times Q') = \bar{S},$$
$$\hat{M}'((Q_1 \cup \ldots \cup Q_k) \times Q', (Q_1 \cup \ldots \cup Q_k) \times \bar{Q}') = \bar{M}',$$
$$\hat{M}'((Q_1 \cup \ldots \cup Q_k) \times \bar{Q}', Q') = \bar{P},$$

where by definition,

$$\bar{M}'((Q_{m_1}, q_1), (Q_{m_2}, \bar{q}_2)) = \delta_{m_1, m_2} \cdot \delta_{q_1, q_2} \otimes z_1^{m_1}, \quad 1 \leq m_1, m_2 \leq k, \ q_1, q_2 \in Q'.$$

Let $\hat{h} : Z_1^* \to \mathfrak{Alg}(A')$ be the multiplicative monoid morphism defined by $\hat{h}((z_1^m)_{q_1, q_2}) = h_m(a_m)_{q_1, q_2}, q_1, q_2 \in Q_m, 1 \leq m \leq k$. Extend $\hat{h}$ to a complete semiring morphism from $\mathbb{N}_A^\infty \langle\!\langle Z_1^* \rangle\!\rangle$ to A. Then we obtain $\hat{h}(\hat{M}') = \hat{M}$ and $\hat{h}(\hat{P}) = \hat{P}$. This implies $\hat{h}(\hat{M}'^* \hat{P}) = \hat{h}(\hat{M}')^* \hat{h}(\hat{P}) = \hat{M}^* \hat{P}$. In particular, we infer $\|\hat{\mathfrak{A}}\| = \hat{S} \hat{M}^* \hat{P} = (\hat{M}^* \hat{P})_{q_0} = \hat{h}(\hat{M}'^* \hat{P})_{q_0}$.

Consider now the $\mathfrak{Rat}(A')$-algebraic system

$$z = \hat{M}'z + \hat{P}, \quad z_i^m = h_m(p_i^m), \quad 1 \leq i \leq n_m, \ 1 \leq m \leq k. \tag{$*$}$$

Here z is a column vector of variables indexed by $\hat{Q}$, whose q-entry, $q \in \hat{Q}$, is the variable z_q. Hence, the variables of the $\mathfrak{Rat}(A')$-algebraic system $(*)$ are z_q, $q \in \hat{Q}$, and $(z_i^m)_{q_1, q_2}, q_1, q_2 \in Q_m, 1 \leq i \leq n_m, 1 \leq m \leq k$. Observe that the entries of $\hat{M}'z$ and $\hat{P}$ are polynomials in the polynomial semiring over A with variables in $\{z_q \mid q \in \hat{Q}\} \cup \{(z_1^m)_{q_1, q_2} \mid q_1, q_2 \in Q_m, 1 \leq m \leq k\}$ and coefficients in $\mathfrak{Rat}(A')$.

We claim that $a \in \mathfrak{L}$, the element we started with, is the z_{q_0}-component of the least solution of the $\mathfrak{Rat}(A')$-algebraic system $(*)$. Since, for each $1 \leq m \leq k$, the $\mathfrak{Rat}(A')$-algebraic system $z_i^m = h_m(p_i^m)$ is an independent subsystem of $(*)$, the $(z_1^m)_{q_1, q_2}$-entry of the least solution of $(*)$ is given by $h_m(a_m)_{q_1, q_2}$, $q_1, q_2 \in Q_m$. This implies that the z_q-component of the least solution of $(*)$, $q \in \hat{Q}$, is given by the z_q-component of the least solution of the $\mathfrak{Alg}(A')$-linear system $z = \hat{h}(\hat{M}')z + \hat{P}$. The least solution of this $\mathfrak{Alg}(A')$-linear system is given by $\hat{h}(\hat{M}')^* \hat{P} = \hat{M}^* \hat{P}$. Hence, we obtain that the z_{q_0}-component of the least solution of $(*)$ is equal to $\|\hat{\mathfrak{A}}\| = a$.

Theorem 9 *Let $\mathfrak{L}$ be an $\mathfrak{H}$-closed semiring where $A' \subseteq \mathfrak{L} \subseteq \mathfrak{Alg}(A')$ and $\mathfrak{R} \subseteq \mathfrak{Alg}(A')$ such that $\mathfrak{L} = \mathfrak{Rat}(\mathcal{K}(\mathfrak{R}))$. For $a \in \mathfrak{L}$, the z_{q_0}-component of the least solution of the $\mathfrak{Rat}(A')$-algebraic system $(*)$ is equal to a.*

For a complete characterization of $\mathfrak{L} = \mathfrak{Rat}(\mathcal{K}(\mathfrak{R}))$ we need a converse to Theorem 9.

Let I be an index set. By definition, a *family* $\mathfrak{S} = \{y_i^m = p_i^m \mid 1 \leq i \leq n_m, \ m \in I\}$ of $\mathfrak{Rat}(A')$-*algebraic systems together with a set* $Y \subseteq \{y_i^m \mid 1 \leq i \leq n_m, \ m \in I\}$ of distinguished variables *generates the subset of* $\mathfrak{Alg}(A')$

$\{a \mid$ there exists a $j \in I$ and a variable $z \in \{y_i^j \mid 1 \leq i \leq n_j\} \cap Y$ such that the z-component of the least solution of the $\mathfrak{Rat}(A')$-algebraic system $y_i^j = p_i^j, 1 \leq i \leq n_j$, in $\mathfrak{S}$ equals $a\}$.

Let $\mathfrak{S} = \{y_i^m = p_i^m \mid 1 \leq i \leq n_m, \ m \in I\}$ be a family of $\mathfrak{Rat}(A')$-algebraic systems such that the sets of variables $\{y_i^m \mid 1 \leq i \leq n_m\}, m \in I$, are mutually disjoint. Let $\{z_q \mid q \in Q_\infty\}$ be a set of new variables. Then we define $\mathcal{F}(\mathfrak{S})$ to be the collection of all $\mathfrak{Rat}(A')$-algebraic systems of the form

$$z = Mz + P, \quad z_i^m = h_m(p_i^m), \quad 1 \leq i \leq n_m, \, m \in F, \qquad (**)$$

where F is a finite subset of I. Here $h_m : A \to A^{Q_m \times Q_m}$, $m \in F$, is a morphism in $\mathfrak{H}$ and z_i^m is a $Q_m \times Q_m$-matrix of variables with entries $(z_i^m)_{q_1, q_2}$, $q_1, q_2 \in Q_m$, $1 \leq i \leq n_m$, $m \in F$. Moreover, z is a column vector of variables indexed by $Q \subset Q_\infty$, whose q-entry, $q \in Q$, is the variable z_q, M is a matrix in $(\mathbb{N}_A^\infty \langle Z_1 \cup \varepsilon \rangle)^{Q \times Q}$, where $Z_1 = \{(z_1^m)_{q_1, q_2} \mid q_1, q_2 \in Q_m, \, m \in F\}$, and P is a column vector in $(\mathbb{N}_A^\infty)^{Q \times 1}$. Observe that the entries of Mz and P are polynomials in the polynomial semiring over A with variables in $\{z_q \mid q \in Q\} \cup Z_1$ and coefficients in $\mathfrak{Rat}(A')$.

Theorem 10 *Assume that* $\mathfrak{R} \subseteq \mathfrak{Alg}(A')$ *with* $A' \subseteq \mathfrak{R}$. *Let* $\mathfrak{S} = \{y_i^m = p_i^m \mid 1 \leq i \leq n_m, \, m \in I\}$ *be a family of* A'-*algebraic systems with mutually disjoint sets of variables that together with* $\{y_1^m \mid m \in I\}$ *generates* $\mathfrak{R}$. *Then the family* $\mathcal{F}(\mathfrak{S})$ *of* $\mathfrak{Rat}(A')$-*algebraic systems together with* $\{z_q \mid q \in Q_\infty\}$ *generates the* $\mathfrak{H}$-*closed semiring* $\mathfrak{L} = \mathfrak{Rat}(\mathcal{K}(\mathfrak{R}))$.

Corollary 11 *Let* $A' \subseteq \mathfrak{L} \subseteq \mathfrak{Alg}(A')$. *Then* $\mathfrak{L}$ *is an* $\mathfrak{H}$-*closed semiring iff there exists a family* $\mathfrak{S}$ *of* A'-*algebraic systems such that* $\mathcal{F}(\mathfrak{S})$ *together with* $\{z_q \mid q \in Q_\infty\}$ *generates* $\mathfrak{L}$.

An $\mathfrak{H}$-closed semiring $\mathfrak{L}$ is called *principal* if there exists an $a \in A$ such that $\mathfrak{L} = \mathfrak{Rat}(\mathcal{K}(\{a\}))$.

Corollary 12 *Let* $A' \subseteq \mathfrak{L} \subseteq \mathfrak{Alg}(A')$. *Then* $\mathfrak{L}$ *is a principal* $\mathfrak{H}$-*closed semiring iff there exists a family* $\mathfrak{S}$ *of* A'-*algebraic systems containing just one* A'-*algebraic system such that* $\mathcal{F}(\mathfrak{S})$ *together with* $\{z_q \mid q \in Q_\infty\}$ *generates* $\mathfrak{L}$.

Hence, in this case $\mathcal{F}(\mathfrak{S})$ is a family of $\mathfrak{Rat}(A')$-algebraic systems of the form $z = Mz + P$, $z_i = h(p_i)$, $1 \leq i \leq n$, where the specifications for z, M, P, h, z_i and p_i, $1 \leq i \leq n$, are analogous to $(**)$.

4   Applications to formal power series and formal languages

We first deal with formal power series. The basic semiring is now the semiring of formal power series $A \langle\langle \Sigma_\infty^* \rangle\rangle$, A commutative and continuous, over an infinite alphabet Σ_∞, and $A' = A\{\Sigma_\infty \cup \varepsilon\}$. This implies that $\mathfrak{Rat}(A') = A^{\mathrm{rat}}\{\{\Sigma_\infty^*\}\}$ and $\mathfrak{Alg}(A') = A^{\mathrm{alg}}\{\{\Sigma_\infty^*\}\}$.

If we choose $\mathfrak{H}$ to be the family of rational representations then the $\mathfrak{H}$-closed semirings that are subsemirings of $A\{\{\Sigma_\infty^*\}\}$ and contain $A\{\Sigma_\infty \cup \varepsilon\}$ are just the *full* abstract families of power series (full AFPs) in the sense of Kuich [8].

If we choose $\mathfrak{H}$ to be the family of *regulated* rational representations then the $\mathfrak{H}$-closed semirings that are subsemirings of $A\{\{\Sigma_\infty^*\}\}$ and contain $A\{\Sigma_\infty \cup \varepsilon\}$

are just the abstract families of power series (AFPs) in the sense of Kuich, Salomaa [9].

To characterize full AFPs and AFPs we have to consider families $\mathfrak{S} = \{y_i^m = p_i^m \mid 1 \leq i \leq n_m, \ m \in I\}$ of algebraic systems. This yields the following specialization of Corollaries 11 and 12.

Theorem 13 *Let $\mathfrak{L}$ be a family of algebraic power series containing a symbol of Σ_∞ and assume that $\mathfrak{H}$ is the collection of rational representations (resp. regulated rational representations). Then $\mathfrak{L}$ is a full AFP (resp. AFP) iff there exists a family $\mathfrak{S}$ of algebraic systems such that $\mathcal{F}(\mathfrak{S})$ together with $\{z_q \mid q \in Q_\infty\}$ generates $\mathfrak{L}$.*

Corollary 14 *Let $\mathfrak{L}$ be a family of algebraic power series containing a symbol of Σ_∞ and assume that $\mathfrak{H}$ is the collection of rational representations (resp. regulated rational representations). Then $\mathfrak{L}$ is a principal full AFP (resp. principal AFP) iff there exists a family $\mathfrak{S}$ containing just one algebraic system such that $\mathcal{F}(\mathfrak{S})$ together with $\{z_q \mid q \in Q_\infty\}$ generates $\mathfrak{L}$.*

Observe the relation of Corollary 14 with Theorems 2.3 and 2.5 of Kuich [6].

We turn to formal languages. The basic semiring is now $\mathfrak{P}(\Sigma_\infty^*)$, the power set of Σ_∞^*, where Σ_∞ is an infinite alphabet. Formal languages are subsets of Σ^*, where $\Sigma \subset \Sigma_\infty$ is a finite alphabet. By the isomorphisms of $\mathfrak{P}(\Sigma_\infty^*)$ with $\mathbb{B}\langle\!\langle \Sigma_\infty^* \rangle\!\rangle$ and the family of formal languages with $\mathbb{B}\{\!\{\Sigma_\infty^*\}\!\}$, specializations of Theorem 13 and Corollary 14 yield characterizations of abstract families of languages (AFLs), full AFLs, principal AFLs and principal full AFLs.

Example 1. We consider one counter languages (see Berstel [1]) and their generalization to power series, the power series in $A^{oc}\{\!\{\Sigma_\infty^*\}\!\}$ (see Kuich, Salomaa [9]). By Theorems 13.26 and 13.27 of Kuich, Salomaa [9], $A^{oc}\{\!\{\Sigma_\infty^*\}\!\}$ is a principal AFP generated by δ_1, where δ_1 is the (unique) solution of the algebraic system (with variable y_1) $y_1 = xy_1\bar{x}y_1 + \varepsilon$.

Corollary 14 yields now a "normal form" for algebraic systems for "one counter power series". Given a power series r in $A^{oc}\{\!\{\Sigma_\infty^*\}\!\}$, there exists an algebraic system of the form $(**)$

$$z = Mz + P, \qquad y = Xy\bar{X}y + E,$$

such that r is a component of the least solution of this system. Here y is a $Q_1 \times Q_1$-matrix of variables y_{q_1,q_2}, $q_1, q_2 \in Q_1$ and $X, \bar{X}$ are matrices in $(A^{rat}\langle\!\langle \Sigma^* \rangle\!\rangle)^{Q_1 \times Q_1}$, $\Sigma \subset \Sigma_\infty$ a finite alphabet, with the following additional property: there exists a $k \geq 1$ such that each product of at least k of the matrices (X, ε) and $(\bar{X}, \varepsilon)$ vanishes. (This additional property originates in the fact that we consider only *regulated* rational representations.) E is the matrix of unity, M is a $Q \times Q$-matrix whose entries are in $A^{rat}\langle\!\langle (\Sigma \cup \{y_{q_1,q_2} \mid q_1, q_2 \in Q_1\})^* \rangle\!\rangle$, P is a $Q \times 1$-column vector whose entries are in $A^{rat}\langle\!\langle \Sigma^* \rangle\!\rangle$ and z is a $Q \times 1$-column vector of variables. (Compare with Example 2.6 of Kuich [6].) $\qquad\square$

Example 2. In language theory, the most important examples of principal AFLs are that of the families of regular and context-free languages. The family of regular languages is generated by $\{x\}$, where x is an arbitrary symbol in Σ_∞, i. e., the family of regular languages coincides with $\mathfrak{Rat}(\mathcal{K}(\{x\}))$.

Let $G = (\{y_2\}, \{x_1, \bar{x}_1, x_2, \bar{x}_2\}, P, y_2)$ be a context-free grammar, where the set P of productions is given by $\{y_2 \to x_1 y_2 \bar{x}_1 y_2, y_2 \to x_2 y_2 \bar{x}_2 y_2, y_2 \to \varepsilon\}$. The language generated by G is called *restricted Dyck language* $D_2'^*$. The family of context-free languages is generated by $D_2'^*$, i. e., the family of context-free languages coincides with $\mathfrak{Rat}(\mathcal{K}(D_2'^*))$. (See Berstel [1].)

Turning to the theory of formal power series, the most important examples of principal AFPs are that of the families of rational and algebraic power series, $A^{\mathrm{rat}}\{\{\Sigma_\infty^*\}\}$ and $A^{\mathrm{alg}}\{\{\Sigma_\infty^*\}\}$, respectively.

We have $A^{\mathrm{rat}}\{\{\Sigma_\infty^*\}\} = \mathfrak{Rat}(\mathcal{K}(\{x\}))$, where x is an arbitrary symbol in Σ_∞, and $A^{\mathrm{alg}}\{\{\Sigma_\infty^*\}\} = \mathfrak{Rat}(\mathcal{K}(\{\delta_2\}))$, where δ_2 is the (unique) solution of the algebraic system (with variable y_2) $y_2 = x_1 y_2 \bar{x}_1 y_2 + x_2 y_2 \bar{x}_2 y_2 + \varepsilon$. (See Kuich, Salomaa [9].)

Corollary 14 yields now a "normal form" for algebraic systems. Since $A^{\mathrm{alg}}\{\{\Sigma_\infty^*\}\}$ is even a principal cone, Example 2.8 of Kuich [6] yields a simpler "normal form". $\qquad\qquad\square$

References

1. Berstel, J.: Transductions and Context-Free Languages. Teubner, 1979.
2. Ginsburg, S.: Algebraic and Automata-Theoretic Properties of Formal Languages. North-Holland, 1975.
3. Goldstern, M.: Vervollständigung von Halbringen. Diplomarbeit, Technische Universität Wien, 1985.
4. Karner, G.: On limits in complete semirings. Semigroup Forum 45(1992) 148–165.
5. Karner, G., Kuich, W.: On abstract families of languages, power series, and elements. Lect. Notes Comput. Sci., to appear.
6. Kuich, W.: Matrix systems and principal cones of algebraic power series. MFCS 86, Lect. Notes Comput. Sci. 233(1986) 512–517 and Theoret. Comput. Sci. 57(1988) 147–152.
7. Kuich, W.: The algebraic equivalent of AFL theory. ICALP95, Lect. Notes Comput. Sci. 944(1995) 39–50.
8. Kuich, W.: Semirings and formal power series: Their relevance to formal languages and automata theory. In: G. Rozenberg and A. Salomaa, eds., Handbook of Formal Languages. Springer, 1997, 609–677.
9. Kuich, W., Salomaa, A.: Semirings, Automata, Languages. EATCS Monographs on Theoretical Computer Science, Vol. 5. Springer, 1986.
10. Lausch, H., Nöbauer, W.: Algebra of Polynomials. North Holland, 1973.
11. Sakarovitch, J.: Kleene's theorem revisited. Lect. Notes Comput. Sci. 281(1987) 39–50.

Repetitiveness of D0L-Languages Is Decidable in Polynomial Time [*]

Yuji Kobayashi[1] and Friedrich Otto[2]

[1] Department of Information Science, Faculty of Science,
Toho University, Funabashi 274, Japan.
Internet: kobayasi@is.sci.toho-u.ac.jp
[2] Fachbereich Mathematik/Informatik, Universität Kassel,
D-34109 Kassel, Germany.
Internet: otto@theory.informatik.uni-kassel.de

Abstract. We study the repetition of subwords in languages generated by morphisms. First we give a new proof for the fact that such a language is repetitive if and only if it is strongly repetitive (Ehrenfeucht and Rozenberg, 1983). Central to our proof is the notion of *quasi-repetitive* elements for morphisms. Then, from this proof, we derive a structurally simple polynomial-time algorithm for deciding whether a D0L-language is repetitive.

1 Introduction

An important part of formal language theory is concerned with the combinatorial structure of languages. One of the most basic combinatorial properties of a language is the repetition of subwords. Accordingly, a language L is called *repetitive* if, for each positive integer n, there exists a word $w \in L$ that contains a subword of the form x^n for some non-empty word x.

For context-free languages repetition of subwords is a very natural property. Indeed a context-free language is not repetitive if and only if it is finite. Actually, an infinite context-free language L is not only repetitive, but it is even *strongly repetitive*, that is, there exists a non-empty word x such that x^n is a subword of L for all positive integers n. Hence, a context-free language is repetitive if and only if it is strongly repetitive. This equivalence is not true in general as is shown by a simple example in [ER83].

In the theory of L-systems morphisms are used to describe and define languages. The simplest type of L-system is the D0L-*system*, where a language L is generated from a given word w by iterating a given morphism f, that is, $L = \{f^n(w) \mid n \geq 0\}$. See the monograph by G. Rozenberg and A. Salomaa for an introduction to the theory of L-systems [RS80], and see the collection [RS92]

[*] **Acknowledgement:** The results presented here were obtained while the second author was visiting at Toho University. He gratefully acknowledges the hospitality of the Faculty of Science of Toho University and the support by the Deutsche Forschungsgemeinschaft.

edited by G. Rozenberg and A. Salomaa for a detailed survey on recent developments concerning this theory and its impacts on theoretical computer science, computer graphics, and developmental biology.

In [ER81] A. Ehrenfeucht and G. Rozenberg investigate the subword complexity of square-free D0L-languages. In a later paper the same authors prove that a D0L-language is repetitive if and only if it is strongly repetitive [ER83]. In addition, they show that it is decidable whether a given D0L-system generates a repetitive language. Both these results are proved in [ER83] by reducing the general case of an arbitrary D0L-system to that of a very special D0L-system. But even for these special systems the given proof is still fairly complicated.

In [MS93] F. Mignosi and P. Séébold investigate a related problem. Call a language L k-power free if it does not contain any subword that is a non-empty k-th power. F. Mignosi and P. Séébold show that there exists a recursive function ι such that $x^{\iota(\Sigma, f, w)} \in S(L)$ implies that $x^+ \subseteq S(L)$, where (Σ, f, w) is a D0L-system generating the language L. Here S denotes the subword operator. Hence, it follows immediately that a D0L-language is strongly repetitive if it is repetitive. Since it is decidable whether a D0L-language is k-power free [MS93], this leads to another algorithm for deciding whether a given D0L-language is repetitive. However, as with the algorithm of A. Ehrenfeucht and G. Rozenberg [ER83] the exact degree of complexity of this algorithm is not known.

Here we address the repetitiveness of languages generated by morphisms in a different way. Central to our approach is the notion of a *quasi-repetitive* element for a morphism f. A non-empty word v is a quasi-repetitive element for f, if there exist integers $n > 0$ and $p > 1$ such that $f^n(v) = v_1^p$, where v_1 is a conjugate of v. The quasi-repetitive elements for the morphism f can be seen as a generalization of those words w for which the D0L-system (Σ, f, w) is *periodic* [HL86a]. It follows from a result of T. Harju and M. Linna that it is decidable whether a morphism f admits a quasi-repetitive element [HL86]. It is rather straightforward to verify that the D0L-language L is strongly repetitive if it contains a subword that is quasi-repetitive for the generating morphism. On the other hand, we will show that the D0L-language $L := \{f^n(w) \mid n \geq 0\}$ contains a subword that is a quasi-repetitive element for f, if the language L is repetitive. Thus, L is repetitive if and only if it is strongly repetitive. In addition, we obtain a bound on the length of a shortest quasi-repetitive element for f and on its quasi-period n, which yields an algorithm for deciding whether a D0L-language is repetitive. In fact, we will see that this algorithm runs in polynomial time, provided the size of the underlying alphabet is fixed.

This paper is structured as follows. In Section 2 we define the basic notions used throughout the paper. In particular, we consider the notion of *simplification* of a morphism, a notion that was introduced by A. Ehrenfeucht and G. Rozenberg in [ER78]. In Section 3 we define the injective simplification g of a morphism f, showing that the language $L(f)$ is (strongly) repetitive if and only if $L(g)$ is. In Section 4 we derive some technical results on quasi-repetitive elements, which are then used to prove the two results mentioned above. We close with some remarks on possible directions for future research. In order to not exceed the page limit proofs are mostly omitted. They can be found in [KO97].

2   Repetitiveness and simplifications of morphisms

We assume that the reader is familiar with the basics of formal language theory and the theory of D0L-systems. As our main references we use the monograph by J. Hopcroft and J. Ullman [HU79] for the former and that by G. Rozenberg and A. Salomaa [RS80] for the latter.

Let Σ be a finite alphabet. If $f : \Sigma^* \to \Sigma^*$ is a morphism, then $L(f)$ denotes the language $L(f) := \{f^n(a) \mid a \in \Sigma, n \in \mathbb{N}\}$. We call $L(f)$ the *language generated* by f. A D0L-*system* is a triple $G = (\Sigma, f, w)$, where $w \in \Sigma^*$ is called the *axiom* of G, and $f : \Sigma^* \to \Sigma^*$ is a morphism. The language $L(G) := \{f^n(w) \mid n \in \mathbb{N}\}$ is the *language generated* by G.

For a language $L \subseteq \Sigma^*$, $S(L)$ denotes the language consisting of all subwords of L. To simplify the notation we write $SL(f)$ and $SL(G)$ for $S(L(f))$ and $S(L(G))$, respectively. Obviously, we have $L(f) = \bigcup_{a \in \Sigma} L((\Sigma, f, a))$.

A language L is called *repetitive*, if, for each integer $n > 0$, there exists a non-empty word x such that $x^n \in S(L)$. The language L is called *strongly repetitive* if $x^* \subseteq S(L)$ holds for some non-empty word x.

For D0L-languages the following result holds, where alph(w) denotes the set of letters that actually have occurrences in the word w.

Proposition 1. *For a* D0L-*system* $G = (\Sigma, f, w)$ *the following two statements are equivalent:*

(a) the language $L(G)$ is (strongly) repetitive;

(b) for some letter $a \in$ alph(w), the language $L((\Sigma, f, a))$ is (strongly) repetitive.

For a D0L-system $G = (\Sigma, f, w)$, let $\Sigma_G := \text{alph}(L(G))$. It is easily seen that Σ_G can be determined in polynomial time from G. Further, let f_G denote the restriction of the morphism f to the subalphabet Σ_G. From Proposition 1 we obtain the following consequence.

Corollary 2. *Let $G = (\Sigma, f, w)$ be a* D0L-*system. Then the following statements are equivalent:*

(a) $L(G)$ is (strongly) repetitive;

(b) $L(f_G)$ is (strongly) repetitive.

Thus, instead of looking at D0L-systems, we only need to consider morphisms. For the intended characterization of repetitive languages generated by morphisms the following notions will be crucial.

A word v is called *bounded* for f, if the D0L-language $L(\Sigma, f, v)$ is finite, otherwise, v is called *unbounded* for f. The morphism f is called *bounded* if all letters $a \in \Sigma$ are bounded for f. Obviously, f is bounded if and only if the language $L(f)$ is finite. If no letter is bounded for f, then f is called *growing*.

A word $v \in \Sigma^+$ is called a *quasi-repetitive element* for a morphism f, if there exist integers $n > 0$ and $p > 1$ such that $f^n(v) = v_1^p$ holds for some conjugate v_1 of v. The number n is called a *quasi-period* of v, and the corresponding number p is called a *quasi-multiplicity* of v. For future reference we note that, if v is a quasi-repetitive element for f, then each conjugate v_1 of v is also a quasi-repetitive element for f.

The following proposition states the basic property of quasi-repetitive elements which motivates our interest in these elements.

Proposition 3. *If a morphism f has a quasi-repetitive element v, then $v^* \subseteq SL((\Sigma, f, a))$ for each unbounded letter $a \in$ alph(v). In particular, $L(f)$ is strongly repetitive.*

Let Σ and Δ be two finite alphabets, and let $f : \Sigma^* \to \Sigma^*$ and $g : \Delta^* \to \Delta^*$ be morphisms. We say that f and g are *twined*, if there exist morphisms $h : \Sigma^* \to \Delta^*$ and $k : \Delta^* \to \Sigma^*$ satisfying the equalities $k \circ h = f$ and $h \circ k = g$. If $|\Delta| < |\Sigma|$ and f and g are twined, then g is called a *simplification* of f.

If f and g are twined with respect to the morphisms (h, k), then the following statements hold for all integers $n \geq 0$:

$$(2.1) \ f^n \circ k = k \circ g^n \text{ and } h \circ f^n = g^n \circ h.$$
$$(2.2) \ f^{n+1} = k \circ g^n \circ h \text{ and } g^{n+1} = h \circ f^n \circ k.$$

Based on this observation the following proposition relating properties of D0L-languages generated by f to those of D0L-languages generated by g can be derived.

Proposition 4. *Let f and g be morphisms that are twined with respect to (h, k).*

(a) If k is non-erasing and $L((\Delta, g, h(w)))$ is (strongly) repetitive, then so is the language $L((\Sigma, f, w))$.

(b) If h and k are both non-erasing, then $L((\Sigma, f, w))$ is (strongly) repetitive if and only if $L((\Delta, g, h(w)))$ is.

(c) If $v \in \Delta^+$ is (quasi-) repetitive for g and $k(v) \neq \varepsilon$, then $k(v)$ is (quasi-) repetitive for f, and each (quasi-) period of v is a (quasi-) period of $k(v)$.

From Proposition 1 and Proposition 4 we obtain the following conclusion.

Corollary 5. *Let f and g be morphisms that are twined with respect to (h, k).*
(a) If k is non-erasing, then $L(f)$ is (strongly) repetitive, if $L(g)$ is.
(b) If h and k are both non-erasing, then $L(f)$ is (strongly) repetitive if and only if $L(g)$ is.

3 Injective simplifications of morphisms

Let $\Gamma := \{a \in \Sigma \mid \exists n \geq 1 : f^n(a) = \varepsilon\}$, that is, Γ contains those letters of Σ that ultimately disappear. To determine Γ we construct a sequence of subalphabets of Σ inductively as follows:

$$\Gamma_1 := \{a \in \Sigma \mid f(a) = \varepsilon\}, \ \Gamma_{i+1} := \Gamma_i \cup \{a \in \Sigma \setminus \Gamma_i \mid f(a) \in \Gamma_i^+\}, \ i \geq 1.$$

Then $\Gamma_1 \subseteq \Gamma_2 \subseteq \ldots \subseteq \bigcup_{i \geq 1} \Gamma_i = \Gamma \subseteq \Sigma$, and $\Gamma_j = \Gamma_{j+1}$ implies $\Gamma_j = \Gamma$.

Let $\ell := \min\{n \mid \forall a \in \Gamma : f^n(a) = \varepsilon\}$. Then $\ell \leq |\Sigma|$ and $\Gamma = \Gamma_\ell$. Further, let $\Sigma_0 := \Sigma \smallsetminus \Gamma$, and define a morphism $\bar{f} : \Sigma_0^* \to \Sigma_0^*$ as follows: $\bar{f}(a) := \pi_{\Sigma_0}(f(a))$, $a \in \Sigma_0$, where $\pi_{\Sigma_0} : \Sigma^* \to \Sigma_0^*$ denotes the canonical projection. Observe that $\bar{f}(a) \neq \varepsilon$ for all $a \in \Sigma_0$.

If $\Gamma = \Sigma$, then $\Sigma_0 = \emptyset$ and $f^\ell(w) = \varepsilon$ for all $w \in \Sigma^*$, and if $\Gamma = \emptyset$, then $\bar{f} = f$. To exclude these trivial cases we assume for the following considerations that Γ is a proper subalphabet of Σ, that is, $\Sigma_0 \neq \emptyset \neq \Gamma$. Define a morphism $\bar{k} : \Sigma_0^* \to \Sigma^*$ through $\bar{k}(a) := f^\ell(a)$, $a \in \Sigma_0$. Then the following result can easily be established.

Lemma 6. *The morphisms $f^\ell : \Sigma^* \to \Sigma^*$ and $\bar{f}^\ell : \Sigma_0^* \to \Sigma_0^*$ are twined with respect to $(\pi_{\Sigma_0}, \bar{k})$. In particular, $\bar{f}^\ell$ is a non-erasing simplification of f^ℓ.*

Since the projection π_{Σ_0} is erasing, Proposition 4(b) is not applicable to f and $\bar{f}$. Nevertheless, the following analogous result holds.

Proposition 7.

(a) For $w \in \Sigma^*$, the language $L((\Sigma, f, w))$ is (strongly) repetitive if and only if $L((\Sigma_0, \bar{f}, \pi_{\Sigma_0}(w)))$ is.
(b) $L(f)$ is (strongly) repetitive if and only if $L(\bar{f})$ is.
(c) If a word $v \in \Sigma^+$ is (quasi-) repetitive for f with (quasi-) period $n > 0$, then $\pi_{\Sigma_0}(v)$ is (quasi-) repetitive for $\bar{f}$ with (quasi-) period n.
(d) If a word $v \in \Sigma_0^+$ is (quasi-) repetitive for $\bar{f}$ with (quasi-) period $n > 0$, then $f^\ell(v)$ is (quasi-) repetitive for f with (quasi-) period n.

Proposition 7 shows that in our investigations we can restrict our attention to non-erasing morphisms. Thus, in the following we assume that the morphisms considered are non-erasing.

Let $f : \Sigma^* \to \Sigma^*$ be a non-erasing morphism. If f is not injective, then the set $\{f(a) \mid a \in \Sigma\}$ is not a code. Hence, by the defect theorem [Lot82] there exists a code $C \subseteq \Sigma^+$ such that $|C| < |\Sigma|$ and $f(\Sigma) \subseteq C^+$. Let Δ be an alphabet in 1-to-1 correspondence to C, and let $k : \Delta^* \to \Sigma^*$ be the morphism that is induced by the bijection from Δ onto C. For each $a \in \Sigma$, there exists a unique word $w_a \in C^+$ such that $f(a) = w_a$. Hence, we can define a morphism $h : \Sigma^* \to \Delta^*$ through $h(a) := k^{-1}(f(a))$. Further, let $g : \Delta^* \to \Delta^*$ denote the morphism $g := h \circ k$. Then $f = k \circ h$, and thus, f and g are twined with respect to (h, k). Since $|\Delta| = |C| < |\Sigma|$, g is a simplification of f. Since f is non-erasing, h and g are non-erasing, too.

If g is not injective, either, we can repeat the above construction. Thus, we obtain a finite sequence of morphisms

$$f_0 := f : \Delta_0^* \to \Delta_0^*, \; f_1 := g : \Delta_1^* \to \Delta_1^*, f_2 : \Delta_2^* \to \Delta_2^*, \ldots, f_t : \Delta_t^* \to \Delta_t^*,$$

where $\Delta_0 := \Sigma$ and $\Delta_1 := \Delta$, such that

(1.) f_i and f_{i+1} are twined with respect to the non-erasing morphisms $h_i : \Delta_i^* \to \Delta_{i+1}^*$ and $k_i : \Delta_{i+1}^* \to \Delta_i^*$, $i = 0, 1, \ldots, t-1$,

(2.) $|\Delta_i| > |\Delta_{i+1}|$, $i = 0, 1, \ldots, t-1$, and

(3.) f_t is injective.

The morphism f_t is called an *injective simplification* of f. Let $h := h_{t-1} \circ \ldots \circ h_1 \circ h_0$ and $k := k_0 \circ k_1 \circ \ldots \circ k_{t-1}$. Then $k \circ h = f^t$.

Proposition 8. *Let $f : \Sigma^* \to \Sigma^*$ be a non-erasing morphism, and let $f_t : \Delta_t^* \to \Delta_t^*$ be an injective simplification of f.*

(a) $L(f)$ is (strongly) repetitive if and only if $L(f_t)$ is.

(b) If a word $v \in \Sigma^+$ is (quasi-) repetitive for f with (quasi-) period n, then $h(v)$ is (quasi-) repetitive for f_t with (quasi-) period n.

(c) If a word $v \in \Delta_t^+$ is (quasi-) repetitive for f_t with (quasi-) period n, then $k(v)$ is (quasi-) repetitive for f with (quasi-) period n.

4 Repetitiveness and quasi-repetitive elements

In Proposition 3 we have seen that the language $L(f)$ is strongly repetitive, if the morphism f has a quasi-repetitive element. Now we will see that f has such an element, if the language $L(f)$ is repetitive.

For the following considerations let $f : \Sigma^* \to \Sigma^*$ be a morphism that is non-erasing, let $r := |\Sigma|$, and let $\mu := \max\{|f(a)| \mid a \in \Sigma\}$.

Lemma 9. *Let $w, v \in \Sigma^+$ and $k, n \in \mathbb{N}$. If v^p is a subword of $f^n(w)$ for some $p \geq k \cdot \mu'(w)$, where $\mu'(w) := \max\{\mu, |w|\}$, then there exist an integer $m \in \{0, 1, \ldots, n\}$, a letter $a \in \mathrm{alph}(f^m(w))$, and a conjugate v_1 of v such that v_1^k is a prefix or a suffix of $f^{n-m}(a)$.*

Recall that a word v is called *unbounded* for f, if the language $L((\Sigma, f, v))$ is infinite.

Lemma 10. *Let $v \in \Sigma^+$ be a primitive word that is unbounded for f. If $f^n(a)$ has a prefix v^p, where $a \in \Sigma$ and $n, p \in \mathbb{N}$ are such that $p \geq 2 \cdot \mu^r$, then v is a quasi-repetitive element of quasi-period q for f for some integer $q \leq r$.*

An analogous result holds if v^p is a suffix of $f^n(a)$. Based on this lemma we now derive the following important technical result.

Theorem 11. *Let $f : \Sigma^* \to \Sigma^*$ be a non-erasing morphism, $r := |\Sigma|$, $\mu := \max\{|f(a)| \mid a \in \Sigma\}$, and $v \in \Sigma^+$ be a primitive word that is unbounded for f. If there exists an integer $p \geq 2 \cdot \mu^r \cdot \max\{\mu, |w|\}$ such that $v^p \in SL((\Sigma, f, w))$, then v is a quasi-repetitive element for f of quasi-period $q \leq r$.*

Proof. Assume that v^p is a subword of $f^n(w)$. Hence, by Lemma 9 there exist an integer $m \in \{0, 1, \ldots, n\}$ and a letter $a \in \text{alph}(f^m(w))$ such that $v_1^{2 \cdot \mu^r}$ is a prefix or a suffix of $f^{n-m}(a)$ for some conjugate v_1 of v. By Lemma 10 this implies that v_1, and hence v, is a quasi-repetitive element for f with quasi-period $q \le r$. □

For the non-erasing morphism $f : \Sigma^* \to \Sigma^*$, let $\Delta := \{a \in \Sigma \mid a \text{ is bounded for } f\}$. A D0L-system $G = (\Sigma, f, w)$ is called *pushy* if the language $SL(G) \cap \Delta^*$ is infinite. The morphism f is called *pushy* if the D0L-system (Σ, f, a) is pushy for at least one letter $a \in \Sigma$. This notion has been coined by A. Ehrenfeucht and R. Rozenberg in [ER83], where it is observed that a D0L-system is strongly repetitive, if it is pushy.

For a morphism f that is not pushy, let $q(f) := \max\{|x| \mid x \in SL(f) \cap \Delta^*\}$.

Corollary 12. *Let f be a morphism that is non-erasing and not pushy, and let $v \in \Sigma^+$ be a primitive word. If $v^p \in SL(f)$ for some $p \ge \max\{2 \cdot \mu^{r+1}, 2 \cdot \mu^r \cdot |w|, (q(f) + 1)/|v|\}$, then v is a quasi-repetitive element for f of quasi-period $q \le r$.*

Proof. Since $|v^p| = p \cdot |v| \ge q(f) + 1$, v^p contains a letter that is unbounded for f, and so v is unbounded for f. Hence, by Theorem 11 v is a quasi-repetitive element for f with quasi-period $q \le r$. □

From this observation we get the following characterization of repetitive morphisms that are not pushy.

Corollary 13. *Let f be a morphism that is non-erasing and not pushy. Then the following three statements are equivalent:*

(a) $L(f)$ is repetitive;
(b) $L(f)$ is strongly repetitive; and
(c) there exists a quasi-repetitive element $v \in SL(f)$ for f.

Hence, we obtain the following result of [ER83].

Corollary 14. *Let f be a morphism. Then $L(f)$ is repetitive if and only if it is strongly repetitive.*

Proof. Let $f : \Sigma^* \to \Sigma^*$ be a morphism, and let $\bar{f} : (\Sigma \setminus \Gamma)^* \to (\Sigma \setminus \Gamma)^*$ be the non-erasing simplification of f. By Proposition 7 $L(f)$ is (strongly) repetitive if and only if $L(\bar{f})$ is. Now if $\bar{f}$ is pushy, then $L(\bar{f})$ is strongly repetitive, and if $\bar{f}$ is not pushy, then Corollary 13 applies. □

For D0L-languages the analogous result follows from Corollary 2. Observe that Corollary 13 characterizes the repetitiveness of a morphism through the existence of a quasi-repetitive element. In the rest of this section we will derive an algorithm for deciding whether a morphism is repetitive that is based on this characterization. In order to do so, however, we need some additional information on quasi-repetitive elements.

Lemma 15. *Let $f : \Sigma^* \to \Sigma^*$ be an injective morphism, and let $v \in \Sigma^*$ be a primitive word that is quasi-repetitive for f. Then no unbounded letter $a \in \Sigma \setminus \Delta$ occurs more than once in v.*

This lemma implies that, if v is a primitive word that is quasi-repetitive for an injective and growing morphism, then no letter occurs more than once in v, and hence, $|v| \leq |\Sigma|$.

Theorem 16. *Let f be a morphism that is injective, but not pushy. If $L(f)$ is repetitive, then $SL(f)$ contains a quasi-repetitive element v for f that has length $|v| \leq (r_1 + 1) \cdot q(f) + r_1$ and quasi-period at most r, where $r_1 := |\Sigma \setminus \Delta|$.*

If the non-erasing morphism f is not injective, then using the construction of Section 3 we obtain an injective simplification g of f. For all $w \in \Sigma^*$, $L((\Sigma, f, w))$ is repetitive if and only if $L((\Delta_t, g, h(w)))$ is repetitive by Proposition 4(b). Further, if the language $L((\Sigma, f, w))$ is not pushy, then neither is the language $L((\Delta_t, g, h(w)))$. Hence, Theorem 16 implies that the language $SL(g)$ contains a quasi-repetitive element v for g that has length $|v| \leq (r_1 + 1) \cdot q(g) + r_1$ and quasi-period at most $|\Delta_t|$, where r_1 is the number of unbounded letters in Δ_t (with respect to the morphism g). Hence, by Proposition 8(c) $k(v) \in SL(f)$ is a quasi-repetitive element for f with quasi-period at most $|\Delta_t| \leq |\Sigma|$. Since $k \circ h = f^t$, we see that $|k(v)| \leq \mu^t \cdot |v|$. Since $t \leq |\Sigma| - 1$, we obtain the following result.

Corollary 17. *Let f be a morphism that is non-erasing and repetitive, but not pushy. Then the language $SL(f)$ contains a quasi-repetitive element v for f that has length at most $\mu^{|\Sigma|-1} \cdot ((|\Sigma| + 1) \cdot q(f) + |\Sigma|)$ and quasi-period at most $|\Sigma|$.*

Assume that v is a primitive word that is a quasi-repetitive element for f of quasi-period $n \leq |\Sigma|$. Let i be chosen such that $p^i \geq \max\{|v|, \mu\}$, where $p > 1$ is the quasi-multiplicity of v corresponding to the quasi-period n. Then $f^{n \cdot i}(v) = v_1^{p^i}$ for some conjugate v_1 of v. Hence, by Lemma 9 there exist an integer $m \leq n \cdot i$, a letter $a \in \text{alph}(f^m(v))$, and a conjugate v_2 of v such that v_2 is a prefix or a suffix of $f^{n \cdot i - m}(a)$. Obviously, a is an unbounded letter. Thus, in order to check whether there exists a quasi-repetitive element for f it suffices to consider all prefixes and suffixes of $f^j(a)$, where a is an unbounded letter and $j \leq n \cdot i \leq |\Sigma| \cdot \log_p(\max\{\mu, |v|\}) \leq |\Sigma| \cdot \log_2(\mu^{|\Sigma|-1} \cdot ((|\Sigma| + 1) \cdot q(f) + |\Sigma|)) \leq |\Sigma| \cdot (|\Sigma| \cdot \log_2 \mu + \log_2(|\Sigma| + 1) + \log_2(q(f) + 1))$. Thus, we get the following improvement of Corollary 17.

Corollary 18. *Let f be a morphism that is non-erasing and repetitive, but not pushy. Then there exist an unbounded letter $a \in \Sigma$ and an integer $m \leq |\Sigma| \cdot (|\Sigma| \cdot \log_2 \mu + \log_2(|\Sigma| + 1) + \log_2(q(f) + 1))$ such that $f^m(a)$ has a prefix or a suffix that is a quasi-repetitive element for f of quasi-period at most $|\Sigma|$.*

In order to be able to exploit the above corollary we need a simple bound for the constant $q(f)$. Recall that $q(f) = \max\{|x| \mid x \in SL(f) \cap \Delta^*\}$, where

$\Delta = \{a \in \Sigma \mid a \text{ is bounded for } f\}$. As observed in [KO97] Δ can be determined in polynomial time from f. To simplify the notation we define $\Sigma_u := \Sigma \smallsetminus \Delta$. Further, we introduce the following constants:

(i) $\alpha := |\Delta|$, $\beta := |\Sigma_u|$, and hence, $|\Sigma| = \alpha + \beta$,
(ii) $\gamma := \max\{|f(a)| \mid a \in \Delta\}$, and
(iii) $\delta := \max\{|x| \mid x \in \Delta^* \text{ is a syllable of } f(bc), b, c \in \Sigma_u\}$.

Then the following bound can be established. Actually, it can be shown that this bound is sharp [KO97].

Theorem 19. *If f is non-erasing and not pushy, then $q(f) \leq \delta \cdot \beta \cdot \gamma^{\alpha-1}$.*

Hence, the number $q(f)$ is bounded from above by an exponential function of $|\Sigma|$. However, if the alphabet is fixed, then $\delta \cdot \beta \cdot \gamma^{|\Sigma|}$ is a bound for $q(f)$ that is polynomial in (the size of the description of the morphism) f.

Corollary 18 and Theorem 19 now yield a structurally simple algorithm for deciding whether or not a given morphism f is repetitive. Recall from Section 3 that it suffices to consider morphisms that are non-erasing.

Algorithm 20. INPUT: A non-erasing morphism $f : \Sigma^* \to \Sigma^*$;
               OUTPUT: 'yes', if f is repetitive, 'no', otherwise.

begin (1.) **if** $L(f)$ is finite **then** (* $L(f)$ is not repetitive *)
         {OUTPUT: 'no'; STOP};
     (2.) **if** $L(f)$ is pushy **then** (* $L(f)$ is repetitive [ER83] *)
         {OUTPUT: 'yes'; STOP};
     (3.) $\Delta := \{a \in \Sigma \mid a \text{ is bounded for } f\}$;
         $\alpha := |\Delta|$; $\beta := |\Sigma \smallsetminus \Delta|$;
         $\mu := \max\{|f(a)| \mid a \in \Sigma\}$; $\gamma := \max\{|f(a)| \mid a \in \Delta\}$;
         $q := 2 \cdot \mu \cdot \beta \cdot \gamma^{\alpha-1}$; (* q is an upper bound for $q(f)$ *)
         mmax $:= |\Sigma| \cdot (|\Sigma| \cdot \log_2 \mu + \log_2(|\Sigma| + 1) + \log_2(q + 1))$;
     (4.) **for** all $a \in \Sigma \smallsetminus \Delta$ **do**
        for $i = 1$ **to** mmax **do**
        if there is a prefix or a suffix v of $f^i(a)$ such that v is primitive
            and $|v| \leq \mu^{|\Sigma|-1} \cdot ((|\Sigma| + 1) \cdot q + |\Sigma|)$
            and $\exists \ell \in \{1, 2, \ldots, |\Sigma|\} \, \exists p > 1 \, \exists v_1 \sim v : f^\ell(v) = v_1^p$
        then (* $L(f)$ is repetitive by Prop. 3 *)
         {OUTPUT: 'yes'; STOP};
     (5.) OUTPUT: 'no'
end.

Theorem 21. *The algorithm above decides whether or not the given morphism is repetitive. If the alphabet Σ is fixed, then this algorithm runs in polynomial time.*

Proof. The correctness follows from Corollary 13, Corollary 18, and Theorem 19. So let us consider the running time. The test in (1.) can be performed in polynomial time, and also the subalphabet Δ can be determined in polynomial time from Σ and f [KO97]. According to the proof of Lemma 2.1 of [ER83] $L(f)$ is pushy if and only if it satisfies the so-called 'edge condition'. This, however, can be checked in polynomial time. Also the constants q, μ, and $mmax$ can be determined in polynomial time. Finally, the **for**-loops in (4.) are executed only a polynomial number of times, and each iteration only takes polynomial time, if Σ is fixed. □

Structurally Algorithm 20 is much simpler than the algorithm of Ehrenfeucht and Rozenberg given in [ER83]. Also it is not at all clear whether their algorithm can be made to run in polynomial time. For the algorithm that is presented by F. Mignosi and P. Séébold in [MS93] this is not clear, either.

5 Concluding remarks

Is there a combinatorial characterization of all repetitive morphisms? Such a characterization could be seen as the ultimate extension of the decidability result presented in the present paper. At least for the case of a two-letter alphabet such a characterization is given in [KOS97].

References

[ER78] A. Ehrenfeucht and G. Rozenberg. Simplifications of homomorphisms. *Information and Control*, 38:298–309, 1978.

[ER81] A. Ehrenfeucht and G. Rozenberg. On the subword complexity of square-free D0L languages. *Theoretical Computer Science*, 16:25–32, 1981.

[ER83] A. Ehrenfeucht and G. Rozenberg. Repetition of subwords in D0L languages. *Information and Control*, 59:13–35, 1983.

[HL86] T. Harju and M. Linna. On the periodicity of morphisms on free monoids. *RAIRO Informatique Théorique et Applications*, 20:47–54, 1986.

[HL86a] T. Head and B. Lando. Periodic D0L languages. *Theoretical Computer Science*, 46:83–89, 1986.

[HU79] J.E. Hopcroft and J.D. Ullman. *Introduction to Automata Theory, Languages, and Computation*. Addison-Wesley, Reading, M.A., 1979.

[KO97] Y. Kobayashi and F. Otto. Repetitiveness of languages generated by morphisms. Preprint No. 2/97, Fachbereich 17, Universität Kassel, 1997.

[KOS97] Y. Kobayashi, F. Otto, and P. Séébold. A complete characterization of repetitive morphisms over the two-letter alphabet. *Proceedings of COCOON'97*, Lecture Notes in Computer Science, Springer-Verlag, Berlin, 1997, to appear.

[Lot82] M. Lothaire. *Combinatorics on Words*. Addison-Wesley, Mass., 1982.

[MS93] F. Mignosi and P. Séébold. If a D0L language is k-power free then it is circular. In *Automata, Languages and Programming, Proceedings of ICALP'93*, Lecture Notes in Computer Science 700, pages 507–518. Springer-Verlag, Berlin, 1993.

[RS80] G. Rozenberg and A. Salomaa. *The Mathematical Theory of L Systems*. Academic Press, New York, 1980.

[RS92] G. Rozenberg and A. Salomaa, editors. *Lindenmayer Systems*. Springer-Verlag, Berlin, 1992.

Minimal Letter Frequency
in n-th Power-Free Binary Words

Roman Kolpakov[1] and Gregory Kucherov[2]

[1] French-Russian Institute for Informatics and Applied Mathematics, Moscow
University, 119899 Moscow, Russia, e-mail: roman@vertex.inria.msu.ru
[2] INRIA-Lorraine & CRIN, 615, rue du Jardin Botanique, B.P. 101, 54602
Villers-lès-Nancy France, e-mail: kucherov@loria.fr

Abstract. We show that the minimal proportion of one letter in an
n-th power-free binary word is asymptotically $1/n$. We also consider a
generalization of n-th power-free words defined through the notion of
exponent: a word is x-th power-free for a real x, if it does not contain
subwords of exponent x or more. We study the minimal proportion of
one letter in an x-th power-free binary word as a function of x and prove,
in particular, that this function is discontinuous.

1 Introduction

One of classical topics of formal language theory and word combinatorics is the
construction of infinite words verifying certain restrictions. A typical restriction
is the requirement that the word does not contain a subword of the form specified
by some general pattern. Results of this kind find their applications in different
areas such as algebra, number theory, game theory (see [12,16]).

The oldest results of this kind, dating back to the beginning of the century,
are Thue's famous constructions of infinite square-free and (strongly) cube-free
words over alphabets of three and two letters respectively [17,18] (see also [5]).
A word is *square-free* (respectively *cube-free*, *strongly cube-free*) if it does not
contain a subword uu (respectively uuu, uua), where u is a non-empty word and
a is the first letter of u.

During the last two decades, different generalizations of Thue's results have
been studied. A natural generalization is to consider, instead of squares or cubes,
any n-th power, or, yet more generally, any *pattern* (a word over some alphabet
of variables). Works [4,1] introduce a general property of *avoidability* of a pattern
and propose an algorithm to test it. A pattern is avoidable iff for some k, there
is an infinite word over k letters that does not contain a subword which is an
instance of the pattern. If k is fixed, the pattern is called k-avoidable.[1] In this
terminology, Thue's results state that the pattern xx is 3-avoidable, and the
pattern xxx and, more strongly, the pattern $xyxyx$ are 2-avoidable. A good

---

[1] The difference between avoidability and k-avoidability is important. While avoidability was shown to be decidable in [4,1], decidability of k-avoidability is a long-standing open problem (see [9]).

account of the area of pattern avoidability, together with some new results, is contained in [7]. We would like also to point out work [2].

Many results on avoidability establish some threshold values or some "borderline conditions". As an example, let us mention the result of Roth [14] showing that every pattern over two variables of length six is 2-avoidable. Six is the best possible value, as there are patterns of length five that are not 2-avoidable (e.g. $xxyxx$).

As another example, Dejean [10] strengthens the Thue construction of a square-free word by constructing an infinite word over three letters such that any two occurrences of a non-empty word u are separated by at least $|u|/3$ letters, and she shows that this bound is optimal. There is another formulation of this result: There is an infinite word over three letters that not only avoids repetitions (subwords uu), but does not admit subwords uv, where v is a prefix of u of length more than $3|u|/4$. Generalizations of this result for bigger alphabets have been obtained (see [5] for more references; see also [8] for a related result).

In this paper that fits into this general research direction, we address the following general problem which, to the best of our knowledge, has not been studied so far. Assume that each letter has some weight, and we try to minimize the total weight of a word of given length avoiding the pattern. For example, if one letter is much "heavier" than the others, this leads to the following problem: Assume that a pattern is k-avoidable but not $(k-1)$-avoidable, then what is the minimal proportion of the k-th letter in an infinite word avoiding the pattern?

In this paper we solve this problem for the case of binary alphabet ($k = 2$), and patterns x^n (n-th power) for $n > 2$. In our main result we show that the minimal proportion of one letter in an n-th power-free binary word is asymptotically $1/n$. As for strongly cube-free words, this proportion is asymptotically $1/2$, i.e. it is not possible in this case to reduce by any factor the number of occurrences of one letter with respect to the other. Both these results can be expressed uniformly through the generalized minimal frequency function based on the notion of exponent of a word. In this way, we can generalize n-th power-free words to "x-th power-free words" (more precisely, words without subwords of exponent x) for any real $x > 2$. We study the properties of the generalized minimal frequency function and prove, in particular, that it is discontinuous.

In Section 2 we give some preliminary definitions and results. Section 3 is devoted to the proof of the main theorem. In Section 4 we study the generalized minimal frequency function. We conclude in Section 5 with possible directions for future work.

2   Definitions and preliminary results

As usual, A^* denotes the free monoid over an alphabet A. ε denotes the empty word, and $A^+ = A^* \setminus \{\varepsilon\}$. $u \in A^+$ is a *subword* of $w \in A^*$ if w can be written as $u_1 u u_2$ for some $u_1 u_2 \in A^*$. $|u|$ stands for the length of $u \in A^*$. A *position* in a word w is associated with a natural number between 0 and $|w|$. If $w = u_1 u u_2$, then $|u_1|$ (resp. $|u_1 u|$) is called the *start position* (resp. *end position*) of u in w.

Throughout this paper, we consider the binary alphabet $\{0,1\}$. The word w obtained by concatenating n copies of a word v is called the n-th power of v and denoted v^n. For a natural $n > 1$, a word w is called n-*th power-free* iff it does not contain a subword which is the n-th power of some non-empty word. For $n = 2$ (resp. $n = 3$), such w is called *square-free* (resp. *cube-free*). If w does not contain a subword uua, where u is a non-empty word and a is the first letter of u, then w is called *strongly cube-free*. An equivalent property (see [15]) is overlap-freeness – w is *overlap-free* if it does not contain two overlapping occurrences of a non-empty word u.

Let $c(w)$ be the number of 1's in a word w. For $n, l \in \mathbb{N}$, let $B[n, l]$ be the set of all n-th power-free words of length l. Since the famous Thue's construction of an infinite (strongly) cube-free binary word (see e.g. [15]), it is well known that for $n \geq 3$, $B[n, l]$ is not empty for every $l \in \mathbb{N}$. Therefore, for every $n, l \in \mathbb{N}$, $n \geq 3$, we define $c[n, l] = \frac{1}{l} \min_{w \in B[n,l]} c(w)$ and $c[n] = \underline{\lim}_{l \to \infty} c[n, l]$. (Note that all numbers $c[n, l]$ belong to $[0, 1]$ and therefore $\underline{\lim}_{l \to \infty} c[n, l]$ belongs to $[0, 1]$ too.)

The following two propositions clarify the behaviour of the sequence $\{c[n, l]\}_{l=1}^{\infty}$ with respect to $c[n]$.

Proposition 1. *For every $l \in \mathbb{N}$, $c[n, l] \leq c[n]$.*

Proof. Take any $l \in \mathbb{N}$ and assume that $\{c[n, l_i]\}_{i=1}^{\infty}$ is a subsequence converging to $c[n]$. Take some $l_i > l$. By definition of $c[n, l_i]$, there exists an n-th power-free word w_i of length l_i such that $c(w_i) = l_i c[n, l_i]$. Consider $\lfloor l_i / l \rfloor$ non-overlapping subwords of w_i of length l. Each of these subwords is n-th power-free and then contains at least $l c[n, l]$ 1's. Therefore, w_i contains at least $\lfloor l_i / l \rfloor l c[n, l]$ 1's, that is $c(w_i) \geq \lfloor l_i / l \rfloor l c[n, l]$. We obtain that $c[n, l_i] \geq \lfloor l_i / l \rfloor l c[n, l] / l_i > ((l_i / l) - 1) l c[n, l] / l_i = (1 - (l / l_i)) c[n, l]$. By taking the limit for $i \to \infty$, we conclude that $c[n] = \lim_{i \to \infty} c[n, l_i] \geq \lim_{i \to \infty} (1 - (l / l_i)) c[n, l] = c[n, l]$.

Proposition 2. *For every $n \geq 3$, $c[n] = \lim_{l \to \infty} c[n, l]$.*

Proof. By Proposition 1, $c[n, l] \leq c[n]$ for every l, and then $\overline{\lim}_{i \to \infty} c[n, l] \leq c[n] = \underline{\lim}_{i \to \infty} c[n, l]$.

3 Main result

A central result of this work is the following theorem. Its proof will occupy the whole rest of this section.

Theorem 3.

$$c[n] = \frac{1}{n} + \mathcal{O}\left(\frac{1}{n^2}\right).$$

The lower bound $\frac{1}{n}$ for $c[n]$ can be easily established.

Lemma 4. *For every $n \geq 3$,*

$$c[n] \geq \frac{1}{n}.$$

Proof. It is easily seen that $c[n, n] = 1/n$. By Proposition 1, we have $c[n] \geq 1/n$.

Let us turn to the more difficult part – proving that $\frac{1}{n}$ is an asymptotic upper bound for $c[n]$. We will effectively construct a sequence of words $\{w_k\}_{k=1}^{\infty}$ for which we show that $\lim_{k \to \infty} \frac{c(w_k)}{|w_k|} = \frac{1}{\sqrt{n^2-n}}$. To construct this sequence, we use the technique of morphism iteration, known in a more general setting as DOL-systems. Despite that DOL-systems are a common tool for constructing infinite or arbitrarily long words with required properties (this is the case for most works referred to in the bibliography section), finding a suitable morphism is often a difficult task.

Let $n \geq 3$. Consider the morphism $h : \{0, 1\}^* \longrightarrow \{0, 1\}^*$ defined as follows.

$$h(0) = 0^{n-1}1,$$
$$h(1) = 0^{n-2}1(0^{n-1}1)^{n-1}1.$$

No word of $\{h(0), h(1)\}$ is a prefix or a suffix of the other. In other terms, h defines a biprefix code (see [15]). This implies immediately a number of good properties of h. The following proposition gives two of them that will be used in the sequel.

Proposition 5. *(a) Morphism h is injective,*
(b) If $h(v_1)$ is a prefix (resp. suffix) of $h(v_2)$, then v_1 is a prefix (resp. suffix) of v_2.

For all $k \in \mathbb{N}$, define $w_k = h^k(0)$. Note that every w_k is a concatenation of some number of $h(0)$ and $h(1)$. Since h is a code, this decomposition is uniquely defined. Every occurrence of $h(0), h(1)$ in this decomposition will be called a *coding occurrence*. Positions in w_k delimiting the coding occurrences will be called *landmarks*. This definition implies the following proposition which will often be used below.

Proposition 6. *Let $w_k = u_1 u_0 u_2$ and the start and the end positions of u_0 are landmarks of w_k. Then there exists a decomposition $w_{k-1} = v_1 v_0 v_2$ such that $u_0 = h(v_0)$, $u_1 = h(v_1)$, $u_2 = h(v_2)$.*

The conclusion of the proposition above can be stated slightly differently by saying that $h^{-1}(u_0)$ exists and is a subword of w_{k-1}.

Note that both $h(0)$ and $h(1)$ start from 0 and end up with 1. Therefore, a landmark of w_k cannot occur inside a subword of w_k consisting only of 0's or only of 1's, and such a subword must be a subword of either $h(0)$ or $h(1)$. This implies the following lemma.

Lemma 7. *For every $k \in \mathbb{N}$, w_k contains no more than $n - 1$ contiguous occurrences of 0 and no more than two contiguous occurrences of 1.*

Define now

$$\alpha = 0^{n-1}1, \ \beta = 0^{n-2}1, \ \gamma = 1.$$

We call α, β, γ *elementary words*. Note that $h(0) = \alpha$, $h(1) = \beta\alpha^{n-1}\gamma$, and then every w_k can be represented as a concatenation of α, β, γ. The occurrences of α, β, γ in this representation are called *basic occurrences*. Note that the basic occurrences are delimited by the occurrences of 1.

Further useful properties of basic occurrences are summarized in the following lemma. They follow immediately from properties of h and the definition of $h(0), h(1)$ in terms of α, β, γ.

Lemma 8. *(a) Between any two basic occurrences of β in w_k, there is a basic occurrence of γ,*
(b) Between any two basic occurrences of γ in w_k, there is a basic occurrence of β.

Concerning α, the following result holds.

Lemma 9. *For every $k \in \mathbb{N}$, w_k does not contain n contiguous basic occurrences of α.*

Proof. Assume that for some k, w_k contains n contiguous basic occurrences of α. Note that each of these occurrences is either a part of an occurrence of $h(1)$, or an occurrence of $h(0)$. The first case is impossible since $h(1)$ generates exactly $n - 1$ contiguous basic occurrences of α bounded by basic occurrences of β and γ. The second case is impossible either since by Proposition 6 w_{k-1} would then contain 0^n which contradicts to Lemma 7.

Now we show that w_k's are n-th power-free words.

Lemma 10. *For every $k \in \mathbb{N}$, w_k is n-th power-free.*

Proof. We use induction over k. Clearly, w_1 is an n-th power-free word. Assume by contradiction that for some $k \geq 2$, w_{k-1} is n-th power free but w_k is not. Then w_k contains a subword u^n for some non-empty word u. By Lemma 7, u contains at least one symbol 1. We proceed by case analysis on the occurrences of 1 in u.

CASE 1: Assume that u contains a single occurrence of 1. Then u^n contains $(n - 1)$ consecutive equal basic occurrences. By Lemma 8, these are occurrences of α. This implies further that $|u| = n$. Consider all possible positions of 1 in u.

CASE 1.1: 1 is the first letter of u, that is $u^n = (10^{n-1})^n = 1\alpha^{n-1}0^{n-1}$. Since u^n has the suffix 0^{n-1}, by Lemma 7, the occurrence of u^n must be followed by 1 in w_k. This implies that w_k contains n contiguous basic occurrences of α which contradicts Lemma 9.

CASE 1.2: 1 is the last letter of u, that is $u^n = (0^{n-1}1)^n = \alpha^n$. u^n must either be preceded by 1, or be a prefix of w_k, and this is again a contradiction to Lemma 9.

CASE 1.3: 1 occurs in u at some inner position, that is $u = 0^{j-1}10^{n-j}$ for some $1 < j < n$. Then $u^n = 0^{j-1}1\alpha^{n-1}0^{n-j}$ which shows that w_k contains $n-1$ basic occurrences of α followed by and preceeded by some non-empty words. Consider the basic occurrences immediately before and after these occurrences of α. By Lemma 9, neither of them can be an occurrence of α and by Lemma 8, at least one of them must be γ. But this would mean that the occurrence of 1 in u is followed or preceded by another occurrence of 1 which contradicts to the assumption that u has a single occurrence of 1.

CASE 2: u has at least two occurrences of 1. Recall that since w_k can be written as a concatenation of α, β, γ, any two neighbouring 1's (i.e. those without 1's in between) are separated by either $n-1$, or $n-2$, or zero 0's. We show first that u contains a subword $\beta' = 1\beta$ or a subword $\gamma' = 1\gamma$, i.e. u contains two occurences of 1 which are either contiguous or separated by exactly $n-2$ 0's. Assume that this is not the case, i.e. any two neighbouring occurrences of 1 are separated by $n-1$ 0's. Then $u = 0^i1\alpha^j0^l$ for some $j \geq 1$, $i, l \geq 0$, and $u^n = 0^i1(\alpha^j0^{i+l}1)^{n-1}0^l$. Clearly, $i+l = n-1$ contradicts immediately to Lemma 9. Therefore, $i+l = n-2$ or $i+l = 0$. If $i+l = n-2$, we obtain at least two basic occurrences of β without basic occurences of γ between them. This contradicts to Lemma 8. Similarly, if $i+l = 0$, we have at least two occurrences of γ without occurences of β in between. We conclude that u contains β' or γ'.

Note that since $\beta' = 1\beta$, and β identifies the beginning of $h(1)$, an occurrence of β' in w_k contains a landmark at its position 1. Similarly, $\gamma' = 1\gamma$, and the end position of an occurrence of γ' in w_k is the end position of an occurrence of $h(1)$ and hence a landmark.

Consider an occurrence of u in w_k. Since u contains either β' or γ', the occurrence of u contains a landmark at some position of u. It is important to note that the position of this landmark in u is the same for all occurrences of u in w_k. From Propositions 5,6 it follows readily that all other landmarks in this occurrence of u are uniquely defined, and these landmarks have the same positions in all other occurrences of u in w_k. Consider the leftmost and rightmost landmarks in u and let $u = u_1u_0u_2$ where u_0 is delimited by these landmarks. By Proposition 6, there exists $v = h^{-1}(u_0)$ and v is a subword of w_{k-1}. Since u_0 is delimited by the outermost landmarks in u, u_1 (resp. u_2) is a suffix (resp. prefix) of some coding occurrence.

Consider now an occurrence of u^n in w_k, that is n contiguous occurrences of u. For each occurrence of u, consider the above decomposition $u = u_1h(v)u_2$ and write $u^n = u_1(h(v)u_2u_1)^{n-1}h(v)u_2$. If u_1 and u_2 are both empty, then by Proposition 6, w_{k-1} has to contain v^n which contradicts to the assumption about w_{k-1}. If only one of them is empty, then the other must form a coding occurrence, which contradicts to the choice of outermost landmarks. Let $u_1 \neq \varepsilon$, $u_2 \neq \varepsilon$. Then $u_2u_1 = h(0)$ or $u_2u_1 = h(1)$. Since $|u_2u_1| \geq \min\{|h(0)|, |h(1)|\} = n$, then $|u_1| \geq 2$ or $|u_2| \geq n-1$. Assume that $|u_1| \geq 2$. Since $h(0)$ and $h(1)$ have distinct suffixes of length 2, u_1 uniquely identifies $a \in \{0, 1\}$ such that $h(a) = u_2u_1$. Therefore, the first occurrence of u_1 in u^n is the suffix of a coding occurrence of u_2u_1, and the n-th power $(u_2u_1h(v))^n = (h(a)h(v))^n$ occurs in w_k. This implies that $(av)^n$ occurs in w_{k-1} which contradicts to the assumption that w_{k-1} is n-th power-free. Similarly, if $|u_2| \geq n - 1$, u_2 uniquely identifies $a \in \{0, 1\}$ such that $h(a) = u_2 u_1$. Therefore, w_k contains $(h(v)u_2u_1)^n = (h(v)h(a))^n$ which implies that w_{k-1} contains $(va)^n$. This completes the proof of Case 2 and Lemma 10.

The next step is to compute an upper bound for $c[n]$ by counting the number of 1's in w_k's. This is done in the following lemma.

Lemma 11. *For every $n \geq 3$, $c[n] \leq \dfrac{1}{\sqrt{n(n-1)}}$.*

Proof. First, we count the length of w_k and the number of 1's in it. With every w_k we associate a vector $e_k \in \mathbb{R}^2$, such that its first and second components are the number of 0's and of 1's respectively. Formally,

$$e_k = \begin{pmatrix} |w_k| - c(w_k) \\ c(w_k) \end{pmatrix}.$$

Applying h to a word replaces every occurrence of 0 by $(n-1)$ 0's and one 1 and every occurrence of 1 by $(n^2 - n - 1)$ 0's and $(n+1)$ 1's. As $w_1 = h(0)$ and $w_k = h(w_{k-1})$, we get for every $k \in \mathbb{N}$, $e_k = Ae_{k-1}$, where

$$A = \begin{pmatrix} n-1 & n^2 - n - 1 \\ 1 & n+1 \end{pmatrix}, \qquad e_0 = \begin{pmatrix} 1 \\ 0 \end{pmatrix}.$$

Clearly, $e_k = A^k e_0$ for all $k \in \mathbb{N}$. The eigenvalues of matrix A are $\lambda_1 = n + \sqrt{n^2 - n}$ and $\lambda_2 = n - \sqrt{n^2 - n}$. Note that $\lambda_1 > \lambda_2 > 0$. The eigenvectors corresponding to λ_1 and λ_2 are respectively

$$f_1 = \begin{pmatrix} \sqrt{n^2 - n} - 1 \\ 1 \end{pmatrix}, \qquad f_2 = \begin{pmatrix} \sqrt{n^2 - n} + 1 \\ -1 \end{pmatrix}.$$

Representing e_0 as a linear combination of f_1 and f_2, we get $e_0 = \dfrac{1}{2\sqrt{n^2-n}}(f_1 + f_2)$. Therefore,

$$e_k = A^k e_0 = \frac{1}{2\sqrt{n^2 - n}} \left(A^k f_1 + A^k f_2 \right) = \frac{1}{2\sqrt{n^2 - n}} \left(\lambda_1^k f_1 + \lambda_2^k f_2 \right),$$

and then

$$\begin{pmatrix} |w_k| - c(w_k) \\ c(w_k) \end{pmatrix} = \frac{1}{2\sqrt{n^2 - n}} \begin{pmatrix} \lambda_1^k(\sqrt{n^2 - n} - 1) + \lambda_2^k(\sqrt{n^2 - n} + 1) \\ \lambda_1^k - \lambda_2^k \end{pmatrix}.$$

We obtain that

$$c(w_k) = \frac{1}{2\sqrt{n^2 - n}}(\lambda_1^k - \lambda_2^k), \quad |w_k| = \frac{1}{2}(\lambda_1^k + \lambda_2^k).$$

Clearly, $|w_k| \to \infty$ as $k \to \infty$, and by Proposition 2, $c[n] = \lim_{k \to \infty} c[n, |w_k|]$. Since by Lemma 10, w_k is n-th power-free, we have

$$c[n, |w_k|] = \frac{1}{|w_k|} \min_{w \in B[n, |w_k|]} c(w) \leq \frac{c(w_k)}{|w_k|} =$$
$$= \frac{1}{2\sqrt{n^2-n}}(\lambda_1^k - \lambda_2^k) \bigg/ \frac{\lambda_1^k + \lambda_2^k}{2} = \frac{1}{\sqrt{n^2-n}} \frac{\lambda_1^k - \lambda_2^k}{\lambda_1^k + \lambda_2^k}.$$

We conclude that

$$c[n] = \lim_{k \to \infty} c[n, |w_k|] \leq \lim_{k \to \infty} \frac{1}{\sqrt{n^2-n}} \frac{\lambda_1^k - \lambda_2^k}{\lambda_1^k + \lambda_2^k} =$$

$$= \frac{1}{\sqrt{n^2-n}} \lim_{k \to \infty} \frac{1-(\lambda_1/\lambda_2)^k}{1+(\lambda_1/\lambda_2)^k} = \frac{1}{\sqrt{n^2-n}}.$$

The lemma is proved.

Since $\frac{1}{\sqrt{n^2-n}} = \frac{1}{n} + \mathcal{O}(\frac{1}{n^2})$, Theorem 3 follows from Lemmas 10 and 11.

4 Generalized minimal frequency function

Function $c[n]$ admits an interesting generalization to real arguments through the notion of *exponent* (see [10,6,8]) that generalizes the notion of n-th power. The exponent of a word w is the ratio $\frac{|w|}{\min |v|}$, where the minimum is taken over all periods v of w.[2] Using periods, the function $c[n]$ can be defined on real numbers in the following way. For every real number $x > 2$, define $B[x, l]$ (resp. $B[x + \varepsilon, l]$) to be the set of binary words of length l that do not contain a subword of exponent greater than or equal to (resp. strictly greater than) x. Similar to the case of natural argument, for every real $x > 2$, we can define $c[x, l] = \frac{1}{l} \min_{w \in B[x,l]} c(w)$, $c[x + \varepsilon, l] = \frac{1}{l} \min_{w \in B[x+\varepsilon,l]} c(w)$, and then $c[x] = \underline{\lim}_{l \to \infty} c[n, l]$, $c[x + \varepsilon] = \underline{\lim}_{l \to \infty} c[n + \varepsilon, l]$. Furthermore, Propositions 1 and 2 can be directly generalized too:

Proposition 12. *For every $x > 2$, $l \in \mathbb{N}$, $c[x, l] \leq c[x]$, and $c[x+\varepsilon, l] \leq c[x+\varepsilon]$.*

Proposition 13. *For every $x > 2$, $c[x] = \lim_{l \to \infty} c[x, l]$, and $c[x + \varepsilon] = \lim_{l \to \infty} c[x + \varepsilon, l]$.*

In this section we study some properties of function $c[x]$ and prove, in particular, that it is discontinuous.

Functions $c[x]$, $c[x+\varepsilon]$ are non-increasing with values from $[0, \frac{1}{2}]$. Therefore, at every $x > 2$ there exists a right limit, denoted $c[x+0]$, and $c[x+0] = \sup_{y>x} c[y]$. The following lemma is useful.

Lemma 14. *For every $x > 2$, $c[x + 0] = c[x + \varepsilon]$.*

Proof. Clearly, for every $y > x$, $c[y] \leq c[x + \varepsilon]$, and therefore $c[x + 0] = \sup_{y>x} c[y] \leq c[x + \varepsilon]$. Assume that $c[x + 0] < c[x + \varepsilon]$. Then by Proposition 13, for some l, $c[x + \varepsilon, l] > c[x + 0]$. The exponents of subwords of words of length l can take finitely many possible values. Let $\hat{x}_l$ be the smallest such value strictly greater than x. Then $c[x + \varepsilon, l] = c[\hat{x}_l, l] > c[x + 0]$. By Proposition 12, $c[\hat{x}_l] \geq c[\hat{x}_l, l]$ and then $c[\hat{x}_l] > c[x + 0]$. This contradicts to the fact that $c[x + 0] = \sup_{y>x} c[y]$.

---

[2] v is a period of w iff w is a subword of v^n for some n.

Let us now compute the value $c[2 + \varepsilon]$. Observe that $B[2 + \varepsilon, l]$ is exactly the set of *strongly cube-free* (overlap-free) binary words of length l. These words have been extensively studied (see [13,11]) and their structure has been thoroughly characterized. In particular, it is known that every strongly cube-free word can be decomposed as $v_1 v v_2$, where $|v_1| \leq 2$, $|v_2| \leq 2$ and $v \in \{01, 10\}^*$ (see Lemma 2.2 of [11]). This implies immediately that $c[2 + \varepsilon] = \frac{1}{2}$. However, a stronger result can be stated.

Lemma 15. *For all $x \in (2, \frac{7}{3}]$, $c[x] = \frac{1}{2}$.*

Proof. To prove that every strongly cube-free word w can be decomposed as above it is sufficient to assume that w does not contain a subword vva where $|v| \leq 3$ and a is the first letter of v. We refer the reader to the proof of Lemma 2.1 in [11] to check this out.

We now prove the main result of this section showing that $c[x]$ is discontinuous at $x = 7/3$. Specifically, we prove

Theorem 16.
$$c[\frac{7}{3} + \varepsilon] \leq \frac{10}{21}.$$

Consider the morphism h defined by

$$h(0) = 011010011001001101001,$$
$$h(1) = 100101100100110010110.$$

Similar to the morphism used in the proof of Theorem 3, h defines a biprefix code and verifies Proposition 5. We call $h(0), h(1)$ *coding words*. Note that a coding word is uniquely determined by its first (or last) letter. Consider a word $w \in \{0,1\}^*$ and its image $h(w)$. As in the proof of Theorem 3, an occurrence of $h(a)$, $a \in \{0,1\}$, in $h(w)$, corresponding to the image of a in w, is called a *coding occurrence*. Consider an occurrence in $h(w)$ of some subword u, that is $h(w) = u_1 u u_2$ for some $u_1, u_2 \in \{0,1\}^*$. Consider the minimal subword w' of w such that u is covered by $h(w')$, that is $h(w) = h(w_1) h(w') h(w_2)$, $h(w') = \delta_1 u \delta_2$, $h(w_1)\delta_1 = u_1$, $\delta_2 h(w_2) = u_2$, and δ_1 (resp. δ_2) is a proper prefix (resp. suffix) of a coding occurrence. We call δ_1 the *precursor* of this occurrence of u.

We show that h preserves the property of absence of subwords of exponent greater than $7/3$.

Lemma 17. *For every $w \in \{0,1\}^*$, if w does not contain subwords of exponent greater than $7/3$, then neither does $h(w)$.*

Proof. Assume that w does not contain subwords of exponent greater than $7/3$. First show that $h(w)$ does not contain a subword of exponent greater than $7/3$ and with a period less than or equal to 15. If such a subword exists, there exists another one, say v, of length at most 36, with the same period and of exponent greater than $7/3$. Since $|h(0)| = |h(1)| = 21$, v is covered by three contiguous coding occurrences. Since w does not contain 000 or 111, v occurs in one of $h(001), h(010), h(011), h(100), h(101), h(110)$. A direct exhaustive check shows that neither of these words contains a subword of exponent greater than $7/3$ and with a period at most 15.

Now assume that $h(w)$ contains a subword $v = v_1 v_1 v_2$, where $|v_1| \geq 16$, v_2 is a prefix of v_1, and $|v_2| > |v_1|/3$. Let w' be the shortest subword of w such that $h(w')$ contains v, that is $h(w') = \delta_1 v \delta_2$, where $\delta_1, \delta_2 \in \{0,1\}^*$, and δ_1 is the precursor of the considered occurrence of v.

We now observe that if u is a subword of $h(w)$ of length 16 or more, then the precursor of u is uniquely defined for all occurrences of u. Since every subword of length 16 is located within two coding occurrences, this can be shown by checking this property for all subwords of length 16 occurring in words $h(00), h(01), h(10), h(11)$.[3]

By applying this argument to the two occurrences of v_1 in v and by using properties of h, we can rewrite $h(w') = h(a)v_1'h(a)v_1'h(a)v_2'\delta_2$, $a \in \{0,1\}$. Now observe that

(1) v_1' is non-empty (otherwise w would contain aaa),
(2) $|h(a)v_1'| = |v_1|$, $|h(a)v_2'| \geq |v_2|$,
(3) $v_2'\delta_2$ is a prefix of $v_1'h(a)$,
(4) $v_1' = h(w_1)$ and $v_2'\delta_2 = h(w_2)$ for some $w_1, w_2 \in \{0,1\}$.

By taking the inverse image of $h(w')$, we get $w' = aw_1aw_1aw_2$, where w_2 is a prefix of w_1a, and

$$|aw_2| = \frac{|h(a)v_2'| + |\delta_2|}{21} \geq \frac{|v_2|}{21} > \frac{1}{3} \cdot \frac{|v_1|}{21} = \frac{1}{3}|aw_1|.$$

We conclude that w' is a word of exponent greater than $7/3$, which is a contradiction.

Theorem 16 follows from Lemma 17. Consider words $h(0), h^2(0), \ldots, h^k(0), \ldots$. By Lemma 17, these words don't contain subwords of exponent greater than $7/3$. On the other hand, since both $h(0), h(1)$ are of length 21 and contain ten 1's, then $c[h^k(0)] = \frac{10}{21}|h^k(0)|$. By Proposition 13, we conclude that $c[7/3 + \varepsilon] \leq \frac{10}{21}$. By Lemma 14, $c[7/3 + 0] \leq \frac{10}{21}$ which proves that $c[x]$ has a jump at $x = 7/3$.

5 Concluding remarks

A possible direction of generalizing the results of this paper is to consider the general notion of k-avoidability of a pattern (see Introduction). The general question is: If a pattern p is not k-avoidable but is $(k+1)$-avoidable, what is the minimal frequency of a letter in an infinite word over $(k+1)$ letters avoiding p? For example, what is the minimal frequency of a letter in an infinite ternary square-free word? A pattern which is 4-avoidable but not 3-avoidable is given in [3]. What is the minimal proportion of the 4th letter needed to avoid that pattern?

---

[3] This does not hold for subwords of length 15. For example, 010011001011001 occurs in $h(01)$ as well as in $h(10)$, and these occurrences have different precursors.

Acknowledgements This work was supported by the French-Russian A.M.Liapunov Institut of Applied Mathematics and Informatics at Moscow University. The first author was also supported by the Russian Foundation of Fundamental Research (grant 96–01–01068). We are indebted to Alexandre Ugolnikov for his contributions to this work, both scientific and organizing.

References

1. А.И. Зимин. Блокирующие множества термов. *Математический Сборник*, 119(3):363–375, 1982. English Translation: A.I.Zimin, Blocking sets of terms, Math. USSR Sbornik 47 (1984), 353-364.

2. А.А. Евдокимов. Полные множества слов и их числовые характеристики. In *Методы дискретного анализа в исследовании экстремальных структур*, volume 39, pages 7–19, Новосибирск, 1983.

3. K. Baker, G. McNulty, and W. Taylor. Growth problems for avoidable words. *Theoret. Comp. Sci.*, 69:319–345, 1989.

4. D. Bean, A. Ehrenfeucht, and G. McNulty. Avoidable patterns in strings of symbols. *Pacific J. Math.*, 85(2):261–294, 1979.

5. J. Berstel. Axel thue's work on repetitions in words. Invited Lecture at the 4th Conference on Formal Power Series and Algebraic Combinatorics, Montreal, 1992, June 1992. accessible at http://www-litp.ibp.fr:80/berstel/.

6. J. Berstel and D. Perrin. *Theory of codes*. Academic Press, 1985.

7. J. Cassaigne. *Motifs évitables et régularités dans les mots*. Thèse de doctorat, Université Paris VI, 1994.

8. M. Crochemore and P. Goralcik. Mutually avoiding ternary words of small exponent. *International Journal of Algebra and Computation*, 1(4):407–410, 1991.

9. J. Currie. Open problems in pattern avoidance. *American Mathematical Monthly*, 100:790–793, 1993.

10. F. Dejean. Sur un théorème de Thue. *J. Combinatorial Th. (A)*, 13:90–99, 1972.

11. A. Kfoury. A linear time algorithm testing whether a word contains an overlap. *RAIRO Inf. Th.*, 22:135–145, 1988.

12. M. Lothaire. *Combinatorics on Words*, volume 17 of *Encyclopedia of Mathematics and Its Applications*. Addison Wesley, 1983.

13. A. Restivo and S. Salemi. On weakly square free words. *Bull. of the EATCS*, 21:49–56, 1983.

14. P. Roth. Every binary pattern of length six is avoidable on the two-letter alphabet. *Acta Informatica*, 29:95–106, 1992.

15. A. Salomaa. *Jewels of formal language theory*. Computer Science Press, 1986.

16. M. Sapir. Combinatorics on words with applications, December 1993. accessible at http://www.math.unl.edu/~msapir/ftp/course.

17. A. Thue. Über unendliche Zeichenreihen. *Norske Vid. Selsk. Skr. I. Mat. Nat. Kl. Christiania*, 7:1–22, 1906.

18. A. Thue. Über die gegenseitige Lage gleicher Teile gewisser Zeichenreihen. *Norske Vid. Selsk. Skr. I. Mat. Nat. Kl. Christiania*, 10:1–67, 1912.

Real-Time Generation of Primes
by a One-Dimensional Cellular Automaton
with 11 States

IVAN KOREC

Mathematical Institute, Slovak Academy of Sciences,
Štefánikova 49, 814 73 Bratislava, Slovakia

ABSTRACT. A 11-state one-dimensional cellular automaton is constructed which generates the primes in the following sense: The content of the 0-th cell at time t is equal to '1' if t is a prime, and is equal to '0' otherwise. The neighbourhood type of this CA is $(-1, 0, 1)$, i.e. the most usual one. At time $t = 0$ only the 0-th cell is in the non-quiescent state (here '0' is not the quiescent state). Further, a one-dimensional CA is constructed with the radius 12 but with two states only which also generates the primes. (At time $t = 0$ only the 1-st cell is in non-quiescent state.) Also a generalized Pascal triangle with 83 distinct elements is constructed which generates the odd primes in a similar sense. Hence the primes can be real-time generated also by a 83-state one-dimensional CA with the neighborhood type $(-1, 1)$.

1. Introduction

By P. C. Fischer [Fi], the primes can be generated by a one-dimensional cellular automaton (abbreviation: 1D CA) in real time in the following sense: The content of the 0-th cell of the 1D CA at time t is equal to '1' if t is a prime, and it is equal to '0' otherwise. Moreover, all cells except the 0-th one are in the quiescent state at time $t = 0$, and all cells on the left from the 0-th one are in the quiescent state forever. The idea of sieve of Eratosthenes is used. (To be quite precise, one-dimensional iterative arrays are considered in [Fi], but the difference in terminology is not substantial.) The 1D CA presented in [Fi] uses $37^3 = 50653$ (after reduction ≈ 30000) states. The main result of the present contribution is:

Claim 1. *There is a 11-state one-dimensional cellular automaton (with radius 1) which generates the primes in real time.*

(Three *claims* of the Introduction are repeated more formally as initial *theorems* in the Sections 3–5, Claim 4 as a *corollary* of Theorem 3. Necessary definitions are given in the next section.) The sieve of Eratosthenes is again used. The initial configuration remains the same as above, i.e. as simplest as possible. From Claim 1 we shall derive:

This research was partially supported by GA ČR No 201/95/0976 "HypercompleX" and by Grant 2/4034/97 of VEGA (SAV)

Claim 2. *There is a 2-state one-dimensional cellular automaton with radius 12 which generates the primes in real time.*

Here two states '0', '1' will be used, and '0' is the quiescent state. Again, the content of the 0-th cell at time t is '1' if t is prime and '0' otherwise, and at the time $t = 0$ only a unique cell is in a non-quiescent state (i.e. '1' in this case). However, it cannot be the 0-th cell because its content must be '0'; the 1-st cell can be used for this purpose.

Further, we obtain a similar result for generalized Pascal triangles, abbreviation: GPT. They correspond to some computations of 1D CA with the neighborhood type $(-1, 1)$. (GPT are constructed similarly as the classical Pascal's triangle, but arbitrary binary operation is used instead of addition.)

Claim 3. *There is a generalized Pascal triangle with 83 distinct elements which generates the odd primes in real time.*

We have to consider here only odd primes because a column of a GPT has common elements either with its odd rows or with its even rows. (Various modifications with two distinguished columns are possible, but such complication seems to be unnecessary here.) We can construct another (rather trivial) GPT which recognizes $\{2\}$, and obtain:

Claim 4. *There is a 83-state one-dimensional cellular automaton with neighbourhood type $(-1, 1)$ which generates the primes in real time.*

All four claims will be proved by explicit construction of 1D CA, resp. GPT. At the end some open problems are formulated.

2. Notation and technicalities

The set of nonnegative integers will be denoted by $\mathbb{N}$ and the set of all integers by $\mathbb{Z}$. The set of all nonempty words in an alphabet Σ will be denoted by Σ^+. The i-th symbol of a word w will be denoted by $w(i)$; the counting starts with 0. Length of a word w will be denoted by $|w|$. We shall use single quotation marks ' ' around letters and words. It is suitable because to obtain more transparent figures we use also '.' and '/' as letters (or states).

Definition 1. *A one-dimensional CA is an ordered quadruple*

$$C = (S, N, f, q), \tag{2.1}$$

where

> S *is a finite set of states;*
> $N = (a_1, \ldots, a_n)$, *the neighbourhood vector, is a finite sequence of pairwise distinct integers;*
> $f : S^n \to S$ *is a local transition function;*
> $q \in S$, *the quiescent state, satisfies* $f(q, \ldots, q) = q$.

A computation of one-dimensional CA C is a function $F : \mathbb{Z} \times \mathbb{N} \to S$ such that

$$F(z, t+1) = f(F(z + a_1, t), \ldots, F(z + a_n, t)) \tag{2.2}$$

for all $z \in \mathbb{Z}, t \in \mathbb{N}$.

The restrictions of F to sets $\mathbb{Z} \times \{t\}, t \in \mathbb{N}$ will be called configurations, for $t = 0$ the initial configuration. A configuration (at time t) is called finite if $F(z, t) = q$ for all but finitely $z \in \mathbb{Z}$.

We shall say that C is a one-dimensional CA of radius r if $|a_i| \leq r$ for all $i = 1, \ldots n$.

Notice that usually configurations of CA are defined at first, and then a computation of CA is defined as a sequence of configurations which satisfies some conditions ((2.2) in essential).

If a computation F of a 1D CA is given we can try to reconstruct the original 1D CA. It never can be done uniquely because, e.g., we can add unnecessary states and enlarge the neighborhood type. Even if we know S, N and q it may happen that (2.2) determines the local rule f only partially. However, if we are interested only in the computation F (and not in other computations of the CA) we can complete f arbitrarily. This will be our situation below. (Here we do not touch *algorithmic* problems like: how large piece of F must be used.)

Now we shall introduce generalized Pascal triangles (GPT). (The notion of Pascal's triangle was generalized in many ways; for example, many arithmetical ones are considered in [Bo]. However, we follow only [K1] and [K2], where a notion strongly related to 1D CA is given.) GPT are mappings of some subsets of $\mathbb{N} \times \mathbb{N}$ into finite sets. They are associated to some finite algebras and nonempty words analogously as the classical Pascal triangle can be associated to the (infinite) algebra $\langle \mathbb{N}; +, 0 \rangle$ and "the word" 1 (of length 1). More formally:

Definition 2. (i) *To every algebra $\mathcal{A} = (\mathbf{A}; *, \mathrm{o})$ such that $\mathrm{o} * \mathrm{o} = \mathrm{o}$ and to every word $w \in \mathbf{A}^+$ the partial function $G = \mathrm{GPT}(\mathcal{A}, w) : \mathbb{N} \times \mathbb{N} \to \mathbf{A}$ will be associated by the formula:*

$$
G(x, y) = \begin{cases}
\text{undefined} & \text{if } x + y < |w| - 1, \\
w(x) & \text{if } x + y = |w| - 1, \\
\mathrm{o} * G(0, y - 1) & \text{if } x = 0, \ y \geq |w|, \\
G(x - 1, 0) * \mathrm{o} & \text{if } y = 0, \ x \geq |w|, \\
G(x - 1, y) * G(x, y - 1) & \text{if } x + y \geq |w|, \ x > 0, \ y > 0.
\end{cases}
$$

(ii) *A partial function G will be called GPT if it can be represented in the form $\mathrm{GPT}(\mathcal{A}, w)$ for a finite algebra $\mathcal{A}$ and $w \in \mathbf{A}^+$.*

(iii) *The i-th row of GPT G consists of all $G(x, y)$ with $x + y = i$ (it is a word). The i-th column of a GPT G consists of all $G(x, y)$ with $x - y = i$ (it is an infinite sequence). In both cases we consider the values in the order given by x.*

We shall explain the relationship between GPT and computations of 1D CA in its the simplest case only, for the neighborhood type $(-1, 1)$ and computations starting from configurations with non-quiescent state only in the 0-th cell. (We shall need exactly this case below.)

Computations of 1D CA can be considered as functions defined on $(1 + 1)$-dimensional discrete space-time $\mathbb{Z} \times \mathbb{N}$. For displaying computations of 1D CA we use the coordinate system with the space coordinate z horizontal, and the time coordinate t vertical, oriented downwards. For displaying GPT, the system of coordinates is chosen so that the whole GPT lies in the first (i.e., "positive") quadrant of the plane. The axis x is oriented right-downward, and the axis y is oriented left-downward. (Then point (iii) of Definition 2.2 is natural.) The length unit are chosen so that $x + y$ corresponds to the time coordinate t and $x - y$ to the space coordinate z mentioned above.

Then the domain of a GPT will be considered as a "light cone", which contains the whole interesting part of the CA computation; all positions outside of it are in the quiescent state q (the constant o).

3. A 11-state 1D cellular automaton

Theorem 1. *There is a 11-state 1D CA which has a computation F with the following properties:*

(1) $F(0,0) = $ '0' and $F(x,0) = $ '.' for all $x \neq 0$.

(2) For all $x < 0$ and $t \geq 0$ it holds $F(x,t) = $ '.'.

(3) For all $t \geq 0$ we have $F(0,t) = $ '1' if t is a prime, $F(0,t) = $ '0' otherwise.

Proof. The computation F is presented in Figure 1.

The alphabet of states of the constructed 1D CA will be

$$\{ `.`, `/`, `0`, `1`, `L`, `R`, `r`, `V`, `v`, `A`, `B` \},$$

where '.' is used as the quiescent state. The alphabet is chosen to obtain the figures as transparent as possible. For example, the quiescent state '.' is chosen as similar to empty space. Capitals 'L', 'R', 'V' remind "left", "right", "vertical", respectively. The symbol '/' runs to the left, and $F(z,t) = $ '/' denotes that $z + t$ is composed. Small letters 'r', 'v' can be considered as the corresponding capitals joined with '/' (however, 'L' is considered as already containing '/'). It is also suitable to imagine that '0' is '1' with '/'.

The upper part of F contains some irregularities, and ought to be considered separately. It concerns mainly the rows up to the 15-th; they ought to be verified ad hoc.

The further part of F contains several kinds of signals which spread with distinct speeds. If "speed of light" (i. e. the maximal possible one) is ± 1 (where the sign $+$ determines direction to the right) then the main signals and their speeds are:

(1) speed -1: '/' and 'L'

(2) speed 0: 'V' and '1' (in the 0-th column; the right neighbour is non-quiescent)

(3) speed $\frac{1}{3}$: 'A'–'B'–'BL'

(4) speed 1: 'R' and '1' (when '.' are on the right).

The signals behave in various ways at their crossings; they may disappear, form another signals, etc.

The most important signals are '/' ("composed"); when such signal reaches the 0-th column it temporarily changes '1' to '0', and then disappears. (Of course, '1' is changed to '0' very often.) Most other signals are mainly used to produce '/'.

To every odd integer $d > 3$ a segment between two consecutive 'V' is associated (the segments are labelled during the computation). In this segment multiples of d (starting from $\frac{d.(d+5)}{2}$) are recognized with help of the signals 'L' and 'R'. The divisors $d = 2$ and $d = 3$ are considered separately in the second and the first cell, respectively.

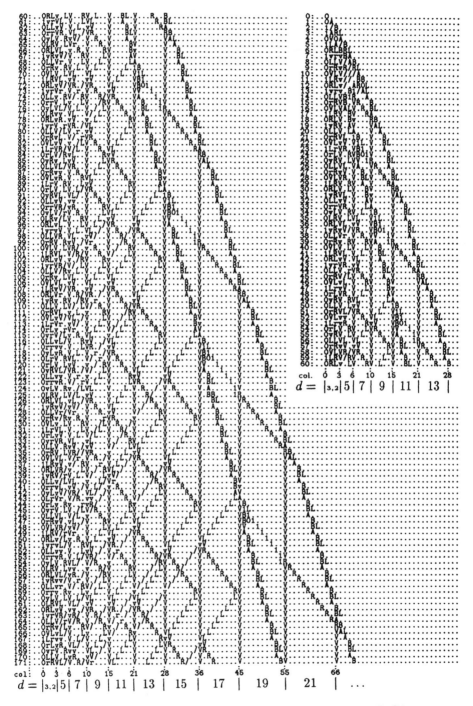

Figure 1. Real-time generation of primes by a 11-state 1D CA.

The symbols 'V' which separate the segments for divisors $d > 3$ are displaced using signals 'A'–'B'–'BL', '1' and 'R'. There are two instances of the signal 'A'–'B'–'BL'. In the crossing of the lower one with 'V' also the signal 'L' arises which starts to recognize multiples of d.

We shall not write the local transition rule f explicitly. Most of the necessary values of f can be read from Figure 1. Several further values can be obtained when temporary changes caused by crossing with '/' are considered. For example, in the row 494 we need $f(\text{'r'}, \text{'/'}, \text{'/'}) = \text{'r'}$, which can be derived from $f(\text{'R'}, \text{'.'}, \text{'.'}) = \text{'R'}$ in the row 100 (but also already in the row 15) or from $f(\text{'r'}, \text{'.'}, \text{'/'}) = \text{'r'}$ in the row 119. (The numbers of the rows with *values* of f are shown.) Together values of f for 323 triples of arguments are necessary; the other $11^3 - 323$ triples never occur (as subwords of the rows of F). Notice that these numerical observations are not necessary for the proof. □

We shall present without giving all details also a 14-state solution which is (better structured and) simpler in the sense that

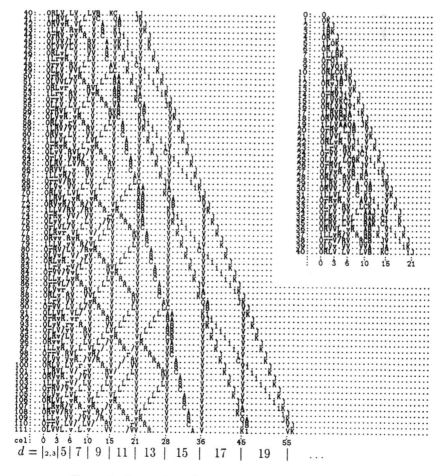

Figure 2. Generation of primes by a 14-state 1D CA.

(1) creating of 'L' signals is separated from creating of 'V' signals,
(2) the structure of signals is the simplest possible, and
(3) the signals used to form new 'V' are sent always in the same time. (In the solution above the right-hand signal is sent sooner.)

The computation is displayed in in Figure 2.

4. A 2-state 1D CA with radius 12

Theorem 2. *There is a 2-state 1D CA with radius 12 which has a computation H with the following properties:*

(1) $H(1,0) = $ '1' *and* $H(z,0) = $ '0' *for all* $z \neq 1$.
(2) *For all* $z < 0$ *and* $t \geq 0$ *it holds* $H(z,t) = $ '0' .
(3) *For all* $t \geq 0$ *we have* $H(0,t) = $ '1' *if* t *is a prime, and* $H(0,t) = $ '0' *otherwise.*

Proof. Let us denote

$$U = \{ \text{'0000000', '0010100', '0001000', '0011110', '0001100', '0110110',}$$
$$\text{'0101010', '0010110', 0011010, '0101110', '0111010'} \}.$$

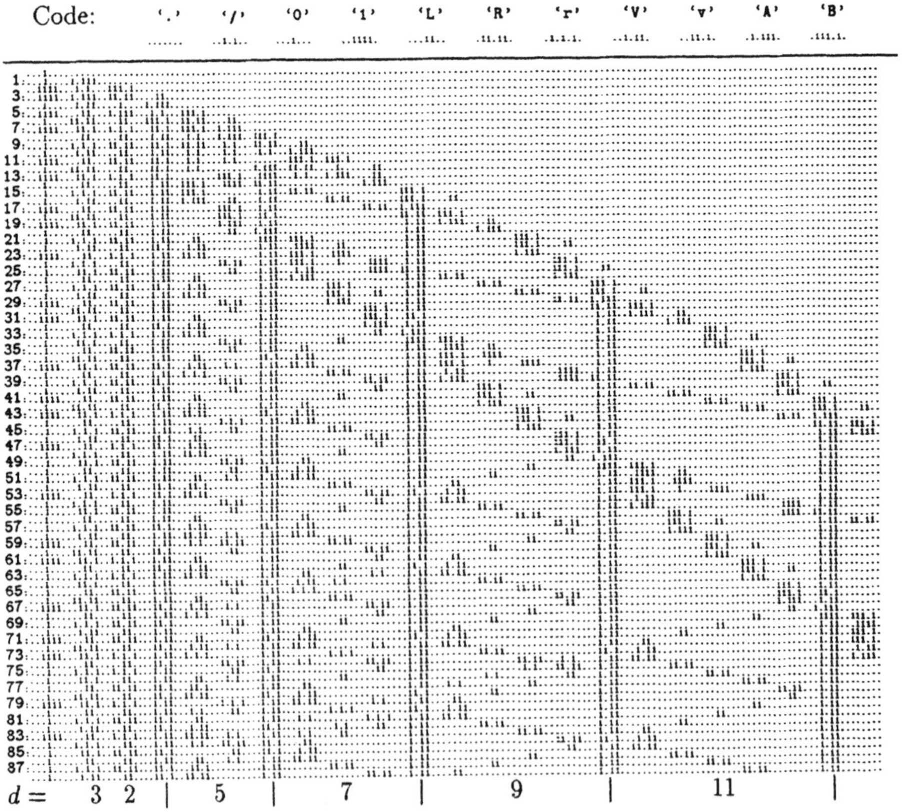

Figure 3. Generation of primes by 2-state 1D CA with $r = 12$.

The following statement can be proved:

(∗) *Let x be a subword of a word $w \in U^+$, and let some $u \in U$, $u \neq$ '0000000', be a subword of x. Then there are uniquely determined words x_1, x_2, x_3 such that $x = x_1 x_2 x_3$, $|x_1| < 7$, $x_2 \in U^+$, and $|x_3| < 7$.*

Let us associate '.' ↦ '0000000', '0' ↦ '0001000', '1' ↦ '0011110', and the other symbols from F in Figure 1 to the other elements of U (see also the upper part of Figure 3). Let us replace every value of F by the associated word. So we obtain the computation H. It is displayed in Figure 3 (where every '0' is again replaced by '.' to obtain a more transparent figure). Of course, seven columns of H are obtained from one column of F. We shall enumerate the columns of H so that the leftmost column which contains some '1' obtains the number 0.

To prove that H is a computation of a 1D CA with radius 12 we must show that the symbol $x = H(z, t+1)$ can be determined from the word w consisting of the symbols $H(i, t)$, $i = z - 12, z - 11, \ldots, z + 12$. (However, z, t are not known.) If all these symbols are '0' then $x =$ '0'. Otherwise we can apply the statement (∗), and reconstruct three consecutive elements of the t-th row of F. (They are contained in w, up to at most one '0' at the margin.) So we can determine $y \in U$ (in the $(t+1)$-th row of F) which contains x, and then x itself, too. □

5. A generalized Pascal triangle

Theorem 3. *There is an algebra $\mathcal{A} = (A; *, '.')$ such that '.' is its idempotent, $\mathrm{card}(A) = 83$ and for some $w \in A$ the function $G = \mathrm{GPT}(\mathcal{A}, w)$ has the following properties:*

(1) $G(0, 0) \neq$ '.'.
(2) *For all $x, y \in \mathbb{N}$, if $x - y < -1$ then $G(x, y) =$ '.' .*
(3) *For all $t \in \mathbb{N}$ we have $G(t, t+1) =$ '1' if $2t + 1$ is a prime, and $G(t, t+1) =$ '0' otherwise.*

Proof. The requested GPT G is displayed in Figure 4. However, the elements '.', '0', '1' from the theorem are replaced by '..', '.0', '.1'. (A "leading dot" is used similarly as sometimes leading zeros are used in writing integers.) Figure 4 is obtained simply by suitable gluing the elements of F from Figure 1 into pairs. (Technically, spaces are made between not glued elements, and ' ' are not used.) Therefore G clearly contains at most 11^2 distinct elements. However, some pairs never occur. The computer simulation found only 83 pairs in G, and it can be proved that no new pairs occur later.

The 0-th column of F is now distributed into the 0-th column of G (even rows; the first components) and the (-1)-st (odd rows; the second components). So it is guarranted that (3) holds. □

As a consequence of Theorem 3 we can obtain:

Corollary 4. *There is a 83-state one-dimensional cellular automaton with neighbourhood type $(-1, 1)$ and its computation K such that:*

(1) $K(z, 0) =$ '0' for all $z \in \mathbb{Z} \setminus \{0, 1\}$.
(2) *For all $z < 0$ and $t \geq 0$ it holds $K(z, t) =$ '0' .*
(3) *For all $t \geq 0$ we have $K(0, t) =$ '1' if t is a prime, and $K(0, t) =$ '0' otherwise.*

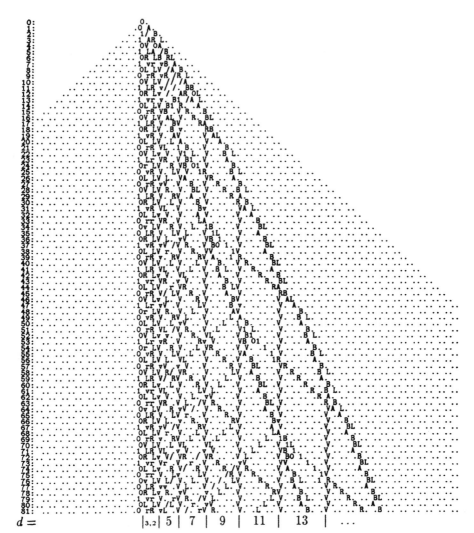

Figure 4. Generation of odd primes by a GPT.

Proof. The computation K can be composed from two GPT, a trivial one which only generates $\{2\}$, and G from Theorem 3. The elements of any of them will fill the gaps between neighbour elements of the other (like black and white fields in the chessboard); the area outside GPT will be filled by the quiescent state. □

Remarks. 1. The corollary can be reformulated (with the same constant 83) also to one-way CA, i.e. 1D CA with neighbourhood type $(0, 1)$ or $(-1, 0)$; see [CHY] (and [Dy]). However, in these cases the result cannot be found in a fixed cell; it ought to be read in the leftmost non-quiescent cell. Further, several states will correspond to composed integers.

2. The bound 83 (in Theorem 3 as well as in the Corollary) can be diminished by a modification of G.

6. Remarks and problems

It would be interesting to diminish the number of states in Theorem 1, resp. the radius in Theorem 2. One possibility in Theorem 1 would be to spare symbols in the signal with the speed $\frac{1}{3}$; maybe '11'–'11L'–'11/' would be suitable. (However, problems with crossings, resp. reflections of signals must be solved.)

Conjecture 1. *There is a one-dimensional cellular automaton (with radius 1) which generates the primes in real time and which uses only 9 states for t greater than a suitable t_0.*

It would be also interesting to spare substantially the space. Let us denote $\text{SPACE}_F(t) = \max\{|z| \mid F(t,z) \neq \text{'.'}\}$ and analogously for other computations of 1D CA (with necessary modification if the quiescent state is changed). The 1D CA from Theorem 1 works in linear space, more precisely $\text{SPACE}_F(t) = \frac{1}{3}t + O(\sqrt{t})$. Can this space be substantially diminished? (Of course, the bound 11 for the number of states need not be preserved.)

Conjecture 2. *The bound $\text{SPACE}_F(t) = O(\sqrt{t} \cdot \log t)$ can be achieved in real-time generation of the primes by 1D CA.*

Here the idea of sieve of Erasthotenes probably again can be applied; counting up to d must be performed in the space $O(\log d)$.

Problem 1. *Does the bound $\text{SPACE}_F(t) = O((\log t)^k)$ for some fixed k suffice for real-time generation of the primes by 1D CA?*

Problem 2. *Is there a Turing machine which generates the primes in real time?*

Here we ask the signals to appear in a fixed position of a fixed tape; another possibility is a special (real time) output tape. The problem can be stated for various kinds of Turing machines (and also for linear time instead of real time).

References

[Bo]   Bondarenko, B. A., *Generalized Pascal triangles and pyramids, their fractals, graphs and applications* (in Russian), "Fan", Tashkent, 1990.

[CHY]   Culik, K. II — Hurd, L. P. — Yu, S., *Computation theoretic aspects of cellular automata*, Physica D **45** (1990), 357-378.

[Dy]   Dyer, C. R., *One-way bounded cellular automata*, Information and Control **44** (1980), no. 3, 261-281.

[Fi]   Fischer, P. C., *Generation of primes by a one-dimensional real-time iterative array*, J. ACM **12** (1965), no. 3, 388-394.

[K1]   Korec, I., *Generalized Pascal triangles*, In: K. Halkowska and S. Stawski, ed.: Proceedings of the V Universal Algebra Symposium, Turawa, Poland, May 1988, World Scientific, Singapore, 1989, pp. 198-218.

[K2]   ———, *Generalized Pascal triangles, their relation to cellular automata and their elementary theories*, J. Dassow, A. Kelemenová (Eds.): Development in theoretical computer science, Proceedings of the 7th IMYCS, Smolenice 92, November 16-20, 1992, Gordon and Breach Science Publishers, 1994, pp. 59-70.

Optimal Algorithms for Complete Linkage Clustering in d Dimensions

Drago Krznaric* and Christos Levcopoulos*

Department of Computer Science, Lund University, Box 118, S-221 00 Lund, Sweden

Abstract. It is shown that the complete linkage clustering of a set of n points in $\mathbb{R}^d$, where $d \geq 1$ is a constant, can be computed in optimal $O(n \log n)$ time and linear space, under the L_1 and L_∞-metric. Furthermore, it is shown that, for every other fixed L_t-metric, it can be approximated within an arbitrarily small constant factor in $O(n \log n)$ time using linear space.

1 Introduction

Given a set S of n objects, the complete linkage (c-link) method produces a hierarchy of clusters as follows. Initially each object in S constitutes a cluster. Then, as long as there is more than one cluster, a closest pair of clusters is merged into a single cluster. The c-link distance of two clusters C and C', $\delta(C, C')$, is defined as $\max\{d(x, y) \mid x \in C \text{ and } y \in C'\}$, where $d(x, y)$ denotes the distance (dissimilarity) between objects x and y.

Hierarchical clustering algorithms are of great importance for structuring and interpreting data in domains such as biology, medicine, and image processing. Among the different methods for producing a hierarchy of clusters, the c-link clustering is one of the most well-known, and has thus been used for applications [1].

The obvious algorithm for computing a c-link hierarchy takes cubic time. Using priority queues, Day and Edelsbrunner [4] showed that it can be obtained in $O(n^2 \log n)$ time. A quadratic-time algorithm was described by Murtagh [12]. Recently, Křivánek [9] developed a quadratic-time algorithm based on the (a, b)-tree data structure. (These three algorithms work also for some other clustering methods.) A quadratic-time algorithm which uses only linear space was proposed by Defays [5]; however, it only approximates the hierarchy since its output depends on a certain insertion order [2]. Parallel algorithms for the c-link clustering have also been developed [11,8], but asymptotically the total work was still at least quadratic.

If the input does not need to explicitly contain the distance of each of the quadratic number of pairs of objects (for example, when objects correspond to points in some metric space) then faster algorithms can be found. Recently, we showed that for n points in the Euclidean plane, the c-link clustering can be

---

* {drago,christos}@dna.lth.se

computed in $O(n \log^2 n)$ time using linear space [7]. In addition, we developed an $O(n(\log n + \log^2 \epsilon))$ algorithm for constructing a c-link ϵ-approximation. By a *c-link ϵ-approximation* we mean a hierarchy that can be produced like a c-link hierarchy, except that the following holds at each merging: if P is the pair of clusters next merged and P' is a closest pair of clusters, then $\delta(P) \leq (1 + \epsilon)\delta(P')$. However, considering only the 2-dimensional case is too restrictive for real applications. Therefore, the results of this paper are especially important since, in addition to reducing the time complexity from $O(n \log^2 n)$ to $O(n \log n)$ for the L_1 and L_∞-metric, they hold also for the multi-dimensional case. (In [7] we used some data structures that are not known to work efficiently in higher dimensions. In this paper we dispense with those data structures by making a non-trivial use of an efficient data structure for dynamic closest point queries [3]. This gives, at the same time, a simplification of the methods in [7].)

We assume that the input set S corresponds to n distinct points in d-dimensional real space, where $d \geq 1$ is an integer constant, and that we are given a positive real constant ϵ and a fixed Minkowsky metric L_t. We use $d(x, y)$ to denote the distance between points $x = (x_1, x_2, \ldots, x_d)$ and $y = (y_1, y_2, \ldots, y_d)$, that is,

$$d(x, y) = \left(\sum_{i=1}^{d} |x_i - y_i|^t \right)^{1/t}.$$

(Note that the L_∞-distance is given by $\max\{|x_i - y_i| \mid 1 \leq i \leq d\}$.) Under these assumptions, we show that a c-link ϵ-approximation of S can be obtained in $O(n \log n)$ time using $O(n)$ space. Moreover, for $t = 1$ and $t = \infty$, we show that the complete linkage clustering of S can be computed in optimal $O(n \log n)$ time and $O(n)$ space.

The rest of the paper is organized as follows. In the next section we describe how c-link distances can be approximated in order to compute a c-link ϵ-approximation. In Section 3 we give a lemma that we use in order to keep track of relevant pairs of clusters during the merging process. Then, in Section 4 we state the algorithm, and in Section 5 we make a run time analysis. In Section 6 we note that the algorithm fits in the algebraic computation tree model, and we show that it can compute c-link distances exactly under the L_1 and L_∞-metric, thus concluding that it is optimal up to a constant factor in these two cases. Finally, in Section 7, we make some remarks concerning other hierarchical clustering methods.

2 Approximating C-Link Distances

For each cluster we keep track of k points of the cluster where k is a constant $\geq 2d$. These so-called *k-extremes* are used in order to approximate c-link distances of pairs of clusters. The k-extremes of a cluster C are determined by a set $\mathcal{V}$ of k d-dimensional vectors, each vector having its tail at the origin ($\mathcal{V}$ will be defined in a moment). If $v_1, v_2, \ldots, v_k$ are the vectors in $\mathcal{V}$, then the k-extremes of C consist of points $p_1, p_2, \ldots, p_k$ such that each p_i is an extreme point of C

in the direction given by vector v_i; that is, there is no point p in C such that $(p, v_i) > (p_i, v_i)$, where $(\cdot, \cdot)$ stands here for the inner product of two vectors. (Note that the k-extremes are not necessarily distinct and that they are not uniquely defined.) To define the set $\mathcal{V}$, let w be an integer ≥ 1 and let

$$W = \{i/w \mid i \text{ is an integer and } -w < i < w\}.$$

Next, define V_i^+ as the set of all possible vectors $(x_1, x_2, \ldots, x_d)$ such that $x_i = 1$ and $x_j \in W$ for every $j \neq i$. The set V_i^- is defined similarly except that it consists of vectors having -1 in position i. Then,

$$\mathcal{V} = \bigcup_{i=1}^{d} (V_i^+ \cup V_i^-).$$

So there are $k = 2d(2w-1)^{d-1}$ vectors in $\mathcal{V}$ and they are almost evenly spread around the origin.

To give a more intuitive picture of the k-extremes, let us examine the 3-dimensional case. Consider a cube centered at the origin such that its edges are parallel with the coordinate axes and have length 2. Partition each side of this cube into $(2w)^2$ subsquares. Clearly, the vectors in $\mathcal{V}$ correspond to those vertices of these subsquares that do not lie on an edge of the cube. Now, consider k planes that come from infinity and move toward a cluster, where each plane is orthogonal to a distinct vector in $\mathcal{V}$ and comes from the direction pointed by that vector. For each of these planes, select a point of the cluster such that the plane hits this point first (if more than one points are hit simultaneously, select one arbitrarily). Then, the union of the selected points, considering all k planes, constitutes the k-extremes of the cluster.

We define the k-*distance* of two clusters C and C', denoted by $\delta_k(C, C')$, as the distance of two k-extremes, one from C and the other from C', that are farthest apart (according to the L_t-metric). It is not hard to realize that for any $\epsilon > 0$ there exists a constant k, depending on ϵ and d, such that $\delta_k(P) \leq (1 + \epsilon)\delta(P)$ for any pair P of clusters. In the remainder of this paper, let k be such a constant. (In Section 6 it is shown that for the L_1 or L_∞-metric we can define the k-extremes so that $\delta_k(P) = \delta(P)$ for any pair P of clusters.)

When a new cluster C is created by merging a pair P of clusters, we compute the k-extremes of C by selecting points among the k-extremes of the clusters of P. More precisely, for each of the k directions, we compare the k-extreme for that direction of one of the clusters of P with the k-extreme for that direction of the other cluster of P, and select one which is most extreme in that direction.

3   Representing Clusters

For each cluster we have a what we call *leader*, which is simply a k-extreme of the cluster (we could actually choose any point of the cluster). All leaders are stored in a data structure for dynamic closest pair queries. Henceforth, that data

structure will be referred to as the *DCP-structure*. The DCP-structure supports
the following three operations: (i) find a closest pair of points, (ii) delete a
point, and (iii) insert a new point. We assume that the DCP-structure uses
linear space and makes it possible to carry out each of these three operations in
logarithmic worst-case time (although logarithmic amortized time per operation
would suffice), and that it fits in the algebraic computation tree model. These
requirements can be met by using the data structure of Bespamyatnikh [3].

Initially the points of S are inserted into the DCP-structure. Then, each time
a new cluster C is created by merging a pair P of clusters, the leader for one of
the clusters of P is deleted from the DCP-structure, and the leader for the other
cluster of P becomes the leader for C.

As it is described in the next section, our algorithm works in phases where
each phase uses the DCP-structure to find relevant pairs of clusters. The following lemma states an upper bound for how fast these pairs can be found.

Lemma 1. *Let S be a set of n points in $\mathbb{R}^d$, let l be a positive real, and let κ
be an integer such that for each point of S there are at most $\kappa - 1$ other points
in S within distance less than l from it (distances are according to some fixed
but arbitrary L_t-metric, and $d \geq 1$ is an integer constant). Then, if the points
of S are stored in the DCP-structure, the set $\{(x, y) \mid d(x, y) < l \text{ and } x, y \in S\}$
can be found in $O(\kappa^d \eta \log n)$ time, where η denotes the cardinality of the set
$\{(x, y) \mid d(x, y) < 3l \text{ and } x, y \in S\}$.*

Proof. For each point x we have a set S_x where $S_x = \{x\}$ initially. First we
repeat the following four steps until the condition at Step 2 holds:

1. Find a closest pair (x, y) of points in the DCP-structure.
2. If $d(x, y) > l/(|S_x| + |S_y|)$ then halt.
3. Let S_x equal $S_x \cup S_y$.
4. Delete y from the DCP-structure.

After this procedure, let l' be the distance of a closest pair of points in the
DCP-structure, and consider an arbitrary point x in the DCP-structure. If p_1
and p_m are any two points in S_x, then there must exist points $p_1, p_2, \ldots, p_m$
in S_x such that $d(p_i, p_{i+1}) \leq l/|S_x|$ for all $1 \leq i < m$. Thus we can conclude
that $d(p_1, p_m) < l$. This means that there are at most κ points in S_x, and that
$l' > l/(2\kappa)$. Clearly, the total time for the procedure is $O(\eta \log n)$.

Next, as long as the distance of a closest pair of points in the DCP-structure
is $< 3l$, we repeat the following five steps:

(i) find a closest pair (x, y) of points in the DCP-structure,
(ii) find every point z in the DCP-structure such that $d(x, z) < 3l$,
(iii) for each found z output every (p_x, p_z) such that $p_x \in S_x, p_z \in S_z$ and
  $d(p_x, p_z) < l$,
(iv) output each pair of points in S_x, and
(v) delete x from the DCP-structure.

It is not hard to see that this procedure indeed outputs the set $\{(x,y) \mid d(x,y) < l \text{ and } x,y \in S\}$. So it remains only to show that it takes $O(\kappa^d \eta \log n)$ time. To do this, we need the following observation.

Observation 2. *One iteration of Step (ii) takes $O(\kappa^d \log n)$ time.*

Proof. Let h be a d-dimensional cube centered at x such that its edges are parallel with the coordinate axes and have length $6l$. Clearly, every point z that we are looking for lies in h. To find the points in h, partition h into $(12ld/l')^d$ subcubes with edges of length $l'/(2d)$. Note that the L_t-diameter of a subcube is at most equal to its L_1-diameter, which is equal to $l'/2$.

Now, let h_i be one of the subcubes, let c_i be the point at which h_i is centered, and suppose that there is a point z of the DCP-structure that lies in h_i. Then, if we insert c_i into the DCP-structure, (c_i, z) becomes the closest pair of points in the DCP-structure. This is so because $d(c_i, z) < l'/2$ whereas $d(z, z') \geq l'$ for every $z' \neq c_i$ in the DCP-structure. Hence, we can find all points in h by using one insertion, query, and deletion per subcube, which totally takes time $O((24d\kappa)^d \log n)$ (recall that $l' > l/(2\kappa)$ so the number of subcubes is $< (24d\kappa)^d$). ☐

In Step (iii) we can compute the distance from each point in S_x to each point in S_z for every z found at Step (ii). But, if we in this way compute the distance from a point in S_x to a point in S_z, then that point in S_z must be within distance $< 4l$ from x. By partitioning the cube with edges of length $8l$ and centered at x into $(8d)^d$ subcubes, each subcube having L_1-diameter l, we realize that for each point in S_x we compute at most $\kappa(8d)^d$ distances. Consequently, one iteration of Step (iii) takes $O(\kappa^2)$ time. We can now conclude that one iteration of Steps (i) through (v) takes $O(\kappa^d \log n)$ time, and since we do at most η iterations, the total time used by the algorithm is $O(\kappa^d \eta \log n)$. ☐

4 The Algorithm

We compute a c-link ϵ-approximation in a sequence $p_1, p_2, \ldots$ of phases. The objective of a phase p_i is to merge every pair P of clusters such that $\delta_k(P) \in [l_i, 2l_i)$, where the parameter l_i is defined as follows. At phase p_1, l_1 equals the distance of a closest pair of points in S. For $i > 1$, let l be the distance of a closest pair of leaders in the DCP-structure immediately after phase p_{i-1}. So $\delta_k(P) \geq l$ for every pair P of clusters. Therefore, if $l > 2l_{i-1}$ then we set l_i to l, otherwise we set l_i to $2l_{i-1}$.

Using a priority queue and Lemma 1, a phase p_i is rather straightforward to compute. First, we do the following two operations:

(a) Find each pair of clusters such that the distance of its corresponding pair of leaders is $< 2l_i$.
(b) Insert each such pair of clusters into a priority queue, initially empty, according to the k-distance of the pair.

Then, as long as the k-distance of a closest (according to the k-distance) pair of clusters in the priority queue is $< 2l_i$, repeat the following three steps:

1. Remove a closest pair P of clusters from the priority queue.
2. Merge P into a single cluster C_P.
3. For each pair (C, C') in the priority queue such that C' is a cluster of P do the following:
   - Remove (C, C') from the priority queue.
   - If $\delta_k(C, C_P) < 2l_i$, then insert (C, C_P) into the priority queue.

5 Run Time Analysis

As our algorithm works, at phases $p_1, p_2, \ldots, p_{i-1}$ we do not merge a pair P of clusters for which $\delta_k(P) \geq 2l_{i-1}$. Hence, at the beginning of phase p_i, there must be at least one pair of clusters of k-distance $\leq 3l_i$. This means that if no merging occurs at phase p_i, then at least one merging must occur at phase p_{i+1}. Thus the total number of phases is no more than $2n - 2$ (the total number of mergings is $n - 1$).

To proceed with the analysis we need a couple of notations. Let $\mathcal{P}_i$ be the set consisting of every pair of clusters such that the pair exists at some time during phase p_i and the distance of its corresponding pair of leaders is $< 6l_i$. Further, let η_i denote the cardinality of $\mathcal{P}_i$. Finally, let κ be the smallest integer such that the following holds at any time during any phase p_i: for each leader there are at most $\kappa - 1$ other leaders within distance $< 6l_i$ from it. (In Lemma 3 below we show that κ is never greater than some constant.)

By Lemma 1, Operation (a) takes $O(\kappa^d \eta_i \log n)$ time. Clearly, (b) takes $O(\eta_i \log \eta_i)$ time. Observe that only a pair in $\mathcal{P}_i$ may be considered at Step 3, and that each pair in $\mathcal{P}_i$ is considered at most once at Step 3. So the total time for Step 3 is $O(\eta_i \log \eta_i)$. Next, observe that each time we iterate Steps 1 through 3, a pair of clusters is merged into a single cluster. Consequently, the total time for Steps 1 and 2 is $O(m \log \eta_i)$, where m denotes the total number of mergings performed during phase p_i. We can thus conclude that the total time used by phase p_i is $O(\kappa^d \eta_i \log n + \eta_i \log \eta_i + m \log \eta_i)$.

Now, consider a pair P in $\mathcal{P}_i$. It holds that $\delta_k(P) < 10l_i$. Therefore, one of the clusters of P must participate in a merging at one of the phases p_i, p_{i+1} or p_{i+2}. Let us associate each pair in $\mathcal{P}_i$ with the first merging in which one of the clusters of the pair participate. If we do this for each set $\mathcal{P}_i$, we associate at most $3(\kappa - 1)$ pairs to each merging. Hence, the total sum of the η_i's is $O(\kappa n)$, which implies that the total time used by the algorithm is $O(\kappa^{d+1} n \log(\kappa n))$.

From the following lemma we can conclude that our algorithm actually runs in $O(n \log n)$ time. (A special case of this lemma handling only the Euclidean plane was stated but not proved in [7].)

Lemma 3. *There exists a constant (depending on d and k) greater than κ.*

Proof. Throughout the proof, by a *cube* we mean a d-dimensional cube with edges parallel to the coordinate axes. The proof is by induction on i. The induction hypothesis is as follows: at the beginning of phase p_i (before any clusters are merged) any cube having edges of length l_i contains at most $\lambda(d+1)^d$ leaders, where λ is a constant ≥ 1 that we specify later. Recall, at the beginning of phase p_1, each leader corresponds to a single point in S, and l_1 equals the distance of a closest pair of points in S. It is not hard to show that any cube having edges of length l_1 contains at most $(d+1)^d$ points of S. Thus the statement is true for $i = 1$. Let h be an arbitrary cube with edges of length l_{i+1}. To complete the proof it suffices to show that h contains at most $\lambda(d+1)^d$ leaders at the beginning of phase p_{i+1}.

From the definition of the parameter l_i it follows that if $l_{i+1} \neq 2l_i$ then, at the beginning of phase p_{i+1}, the distance of a closest pair of leaders equals l_{i+1}. So, similarly as for $i = 1$, h contains at most $(d+1)^d$ leaders in this case. Therefore, in the continuation we can (and will) assume that $l_{i+1} = 2l_i$.

From the set of clusters that exist at the beginning of phase p_i we extract two subsets H and H' as follows. The set H consists of every cluster that has a k-extreme in h. The set H' consists of every cluster that, at the end of phase p_i, is included in a cluster containing a cluster of H. Note that $H \subseteq H'$ and that the clusters in H' are merged only with each other during phase p_i. Let n' be the number of clusters in H'. We aim to show that sufficiently many mergings of clusters in H' occur during phase p_i. But first we need some more definitions.

Let h' be the cube having edges of length $6l_i$ and being concentric with h. It is not hard to see that each cluster in H' has all of its k-extremes in h'. Define the *triggers* as the r^d cubes that partition h' into subcubes with edges of length $6l_i/r$, where $r = 6d(1 + (d+1)\log_2 6)$. Now, consider a cluster with all its k-extremes in h'. For each k-extreme of the cluster we select the trigger in which it is contained, thus selecting k triggers $t_1, t_2, \ldots, t_k$. We say that the cluster is of *type* τ, where τ is the unordered k-tuple $(t_1, t_2, \ldots, t_k)$. We are now in position to set the constant λ, namely, λ equals the maximum number of distinct types of clusters. By traditional combinatorics, $\lambda = \binom{r^d+k-1}{k}$.

Next, define the k-diameter of a cluster as the distance of its two k-extremes that are farthest apart. It is easily seen that if C and C' are any two clusters produced by our algorithm, then the k-diameter of $C \cup C'$ equals $\delta_k(C, C')$.

Observation 4. *Let C and C' be two clusters with all their k-extremes in h' such that each of them is of type τ and has k-diameter less than l. Then the k-diameter of $C \cup C'$ is less than $l + 6l_i d/r$.*

Proof. Since C and C' are of the same type, there are k triggers (not necessarily distinct) such that each of them contains two k-extremes, one from C and the other from C'. Let p and p' be two k-extremes of $C \cup C'$ such that the distance between them is maximized. So the k-diameter of $C \cup C'$ equals $d(p, p')$. We can assume that p is a k-extreme of C and p' is a k-extreme of C', because $d(p, p')$ would otherwise be $< l$. Let t_p be the trigger containing p. As mentioned above, t_p also contains a k-extreme q' of C', and we know that $d(p', q') < l$. But $d(p, q')$

is at most the L_1-diameter of t_p, which is equal to $6l_id/r$. Hence, by triangle inequality, $d(p, p') < l + 6l_id/r$. □

Suppose that there is a subset T of H' consisting only of (at least two) clusters that are of the same type. Let C and C' be two clusters in T. First we observe that both C and C' have k-diameter $< l_i$, because they were created at some phase before p_i. By Observation 4, the k-diameter of $C \cup C'$ is $< l_i + 6l_id/r$. Consequently, during phase p_i at least one of C and C', let us say C, will be merged with a cluster C'' such that $C \cup C''$ has k-diameter $< l_i + 6l_id/r$ (C'' and C' might be the same cluster). Thus we realize that each cluster in T except at most one will participate in a merging during phase p_i, in such a way that the new clusters resulting from these mergings have k-diameters $< l_i + 6l_id/r$.

In the remainder we only consider the clusters in H' and those clusters that are created during phase p_i by merging two or more clusters of H'. As indicated in the previous paragraph, at least $n' - \lambda$ clusters will participate in a merging during phase p_i, and the new clusters resulting from these mergings have k-diameters $< l_i + 6l_id/r$. The number of clusters that remain after these mergings is at most $(n' - \lambda)/2 + \lambda$. We can repeat the scenario for these clusters. After having done that we are left with at most $(n' - \lambda)/2^2 + \lambda$ clusters, each cluster having k-diameter $< l_i + 2 \cdot 6l_id/r$. Indeed, we can repeat the scenario as long as we do not merge two clusters whose union has k-diameter $\geq 2l_i$ (we may assume that there are after each repetition sufficiently many clusters left for the next repetition to work).

Now, if we repeat the scenario j times, we are left with at most

$$\frac{n' - \lambda}{2^j} + \lambda < \frac{n'}{2^j} + \lambda$$

clusters, each cluster having k-diameter

$$< l_i + j \cdot 6l_id/r,$$

which is $< 2l_i$ for $j = \lfloor 1 + (d+1)\log_2 6 \rfloor > (d+1)\log_2 6$. But since h' can be partitioned into 6^d subcubes with edges of length l_i, we have by our induction hypothesis that $n' \leq 6^d \lambda (d+1)^d$. Hence, at the end of phase p_i, the number of clusters that have a k-extreme in h is at most

$$\frac{6^d \lambda (d+1)^d}{2^{(d+1)\log_2 6}} + \lambda = \frac{\lambda (d+1)^d}{6} + \lambda < \lambda (d+1)^d,$$

which completes the proof of Lemma 3. □

We can summarize this section by the following theorem.

Theorem 5. *Let $d \geq 1$ be an integer constant, and $\epsilon > 0$ a real constant. Then, under any fixed L_t-metric, a c-link ϵ-approximation of any set of n points in $\mathbb{R}^d$ can be computed in $O(n \log n)$ time using $O(n)$ space.*

6 Complete Linkage Clustering in the L_1 and L_∞-Metric

Under the L_∞-metric it is possible to define the k-extremes so that $\delta_k(P) = \delta(P)$ for any pair P of clusters. To see this, consider a cluster and a point x. All points within L_∞-distance $\leq l$ from x, for some $l > 0$, comprise a d-dimensional cube centered at x such that its edges are parallel to the coordinate axes and have length $2l$. Hence, a point of the cluster which is farthest away from x must be an extreme point of the cluster in one of the $2d$ coordinate directions. Therefore, under the L_∞-metric, for every cluster we need only to keep track of an extreme point of the cluster in each of the $2d$ coordinate directions. (This corresponds to the definition of k-extremes given in Section 2 if the set $\mathcal{V}$ of vectors is defined for $w = 1$.)

A similar observation can be made for the L_1-metric. In this case, all points within L_1-distance $\leq l$ from x comprise a d-dimensional cross-polytope where each edge has length $\sqrt{2}\,l$, that is, a regular polytope with d diagonals (a straight-line segment connecting two vertices of the polytope such that its interior does not intersect the boundary of the polytope) each diagonal being parallel to one of the coordinate axes. (In $\mathbb{R}^3$ it is a octahedron with diagonals parallel to the coordinate axes.) It is not hard to realize that this polytope is bounded by 2^d planes. Therefore, under the L_1-metric, for every cluster we need only to keep track of an extreme point of the cluster in each of these 2^d directions. The set $\mathcal{V}$ of vectors from Section 2 has to be defined in a slightly different way in order to correspond to these 2^d directions. Namely, in this case we define $\mathcal{V}$ so that it consists of all possible vectors $(x_1, x_2, \ldots, x_d)$ such that each $x_i \in \{1, -1\}$. So the vectors in $\mathcal{V}$ correspond to the binary representation of $0, 1, \ldots, d$ where each 0 is replaced by -1.

It is not hard to see that our algorithm can be implemented using only operations allowed by the algebraic computation tree model. Moreover, by reduction to the static closest pair problem it is easy to conclude that $\Omega(n \log n)$ is a lower bound for the c-link clustering, even if we restrict ourselves to 1-dimensional space, in the algebraic computation tree model (see, e.g., [13]). Thus our algorithm is optimal in that model of computation.

This section is summarized in the following theorem.

Theorem 6. *Let S be a set of n points in $\mathbb{R}^d$ where $d \geq 1$ is an integer constant. Then, under the L_1 and L_∞-metric, the complete linkage clustering of S can be computed in $O(n \log n)$ time using $O(n)$ space, which is optimal in the algebraic computation tree model.*

7 Final Remarks

The complete linkage clustering belongs to a family of clustering methods known as SAHN methods (sequential, agglomerative, hierarchical, and nonoverlapping). Two other methods in this family are the centroid [4,6,14] and the median [4,6,10] method. Given a set of n points in $\mathbb{R}^d$, these two methods work by repeatedly replacing a closest pair of points with a single (centroid respectively median) point. Hence, using the DCP-structure of Bespamyatnikh [3], these two clustering methods can trivially be computed in $O(n \log n)$ time, under any fixed L_t-metric.

References

1. P. Arabie, L. J. Hubert, and G. De Soete, editors. *Clustering and Classification*. World Scientific, 1996.
2. F. Aurenhammer and R. Klein. Voronoi diagrams. Technical Report 198-5, Informatik, FernUniversität, Hagen, Germany, 1996.
3. S. N. Bespamyatnikh. An optimal algorithm for closest pair maintenance. In *Proceedings of the 11th Annual ACM Symposium on Computational Geometry*, pages 152–161, 1995.
4. W. H. E. Day and H. Edelsbrunner. Efficient algorithms for agglomerative hierarchical clustering methods. *Journal of Classification*, 1(1):7–24, 1984.
5. D. Defays. An efficient algorithm for a complete link method. *Computer Journal*, 20:364–366, 1977.
6. J. C. Gower. A comparison of some methods of cluster analysis. *Biometrics*, 23:623–638, 1967.
7. D. Krznaric and C. Levcopoulos. The first subquadratic algorithm for complete linkage clustering. In *Proceedings of ISAAC '95*, LNCS 1004, pages 392–401. Springer, 1995.
8. T. Kurita. An efficient agglomerative clustering algorithm using a heap. *Pattern Recognition*, 24(3):205–209, 1991.
9. Křivánek. Connected admissible hierarchical clustering. Paper presented at the DIANA III conference, Bechyne, Czechoslovakia, June 1990.
10. G. N. Lance and W. T. Williams. A generalised sorting strategy for computer classifications. *Nature*, 212:218, 1966.
11. X. Li. Parallel algorithms for hierarchical clustering and cluster validity. *IEEE Transactions on Pattern Analysis and Machine Intelligence*, 12(11):1088–1092, 1990.
12. F. Murtagh. Complexities of hierarchic clustering algorithms: State of the art. *Computational Statistics Quarterly*, 1:101–113, 1984.
13. F. P. Preparata and M. I. Shamos. *Computational Geometry: An Introduction*. Springer-Verlag, New York, 1985.
14. R. R. Sokal and C. D. Michener. A statistical method for evaluating systematic relationships. *University of Kansas Science Bulletin*, 38:1409–1438, 1958.

Invertible Linear Cellular Automata over $\mathbf{Z}_m$: Algorithmic and Dynamical Aspects

Giovanni Manzini[1,2], Luciano Margara[3]

[1] Dipartimento di Scienze e Tecnologie Avanzate, Università di Torino, Via Cavour 84, 15100 Alessandria, Italy.

[2] Istituto di Matematica Computazionale, Via S. Maria, 46, 56126 Pisa, Italy.

[3] Dipartimento di Scienze dell'Informazione, Università di Bologna, Mura Anteo Zamboni 7, 40127 Bologna, Italy.

Abstract. We give an explicit and efficiently computable formula for the inverse of D-dimensional linear cellular automata over $\mathbf{Z}_m$ ($D \geq 1$, $m \geq 2$). We use this formula to get an easy-to-check necessary and sufficient condition for an invertible 1-dimensional linear CA to be expansive.

1 Introduction

Cellular Automata (CA) are dynamical systems consisting of a D-dimensional lattice of variables which can take a finite number of discrete values. The *global state* of the CA, specified by the values of all the variables at a given time, evolves in synchronous discrete time steps according to a given *local rule* which acts on the value of each single variable.

A CA is *invertible* if its global transition map, obtained by applying the local rule to all sites of the lattice, is invertible, i.e., if every global state which, by definition, has exactly one successor, also has one predecessor. Invertibility was explicitly addressed for the first time in two seminal papers [1, 10]. In particular, in [10] Richardson proved that if a CA is invertible, then its inverse is still a CA. After that, theoretical works on invertible CA proliferated (for a review on invertible CA see [14]). In particular, many efforts were made for (i) understanding which properties of the local rule of a given CA make it invertible and (ii) designing algorithms for computing the inverse of a CA. Unfortunately, for general CA these problems are known to be very hard. For example, in [7, 8] the authors prove that is algorithmically unsolvable the problem of deciding if a given 2-dimensional CA is invertible.

In this paper we give an explicit formula for the computation of the inverse of a *linear* CA over $\mathbf{Z}_m$, that is, of a CA based on a linear local rule. Linear CA have received considerable attention in the last few years [2, 4, 5, 6, 11, 12], and in many cases it has been possible to characterize set theoretic and topological properties of the global map in terms of the coefficients of the linear local rule. The problem of determining whether a linear CA over $\mathbf{Z}_m$ is invertible has been solved by Ito *et al.* in [6]. They give a necessary and sufficient condition for invertibility based on the coefficients of the local map and the prime factors of m. The problem of computing the inverse of a linear CA over $\mathbf{Z}_m$ has been addressed in [12] where the author gives a formula for the inversion of *group structured* linear CA. Although not every linear CA is group structured, the author proves that for any linear CA there exists an integer n such that the n-th iteration of the CA is group structured.

The formula we give in this paper can be applied to any linear D-dimensional CA over $\mathbf{Z}_m$ ($D \geq 1, m \geq 2$). In addition, our formula provides valuable information on the structure of the inverse CA. We use this information to study one of the most important topological property of invertible maps, namely *expansivity*.

The main results of this paper can be summarized as follows.

(a) We provide an explicit formula for inverting D-dimensional linear CA over $\mathbf{Z}_m$, for every $m \geq 2$ and $D \geq 1$ (Theorem 2).

(b) We present an alternative procedure for the computation of the inverse which has a much smaller computational cost since it avoids integer factorization (Theorem 6).

(c) We give an easy-to-check necessary and sufficient condition for an invertible linear CA over $\mathbf{Z}_m$ to be expansive (Theorem 8).

2 Basic definitions

Let $\mathbf{Z}_m = \{0, 1, \ldots, m-1\}$ be a finite alphabet of cardinality $m \geq 2$. We consider the *space of configurations*

$$\mathbf{Z}_m^{\mathbf{Z}^D} = \left\{ c \mid c : \mathbf{Z}^D \to \mathbf{Z}_m \right\}.$$

Each element of $\mathbf{Z}_m^{\mathbf{Z}^D}$ can be visualized as an infinite D-dimensional lattice in which each cell contains an element of $\mathbf{Z}_m$. Let $s \geq 1$ and $f : \mathbf{Z}_m^s \to \mathbf{Z}_m$ be any map. A *neighborhood frame* of size s is an ordered set of D-dimensional vectors $\mathbf{u}_1, \mathbf{u}_2, \ldots, \mathbf{u}_s$. A D-dimensional CA based on a *local rule* f is a pair $(\mathbf{Z}_m^{\mathbf{Z}^D}, F)$, where $F : \mathbf{Z}_m^{\mathbf{Z}^D} \to \mathbf{Z}_m^{\mathbf{Z}^D}$, is the *global transition map* defined as follows. For every $c \in \mathbf{Z}_m^{\mathbf{Z}^D}$ the configuration $F(c)$ is such that for every $\mathbf{v} \in \mathbf{Z}^D$

$$[F(c)](\mathbf{v}) = f\left(c(\mathbf{v} + \mathbf{u}_1), \ldots, c(\mathbf{v} + \mathbf{u}_s)\right), \tag{1}$$

In other words, the content of cell $\mathbf{v}$ in the configuration $F(c)$ is a function of the content of cells $\mathbf{v} + \mathbf{u}_1, \ldots, \mathbf{v} + \mathbf{u}_s$ in the configuration c. Note that the local rule f and the neighborhood frame completely determine F. In this paper we consider *linear* CA, that is, CA which have a local rule of the form $f(x_1, \ldots, x_s) = \sum_{i=1}^s \lambda_i x_i \mod m$. Note that for a linear D-dimensional CA, equation (1) becomes

$$[F(c)](\mathbf{v}) = \sum_{i=1}^s \lambda_i c(\mathbf{v} + \mathbf{u}_i) \mod m. \tag{2}$$

For linear 1-dimensional CA we use a simplified notation. A local rule of *radius r* is written as $f(x_{-r}, \ldots, x_r) = \sum_{i=-r}^{r} a_i x_i \bmod m$, where at least one between a_{-r} and a_r is nonzero. Using this notation, the global map F becomes

$$[F(c)](i) = \sum_{j=-r}^{r} a_j c(i+j) \mod m, \qquad c \in \mathbf{Z}_m^{\mathbf{Z}}, \ i \in \mathbf{Z}.$$

A convenient notation for the study of linear CA, is the *Formal Power Series* (fps) representation of the configuration space $\mathbf{Z}_m^{\mathbf{Z}^D}$ (see [6, Sec. 3] for details). For example, for $D = 1$, to each configuration $c \in \mathbf{Z}_m^{\mathbf{Z}}$ we associate the fps

$$P_c(X) = \sum_{i \in \mathbf{Z}} c(i) X^i.$$

The advantage of this representation is that the computation of a linear map is equivalent to power series multiplication. Let $F: \mathbf{Z}_m^{\mathbf{Z}} \to \mathbf{Z}_m^{\mathbf{Z}}$ be a linear map with local rule $f(x_{-r}, \ldots, x_r) = \sum_{i=-r}^{r} a_i x_i$. We associate to F the finite fps $A_f(X) = \sum_{i=-r}^{r} a_i X^{-i}$. Then, for any $c \in \mathbf{Z}_m^{\mathbf{Z}}$ we have

$$P_{F(c)}(X) = P_c(X) A_f(X) \mod m.$$

Note that each coefficient of $P_{F(c)}(X)$ is well defined since $A_f(X)$ has only finitely many nonzero coefficients. Note also that the finite fps associated to F^n is $A_f^n(X)$. More in general, we associate to each configuration $c \in \mathbf{Z}_m^{\mathbf{Z}^D}$ the formal power series

$$P_c(X_1, \ldots, X_D) = \sum_{i_1, \ldots, i_D \in \mathbf{Z}} c(i_1, \ldots, i_D) X_1^{i_1} \cdots X_D^{i_D}.$$

The computation of a linear map F over $\mathbf{Z}_m^{\mathbf{Z}^D}$ is equivalent to the multiplication by a fps $A(X_1, \ldots, X_D)$ which can be easily obtained by the local rule f and the neighborhood frame $\mathbf{u}_1, \ldots, \mathbf{u}_s$. The finite fps associated to the map F defined by (2) is

$$A(X_1, \ldots, X_D) = \sum_{i=1}^{s} \lambda_i X_1^{-\mathbf{u}_i(1)} \cdots X_D^{-\mathbf{u}_i(D)}$$

where $\mathbf{u}_i(j)$ denotes the j-th component of vector $\mathbf{u}_i$.

In this paper we consider the problem of computing the inverse of a linear CA over $\mathbf{Z}_m$. Our starting point is a result by Ito *et al.* [6] which characterizes the invertible D-dimensional linear CA in terms of the coefficients $\lambda_1, \ldots, \lambda_s$ of the local rule. Their result establishes that the global map F is invertible if and only if, for each prime factor p of m there exists a unique coefficient λ_j such that $p \nmid \lambda_j$. In other words, if $\mathcal{P}$ denotes the set of prime factors of m, F is invertible if and only if

$$\text{for each } p \in \mathcal{P} \quad \exists j \colon p \nmid \lambda_j \text{ and } p \mid \lambda_i \text{ for } i \neq j. \tag{3}$$

Throughout the paper, $F(c)$ will denote the result of the application of the map F to the configuration c, and $c(\mathbf{v})$ will denote the value assumed by c in $\mathbf{v}$. For $n \geq 0$, we recursively define $F^n(c)$ by $F^n(c) = F(F^{n-1}(c))$, where $F^0(c) = c$. If F is invertible, F^n with $n < 0$ will denote $(F^{-1})^{|n|}$.

3 An inversion formula for linear CA

In this section we consider the problem of finding the inverse of a linear CA over $\mathbf{Z}_m$. By the discussion of the previous section we know that this is equivalent to the problem of finding the inverse of a finite fps. In other words, given a finite fps $F(X_1, \ldots, X_D)$ we want to find a finite fps $G(X_1, \ldots, X_D)$ such that $F(X_1, \ldots, X_D)G(X_1, \ldots, X_D) \equiv 1 \pmod{m}$.

In the following $F(X_1, \ldots, X_D) = \sum_{i=1}^{s} \lambda_i X_1^{i_1} \cdots X_D^{i_D}$ will denote a generic finite fps, where for $i = 1, \ldots, s$, the vector $\langle i_1, \ldots, i_D \rangle$ belongs to $\mathbf{Z}^D$. We assume that for $i \neq j$ we have $\langle i_1, \ldots, i_D \rangle \neq \langle j_1, \ldots, j_D \rangle$. We say that $F(X_1, \ldots, X_D)$ is invertible if the associated CA over $\mathbf{Z}_m$ is invertible, that is, if the coefficients λ_i's satisfy (3). To simplify the exposition, in this section we will often use the same letter F to denote both a finite fps and the associated CA.

Our first observation is that it suffices to find an inverse of F modulo the prime powers which compose m. This fact is a consequence of the Chinese Remainder theorem and it is proven by the following lemma.

Lemma 1. *Let* $m = p_1^{k_1} p_2^{k_2} \ldots p_h^{k_h}$ *and* $F(X_1, \ldots, X_D)$ *be a fps over* $\mathbf{Z}_m$. *Given* h *finite fps's* $G_1, \ldots, G_h$ *such that* $FG_i \equiv 1 \pmod{p_i^{k_i}}$, *we can find a finite fps* G *such that* $F(X_1, \ldots, X_D)G(X_1, \ldots, X_D) \equiv 1 \pmod{m}$.

Proof. Let $\alpha_i = m/p_i^{k_i}$. Since $\gcd(\alpha_i, p_i) = 1$ we can find β_i such that $\beta_i \alpha_i \equiv 1 \pmod{p_i^{k_i}}$. Let

$$G(X_1, \ldots, X_D) = \sum_{i=1}^{h} \beta_i \alpha_i G_i(X_1, \ldots, X_D),$$

we claim that G is the inverse of F. Since $G \equiv G_i \pmod{p_i^{k_i}}$, for $i = 1, \ldots, h$ we have $FG \equiv 1 \pmod{p_i^{k_i}}$. Hence $FG - 1$ is a multiple of every $p_i^{k_i}$, and therefore it is a multiple of m. $\qquad\square$

We now concentrate on the problem of inverting a finite fps modulo a prime power. For $m = p^k$, (3) tells us that F is invertible iff there exists a unique coefficient λ_j such that $p \nmid \lambda_j$. Hence, F can be written as

$$F(X_1, \ldots, X_D) = \lambda_j X_1^{j_1} \cdots X_D^{j_D} + pH(X_1, \ldots, X_D), \tag{4}$$

where $\gcd(\lambda_j, p) = 1$.

Theorem 2. *Let* $F(X_1, \ldots, X_D)$ *denote an invertible finite fps over* $\mathbf{Z}_{p^k}$, *and let* λ_j *and* H *be defined as in* (4). *Let* λ_j^{-1} *be such that* $\lambda_j^{-1} \lambda_j \equiv 1 \pmod{p^k}$. *Then, the inverse of* F *is given by*

$$G(X_1, \ldots, X_D) = \lambda_j^{-1} X_1^{-j_1} \cdots X_D^{-j_D} (1 + p\tilde{H} + p^2 \tilde{H}^2 + \cdots + p^{k-1}\tilde{H}^{k-1}),$$

where $\tilde{H}(X_1, \ldots, X_D) = -\lambda_j^{-1} X_1^{-j_1} \cdots X_D^{-j_D} H(X_1, \ldots, X_D)$.

Proof. We can rewrite $F(X_1, \ldots, X_D)$ as

$$F(X_1, \ldots, X_D) = \lambda_j X_1^{j_1} \cdots X_D^{j_D} (1 - p\tilde{H}(X_1, \ldots, X_D)).$$

Hence

$$
\begin{aligned}
FG &\equiv (1 - p\tilde{H})(1 + p\tilde{H} + p^2\tilde{H}^2 + \cdots + p^{k-1}\tilde{H}^{k-1}) \pmod{p^k} \\
&\equiv (1 - p^k\tilde{H}^k) \pmod{p^k} \\
&\equiv 1 \pmod{p^k}.
\end{aligned}
$$

$\square$

4 Fast computation of the inverse

If we look at the procedure given in Sect. 3 for the computation of the inverse of a CA, we can see that the computation takes time polynomial in $\log m$, D, and s (the size of the neighborhood frame). In fact, additions and products modulo n, the computation of a multiplicative inverse modulo n, and the computation of $\gcd(a, b)$ with $0 \leq a, b \leq n$ take time polynomial in $\log n$. However, in Sect. 3 we assume to know the factorization of m. If this is not the case, we can no longer compute the inverse in polylogarithmic time since the fastest known factorization algorithms require time exponential in $\log m$ (see for example [15]). To overcome this problem, in this section we show how to compute the inverse of a finite fps over $\mathbf{Z}_m$ without knowing the factorization of m. Our first step is a procedure for recognizing if a finite fps is invertible without knowing the factorization of m.

Theorem 3. *Let $F(X_1, \ldots, X_D) = \sum_{i=1}^{s} \lambda_i X_1^{i_1} \cdots X_D^{i_D}$ be a finite fps over $\mathbf{Z}_m$, and let $k = \lfloor \log_2 m \rfloor$. For $i = 1, \ldots, s$, define*

$$z_i = [\gcd(\lambda_i, m)]^k, \qquad \text{and} \qquad q_i = \frac{m}{\gcd(m, z_i)}.$$

Then, $F(X_1, \ldots, X_D)$ is invertible if and only if $q_1 q_2 \cdots q_s = m$.

Proof. Let $m = p_1^{k_1} \cdots p_h^{k_h}$. Since $k_j \leq k$, z_i contains all prime factors p_j's of m which are in λ_i, taken with exponent at least k_j. Hence, $\gcd(m, z_i)$ contains all primes p_j's which are in λ_j with exponent exactly k_j. As a consequence, the coefficients $q_1, \ldots, q_s$ are related to $\lambda_1, \ldots, \lambda_s$ as follows

$$p_j \nmid \lambda_i \implies p_j^{k_j} | q_i, \qquad p_j | \lambda_i \implies p_j \nmid q_i. \tag{5}$$

Hence, if p_j does not divide t_j coefficients λ_i's, $p_j^{k_j}$ divides t_j coefficients q_i and the product $q_1 \cdots q_s$ contains exactly the factor $p_j^{k_j t_j}$. The condition $q_1 \cdots q_s = m$ implies $t_j = 1$, for $j = 1, \ldots, h$, which is equivalent to (3). $\square$

Theorem 3 not only provides a procedure for testing if a finite fps is invertible, but also suggests an alternative way of computing the inverse using the Chinese Remainder theorem. In fact the following corollary holds.

Corollary 4. *Let $F(X_1, \ldots, X_D)$ be an invertible finite fps and let $q_1, \ldots, q_s$ be defined as in Theorem 3. For $1 \leq i, j \leq s$, $i \neq j$ we have $\gcd(q_i, q_j) = 1$.*

Proof. Assume by contradiction that that there exists a prime p such that $p \mid \gcd(q_i, q_j)$. Since each q_i divides m, we have $p \mid m$ and, by (5), $p \nmid \lambda_i$ and $p \nmid \lambda_j$. This is impossible since F is invertible and (3) must hold. $\qquad\square$

In view of Corollary 4 we have that in order to compute the inverse of F modulo m is suffices to compute the inverse G_i of F modulo q_i, for $i = 1, \ldots, s$ (obviously we can ignore those q_i's which are equal to 1). In fact, since $q_1 \cdots q_s = m$ and $\gcd(q_i, q_j) = 1$, given $G_1, \ldots, G_s$ we can obtain F^{-1} using the same procedure described in Lemma 1. We are therefore left with the problem of computing the inverse of F modulo q_i, for all q_i's which are greater than 1. To show how to do this without knowing the factorization of q_i we need a preliminary lemma.

Lemma 5. *Let $F(X_1, \ldots, X_D)$ be an invertible finite fps, $q_1, \ldots, q_s$ be defined as in Theorem 3, and $k = \lfloor \log_2 m \rfloor$. For $i = 1, \ldots, s$, we have*

$$\gcd(q_i, \lambda_i) = 1, \quad and \quad [\gcd(\lambda_1, \ldots, \lambda_{i-1}, \lambda_{i+1}, \ldots, \lambda_s)]^k \equiv 0 \pmod{q_i}.$$

Proof. Let $m = p_1^{k_1} \cdots p_h^{k_h}$. Since $q_i \mid m$, we have $q_i = p_1^{l_1} \cdots p_h^{l_h}$, with $0 \leq l_j \leq k_j$. By (5), we have $p_j \mid \lambda_i \implies p_j \nmid q_i$ hence $\gcd(q_i, \lambda_i) = 1$ as claimed. In addition, by (5) and (3) we have

$$p_j \mid q_i \implies p_j \nmid \lambda_i \implies p_j \mid \lambda_t \; \forall t \neq i.$$

Hence, $\gcd(\lambda_1, \ldots, \lambda_{i-1}, \lambda_{i+1}, \ldots, \lambda_s)$ contains all prime factors of q_i. Since, for $j = 1, \ldots, h$, $k \geq l_j$ we have that $[\gcd(\lambda_1, \ldots, \lambda_{i-1}, \lambda_{i+1}, \ldots, \lambda_s)]^k$ is a multiple of q_i as claimed. $\qquad\square$

Let $w_i = \gcd(\lambda_1, \ldots, \lambda_{i-1}, \lambda_{i+1}, \ldots, \lambda_s)$. We rewrite F as

$$F(X_1, \ldots, X_D) = \lambda_i X_1^{i_1} \cdots X_D^{i_D} + w_i H(X_1, \ldots, X_D).$$

Since $\gcd(\lambda_i, q_i) = 1$, there exists λ_i^{-1} such that $\lambda_i \lambda_i^{-1} \equiv 1 \pmod{q_i}$. By setting

$$\tilde{H}(X_1, \ldots, X_D) = -\lambda_i^{-1} X_1^{-i_1} \cdots X_D^{-i_D} H(X_1, \ldots, X_D)$$

we have

$$F(X_1, \ldots, X_D) \equiv \lambda_i X_1^{i_1} \cdots X_D^{i_D} [1 - w_i \tilde{H}(X_1, \ldots, X_D)] \pmod{q_i}.$$

Since by Lemma 5 we know that $w_i^k \equiv 0 \pmod{q_i}$, reasoning as in the proof of Lemma 2 we have that the inverse of F modulo q_i is given by

$$G_i(X_1, \ldots, X_D) = \lambda_i^{-1} X_1^{-i_1} \cdots X_D^{-i_D} (1 + w_i \tilde{H} + w_i^2 \tilde{H}^2 + \cdots + w_i^{k-1} \tilde{H}^{k-1}).$$

We have therefore proven the following result.

Theorem 6. *Given the finite fps $F(X_1, \ldots, X_D) = \sum_{i=1}^s \lambda_i X_1^{i_1} \cdots X_D^{i_D}$ we can test if F is invertible over $\mathbf{Z}_m$ and compute the inverse of F in time polynomial in $\log m$, D, and s.* $\qquad\square$

5 Characterization of expansive linear CA

A powerful tool for the study of CA, and of discrete time dynamical systems in general, is the analysis of their *topological properties*. These properties, e.g. sensitivity, transitivity, expansivity, denseness of periodic orbits, *etc.*, determine the qualitative behavior of the system under iteration of the global map F. In [3] and [9] the authors present necessary and sufficient criteria for determining whether a linear CA is transitive, sensitive to initial conditions, positively expansive, or strongly transitive. In this section we recall the definition of an important topological property of invertible dynamical systems, namely expansivity, and we give a complete characterization of expansive linear CA.

Definition 7 (Expansivity). Let X denote a set equipped with a distance d, and let $F: X \to X$ be an invertible map which is continuous with respect to the topology induced by d. We say that the dynamical system (X, F) is expansive if and only if there exists $\delta > 0$ such that for every $x, y \in X$, with $x \neq y$, there exists $n \in \mathbf{Z}$ such that $d(F^n(x), F^n(y)) > \delta$.

The topological properties of CA are usually defined with respect to the metric topology induced by the *Tychonoff distance*. Let $\Delta: \mathbf{Z}_m \times \mathbf{Z}_m \to \{0, 1\}$ given by

$$\Delta(i, j) = \begin{cases} 0, & \text{if } i = j, \\ 1, & \text{if } i \neq j. \end{cases}$$

For any pair $a, b \in \mathbf{Z}_m^{\mathbf{Z}^D}$ the Tychonoff distance $d(a, b)$ is defined by

$$d(a, b) = \sum_{\mathbf{v} \in \mathbf{Z}^D} \frac{\Delta(a(\mathbf{v}), b(\mathbf{v}))}{2^{\|\mathbf{v}\|_\infty}}, \tag{6}$$

where $\|\mathbf{v}\|_\infty$ denotes the maximum of the absolute value of the components of $\mathbf{v}$. It is easy to verify that d is a metric on $\mathbf{Z}_m^{\mathbf{Z}^D}$ and that the topology induced by d coincides with the product topology induced by the discrete topology of $\mathbf{Z}_m$.

Given two configurations $a, b \in \mathbf{Z}_m^{\mathbf{Z}^D}$ we define their sum $a + b$ by the rule $(a + b)(\mathbf{v}) = a(\mathbf{v}) + b(\mathbf{v}) \bmod m$. Note that, with respect to this sum, the Tychonoff distance is translation invariant, that is, $d(a, b) = d(a + c, b + c)$. In addition, if F is linear we have $F(a + b) = F(a) + F(b)$. A special configuration is the *null* configuration $\mathbf{0}$ which has the property that $\mathbf{0}(\mathbf{v}) = 0$ for all $\mathbf{v} \in \mathbf{Z}^D$.

Since expansive CA, whether linear or not, do not exist in any dimension greater than 1 (see [13, Theorem 2]), in order to characterize expansive linear CA, we can restrict ourself to the 1-dimensional case. Let $F: \mathbf{Z}_m^{\mathbf{Z}} \to \mathbf{Z}_m^{\mathbf{Z}}$ denote the global transition map of a 1-dimensional invertible linear CA. It is straightforward to verify that F is expansive if and only if there exists $\delta > 0$ such that for any configuration $c \in \mathbf{Z}_m^{\mathbf{Z}}$ we have

$$\forall c' \in \mathbf{Z}_m^{\mathbf{Z}} \quad c' \neq \mathbf{0} \quad \exists n \in \mathbf{Z}: \quad d(F^n(c + c'), F^n(c)) > \delta.$$

Since d is translation invariant and both F and F^{-1} are linear, we can get rid of the particular configuration c. We have

$$d(F^n(c+c'), F^n(c)) = d(F^n(c) + F^n(c'), F^n(c)) = d(F^n(c'), \mathbf{0}).$$

Hence, F is expansive if and only if for any $c' \neq \mathbf{0}$ we have $d(F^n(c'), \mathbf{0}) > \delta$ for some $n \in \mathbf{Z}$. By (6), this is equivalent to assuming that there exists $M > 0$ such that

$$\forall c' \in \mathbf{Z}_m^{\mathbf{Z}} \quad c' \neq \mathbf{0} \quad \exists n \in \mathbf{Z}: \quad [F^n(c')](i) \neq 0 \text{ for some } i \text{ with } |i| < M.$$

For any integer $k > 0$, let W_k denote the set of configurations $c \in \mathbf{Z}_m^{\mathbf{Z}}$ such that $c(i) = 0$ for $|i| < k$ and at least one between $c(k)$ and $c(-k)$ is different from zero. Since δ can be chosen arbitrarily, we have that F is expansive if and only if $\exists \tilde{k}$ such that for all $k > \tilde{k}$

$$\forall c' \in W_k \quad \exists n \in \mathbf{Z}: \quad [F^n(c')](i) \neq 0 \text{ for some } i \text{ with } |i| < M. \tag{7}$$

If we visualize each configuration as a bi-infinite array, (7) tells us that the essential feature of expansive linear maps is that any pattern of nonzero values can "propagate" from positions arbitrarily away from 0 up to a position i with $|i| < M$. Informally, we say that any nonzero pattern can propagate for an arbitrarily large distance. Note that this propagation can be seen "in the future", when we consider F^n with $n > 0$, or "in the past" (in the future for F^{-1}), when we consider F^n with $n < 0$.

Theorem 8. *Let F denote the global transition map of an invertible linear 1-dimensional CA over $\mathbf{Z}_m$ with local rule $f(x_{-r}, \ldots, x_r) = \sum_{i=-r}^{r} a_i x_i \bmod m$. The map F is expansive if and only if*

$$\gcd(m, a_{-r}, \ldots, a_{-1}, a_1, \ldots, a_r) = 1 \tag{8}$$

Proof. Let $A(X)$ denote the finite fps associated to F. Since F is invertible, by (3) we know that for each prime factor p of m, there exists a unique coefficient a_j such that $p \nmid a_j$. Hence $A(X)$ can be written as

$$A(X) = a_j X^{-j} + p H(X). \tag{9}$$

Consider now the the finite fps $B(X)$ corresponding to F^{-1}. By the proofs of Theorems 1 and 2 we know that $B(x)$ has the form

$$B(X) = b_j X^j + p K(X). \tag{10}$$

where $b_j a_j \equiv 1 \pmod{p}$. Note that (8) implies that (9) and (10) hold with $j \neq 0$. Vice versa, $\gcd(m, a_{-r}, \ldots, a_{-1}, a_1, \ldots, a_r) > 1$ implies that there exists at least a prime p for which (9) and (10) hold for $j = 0$.

We are now ready to prove that (8) is a necessary condition for expansivity. Let p denote a prime for which (9) and (10) hold for $j = 0$ and let $q = m/p$. For any integer $k > 0$ let $\tilde{c} \in W_k$ denote the configuration defined by $\tilde{c}(i) = q$ if $i = k$ and $\tilde{c}(i) = 0$ otherwise. We show that for every $n \in \mathbf{Z}$ and $i \neq k$ we have

$[F^n(\tilde{c})](i) = 0$ which implies that F is not expansive. Let $P_{\tilde{c},n}(X)$ denote the fps associated to $F^n(\tilde{c})$. Since the fps associated to $\tilde{c}$ is qX^k, for $n \geq 0$ we have

$$P_{\tilde{c},n}(X) = qX^k A^n(X) = qX^k(a_0 + pH(X))^n = qX^k(a_0^n + pH'(X)),$$

where $pH'(X) = \sum_{i=1}^{n} \binom{n}{i} p^i H^i(X) a_0^{n-i}$. Since $pq = m$ we have $P_{\tilde{c},n}(X) = a_0^n qX^k$. With the same argument we can prove that for $n < 0$

$$P_{\tilde{c},n}(X) = qX^k B^n(X) = b_0^n qX^k.$$

Hence, for $i \neq k$, we have $[F^n(\tilde{c})](i) = 0$ for all $n \in \mathbf{Z}$ and F cannot be expansive.

Now we prove that condition (8) implies expansivity. Assume that for each prime p (9) and (10) hold with $j \neq 0$ and let c any configuration in $\mathcal{W}_k$. Let $C(X) = \sum_i c_i X^i$ be the fps associated to c. At least one between c_k and c_{-k} is nonzero. In the following we assume $c_k \neq 0$, the case $c_k = 0$ being analogous. Since $(c_k \bmod m) \neq 0$ there exists a prime p and a positive integer h such that $p^h | m$ and $p^h \not| c_k$. Let n be any positive multiple of $p^h h!$. The fps associated to $F^n(c)$ and $F^{-n}(c)$ are $A^n(X)C(X)$ and $B^n(X)C(X)$ respectively. We have

$$A^n(X) = (a_j X^{-j} + pH(X))^n$$

$$= a_j^n X^{-nj} + \sum_{i=1}^{n} \binom{n}{i} p^i H^i(X) a_j^{n-i}$$

$$= a_j^n X^{-nj} + \left(\sum_{i=1}^{h-1} \binom{n}{i} p^i H^i(X) a_j^{n-i} \right) + p^h \left(\sum_{i=h}^{n} \binom{n}{i} p^{i-h} H^i(X) a_j^{n-i} \right)$$

$$= a_j^n X^{-nj} + p^h H'(X),$$

where the last equality holds since, for $i < h$, $\binom{n}{i}$ is a multiple of p^h. Similarly, we have $B^n(X) = b_j^n X^{nj} + p^h K'(X)$. Hence, working $\bmod p^h$ we get

$$A^n(X)C(X) = a_j^n \sum_i c_i X^{i-nj}, \quad B^n(X)C(X) = b_j^n \sum_i c_i X^{i+nj},$$

Consider the term $i = k$ of the above series. Since $p^h \not| c_k$, $p \not| a_j^n$, and $p \not| b_j^n$, working $\bmod p^h$ we have

$$[F^n(c)](k - nj) = a_j^n c_k \neq 0, \quad [F^{-n}(c)](k + nj) \equiv b_j^n c_k \neq 0,$$

which implies

$$[F^n(c)](k - nj) \not\equiv 0 \pmod{m}, \quad [F^{-n}(c)](k + nj) \not\equiv 0 \pmod{m}. \quad (11)$$

Thus, by considering both F^n and F^{-n}, we can move the nonzero value from position k to positions $k \pm nj$, where n is a multiple of $p^h h!$, and $j \neq 0$. Define $z = rp^h h!$, where r is the radius of the CA. By (11), it follows that we can find $n \in \mathbf{Z}$ such that $[F^n(c)](i) \neq 0$ for some i with $|i| \leq z$. Since the same reasoning holds for any $c \in \mathcal{W}_k$ and for all $k > 0$, by (7) we have that F is expansive. This completes the proof. $\qquad \square$

References

1. S. Amoroso and Y. N. Patt. Decision procedures for surjectivity and injectivity of parallel maps for tesselation structures. *Journal of Computer and System Sciences*, 6:448–464, 1972.
2. H. Aso and N. Honda. Dynamical characteristics of linear cellular automata. *Journal of Computer and System Sciences*, 30:291–317, 1985.
3. G. Cattaneo, E. Formenti, G. Manzini, and L. Margara. On ergodic linear cellular automata over Z_m. In *14th Annual Symposium on Theoretical Aspects of Computer Science (STACS '97)*, volume 1200 of *LNCS*, pages 427–438. Springer Verlag, 1997.
4. P. Favati, G. Lotti, and L. Margara. One dimensional additive cellular automata are chaotic according to Devaney's definition of chaos. *Theoretical Computer Science*, 174:157–170, 1997.
5. P. Guan and Y. He. Exacts results for deterministic cellular automata with additive rules. *Jour. Stat. Physics*, 43:463–478, 1986.
6. M. Ito, N. Osato, and M. Nasu. Linear cellular automata over Z_m. *Journal of Computer and System Sciences*, 27:125–140, 1983.
7. J. Kari. Reversability of 2D cellular automata is undecidable. *Physica*, D 45:379–385, 1990.
8. J. Kari. Reversibility and surjectivity problems of cellular automata. *Journal of Computer and System Sciences*, 48(1):149–182, 1994.
9. G. Manzini and L. Margara. A complete and efficiently computable topological classification of D-dimensional linear cellular automata over Z_m. Technical Report B4-96-18, Istituto di Matematica Computazionale, CNR, Pisa, Italy, 1996.
10. D. Richardson. Tessellation with local transformations. *Journal of Computer and System Sciences*, 6:373–388, 1972.
11. T. Sato. Decidability for some problems of linear cellular automata over finite commutative rings. *Information Processing Letters*, 46(3):151–155, 1993.
12. T. Sato. Group structured linear cellular automata over Z_m. *Journal of Computer and System Sciences*, 49(1):18–23, 1994.
13. M. A. Shereshevsky. Expansiveness, entropy and polynomial growth for groups acting on subshifts by automorphisms. *Indag. Mathem. N.S.*, 4:203–210, 1993.
14. T. Toffoli and H. Margolus. Invertible cellular automata: A review. *Physica*, D 45:229–253, 1990.
15. J. van Leeuwen (Managing Editor). *Handbook of Theoretical Computer Science. Volume A: Algorithms and complexity.* The MIT press/Elsevier, 1990.

Two-Level Contextual Grammars:
The Internal Case

Carlos MARTIN-VIDE[a], Joan MIQUEL-VERGES[a], Gheorghe PĂUN[b]

[a]Universitat Rovira i Virgili, Grup de Recerca en Linguistica Matematica
Pl. Imperial Tarraco, 1, 43005 Tarragona, Spain
[b]Institute of Mathematics of the Romanian Academy
PO Box 1 – 764, 70700 Bucureşti, Romania

Abstract. We consider generative mechanisms similar to contextual grammars with infinite sets of contexts, such that the contexts are also generated in a contextual way (by adjoining contexts, depending on given selectors). In this way one generalizes a proposal from [6], where grammars generating sets of contexts are suggested (here we also use the produced contexts in order to generate a language). We investigate such two-level contextual grammars with finite or regular choice or without choice, at each of the two levels. All the relations between the nine families obtained in this way (only five are distinct), and between them and Chomsky families or usual contextual language families are settled. Two further variants are then proposed; for one of them, the relationships between the corresponding families are settled (all the nine families are now distinct).

1. Introduction

The contextual grammars are generative mechanisms based on the operation of adjoining contexts (pairs of strings) to associated strings. One starts from a finite set of given *axioms* and, by iterated adjoining of contexts, one obtains a language. In [5], where these grammars were introduced, the contexts are adjoined at the ends of the current string (external grammars); the internal variant was considered first in [12]. Several variants were investigated, in order to better capture this basic aspect in descriptive linguistics which is the acception of a context by a string, with respect to a language (see, e.g., [4]), and to cover (non-context-free) constructions specific to natural languages, such as duplication, multiple or crossed agreements, etc. From this point of view, internal contextual grammars with maximal use of selectors, introduced in [8], seem to be the most promising (see [7]). Details about the early development of the theory of contextual grammars can be found in [9], recent results and references can be found in [11] and in [2].

One idea of enlarging the power of contextual grammars is to use an infinite number of contexts, with some restrictions about their forms or about their pairing with the selecting strings. This has been explored in [10] (the contexts are given and one imposes the restriction to have only finitely many contexts associated to each string).

Here we investigate another approach, suggested in [6]: the contexts we use are also produced in a contextual manner, by adjoining contexts. More specifically, because a context is a pair of strings and these strings are supposed to grow in a related way, we use pairs of contexts (hence quadruples) in order to produce contexts. In this way we get a sort of contextual grammar with two levels, one generating contexts, the other one using these contexts in the classic manner; both levels use selection mappings as usual in contextual grammars. Depending on the type of selectors defined by these mappings (sets of strings associated to sets of contexts or to pairs of contexts), we obtain classifications of our mechanisms.

We shall examine here two-level contextual grammars with finite or regular choice, or without choice, these variants being considered at each of the two levels; always we use only the internal derivation. In this way we obtain nine families of languages. We settle all generative capacity questions about them: mutual relationships (only five families are distinct), relationships with Chomsky families or with families of usual internal contextual languages.

Two further variants of two-level contextual grammars are proposed, which looks more apropriate; for instance, one of they leads to nine distinct families, because it does not allow the simulation techniques which make some of the families of languages generated by initial two-level grammars to coincide (which means that the initial definition is not sensitive enough).

2. Contextual Grammars; Basic Notions and Results

As usual, we use the following general notations: $V^* = $ the free monoid generated by the alphabet V under the operation of concatenation, $\lambda = $ the empty string, $V^+ = V^* - \{\lambda\}$, $|x| = $ the length of $x \in V^*$, $|x|_a = $ the number of occurrences of the symbol $a \in V$ in $x \in V^*$, FIN, REG, CF, CF, RE = the families of finite, regular, context-free, context-sensitive, and recursively enumerable languages, respectively. The Parikh mapping, Ψ_V, associated to an alphabet $V = \{a_1, \ldots, a_n\}$ (with this ordering of the symbols) is defined by $\Psi_V(x) = (|x|_{a_1}, \ldots, |x|_{a_n})$, and it is extended to languages $L \subseteq V^*$ in the natural way, $\Psi_V(L) = \{\Psi_V(x) \mid x \in L\}$.

A Chomsky grammar is denoted by $G = (N, T, S, P)$, where N is the nonterminal alphabet, T is the terminal alphabet, $S \in N$ is the axiom, and P is the set of productions (written in the form $u \to v$).

A *contextual grammar* is a construct $G = (V, A, C, \varphi)$, where V is an alphabet, $A \subseteq V^*$ is a finite set (of *axioms*), $C \subseteq V^* \times V^*$ is a finite set (of *contexts*) and $\varphi : V^* \longrightarrow 2^C$ (the *selection* mapping).

With respect to such a grammar, for $x, y \in V^*$ we define:

$$x \Longrightarrow y \text{ iff } x = x_1 x_2 x_3, y = x_1 u x_2 v x_3, \text{ for some } (u, v) \in \varphi(x_2), x_1, x_2, x_3 \in V^*.$$

The language generated by G (in the internal mode) is

$$L(G) = \{x \in V^* \mid w \Longrightarrow^* x, w \in A\}.$$

For $C' \subseteq C$ we define

$$\varphi^{-1}(C') = \{x \in V^* \mid \varphi(x) = C'\}.$$

If $\varphi^{-1}(C') \in F$ for each $C' \subseteq C$, where F is a given family of languages, then we say that G is a grammar *with F choice*. When $\varphi(x) = C$ for all $x \in V^*$, then the grammar is *without choice* (and the mapping φ can be omitted). Clearly, a grammar without choice has regular choice.

We denote by $ICC(F)$ the family of languages $L(G)$ generated (in the internal mode) by grammars with F choice; by IC we denote the family of languages generated by grammars without choice.

In a derivation $x_1x_2x_3 \Longrightarrow x_1ux_2vx_3$, the adjoined context (u,v) is selected depending on x_2 only. A generalization appears in the *total* contextual grammars, which are constructs of the form $G = (V,A,C,\varphi)$, with V,A,C as above and $\varphi : V^* \times V^* \times V^* \longrightarrow 2^C$. The derivation relation is defined by

$$x \Longrightarrow y \text{ iff } x = x_1x_2x_3, y = x_1ux_2vx_3, \text{ for } (u,v) \in \varphi(x_1,x_2,x_3), \ x_1,x_2,x_3 \in V^*.$$

We denote by TC_c the family of languages generated by total contextual grammars with *computable* selection mappings; when arbitrary mappings are used we write TC.

The following results are known:

1. $(IC \cup ICC(FIN)) \subset ICC(REG) \subset TC_c \subset RE, \ ICC(REG) \subset CS$; IC and $ICC(FIN)$ are incomparable;

2. $REG \subset ICC(FIN), \ CF \subset TC_c$;

3. REG is incomparable with IC; CF is incomparable with $IC, ICC(FIN)$, $ICC(REG)$; CS is incomparable with TC_c.

Let us also recall two necessary conditions, useful below:

Lemma 1. *For every language* $L \subseteq V^*, L \in ICC(F)$, *with arbitrary* F, *there is a constant* p *such that every* $x \in L, |x| > p$, *can be written in the form* $x = x_1ux_2vx_3$, *for some* $x_1,x_2,x_3 \in V^*, u,v \in V^*$, *with* $|uv| \le p$, *such that* $x_1u^nx_2v^nx_3 \in L$ *for all* $n \ge 0$.

We say that a language $L \subseteq V^*$ has the *internal bounded step* (IBS) property if there is a constant p such that for each $x \in L, |x| > p$, there is $y \in L$ such that $x = x_1ux_2vx_3, y = x_1x_2x_3$, for some $x_1,x_2,x_3,u,v \in V^*$ with $0 < |uv| \le p$.

Lemma 2. *A language* L *is in the family* TC *if and only if it has the IBS property; if* L *is recursive (its membership question is algorithmically solvable), then* $L \in TC_c$ *iff it has the IBS property.*

Together with the known pumping lemmas for context-free languages (they imply the IBS property) we obtain in this way the inclusion $CF \subseteq TC_c$. A proof of the inclusion $REG \subseteq ICC(FIN)$ can be found in [1], as well as in [2].

3. Two-level Contextual Grammars

A *two-level contextual grammar* with choice is a construct

$$\gamma = (V,A,C,D,\delta,\varphi),$$

where V is an alphabet, $A \subseteq V^*$ is a finite language (of axioms), $C \subseteq V^* \times V^*$ is a finite set (of contexts), $D \subseteq (V^* \times V^*) \times (V^* \times V^*)$ is a finite set (of *double-contexts*), $\delta : V^* \times V^* \longrightarrow 2^D$ and $\varphi : V^* \longrightarrow 2^{K(C,D,\delta)}$, where $K(C,D,\delta)$ is defined as follows: We define the relation $\Longrightarrow$ on $V^* \times V^*$ by

$$(x,y) \Longrightarrow (x',y') \text{ iff } x = x_1x_2x_3, y = y_1y_2y_3, x' = x_1ux_2vx_3, y' = y_1u'y_2v'y_3,$$
$$\text{for } ((u,v),(u',v')) \in \delta(x_2,y_2), x_1,x_2,x_3,y_1,y_2,y_3 \in V^*.$$

(In words, the context (x, y) is derived into (x', y') by simultaneously adjoining a context (u, v) to x and a context (u', v') to y, as selected by the subwords x_2, y_2 according to the mapping δ.)

Then, denoting by $\Longrightarrow^*$ the reflexive and transitive closure of the relation $\Longrightarrow$, we define

$$K(C, D, \delta) = \{(x, y) \in V^* \times V^* \mid (w, z) \Longrightarrow^* (x, y), (w, z) \in C\}.$$

This is the set of contexts generated by the components C, D, δ of γ. Then, for $x, y \in V^*$ we define

$$x \Longrightarrow y \text{ iff } x = x_1 x_2 x_3, y = x_1 u x_2 v x_3, \text{ for some } (u, v) \in \varphi(x_2), x_1, x_2, x_3 \in V^*.$$

Then, the language generated by γ is

$$L(\gamma) = \{x \in V^* \mid w \Longrightarrow^* x, w \in A\}.$$

This is the usual (internally) generated language of the contextual grammar $(V, A, K(C, D, \delta), \varphi)$, hence possibly having infinitely many contexts.

Several specific two-level contextual grammars will appear in the subsequent sections, hence we do not present an example here.

Take a grammar $\gamma = (V, A, C, D, \delta, \varphi)$. For each $D' \subseteq D$ we define the language

$$\delta^{-1}(D') = \{x_1 \# x_2 \mid x_1, x_2 \in V^*, \delta(x_1, x_2) = D'\},$$

where $\#$ is a symbol not in V. We say that δ is of type F, for a given family F of languages, if $\delta^{-1}(D') \in F$ for all $D' \subseteq D$.

The type of φ is defined as for usual contextual grammars, according to the languages $\varphi^{-1}(C') = \{x \in V^* \mid \varphi(x) = C'\}$, for $C' \subseteq K(C, D, \delta)$.

Thus, a two-level contextual grammar $\gamma = (V, A, C, D, \delta, \varphi)$ has (F_1, F_2) choice, for two given families of languages F_1, F_2, if φ is of F_1 type and δ is of F_2 type. We consider here only the cases when F_1, F_2 are one of FIN, REG, as well as the case of φ and/or δ imposing no choice. The corresponding families of languages are denoted by $TLC(F_1, F_2)$, with $F_1, F_2 \in \{FIN, REG, NO\}$, where NO indicates the case of no choice; by convention, $NO \subseteq REG$. When presenting a grammar $\gamma = (V, A, C, D, \delta, \varphi)$ with δ and/or φ imposing no choice, we replace δ and/or φ by $\emptyset$, respectively.

We have obtained nine families of languages. Their size (comparing them to each other, with families in the Chomsky hierarchy and with $IC, ICC(FIN), ICC(REG)$) will be investigated in the following sections.

4. Preliminary Results

First, we give some necessary conditions for a language to be in a family $TLC(F_1, F_2)$ as above.

We say that a language $L \subseteq V^*$ has the *bounded length increase* property if there is a constant p such that for each $x \in L$, $|x| > p$, there is $y \in L$ with $0 < |x| - |y| \leq p$. (Note that the IBS property is stronger: it also imposes the fact that x differs from y by a context, which in turn is of bounded length.) The proof of the next lemma is straightforward.

Lemma 3. *Each language $L \in TLC(F_1, F_2)$, for all F_1, F_2, has the bounded length increase property.*

Of course, the bounded length increase property is also a consequence of the semi-linearity. However, we have the following result, whose proof is omitted (an extension of the proof in [3] can be used):

Theorem 1. *All families $TLC(F_1, F_2), F_1, F_2 \in \{FIN, REG, NO\}, (F_1, F_2) \neq (NO, NO)$, contain non-semilinear languages.*

The following inclusions either follow from definitions or can be proved in an easy way.

Lemma 4. $TLC(F_1, F_2) \subseteq TLC(F_1', F_2')$, *for all $F_1 \subseteq F_1', F_2 \subseteq F_2'$.*

Lemma 5. $ICC(F_1) \subseteq TLC(F_1, F_2)$, $IC \subseteq TLC(NO, F_2)$, *for all F_1 and F_2.*

Lemma 6. $TLC(REG, REG) \subset CS$.

We synthesize the inclusions in Lemma 4, 5, 6, as well as the known relationships between Chomsky families and families of contextual languages in the diagram in Figure 1; an arrow from a family F_1 to a family F_2 indicates the inclusions $F_1 \subseteq F_2$. For saving space, the families $TLC(F_1, F_2)$ are indicated by F_1F_2, abbreviating FIN, REG, NO by F, R, N, respectively.

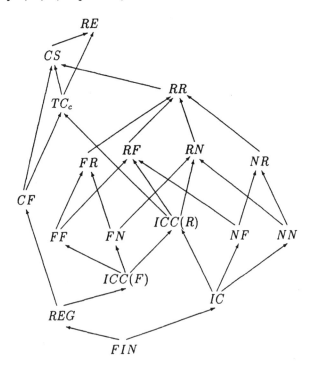

Figure 1

In contrast to Theorem 1, for the remaining family $TLC(NO, NO)$ we have

Theorem 2. *Each language in the family $TLC(NO, NO)$ is semilinear.*

Proof. Consider a two-level contextual grammar $\gamma = (V, A, C, D, \emptyset, \emptyset)$ and construct the right-linear grammar $G = (\{X\}, V, X, P)$ with

$$P = \{X \rightarrow u_1 v_1 u_2 v_2 X \mid ((u_1, v_1), (u_2, v_2)) \in D\}$$
$$\cup \{X \rightarrow uvX \mid (u, v) \in C\} \cup \{X \rightarrow x \mid x \in A\}.$$

The equality $\Psi_V(L(\gamma)) = \Psi_V(L(G))$ is obvious: each string in $L(G)$ corresponds to a string in $L(\gamma)$, in the sense that the same string in A, contexts in C and double-contexts in D are used when building the two strings, but with the symbols in possibly changed positions. Consequently, $L(\gamma)$ is semilinear. $\square$

5. The Generative Capacity

We start by a somewhat surprising result, which however can be explained by the large power of the selection mapping φ.

Theorem 3. $TLC(F_1, FIN) = TLC(F_1, REG) = TLC(F_1, NO)$, $F_1 \in \{FIN, REG\}$.

Proof. Consider a two-level contextual grammar $\gamma = (V, A, C, D, \delta, \varphi)$ with regularly defined δ and construct the grammars

$$\gamma' = (V, A, C, D, \emptyset, \varphi),$$
$$\gamma'' = (V, A, C, D', \delta', \varphi),$$
$$D' = \{((\lambda, a), (\lambda, \lambda)), ((\lambda, \lambda), (\lambda, a)) \mid a \in V\},$$
$$\delta'(\lambda, \lambda) = D'.$$

Obviously, we have $K(C, D, \delta) \subseteq K(C, D, \emptyset)$ and $K(C, D, \delta) \subseteq K(C, D', \delta')$. The mapping φ uses from the larger sets $K(C, D, \emptyset), K(C, D', \delta')$ only the contexts in $K(C, D, \delta)$, hence the equalities $L(\gamma) = L(\gamma') = L(\gamma'')$ follow. This implies the inclusions $TLC(F_1, REG) \subseteq TLC(F_1, F_2)$, for $F_2 \in \{FIN, NO\}$, $F_1 \in \{FIN, REG\}$. The converse inclusions follow by the definitions. $\square$

Consequently, the diagram in Figure 1 can be redrawn as in Figure 2.

All the remaining families are distinct. More precisely, we have

Theorem 4. *All the inclusions in the diagram in Figure 2 are proper and all families which are not linked by a path in this diagram are incomparable.*

Some of the assertions (about pairs involving families $TLC(F_1, F_2)$, the other relations are known) were proved in the previous section, the others follow from the lemmas below.

Lemma 7. *All families $TLC(F_1, F_2), F_1, F_2 \in \{FIN, NO\}$, contain languages which are not in TC.*

Proof. Let us consider the two-level contextual grammar

$$\gamma = (\{a, b, c, d, e, f, h\}, \{\lambda\}, C, D, \emptyset, \emptyset),$$
$$C = \{(\lambda, \lambda)\},$$
$$D = \{((ab, cd), (ef, gh))\}.$$

Because all contexts $(a^n b^n c^n d^n, e^n f^n g^n h^n), n \geq 1$, are in $K(C, D, \emptyset)$, it follows that all strings $a^n b^n c^n d^n e^n f^n g^n h^n, n \geq 1$, are in $L(\gamma)$. No bounded context can be removed from such a string of large enough length, hence $L(\gamma)$ does not have the IBS property. In view of Lemma 2, this language is not in the family TC.

If we put $\delta(\lambda, \lambda) = D$ and/or $\varphi(\lambda) = K(C, D, \delta)$, then we obtain a language $L(\gamma)$ still containing all strings $a^n b^n c^n d^n e^n f^n g^n h^n, n \geq 1$, hence the language is not in the family TC. □

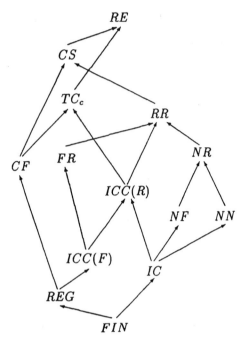

Figure 2.

Lemma 8. $CF - TLC(REG, REG) \neq \emptyset$.

Proof. The linear language $L = \{a^n b^n \mid n \geq 1\} \cup a^+$ is not in $TLC(REG, REG)$. Assume the contrary and take $\gamma = (\{a, b\}, A, C, D, \delta, \varphi)$ such that $L(\gamma) = L$. Each string a^n is in L, hence there is a context $(a^i, a^j) \in K(C, D, \delta)$ and a string a^k such that $(a^i, a^j) \in \varphi(a^k), k \geq 0, i + j \geq 1$. Then we can derive $a^{k+1} b^{k+1} \implies a^{k+i+j+1} b^{k+1}$, which is not in L, a contradiction. □

Lemma 9. $IC - TLC(FIN, REG) \neq \emptyset$.

Proof. Consider the usual contextual grammar without choice

$$G = (\{a, b, c, d, e\}, \{e^5\}, \{(a, c), (b, d)\}).$$

All strings in the language $L(G)$ contain five occurrences of the symbol e and all strings of the form $ea^n eb^m ec^n ed^m e$, for $n, m \geq 1$, are in $L(G)$.

Assume now that $L(G) = L(\gamma)$ for some $\gamma = (\{a, b, c, d, e\}, A, C, D, \delta, \varphi)$ with finitely defined φ. From the fact that $|x|_e = 5$ for all $x \in L(\gamma)$, it follows that $|x|_e = 5$ for all $x \in A$, hence $|uv|_e = 0$ for all $(u, v) \in K(C, D, \varphi)$. In order to generate strings $ea^n eb^m ec^n ed^m e$ with arbitrarily large n and m, we need contexts

$(a^i, c^i), (b^j, d^j), i, j \geq 1$. Due to the finiteness of the domain of φ, only strings as above with at least one of n, m bounded can be generated. As $L(G)$ contains such strings with both n, m arbitrarily large, it follows that $L(G) = L(\gamma)$ cannot hold, a contradiction. □

Lemma 10. $TLC(FIN, FIN) - TLC(NO, REG) \neq \emptyset$.

Proof. The assertion is proved by the language

$$L(\gamma) = \{a^{n_k} cb^{n_k} \ldots a^{n_1} cb^{n_1} da^{n_1} cb^{n_1} \ldots a^{n_k} cb^{n_k} \mid k \geq 0, n_i \geq 0, 1 \leq i \leq k\}.$$

The details are left to the reader. □

Lemma 11. *The families* $TLC(NO, FIN), TLC(NO, NO)$ *are incomparable.*

Proof. Consider the two-level contextual grammar

$$\begin{aligned}
\gamma \ = \ &(\{a, b, c, d, e, f\}, \{\lambda\}, C, D, \delta, \emptyset), \\
&C = \{(c, c), (d, d)\}, \\
&D = \{((a, b), (a, b)), ((e, f), (e, f))\}, \\
&\delta(c, c) = \{((a, b), (a, b))\}, \\
&\delta(d, d) = \{((e, f), (e, f))\}.
\end{aligned}$$

We obtain

$$K(C, D, \delta) = \{(a^n cb^n, a^n cb^n), (e^n df^n, e^n df^n) \mid n \geq 1\}.$$

Therefore, $L(\gamma)$ contains all strings of the form $a^n cb^n e^m df^m a^n cb^n e^m df^m, n, m \geq 1$. On this basis, one can see that $L(\gamma) \notin TLC(NO, NO)$.

Conversely, consider the two-level grammar

$$\begin{aligned}
\gamma \ = \ &(\{a, b, c, d, e, f, g, h, k\}, \{k\}, C, D, \emptyset, \emptyset), \\
&C = \{(abcd, efgh)\}, \\
&D = \{((a, c), (e, g)), ((b, d), (f, h))\}.
\end{aligned}$$

All contexts of the form $(a^n b^m c^n d^m, e^n f^m g^n h^m), n, m \geq 1$, are in $K(C, D, \emptyset)$, hence all strings $a^n b^m c^n d^m k e^n f^m g^n h^m, n, m \geq 1$, are in $L(\gamma)$.

If $L(\gamma) = L(\gamma')$ for some $\gamma' = (\{a, b, c, d, e, f, g, h, k\}, A, C', D', \delta, \emptyset)$, then $(a^n b^m c^n d^m, e^n f^m g^n h^m)$ should be in $K(C', D', \delta)$ for arbitrarily many values of n, m, both of them arbitrarily large. This is not possible when δ has a finite domain, hence $L(\gamma) \notin TLC(NO, FIN)$. □

6. Two Variants

Theorem 3 and, mainly, its proof show the fact that the definition of a two-level contextual grammar as considered in Section 3 is too general, it allows that $K(C, D, \delta)$ can be rather unrestricted but, because φ selects from it certain contexts only, very complex languages can be generated even when $\delta = \emptyset$. At the limit, $K(C, D, \delta)$ can be $V^* \times V^*$ and still no loss of power is entailed due to the ability of φ to select contexts. (This is not possible in the case when φ imposes no selection, hence the families $TLC(NO, F_2), F_2 \in \{FIN, REG, NO\}$, were distinct. In fact, such a case is suggested in [6], without choice.)

A natural idea which seems to eliminate this drawback is to impose that all contexts in $K(C, D, \delta)$ are used in the definition of φ:

$$\bigcup_{x \in Sub(L(\gamma))} \varphi(x) = K(C, D, \delta). \qquad (*)$$

In this way, the definition of a two-level contextual grammar becomes somewhat circular: the generated language appears in the very definition of the grammar generating it. However, we keep condition $(*)$, but we use it in *a posteriori* manner: we consider a grammar *acceptable* only when condition $(*)$ is satisfied.

Let us denote by $ATLC(F_1, F_2)$ the family of languages $L(\gamma)$ such that γ is an acceptable two-level contextual grammar of type (F_1, F_2).

Another attractive idea is to construct the contexts directly associated with the string selectors, in the following way.

A *localized two-level contextual grammar* is a system

$$\gamma = (V, A, (S_1, C_1, D_1, \delta_1), \ldots, (S_n, C_n, D_n, \delta_n)), n \geq 1,$$

where V is an alphabet, $A \subseteq V^*$ is a finite language (of axioms), $S_i \subseteq V^*$ (selectors), $C_i \subseteq V^* \times V^*$ are finite sets (of contexts), $D_i \subseteq (V^* \times V^*) \times (V^* \times V^*)$ are finite sets (of double-contexts), and $\delta_i : V^* \times V^* \longrightarrow 2^D, 1 \leq i \leq n$. As in the previous sections, each triple C_i, D_i, δ_i defines a set of contexts, $K(C_i, D_i, \delta_i)$, which can be adjoined to the strings in S_i. Thus, we implicitly have a mapping φ, defined by

$$\varphi(x) = \bigcup_i K(C_i, D_i, \delta_i),$$

the union being taken over all $i, 1 \leq i \leq n$, such that $x \in S_i$.

No condition is imposed to the selectors (they are not necessarily disjoint, not even necessarily distinct). The grammar Γ is said to be of type (F_1, F_2) if $S_i \in F_1$ for all $i, 1 \leq i \leq n$, and δ_i are of type F_2, in the sense defined in Section 3: $\delta_i^{-1}(D) \in F_2$ for all $D \subseteq D_i, 1 \leq i \leq n$. A grammar is without choice when all its selectors are equal to V^*.

We denote by $LTLC(F_1, F_2)$ the family of languages generated by localized two-level contextual grammars of type (F_1, F_2).

Note that the localized grammars are not necessarily acceptable grammars: the selectors S_i can contain strings which are never used in derivations.

By the definitions, $ATLC(F_1, F_2) \subseteq TLC(F_1, F_2)$ and $LTLC(F_1, F_2) \subseteq TLC(F_1, F_2)$, for all F_1, F_2.

Moreover, $ICC(FIN) \subseteq ATLC(FIN, F_2)$, $IC \subseteq ATLC(NO, F_2)$ and $ICC(REG) \subseteq ATLC(REG, F_2)$, $F_2 \in \{FIN, NO\}$. Similar relations are true for the familes $LTLC(F_1, F_2)$.

Using these relations and those stated in Theorem 4, many relations concerning the families $ATLC(F_1, F_2)$ and $LTLC(F_1, F_2)$ are obtained.

Moreover, the grammar in the proof of Lemma 10 is acceptable and it also can be obviously rewritten as a localized one (with only one selector, $\{d\}$). Thus, we have

Lemma 12. $ATLC(FIN, FIN) - TLC(NO, REG) \neq \emptyset$, $LTLC(FIN, FIN) - TLC(NO, REG) \neq \emptyset$.

The proofs of the next two lemmas are left to the reader.

Lemma 13. $LTLC(FIN, FIN) - LTLC(REG, NO) \neq \emptyset$.

Lemma 14. $LTLC(FIN, NO) - (LTLC(REG, FIN) \cup LTLC(NO, REG)) \neq \emptyset$.
Synthesizing these results, we obtain

Theorem 5. *A diagram like that in Figure 1 holds true for families $LTLC(F_1, F_2)$ such that all inclusions are proper and all two families which are not linked by a path in this diagram are incomparable.*

References

[1] A. Ehrenfeucht, Gh. Păun, G. Rozenberg, On representing recursively enumerable languages by internal contextual languages, *Th. Computer Sci.*, 182 (1997).

[2] A. Ehrenfeucht, Gh. Păun, G. Rozenberg, Contextual grammars and formal languages, in *Handbook of Formal Languages* (G. Rozenberg, A. Salomaa, eds.), Springer-Verlag, Berlin, Heidelberg, 1997.

[3] L. Ilie, A non-semilinear language generated by an internal contextual grammar with finite choice, *Ann. Univ. Buc., Matem.-Inform. Series*, 45, 1 (1996), 63 – 70.

[4] S. Marcus, *Algebraic Linguistics. Analytical Models*, Academic Press, New York, 1967.

[5] S. Marcus, Contextual grammars, *Rev. Roum. Math. Pures Appl.*, 14 (1969), 1525 – 1534.

[6] S. Marcus, Contextual grammars and natural languages, in *Handbook of Formal Languages* (G. Rozenberg, A. Salomaa, eds.), Springer-Verlag, Berlin, Heidelberg, 1997.

[7] S. Marcus, C. Martin-Vide, Gh. Păun, Contextual grammars as generative models of natural languages, *Fourth Meeting on Mathematics of Language, MOL 4*, Philadelphia, 1995.

[8] C. Martin-Vide, A. Mateescu, J. Miquel-Verges, Gh. Păun, Internal contextual grammars with minimal, maximal and scattered use of contexts, *Proc. of the Fourth Bar-Ilan Symp. on Found. of AI, BISFAI 95* (M. Kappel, E. Shamir, eds.), Jerusalem, 1995, 132 – 142.

[9] Gh. Păun, *Contextual Grammars*, The Publ. House of the Romanian Academy, Bucharest, 1982 (in Romanian).

[10] Gh. Păun, A new class of contextual grammars, *Rev. Roum. Ling.*, 33 (1988), 167 – 171.

[11] Gh. Păun, Marcus contextual grammars. After 25 years, *Bulletin of the EATCS*, 52 (1994), 263 – 273.

[12] Gh. Păun, X. M. Nguyen, On the inner contextual grammars, *Rev. Roum. Math. Pures Appl.*, 25 (1980), 641 – 651.

Counting Problems over the Reals

Klaus Meer

RWTH Aachen
Lehrstuhl C für Mathematik
Templergraben 55
D-52062 Aachen, Germany
meer@rwth-aachen.de

Abstract. In this paper we introduce a complexity theoretic notion of counting problems over the real numbers. We follow the approaches of Blum, Shub, and Smale [4] for computability over $\mathbb{R}$ and of Grädel and Meer [9] for descriptive complexity theory in this setting and give a complete characterization of such problems by logical means.

1 Introduction

Many important problems in mathematics and computer science appear in the form of counting problems. As a classical example let us mention the question of counting the (real) zeros of polynomials, analyzed already in the last century (f.e. theorems by Bezout and Sturm, cf. [2]) or counting the dimension of a (semi-) algebraic set. Another example closely related to real zero counting is given by evaluating the number of sign changes in a sequence of reals. The latter has caused increasing attention again during the last years because of its extreme importance for dealing with the existential theory over real closed fields (see for example [14],[13]).

Even though an extensive theory of counting problems on finite domains has been developed in discrete computer science along the counting class #P introduced by Valiant [17], for the above mentioned problems (and many more) this approach fails; it refers to the Turing machine as underlying computational device and thus cannot deal appropriately with problems involving real (or complex) input data. In order to develop a similar theory but for real number problems we are led to a different machine model. Such a model was introduced in 1989 by Blum, Shub, and Smale [4]. It turns out to provide the right framework for dealing with counting problems as those mentioned above.

Recall that over finite alphabets #P is defined as class of problems for which the counting function is given by the number of accepting paths of a nondeterministic polynomial time Turing machine. Equivalently, this class could be modelled also by counting the number of guesses leading a NP-machine into an accepting state. This observation causes an ambiguity concerning the task to define #P within the Blum-Shub-Smale (henceforth: BSS) model. Counting accepting paths will always give a finite number whereas counting guesses can also yield infinitely many answers. Having in mind the zero-counting problem we consider the second approach the more natural. In section 2 we introduce the real analogue $\#P_{\mathbb{R}}$ of #P. This definition is substantiated because all $\#P_{\mathbb{R}}$-functions turn out to be computable.

The main scope of this paper is to give a logical characterization of $\#P_{\mathbb{R}}$. Starting point for such a characterization is a branch in complexity theory called descriptive complexity. It aims to capture complexity classes by logical means only. A cornerstone in this area with respect to the Turing model was a paper by Fagin [8]. There he described the class of NP problems as those sets of finite structures being definable in existential second order logic. Along these lines Saluja, Subrahmanyam, and Thakur [15] where able to classify #P-functions on ordered structures into a hierarchy of five distinct levels. Compton and

Grädel [5] extended the former work to arbitrary structures. A descriptive complexity theory for the BSS model was developed in [9] and further investigated in [6]. Most real number complexity classes can be captured by certain logics defined over so-called IR-structures. In section 3 we make available the basic ingredients of this area needed later on and express #P_{IR} functions via first-order logic on IR-structures. The main results are then presented in section 4. We show that on ordered IR-structures #P_{IR} again can be decomposed into five distinct levels. Each of these levels is given by a certain restriction on the number and type of quantifiers appearing in first-order formulas. Using the main theorem of [9] the most powerful class of formulas with respect to counting is already obtained by considering those of type $\exists x \forall y \psi$, where ψ is quantifier free first-order.

Contrary to the discrete situation where only the first level of the according hierarchy is known to be included in the class of polynomial time computable functions we show furthermore that over the reals the latter holds true for the two basic levels.

2 The class #P_{IR}

We assume the reader to be familiar with the computational model of Blum, Shub, and Smale and the notion of NP_{IR}-completeness (see [3],[4],[12]). We denote by P_{IR} (resp. NP_{IR}) the class of decision problems being decidable (resp. verifiable) in deterministic (resp. nondeterministic) polynomial time by a BSS machine. The class of functions from $IR^{\infty} := \bigoplus_{k \in IN} IR$ to IR^{∞} being (BSS-) computable in polynomial time is denoted by FP_{IR}. An example of a NP_{IR}-complete problem is given by 4-$FEAS$ (see [4]): given a (multivariate) polynomial f with real coefficients, $deg(f) \leq 4$, does there exist a real zero of f?

We want to define the class #P_{IR} of real counting problems. As already mentioned in the introduction for NP_{IR}-machines it is a difference whether to count accepting paths or accepting guesses. In order to deal with functions having result ∞ for certain counting tasks we prefer the second approach.

Definition 1. The class #P_{IR} is given by all functions $f : IR^{\infty} \to IN_0 \cup \{\infty\}$ such that there exists a NP_{IR}-machine M working on $IR^{\infty} \times IR^{\infty}$ and a polynomial q satisfying

$$f(y) = |\{z \in IR^{q(size(y))} | M(y,z) \text{ is an accepting computation}\}| \quad \forall y \in IR^{\infty}.$$

Remark. The dependence on q is used to avoid counting all $z \in IR^{\infty}$ with arbitrary size. On inputs of dimension $n \in IN$ machine M only takes into account that part of a guess which is bounded by a certain dimension $m := q(n)$, where q denotes some polynomial only depending on M. Without limiting the dimension of z functions in #P_{IR} would take values in $\{0, \infty\}$ only.

The following counting problems can easily be seen to belong to #P_{IR}: The "natural" nondeterministic algorithms for establishing the corresponding decision problems (i.e. deciding whether the result is ≥ 1) to be in NP_{IR} guess exactly those objects which are to be counted in the problem formulation (f.e. zeros of polynomials). We collect the problems in the following definition because they will be important later on in section 4 to separate the classes within the logical hierarchy for #P_{IR}.

Definition 2. 1) #4-$FEAS$: given a (multivariate) polynomial f with real coefficients, $deg(f) \leq 4$, count the number of real zeros of f.

2) #SC : given an ordered sequence $x_1, \ldots, x_n$ of real numbers, count the number of sign changes in the sequence, i.e. the number of pairs $(i,j) \in \{1, \ldots, n\}^2$ s.t. $i < j$, $sign(x_i) \cdot sign(x_j) < 0$ and $x_k = 0 \ \forall i < k < j$.

3) $\#SC_{red}$: this is a reduced form of problem $\#SC$. Again given an ordered sequence $x_1, \ldots, x_n$ of real numbers, count the number of indices $i \in \{1, \ldots, n\}$ such that there is at least one more sign change with respect to x_i, i.e. there is at least one $j > i$ such that $\text{sign}(x_i) \cdot \text{sign}(x_j) = -1$.
Note that for all sequences we have $\#SC \leq \#SC_{red}$.

As already mentioned above these counting tasks provide fundamental problems within the area of (semi-) algebraic geometry. Knowing the number of zeros of a (nondegenerated) polynomial system for example is crucial for homotopy methods used to compute approximations of zeros (see [1],[16]).

The following theorem substantiates the introduction of $\#P_{\mathbb{R}}$:

Theorem 3. *Every function in $\#P_{\mathbb{R}}$ is (BSS-) computable in single exponential time.* □

Clearly, if $P_{\mathbb{R}} \neq NP_{\mathbb{R}}$ we have $\#P_{\mathbb{R}} \not\subset FP_{\mathbb{R}}$. On the other hand we can ask whether any function $f : \mathbb{R}^\infty \to \mathbb{N}_0 \cup \{\infty\}$ in $FP_{\mathbb{R}}$ also belongs to $\#P_{\mathbb{R}}$. The problem here is that $FP_{\mathbb{R}}$ functions in polynomial time can compute numbers of exponential bit-size. At least as long as such an f is only moderately increasing we have the

Lemma 4. *Let $f : \mathbb{R}^\infty \to \mathbb{N}_0 \cup \{\infty\}$ be a function in $FP_{\mathbb{R}}$ such that there exists a polynomial q satisfying*

$$f(x) \leq 2^{q(n)} \ \forall x \in \mathbb{R}^n .$$

Then $f \in \#P_{\mathbb{R}}$. □

3 $\mathbb{R}$-structures and counting

In this section we first recall basic notions of $\mathbb{R}$-structures and their logics. This concept was first introduced in [9]; we refer to this paper as well as [6] for a closer study of descriptive complexity in the BSS model.
    Then we relate it to counting problems.

3.1 $\mathbb{R}$-structures

We suppose the reader familiar with the main terminology of logic as well as with the concepts of vocabulary, first-order formula or sentence, interpretation and structure (see for example [7]).

Definition 5. *Let L_s, L_f be finite vocabularies where L_s may contain relation and function symbols, and L_f contains function symbols only. A $\mathbb{R}$-structure of signature $\sigma = (L_s, L_f)$ is a pair $\mathfrak{D} = (\mathcal{A}, \mathcal{F})$ consisting of*

(i) *a finite structure $\mathcal{A}$ of vocabulary L_s, called the* skeleton *of $\mathfrak{D}$, whose universe A will also be said to be the* universe of $\mathfrak{D}$, *and*
(ii) *a finite set $\mathcal{F}$ of functions $X : A^k \to \mathbb{R}$ interpreting the function symbols in L_f.*

Definition 6. *Let $\mathfrak{D}$ be a $\mathbb{R}$-structure with skeleton $\mathcal{A}$. We denote by $|A|$ and also by $|\mathfrak{D}|$ resp. the cardinality of the universe A of $\mathcal{A}$. This number is called the* size *of the structure $\mathfrak{D}$. A $\mathbb{R}$-structure $\mathfrak{D} = (\mathcal{A}, \mathcal{F})$ is* ranked *if there is a unary function symbol $r \in L_f$ whose interpretation ρ in $\mathcal{F}$ bijects A with $\{0, 1, \ldots, |A| - 1\}$. The function ρ is called* ranking.

We will write $i < j$ for $i, j \in A$ iff $r(i) < r(j)$. A k-ranking on A is a bijection between A^k and $\{0, 1, \ldots, |A|^k - 1\}$. It can easily be defined if a ranking is available. We denote by ρ^k the interpretation of the k-ranking induced by ρ.

Throughout this paper we suppose all $\mathbb{R}$-structures to be ranked. We therefore notationally suppress the symbol r in the sets $\mathcal{F}$ considered. The basic logic important for our work is first-order logic which we are going to define now.

Fix a countable set $V = \{v_0, v_1, \ldots, \}$ of variables. These variables range only over the skeleton; we do not use element variables taking values in $\mathbb{R}$.

Definition 7. The language $\mathrm{FO}_{\mathbb{R}}$ contains, for each signature $\sigma = (L_s, L_f)$ a set of formulas and terms. Each term t takes, when interpreted in some $\mathbb{R}$-structure, values in either the skeleton, in which case we call it an *index term*, or in $\mathbb{R}$, in which case we call it a *number term*. Terms are defined inductively as follows

(i) The set of index terms is the closure of the set V of variables under applications of function symbols of L_s.

(ii) Any real number is a number term.

(iii) If $h_1, \ldots, h_k$ are index terms and X is a k-ary function symbol of L_f then $X(h_1, \ldots, h_k)$ is a number term.

(iv) If t, t' are number terms, then so are $t + t'$, $t - t'$, $t \times t'$, t/t' and sign(t).

Atomic formulas are equalities $h_1 = h_2$ of index terms, equalities $t_1 = t_2$ and inequalities $t_1 < t_2$ of number terms, and expressions $P(h_1, \ldots, h_k)$ where P is a k-ary predicate symbol in L_s and $h_1, \ldots, h_k$ are index terms.

The set of formulas of $\mathrm{FO}_{\mathbb{R}}$ is the smallest set containing all atomic formulas and which is closed under Boolean connectives and quantification $(\exists v)\psi$ and $(\forall v)\psi$. Note that we do *not* consider formulas $(\exists x)\psi$ where x ranges over $\mathbb{R}$.

Example 1 (cf. [6],[9]). Let L_s be the empty set and L_f be $\{r, X\}$ where both function symbols have arity 1. Then, a simple class of ranked $\mathbb{R}$-structures with signature (L_s, L_f) is obtained by letting A be a finite set A, $r^{\mathfrak{D}}$ any ranking on A and $X^{\mathfrak{D}}$ any unary function $X^{\mathfrak{D}} : A \to \mathbb{R}$. Since $r^{\mathfrak{D}}$ bijects A with $\{0, 1, \ldots, n - 1\}$ where $n = |A|$, this $\mathbb{R}$-structure is a point $x_{\mathfrak{D}}$ in $\mathbb{R}^\infty$. Conversely, for each point $x \in \mathbb{R}^\infty$ there is an $\mathbb{R}$-structure $\mathfrak{D}$ such that $x = x_{\mathfrak{D}}$. Thus, this class of structures models $\mathbb{R}^\infty$. On the other hand any $\mathbb{R}$-structure $\mathfrak{D} = (A, \mathcal{F})$ can be identified with a vector $e(\mathfrak{D}) \in \mathbb{R}^\infty$ using a natural encoding ([9]).

Example 1 allows us to speak about complexity classes among $\mathbb{R}$-structures. If S is a set of $\mathbb{R}$-structures closed under isomorphisms, we say that S belongs to a complexity class $\mathcal{C}$ over the reals if the set $\{e(\mathfrak{D}) \mid \mathfrak{D} \in S\}$ belongs to $\mathcal{C}$.

Remark. If ρ is a ranking on A and $|A| = n$ then, there are elements $o, 1 \in A$ such that $\rho(o) = 0$ and $\rho(1) = n - 1$. Note that these two elements are first-order definable.

In order to describe the class $\mathrm{NP}_{\mathbb{R}}$ it turns out to be fruitful also considering an extension of first-order logic.

Definition 8. We say that ψ is an *existential second-order sentence* (of signature $\sigma = (L_s, L_f)$) if $\psi = \exists Y_1 \ldots \exists Y_r \phi$ where ϕ is a first-order sentence in $\mathrm{FO}_{\mathbb{R}}$ of signature $(L_s, L_f \cup \{Y_1, \ldots, Y_r\})$. The symbols $Y_1, \ldots, Y_r$ will be called *function variables*. The sentence ψ is true in a $\mathbb{R}$-structure $\mathfrak{D}$ of signature σ when there exist interpretations of $Y_1, \ldots, Y_r$ such that ϕ holds true on $\mathfrak{D}$. The set of existential second-order sentences will be denoted by $\exists \mathrm{SO}_{\mathbb{R}}$. Together with the interpretation above it constitutes *existential second-order logic*.

The following example already gives the right idea to relate counting problems with logics on $\mathbb{R}$-structures.

Example 2 [9]. Let us see how to describe 4-*FEAS* with an existential second-order sentence. Consider the signature $(\emptyset, \{r, c\})$ where the arities of r and c are 1 and 4 respectively, and require that r is interpreted as a ranking.

Let $\mathfrak{D} = (A, \mathcal{F})$ be any $\mathbb{R}$-structure where $\mathcal{F}$ consists of interpretations $C : A^4 \to \mathbb{R}$ and $\rho : A \to \mathbb{R}$ of c and r. Let $n = |A| - 1$ so that ρ bijects A with $\{0, 1, \ldots, n\}$. Then $\mathfrak{D}$ defines a homogeneous polynomial $\widehat{g} \in \mathbb{R}[X_0, \ldots, X_n]$ of degree four, namely

$$\widehat{g} = \sum_{(i,j,k,\ell) \in A^4} C(i, j, k, \ell) X_i X_j X_k X_\ell.$$

We obtain an arbitrary, that is, not necessarily homogeneous, polynomial $g \in \mathbb{R}[X_1, \ldots, X_n]$ of degree four by setting $X_0 = 1$ in $\widehat{g}$. We also say that $\mathfrak{D}$ defines g. Notice that for every polynomial g of degree four in n variables there is a $\mathbb{R}$-structure $\mathfrak{D}$ of size $n + 1$ such that $\mathfrak{D}$ defines g.

Denote by o, 1, ō and ī the first and last elements of A and A^4 with respect to ρ and ρ^4 respectively. The following sentence quantifies two functions $X : A \to \mathbb{R}$ and $Y : A^4 \to \mathbb{R}$

$$\psi \equiv (\exists X)(\exists Y) \ \Big(Y(\bar{o}) = C(\bar{o}) \ \& \ Y(\bar{1}) = 0 \ \& \ X(o) = 1 \ \&$$
$$\& \ \forall u_1 \ldots \forall u_4 \ [u \neq \bar{o} \Rightarrow \exists v_1 \ldots \exists v_4 \ (\rho^4(u) = \rho^4(v) + 1)$$
$$\& \ Y(u) = Y(v) + C(u) X(u_1) X(u_2) X(u_3) X(u_4)] \Big).$$

Here, if $a_i = \rho^{-1}(i)$ for $i = 1, \ldots, n$ then, $(X(a_1), \ldots, X(a_n)) \in \mathbb{R}^n$ describes the zero of g and $Y(u)$ is the partial sum of all its monomials up to $u = (u_1, \ldots, u_4) \in A^4$ evaluated at the point $(X(a_1), \ldots, X(a_n))$.

The sentence ψ describes 4-*FEAS* in the sense that for any $\mathbb{R}$-structure $\mathfrak{D}$ it holds $\mathfrak{D} \models \psi$ if and only if the polynomial g of degree four defined by $\mathfrak{D}$ has a real zero.

The fact that existential second order logic describes a $NP_{\mathbb{R}}$-complete problem is not fortuitous.

Theorem 9 [9]. $\exists SO_{\mathbb{R}} = NP_{\mathbb{R}}$. □

Here two remarks are essential for what will follow: The above example shows that the number of choices for (X, Y) which cause the sentence ψ to hold true equals the number of real zeros of the given polynomial (note that Y is unique as soon as X is determined). This leads to the idea of counting the satisfying assignments for first-order formulas. On the other hand, a closer analysis of the proof of Theorem 9 in [9] shows that first-order formula with two quantifier alternations suffice to represent a $NP_{\mathbb{R}}$ computation - just as it is the case with Fagin's theorem (see [8]). Thus the hierarchy we will define collapses after at most five levels.

Definition 10. a) Let $\sigma = (L_s, L_f)$ be a vocabulary and let $\mathbf{D}$ be a family of $\mathbb{R}$-structures over vocabulary σ. A function $f : \mathbf{D} \to \mathbb{N}_0 \cup \{\infty\}$ belongs to class $\#FO_{\mathbb{R}}$ if the following holds: there exists a first order formula $\psi_{\mathbf{D}}$ over vocabulary $(L_s, L_f \cup \{X_1, \ldots, X_l\})$ such that

$$f(\mathfrak{D}) = |\{(X_1, \ldots, X_l, z_1, \ldots, z_m) : \mathfrak{D} \models \psi_{\mathbf{D}}(X, z)\}| \quad \forall \mathfrak{D} \in \mathbf{D}$$

Here $X := (X_1, \ldots, X_l)$ denotes a sequence of functions from some A^{k_i} to $\mathbb{R}$, $1 \leq i \leq l$, and $z := (z_1, \ldots, z_m)$ denotes a sequence of first order variables. Moreover, l as well as m are natural numbers with $l + m > 0$ depending on ψ_D only.

b) The subclasses $\#\Sigma_{n\mathbb{R}}$ and $\#\Pi_{n\mathbb{R}}, n \in \mathbb{N}$ of $\#FO_\mathbb{R}$ are defined similarly by restricting ψ_D to be a $\Sigma_{n\mathbb{R}}$ resp. $\Pi_{n\mathbb{R}}$ formula. Here $\Sigma_{n\mathbb{R}}$ resp. $\Pi_{n\mathbb{R}}$ consist of those formulas in prenex normal form with n alternating blocks of quantifiers beginning with $\exists$ resp. $\forall$.

In the following example the problems of definition 2 are classified into some of these classes. To this aim we use straightforward representations of the according problem instances as $\mathbb{R}$-structures.

Example 3. 1) $\#4$-$FEAS$: We have already seen how to describe the 4-$FEAS$ problem via an existential second order formula in example 2. The according counting function is then given by $|\{(X, Y) | \psi_{4\text{-}FEAS}(X, Y)\}|$, where

$$\psi_{4\text{-}FEAS}(X, Y) \equiv \Big(Y(\bar{0}) = C(\bar{0}) \ \& \ Y(\bar{1}) = 0 \ \& \ X(o) = 1 \ \&$$
$$\& \ \forall u \in A^4 \ \forall v \in A^4 \ \exists w \in A^4 \ [u \neq \bar{0} \Rightarrow$$
$$\{(u > v \wedge w > v) \rightarrow w \geq u\} \ \&$$
$$\& \ Y(u) = Y(v) + C(u)X(u_1)X(u_2)X(u_3)X(u_4)] \Big),$$

which shows $\#4$-$FEAS$ to belong to $\#\Pi_{2\mathbb{R}}$.

2) $\#SC$: A sequence of real numbers is represented as $\mathbb{R}$-structure $\mathfrak{D} = (A, \mathfrak{S})$, where $\mathfrak{S} : A \rightarrow \mathbb{R}$. The counting function for $\#SC$ is given as

$$|\{(i, j) \in A^2 \mid \mathfrak{D} \models i < j \wedge \text{sign} \, \mathfrak{S}(i) \cdot \text{sign} \, \mathfrak{S}(j) = -1 \wedge \forall k \, (i < k < j \Rightarrow \mathfrak{S}(k) = 0) \ .$$

We thus have $\#SC \in \#\Pi_{1\mathbb{R}}$.

3) $\#SC_{red}$: For the reduced sign change problem we consider the same structure as in 2). The corresponding counting function now is

$$|\{i \in A \mid \mathfrak{D} \models \exists j \in A \ j > i \wedge \text{sign} \, \mathfrak{S}(i) \cdot \text{sign} \, \mathfrak{S}(j) = -1\}|,$$

showing $\#SC_{red} \in \#\Sigma_{1\mathbb{R}}$.

The above problems are not chosen arbitrarily. In the next section they serve to show the distinctness of the corresponding classes they belong to.

The justification for defining class $\#FO_\mathbb{R}$ with respect to $\#P_\mathbb{R}$ is given in the following

Theorem 11. $\#P_\mathbb{R} = \#FO_\mathbb{R} = \#\Pi_{2\mathbb{R}}$.

Proof. The proof is straightforward from the corresponding result in [15] for the discrete classes and the proof of Theorem 9 in [9]. The latter shows $\#FO_\mathbb{R} \subset \#\Pi_{2\mathbb{R}}$ and hence equality (cf. example 3). Any problem in $\#FO_\mathbb{R}$ can be proven to belong to $\#P_\mathbb{R}$ by taking a $NP_\mathbb{R}$ machine which guesses a satisfying assignment. The reverse inclusion follows by noting that in the proof of Theorem 9 every accepting computation on a $NP_\mathbb{R}$ machine corresponds to a satisfying assignment of the related existential second order formula. $\square$

4   The five distinct levels

It is known that for ordered finite structures in the discrete setting the first level $\#\Sigma_0$ of the hierarchy for $\#P$ consists of polynomial time computable functions only, whereas problems in the second level $\#\Sigma_1$ at least allow a fully polynomial time randomized approximation scheme (see [15]). Over the reals we can say more because of the possibly uncountable number of satisfying assignments for a function from A^k to $\mathbb{R}$.

Theorem 12. $\#\Sigma_{1\mathbb{R}} \subset FP_{\mathbb{R}}$.

Proof. Let a function $f : \mathbf{D} \to \mathbb{N}_0 \cup \{\infty\}$ in $\#\Sigma_{1\mathbb{R}}$ be given by

$$f(\mathfrak{D}) = |\{(X_1, \ldots, X_l, z_1 \ldots, z_m) | \mathfrak{D} \models \exists x \in A^k \ \psi(X, x, z)\}| \quad \forall \mathfrak{D} \in \mathbf{D},$$

where ψ is quantifier free. Fix a $z^* \in A^m$ and an $x^* \in A^k$. Consider all terms $X_i(y)$ appearing in this formula (where $X = (X_1, \ldots, X_l)$ as usual). For each one among them (i.e. different i or different argument y) we introduce a real variable. Note that there will be at most a constant number s of variables where s only depends on ψ but not on the size n of $\mathfrak{D}$. (Up to this point the proof closely followed the one in [15] for showing $\#\Sigma_0 \in FP$). Thus for every fixed (x^*, z^*) we obtain a first order formula $\rho(t_1, \ldots, t_s)$ with s free real variables. ρ can be computed uniformly in polynomial time for given $\mathfrak{D}$ and given (x^*, z^*). Now we consider the existential problem

$$\exists t_1, \ldots, t_s \in \mathbb{R} \ \rho(t_1, \ldots, t_s) \ .$$

There are three cases to be analyzed: First assume at least one of the X_i to have an arity ≥ 1. There exists a dimension n_0 such that for every structure $\mathfrak{D}$ of size $n \geq n_0$ the satisfyability of the above formula in $\mathfrak{D}$ implies the existence of infinitely many X s.t. $\psi(X, x^*, z^*)$ holds. This is true because in that case X_i must be defined on more than s arguments showing that for large enough structures some of these arguments do not appear as arguments of X_i in ψ. Hence their values do not affect validity of ψ. Now if the structure $\mathfrak{D}$ has size at least n_0 apply quantifier elimination to $\exists t_1, \ldots, t_s \in \mathbb{R} \ \rho(t_1, \ldots, t_s)$ which can be performed in constant time with respect to n ([10],[14]). If the obtained sentence holds true we have result ∞ for the underlying counting problem, otherwise 0. For structures of dimension less than n_0 we can compute the counting function using a brute algorithm (see Theorem 3) without affecting the overall polynomial time performance. Due to the fact that there are at most polynomially many different tuples (x^*, z^*) the first case is solvable in polynomial time.

Secondly, assume all X_i to be nullary functions. Then the corresponding first order formula over the reals has the fixed number l of free real variables (not depending on $|\mathfrak{D}|$). Applying the *4-FEAS* completeness proof in [4] as well as Theorem 3 we can reduce the problem in constant time to counting the zeros of a polynomial which has constant degree and a constant number of variables (with respect to n). Again this can be done in constant time.

Finally, if ψ does not depend on X (i.e. we only count z) there are only polynomially many assignments for z which have to be checked. $\qquad\square$

We have already seen in the last section that $\#P_{\mathbb{R}}$ splits into at most five levels. The main goal of this section is to show that all these levels are distinct. To this aim we will further exploit the properties of the examples introduced in definition 2.

Theorem 13. *The class $\#P_{\mathbb{R}}$ splits into five different levels*

$$\#\Sigma_{0\mathbb{R}} = \#\Pi_{0\mathbb{R}} \subsetneq \#\Sigma_{1\mathbb{R}} \subsetneq \#\Pi_{1\mathbb{R}} \subsetneq \#\Sigma_{2\mathbb{R}} \subsetneq \#\Pi_{2\mathbb{R}} = \#P_{\mathbb{R}} \ .$$

Proof. The proof is done by showing that every problem of definition 2 will separate the class it belongs to from the next lower one (see example 3).

$\#\Sigma_{0\mathbb{R}} \subsetneq \#\Sigma_{1\mathbb{R}}$: Assume $\#SC_{red} \in \#\Sigma_{0\mathbb{R}}$; let ψ be the corresponding $\Sigma_{0\mathbb{R}}$ formula, i.e.

$$|\{(X, z)|\mathfrak{D} \models \psi(X, z)\}|$$

counts the number of indices causing another sign change in the sequence given by structure $\mathfrak{D}$.

Because of the proof of Theorem 12 the above ψ must be independent of X_i's having arity ≥ 1. Thus we count $|\{(X, z)|\mathfrak{D} \models \psi(X, z)\}|$, where all X_i in X are nullary. Let m be the arity of z; consider a family of $\mathbb{R}$-structures

$$\mathfrak{D}^{(i)} = (A, \mathfrak{S}^{(i)}) \,, \quad \mathfrak{S}^{(i)} = (1, 0, \ldots, 0, \underbrace{-1}_{i}, 0, \ldots, 0)$$

for i varying between 2 and $n = |A|$. Here the size n of A is assumed to be large enough to satisfy the conditions necessary below. For all $2 \leq i \leq n$ it is $\#SC_{red}(\mathfrak{D}^{(i)}) = 1$. Let $(X^{(i)}, z^{(i)})$ be the unique assignment with $\mathfrak{D}^{(i)} \models \psi(X^{(i)}, z^{(i)})$. Obviously, $z^{(i)}$ must depend on i. If n is large enough there exist i_0 and j_0, $i_0 < j_0$ such that neither $z^{(i_0)}$ depends on j_0 nor $z^{(j_0)}$ depends on i_0. Consider the new structure $\mathfrak{D}^{(i_0, j_0)} = (A, \mathfrak{S}^{(i_0, j_0)})$ where

$$\mathfrak{S}^{(i_0, j_0)} = (1, 0, \ldots, 0, \underbrace{-1}_{i_0}, 0, \ldots, 0, \underbrace{-1}_{j_0}, 0, \ldots, 0) \,.$$

The above conditions on i_0, j_0 imply $\mathfrak{D}^{(i_0, j_0)} \models \psi(X^{(i_0)}, z^{(i_0)})$ as well as $\mathfrak{D}^{(i_0, j_0)} \models \psi(X^{(j_0)}, z^{(j_0)})$ (note that $X^{(i_0)}$ and $X^{(j_0)}$ do not depend on any element of A). However, $z^{(i_0)} \neq z^{(j_0)}$ and $\#SC_{red}(\mathfrak{D}^{(i_0, j_0)}) = 1$ gives a contradiction.

$\#\Sigma_{1\mathbb{R}} \subsetneq \#\Pi_{1\mathbb{R}}$: The inclusion $\#\Sigma_{1\mathbb{R}} \subset \#\Pi_{1\mathbb{R}}$ follows exactly as in [15], noting that instead of counting tuples (X, z) satisfying a sentence $\exists x\ \psi(X, x, z)$ it suffices to count those tuples (X, x^*, z) satisfying $\psi(X, x^*, z)$, where x^* denotes the lexicographic smallest among those x with $\psi(X, x, z)$. The latter condition is $\Pi_{1\mathbb{R}}$ definable.

To show distinctness of both classes assume $|\{(X, z)|\mathfrak{D} \models \exists x\ \psi(X, x, z)\}|$ to be a $\Sigma_{1\mathbb{R}}$ formula for computing $\#SC(\mathfrak{D})$ for according problem instances $\mathfrak{D}$. We proceed exactly as in the foregoing part. Choose the same structures $\mathfrak{D}^{(i)}$ and note that according to Theorem 12 again all components X_i in X must be nullary functions. Then for every $2 \leq i \leq n$ there exists an $x^{(i)}$ such that there is a unique satisfying assignment $(X^{(i)}, z^{(i)})$ for $\psi(X, x^{(i)}, z)$. Again taking n large enough we can find i_0, j_0 with $(x^{(i_0)}, z^{(i_0)})$ not depending on j_0 and vice versa. The structure $\mathfrak{D}^{(i_0, j_0)}$ introduced above yields a contradiction once more.

$\#\Sigma_{2\mathbb{R}} \subsetneq \#\Pi_{2\mathbb{R}}$: Suppose $\#4\text{-}FEAS \in \#\Sigma_{2\mathbb{R}}$ and let

$$|\{(X, z)|\mathfrak{D} \models \exists x\ \forall y\ \psi(X, x, y, z)\}| \,, \forall \mathfrak{D} \in \mathbf{D}$$

be the according counting formula in $\Sigma_{2\mathbb{R}}$. Here $\mathbf{D}$ is the set of $\mathbb{R}$-structures representing a degree 4 polynomial as in example 2. Consider a particular polynomial f in n variables given by

$$f(x_1, \ldots, x_n) = \sum_{i=1}^{n} (a_i \cdot x_i - 1)^2 \ .$$

Here the a_i are taken as non-vanishing real coefficients. Thus f has exactly one real zero. Let (X^*, z^*) be the unique assignment such that

$$\mathfrak{D} \models \exists x \ \forall y \ \psi(X^*, x, y, z^*)$$

holds for the structure $\mathfrak{D}$ representing f. Moreover, fix an x^* with

$$\mathfrak{D} \models \forall y \ \psi(X^*, x^*, y, z^*) \ .$$

Now, if in the above situation n is sufficiently large we can find an $i_0 \in A$ such that both x^* and z^* do not depend on i_0. Note that according to our representation of f as $\mathbb{R}$-structure this implies the validity of $\forall y \ \psi(X^*, x^*, y, z^*)$ to be independent of the value for a_{i_0}. Consider a substructure $\tilde{\mathfrak{D}}$ of $\mathfrak{D}$ by taking $\tilde{A} = A \setminus \{i_0\}$ and $\tilde{X}$ being obtained from X^* by deleting all occurences of those tuples involving i_0. The polynomial $\tilde{f}$ related with $\tilde{\mathfrak{D}}$ is of the form

$$\tilde{f} = 1 + \sum_{\substack{j=1 \\ j \neq i_0}}^{n} (a_j \cdot x_j - 1)^2$$

and thus has no zero. On the other hand $\tilde{\mathfrak{D}} \models \forall y \ \psi(\tilde{X}, x^*, y, z^*)$ still holds because x^* and z^* do not depend on i_0 and universal formulas remain valid by passing to substructures. Hence the above $\Sigma_{2\mathbb{R}}$ formula does not count the zeros of $\tilde{f}$ which gives the claim.

The missing inclusion $\#\Pi_{1\mathbb{R}} \subsetneq \#\Sigma_{2\mathbb{R}}$ can be shown in a similar fashion. Because of lack of space the proof will be postponed to the full paper version. $\square$

We want to finish this paper mentioning that from class $\#\Pi_{1\mathbb{R}}$ on not much can be said with respect to complexity. There are counting problems in $\#\Pi_{1\mathbb{R}}$ the according decision problems of which are $NP_{\mathbb{R}}$-complete. And there are problems in $\#\Pi_{2\mathbb{R}} \setminus \#\Sigma_{2\mathbb{R}}$ related to polynomial time solvable decision problems: the above proof similarly could be used for say counting the solutions of a system of linear equations.

References

1. Allgower, E.L., Georg, K.: Continuation and Path Following. Acta Numerica (1992) 1 - 64
2. Benedetti, R., Risler, J.J.: Real algebraic and semi-algebraic sets. Hermann (1990)
3. Blum, L., Cucker, F., Shub, M., Smale, S.: Complexity and Real Computation. Springer-Verlag (to appear)
4. Blum, L., Shub, M., Smale, S.: On a theory of computation and complexity over the real numbers: NP-completeness, recursive functions and universal machines. Bulletin American Mathematical Society **21** (1989) 1–46
5. Compton, K.J., Grädel, E.: Logical Definability of Counting Functions Proceedings of IEEE Conference on Structure in Complexity Theory (1994) 255–266
6. Cucker, F.: Meer, K.: Logics which capture complexity classes over the reals. Extended abstract to appear in: Proc. 11th International Symposium on Fundamentals of Computation Theory, Krakow, Lecture Notes in Computer Science, Springer 1997
7. Ebbinghaus, H.D., Flum, J.: Finite Model Theory. Springer-Verlag (1995)
8. Fagin, R.: Generalized first-order spectra and polynomial-time recognizable sets. SIAM-AMS Proc. **7** (1974) 43–73

9. Grädel, E., Meer, K.: Descriptive complexity theory over the real numbers. In J. Renegar, M. Shub, and S. Smale (editors): The Mathematics of Numerical Analysis, Lectures in Applied Mathematics 32, American Mathematical Society (1996) 381–404

10. Heintz, J., Roy, M.F., Solerno, P.: On the complexity of semialgebraic sets. Proceedings IFIP 1989, San Francisco, North-Holland (1989) 293–298

11. Heintz, J., Krick, T., Roy, M.F., Solerno, P.: Geometric Problems solvable in single exponential time. In: Proc. 8th Conference AAECC, LNCS 508 (1990) 11–23

12. Meer, K., Michaux, C.: A survey on real structural complexity theory. Bulletin of the Belgian Math. Soc. Simon Stevin 4 (1997) 113–148

13. Pedersen, P.: Multivariate Sturm theory. In: Proc. 9th Conference AAECC, LNCS 539 (1991) 318–331

14. Renegar, J.: On the computational Complexity and Geometry of the first-order Theory of the Reals , I - III. Journal of Symbolic Computation 13 (1992) 255–352

15. Saluja, S., Subrahmanyam, K.V., Thakur, M.N.: Descriptive Complexity of #P Functions. In Proc. 7th IEEE Symposium on Structure in Complexity Theory (1992) 169–184

16. Shub, M., Smale, S.: Complexity of Bezout's Theorem I : Geometric aspects. Journal of the AMS 6 (1993) 459–501

17. Valiant, L.: The complexity of computing the permanent. Theoretical Computer Science 8 (1979) 189–201

On the Influence of the State Encoding on OBDD-Representations of Finite State Machines

Christoph Meinel, Thorsten Theobald*

FB IV – Informatik, Universität Trier, D–54286 Trier, Germany
{meinel,theobald}@ti.uni-trier.de

Abstract. Ordered binary decision diagrams are an important data structure for the representation of Boolean functions. Typically, the underlying variable ordering is used as an optimization parameter. When finite state machines are represented by OBDDs the state encoding can be used as an additional optimization parameter. In this paper, we analyze the influence of the state encoding on the OBDD-representations of counter-type finite state machines. In particular, we prove lower bounds, derive exact sizes for important encodings and construct a worst-case encoding which leads to exponential-size OBDDs.

1 Introduction

Ordered binary decision diagrams (OBDDs) introduced by Bryant [Bry86] provide an efficient graph-based data structure for Boolean functions (for a survey see [Bry92] or [Weg94]). The main optimization parameter of OBDDs is the underlying variable ordering. Many research efforts have tried to characterize the complexity of the relevant variable ordering problems [THY93, MS94, BW96] and to come up with efficient optimization algorithms for obtaining large size reductions without aiming at the global minimum [Rud93, BMS95]. Unfortunately, there are many important applications, in particular in the analysis of finite automata/finite state machines [CBM89, CM95], where this optimization technique reaches its limits.

Quite recently, in [MT96] and [MST97] it was shown that in the context of finite state machines the state encoding can be used as an additional optimization parameter: In particular, the optimization techniques for finding suitable state encodings can be well integrated into existing algorithms for finding good variable orderings.

These optimization techniques immediately raise the question *in which extent* the choice of the state encoding can influence the OBDD-size at all. We will consider classes of counters which have a simple structure but which appear in numerous practical examples. For these classes, we analyze the relationship between the state encoding and the OBDD-size from a combinatorial point of view and give some precise answers concerning this relationship.

---

* Supported by DFG-Graduiertenkolleg "Mathematische Optimierung".

In particular, we consider the autonomous counter and the loop counter shown in Figure 1 [GDN92].

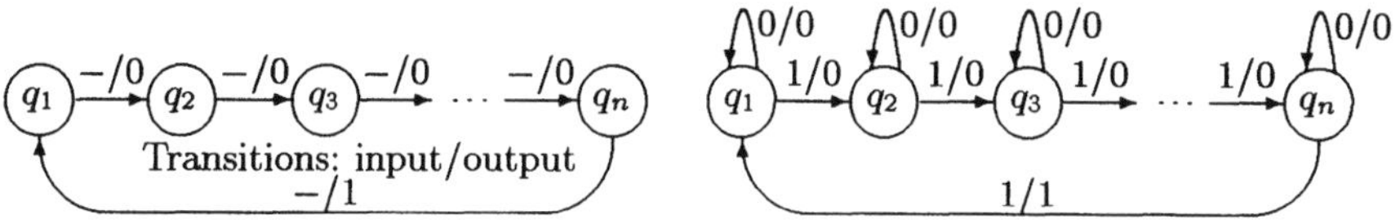

(a) Autonomous counter   (b) Loop counter

Fig. 1. Counter types

In Figure 1 (a), the input symbol '−' means that this transition takes place for both input 1 and 0. Furthermore we analyze an acyclic counter which can be constructed out of the autonomous counter by deleting the "backward" edge from the last state to the first state. The results, although derived for a quite specific class of finite state machines, serve as reference examples for the task of finding re-encodings. The main contributions of this work are:

1. When fixing the variable ordering in a reasonable way, we derive the exact OBDD-sizes for the counters under some important encodings.
2. We present lower bounds for the OBDD-sizes of counter encodings which are very close to the derived OBDD-sizes for the standard encoding. These bounds underline the suitability of the standard encoding in this context.
3. We construct worst-case encodings which lead to exponential-size OBDDs and hence demonstrate the sensitivity in choosing the appropriate state encoding.
4. In general, we give some ideas for the analysis of an important and still growing topic, in which most of the previously known results are based on experimental work.

2 Preliminaries

2.1 Ordered binary decision diagrams

An *ordered binary decision diagram* (OBDD) is defined as a rooted directed acyclic graph with two sink nodes which are labeled 1 and 0. Each internal (= non-sink) node is labeled by an input variable x_i and has two outgoing edges, labeled 1 and 0 (in the diagrams the 0-edge is indicated by a dashed line). A linear variable ordering π is placed on the input variables. The variable occurrences on each OBDD-path have to be consistent with this ordering. An OBDD computes a Boolean function $f : \{0, 1\}^n \to \{0, 1\}$ in a natural manner: each assignment to the input variables x_i defines a unique path through the graph from the root to the sinks. The label of the sink gives the value of the function on that input.

The OBDD is called *reduced* if it does not contain any vertex v such that the 0-edge and the 1-edge of v leads to the same node, and it does not contain any distinct vertices v and v' such that the subgraphs rooted in v and v' are isomorphic. It is well-known that reduced OBDDs are a unique representation of Boolean functions $f : \{0,1\}^n \to \{0,1\}$ w.r.t. a given variable ordering [Bry86]. The *size* of an OBDD is the number of its internal nodes.

2.2 Finite state machines

Let $M = (Q, I, O, \delta, \lambda, Q_0)$ be a finite state machine, where Q is the set of states, I the input alphabet, O the output alphabet, $\delta : Q \times I \to Q$ the next-state function, $\lambda : Q \times I \to O$ the output function and Q_0 the set of initial states. As usual in VLSI design, all components of the state machine are assumed to be binary encoded. Let p be the number of input bits and n be the number of state bits. In particular, with $B = \{0,1\}$, δ is a function $B^n \times B^p \to B^n$. For a finite state machine M, the characteristic function of its *transition relation* is defined by

$$T(x_1, \ldots, x_n, y_1, \ldots, y_n, e_1, \ldots, e_p) = T(x, y, e) = \prod_{1 \le i \le n} (y_i \equiv \delta_i(x, e)).$$

Hence, the function T computes the value 1 for a triple (x, y, e) if and only if the state machine in state x and input e enters the state y.

In [BCL+94] and [CBM89] it was shown that the representation of a finite state machine by means of its transition relation goes well together with typical tasks for analyzing finite state machines like checking equivalence. All these applications are based on a reachability analysis. Hence, we can consider equivalently the transition relation of the underlying non-deterministic machine in which the inputs have been eliminated. In terms of Boolean manipulation this corresponds to an existential quantification over the inputs.

We want to remark that the most efficient implementations of this general concept work with a *partitioned* transition relation [BCL+94].

2.3 Variable orderings

For a finite state machine M, we derive the reduced OBDD-size for the characteristic function of its transition relation. This size is shortly called *OBDD-size of M*. The OBDD-size crucially depends on the chosen variable ordering. There are two variable orderings which often appear in connection with finite state machines: the *separated* ordering $x_1, \ldots, x_n, y_1, \ldots, y_n$ and the *interleaved* ordering $x_1, y_1, x_2, y_2, \ldots, x_n, y_n$ [ATB94].

For practical applications, the interleaved variable ordering is often superior to the separated ordering. If one considers for example a deterministic autonomous (i.e. input-independent) machine with a bijective next-state function, then the OBDD w.r.t. the separated ordering has exponential size. The reason is that after reading the variables $x_1, \ldots, x_n$ all induced subfunctions are different. The restriction to fix the variable ordering is reasonable in our context, as we want to analyze the effect of different state encodings.

3 The Lower Bound

We will investigate lower bounds for the OBDD-size of an autonomous counter with 2^n states where we keep the interleaved variable ordering fixed and vary over all $(2^n)!$ possible n-bit state encodings. For the first lower bound we use that the next-state function of the autonomous counter is bijective.

Theorem 1. *Let M_{2^n} be an autonomous finite state machine with 2^n states, n encoding bits and a bijective next-state function. The OBDD-size of M_{2^n} w.r.t. the interleaved variable ordering is at least $3n$.*

Proof. We show that for each $1 \leq i \leq n$, there are at least 3 nodes with label x_i or y_i. As the next-state function is bijective, each of the $2n$ variables appears on every path from the root to the 1-sink. There exists a node A labeled by x_i whose 1-edge leads to a sub-OBDD which does not represent the constant 0 (otherwise the next-state function cannot be a total function). In this sub-OBDD there exists a path from the root to the sink, and therefore the root must be labeled by y_i (see Figure 2 (a)). Analogously, there exists a node B labeled by x_i whose 0-edge leads to a sub-OBDD with root label y_i.

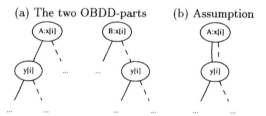

(a) The two OBDD-parts   (b) Assumption

Fig. 2. Lower bound

If we assume that A and B are identical and that the two roots of the sub-OBDDs are identical, then node A can be reduced, see Figure 2 (b). The resulting OBDD has a path from the root to the 1-sink on which the variable x_i does not appear, a contradiction. □

The next theorem improves the lower bound by explicitly using the properties of a counter.

Theorem 2. *The OBDD-size of an autonomous counter with 2^n states, n encoding bits and interleaved variable ordering is at least $4n - 1$.*

Proof. First, we show that every variable $x_2, \ldots, x_n$ must appear at least twice in the OBDD. Assume for a contradiction that x_i appears only once. As the next-state function is bijective each 1-path in the OBDD goes through the vertex labeled by x_i. Every assignment $(x_1, y_1, \ldots, x_{i-1}, y_{i-1}) \in \{0, 1\}^{2i-2}$ that leads to the x_i-node can be combined with every assignment $(x_i, y_i, \ldots, x_n, y_n) \in \{0, 1\}^{2n-2i+2}$ that leads from the x_i-node to the 1-sink in order to construct a transition of the counter. Intuitively, these two groups of variables act independently. For each $(x_i, \ldots, x_n)$ there exists a vector $(y_i, \ldots, y_n)$ such that the assignment $x_i, y_i, \ldots, x_n, y_n$ leads from the x_i-node to the 1-sink. After exactly 2^{n-i} transitions of the form $x_i \ldots x_n \rightarrow y_i \ldots y_n$ the cycle w.r.t. the last $n-i+1$ variables is finished. Analogously, the first $i-1$ variables form a cycle of length 2^{i-1}. The resulting cycle is of length 2^n if and only if the least common multiple of 2^{i-1} and 2^{n-i+1} is 2^n which is impossible for $2 \leq i \leq n$. Analogously, we can show that each variable y_i must appear at least twice in the OBDD. Altogether there are at least $4n-1$ internal nodes. □

4 The Behavior of Important Encodings

4.1 The Standard Minimum-Length Encoding

First, we consider the standard encoding where 2^n states are represented by n bits, and the encoding of a state q_i is the binary representation of i. Let M_{2^n} be the autonomous counter with 2^n states under this encoding.

The *MSB-first* (most significant bit first) variable ordering is the interleaved variable ordering which reads the bits in decreasing significance. In contrast, the *LSB-first* variable ordering reads the bits in increasing significance.

Lemma 3. *For $n \geq 2$, the reduced OBDD for M_{2^n} w.r.t. the MSB-first variable ordering has $5n-3$ internal nodes.*

Proof. The idea is to use the OBDD for $M_{2^{n-1}}$ in order to construct the OBDD for M_{2^n}. Formally, this leads to a proof by induction. We show: The reduced OBDD is of the form like in Figure 3 (a) (in the sense that the shown nodes exist and are not pairwise isomorphic) with sub-OBDDs A and B and has exactly $5n-3$ nodes. The case $n=2$ can easily be checked.

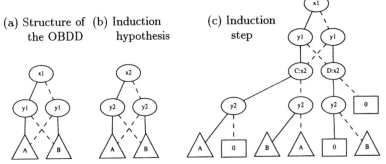

(a) Structure of the OBDD (b) Induction hypothesis (c) Induction step

Fig. 3. MSB-first

Induction step: The OBDD for the $n-1$ bits $x_2, \ldots, x_n$ has the form like in Figure 3 (b). Let $|x|$ be the binary number that is represented by a bit-string $x \in \{0,1\}^+$. With this notation we have
A: leads to the 1-sink if and only if $|y_3 \ldots y_n| = |x_3 \ldots x_n|+1$, $x_3 \ldots x_n \neq 11 \ldots 1$.
B: leads to the 1-sink if and only if $x_3 = \ldots = x_n = 1$, $y_3 = \ldots = y_n = 0$.

We construct the reduced OBDD for M_{2^n} like in Figure 3 (c). The subfunctions rooted in C and D have the following meanings:

C: leads to the 1-sink if and only if $|y_2 \ldots y_n| = |x_2 \ldots x_n|+1$, $x_2 \ldots x_n \neq 11 \ldots 1$.
D: leads to the 1-sink if and only if $x_2 = \ldots = x_n = 1$, $y_2 = \ldots = y_n = 0$.

It can easily be checked that all the subfunctions rooted in the new invented nodes are pairwise different. Therefore $\text{size}(M_{2^n}) = \text{size}(M_{2^{n-1}}) - 3 + 8 = \text{size}(M_{2^{n-1}}) + 5$. □

Analogously, it can be shown that for $n \geq 2$, the OBDDs for the LSB-first ordering lead to the same number of nodes as for MSB-first although the OBDDs are not isomorphic. The main reason for this equality is the fact that in both OBDDs there is only one bit of information that has to be passed from the level of y_{i-1} to the level of x_i for $2 \leq i \leq n$.

It is quite remarkable that these OBDD-sizes for the standard encoding nearly meet the lower bound of $4n - 1$. We conjecture that the standard encoding is even optimal. In order to prove a better lower bound, one might have to establish much more sophisticated communication complexity arguments which connect local OBDD-properties with the global single-cycle-property of a counter.

4.2 The Gray Encoding

Another important minimum-length encoding is the one where the encoding of state i is the Gray code representation of i. The Gray code has the property that all successive code words differ in only one bit. The n-bit Gray code can be constructed by reflecting the $(n-1)$-bit Gray code. To all the new codewords, a leading 1 is added.

By analogous constructions like in the previous subsection, it can be verified that for $n \geq 2$ the OBDD-size of the Gray-encoded autonomous counter with MSB- or LSB-first variable ordering is $10n - 11$. However, the OBDDs are not isomorphic. The constructions in both cases are based on the reflecting property of the Gray code. The essential reason why the Gray encoding has a bigger factor than the standard encoding is the reflecting property of the Gray code which does not allow an immediate use of the OBDD for $M_{2^{n-1}}^G$ when constructing the OBDD for $M_{2^n}^G$.

5 A Worst-Case Encoding

Of course, it is easy to construct autonomous finite state machines with bijective next-state function together with an encoding which have exponential OBDD-size w.r.t. the interleaved variable ordering. Things become more difficult if one wants to construct the encoding of an autonomous *counter* that leads to exponential OBDD-size. In the following we construct an encoding in which the first nodes in the OBDD labeled by $x_1, y_1, \ldots, x_{n/2}, y_{n/2}$ and their outgoing edges lead to a complete tree. The number of nodes with label $x_{n/2+1}$ will therefore be 2^n and the OBDD will have more than 2^{n+1} nodes.

There are 2^n different assignments to the *leading variables* $x_1, y_1, \ldots,$ $x_{n/2}, y_{n/2}$ and 2^n transitions in the finite state machine. To ensure that the leading variables in the OBDD generate a complete tree, we construct the next-state function in the following way:

1. For each assignment to the leading variables, there exists exactly one assignment to the *tail variables* $x_{n/2+1}, y_{n/2+1}, \ldots, x_n, y_n$ which leads to the 1-sink in the OBDD.

2. Each of the 2^n assignments to the tail variables appears exactly once in the construction.

For the construction of the worst-case counter, we build up two tables, the *transition table* and the *tail table*.

Transition table: This table consists of 2^n rows. Each row describes one transition $x_1 \ldots x_n \rightarrow y_1 \ldots y_n$ of the counter. Initially, some of the bits are already set to 1 or 0: All the 2^n assignments for the leading variables $x_1, y_1, \ldots, x_{n/2}, y_{n/2}$ are inserted in the table in the following way: In row i of the table, $0 \leq i < 2^n$, the integer i is represented in binary by the bits $x_1 \ldots x_{n/2} y_1 \ldots y_{n/2}$. The bits of the tail variables are not affected by this initial construction – these entries are marked by stars and will be filled during the construction. The transition table after this initial step is shown in Figure 4.

There is an additional entry called *visited* in each row which helps to keep track which rows of the table have already been filled during the construction algorithm. Initially, every entry in the *visited* column is set to 'FALSE'.

$x_1 \ldots x_{n/2}$	$x_{n/2+1} \ldots x_n$	$y_1 \ldots y_{n/2}$	$y_{n/2+1} \ldots y_n$	*visited*
$00 \ldots 00$	$\star$	$00 \ldots 00$	$\star$	FALSE
$00 \ldots 00$	$\star$	$00 \ldots 01$	$\star$	FALSE
$\vdots$	$\vdots$	$\vdots$	$\vdots$	$\vdots$
$00 \ldots 00$	$\star$	$11 \ldots 11$	$\star$	FALSE
$00 \ldots 01$	$\star$	$00 \ldots 00$	$\star$	FALSE
$\vdots$	$\vdots$	$\vdots$	$\vdots$	$\vdots$
$11 \ldots 11$	$\star$	$11 \ldots 11$	$\star$	FALSE

Fig. 4. Initial structure of the transition table

Tail table: This table consists of 2^n rows which contain the 2^n different assignments for the tail variables $x_{n/2+1}, \ldots, x_n, y_{n/2+1}, \ldots, y_n$. In each row there is an additional entry called *used* which helps to ensure that each assignment is only used once during the construction. Initially, every entry in the *used* column is set to 'FALSE'. The tail table is shown in Figure 5.

$x_{n/2+1} \ldots x_n$	$y_{n/2+1} \ldots y_n$	used
$00 \ldots 00$	$00 \ldots 00$	FALSE
$00 \ldots 00$	$00 \ldots 01$	FALSE
$\vdots$	$\vdots$	$\vdots$
$11 \ldots 11$	$11 \ldots 11$	FALSE

Fig. 5. Initial structure of the tail table

The entries of the transition table which are marked with a star are filled during the construction. The task is to put each assignment for the tail variables into one row of the transition table in such a way that the induced transitions $x_1 \ldots x_n \rightarrow y_1 \ldots y_n$ form a cycle of length 2^n. Note that this construction guarantees that the two properties are satisfied. We use the algorithm of Figure 6.

---

1. For each row of the transition table: Set $x_{n/2+1} \ldots x_n := y_1 \ldots y_{n/2}$.
2. Set the present row to the top row of the transition table.
3. While *visited* in the present row is set to 'FALSE'
   (a) Set the *visited* entry in the present row of the transition table to 'TRUE'.
   (b) Choose the maximal (w.r.t. the represented binary number) assignment for the tail variables from the tail table which matches the assignment for $x_{n/2+1} \ldots x_n$ in the present row and whose *used* entry is set to 'FALSE'.
   (c) Set the *used* entry for the chosen assignment in the tail table to 'TRUE'.
   (d) Set $y_{n/2+1} \ldots y_n$ in the transition table to the chosen assignment for $y_{n/2+1} \ldots y_n$.
   (e) Let R be the row in the transition table in which the assignment for $x_1 \ldots x_n$ is identical with the assignment for $y_1 \ldots y_n$ in the present row.
   (f) Set the present row to row R.

Fig. 6. Construction of the worst-case counter

Claim. The finite state machine that is constructed by the algorithm is a counter, i.e. the cycle has length 2^n.

The claim follows from three statements:

1. In step 3.(a) there exists an assignment in the tail table with the desired properties.
2. If the while-condition in step 3 is not satisfied (i.e. the while-loop terminates), then $x_1 \ldots x_n = 00 \ldots 0$ in the present row of the transition table.
3. When the while-loop terminates, all 2^n assignments for the tail variables have been marked as used.

Before proving the statements we show why the claim follows from them: From statement 1 it follows that step 3.(b) is well-defined in each processing of the loop body. Due to the finiteness of the tail table and step 3.(c), the algorithm terminates. Statements 2 and 3 guarantee that the construction builds a cycle of length 2^n.

Proofs of the statements:
1. The *visited* column of the transition table guarantees that each row of the transition table is at most once the present row – otherwise, the while-loop immediately terminates. For a *fixed* assignment X to $x_{n/2+1} \ldots x_n$ the number of rows in the transition table in which the assignment X appears is $2^{n/2}$. Therefore there are at most $2^{n/2}$ situations during the run of the algorithm in which an assignment is needed in step 3.(b) that extends X. On the other hand there are also $2^{n/2}$ assignments for the tail variables in the tail table which extend X.

Therefore in each processing of step 3.(b), there is at least one previously unused suitable assignment left.

2. Consider the assignment for $y_1 \ldots y_n$ in the present row after step 3.(d). Due to step 1 of the algorithm this assignment is equal to $x_{n/2+1} \ldots x_n y_{n/2+1} \ldots y_n$ in the present row. Due to step 3.(d) and (e) this assignment determines the assignment for $x_1 \ldots x_n$ in the new present row. It follows: Whenever a row with a given bit sequence for $x_1 \ldots x_n$ becomes the present row in step 3.(f), this bit sequence has just been marked as used in the tail table in step 3.(c). This makes it impossible that a row becomes the present row in step 3.(f) more than once.

After the first processing of the loop body, the top row is marked as visited, but the bit sequence $00 \ldots 0$ has not been marked as used in the tail table. Therefore the only possibility to enter a row whose *visited* entry is already set to 'TRUE' is to enter the top row.

3. Due to statement 2, the last chosen assignment for the tail variables is $00 \ldots 0$. As in step 3.(b) the maximal assignment is chosen, all $2^{n/2}$ assignments of the form $00 \ldots 0 \ldots \in 0^{n/2}\{0,1\}^{n/2}$ must have already been used now. Therefore, due to the bit shifting in step 1 and step 3.(e), all assignments $\ldots 00 \ldots 0 \in \{0,1\}^{n/2}0^{n/2}$ in the tail table must have already been used now. Using again that in 3.(b) the maximal assignment is chosen, it follows that all 2^n assignments for the tail variables have been used when the while-loop terminates. □

Remark. The first paragraph of this section implies that the constructed OBDD has at least 2^n subfunctions of the form $f_{x_1=a_1,\ldots,x_{n/2}=a_{n/2},y_1=b_1,\ldots,y_{n/2}=b_{n/2}}$ for $a_1, \ldots, a_{n/2}, b_1, \ldots, b_{n/2} \in \{0,1\}$. Hence, the worst-case construction also holds for a bigger class of variable orderings than only the fixed interleaved ordering $x_1, y_1, \ldots, x_n, y_n$: Namely, it also holds for all variable orderings in which all variables from the upper half come from the set $\{x_1, \ldots, x_{n/2}, y_1, \ldots, y_{n/2}\}$ and all variables from the lower half come from the set $\{x_{n/2+1}, \ldots, x_n, y_{n/2+1}, \ldots, y_n\}$.

6 Related Topologies

With similar techniques most of the results that have been proven for the autonomous counter can also be established for the loop counter and the acyclic counter. The lower bound of $4n - 1$ can be transferred to the acyclic counter. The construction can be slightly modified to prove a $4n - 2$ lower bound for the loop counter which becomes non-deterministic after the elimination of the inputs. The worst-case construction for the autonomous counter can also be used to construct a worst-case encoding for the acyclic counter.

7 Conclusion and Open Questions

We have given some precise results on the relation between state encodings and the size of OBDD-representations. These results show a strong dependence and therefore underline the importance of the OBDD-optimization by re-encoding techniques [MT96].

The general open problems are to analyze non-linear topologies of finite state machines, more general variable orderings and the behavior during reachability analysis in the context of state encodings. For these tasks, the presented results and analysis techniques form the basic ingredients.

References

[ATB94] A. Aziz, S. Taziran, and R. K. Brayton. BDD variable ordering for interacting finite state machines. In *Proc. 31st ACM/IEEE Design Automation Conference*, 283–288, 1994.

[BCL⁺94] J. R. Burch, E. M. Clarke, D. E. Long, K. L. McMillan, and D. L. Dill. Symbolic model checking for sequential circuit verification. *IEEE Transactions on CAD*, 13:401–424, 1994.

[BMS95] J. Bern, Ch. Meinel, and A. Slobodová. Global rebuilding of OBDDs - avoiding memory requirement maxima. In *Proc. Computer-Aided Verification*, *Lecture Notes in Computer Science* 939, 4–15. Springer, 1995.

[Bry86] R. E. Bryant. Graph-based algorithms for Boolean function manipulation. *IEEE Transactions on Computers*, C–35:677–691, 1986.

[Bry92] R. E. Bryant. Symbolic Boolean manipulation with ordered binary decision diagrams. *ACM Computing Surveys*, 24(3):293–318, 1992.

[BW96] B. Bollig and I. Wegener. Improving the variable ordering of OBDDs is NP-complete. *IEEE Transactions on Computers*, 45:993–1002, 1996.

[CBM89] O. Coudert, C. Berthet, and J. C. Madre. Verification of synchronous sequential machines using symbolic execution. In *Proc. Workshop on Automatic Verification Methods for Finite State Machines, Lecture Notes in Computer Science* 407, 365–373. Springer, 1989.

[CM95] O. Coudert and J. C. Madre. The implicit set paradigm: A new approach to finite state system verification. *Formal Methods in System Design*, 6(2):133–145, 1995.

[GDN92] A. Ghosh, S. Devadas, and A. R. Newton. *Sequential Logic Testing and Verification*. Kluwer Academic Publishers, 1992.

[MS94] Ch. Meinel and A. Slobodová. On the complexity of constructing optimal ordered binary decision diagrams. In *Proc. Mathematical Foundations in Computer Science, Lecture Notes in Computer Science* 841, 515–524, 1994.

[MST97] Ch. Meinel, F. Somenzi, and T. Theobald. Linear sifting of decision diagrams. To appear at 34th ACM/IEEE Design Automation Conference (Anaheim, CA), 1997.

[MT96] Ch. Meinel and T. Theobald. Local encoding transformations for optimizing OBDD-representations of finite state machines. In *Proc. International Conference on Formal Methods in CAD (Palo Alto, CA), Lecture Notes in Computer Science* 1166, 404–418. Springer, 1996.

[Rud93] R. Rudell. Dynamic variable ordering for ordered binary decision diagrams. In *Proc. IEEE International Conference on CAD*, 42–47, 1993.

[THY93] S. Tani, K. Hamaguchi, and S. Yajima. The complexity of the optimal variable ordering problems of shared binary decision diagrams. In *Proc. International Symposium on Algorithms and Computation '93, Lecture Notes in Computer Science* 762, 389–398. Springer, 1993.

[Weg94] I. Wegener. Efficient data structures for Boolean functions. *Discrete Mathematics*, 136:347–372, 1994.

Decomposition of TrPTL Formulas

Raphaël Meyer and Antoine Petit

LSV, URA 2236 CNRS, ENS de Cachan,
61, av. du Prés. Wilson, F-94235 Cachan Cedex

Abstract. Partial orders based verifications methods are now well developed. In this framework, several suitable logics have already been defined. We focus on this paper on the logic $TrPTL$, as defined by Thiagarajan, for which models are the well known (infinite) Mazurkiewicz traces. We study the case where the alphabet is not connected. Our main theoretical result is that any $TrPTL$ formula can be decomposed in an effective way as the disjunction of formulas on the connected components. Note that this result can be viewed as a direct logical counterpart of the famous Mezei's theorem on recognizable sets in a direct product of free monoids.

Finally, we show that our result can also be of practical interest. Precisely, we exhibit families of formulas for which the use of our decomposition procedure decreases the complexity of the decision procedure of satisfiability.

1 Introduction

The industrial and economic need of correct software has increased the interest on researches on specification and verification of sequential and distributed programs. In order to express properties of these programs, several logics have been defined, among which the famous Propositional linear time Temporal Logic (PTL) of Pnueli has to be mentioned [Pnu77]. These logics have been interpreted for a long time on infinite sequences describing the program behaviors. In the case of distributed programs, techniques allowing to verify a property for just one representative sequential behavior of each partially ordered computation is a subject of active research (see e.g. [KP92,GW94]).

An alternative way to treat the distributed programs is to represent theirs behaviors directly by partial order based models. Among the possible models, Mazurkiewicz traces [Maz77] play a central role. Indeed, the theory of traces is very well developed (see e.g [Die90,DR95]) and strongly related to other partial order based formalisms such as Petri nets or event structures. Moreover recognizable languages of infinite traces have been characterized from algebraic, automata and logical points of view [GP92,EM93,GPZ94], extending the classical theory of infinite words [Tho90].

In a natural way, several logics directly interpreted over traces or partial order based models have been proposed [Pen88,LRT92,MT92,PK95]. Unfortunately, these logics do not have natural automata counterparts. This motivated the work of Thiagarajan to define the logic $TrPTL$ as a natural extension of PTL to be interpreted on infinite traces. Using automata techniques, the satisfiability problem for $TrPTL$ turns out to be decidable [Thi94,MT96]. A major open question on this logic is to know whether it is as expressive as the extension of the first-order logic $FO(<)$ to traces proposed by Thomas [Tho89] and studied also in [EM93].

We focus in this paper of the decomposition of $TrPTL$ formulas in the case where the underlying dependent alphabet is not connected. Our main theoretical result claims that any $TrPTL$ formula can be decomposed in an effective way as the disjunction of formulas on the connected components. Note that this result can be seen as a direct logical counterpart of the famous Mezei's theorem [Ber79] on recognizable sets in a direct product of free monoids. However, it is not a consequence since, as recalled above, the equivalence between $TrPTL$ and first-order logic $FO(<)$ is still an open problem.

Finally, we show that our result can also be of practical interest. Precisely, we exhibit families of formulas for which the use of our decomposition procedure decreases the complexity of the decision procedure of satisfiability. Note that, obviously, we can not expect to decrease time in the worst case with any such decomposition.

The paper is organized as follows. In Section 2, we briefly recall the basis on dependence graphs and $TrPTL$ logic. We give our main decomposition theorem in Section 3. In order to prove this theorem, we introduce a new operator which can be viewed as a "reset" of the configuration. When dealing with initial equivalence of formulas, we prove that this operator does not increase the power of $TrPTL$. Section 4 is dedicated to the proof of the main theorem. Finally, we show in Section 5 how our theoretical result can be of practical interest for decreasing the complexity procedure of satisfiability in some suitable cases, and give our conclusion in Section 6.

2   TrPTL

We recall in this section the needed bases on traces and on the $TrPTL$ logic (see [DR95] and [Thi94,MT96] for more complete presentations on these subjects).

Let $P = \{1, \ldots, k\}$ be a set of processes, each process $i \in P$ having a local alphabet Σ_i of actions. We may have $\Sigma_i \cap \Sigma_j \neq \emptyset$, some actions involving different processes. Let $\Sigma = \bigcup_{i=1}^{k} \Sigma_i$ be the global alphabet of actions. Each action $a \in \Sigma$ can be mapped to the set $pr(a) = \{i \in P \mid a \in \Sigma_i\}$ of the process it involves. Upon the alphabet Σ, a reflexive and symmetrical relation $D \subseteq \Sigma \times \Sigma$, called the dependence relation, is defined in the natural following way: $D = \{(a, b) \in \Sigma \times \Sigma \mid pr(a) \cap pr(b) \neq \emptyset\}$. This relation expresses the fact that two actions which involve a common process can not be executed simultaneously, and are therefore dependent. The complementary relation $I = (\Sigma \times \Sigma) \setminus D$ is called the independence relation induced by D upon Σ.

2.1 Infinite traces

An infinite trace is a (equivalence class up to isomorphism of) dependence graph(s), that is to say a Σ-labelled partially-ordered graph $F = (E, \leq, \lambda)$ satisfying the following conditions:

- E is a countable set of vertices;
- $\leq$ is a partial order relation on E; we shall write $e \lessdot e'$ (covering relation) iff $e < e'$ and $\forall e''$, $e \leq e'' \leq e'$ implies $e = e''$ or $e'' = e'$;
- λ is an (labelling) application from E to Σ;
- $\forall e \in E$, $\downarrow e = \{e' \in E \mid e' \leq e\}$ is a finite set;
- $\forall e, e' \in E$, $(\lambda(e), \lambda(e')) \in D$ implies $e \leq e'$ or $e' \leq e$;
- $\forall e, e' \in E$, $e \lessdot e'$ implies $(\lambda(e), \lambda(e')) \in D$.

Let $F = (E, \leq, \lambda)$ be a dependence graph. The set E is called the set of events of F, and the partial order relation $\leq$ is called the causality relation ($e \leq e'$ meaning that event e occurs before event e' in F). Let $E_i = \{e \in E \mid \lambda(e) \in \Sigma_i\}$ be the set of events relative to process i. A configuration of a dependence graph F is a finite set of events which respects the causality relation. More formally, c is a configuration of F iff c is a finite subset of E such that $\downarrow c = c$, where $\downarrow c = \bigcup_{e \in c} \downarrow e$. Let C_F be the set of all configurations of F. The notion of configuration is a set-theoretical translation of the one of prefix. For instance, the null prefix of a trace corresponds to the $\emptyset$ configuration.

A transition relation between configurations of a dependence graph is defined in the following way:

$$\forall c, c' \in C_F, \ \forall e \in E, \ (c \Longrightarrow^e_F c') \Leftrightarrow (c' = c \cup \{e\} \text{ and } e \notin c)$$

For every configuration c of F, one can define the i-view of c, denoted by $\downarrow^i(c)$: $\downarrow^i(c) = \downarrow(c \cap E_i)$. Obviously, $\downarrow^i(c)$ is also a configuration.

2.2 Syntax and semantics of TrPTL

In order to express and verify properties on traces, we shall use the $TrPTL$ logic defined by Thiagarajan [Thi94]. This logic $TrPTL$ is built up from a countable set $AP = \{p, q, \ldots\}$ of atomic propositions indexed by the processes, the boolean connectives $\vee$ and $\neg$, and two temporal operators O_i (which is local "next-time" operator), and U_i (which is a local "until" operator).

The syntax of $TrPTL$ is defined in the following way:

$$\Phi_{TrPTL} ::= p(i) \mid \neg\alpha \mid \alpha \vee \beta \mid O_i\alpha \mid \alpha U_i\beta \text{ where } p \in AP \text{ and } i \in P.$$

A model for $TrPTL$ is a pair $M = (F, V)$, where F is a dependence graph and V an evaluation function from C_F into $(\wp(AP))^k$, k being the number of processes. Intuitively, for any configuration c and any process i, $V(c)[i]$ is the set of atomic propositions verified by process i at configuration c. In order to keep the distributive aspect of the model, we assume that the following locality condition is verified:

$$c \Longrightarrow^e_F c' \wedge \lambda(e) = a \Rightarrow \forall i \notin pr(a) \,, \; V(c)[i] = V(c')[i]$$

That is to say, when an action a is being executed, the values of atomic propositions concerning processes not involved in this action can not be modified.

Let $M = (F, V)$ be a model and c a configuration of this model. The satisfiability of a formula α of $TrPTL$ at c in model M, denoted by $c \models_M \alpha$, is defined inductively as follows:

- $c \models_M p(i)$ iff $p \in V(c)[i]$;
- $c \models_M \neg\alpha$ iff $c \not\models_M \alpha$;
- $c \models_M \alpha \vee \beta$ iff $c \models_M \alpha$ or $c \models_M \beta$;
- $c \models_M O_i\alpha$ iff $\exists e \in E_i \mid \downarrow e \models_M \alpha$ and $(c \cap E_i) \subset \downarrow e \cap E_i = (c \cap E_i) \cup \{e\}$;
- $c \models_M \alpha U_i\beta$ iff $\exists c' \in C_F \mid c \subseteq c'$ and $\downarrow^i(c') \models_M \beta$, and $\forall c'' \in C_F$, if $\downarrow^i(c) \subseteq \downarrow^i(c'') \subset \downarrow^i(c)$ then $\downarrow^i(c'') \models_M \alpha$.

The formula $p(i)$, read "p at i", is satisfied if p is part of the atomic propositions provided by V for process i in configuration c. We denote by $\alpha \wedge \beta$ the formula $\neg(\neg\alpha \vee \neg\beta)$. We shall also use $\top = p \vee \neg p$, and $\bot = \neg\top$. The O_i operator is a local "next time" operator. The semantics of $O_i\,\alpha$ is that there exists a next i-view and that, at this i-view, α is satisfied. The U_i operator is an "until" operator restricted to events of E_i.

A formula α is said *satisfiable* if there exists a model $M = (F, V)$, and a configuration $c \in C_F$ such that $c \models_M \alpha$. A formula α is said *root-satisfiable* if there exists a model M such that $\emptyset \models_M \alpha$.

The interest of the $TrPTL$ logic lies mainly on the following result due to Thiagarajan [Thi94].

Theorem 1. *The satisfiability problem for $TrPTL$ formulas is decidable. It can be decided in time $2^{O(max(m^2 log(m),n)m)}$, where n is the size of the formula, and m the number of different processes mentioned in it.*

3 Decomposition theorem

On this paper, we focus on the case where the dependence alphabet is not connected. In this case, the dependence alphabet (Σ, D) is the disjoint union of two dependence alphabets (Σ_A, D_A) and (Σ_B, D_B) that is to say $\Sigma = \Sigma_A \cup \Sigma_B$, $\Sigma_A \cap \Sigma_B = \emptyset$ and $D = D_A \cup D_B$. We will say in the sequel that Σ_A and Σ_B are *components* of Σ. A component Σ_A is said *connected* if the graph of (Σ_A, D_A) is connected. It is easy to show that for every process i, all the actions of Σ_i belong to the same component of Σ. We can thus define $P_A = \{i \in P \mid \Sigma_i \subseteq \Sigma_A\}$ and $P_B = \{i \in P \mid \Sigma_i \subseteq \Sigma_B\}$.

For any subset Q of P (and in particular for P_A and P_B), we can now define in a natural way a corresponding subset of $TrPTL$ formulas, using syntactical restrictions, in the following way:

Definition 2. Let $Q \subseteq P$. We denote by $TrPTL(Q)$ the least subset of $TrPTL$ formulas built from elementary formulas $p(i)$, the connectives $\neg$ and $\vee$, and the operators O_i and U_i with the constraint $i \in Q$.

Two formulas ϕ and ψ are said *equivalent*, denoted by $\phi \equiv \psi$, if for every model M and configuration c of that model, it holds: $c \models_M \phi$ iff $c \models_M \psi$. In a similar way, ϕ and ψ are said *initially equivalent*, denoted by $\phi \equiv_i \psi$, if for every model M, it holds: $\emptyset \models_M \phi$ iff $\emptyset \models_M \psi$.

Our main result is that every $TrPTL$ formula can be effectively decomposed into an initially equivalent boolean combination of formulas of $TrPTL(P_A)$ and formulas of $TrPTL(P_B)$. As a first step we study the case of the equivalence. To this purpose, we need to define a new temporal operator, denoted by R for *Reset*, which semantics is: $c \models_M R\phi$ iff $\emptyset \models_M \phi$. Intuitively, this Reset operator allows us to go back in time by getting rid of the current configuration and continuing the semantic interpretation from the initial configuration.

With the help on this new operator, we can define the following class of $TrPTL$ formulas:

Definition 3. A $TrPTL$ formula γ is called a separated formula if there exist formulas γ_A and γ'_A of $TrPTL(P_A)$, and formulas γ_B and γ'_B of $TrPTL(P_B)$ such that
$$\gamma = \gamma_A \wedge R\gamma'_A \wedge \gamma_B \wedge R\gamma'_B$$

We can now state the decomposition theorem related to equivalence of $TrPTL$ formulas.

Theorem 4. *Let (Σ, D) be a non connected dependence alphabet such that $\Sigma = \Sigma_A \cup \Sigma_B$, Σ_A and Σ_B being distinct components of Σ. Every $TrPTL$ formula γ on (Σ, D) is equivalent to a disjunction of separated formulas. Moreover, these formulas can effectively be constructed from γ.*

As a direct corollary of this result, using the fact that Reset operators disappear when dealing with initial equivalence, we obtain a decomposition theorem related to initial equivalence:

Theorem 5. *Let (Σ, D) be a non connected dependence alphabet such that $\Sigma = \Sigma_A \cup \Sigma_B$, Σ_A and Σ_B being distinct components of Σ. Every $TrPTL$ formula γ on (Σ, D) is initially equivalent to a disjunction of formulas $\gamma_A \wedge \gamma_B$, where γ_A belongs to $TrPTL(P_A)$, and γ_B belongs to $TrPTL(P_B)$.*

By an immediate induction, we get the following corollary when dealing with the connected components of the alphabet (Σ, D):

Corollary 6. *Let (Σ, D) be a non connected dependence alphabet such that $\Sigma = \bigcup_{j=1}^l \Sigma^j$, the Σ^j being the distinct connected components of Σ. Denote by P_j the set of processes i such that $\Sigma_i \subseteq \Sigma^j$. Every $TrPTL$ formula γ on (Σ, D) is initially equivalent to a disjunction of formulas $\gamma_1 \wedge \gamma_2 \wedge \ldots \wedge \gamma_l$, where for all j, γ_j belongs to $TrPTL(P_j)$.*

Note that this result can be seen as a direct logical counterpart of the famous Mezei's theorem [Ber79] on recognizable sets of product of free monoids since, as we recalled in the introduction, the equivalence between $TrPTL$ and first-order logic $FO(<)$ is still an open problem.

The following section is dedicated to the proof of Theorem 4.

4   Proofs

In order to prove Theorem 4, we will first focus on the distributivity of the temporal operators upon the boolean connectives. The following equivalences follow easily from the definitions of the temporal operators.

Proposition 7. *Let* γ, ψ *and* χ *be* $TrPTL$ *formulas, and let* $i \in P$, *the following equivalences hold:*

$$O_i(\gamma \vee \psi) \equiv O_i(\gamma) \vee O_i(\psi)$$

$$O_i(\gamma \wedge \psi) \equiv O_i(\gamma) \wedge O_i(\psi)$$

$$(\gamma \wedge \psi)\, U_i\, \chi \equiv (\gamma\, U_i\, \chi) \wedge (\psi\, U_i\, \chi)$$

$$\gamma\, U_i\, (\psi \vee \chi) \equiv (\gamma\, U_i\, \psi) \vee (\gamma\, U_i\, \chi)$$

Now we shall focus on the non-connected aspect of the alphabet Σ. Let $E_A = \{e \in E \mid \lambda(e) \in \Sigma_A\}$ and $E_B = \{e \in E \mid \lambda(e) \in \Sigma_B\}$. Every dependence graph $F = (E, \leq, \lambda)$ on (Σ, D) can thus be decomposed in two dependence graphs $F_A = (E_A, \leq, \lambda)$ and $F_B = (E_B, \leq, \lambda)$ on (Σ_A, D_A) and (Σ_B, D_B) respectively. For every configuration c of C_F, we define the configurations c_A and c_B induced on the two components, and then it holds: $c = c_A \cup c_B$, where the union is to be taken over graphs. Remark that c_A and c_B are configurations of F.

Proposition 8. *Let* $i \in P_A$ *and let* c *be a configuration of a model* M, *then:*

- $c \models_M p(i) \Leftrightarrow c_A \models_M p(i)$
- $c \models_M O_i \alpha \Leftrightarrow c_A \models_M O_i \alpha$
- $c \models_M \alpha U_i \beta \Leftrightarrow c_A \models_M \alpha U_i \beta$

Proof. It suffices to notice that since $i \in P_A$, we have $E_i \subseteq E_A$, and hence $c \cap E_i = c_A \cap E_i$. $\qquad\square$

With the help of the previous proposition, we can extend the distributivity of the temporal operators to some particular cases:

Proposition 9. *Let* $i \in P_A$, *let* α *and* β *be* $TrPTL$ *formulas, let* α_A *and* α_B *be* $TrPTL(P_A)$ *formulas, and let* α_B *and* β_B *be* $TrPTL(P_B)$ *formulas. Then the following equivalences hold:*

1. $O_i\, \alpha_B \equiv O_i\, \top \wedge R\, \alpha_B$
2. $O_i\, (R\, \alpha) \equiv O_i\, \top \wedge R\, \alpha$

3. $\alpha\, U_i\, (\beta_A \wedge \beta_B \wedge R\,\beta) \equiv \alpha\, U_i\, \beta_A \wedge R\,(\beta_B \wedge \beta)$

4. $(\alpha_A \vee \alpha_B)\, U_i\, \beta_A \equiv (\alpha_A\, U_i\, \beta_A) \vee (R\,\alpha_B \wedge \top\, U_i\, \beta_A)$

Proof. We shall only focus on the proof of equivalence 4, which gives a good idea of what the other proofs could look like. Let c be a configuration of a model M. We use the following notation: $S_i(max(c \cap E_i))$ is the set of all E_i-events that are greater than or equal to the greatest E_i-event appearing in configuration c. Now, $c \models_M (\alpha_A \vee \alpha_B)\, U_i\, \beta_A$ iff there exists $e \in S_i(max(c \cap E_i))$, such that $\downarrow e \models_M \beta_A$, and for all $e' \in S_i(max(c \cap E_i))$, if $e' < e$ then $\downarrow e' \models_M \alpha_A \vee \alpha_B$. Since $\downarrow e' \subset E_A$, Proposition 8 shows that $\downarrow e' \models_M \alpha_A \vee \alpha_B$ iff $\downarrow e' \models_M \alpha_A$ or $\emptyset \models_M \alpha_B$, which is equivalent to $c \models_M R\,\alpha_B$. Thus, if $c \models_M R\,\alpha_B$ then we just need to check that $c \models_M \top\, U_i\, \beta_A$; otherwise, we have to check that $c \models_M \alpha_A\, U_i\, \beta_A$. This leads to the equivalence between $c \models_M (\alpha_A \vee \alpha_B)\, U_i\, \beta_A$ and $c \models_M (\alpha_A\, U_i\, \beta_A) \vee (R\,\alpha_B \wedge \top\, U_i\, \beta_A)$. □

We can now prove our main theorem, that is the decomposition of $TrPTL$ formulas:

Proof (of Theorem 4). We prove the theorem by induction on the length of the formula γ.

- If $|\gamma| = 1$, then $\gamma = p(i)$. We can assume, without loss of generality, that $i \in P_A$. Then it suffices to write γ as $p(i) \wedge R\,\top \wedge \top \wedge R\,\top$, since $R\,\top \equiv \top$.
- If $|\gamma| > 1$, several cases occur, depending on the nature of γ:
  1. If $\gamma = \alpha \vee \beta$, the conclusion is trivial: it suffices to use the induction hypothesis on α and β.
  2. If $\gamma = \neg\alpha$, we use the induction hypothesis, then we make intensive use of some easy combinatorial properties and of the fact that the operators $\neg$ and R commute to put things in the right form.
  3. If $\gamma = O_i\,\alpha$, we use the induction hypothesis, then we use the distributivity of O_i upon $\vee$ and $\wedge$, and the equivalences 1 and 2 of Proposition 9.
  4. If $\gamma = \alpha\, U_i\, \beta$, we can also assume that $i \in P_A$. Using the induction hypothesis, β is shown to be equivalent to a disjunction of separated formulas. Using the distributivity of U_i upon $\vee$, γ can be written as an equivalent disjunction of formulas like $\alpha\, U_i\, (\beta_A \wedge R\,\beta'_A \wedge \beta_B \wedge R\,\beta'_B)$. Denote by $\alpha\, U_i\, \beta_{sep}$ this last formula. Using equivalence 3 of Proposition 9, we show the equivalence between $\alpha\, U_i\, \beta_{sep}$ and $\alpha\, U_i\, \beta_A \wedge R\,(\beta'_A) \wedge R\,(\beta_B \wedge \beta'_B)$. Now, using the induction hypothesis on α, we have: $\alpha \equiv \bigvee_{k=1}^{d_\alpha} \alpha_{A,k} \wedge R\,\alpha'_{A,k} \wedge \alpha_{B,k} \wedge R\,\alpha'_{B,k}$. This can be written as $\alpha \equiv \bigwedge_{K \subseteq \{1,\ldots,d_\alpha\}} (\alpha_{A,K} \vee \alpha_{B,K})$, where $\alpha_{A,K} = \bigvee_{k \in K} (\alpha_{A,k} \wedge R\,\alpha'_{A,k})$. Hence $\alpha\, U_i\, \beta_A$ is equivalent to $\bigwedge_{K \subseteq \{1,\ldots,d_\alpha\}} ((\alpha_{A,K} \vee \alpha_{B,K})\, U_i\, \beta_A)$. Using equivalence 4 of Proposition 9, we show that $c \models_M (\alpha_{A,K} \vee \alpha_{B,K})\, U_i\, \beta_A$ iff $c \models_M ((\top\, U_i\, \beta_A) \wedge R\,\alpha_{B,k}) \vee \alpha_{A,K}\, U_i\, \beta_A$. By definition $\alpha_{A,K}$ is equal to $\bigvee_{k \in K} (\alpha_{A,k} \wedge R\,\alpha'_{A,k})$, which is equivalent to $\bigwedge_{L \subseteq K} (\alpha_{A,L} \vee R\,\alpha'_{A,L})$, where $\alpha_{A,L} = \bigvee_{l \in L} \alpha_{A,l}$ and $\alpha'_{A,L} = \bigvee_{l \notin L} \alpha'_{A,l}$, using the distributivity of I upon $\vee$. Thus, $\alpha_{A,K}\, U_i\, \beta_A$ is equivalent to $\bigwedge_{L \subseteq K} (\alpha_{A,L} \vee R\,\alpha'_{A,L})\, U_i\, \beta_A$,

also equivalent to $\bigwedge_{L \subseteq K}((\alpha_{A,L} U_i \beta_A) \vee (\top U_i \beta_A \wedge R\alpha'_{A,L}))$. Replacing the previous results in $\alpha \bar{U}_i \beta_A$, one can rewrite this formula as a big boolean combination of separated formulas. But disjunctions of separated formulas are stable under negation, and hence under boolean combination. Thus $\alpha U_i \beta_A$, and then $\alpha U_i \beta$, can be written as equivalent disjunctions of separated formulas, which concludes the proof of the theorem. □

5 Application to verification of TrPTL formulas

The proof of Theorem 4 has shown that decomposing a $TrPTL$ formula can lead to combinatorial explosions, especially when dealing with negations and U_i operators. In order to avoid this explosion, we can define a subclass of $TrPTL$ formulas for which the separation procedure is actually efficient.

Recall that if a formula is of size n and involves m different process, then the satisfiability problem for this formula is in time $2^{O(max(m^2 log(m),n)m)}$.

Let Φ_s be the least set of $TrPTL$ formulas satisfying the following conditions:

1. $p(i) \in \Phi_s$ for all $p \in PA$ and all $i \in P$.
2. If $\alpha, \beta \in \Phi_s$, then $\alpha \vee \beta \in \Phi_s$.
3. If $\alpha \in \Phi_s$, then for all $i \in P$, $O_i \alpha \in \Phi_s$.
4. If $\alpha \in TrPTL(P_A)$, and $\beta \in \Phi_s$, then $\alpha U_i \beta \in \Phi_s$, for all $i \in P_A$.
5. If $\alpha \in TrPTL(P_B)$, and $\beta \in \Phi_s$, then $\alpha U_i \beta \in \Phi_s$, for all $i \in P_B$.

Note that the formulas $\neg p(i)$ have been excluded only for sake of simplicity, since they can be simulated by defining new atomic propositions.

The Φ_s formulas, called separable formulas, can be decomposed as a disjunction of separated formulas, without any risk of combinatorial explosion.

Proposition 10. *Let γ be a Φ_s formula of size n containing d disjunction operators, then γ can effectively be written as an equivalent conjunction of d formulas of size at most $(n - d)$ not containing any $\vee$ operator.*

Proof. Since it holds that $O_i(\alpha \vee \beta) \equiv O_i(\alpha) \vee O_i(\beta)$, and $\alpha U_i(\beta \vee \theta) \equiv \alpha U_i \beta \vee \alpha U_i \theta$, we can move all the $\vee$ operators out of the temporal ones, which leads immediatly to the conclusion. □

Now that we are dealing with formulas without $\vee$ operators, we consider the following function defined inductively on these formulas (we write $(i, j) \in C$ whenever i and j belong to the same connected component):

$$\forall i \in P, \ \delta(p(i)) = p(i)$$

If $(i, j) \in C$ then:
$\delta(O_i p(j)) = O_i p(j)$
$\delta(\alpha U_i p(j)) = \alpha U_i p(j)$
$\delta(O_i O_j \alpha) = O_i \delta(O_j \alpha)$
$\delta(O_i(\alpha U_i \beta) = O_i \delta(\alpha U_i \beta)$
$\delta(\alpha U_i(O_j \beta)) = \alpha U_i(\delta(O_j \beta))$
$\delta(\alpha U_i(\beta U_j \theta) = \alpha U_i(\delta(\beta U_j \theta))$

If $(i, j) \notin C$ then:
$\delta(O_i p(j)) = O_i \top \wedge Rp(j)$
$\delta(\alpha U_i p(j)) = \alpha \wedge Rp(j)$
$\delta(O_i O_j \alpha) = O_i \top \wedge \delta(O_j \alpha)$
$\delta(O_i(\alpha U_i \beta)) = O_i \top \wedge \delta(\alpha U_i \beta)$
$\delta(\alpha U_i(O_j \beta)) = \alpha \wedge R(\delta(O_j \beta))$
$\delta(\alpha U_i(\beta U_j \theta)) = \alpha \wedge R(\delta(\beta U_j \theta))$

Then it is easy to prove (by induction) the following:

Proposition 11. *Let γ be a Φ_s formula without any $\vee$ operator, then $\gamma \equiv \delta(\gamma)$.*

Example 12. Let $\gamma = p(1)U_1(O_1(p(2)U_2 q(1)))$ with $1 \in P_A$ and $2 \in P_B$. Then $\delta(\gamma) = p(1)U_1(O_1 \top \wedge R(p(2) \wedge Rq(1)))$.

Now, using $\alpha U_i(\beta \wedge R\theta) \equiv \alpha U_i \beta \wedge R\theta$, $O_i(\alpha \wedge R\beta) \equiv O_i \alpha \wedge R\beta$, and $R(\alpha \wedge R\beta) \equiv R\alpha \wedge R\beta$, we can easily transform $\delta(\gamma)$ into a separated formula.

Example 13. With the same formula γ as above, it holds that $\delta(\gamma) \equiv p(1)U_1(O_1 \top) \wedge Rp(2) \wedge Rq(1)$.

When dealing with initial equivalence, we can then remove all the R operators from $\delta(\gamma)$. This gives us a formula that is initially equivalent to γ. Intuitively, this formula is built from γ by cutting γ in different pieces whenever two temporal operators relating to different components occur successively.

Obviously no progress will be made in the worst case, for instance if we start with a formula already in $TrPTL(P_A)$. But, in the best case, one can hope to bound by the size of the subformulas by a constant, while the number of processes involved can be divided by 2. Testing the root-satifiability problem for the subformulas can thus sometimes be done dramatically faster than the initial problem for γ, depending on the structure of γ.

Example 14. Let $\gamma = O_1 O_2 O_1 O_2 \ldots O_1 O_2 p(2)$, with $1 \in P_A$ and $2 \in P_B$. Then γ can be decomposed as the initially equivalent formula $O_1 \top \wedge O_2 p(2)$, regardless of the size of γ.

6   Conclusion

By the time we had submitted this work, Thiagarajan and Walukiewicz [TW97] have exhibed a new logic on traces, called $LTrL$, that is actually expressively complete (i.e. equivalent to the first-order logic $FO(<)$ on traces), but does not have yet an elementary decision procedure, as opposed to $TrPTL$. Therefore Mezei's theorem holds for $LTrL$-definable languages. Nevertheless, [TW97] does not give any way to obtain an effective decomposition of $LTrL$ formulas. Thus, we shall extend the results of the present paper to this new logic $LTrL$ in a future version of this work.

References

[Ber79] J. Berstel. *Transductions and context-free languages.* Teubner Studienbücher, 1979.

[Die90] V. Diekert. *Combinatorics on Traces.* Number 454 in LNCS. Springer, 1990.

[DR95] V. Diekert and G. Rozenberg, editors. *The Book of Traces.* World Scientific, Singapore, 1995.

[EM93] W. Ebinger and A. Muscholl. Logical definability on infinite traces. In A. Lingas, R. Karlsson, and S. Carlsson, editors, *Proc. of the 20th ICALP, Lund (Sweden) 1993*, number 700 in LNCS, pages 335–346. Springer, 1993.

[GP92] P. Gastin and A. Petit. Asynchronous automata for infinite traces. In W. Kuich, editor, *Proc. of the 19th ICALP, Vienna (Austria) 1992*, number 623 in LNCS, pages 583–594. Springer, 1992.

[GPZ94] P. Gastin, A. Petit, and W. Zielonka. An extension of Kleene's and Ochmański's theorems to infinite traces. *Theoret. Comp. Sci.*, 125:167–204, 1994. A preliminary version was presented at ICALP'91, LNCS 510 (1991).

[GW94] P. Godefroid and P. Wolper. A partial approach to model checking. *Inform. and Comp.*, 110:305–326, 1994.

[KP92] S. Katz and D. Peled. Interleaving set temporal logic. *Theoret. Comp. Sci.*, 75:21–43, 1992.

[LRT92] K. Lodaya, R. Ramajunam, and P.S. Thiagarajan. Temporal logics for communicating sequential agents:I. *Int. J. of Found. of Comp. Sci.*, 3(2):117–159, 1992.

[Maz77] A. Mazurkiewicz. Concurrent program schemes and their interpretations. DAIMI Rep. PB 78, Aarhus University, Aarhus, 1977.

[MT92] M. Mukund and P.S. Thiagarajan. A logical characterization of well branching event structures. *Theoret. Comp. Sci.*, 96:35–72, 1992.

[MT96] M. Mukund and P.S. Thiagarajan. Linear time temporal logics over Mazurkiewicz traces. In *Proc. of the 21th MFCS, 1996*, number 1113 in LNCS, pages 62–92. Springer, 1996.

[Pen88] W. Penczek. A temporal logic for event structures. *Fundamenta Informaticae*, XI:297–326, 1988.

[PK95] W. Penczek and R. Kuiper. Traces and logic. In V. Diekert and G. Rozenberg, editors, *The book of Traces*, pages 307–381, 1995.

[Pnu77] A. Pnueli. The temporal logics of programs. In *Proc. of the 18th IEEE FOCS, 1977*, pages 46–57, 1977.

[Thi94] P.S. Thiagarajan. A trace based extension of linear time temporal logic. In *Proc. of the 9th LICS, 1994*, pages 438–447, 1994.

[Tho89] W. Thomas. On logical definability of trace languages. In V. Diekert, editor, *Proc. an ASMICS workshop, Kochel am See 1989*, Report TUM-I9002, Technical University of Munich, pages 172–182, 1989.

[Tho90] W. Thomas. Automata on infinite objects. In J. v. Leeuwen, editor, *Handbook of Theoretical Computer Science*, pages 133–191. Elsevier Science Publishers, 1990.

[TW97] P.S. Thiagarajan and I. Walukiewicz. An expressively complete linear time temporal logic for Mazurkiewicz traces. In *Proc. of LICS'97 (to appear)*, 1997.

NP-Hard Sets Have Many Hard Instances

Martin Mundhenk [*]

Universität Trier, FB IV - Informatik, D-54286 Trier, Germany

Abstract. The notion of *instance complexity* was introduced by Ko, Orponen, Schöning, and Watanabe [9] as a measure of the complexity of individual instances of a decision problem. Comparing instance complexity to Kolmogorov complexity, they stated the "instance complexity conjecture," that every set not in P has p-hard instances. Whereas this conjecture is still unsettled, Buhrman and Orponen [4] showed that E-complete sets have exponentially dense hard instances, and Fortnow and Kummer [5] proved that NP-hard sets have p-hard instances unless P = NP. They left open whether the p-hard instances of NP-hard sets must be dense. In this work, we introduce a slightly weaker notion of hard instances and obtain a superpolynomial lower bound on the density of hard instances in the case of NP-hard sets. We additionally show that NP-hard sets cannot consist of hard instances only, unless P = NP. Kummer [10] proved that the class of recursive sets cannot be characterized by a respective version of the instance complexity conjecture, i.e. there exist nonrecursive sets without hard instances. We give a complete characterization of the class of recursive sets comparing the instance complexity to a relativized Kolmogorov complexity of strings. A set A is shown to be recursive iff $\mathrm{ic}(x : A) \leq C^{K_0 \oplus A}(x)$ for almost all x. This translates to a characterization of P.

1 Introduction

Ko, Orponen, Schöning, and Watanabe [9] introduced instance complexity as a measure of the complexity of individual instances of a decision problem. The t-time bounded instance complexity $\mathrm{ic}^t(x : A)$ of x w.r.t. A is the length of the shortest program which correctly computes $A(x)$ in time $t(|x|)$ and is consistent with A (i.e. on every input y it either outputs $A(y)$ in time $t(|y|)$ or "says" that it is unable to make a decision). In [9] it is shown that P is the class of sets with polynomial-time instance complexity bounded by a constant, formally $A \in P \Leftrightarrow \exists$ polynomial p, constant $c \; \forall^\infty x \; : \; \mathrm{ic}^p(x : A) < c$. We show that the constant c can be replaced by the unbounded function $C^A(x)$, which is the Kolmogorov complexity of x relative to A. Because for every c and almost every x, $c < C^A(x)$, this yields a strong notion of hard instances for a set A, namely the set of hard instances $H(A) = \{x \mid \forall$ polynomial $p : \mathrm{ic}^p(x : A) >$

[*] Supported in part by the Office of the Vice Chancellor for Research and Graduate Studies at the University of Kentucky, and by the Deutsche Forschungsgemeinschaft (DFG), grant Mu 1226/2-1. Parts of the work done at University of Kentucky.

$\mathrm{C}^A(x)\}$. In fact, this set is an infinite complexity core (as defined by Lynch [12]) for A, if A is not in P. We investigate its structural properties and show that if any of its "slices" $H_p(A) = \{x \mid \mathrm{ic}^p(x : A) > \mathrm{C}^A(x)\}$ is sparse, then A reduces to a sparse set via a composition of a polynomial-time 2-truth-table and a conjunctive reducibility. Consequently, we show that no $\leq_m^p$-hard set for NP has a sparse set of hard instances unless P = NP. This result has a flavour similar to the characterization of sets having only sparse complexity cores by *almost* -P (see [14]). The latter is contained in the closure of sparse sets under polynomial-time 1-truth-table reductions. We leave open whether every complexity core for A is almost everywhere contained in $H(A)$, and conjecture that this is not the case.

From this, we turn to the instance complexity conjecture, stated in [9]. It says that for every set $A \notin P$ and every polynomial q, there exists a polynomial q' and a constant c such that the set $\{x \mid \mathrm{ic}^q(x : A) > \mathrm{C}^{q'}(x) - c\}$ is infinite, or "A has p-hard instances" for short. Buhrman and Orponen [4] showed that every $\leq_m^p$-complete set for E has exponentially dense p-hard instances. Later, Fortnow and Kummer [5] settled the conjecture for all decidable tally sets and for decidable honest Turing NP-hard sets (unless P = NP in the latter case). In their proofs, they construct at most n p-hard instances for each length n. They left open whether NP-hard sets must have non-sparse sets of p-hard instances. As an application of the above results, we answer a slightly weaker variant of this question positively, and show that every decidable $\leq_m^p$-hard set for NP has superpolynomially many hard instances, unless P = NP.

Whereas the instance complexity conjecture is still open, Kummer [10] proved that its respective formulation for recursive sets does not hold. Therefore, we consider how instance complexity and Kolmogorov complexity are related for recursive sets. We prove that each recursive set A has (unbounded) instance complexity bounded by the Kolmogorov complexity relative to the Halting Problem K_0 and A itself. Attaching polynomial time bounds and replacing K_0 by the NP-complete set SAT, we get an exact characterization of P, namely a set A is in P iff there exist polynomials p, p' such that for almost every x, $\mathrm{ic}^p(x : A) \leq \mathrm{C}^{\mathrm{SAT}\oplus A, p'}(x)$. Compared to the characterization of P from [9] cited above, the right-hand side of our inequality has no limit inferior.

Finally, we consider sets with maximal density of hard instances. Ko [8] showed a lower bound on their instance complexity and proved, that Martin-Löf random sets have this property. Book and Lutz [3] showed that random sets cannot be NP-hard unless P = NP. We obtain a similar result for sets consisting only of very hard instances.

2 Notation and Definitions

We consider strings over the alphabet $\Sigma = \{0, 1\}$. The empty string is denoted ε. The length of a string x is denoted by $|x|$. Sets are considered to be subsets of Σ^*. A set A is called *tally*, if $A \subseteq 0^*$. A set is *sparse*, if its density is bounded by a polynomial. TALLY denotes the class of all tally sets, SPARSE that of sparse sets. $A \oplus B$ denotes the marked union of sets $\{0x \mid x \in A\} \cup \{1x \mid x \in B\}$.

Complexity classes and languages are defined in the standard manner [2]. Reduction classes are denoted $R_\alpha^p(\mathcal{C}) = \{A \mid \exists B \in \mathcal{C} : A \leq_\alpha^p B\}$. The polynomial-time reducibilities used here are many-one $\leq_m^p$, conjunctive $\leq_c^p$, 2-truth table $\leq_{2\text{-}tt}^p$, and Turing $\leq_T^p$.

Our model of computation is the deterministic (oracle) Turing machine. Each program $\pi \in \Sigma^*$ computes a partial function $\Sigma^* \to \Sigma^*$. For any string x, let $\pi(x)$ denote the output of π on input x.

We will give a short review of necessary definitions for Kolmogorov and instance complexity. We leave out details like the choice of a universal Turing machine with respect to which the size and the computation time of the programs are measured, and suppose that an "optimal interpreter" is used to run the programs. The existence, robustness, and invariance of such an interpreter and more details can be found in [6, 9, 11]. For simplicity and w.l.o.g. we assume that the program $\pi = \varepsilon$ denoted by the empty string halts on every input without any computation and output.

The *Kolmogorov complexity* of a string $x \in \Sigma^*$ relative to oracle A is

$$C^A(x) = \min\{|\pi| \mid \pi^A(\varepsilon) = x\} \quad .$$

We use $C(x)$ to denote $C^\emptyset(x)$. The notion of time-bounded Kolmogorov complexity was introduced by Hartmanis [6]. The *t-time bounded Kolmogorov complexity* is

$$C^{A,t}(x) = \min\{|\pi| \mid \pi^A(\varepsilon) = x \text{ and } \pi^A(\varepsilon) \text{ makes at most } t(|x|) \text{ steps}\} \quad .$$

A set S is in KT[log, poly], if there exists a constant c such that $C^{n^c+c}(x) \leq c \log|x|$ for every $x \in S$.

The characteristic function of a set A is denoted by $A(\cdot)$, where $A(x) = 1$ if $x \in A$, and $A(x) = 0$ if $x \notin A$. For $a, b \in \{0, 1, \bot\}$, we denote $a \simeq b$, if either $a = b$ or $\bot \in \{a, b\}$. Let π be a Turing machine. Then $\pi(x) = \bot$ denotes that either π does not halt on input x, or it halts with output not in $\{0, 1\}$. Machine π is *consistent with A on input x* (denoted $\pi(x) \simeq A(x)$), if $\pi(x) = \bot$ or $\pi(x) = A(x)$. π is called *A-consistent* (denoted $\pi \simeq A$), if $\pi(x) \simeq A(x)$ for every $x \in \Sigma^*$.

The notion of (unbounded and bounded) instance complexity was introduced by Ko, Orponen, Schöning, and Watanabe [9] . The *instance complexity of a string x w.r.t. A* is defined as

$$\text{ic}(x : A) = \min\{|\pi| \mid \pi \simeq A \wedge \pi(x) = A(x)\} \quad .$$

For a function t, let $\pi(x)_t$ be the decision of π on input x after $t(|x|)$ steps, i.e. $\pi(x)_t = \pi(x)$ if π on input x halts after at most $t(|x|)$ steps with output 0 or 1, and $\pi(x)_t = \bot$ otherwise. π is called *t-time bounded A-consistent* (denoted $\pi_t \simeq A$), if $\pi(x)_t \simeq A(x)$ for every $x \in \Sigma^*$. Note that for $\pi = \varepsilon$, by the above convention it holds that $\pi_t \simeq A$ for every A and time bound t.

The *t-time bounded instance complexity of a string x w.r.t. A* is defined as

$$ic^t(x : A) = \min\{|\pi| \mid \pi_t \simeq A \wedge \pi(x)_t = A(x)\} \quad .$$

A set A is said to have *p-hard instances*, if for every polynomial q there exists a polynomial q' and infinitely many x such that $ic^q(x : A) > C^{q'}(x)$. This definition stems from [9], where it was conjectured that every set not in P has p-hard instances. This conjecture is known as the *Instance Complexity Conjecture*.

3 Density of Hard Instances

We say that a set A has *sparse hard instances*, if there exists a sparse set S and a polynomial p, such that the set $H_p(A) = \{x \mid ic^p(x : A) > C^A(x)\}$ is contained in S. The main result shows that sets with sparse hard instances reduce to sparse sets. We then apply this result on sets with sparse p-hard instances (see below for a definition) and conclude that decidable NP-hard sets do not have sparse p-hard instances unless P = NP. This improves the result from [5], where sparse sets of hard instances are shown to exist in NP-hard sets.

Theorem 1. *Let A be a set. If for some polynomial p, the set of hard instances $H_p(A) = \{x \mid ic^p(x : A) > C^A(x)\}$ is sparse, then A is in $R^p_{2\text{-}tt}(R^p_c(SPARSE))$.*

Proof. Let A be a set and p be a polynomial, and assume $|H_p(A) \cap \Sigma^n| \le n^a$ for a constant a and almost every n. The proof idea is as follows. We consider the set of instances which can be consistently decided in polynomial time by programs of logarithmic size. It turns out, that all but a sparse set can. All the instances of logarithmic instance complexity disjunctively reduce to a tally set (essentially an encoding of the small programs), and the sparse rest of the instances many-one reduces to its intersection with A. Putting both reductions together, we get the desired result.

Formally, let $N_{m,k}$ be the following Turing machine, which takes no input, uses oracle A, and computes some output.

$$\Pi := \{\pi \in \Sigma^{\le (a+3)\log m} \mid \forall x \in \Sigma^m : \pi(x)_p \simeq A(x)\}$$
$$T := \{y \in \Sigma^m \mid \text{for all } \pi \in \Pi : \pi(y)_p = \perp\}$$
if $|T| \ge k$
    then output the k-th element of T w.r.t. lex. order
end

$N^A_{m,k}$ halts for all m and k. Let $T(m)$ and $\Pi(m)$ denote the contents of the set variables T and Π when $N^A_{m,k}$ halted. For every m and every $y \in T(m)$ it holds that

$$ic^p(y : A) > (a+3)\log m$$

by the definition of $\Pi(m)$ and $T(m)$.

The size $|N_{m,k}|$ of $N_{m,k}$ depends mainly on the sizes of m and k. I.e. there exists a constant c, such that for all m, k it holds that $|N_{m,k}| = c + |m| + |k|$.

Thus, if y is the output of $N_{m,k}^A$, then $C^A(y) \leq c + |m| + |k|$. Let y be one of the lexicographically first m^{a+1} elements of $T(m)$. Then

$$C^A(y) \leq c + |m| + |m^{a+1}| \leq (a+3) \log m$$

for almost every m. Combining both inequalities, we get

$$\mathrm{ic}^p(y : A) > C^A(y)$$

for almost every m and every y of the lexicographically first m^{a+1} elements of $T(m)$. Thus, each of the lexicographically first m^{a+1} elements of $T(m)$ is in $H_p(A) \cap \Sigma^m$, which contains at most m^a elements. Therefore $|T(m)| \leq m^a$ for almost every m.

Let $S = \bigcup_m T(m) \cap A$. Then S is a sparse set. Let U denote an encoding of $\bigcup_m \Pi(m)$ into a tally set, e.g. $U = \{0^{\langle m,q \rangle} \mid m \geq 0, q \in \Pi(m)\}$.

Then for every $x \in \Sigma^*$,

$$x \in A \Leftrightarrow \exists \pi \in \Pi(|x|) : \pi(x) \text{ accepts in time } p(|x|) \text{, or } x \in S$$
$$\Leftrightarrow \{0^{\langle |x|, \pi \rangle} \mid \pi \in \Sigma^{\leq (a+3) \log |x|}, \pi(x)_p = 1\} \cap U \neq \emptyset \text{, or } x \in S \ .$$

Thus, A can be decided with two queries to different oracles: one query to a set which disjunctively reduces to the tally set U, and one query to the sparse set S. Because the class of sparse sets is closed under marked union, and because $R_d^p(\text{TALLY}) = \text{co-}R_c^p(\text{TALLY})$, this yields that $A \in R_{2\text{-}tt}^p(R_c^p(\text{SPARSE}))$.

Our goal is to get a lower bound on the density of p-hard instances for NP hard sets. Analogously to the notion of sparse hard instances, we say that a set A has *sparse p-hard instances*, if there exists a k and a polynomial q such that for every polynomial q' and every n, $|\{x \mid \mathrm{ic}^q(x : A) > C^{q'}(x)\} \cap \Sigma^n| \leq n^k$. Let the latter set be called $H_q^{q'}$.

Theorem 2. *No decidable $\leq_m^p$-hard set for NP has sparse p-hard instances, unless* P = NP.

Proof. Let A be a decidable $\leq_m^p$-hard set for NP, and let k and polynomial p witness that A has sparse p-hard instances, i.e. $|H_p^{p'} \cap \Sigma^n| \leq n^k$ for every polynomial p' and every n.

Consider any y such that $\mathrm{ic}^p(y : A) < C^{p'}(y)$ holds for all polynomials p'. Let q_y be a polynomial, where $q_y(|y|)$ is greater than the number of steps used to compute y from its shortest description of length $C(y)$. Then $C^{q_y}(y) = C(y)$. Because for every y there exists such a polynomial q_y, it follows that $H_p = \{x \mid \forall p' : \mathrm{ic}^p(x : A) > C^{p'}(x)\}$, and H_p is sparse. Remind that there exists a constant k such that for every x holds $C(x) = C^A(x) + k$. Thus by Theorem 1 it follows that $A \in R_{2\text{-}tt}^p(R_c^p(\text{SPARSE}))$. By a result in [1], no $\leq_m^p$-hard set for NP is in $R_{2\text{-}tt}^p(R_c^p(\text{SPARSE}))$, unless P = NP.

Relaxing from many-one to Turing reductions yields a weaker collapse consequence.

Theorem 3. *No decidable $\leq_T^p$-hard set for* NP *has sparse p-hard instances, unless* PH $= \Sigma_2^p$.

Proof. By the above argument it follows that NP $\subseteq R_T^p$(SPARSE), if a Turing hard set A for NP has sparse p-hard instances. From a result due to Meyer in [7] it then follows that the polynomial-time hierarchy collapses to Σ_2^p.

4 Hard Instances

Kummer [10] showed that the instance complexity conjecture does not hold for recursive sets, e.g. there exists a nonrecursive set A such that $\mathrm{ic}(x : A) \leq C(x)$ for almost every x. We show how to change the right-hand side of the inequality to get an exact characterization of the class of recursive sets. Essentially, the Kolmogorov complexity has to be taken relative to the halting problem K_0 and to the considered set itself.

Theorem 4. *Let A be any subset of Σ^*. A is recursive iff* $\mathrm{ic}(x : A) \leq C^{K_0 \oplus A}(x)$ *for almost every $x \in \Sigma^*$.*

Proof. If A is recursive, then there exists a Turing machine M deciding all instances of A. Thus for every $x \in \Sigma^*$: $\mathrm{ic}(x : A) < |M|$. Because for every c there exists an x_0 such that for all $x > x_0$ it holds that $C^{K_0 \oplus A}(x) > c$, it follows that $\mathrm{ic}(x : A) \leq C^{K_0 \oplus A}(x)$ for almost every $x \in \Sigma^*$.

For the other proof direction, assume that A is not recursive. We show that the following algorithm N with the halting problem and A as oracles computes hard instances.

input 0^n
$\Pi := \Sigma^{\leq n}$
for $m := n, n+1, n+2, \ldots$ **do**
  for all $\pi \in \Pi$ and all $y \in \Sigma^m$ **do**
    if $\pi(y) \not\simeq A(y)$ **then** $\Pi := \Pi - \{\pi\}$ **end**
  end
  if $\exists y \in \Sigma^m \; \forall \pi \in \Pi : \pi(y) = \perp$ **then**
    output the lex first such y and halt
  end
end

Note that Π does not become empty, since the program $\pi = \varepsilon$ will not be removed from Π.

Assume, that $N^{K_0 \oplus A}$ does not halt on some input 0^n. Since every single line of the program can be executed in finite time using oracles K_0 and A, the outer for-loop must then be repeated for infinitely many m. Then the set Π is not changed from some m_0 on. Therefore, Π is a finite set of programs which are almost everywhere consistent with A and almost every instance x is decided by some program in Π, namely all instances of length at least m_0. Thus A is recursive, contradiction the above assumption. Therefore, $N^{K_0 \oplus A}$ must halt for every input 0^n. Let y_n be the output of $N^{K_0 \oplus A}$ on input 0^n. Then $\{y_n \mid n \geq 0\}$ is an infinite set, because y_n exists and $|y_n| \geq n$.

By construction of N, $\mathrm{ic}(y_n : A) > n$ for almost every n. On the other hand, there exists a constant c such that for every n, $\mathrm{C}^{K_0 \oplus A}(y_n) \leq \mathrm{C}^{K_0 \oplus A}(0^n) + c$, where c is essentially the size of N. Because for a constant c' and every n it holds that $\mathrm{C}^{K_0 \oplus A}(0^n) \leq \log n + c'$, and because $\log n + c + c' \leq n$ for almost every n, it follows that $\mathrm{ic}(y_n : A) > \mathrm{C}^{K_0 \oplus A}(y_n)$ for almost every n. Therefore there exist infinitely many x such that $\mathrm{ic}(x : A) > \mathrm{C}^{K_0 \oplus A}(x)$.

Note that the upper bound on the instance complexity of decidable sets is tight in the following sense. For every set A and for every recursive nondecreasing unbounded function f, there exist infinitely many x such that $\mathrm{C}^{K_0 \oplus A}(x) < f(x)$ (see [11, Theorem 2.5]). Thus there are infinitely many x such that $\mathrm{ic}(x : A) > \mathrm{C}^{K_0 \oplus A}(x) - f(x)$. On the other hand, $\mathrm{C}^{K_0 \oplus A}(x)$ can be replaced by any other unbounded nondecreasing recursive function in $\mathrm{C}^{K_0 \oplus A}(x)$.

Corollary 5. *Let A be a set and f be an unbounded nondecreasing recursive function. A is recursive iff for almost every $x \in \Sigma^*$: $\mathrm{ic}(x : A) \leq f(\mathrm{C}^{K_0 \oplus A}(x))$.*

This characterization of decidable sets can be turned into one for sets decidable in polynomial time. The oracle set K_0 will be replaced by the NP complete set SAT, relative to which the search for undecided strings can be performed in polynomial time. To keep the running time polynomial in the length of the output, we have to slow down the removal of non-consistent programs.

Theorem 6. *Let A be any subset of Σ^*. $A \in P$ iff there exists a polynomial p such that for every polynomial p' and almost every $x \in \Sigma^*$: $\mathrm{ic}^p(x : A) \leq \mathrm{C}^{\mathrm{SAT} \oplus A, p'}(x)$.*

Proof. If $A \in P$, then there exist a polynomial p and a constant c such that $\mathrm{ic}^p(x : A) \leq c$ for almost every x [9]. For every polynomial p', there are only finitely many x with $c \leq \mathrm{C}^{\mathrm{SAT} \oplus A, p'}(x)$, proving the "forward" direction of the statement.

For the reverse proof direction, assume that $A \notin P$. Fix any polynomial p. We modify the algorithm N from the proof of Theorem 4 to the following algorithm M.

```
input 0^n
Π := Σ^{≤2 log n}
for m := n, n+1, n+2, ... do                        (* line 3 *)
    for all π ∈ Π and y ∈ Σ^{log m} do              (* line 4 *)
        if π(y)_p ≠ A(y) then Π := Π - {π} end       (* line 5 *)
    end
    if ∃y ∈ Σ^m ∀π ∈ Π : π(x)_p = ⊥ then            (* line 7 *)
        output such a y and halt
    end
end
```

As in the proof of Theorem 4, we can argue that M halts for every input 0^n. We have to argue that M runs in time polynomial in the length of its output. First note that Π has size $2n^2 - 1$ at the beginning of the program. Thus in every repetition of the loop beginning at line 3, for loop-counter m it holds that $|\Pi| \leq 2m^2 - 1$. Line 5 can be computed in time polynomial in $m = |y|$, say time $s(m)$, if A is used as oracle. Line 5 is repeated at most $(2m^2 - 1) \cdot (2m - 1)$ times. Thus the loop at line 4 takes time $O(m^3 \cdot s(m))$. Line 7 can be computed in polynomial time, say $s'(m)$, using the NP set $\{\langle \Pi, z, 1^m, 1^k \rangle \mid z$ is prefix of a string $y \in \Sigma^m \; \forall \pi \in \Pi : \pi(y)_k = \bot\}$ as oracle. Therefore, the execution of the loop beginning at line 3 for loop-counter m using oracles SAT and A takes $O(r_0(m))$ steps for $r_0(m) = m \cdot s(m) + s'(m)$ which can be chosen as nondecreasing polynomial. If eventually an output of length m is produced and the algorithm halts, it has a running time of $O(m \cdot r_0(m))$.

Let y_n be the output of M on input 0^n. Then $\mathrm{ic}^p(y_n : A) > 2\log n$, and $2\log n \geq C^{\mathrm{SAT} \oplus A, p'}(y_n)$ for almost every n, what completes the proof.

We can skip the SAT oracle from the above theorem, if we restrict our attention to decidable sets. As a trade-off, we lose control over the time needed to construct a hard instance. Actually, an exponential time bound applies.

Theorem 7. *Let A be any subset of Σ^*. $A \in P$ iff there exists a polynomial p such that for almost every $x \in \Sigma^*$: $\mathrm{ic}^p(x : A) \leq C^A(x)$.*

If A is decidable, even the oracle A can be omitted.

The above machine M can be modified to print out hard instances of a given length. A set A is called P^B-printable, if there exists a polynomial time-bounded oracle machine which on input 0^n using oracle B outputs $A \cap \Sigma^{=n}$. Printable sets are sparse and polynomial-time isomorphic to subsets of $KT[\log, \mathrm{poly}]$. We say that A has P^B-printable p^B-hard instances, if for every polynomial q there exists a polynomial q' and an infinite P^B-printable set S such that for almost every $x \in S$ it holds that $\mathrm{ic}^q(x : A) > C^{B, q'}(x)$. We show that if a set has hard instances, then some of them are not too hard to find.

Theorem 8. *Let A be a set. A is not in P iff A has $P^{\mathrm{SAT} \oplus A}$-printable $p^{\mathrm{SAT} \oplus A}$-hard instances.*

Proof. If $A \in P$, then by Theorem 6 it follows that A has no $p^{\mathrm{SAT} \oplus A}$-hard instances. For the other proof direction, suppose $A \notin P$. Let $\hat{M}$ be the Turing machine which on input $\langle 0^n, 0^k \rangle$ simulates machine M (from the proof of Theorem 6) on input 0^n, but halts (possibly without output) when the loop-counter m reached value k. Then the following machine prints hard instances.

```
input 0^n
for i := 0 to n do
    simulate M̂(⟨0^i, 0^n⟩)
    output all strings of length n computed by M̂
end
```

Since $\hat{M}$ on input $\langle 0^n, 0^k \rangle$ simulates at most $k \cdot r_0(k)$ of N (where r_0 is the time needed by M to execute one loop), it follows that $\hat{M}$ is polynomial time-bounded. This immediately induces that the above algorithm also is polynomial time-bounded. By the properties of M it follows, that the output strings have the desired hardness. Also by the properties of M it follows that the output set is infinite.

5 Very Hard Sets

Now we consider sets where every instance has maximal complexity. Not all instances of a set can have complexity as high as their Kolmogorov complexity. Ko [8] showed that there exists a set A such that for a constant k and almost every x, $ic(x : A) > C(x) - k \log |x|$. We use a time-bounded version.

Theorem 9. *Let A be a set such that for a constant k and almost every x,*
$$ic^{2^{2^{n+1}}}(x : A) > C(x) - k \log |x|. \text{ Then every } B \in \mathrm{DTIME}(2^n) \text{ which } \leq_m^p\text{-reduces}$$
to A is in $R_m^p(SPARSE)$.

Proof. Let A be as in the statement of the theorem, and let $B \in \mathrm{DTIME}(2^n)$ be a set such that $B \leq_m^p A$ via f. Define $L = \{f(x) \mid x \in B \wedge |f(x)| \geq \log |x|\}$. Then $y \in L$ can be decided by searching through all $x \in B$ up to length $2^{|y|}$ and checking whether $y = f(x)$. Since $f \in \mathrm{FP}$, it follows that L is in $\mathrm{DTIME}(2^{2^{n+1}})$. Each program accepting L can be altered to a program only accepting instances in L and not deciding instances in $\overline{L}$. Because $L \subseteq A$, this altered program is A-consistent, and therefore it follows that for a constant c and for all $x \in L$, $ic^{2^{2^{n+1}}}(x : A) < c$. Using the assumption $C(x) - k \log |x| < ic^{2^{2^{n+1}}}(x : A)$ for almost every x, we get that $C(x) < c + k \log |x|$ for almost every $x \in L$. Because there are at most $2^{c+k \log n}$ many such strings of each length n, L must be sparse.

Define $A' = \{y10^{2^{|y|}} \mid y \in A\}$. Since each string in A' contains only a logarithmic number of significant bits, it follows that A' is sparse too. Let $A' \oplus L = \{0x \mid x \in A'\} \cup \{1y \mid y \in L\}$ be the marked union of A' and L. Clearly, $A' \oplus L$ is sparse, and $B \leq_m^p A' \oplus L$ via the following polynomial-time function g.

$$g(x) = \begin{cases} 0f(x)10^{2^{|f(x)|}}, & \text{if } |f(x)| < \log |x| \\ 1f(x), & \text{otherwise} \end{cases}.$$

Thus, B is in $R_m^p(SPARSE)$.

Using results from [15] and from [13] we finally obtain from the above Theorem

Theorem 10. *Let A be a set such that for a constant k and almost every x,*
$$ic^{2^{2^{n+1}}}(x : A) > C(x) - k \log |x|.$$

1. *A is not $\leq_m^p$-hard for EXPTIME.*
2. *A is not $\leq_m^p$-hard for NP, unless $P = NP$.*

Acknowledgements. The author would like to thank Elvira Mayordomo for discussions which led to the proof of Theorem 9, and for the comments of a referee. The author also likes to thank Judy Goldsmith for her hospitality and the research environment provided during his visit at University of Kentucky.

References

1. V. Arvind, J. Köbler, and M. Mundhenk. On bounded truth-table, conjunctive, and randomized reductions to sparse sets. In *Proceedings 12th Conference on the Foundations of Software Technology & Theoretical Computer Science*, pages 140–151. Lecture Notes in Computer Science #652, Springer-Verlag, 1992.

2. J.L. Balcázar, J. Díaz, and J. Gabarró. *Structural Complexity I/II*. EATCS Monographs on Theoretical Computer Science. Springer Verlag, 1988/1990.

3. R. Book and J. Lutz. On languages with very high space-bounded Kolmogorov complexity. *SIAM Journal on Computing*, 22(2):395–402, 1993.

4. H. Buhrman and P. Orponen. Random strings make hard instances. In *Proc. 9th Structure in Complexity Theory Conference*, pages 217–222. IEEE, 1994.

5. L. Fortnow and M. Kummer. Resource-bounded instance complexity. In *Proceedings of 12th Symposium on Theoretical Aspects of Computer Science*. Lecture Notes in Computer Science #900, Springer-Verlag, 1995.

6. J. Hartmanis. Generalized Kolmogorov complexity and the structure of feasible computations. In *Proc. 24th IEEE Symp. on Foundations of Computer Science*, pages 439–445, 1983.

7. R. Karp and R. Lipton. Some relations between nonuniform and uniform complexity classes. In *Prooceedings of the 12th ACM Symposium on Theory of Computing*, pages 302–309, April 1980.

8. K. Ko. A note on the instance complexity of pseudorandom sets. In *Proc. 7th Structure in Complexity Theory Conference*, pages 327–337. IEEE, 1992.

9. K. Ko, P. Orponen, U. Schöning, and O. Watanabe. Instance complexity. *Journal of the ACM*, 41:96–121, 1994.

10. M. Kummer. The instance complexity conjecture. In *Proc. 10th Structure in Complexity Theory Conference*, pages 111–124. IEEE, 1995.

11. M. Li and P. Vitányi. *An introduction to Kolmogorov complexity and its applications*. Springer-Verlag, 1993.

12. N. Lynch. On reducibility to complex or sparse sets. *Journal of the ACM*, 22:341–345, 1975.

13. S. Mahaney. Sparse complete sets for NP: Solution of a conjecture of Berman and Hartmanis. *Journal of Computer and System Sciences*, 25(2):130–143, 1982.

14. P. Orponen and U. Schöning. The structure of polynomial complexity cores. In *11th Symp. on Mathematical Foundations of Computer Science*. Lecture Notes in Computer Science #176, Springer-Verlag, 1984.

15. O. Watanabe. Polynomial time reducibility to a set of small density. In *Proc. 1987 Structure in Complexity Theory Conference*, pages 138–146. Lecture Notes in Computer Science #223, Springer-Verlag, 1987.

Deciding Verbose Languages with Linear Advice

Arfst Nickelsen

Technische Universität Berlin
Fachbereich Informatik
10623 Berlin, Germany
nicke@cs.tu-berlin.de

Abstract. A language A is verbose if for some k there is a Turing machine M that for every input of k words $w_1, \ldots, w_k$ computes a bitstring of length k that is not the characteristic string $\chi_A(w_1, \ldots, w_k)$. A language A is p-verbose (or $A \in$ P-verb) if M is polynomially time bounded. Linear advice is sufficient to decide p-verbose languages in linear exponential time. Even languages that are linear-exponential time Turing reducible with linearly many queries to a p-verbose language are in E/lin. In [BL97] the special case of Turing reductions with a bounded number of queries to a p-selective language was investigated. Their results are extended to the general case of bounded Turing reductions to p-verbose languages.

In particular, it is shown that $E_{lin-T}(\text{P-verb}) \subset E/\text{lin}$; and EXP/poly is characterized as $\text{EXP}_{poly-T}(\text{P-verb})$. On the other hand for fixed c and k it holds that $E \not\subseteq P_{cn-T}(\text{P-}k\text{-verb})$ and $\text{EXP} \not\subseteq E_{n^c}(\text{P-}k\text{-verb})$.

1 Introduction

Even if a language $A \subseteq \{0, 1\}^*$ is not in P, some knowledge about the characteristic function of A may be computable in polynomial time. The idea is to consider k-tuples $(w_1, \ldots, w_k)$ of words as inputs and compute some partial information about $\chi_A(w_1, \ldots, w_k)$. This has lead (depending on the kind of "partial information" one is interested in) to the definition of various complexity notions, among them as the most prominent ones the notion of p-selectivity and p-verboseness (or non-p-superterseness). Also frequency computations, easily countable sets, cheatability, and multiselectivity fall into this approach.

The weakest type of partial information leads to the notion of verbose languages where only one of the potentially 2^k possible characteristic strings for $(w_1, \ldots, w_k)$ is excluded. A language A is verbose if for some $k \in \mathbb{N}$ there is a Turing machine M that for every input of k words $w_1, \ldots, w_k$ computes a bitstring $b_1 \ldots b_k$ with $\chi_A(w_1, \ldots, w_k) \neq b_1 \ldots b_k$. We are especially interested in the case where M is time bounded. A language A is p-verbose (or $A \in$ P-verb) if M is polynomially time bounded (or $A \in$ P-verb).

The notion of verboseness is most general in the sense that other types of partial information (for fixed tuple length) yield subclasses of the class of verbose languages. E.g., the class P-sel of p-selective languages forms a proper subclass of P-2-verb. A language A is p-selective if there is a polynomial time computable function f that on input of two words w_1, w_2 outputs one of these words that is "more likely to be in A" (see Section 2 for formal definitions).

Because P-verb extends P it is natural to ask wether P-verb is large enough to include standard complexity classes above P. Polynomially verbose sets can be considered "easy" in the sense that they exhibit at least some polynomial time behaviour. It turns out to be very instructive to compare this notion with another notion of "easy" sets, namely sets that can be decided with small advice. One would like to know how much additional information (depending only on the input length) is needed to actually decide different types of verbose languages. Furthermore one hopes to characterize advice classes as the closure of a class of verbose languages under an appropriate reduction. The motivating and best known result in this respect is the characterization of P/poly as the closure of P-sel under polynomial time Turing reductions [Sel79].

We sum up the preceding remarks by formulating the following problems for p-verbose sets:

1. Which uniform complexity classes C are not contained in P-verb?
2. For which uniform complexity class C and which reduction $\leq_r$ it holds that C is not contained in the closure of of P-verb under $\leq_r$-reductions?
3. For which nonuniform complexity class C it holds that P-verb $\subseteq C$?
4. Which nonuniform complexity classes can be characterized as the closure of P-verb under an appropriate reduction?

Question 3 and 4 are only asked for non-uniform classes because it is known that there are non-recursive p-selective sets [Sel82].

Concerning questions 1 and 2 it is known that, unless P = NP, there are languages in NP that are not p-verbose. [BKS94] even showed that, unless P = NP, there is no p-verbose set that is NP-hard under polynomial time $n^{o(1)}$-tt-reductions. There are sets in EXP that are not p-selective. [BL97] showed that even the following holds: For every fixed polynomial n^k there are languages in EXP that are not linear-exponential time Turing reducible to a p-selective language with at most n^k queries. In this paper this result is extended from p-selective sets to p-k-verbose sets. For fixed tuple-length k and and fixed linear functions cn or fixed polynomials n^c we obtain the following separation results:

$$E \not\subseteq P_{cn-T}(\text{P-}k\text{-verb}) \quad \text{and} \quad \text{EXP} \not\subseteq E_{n^c}(\text{P-}k\text{-verb})$$

On the other hand it remains unknown wether there are any sets in EXP that are not polynomial time Turing reducible to a set in P-verb.

Concerning questions 3 and 4 it is known that P-sel $\subseteq P/O(n^2)$ [Ko83] and that P-verb $\subseteq$ P/poly [ABG90]. The closure of P-sel (and thus of P-verb) under polynomial time Turing reductions equals P/poly [Sel79]. In [HT96] it was proven that P-sel $\subseteq$ NP/lin $\cap$ co-NP/lin. It is easy to see that P-sel $\subseteq$ E/lin. In [BL97] Burtschick and Lindner show that $E_{lin-T}(\text{P-sel}) \subset E/\text{lin}$. They also proved that $\text{EXP}_{poly-T}(\text{P-sel}) = \text{EXP}/\text{poly}$, i.e., EXP/poly can be characterized as the closure of P-sel under exponential time Turing reductions with polynomially many queries. We extend these results from the special case of p-selective languages to the general case of p-verbose languages. Particularly we prove the following:

$$\text{P-verb} \subseteq \text{E-verb} \subseteq \text{E/lin}$$

$$\text{EXP-verb} \subseteq \text{EXP/lin}$$

$$\text{E}_{lin-T}(\text{P-verb}) \subset \text{E/lin}$$

$$\text{EXP}_{poly-T}(\text{P-verb}) = \text{EXP/poly}$$

The last result can also be restated in the following way: For exponential time Turing reductions with polynomially many queries p-selective oracles are as powerful as p-verbose oracles.

Because $\text{E} \subseteq \text{P/poly}$ if and only if $\text{EXP} \subseteq \text{P/poly}$ if and only if $\text{EXP/poly} \subseteq \text{P/poly}$ both the inclusion and separation results might shed some light on the still unresolved problem whether $\text{EXP} \subseteq \text{P/poly}$.

2 Definitions

For basic notions and results in complexity theory see, e.g., [BDG88]. A language is a subset of $\{0,1\}^*$. For a language A the characteristic function $\chi_A : \{0,1\}^* \to \{0,1\}$ is defined by $\chi_A(w) = 1 \Leftrightarrow w \in A$. We extend this to tuples of words by setting $\chi_A(w_1, \ldots, w_k) = (\chi_A(w_1), \ldots, \chi_A(w_k))$. A is in $\text{DTIME}(T(n))$ if there is a deterministic Turing machine M with $L(M) = A$ for which the running time on inputs w with $|w| = n$ is in $O(T(n))$. We are especially interested in the complexity classes

$$\text{P} = \bigcup_{c \in \mathbb{N}} \text{DTIME}(n^c), \; \text{E} = \bigcup_{c \in \mathbb{N}} \text{DTIME}(2^{cn}) \text{ and } \text{EXP} = \bigcup_{c \in \mathbb{N}} \text{DTIME}(2^{n^c}).$$

A language A is k-verbose if there is a Turing machine M that on input $(w_1, \ldots, w_k)$ outputs a string $b = b_1 \cdots b_k \in \{0,1\}^k$ such that $\chi_A(w_1, \ldots, w_k) \neq b$. The language A is in $\text{DTIME}(T(n))$-k-verbose if the running time of M on an input with $\max |w_i| = n$ is in $O(T(n))$. In analogy to the complexity classes defined above we can define classes P-k-verb, E-k-verb and EXP-k-verb. E.g., $A \in$ P-k-verb if there is a $c \in \mathbb{N}$ such that $A \in \text{DTIME}(n^c)$-k-verbose. Finally, A is verbose if it is k-verbose for some k. This leads to the definition of the classes P-verb, E-verb and EXP-verb.

An advice Turing machine has in addition to its input tape and work tapes a special advice tape. A language A is decidable with advice h if there is an advice Turing machine M such that for every word w: $w \in A$ iff M started with w as input and $h(|w|)$ on the advice tape accepts. A language A is in $\text{DTIME}(t(n))/f(n)$ if A is decidable with advice h by an advice machine M such that

- the running time on input w is bounded by $t(|w|)$ and
- for all n $|h(n)| \leq f(n)$.

Observe that the running time is measured only relative to $|w| = n$, not relative to $|h(n)|$. Now we are ready to define different advice classes; e.g.,

$$P/poly = \bigcup_{c \in \mathbb{N}} \bigcup_{d \in \mathbb{N}} \text{DTIME}(n^c)/n^d \quad \text{or} \quad E/lin = \bigcup_{c \in \mathbb{N}} \bigcup_{d \in \mathbb{N}} \text{DTIME}(2^{cn})/dn.$$

In the same manner also EXP/lin and EXP/poly can be defined.

A language A is Turing reducible to a language B if there is an oracle Turing machine such that $A = L(M, B)$. A query tree of an oracle Turing machine on input w is a binary tree in which the nodes are labeled with all queries that M can ask on input w; the root is labeled with the first query, and for each internal node correspondig to a query q, the left (resp right) successor corresponds to the next query M asks with a negative (resp positive) answer on q.

For a time bound $t(n)$, a bound for the number of queries $q(n)$ and a class of languages $\mathcal{C}$, let $\text{DTIME}(t(n))_{q(n)-T}(\mathcal{C})$ denote the class of all languages that are Turing reducible to a language in $\mathcal{C}$ by a Turing machine with time bound $t(n)$ that asks at most $q(n)$ queries. We extend this notion to complexity classes and classes of functions bounding the number of queries, e. g., $P_{lin-T}(\mathcal{C})$, $E_{poly-T}(\mathcal{C})$, etc. All time bounds and bounds for the number of queries are assumed to be monotonic increasing and time constructible. Therefore for query trees we can always assume that all branches of the tree have the same length.

3 A Combinatorial Lemma

To prove that verbose sets can be decided with linear advice we will always follow the same scheme. Suppose we want to decide membership in a verbose set A for w with $|w| = n$. Let $w_1, \ldots, w_{2^n}$ be all words of length n. We then use the verboseness of A to precompute a list of bitstrings that contains the correct bitstring $\chi_A(w_1, \ldots, w_{2^n})$. The clue is that this list is not too large. We can then use the advice to choose the correct string out of the precomputed list. To bound the size of the list of bitstrings that appear in proofs of later theorems we need the combinatorial facts stated in Proposition 2 and Lemma 4.

Definition 1. For $t \in \mathbb{N}$ and strings $v, v' \in \{0, 1\}^t$ the Hamming distance $d(v, v')$ is defined as the number of $s \in \{1, \ldots, t\}$ with $v_s \neq v'_s$. For a string $v \in \{0, 1\}^t$ and $1 \leq k \leq t$ let $f(k, t)$ denote the number of strings $v' \in \{0, 1\}^t$ such that $d(v, v') > t - k$. Obviously the value of $f(k, t)$ does not depend on the choice of v.

Proposition 2. For $1 \leq k \leq t$ the following holds:

$$f(k, t) = \sum_{i=0}^{k-1} \binom{t}{i}$$

Proof. Given v, the number of strings v' that differ from v in all but i positions is $\binom{t}{i}$. Add up these values for $i = 0, \ldots, k-1$. □

Remark. Observe that for fixed k $f(k,t)$ is a polynomial; and it is easy to show that $f(k,t) \le t^{k-1} + 1$ for all $1 \le k \le t$.

Definition 3. Consider k, t with $1 \le k \le t$ and let H be a function that maps tuples from $\{(i_1, \ldots, i_k) | 1 \le i_1 < \ldots < i_k \le t\}$ to bitstrings from $\{0,1\}^k$. We say a string $v \in \{0,1\}^t$ is hit by H if there is a tuple $(i_1, \ldots, i_k)$ such that $H(i_1, \ldots, i_k) = v_{i_1} \ldots v_{i_k}$. The set of strings that are hit by H is called $Hit(H)$ and $f(H, k, t)$ denotes the number of strings that are not in $Hit(H)$.

Lemma 4. *For $1 \le k \le t$ and H a function as in Definition 3 it holds that*

$$f(H, k, t) \le f(k, t) .$$

Proof. The idea of the proof is to successively modify H until it has some standard form for which $f(H, k, t) = f(k, t)$. The modification has to be done in such a way that the value of $f(H, k, t)$ does not decrease in each step. This is achieved by successively projecting positions of H to zero.

Let k and t be fixed. For a given function H and a position $s \in \{1, \ldots, t\}$ we say that s is non-zero for H if there exists a tuple $(i_1, \ldots, i_k)$ with $i_l = s$ for some l and $H(i_1, \ldots, i_k)_k \ne 0$. (Note that even if s is non-zero for H there may be tuples $(i_1, \ldots, i_k)$ with $i_l = s$ for some l such that $H(i_1, \ldots, i_k)_l = 0$.) Now the proof is by induction on the number of positions s that are non-zero for H. If there are no such positions then $H(i_1, \ldots, i_k) = 0^k$ for all tuples $(i_1, \ldots, i_k)$. Then the strings that are not hit by H are those that have at most $k - 1$ zeros. These are the strings with $d(0^t, v) > t - k$, and there are $f(k, t)$ such strings.

Now let H be a function such that $n + 1$ positions are non-zero for H. Let s be such a position. Define a function H' that differs from H only in the values at position s as follows:

- For $(i_1, \ldots, i_k)$ where $s = i_l$ for some l
  let $H'(i_1, \ldots, i_k)_l = 0$ and $H'(i_1, \ldots, i_k)_m = H(i_1, \ldots, i_k)_m$ for all $m \ne l$.
- Otherwise let $H'(i_1, \ldots, i_k) = H(i_1, \ldots, i_k)$.

There are n positions left that are non-zero for H'. It remains to show that $f(H', k, t) \ge f(H, k, t)$. We prove this by constructing a mapping π from $Hit(H)$ onto $Hit(H')$. For $v \in Hit(H)$ define $\pi(v)$ as follows:

- If for all tuples of positions $(i_1, \ldots, i_k)$ with $H(i_1, \ldots, i_k) = v_{i_1} \ldots v_{i_k}$ there is an l such that $s = i_l$ then $\pi(v) = v_1 \ldots v_{s-1} 0 v_{s+1} \ldots v_t$.
- Otherwise there is an $(i_1, \ldots, i_k)$ with $H(i_1, \ldots, i_k) = v_{i_1} \ldots v_{i_k}$ such that $s \ne i_l$ for all $l \in \{1, \ldots, k\}$. Then $\pi(v) = v$.

For every $v' \in Hit(H')$ we now have to show that there is a $v \in Hit(H)$ with $\pi(v) = v'$.

- Case 1: For all tuples of positions $(i_1, \ldots, i_k)$ with $H'(i_1, \ldots, i_k) = v'_{i_1} \ldots v'_{i_k}$ there is an l such that $s = i_l$. Fix such a tuple $(i_1, \ldots, i_k)$. If $H(i_1, \ldots, i_k)_l = 0$ then $v = v'_1 \ldots v'_{s-1} 0 v'_{s+1} \ldots v'_t$ is in $Hit(H)$, if $H(i_1, \ldots, i_k)_l = 1$ then $v = v'_1 \ldots v'_{s-1} 1 v'_{s+1} \ldots v'_t$ is in $Hit(H)$. In both cases $\pi(v) = v'$.
- Case 2: There is an $(i_1, \ldots, i_k)$ such that $s \ne i_l$ for all $l \in \{1, \ldots, k\}$ and $H'(i_1, \ldots, i_k) = v'_{i_1} \ldots v'_{i_k}$. Then $v' \in Hit(H)$ and $\pi(v') = v'$. □

4 Deciding Verbose Sets with Linear Advice

In this section Lemma 4 is applied to show how verbose sets can be decided with linear advice. The following definition links functions H that map tuples of natural numbers to bitstrings with Turing machines that witness verboseness by mapping tuples of words to bitstrings.

Definition 5. Let A be a language and M_A a Turing machine such that A is k-verbose via M_A. Let $(w_1, \ldots, w_t)$ be a sequence of t words and $b = b_1 \ldots b_t \in \{0,1\}^t$ a bitstring of length t. We say b is hit by M_A (for $(w_1, \ldots, w_t)$) if there is a tuple of positions $(i_1, \ldots, i_k)$ such that $b_{i_1}, \ldots, b_{i_k} = M_A(w_{i_1}, \ldots, w_{i_k})$.

If we want to know $\chi_A(w_1, \ldots, w_t)$ for a given sequence of words we can remove from the list of 2^t bitstrings that are possible values for $\chi_A(w_1, \ldots, w_t)$ at first hand those strings that are hit by M_A. Since $\chi_A(w_1, \ldots, w_t)$ is not hit by M_A it is not removed. The following lemma formalizes this idea and states that the number of remaining bitstrings is small.

Lemma 6. Let A be a language that is k-verbose via a Turing machine M_A, $T_A(n)$ a bound for the running time of M_A, $t \geq k$ and $(w_1, \ldots, w_t)$ a sequence of t words. The following Turing machine M_{List} computes on input $(w_1, \ldots, w_t)$ a list of bitstrings of length t such that $\chi_A(w_1, \ldots, w_t)$ is in this list.

The list that M_{List} outputs contains at most $f(k,t)$ bitstrings. If $|w_i| \leq n$ for all $i \in \{1, \ldots, t\}$ then the running time of M_{List} is in $O(t^{2k} + t^{k+1}T_A(n))$.

M_{List}: input $(w_1, \ldots, w_t)$
    initialize List (a list of bitstrings) with all bitstrings of length $k-1$
    for all j from k to t do
        for all $b \in List$
            replace b by the two bitstrings $b0$ and $b1$
                    * List now contains bitstrings of length j
                    * the size of List is doubled
        for all tuples $(i_1, \ldots, i_{k-1}, j)$ with $1 \leq i_1 < \ldots i_{k-1} \leq j - 1$
            compute $M_A(w_{i_1}, \ldots, w_{i_{k-1}}, w_j)$
            if $M_A(w_{i_1}, \ldots, w_{i_{k-1}}, w_j) = b_{i_1} \ldots b_{i_{k-1}} b_j$
                then remove b from List
                    * b is removed if b is hit by M_A
    output List

Proof. For each j from k to t, only bitstrings that are hit by M_A for $(w_1, \ldots, w_j)$ are removed from *List*. But for all j and all tuples of positions $(i_1, \ldots, i_{k-1}, j)$ it holds that $\chi_A(w_1, \ldots, w_j) \neq M_A(w_{i_1}, \ldots, w_{i_{k-1}}, w_j)$. Therefore $\chi_A(w_1, \ldots, w_j)$ is never hit by M_A and in the end $\chi_A(w_1, \ldots, w_t)$ is in the output list. For fixed j define H_j for $1 < i_1 < \ldots < i_k \leq j$ by $H_j(i_1, \ldots, i_k) = M_A(w_{i_1}, \ldots, w_{i_k})$. A string b is removed from *List* if $b \in Hit(H_j)$. Therefore, by Lemma 4, the size of *List* after the removals is bounded by $f(k,j)$. Because f is monotonic, the size of *List* never exceeds $2f(k,t)$.

During the computation M_{List} has to simulate $\binom{t}{k}$ computations of M_A; each single simulation takes $T_A(n)$ steps. Thus $O(t^{k+1} T_A(n))$ steps are contributed by the simulations to the running time. M_{List} has to replace all b by $b0$ and $b1$ in the list $t - k + 1$ times; there are at most $f(k, t)$ strings in $List$, each of length at most t. Thus the replacement operations consume $O(t^{k+1})$ steps. The check whether a tuple is hit (and eventually remove it from the list) has to be done for $\binom{t}{k}$ tuples $(i_1, \ldots, i_k)$ for at most $f(k, t)$ strings b, each of length $\leq t$. This can be done in $O(\binom{t}{k} t^{k-1} t) = O(t^{2k})$. Therefore the total running time of M_{List} is in $O(t^{2k} + t^{k+1} T(n))$. □

We now apply Lemma 6 for various time bounds for the machine witnessing verboseness. The small size of the list output by M_{List} yields a small advice.

Theorem 7. *For all $k \in \mathbb{N}$:*

$$P\text{-}k\text{-}verb \subseteq DTIME(2^{2kn})/(k-1)n + 1$$

$$E\text{-}k\text{-}verb \subseteq E/(k-1)n + 1$$

$$EXP\text{-}k\text{-}verb \subseteq EXP/(k-1)n + 1$$

Proof. The proof is the same for all three results. One only has to take care of the different running times $T_A(n)$ of the machine M_A. Therefore only the first result is shown. The case $k = 1$ is easy because $P \subset DTIME(2^{2n})$. Now consider $k \geq 2$. Let $A \in$ P-k-verb via a machine M_A with running time $T_A(n) \in O(n^c)$. We describe a machine M' and an advice $h(n)$ for every input length n such that an input w is accepted by M' with advice $h(n)$ iff $w \in A$.

Consider a fixed input length n. Let $w_1, \ldots, w_{2^n}$ be the sequence of words of length n in lexicographical order. Let $List(n)$ be the list of bitsrings of length 2^n that M_{List} (from Lemma 6) outputs on input $(w_1, \ldots, w_{2^n})$. One of these strings equals $\chi_A(w_1, \ldots, w_{2^n})$. Choose as $h(n)$ the number of this string in $List(n)$ in binary representation. This number is at most $f(k, 2^n) \leq 2^{(k-1)n} + 1$. Therefore its binary representation consumes at most $(k-1)n + 1$ bits.

If M' gets w with $|w| = n$ as an input M' does the following. M' generates the sequence $(w_1, \ldots, w_{2^n})$ and simulates M_{List} on this sequence. M_{List} outputs $List(n)$. M' uses the advice to single out $\chi_A(w_1, \ldots, w_{2^n})$ from the list. Then M' looks up $\chi_A(w)$ in the string $\chi_A(w_1, \ldots, w_{2^n})$. M' accepts if $\chi_A(w) = 1$ and rejects if $\chi_A(w) = 0$. The running time of M' is essentially the running time of M_{List}. Substituting t by 2^n and $T_A(n)$ by n^c in the $O(t^{2k} + t^{k+1} T_A(n))$ bound from Lemma 6 we get the time bound $O(2^{2kn} + 2^{(k+1)n} n^c)$ which is in $O(2^{2kn})$ for $k \geq 2$. □

We restate the above result for the classes obtained by forming the union over all $k \in \mathbb{N}$:

Corollary 8. $P\text{-}verb \subseteq E\text{-}verb \subseteq E/lin$ *and* $EXP\text{-}verb \subseteq EXP/lin$

5   Sets Reducible to Polynomially Verbose Sets

Now we consider sets that are polynomial time Turing reducible to verbose sets where the number of queries is bounded by a function $q(n)$. For p-selective sets it was proven in [BL97] that

$$\mathrm{DTIME}(t(n))_{q(n)-T}(\text{P-sel}) \subseteq \bigcup_{r \geq 1} \mathrm{DTIME}(t(n)^r \, 2^{2q(n)+2n})/q(n) + n + 1 \ .$$

By modifying their proof and applying Lemma 6 we get:

Theorem 9. *For all $k \in \mathbb{N}$*

$$DTIME(t(n))_{q(n)-T}(P\text{-}k\text{-}verb)$$
$$\subseteq \bigcup_{r \geq 1} DTIME(t(n)^r \cdot 2^{2k(q(n)+n+1)})/(k-1)(q(n)+n+1)$$

Proof. Let B be a set Turing reducible to a set $A \in$ P-k-verb via an $O(t(n))$ time-bounded oracle Turing machine M_{red} which asks $q(n)$ queries on every path of the query tree on some input of length n. Let M_A be a Turing machine that witnesses $A \in$ P-k-verb. Fix an input length n and let $w_1, \ldots, w_{2^n}$ be the words of length n in lexicographic order. For each i let $Q(w_i)$ be the queries in the query tree of M_{red} on input w_i (ordered by depth first search). Let $Q(n) = Q(w_1)Q(w_2)\cdots Q(w_{2^n})$ be the concatenation of these lists. Observe that multiple occurences of queries are not removed. $|Q(n)|$, the number of entries in $Q(n)$, is bounded by $2^n(2^{q(n)+1} - 1)$. $Q(n)$ does only depend on n but not on a specific input of this length. Therefore we can use $Q(n)$ to construct an advice $h(n)$ for length n. Run the machine M_{List} from Lemma 6 on $Q(n)$. M_{List} outputs a list $List(n)$ of bistrings that contains $\chi_A(Q(n))$, the characteristic bitstring for $Q(n)$ with respect to A. For the number of entries in $List(n)$ we get

$$|List(n)| \leq f(k, 2^n(2^{q(n)+1} - 1))$$

Because of the remark after Propostion 2 we get

$$|List(n)| \leq 2^{(k-1)n}(2^{q(n)+1} - 1)^{k-1} + 1 \leq 2^{(k-1)(q(n)+n+1)}.$$

As advice $h(n)$ we take the binary representation of the number of the characteristic string $\chi_A(Q(n))$ in $List(n)$. Thus the advice has size $(k-1)(q(n)+n+1)$.

The following algorithm decides $w \in B$. Let $|w| = n$ and let $w = w_i$ in the sequence $(w_1, \ldots, w_{2^n})$ of all words of length n. Generate $Q(n)$ by traversing the query trees depth first for all $w_j \in (w_1, \ldots, w_{2^n})$. Apply M_{List} to $Q(n)$. M_{List} outputs $List(n)$. Use the advice $h(n)$ to single out $\chi_A(Q(n))$ from $List(n)$. Extract $\chi_A(Q(w_i))$ from $\chi_A(Q(n))$. Run M_{red} on w. If M_{red} asks a query look up the answer in $\chi_A(Q(w_i))$. Accept w iff M_{red} accepts w.

Generating $Q(w_j)$ for a $j \in \{1, \ldots, 2^n\}$ takes $O(2^{q(n)} t(n))$ steps. Generating $Q(n)$ therefore takes $O(2^n 2^{q(n)} t(n))$ steps. The size of a single entry in $Q(n)$ is bounded by $t(n)$. Therefore the running time of M_{List} with input $Q(n)$ is in

$$O(|Q(n)|^{2k} + |Q(n)|^{k+1} T_A(t(n))) \ .$$

Plugging in $|Q(n)| \leq 2^n(2^{q(n)+1} - 1)$ and $T_A(n) \in O(n^r)$ we get a bound in

$$O(2^{2k(q(n)+n+1)} + 2^{(q(n)+n+1)(k+1)} t(n)^r) \subseteq O(t(n)^r 2^{2k(q(n)+n+1)}) . \quad \square$$

We now apply Theorem 9 to achieve a characterization of EXP/poly :

Theorem 10. $EXP/poly = EXP_{poly-T}(P\text{-}verb)$.

Proof. In [BL97] it is shown that $\text{EXP}_{poly-T}(\text{P-sel}) = \text{EXP}/\text{poly}$. Since p-selective sets are p-verbose $\text{EXP}_{poly-T}(\text{P-sel})$ is contained in $\text{EXP}_{poly-T}(\text{P-verb})$. It remains to show that $\text{EXP}_{poly-T}(\text{P-verb}) \subseteq \text{EXP}/\text{poly}$. But from Theorem 9 it follows that for all k and for all polynomials $p(n)$ and $q(n)$

$$\text{DTIME}(2^{p(n)})_{q(n)-T}(\text{P-}k\text{-verb})$$

$$\subseteq \bigcup_{r \geq 1} \text{DTIME}(2^{rp(n)} 2^{2k(q(n)+n+1)})/(k-1)(q(n) + n + 1)$$

and the right hand side is included in EXP/poly. $\quad \square$

For E/lin we do not get an analogous characterization but a strict inclusion:

Theorem 11. $E_{lin-T}(P\text{-}verb) \subset E/lin$

Proof. From Theorem 9 it follows that for fixed c and k

$$E_{cn-T}(\text{P-}k\text{-verb}) \subseteq \bigcup_{r \geq 1} \text{DTIME}(2^{nr} 2^{2k(cn+n+1)})/(k-1)(cn+n+1)$$

$$= \bigcup_{r \geq 1} \text{DTIME}(2^{(r+2kc+1)n+1})/(k-1)((c+1)n+1) .$$

Hence $E_{lin-T}(\text{P-verb}) \subseteq \text{E/lin}$. [BL97] showed that the inclusion $E_{lin-T}(\text{P-sel}) \subset \text{E/lin}$ is strict. Their argument involving Kolmogorov random strings applies to the case of P-verb as well. This part of the proof is therefore omitted. $\quad \square$

Now we turn to two non-inclusion results. For the proof of Theorem 13 we need a lemma from [BL97]:

Lemma 12. *Let $f : \mathbb{N} \to \mathbb{N}$ be a function with $n \leq f(n) < 2^n$. Then*

$$DTIME(2^{4f(n)}) \nsubseteq DTIME(2^{f(n)})/f(n) .$$

Theorem 13. *Fix c and $k \in \mathbb{N}$. Then*

$$E \nsubseteq P_{cn-T}(P\text{-}k\text{-}verb) \quad and \quad EXP \nsubseteq E_{n^c}(P\text{-}k\text{-}verb) .$$

Proof. By Theorem 9 for every c, k and $s \in \mathbb{N}$

$$\text{DTIME}(n^s)_{cn-T}(\text{P-}k\text{-verb}) \subseteq \bigcup_{r \geq 1} \text{DTIME}((n^s)^r 2^{2k(cn+n+1)})/(k-1)(cn+n+1) .$$

Therefore, by adjusting the parameters for advice and running time we get:

$$P_{cn-T}(\text{P-}k\text{-verb}) \subseteq \text{DTIME}(2^{2k(cn+n+1)}/2k(cn+n+1) .$$

But $E \nsubseteq \text{DTIME}(2^{2k(cn+n+1)}/2k(cn+n+1)$ by Lemma 12.
The second statement is proved similarly. $\quad \square$

6 Some Open Problems

Apart from the main open problem whether EXP is in P/poly some other questions arise. As the closure under polynomial time Turing reductions of P-verb and P-sel coincide what is the weakest reduction for which this is the case? Can Theorem 13 be extended such that (without fixing a tuple-length k) we get $E \nsubseteq P_{cn-T}(\text{P-verb})$ and $EXP \nsubseteq E_{n^c}(\text{P-verb})$ for fixed c? Is the advive given by the construction preceding Theorem 7 optimal? P-verb and P-sel contain nonrecursive languages. Maybe one should also study in detail closure properties for those subclasses of P-verb that contain only recursive sets.

Acknowledgements: I wish to thank Dieter Hofbauer, Wolfgang Lindner, Birgit Schelm, and Dirk Siefkes for helpful discussions. Special thanks go to Hans-Jörg Burtschick for support, inspiriring discussions and long-time cooperation.

References

[ABG90] A. Amir, R. Beigel, and W. Gasarch. Some connections between bounded query classes and non-uniform complexity. In *Proc. 5th Structure in Complexity Theory*, 1990.

[BDG88] J. Balcázar, J. Díaz, and J. Gabarró. *Structural Complexity I*. Springer, 1988.

[BKS94] R. Beigel, M. Kummer, and F. Stephan. Approximable sets. In *Proc. 9th Structure in Complexity Theory*, 1994.

[BL97] H.-J. Burtschick and W. Lindner. On sets turing reducible to p-selective sets. *Theory of Computing Systems (formerly MST)*, 30:135–143, 1997.

[HT96] L. Hemaspaandra and T. Torenvliet. Optimal advice. *Theoretical Computer Science*, 154:367–377, 1996.

[Ko83] K.-I. Ko. On self-reducibility and weak p-selectivity. *Trans. Amer. Math. Soc.*, 131:420–436, 1983.

[Nic97] A. Nickelsen. On polynomially d-verbose sets. In *Proc. STACS 97*, 1997.

[Sel79] A. Selman. P-selective sets, tally languages and the behaviour of polynomial time reducibilities on np. *Mathematical Systems Theory*, 13:55–65, 1979.

[Sel82] A. Selman. Analogues of semirecursive sets and effective reducibilities to the study of NP complexity. *Information and Control*, 1, 1982.

Homomorphic Images of Sentential Forms and Terminating Grammars (Extended Abstract)

Holger Petersen

Institut für Informatik, Universität Stuttgart
Breitwiesenstr. 20–22, D-70565 Stuttgart
petersen@informatik.uni-stuttgart.de

Abstract. We consider languages generated as homomorphic images of sentential form languages and related grammars. A characterization of these languages in terms of their topological properties is given. Previous hierarchy results are strengthened by constructing separating languages over a single letter alphabet. The relation to languages generated by terminating grammars is investigated.

1 Introduction

In formal language theory the usual way of defining the set of words generated by a grammar is based on the concepts of terminal and nonterminal symbols. Only words over the terminal alphabet are considered to be included in the language defined by the grammar. If however the structural properties of a grammar are important it is useful to investigate all sequences of symbols, terminal or nonterminal, that can be obtained. This set is called the sentential form language and grammars interpreted from this point of view are also called pure [7]. As an example illustrating this difference we mention the well-known constructions transforming context-free grammars into some normal-form. These transformations preserve the terminal language generated by a grammar while in general they will modify the way words are generated and thus influence the sentential form language.

A word from a sentential form language reveals all of its inherent information. We get an intermediate situation if we "hide" some of its structure by applying a homomorphism to the sentential form language. While there are even non-empty regular languages that are not sentential form languages of any context-free grammar [6, Proposition 2.21], every non-empty context-free language L can be obtained as the homomorphic image of the sentential form language generated by a reduced grammar for L [4, Theorem 9](with the restrictions of their definition imposed on context-free grammars the homomorphism may be non-erasing). Very thorough investigations of the relations between the resulting classes of languages have been carried out in [6] for context-free and non-context-free rewriting using various types of grammars and homomorphisms. Further related results are surveyed in [2, Chapter 5].

This article is organized as follows. First some definitions and notations are introduced. Our first results explore the topological properties of the language classes. For images of sentential form languages of phrase-structure grammars under arbitrary homomorphisms we obtain a complete characterization. In the next section we investigate single-letter alphabet languages and are able to establish separations based on these structurally simpler sets. In the last section of our results we consider terminating grammars as introduced in [3]. We close the paper with a short discussion.

2 Definitions and Notation

We denote the set of non-negative integers by $\mathcal{N}$. For a set M its power-set is $\wp(M)$. A *grammar* will be denoted as $G = (V, X, P, M)$, where V is a finite set of *variables*, X is a finite set of *terminal symbols*, P is a finite set of *productions* $\alpha \to \beta$ with $\alpha, \beta \in (V \cup X)^*$, and $M \subseteq (V \cup X)^*$ is a finite set of *axioms*. If M is a singleton set $M = \{S\}$ we write $G = (V, X, P, S)$ instead of $G = (V, X, P, \{S\})$. A *derivation step* according to a production $\alpha \to \beta$ transforms a string $u\alpha v$ into $u\beta v$. If a string w can be transformed by zero or more derivation steps into w', we write $w \stackrel{*}{\Rightarrow} w'$. The *language* generated by G is the set $L(G) = \{w \in X^* \mid S \stackrel{*}{\Rightarrow} w \text{ for some } S \in M\}$. The sentential form language generated by G is $S(G) = \{w \in (V \cup X)^* \mid S \stackrel{*}{\Rightarrow} w \text{ for some } S \in M\}$. Note that for the grammar $G = (\emptyset, \emptyset, \emptyset, \emptyset)$ we have $L(G) = S(G) = \emptyset$. A grammar is *terminating* (t-grammar) if for every $A \in V$ there is a $w \in X^*$ with $A \stackrel{*}{\Rightarrow} w$.

Let $\hat{H}IS$ (HIS) be the class of languages $L = h(S(G))$, where G is a grammar, $G = (V, X, P, M)$, and $h : (V \cup X)^* \to Y^*$ is a homomorphism (non-erasing homomorphism) to some alphabet Y. If the grammar is propagating or monotone and the homomorphism is non-erasing we obtain a language in $HPFIS \subset HIS$. If the homomorphism is the identity and the grammar is arbitrary (has at most one axiom) we get the class FIS (IS). These definitions are taken from [6].

Let T be the class of languages $L(G)$ for terminating grammars G.

By FIN, REG, CF, CS, RE we denote the classes of finite, regular, context-free, context-sensitive, and recursively enumerable languages, respectively.

The following definitions are taken from [3]. Let $L \subseteq X^*$ be a language and define a function $f : X^* \times X^* \to \mathcal{N}$ by

$$f(w, w') = \min\{n \in \mathcal{N} \mid \exists x, y, y', z \in X^* : n = |y| + |y'|, w = xyz, w' = xy'z\}.$$

Now we define a distance $d_L : L \times L \to \mathcal{N}$ by

$$d_L(w, w') = \min \left\{ \sum_{i=1}^{n} f(w_{i-1}, w_i) \mid n \in \mathcal{N}, w_0 = w, w_n = w', \forall i : w_i \in L \right\}.$$

Note that for $L = X^*$ the distance d_L coincides with the edit distance, where deletions and insertions are the allowed operations.

For every $r > 0$ the graph $L(r)$ has L as its set of vertices and $(w, w') \in L \times L$ is an edge in $L(r)$ if and only if $0 < d_L(w, w') \leq r$.

The Turing machine model we use is equipped with a single, semi-infinite tape as defined in [5]. We also use notions from complexity theory.

3 Results

3.1 Topological Properties

The first result of this section is an alternative characterization of the class $\hat{H}IS$ making use of its topological properties. Then we will investigate the subclass $HPFIS$ of HIS generated by propagating grammars from a finite set of axioms. It will be shown that these classes are separated by a decidable language, again exploiting the topological restrictions enforced by the fact that every sentential form contributes to the generated language.

In [3, Theorem 5] it is shown that for any $L \in \hat{H}IS$ an $r > 0$ exists such that the graph $L(r)$ is connected. Note that the distance has to "respect" L in the sense that all intermediate words are in L. Taking the usual edit distance $\{a^n b^n c^n \mid n \geq 0\} \notin \hat{H}IS$ would have a connected graph for distance 3. By adding the property of being recursively enumerable, which trivially holds for languages in $\hat{H}IS$, we show the converse implication (notice that we assume $\emptyset \in \hat{H}IS$).

Theorem 1. *Let L be a language, then $L \in \hat{H}IS$ if and only if L is recursively enumerable and there exists an $r > 0$ such that $L(r)$ is connected.*

Proof. By the preceding remarks it suffices to show that a recursively enumerable language L with $L(r)$ connected can be generated as the homomorphic image of a sentential form language.

We start with an informal exposition of the construction. Every recursively enumerable language L can be generated by some grammar. We construct a new grammar that simulates a derivation of a word w' in the language with the help of sentential forms that have a word w (from L) as their image, where w is "similar" to w'. This process is repeated eventually reaching every word due to the topological property of L. The initial word is the image of the axiom. In order to reach our goal we duplicate the terminal alphabet and let these new symbols commute with the original terminal alphabet. Further we make use of end-markers and three messenger symbols which compare the previous word w and w' (still encoded in the new alphabet). Since these words are not identical the messenger B nondeterministically selects a portion of w (marked by auxiliary symbols "!") and passes control to A. When A reaches the left end-marker it is transformed to R, a messenger moving to the right and replacing the marked portion, thus generating a sentential form that has w' as its image.

If $L = \emptyset$ the conclusion of the theorem trivially holds. Let $L = L(G) \neq \emptyset$ for some grammar $G = (V, X, P, S)$ with a single axiom S (there is no loss of generality in this assumption) and for some $r \geq 1$ the graph $L(r)$ is connected. We introduce a set of new symbols $\bar{X} = \{\bar{a} \mid a \in X\}$. Since $L \neq \emptyset$ we may fix some word $w \in L$. Define the homomorphism $^- : (V \cup X)^* \to (V \cup \bar{X})^*$ by $\overline{A} = A$ for $A \in V$, $\bar{a} = \bar{a}$. Now we define a new grammar $G' = (V', X, P', S')$ with

$$V' = V \cup \bar{X} \cup \{\vdash, \dashv, S', !, B, A, R\}.$$

The productions are defined as follows:

$$P' = \{\bar{\alpha} \to \bar{\beta} \mid \alpha \to \beta \in P\} \cup \{a\bar{b} \to \bar{b}a \mid a, b \in X\} \cup$$
$$\{S' \to \vdash wSB\dashv\} \cup \{a\bar{a}B \to Ba \mid a \in X\} \cup$$
$$\{u\bar{v}B \to A!u\bar{v}! \mid |uv| \le r\} \cup \{a\bar{a}A \to Aa \mid a \in X\} \cup$$
$$\{\vdash A \to \vdash R\} \cup \{Ra \to aR \mid a \in X\} \cup$$
$$\{R!u\bar{v}! \to vR \mid |uv| \le r\} \cup \{R\dashv \to SB\dashv\}$$

Finally we define the homomorphism $h : (V' \cup X)^* \to X^*$ by $h(S') = w$, $h(Y) = \lambda$ for $Y \in V' \setminus \{S'\}$, and $h(a) = a$ for $a \in X$.

Every image $h(u)$ of a sentential form u generated by G' is a word in L. By the connectedness of $L(r)$ there is a path $w_0, w_1, \ldots w_n$ with $f(w_{i-1}, w_i) \le r$ for every $1 \le i \le n$ from the initial word $w = w_0$ to every word $w_n \in L$. The grammar G' can generate these words in turn and thus reach w_n. We conclude that $L = h(S(G'))$. $\qquad\square$

Remark. The connectedness of $L(r)$ does not ensure that L is a recursively enumerable language. Consider a set D of non-negative integers which is not recursively enumerable. Define $L \subseteq \{a\}^*$ by $L = \{a^{2n} \mid n \ge 0\} \cup \{a^{2m+1} \mid m \in D\}$. Then L is not recursively enumerable, but for every $w \in L$ there exists some $w' \in L$ with $d_L(w, w') \le 2$ and hence $L(2)$ is connected.

From [6, Fig. 3] the inclusion $HPFIS \subset HIS$ is known to be proper, a fact following from the observation that all languages in $HPFIS$ are decidable, while HIS contains non-recursive sets. We will strengthen this separation by showing that there are decidable languages in HIS which cannot be generated by monotone grammars. This result will later be of some additional interest when we investigate context-sensitive languages in HIS, see Section 3.3, since it shows that not every context-sensitive language L with $L(r)$ connected for some $r > 0$ is a member of $HPFIS$.

Theorem 2. *Let*

$$L_S = \{a^{n^2}b^m \mid m, n \ge 0\} \cup \{a^m b^{n^3} \mid n \ge 1, (n-1)^2 < m < n^2\}.$$

Then L_S is a decidable language with the property $L_S \in HIS$, but $L_S \notin HPFIS$.

Proof. In principle it would be possible to design a grammar and a homomorphism according to the definition of HIS generating L_S. It will however be more convenient to use a homomorphism that is limited erasing (in fact at most four symbols are erased in every sentential form). By the closure under limited erasing homomorphisms [6, Theorem 3.17] we will thus have shown that $L_S \in HIS$.

The grammar $G = (V, \{a, b\}, P, S)$ and homomorphism h generating the language L_S are given by: $h(X) = \lambda$ for $X \in V$, $h(a) = a$, $h(b) = b$, and

$$V = \{A, B, C, D, E, F, G, H, I, J, K, L, M, N, P, Q, \vdash, !\}.$$
$$P = \{S \to \vdash !A, A \to Bb, aB \to CaD, Da \to aD, Db \to Eb^7,$$

$$aE \to Ea, CE \to B, !B \to F!, aF \to GaH, Ha \to aH, H! \to !H$$
$$Hb \to Ib^4, aI \to Ia, !I \to I!, GI \to F, \vdash F \to \vdash J,$$
$$Ja \to aJ, J! \to !aK, Ka \to LaM, aL \to La, !L \to a!Q,$$
$$!Q \to a!N, Na \to aN, NM \to K, Kb \to Ab, A \to P, P \to Pb, Pb \to P\}$$

We will briefly explain the idea behind this construction. Every generated string containing the symbol A has the form

$$\vdash \overbrace{a \cdots a!}^{n^2+1} \underbrace{a \cdots a}_{n} A \overbrace{b \cdots b}^{n^3}.$$

The productions involving P add or delete any number of b's from this sentential form. With the help of several messenger symbols the grammar passes from n to $n + 1$ by making use of the identities $(n + 1)^2 = n^2 + 2n + 1$ and $(n + 1)^3 = n^3 + 3n^2 + 3n + 1$.

Now we will show that $L_S \notin HPFIS$. For the sake of contradiction suppose we have a monotone grammar $G = (V, X, P, M)$ and a non-erasing homomorphism h with $L_S = h(S(G))$. Let $a = \max\{|w| \mid w \in M\}$, $b = \max\{|\alpha| \mid \{\alpha \to \beta, \beta \to \alpha\} \cap P \neq \emptyset\}$, and $c = \max\{|h(A)| \mid A \in (V \cup X)\}$. Now set $n = \max\{ac, bc\} + 2$ and consider the word $a^{n^2} \in L_S$, $a^{n^2} = h(w)$ with $w \in S(G)$, $|w| \leq n^2$. Any derivation of w necessarily contains a word w' with $h(w') = a^{n^2} b^j$ and $n^3 > j \geq n^3 - n > n^3 - n^2$. To see this fix some derivation of w and consider the longest trailing sequence of sentential forms occurring in the derivation and satisfying $h(u) = a^{n^2} b^j$ with $n^3 - n > j \geq 0$ for each u in the sequence (the sequence is nonempty because $h(w) = a^{n^2} b^0$). None of these sentential forms can be an axiom ($n > ac$), therefore there has to be a sentential form w' occurring in the derivation immediately before the sequence with $h(w') \notin \{a^{n^2} b^j \mid n^3 - n > j \geq 0\}$, but $h(w') \in L_S \subseteq a^*b^*$. The possibility $h(w') = a^i b^j$ for $i \leq (n - 1)^2$ or $i \geq (n + 1)^2$, $j \geq 0$ is ruled out because $n^2 - (n - 1)^2 = 2n - 1 > bc$ and $(n + 1)^2 - n^2 = 2n + 1 > bc$. We conclude that $h(w') = a^{n^2} b^j$ with $n^3 > j \geq n^3 - n > n^3 - n^2$. Therefore $|w'| > ((n^3 - n^2) + n^2)/c = n^3/c > n^2$. But since w' appears in a derivation of w and $|w| = n^2$ this contradicts the monotonicity of G. Thus $L_S \notin HPFIS$. $\square$

3.2 Single Letter Alphabet Languages

The following theorem strengthens the separation of HIS and $\hat{H}IS$ known from [6, Theorem 3.20] (note that $HIS = CIS$ and $\hat{H}IS = \hat{C}IS$, where the classes CIS and $\hat{C}IS$ are defined using codings mapping symbols to symbols instead of arbitrary homomorphisms). We will show that this separation can be achieved with the help of a language over a single letter alphabet. At the same time the proof is based on a straightforward diagonal argument and does not not make use of results from complexity theory.

Theorem 3. *The classes HIS and ĤIS are separated by a language over a single letter alphabet, i.e., there is a language $L_D \subseteq \{a\}^*$ such that $L \notin HIS$ and $L \in \hat{H}IS$.*

Proof (sketch). We describe an iterative process constructing a language L_D word by word, where the construction proceeds from shorter to longer words in $\{a\}^*$. This process will be carried out by a single tape Turing machine M. We will thus view M as a generator, that does not need to stop its generating process, since all configurations contribute to the language L_D.

The Turing machine M will have two special symbols 1 and 2 that are not used in the course of M's normal operation, and none of which is initially written on the tape. A variable n is initialized to 0.

Let G_i, $i \geq 1$, be an enumeration of grammars. Without loss of generality the homomorphisms applied to sentential forms are codings [6, Theorem 3.7]. Therefore they are forced to map every symbol to a and we will ignore them in our discussion. In the ith iteration grammar G_i of this enumeration will be considered by M.

For every grammar G_i of the enumeration Turing machine M writes down a description of G_i and a set K of reachable sentential forms, which is initialized with the axioms of G_i as its only elements. While K keeps growing and the length of no sentential form in K exceeds n, M adds reachable sentential forms to K. Note that this process necessarily terminates, since the set of sentential forms bounded in length by n is finite.

If no sentential form reaches a length exceeding n, the iteration step is finished and M turns to the next grammar.

If a sentential form v is added to K with $|v| > n$ let $m = |v|$. Now M writes down $m - n - 1$ additional symbols 1 onto its tape and one symbol 2 (we suppose that M reserves some portion of its tape for this purpose). Then M updates n to $m + 1$ and continues its construction with the next grammar.

The description of M readily translates into a grammar G_D. A suitable homomorphism h_D is defined by $h_D(1) = a$, $h_D(2) = aa$, and $h_D(x) = \lambda$ for every symbol $x \notin \{1, 2\}$.

By construction $L_D = h_D(S(G_D)) \in \hat{H}IS$. We will argue that $L_D \notin HIS$. First we observe that L_D is infinite, since for every grammar generating an infinite language at least the word a^{m+1} is added to L_D by the above process. Suppose L_D were generated by some G and the coding h mapping every symbol to a. Then a sentential form v with $|v| = m > n$ for the current n will be generated when M simulates G, and $h(v) = a^m$ is a word in the language generated by G. But the symbols 1 and 2 are added in a way that avoids the length m, the ultimate length of the image under h_D is $m + 1$, and later iterations do not decrease the length of the image. We conclude that a^m is not in L_D and derive a contradiction. □

It is clear that $HPFIS$ contains decidable languages only. The following lemma, similar in spirit to [7, Theorem 5.2], shows that this is not the case for HIS, even over a single-letter alphabet, thus separating the classes.

Lemma 4. *For every recursively enumerable set M of non-negative numbers the set $L_M = \{a^{2n} \mid n \geq 0\} \cup \{a^{2m+1} \mid m \in M\}$ is in HIS.*

Proof (sketch). The idea is to simulate a computation on some m with the help of sentential forms that yield only strings of even length. Should $m \in M$ hold, a string of length $2m + 1$ is produced by either adding or deleting two symbols at a time until $2m$ is reached. Then a single symbol is added in order to produce the desired length. In addition all strings a^{2n} for $n \geq 0$ are generated. $\square$

The following result is well known [7, 2] and follows from Büchi's result on regular canonical systems [1], since the rewriting in a string over a single letter alphabet may occur at the right end.

Lemma 5. *The languages in FIS over a single letter alphabet are exactly the regular languages over this alphabet.*

It follows from [6, Lemma 3.24] that $IS \neq FIS$, but the proof is based on a language over the alphabet $\{a, b\}$. We will show the stronger assertion that the classes are separated by a single letter alphabet language.

Lemma 6. *There is a language $L \subseteq \{a\}^*$, such that $L \in FIS$ but $L \notin IS$.*

Proof. Let $L = \{a^{3n} \mid n \geq 0\} \cup \{a^{3n+1} \mid n \geq 0\}$. Suppose that L can be generated from a single axiom a^{3n} for some $n \geq 0$. Then $a^{3n} \overset{*}{\Rightarrow} a^{3n+1}$ and $a^{3n+1} = aa^{3n} \overset{*}{\Rightarrow} aa^{3n+1} = a^{3n+2} \notin L$ which is a contradiction. Now suppose that L can be generated from an axiom a^{3n+1} for some $n \geq 0$. Then $a^{3n+1} \overset{*}{\Rightarrow} a^{3n+3}$ and $a^{3n+1} = aaa^{3n+1} \overset{*}{\Rightarrow} aaa^{3n+3} = a^{3(n+1)+2} \notin L$ which again yields a contradiction. $\square$

The classes RE and $\hat{H}IS$ are separated by $L' = \{a^{2^n} \mid n \geq 0\} \cup \{\lambda\}$ [6, Corollary 3.10], $HPFIS$ and FIS by the non-regular set $\{a\}^* - L'$ [6, Lemma 3.11]. Finally we remark that IS contains every finite set by taking a word w of maximum length as the axiom and including a production $w \to v$ for every $v \in L - \{w\}$.

Proposition 7. *The following relations hold, where inclusions are strict:*

$$FIN \cap \wp(\{a\}^*) \subset IS \cap \wp(\{a\}^*) \subset$$
$$FIS \cap \wp(\{a\}^*) = REG \cap \wp(\{a\}^*) = CF \cap \wp(\{a\}^*) \subset$$
$$HPFIS \cap \wp(\{a\}^*) \subset HIS \cap \wp(\{a\}^*) \subset \hat{H}IS \cap \wp(\{a\}^*) \subset RE \cap \wp(\{a\}^*)$$
$$HPFIS \cap \wp(\{a\}^*) \subset CS \cap \wp(\{a\}^*) \subset RE \cap \wp(\{a\}^*)$$

3.3 Terminating Grammars

First we will address the question, whether it is decidable if a grammar is terminating. As it turns out this is impossible in general. Note that a typical construction showing undecidability of the question whether the initial symbol of a grammar is non-terminating would make use of nonterminals that can be rewritten only in some context. The resulting grammar would be non-terminating by definition.

Terminating grammars were introduced in [3] in order to characterize homo-morphic images of sentential form languages. In the second part of this section we will investigate to what extent unrestricted t-grammars are suitable for this purpose.

Theorem 8. *It is undecidable whether a given grammar is terminating.*

Proof. We will describe a reduction of the membership problem for single tape Turing machines to the problem mentioned in the theorem.

Let $M = (Q, \Sigma, \Gamma, \delta, q_0, B, F)$ be a deterministic single tape Turing machine. Its input word is w. The grammar $G = (V, X, P, S)$ to be constructed will have the set of variables $V = Q \cup \{S\}$ (where $S \notin Q$), the terminal alphabet $X = \Sigma \cup \Gamma \cup \{\vdash, \dashv\}$, and the axiom S. The set of productions P includes $S \to \vdash q_0 w \dashv$, $q \to S$ and $q \dashv \to qB \dashv$ for every $q \in Q$, $q_f \to \lambda$ for every $q_f \in F$. The remaining productions correspond to the move function of M in the natural way:

$$\delta(q, a) = (q', a', R): \quad qa \to a'q' \in P,$$
$$\delta(q, a) = (q', a', L): \quad bqa \to q'ba' \in P \text{ for all } b \in \Sigma \cup \Gamma.$$

If M accepts w, then a derivation from S to some terminal string exists that simulates M's computation until a final state is reached and the single variable can be eliminated. But since for every variable q a production $q \to S$ is in P, the grammar G is terminating.

Conversely, if G is terminating, then fix a terminating derivation for S and consider the last application of $S \to \vdash q_0 w \dashv$. After this step the derivation is forced to simulate a computation of M, since there will be at most one variable in each sentential form and according to the productions it will remain between $\vdash$ and $\dashv$. The only way the derivation can yield a terminal string is by a production $q_f \to \lambda$, which shows that M accepts w. □

In the following paragraphs we will investigate subclasses of $\hat{H}IS$, which are characterized by t-grammars.

By a technique similar to the one employed in the characterization of $\hat{H}IS$ (Theorem 1) it is possible to establish that the families HIS and $\hat{H}IS$ coincide on their context-sensitive members.

Lemma 9. *If for a context-sensitive language L there is an $r > 0$, such that $L(r)$ is connected, then $L \in HIS$.*

Theorem 10. *The families $HIS \cap CS$ and $\hat{H}IS \cap CS$ are equal.*

Proof. By definition $HIS \cap CS \subseteq \hat{H}IS \cap CS$. For any language $L \in \hat{H}IS$ there exists an $r > 0$, such that $L(r)$ is connected [3, Theorem 5]. The theorem follows from the preceding lemma. □

Recall that T is the family of languages generated by t-grammars. Since the inclusions $HIS \subseteq T$ and $T \subseteq \hat{H}IS$ are known [3, Theorem 2, Lemma 4], we can conclude that $T \cap CS$ and $\hat{H}IS \cap CS$ are equal. We are however able to establish a much stronger statement. For a given positive number c define the tower function t_c by $t_c(0) = 1$, $t_c(n) = c^{t_c(n-1)}$ for $n > 0$.

Lemma 11. *If for a language $L \in NSPACE(t_c(n))$ (with $c \geq 1$) there is an $r > 0$, such that $L(r)$ is connected, then $L \in T$.*

Proof (sketch). The construction resembles the one elaborated in the proof of Theorem 1. A sentential form will encode some word w in the language L such that the nonterminals encoding the first and the last symbol of w are marked, while nonterminal symbols that yield the empty word simulate a Turing machine computation in order to accept a word w' that is connected to w in $L(r)$. The set of nonterminals is thus divided in two groups, one of which ensures termination via productions of the form $A \rightarrow a$ with terminal symbol a (non-erasable symbols), and the second one via $B \rightarrow \lambda$ (erasable symbols). Terminal symbols will block the Turing machine simulation.

Let us for the sake of simplicity first assume that $|w| = |w'|$. The idea of the simulation is that at the right side of the ith non-erasable symbol a block of $t_c(i)$ erasable symbols is built up. This is trivial for the first block, its length is c. Assuming the ith block has been built, the $i + 1$st block can be generated by counting in base c on the symbols of the ith block generating an erasable symbol in block $i + 1$ for every counting step. Eventually the the $i + 1$st block contains (at most) $c^{t_c(i)} = t_c(i + 1)$ symbols. The last block is then used for simulating the nondeterministic Turing machine computation on w'. Finally the counting process is reversed and it is checked that no symbol has been lost during any of the three phases described above. To see that this is possible, imagine that at any time an erasable symbol is actually erased. Then look at the leftmost block in which this happens and notice that the missing symbol will necessarily be caught, since at least for block 1 its length can be verified correctly. Finally the encoding of w is transformed into w' and the process is repeated.

Now we discuss the case $|w| \neq |w'|$. By the topological restriction of L the difference in length of two directly connected words is bounded by a constant. Thus it suffices to take as the length of the first block $t_c(k)$ for a suitably chosen k. We omit the detailed description of this simulation, including the generation of a new word and access to the input for the simulated machine. $\square$

Theorem 12. *The families $T \cap NSPACE(t_c(n))$ and $\hat{H}IS \cap NSPACE(t_c(n))$ (for every $c \geq 1$) are equal.*

We conclude that if the families T and $\hat{H}IS$ can be separated, then the separating languages have to be of a high complexity. Now we will take advantage of the previous results in order to separate T and HIS.

Theorem 13. *The inclusion $HIS \subset T$ is proper.*

Proof (sketch). It is possible to adapt the diagonal construction from the proof of Theorem 3 (and then the present result would imply Theorem 3). Instead we give an argument based on the ideas of the proof of [6, Theorem 3.20].

Let $L_1 \subseteq \{a\}^*$ with $L_1 \in NSPACE(n^3)$ and $L_1 \notin NSPACE(n^2)$ [8, Theorem U-1-N]. Consider the sequence of words $w_1, w_2, \ldots \in L_1$ in length increasing order. Form the language $L_2 = \text{pref}(\{w_1\$w_2\$ \cdots \$w_k \mid k \geq 1\})$, where pref is the operator generating all prefixes of words in a language. By the preceding theorem $L_2 \in T$, since a prefix closed language has a connected graph for $r = 1$ and the space bound suffices to check every word possibly in the sequence. Suppose $L_2 \in HIS$, without loss of generality the homomorphism involved in its definition may be a coding [6, Theorem 3.7]. In order to decide whether $a^n \in L_1$ it suffices to simulate the grammar generating L_2 until some string of length at least

$$(n + 1) + \sum_{i=1}^{n} i = \frac{1}{2}n^2 + \frac{3}{2}n + 1$$

is produced. This bound plus a constant holds for the length of the sentential form, thus $L_1 \in \text{NSPACE}(n^2)$ contradicting the choice of L_1. $\square$

4 Discussion

We were able to establish a topological characterization of the class $\hat{H}IS$ and strengthen the known separation results for several sub-families of this class. The intricate structure of terminating grammars is witnessed by the fact that it cannot even be decided whether an arbitrary grammar has this property. For a rich subset of languages, that includes those accepted by nondeterministic Turing machines in exponential space, t-grammars and homomorphic images of sentential form languages are equivalent. It remains open whether in general the equality $T = \hat{H}IS$ holds.

Acknowledgements I thank Volker Diekert for drawing my attention to the present subject. Support by the German-Hungarian project "Formal Languages, Automata and Petri-Nets" BMBF 233.6/TéT Foundation D/102 is gratefully acknowledged.

References

1. J. R. Büchi. Regular canonical systems. *Archiv Math. Logik und Grundlagenforschung*, 6:91–111, 1964.
2. J. Dassow and G. Păun. *Regulated Rewriting in Formal Language Theory*, volume 18 of *EATCS Monographs on Theoretical Computer Science*. Springer, Berlin-Heidelberg-New York, 1989.
3. V. Diekert. Investigations on Hotz groups for arbitrary grammars. *Acta Informatica*, 22:679–698, 1986.
4. A. Ehrenfeucht and G. Rozenberg. Nonterminals versus homomorphisms in defining languages for some classes of rewriting systems. *Acta Informatica*, 3:265–283, 1974.
5. J. E. Hopcroft and J. D. Ullman. *Introduction to Automata Theory, Languages, and Computation*. Addison-Wesley, Reading Mass., 1979.
6. M. Jantzen and M. Kudlek. Homomorphic images of sentential form languages defined by semi-Thue systems. *Theoretical Computer Science*, 33:13–43, 1984.
7. H. A. Maurer, A. Salomaa, and D. Wood. Pure grammars. *Information and Control*, 44:47–72, 1980.
8. J. I. Seiferas. Relating refined space complexity classes. *Journal of Computer and System Sciences*, 14:100–129, 1977.

Simplification Orders for Term Graph Rewriting

Detlef Plump[*]

CWI, Kruislaan 413, 1098 SJ Amsterdam, The Netherlands
E-mail: det@cwi.nl

Abstract. Term graph rewriting differs from term rewriting in that common subexpressions can be shared, improving the efficiency of rewriting in space and time. Moreover, computations by term graph rewriting terminate more often than computations by term rewriting. In this paper, *simplification orders* on term graphs are introduced as a means for proving termination of term graph rewriting. Simplification orders are based on an extension of the homeomorphic embedding relation from trees to term graphs. By generalizing Kruskal's Tree Theorem to term graphs, it is shown that simplification orders are well-founded. Then a *recursive path order* on term graphs is defined by analogy with the well-known order on terms, and is shown to be a simplification order. Examples of termination proofs with the recursive path order are given for rewrite systems that are non-terminating under term rewriting.

1 Introduction

When computations with term rewrite rules are implemented in, for example, interpreters of functional programming languages, symbolic computation systems, or theorem provers, terms are often represented by graph-like data structures. Graphs, in contrast to trees, allow to *share* common subterms. This improves the efficiency of rewriting not only in space but also in time since repeated computations can be avoided.

Term graph rewriting is a computational model in which term rewrite rules operate on graphs that represent terms. The technical setting of the present paper conforms to [8,15,16]. (See [1,2,9] and the collection [17] for some alternative approaches.) In this approach, term graphs can be transformed by both applications of term rewrite rules and so-called collapse steps which enhance the degree of sharing.

Compared with term rewriting, term graph rewriting is not only more efficient but also enjoys termination for a larger class of rewrite systems. For instance, the following non-terminating term rewriting system is given in [4]:

$$f(a, b, x) \to f(x, x, x)$$
$$a \to b$$

---

[*] On leave from Universität Bremen, Germany. Author's research is partially supported by the HCM Network EXPRESS, the ESPRIT Working Group APPLIGRAPH, and the TMR Network GETGRATS.

Non-termination is witnessed by the infinite rewrite sequence $f(a,b,a) \rightarrow f(a,a,a) \rightarrow f(a,b,a) \rightarrow \ldots$ In contrast, the same system does terminate under term graph rewriting. This is because graph rewrite steps with the first rule do not copy the argument x but create a shared subgraph. A terminating computation starting from the tree representing $f(a,b,a)$ looks as follows:

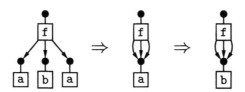

The question arises how to prove termination for systems like the present one. Obviously, the techniques available for term rewriting (see [4] for a survey) are not directly applicable. In this paper, the well-known concept of a *simplification order* [3,12,18] is generalized from terms to term graphs. The main idea is to base simplification orders on precedences of so-called tops, which are graphs containing a single function symbol or variable. By ordering tops instead of function symbols, the homeomorphic embedding relation on trees can be extended to term graphs such that sharing as in the above derivation is reflected.

Consider, for instance, the following precedence (where the three tops in the middle of the first row are all smaller than the left top and greater than the right top):

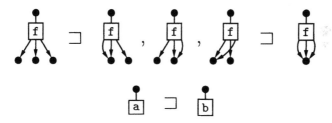

Under this precedence the right term graph of the above derivation is embedded in the left term graph, but the left graph is *not* embedded in the middle graph. In contrast, the left graph (which is a tree) is homeomorphically embedded in the tree corresponding to the middle graph if a is greater than b.

Below it is shown that the embedding relation is a well-quasi-order on term graphs whenever the given precedence is a well-quasi-order on tops. This result extends Kruskal's Tree Theorem [11] to term graphs. Simplification orders are then defined as certain strict orders on term graphs such that "strictly embedded" is a special case of "simpler". These orders are shown to be well-founded whenever the underlying precedence is a well-quasi-order. Subsequently, a *recursive path order* on term graphs is introduced by analogy with the corresponding order on terms and is shown to be a simplification order. In the present example, the recursive path order over the given precedence allows to prove termination of term graph rewriting.

2 Term Graphs

A *signature* Σ is a set of function symbols such that each $f \in \Sigma$ comes with a natural number arity$(f) \geq 0$. Function symbols of arity 0 are called *constants*. For simplicity, it is assumed that Σ contains at least one constant. A set X of *variables* for Σ must satisfy $X \cap \Sigma = \emptyset$. For each variable x, let arity$(x) = 0$.

A *hypergraph* over $\Sigma \cup X$ is a system $G = \langle V_G, E_G, \text{lab}_G, \text{att}_G \rangle$ consisting of two finite sets V_G and E_G of *nodes* and *hyperedges*, a labelling function $\text{lab}_G : E_G \rightarrow \Sigma \cup X$, and an attachment function $\text{att}_G : E_G \rightarrow V_G^*$ which assigns a string of nodes to a hyperedge e such that the length of $\text{att}_G(e)$ is $1 + \text{arity}(\text{lab}_G(e))$. In the following, hypergraphs and hyperedges are simply called graphs and edges.

Given a graph G and an edge e with $\text{att}_G(e) = v\,v_1 \ldots v_n$, node v is the *result node* of e while $v_1, \ldots, v_n$ are the *argument nodes*. The result node is denoted by res(e). For each node v, $G[v]$ is the subgraph consisting of all nodes that are reachable from v and all edges having these nodes as result nodes.

In pictures of graphs, edges are depicted as boxes with inscribed labels, and bullets represent nodes. A line connects each edge with its result node while arrows point to the argument nodes. The order among the argument nodes is given by the left-to-right order of the arrows leaving the box.

Definition 1 (Term graph). A graph G is a *term graph* if

(1) there is a node root_G from which each node is reachable,
(2) G is acyclic, and
(3) each node is the result node of a unique edge.

The set of all term graphs over $\Sigma \cup X$ is denoted by $\mathcal{TG}_{\Sigma,X}$, and $\mathcal{TG}_\Sigma$ stands for the subset of all term graphs without variables; the latter are called *ground term graphs*.

A *graph morphism* $f : G \rightarrow H$ between two graphs G and H consists of two functions $f_V : V_G \rightarrow V_H$ and $f_E : E_G \rightarrow E_H$ that preserve labels and attachment to nodes, that is, $\text{lab}_H \circ f_E = \text{lab}_G$ and $\text{att}_H \circ f_E = f_V^* \circ \text{att}_G$ (where $f_V^* : V_G^* \rightarrow V_H^*$ maps a string $v_1 \ldots v_n$ to $f_V(v_1) \ldots f_V(v_n)$). The morphism f is an *isomorphism* if f_V and f_E are bijective. In this case G and H are *isomorphic*, which is denoted by $G \cong H$.

3 A Well-quasi-order on Term Graphs

In this section, precedences are introduced as orders on certain small graphs. Every precedence induces an embedding relation on term graphs. Recall that a *preorder* (or *quasi-order*) is a reflexive and transitive relation, while a *strict order* is irreflexive and transitive. A strict order $\succ$ on a set A is *well-founded* (or *terminating*) if no infinite sequence $a_1 \succ a_2 \succ \ldots$ over A exists. A preorder $\succeq$ on A is a *well-quasi-order* (wqo for short) if for every infinite sequence $a_1, a_2, \ldots$ over A there are i and j such that $i < j$ and $a_i \preceq a_j$. Note that if A is finite, then every preorder on A is a well-quasi-order.

Definition 2 (Top). Let G be a term graph. The *top* of G, denoted by top_G, is the subgraph consisting of the unique edge e with $\text{res}(e) = \text{root}_G$ and all nodes in $\text{att}_G(e)$. The unique edge label of a top t is denoted by $\text{lab}(t)$, and Tops_Σ is the set of all tops with function symbols from Σ.

Definition 3 (Precedence). Given a signature Σ, a *precedence* is a transitive relation $\sqsupseteq$ on Tops_Σ such that for all $s, t \in \text{Tops}_\Sigma$, $s \cong t$ implies $s \sqsupseteq t$.

Thus, precedences are preorders satisfying a stronger property than reflexivity. The containment of isomorphism guarantees that precedences are well-quasi-orders whenever Σ is finite. (Reflexivity is not sufficient for this as there are infinitely many isomorphic copies of every top.)

Definition 4 (String embedding). Let $\sqsupseteq$ be a preorder on a set A. The *string embedding* relation $\sqsupseteq^{\text{str}}$ on A^* is defined as follows: $a_1 \ldots a_m \sqsupseteq^{\text{str}} b_1 \ldots b_n$ if $b_1 \ldots b_n$ is empty or if there are $j_1, \ldots, j_n$ such that $1 \le j_1 < j_2 \ldots < j_n \le m$ and $a_{j_1} \sqsupseteq b_1, \ldots, a_{j_n} \sqsupseteq b_n$.

Hence, $a \sqsupseteq^{\text{str}} b$ means that b is embedded in a. By Higman's Lemma [7], $\sqsupseteq^{\text{str}}$ is a well-quasi-order on A^* if $\sqsupseteq$ is a well-quasi-order on A.

Definition 5 (Immediate subgraphs). Let G be a term graph and e be the unique edge such that $\text{att}_G(e) = \text{root}_G v_1 \ldots v_n$ for some nodes $v_1, \ldots, v_n$ ($n \ge 0$). Then $G[v_1], \ldots, G[v_n]$ are the *immediate subgraphs* of G and sub_G is the string $G[v_1] \ldots G[v_n]$.

The next definition extends homeomorphic embedding from trees to term graphs (see [4] for a definition of tree embedding).

Definition 6 (Embedding). Let $\sqsupseteq$ be a precedence. The *embedding* relation $\trianglerighteq$ on $\mathcal{TG}_\Sigma$ is defined inductively as follows: $G \trianglerighteq H$ if

(1) $S \trianglerighteq H$ for some immediate subgraph S of G, or
(2) $\text{top}_G \sqsupseteq \text{top}_H$ and $\text{sub}_G \trianglerighteq^{\text{str}} \text{sub}_H$.

It is easy to show that $\trianglerighteq$ is a preorder containing isomorphism of ground term graphs. In order to state Kruskal's Tree Theorem in terms of $\trianglerighteq$, call a term graph G a *tree* if $\text{indegree}(v) = 1$ for each non-root node v.[1]

Theorem 7 (Tree Theorem [11]). *Let $\ge$ be a well-quasi-order on Σ and $\sqsupseteq$ be the precedence $\{\langle s, t \rangle \in \text{Tops}_\Sigma^2 \mid \text{lab}(s) \ge \text{lab}(t)\}$. Then $\trianglerighteq$ is a well-quasi-order on the set of all trees over Σ.*

Note that the above precedence in general contains pairs with tops that are not in tree form. But the restriction of $\trianglerighteq$ to trees is clearly independent of this part of the precedence.

---

[1] Given a node v in a term graph G, $\text{indegree}(v)$ is the total number of occurrences of v in the attachment strings of all edges e with $\text{res}(e) \ne v$.

Definition 8. The relations $\triangleq$ and $\triangleright$ on $\mathcal{TG}_\Sigma$ are defined as follows:
(1) $G \triangleq H$ if $G \trianglerighteq H$ and $H \trianglerighteq G$.
(2) $G \triangleright H$ if $G \trianglerighteq H$ and $H \ntrianglerighteq G$.

Observe that $G \triangleq H$ need not imply that G and H are isomorphic, even with isomorphism as precedence. For example, the following equivalence holds over every precedence:

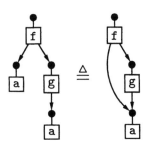

Now the Tree Theorem is extended to term graphs.

Theorem 9. *Let $\sqsupseteq$ be a precedence that is a well-quasi-order on* Tops_Σ. *Then $\trianglerighteq$ is a well-quasi-order on $\mathcal{TG}_\Sigma$.*

The Tree Theorem is a corollary of this result. For if $\geq$ is a well-quasi-order on Σ, the precedence $\{\langle s,t\rangle \in \text{Tops}_\Sigma^2 \mid \text{lab}(s) \geq \text{lab}(t)\}$ is clearly a well-quasi-order on Tops_Σ. With Theorem 9 follows that $\trianglerighteq$ is a well-quasi-order on $\mathcal{TG}_\Sigma$, and hence, in particular, on the set of all trees over Σ.

Theorem 9 can be proved—without difficulties—by the "minimal bad sequence" method used by Nash-Williams for proving the Tree Theorem [13]. Alternatively, Theorem 9 can be proved by the Tree Theorem via an encoding of term graphs as trees. This proof is given below.

Proof of Theorem 9. First, Σ is enlarged to a signature Σ_θ such that there is a bijection between function symbols in Σ_θ and isomorphism classes of tops over Σ. To this end, introduce for every $f \in \Sigma$ and every equivalence relation $\sim$ on $\{1, \ldots, \text{arity}(f)\}$ a function symbol $f_\sim$ with $\text{arity}(f_\sim) = \text{arity}(f)$. Let $\Sigma_\theta = \{f_\sim \mid f \in \Sigma\}$. Now consider any $t \in \text{Tops}_\Sigma$ with $\text{lab}(t) = f$ and string of argument nodes $v_1 \ldots v_n$ $(n \geq 0)$. Define $\theta(t) = f_\sim$, where $\sim$ is the equivalence relation $\{\langle i,j\rangle \mid v_i = v_j\}$ on $\{1, \ldots, n\}$.

Claim: The relation $\geq_\theta = \{\langle \theta(s), \theta(t)\rangle \mid \langle s,t\rangle \in \sqsupseteq\}$ is a wqo on Σ_θ.

Observe first that reflexivity of $\geq_\theta$ follows from reflexivity of $\sqsupseteq$ and surjectivity of the mapping θ. To see that θ is transitive, suppose that $\theta(t_1) \geq_\theta \theta(t_2) = \theta(t_2') \geq_\theta \theta(t_3)$. Then $t_1 \sqsupseteq t_2 \cong t_2' \sqsupseteq t_3$ because θ identifies only isomorphic tops. Hence $t_1 \sqsupseteq t_3$ and $\theta(t_1) \geq_\theta \theta(t_3)$. Finally, since $\sqsupseteq$ is a wqo, surjectivity of θ implies that $\geq_\theta$ is a wqo, too.

Next, θ is extended to a mapping Θ from $\mathcal{TG}_\Sigma$ to the set of trees over Σ_θ as follows: If G is a term graph with $\text{sub}_G = S_1 \ldots S_n$ $(n \geq 0)$, then $\Theta(G)$ is a tree with $\text{lab}(\text{top}_{\Theta(G)}) = \theta(\text{top}_G)$ and $\text{sub}_{\Theta(G)} = \Theta(S_1) \ldots \Theta(S_n)$. (This defines $\Theta(G)$ uniquely up to isomorphism.) Now consider the precedence $\sqsupseteq_\theta = \{\langle s, t \rangle \in \mathrm{Tops}^2_{\Sigma_\theta} \mid \mathrm{lab}(s) \geq_\theta \mathrm{lab}(t)\}$ and its induced embedding relation $\unrhd_\theta$. By the above claim and the Tree Theorem, $\unrhd_\theta$ is a wqo on the set of all trees over Σ_θ. Moreover, an easy induction on the size of (combined) term graphs shows that for all $G, H \in \mathcal{TG}_\Sigma$, $G \unrhd H$ if and only if $\Theta(G) \unrhd_\theta \Theta(H)$. It follows that $\unrhd$ is a wqo, too. $\qquad\square$

The next two lemmas characterize the equivalence $\triangleq$ and the strict part $\rhd$ of $\unrhd$. Given a string $a = a_1 \ldots a_n$, $|a|$ denotes its length n while, for $i = 1, \ldots, n$, $a[i]$ refers to the element a_i. The relations $\equiv$, $\sqsupset$ and $\rhd^{\mathrm{str}}$ are defined as follows: $\equiv = (\sqsupseteq \cap \sqsubseteq)$, $\sqsupset = (\sqsupseteq - \sqsubseteq)$ and $\rhd^{\mathrm{str}} = (\unrhd^{\mathrm{str}} - \lhd^{\mathrm{str}})$.

Lemma 10. *Let $\sqsupseteq$ be a precedence. Then for all term graphs G and H, $G \triangleq H$ if and only if (1) $\mathrm{top}_G \equiv \mathrm{top}_H$, (2) $|\mathrm{sub}_G| = |\mathrm{sub}_H|$, and (3) $\mathrm{sub}_G[i] \triangleq \mathrm{sub}_H[i]$ for $i = 1, \ldots, |\mathrm{sub}_G|$.*

Lemma 11. *Let $\sqsupseteq$ be a precedence. Then for all term graphs G and H, $G \rhd H$ if and only if (1) $S \unrhd H$ for some immediate subgraph S of G, or (2) $\mathrm{top}_G \sqsupset \mathrm{top}_H$ and $\mathrm{sub}_G \unrhd^{\mathrm{str}} \mathrm{sub}_H$, or (3) $\mathrm{top}_G \equiv \mathrm{top}_H$ and $\mathrm{sub}_G \rhd^{\mathrm{str}} \mathrm{sub}_H$.*

4 Simplification Orders

Simplification orders are certain strict orders that contain the strict embedding relation. Theorem 9 guarantees that such orders are well-founded whenever the given precedence is a well-quasi-order.

Definition 12 (Simplification order). Let $\unrhd$ be the embedding relation induced by a precedence that is a well-quasi-order. A transitive relation $\succ$ on $\mathcal{TG}_\Sigma$ is a *simplification order* if it contains $\rhd$ and if for all $G, H \in \mathcal{TG}_\Sigma$, $G \triangleq H$ implies $G \not\succ H$.

Note that simplification orders are irreflexive, in particular.

Theorem 13. *Every simplification order is well-founded.*

Proof. Let $\succ$ be a simplification order. Then, by Theorem 9, the underlying embedding relation $\unrhd$ is a well-quasi-order. Now suppose that there is an infinite sequence $G_1 \succ G_2 \succ \ldots$ Then there are i and j such that $G_i \unlhd G_j$. On the other hand, $G_i \succ G_{i+1} \succ \ldots \succ G_j$ implies $G_i \succ G_j$ by transitivity of $\succ$. Hence, by the definition of simplification orders, $G_i \triangleq G_j$ is impossible. But then $G_i \lhd G_j$ and therefore $G_i \prec G_j$. It follows $G_i \succ G_i$, contradicting the irreflexivity of simplification orders. Thus, $\succ$ is well-founded. $\qquad\square$

In order to introduce a recursive path order on term graphs, the lifting of an order to a multiset order is recalled.

Definition 14 (Multiset extension). Let $\succ$ be a strict order on a set A. The *multiset extension* $\succ^{\mathrm{mul}}$ on the set of finite multisets over A is defined as follows: $M \succ^{\mathrm{mul}} N$ if there are multisets X and Y such that

(1) $\emptyset \neq X \subseteq M$,
(2) $N = (M - X) \cup Y$, and
(3) for all $y \in Y$ there is some $x \in X$ with $x \succ y$.

Lemma 15 (Dershowitz and Manna [6]). *If $\succ$ is a strict order on a set A, then $\succ^{\mathrm{mul}}$ is a strict order on the set of finite multisets over A. If $\succ$ is moreover well-founded, then $\succ^{\mathrm{mul}}$ is well-founded, too.*

The equivalence relation $\approx_{\mathrm{rpo}}$ defined next will be used in the definition of the recursive path order.

Definition 16. Let $\sqsupseteq$ be a precedence. The relation $\approx_{\mathrm{rpo}}$ on $\mathcal{TG}_\Sigma$ is defined inductively as follows: $G \approx_{\mathrm{rpo}} H$ if (1) $\mathrm{top}_G \equiv \mathrm{top}_H$, (2) $|\mathrm{sub}_G| = |\mathrm{sub}_H|$, and (3) there is a bijection π on $\{1, \ldots, |\mathrm{sub}_G|\}$ such that $\mathrm{sub}_G[i] \approx_{\mathrm{rpo}} \mathrm{sub}_H[\pi(i)]$ for $i = 1, \ldots, |\mathrm{sub}_G|$.

The equivalence class of a ground term graph G with respect to $\approx_{\mathrm{rpo}}$ is written $[G]$. Given a strict order $\succ$ on $\mathcal{TG}_\Sigma$ such that $G' \approx_{\mathrm{rpo}} G \succ H \approx_{\mathrm{rpo}} H'$ implies $G' \succ H'$, $\succ$ is lifted to an order on equivalence classes as follows: $[G] \succ [H]$ if $G \succ H$. (See [12] for a similar lifting of preorders.) For $G \in \mathcal{TG}_\Sigma$ with $\mathrm{sub}_G = S_1 \ldots S_n$, the multiset $\{[S_1], \ldots, [S_n]\}$ of equivalence classes of immediate subgraphs is denoted by SUB_G.

Definition 17 (Recursive path order). Let $\sqsupseteq$ be a precedence. The *recursive path order* $\succ_{\mathrm{rpo}}$ on $\mathcal{TG}_\Sigma$ is defined inductively as follows: $G \succ_{\mathrm{rpo}} H$ if

(1) $S \succ_{\mathrm{rpo}} H$ or $S \approx_{\mathrm{rpo}} H$ for some immediate subgraph S of G, or
(2) $\mathrm{top}_G \sqsupset \mathrm{top}_H$ and $G \succ_{\mathrm{rpo}} T$ for all immediate subgraphs T of H, or
(3) $\mathrm{top}_G \equiv \mathrm{top}_H$ and $\mathrm{SUB}_G \succ^{\mathrm{mul}}_{\mathrm{rpo}} \mathrm{SUB}_H$.

Lemma 18. *For all $G', G, H, H' \in \mathcal{TG}_\Sigma$, $G' \approx_{\mathrm{rpo}} G \succ_{\mathrm{rpo}} H \approx_{\mathrm{rpo}} H'$ implies $G' \succ_{\mathrm{rpo}} H'$.*

Theorem 19. *The recursive path order is a simplification order whenever the underlying precedence is a well-quasi-order.*

The proof of this result requires to show the three conditions of Definition 12: (1) transitivity of $\succ_{\mathrm{rpo}}$, (2) $\rhd \subseteq \succ_{\mathrm{rpo}}$, and (3) for all $G, H \in \mathcal{TG}_\Sigma$, $G \triangleq H$ implies $G \not\succ_{\mathrm{rpo}} H$. These properties are shown by induction on the size of term graphs, where the induction steps use case distinctions according to the three cases of Definition 17.

As a corollary of Theorem 19, $\succ_{\mathrm{rpo}}$ is well-founded if the given precedence is a well-quasi-order. This can also be shown by using the corresponding result for the recursive path order on terms [4], exploiting the encoding Θ of term graphs as trees given in the proof of Theorem 9. One has to show that for all $G, H \in \mathcal{TG}_\Sigma$, $G \succ_{\mathrm{rpo}} H$ if and only if $\Theta(G) \succ^\theta_{\mathrm{rpo}} \Theta(H)$, where $\succ^\theta_{\mathrm{rpo}}$ is the recursive path order over the enlarged signature Σ_θ.

5 Termination of Term Graph Rewriting

This section starts with a brief review of the term graph rewriting model investigated in [8,15,16]. In this approach, rewriting includes not only applications of term rewrite rules but also steps for compressing term graphs.

Definition 20 (Collapsing). Given two term graphs G and H, G *collapses* to H if there is a graph morphism $f : G \to H$ mapping root_G to root_H. This is denoted by $G \succeq_c H$. The collapsing is *proper*, denoted by $G \succ_c H$, if f is non-injective.

A *term rewrite rule* $l \to r$ consists of two terms l and r such that l is not a variable and all variables in r occur also in l. A set $\mathcal{R}$ of term rewrite rules is a *term rewriting system.* (See [5,10,14] for surveys of term rewriting.)

For every term t, let $\Diamond t$ be a term graph representing t such that only variables are shared.[2] The graph resulting from a term graph G after removing all edges labelled with variables is denoted by $\underline{G}$.

Definition 21 (Instance). A term graph H is an *instance* of a term graph G if there is graph morphism $\underline{G} \to H$ sending root_G to root_H. An instance that is a ground term graph is a *ground instance.*

Definition 22 (Term graph rewriting). Let G and H be term graphs, $l \to r$ be a rewrite rule and v be a node in G such that $G[v]$ is an instance of $\Diamond l$. Then there is a *proper rewrite step* $G \Rightarrow_{v, l \to r} H$ if H is isomorphic to the term graph G_3 constructed as follows:

(1) $G_1 = G - \{e\}$ is the graph obtained from G by removing the unique edge e satisfying $\text{res}(e) = v$.
(2) G_2 is the graph obtained from the disjoint union $G_1 + \Diamond r$ by
   - identifying v with $\text{root}_{\Diamond r}$,
   - identifying the image of $\text{res}(e_1)$ with $\text{res}(e_2)$, for each pair $\langle e_1, e_2 \rangle \in E_{\Diamond l} \times E_{\Diamond r}$ with $\text{lab}_{\Diamond l}(e_1) = \text{lab}_{\Diamond r}(e_2) \in X$.
(3) $G_3 = G_2[\text{root}_G]$ is the term graph obtained from G_2 by removing all nodes and edges not reachable from root_G ("garbage collection").

Now the term graph rewrite relation $\Rightarrow_{\mathcal{R}}$ on $\mathcal{TG}_{\Sigma, X}$ is defined by adding proper collapse steps: $G \Rightarrow_{\mathcal{R}} H$ if $G \succ_c H$ or $G \Rightarrow_{l \to r} H$ for some rule $l \to r$ in $\mathcal{R}$. The relation $\Rightarrow_{\mathcal{R}}$ is *terminating* if no infinite sequence $G_1 \Rightarrow_{\mathcal{R}} G_2 \Rightarrow_{\mathcal{R}} \ldots$ exists.

Definition 23. A precedence $\sqsupseteq$ is *collapse compatible* if whenever there is a graph morphism $t \to u$ between two tops $t, u \in \text{Tops}_\Sigma$, then $t \sqsupseteq u$. A collapse compatible precedence that is a well-quasi-order is a *well-precedence.*

Lemma 24. *Let $\sqsupseteq$ be a precedence. The embedding relation $\unrhd$ contains the collapse relation $\succeq_c$ if and only if $\sqsupseteq$ is collapse compatible.*

---

[2] That is, $\text{indegree}(\text{res}(e)) \leq 1$ for each edge e with $\text{lab}_{\Diamond t}(e) \notin X$, and $e_1 = e_2$ for all edges e_1, e_2 with $\text{lab}_{\Diamond t}(e_1) = \text{lab}_{\Diamond t}(e_2) \in X$.

Theorem 25. *Let* $\succ_{\text{rpo}}$ *be induced by a well-precedence. Then* $\Rightarrow_{\mathcal{R}}$ *is terminating if* $G \Rightarrow_{l \to r} H$ *implies* $G \succ_{\text{rpo}} H$, *for every rule* $l \to r$ *in* $\mathcal{R}$ *and all ground term graphs* G *and* H.

Proof. It suffices to show the absence of infinite derivations over $\mathcal{TG}_\Sigma$, since all occuring variables can be replaced by a constant. Suppose that there is an infinite sequence $G_1 \Rightarrow_{\mathcal{R}} G_2 \Rightarrow_{\mathcal{R}} \ldots$ over $\mathcal{TG}_\Sigma$. As proper collapsing is terminating, there are $i_1, i_2, \ldots$ such that $1 = i_1 \leq i_2 < i_3 \leq i_4 < \ldots$ and $G_{i_1} \succeq_C G_{i_2} \Rightarrow_{\mathcal{R}} G_{i_3} \succeq_C G_{i_4} \Rightarrow_{\mathcal{R}} \ldots$, where all $\Rightarrow_{\mathcal{R}}$-steps are proper rewrite steps. By the assumption and Lemma 24, this implies $G_{i_1} \trianglerighteq G_{i_2} \succ_{\text{rpo}} G_{i_3} \trianglerighteq G_{i_4} \succ_{\text{rpo}} \ldots$ As $\succ_{\text{rpo}}$ is a simplification order, $\trianglerighteq$ is contained in $\succ_{\text{rpo}} \cup \approx_{\text{rpo}}$. With Lemma 18 follows that there is an infinite subsequence $G_{j_1} \succ_{\text{rpo}} G_{j_2} \succ_{\text{rpo}} \ldots$ of $G_1 \Rightarrow_{\mathcal{R}} G_2 \Rightarrow_{\mathcal{R}} \ldots$ But $\succ_{\text{rpo}}$ is well-founded by Theorems 19 and 13, a contradiction. Thus $\Rightarrow_{\mathcal{R}}$ is terminating. □

Due to a monotonicity property of $\succ_{\text{rpo}}$, the premise of Theorem 25 can be weakened.

Theorem 26. *Let* $\succ_{\text{rpo}}$ *be induced by a well-precedence. Then* $\Rightarrow_{\mathcal{R}}$ *is terminating if* $L \Rightarrow_{\text{root}_L, l \to r} R$ *implies* $L \succ_{\text{rpo}} R$, *for every rule* $l \to r$ *in* $\mathcal{R}$ *and every ground instance* L *of* $\Diamond l$.

Example 27. Consider the following rewrite system $\mathcal{R}$:

$$f(x) \to g(x, x)$$
$$a \to b$$
$$g(a, b) \to f(a)$$

This system is non-terminating under term rewriting because there is an infinite rewrite sequence $f(a) \to g(a, a) \to g(a, b) \to f(a) \to \ldots$ Termination of term graph rewriting can easily be checked by means of Theorem 26, using the following well-precedence:

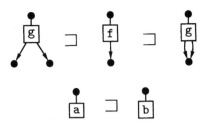

Acknowledgement. The author is grateful to Annegret Habel and Andreas Weiermann, who gave valuable comments on a previous version of this paper.

References

1. Zena M. Ariola and Jan Willem Klop. Equational term graph rewriting. *Fundamenta Informaticae*, 26:207–240, 1996.

2. Andrea Corradini and Francesca Rossi. Hyperedge replacement jungle rewriting for term rewriting systems and logic programming. *Theoretical Computer Science*, 109:7–48, 1993.

3. Nachum Dershowitz. Orderings for term rewriting systems. *Theoretical Computer Science*, 17:279–301, 1982.

4. Nachum Dershowitz. Termination of rewriting. *Journal of Symbolic Computation*, 3:69–116, 1987.

5. Nachum Dershowitz and Jean-Pierre Jouannaud. Rewrite systems. In Jan van Leeuwen, editor, *Handbook of Theoretical Computer Science*, volume B, chapter 6. Elsevier, 1990.

6. Nachum Dershowitz and Zohar Manna. Proving termination with multiset orderings. *Communications of the ACM*, 22(8):465–476, 1979.

7. Graham Higman. Ordering by divisibility in abstract algebras. *Proceedings of the London Mathematical Society*, 3(2):326–336, 1952.

8. Berthold Hoffmann and Detlef Plump. Implementing term rewriting by jungle evaluation. *RAIRO Theoretical Informatics and Applications*, 25(5):445–472, 1991.

9. Richard Kennaway, Jan Willem Klop, Ronan Sleep, and Fer-Jan de Vries. On the adequacy of term graph rewriting for simulating term rewriting. *ACM Transactions on Programming Languages and Systems*, 16(3):493–523, 1994.

10. Jan Willem Klop. Term rewriting systems. In S. Abramsky, Dov M. Gabbay, and T.S.E. Maibaum, editors, *Handbook of Logic in Computer Science*, volume 2, pages 1–116. Oxford University Press, 1992.

11. Joseph B. Kruskal. Well-quasi-ordering, the Tree Theorem, and Vazsonyi's conjecture. *Transactions of the American Mathematical Society*, 95:210–225, 1960.

12. Aart Middeldorp and Hans Zantema. Simple termination revisited. In *Proc. 12th International Conference on Automated Deduction*, volume 814 of *Lecture Notes in Artificial Intelligence*, pages 451–465. Springer-Verlag, 1994.

13. C. St. J. A. Nash-Williams. On well-quasi-ordering finite trees. *Proceedings of the Cambridge Philosophical Society*, 59:833–835, 1963.

14. David A. Plaisted. Equational reasoning and term rewriting systems. In Dov M. Gabbay, C.J. Hogger, and J.A. Robinson, editors, *Handbook of Logic in Artificial Intelligence and Logic Programming*, volume 1, pages 273–364. Clarendon Press, 1993.

15. Detlef Plump. Collapsed tree rewriting: Completeness, confluence, and modularity. In *Proc. Conditional Term Rewriting Systems*, volume 656 of *Lecture Notes in Computer Science*, pages 97–112. Springer-Verlag, 1993.

16. Detlef Plump. Evaluation of functional expressions by hypergraph rewriting. Dissertation, Universität Bremen, Fachbereich Mathematik und Informatik, 1993.

17. Ronan Sleep, Rinus Plasmeijer, and Marko van Eekelen, editors. *Term Graph Rewriting: Theory and Practice*. John Wiley, 1993.

18. Joachim Steinbach. Simplification orderings — history of results. *Fundamenta Informaticae*, 24:47–87, 1995.

Dependency-Based Action Refinement

Arend Rensink and Heike Wehrheim*

Institut für Informatik, University of Hildesheim
Postfach 101363, D–31113 Hildesheim, Germany
{rensink,wehrheim}@informatik.uni-hildesheim.de

Abstract. *Action refinement* in process algebras has been widely studied in the last few years as a means to support top-down design of systems. A specific notion of refinement arises when a *dependency relation*
on the actions (in the Mazurkiewicz sense) is used to control the inheritance of orderings from the abstract level. In this paper we present a
rather simple *operational semantics* for dependency-based action refinement. We show the consistency of the operational with a (previously
published) denotational semantics. We moreover show that bisimulation
is a *congruence* for dependency-based refinement. Finally, we give an
illustrative example.

1 Introduction

Action refinement in process algebras [1, 2, 18, 11, 8, 9] has been proposed to
support hierarchical design of reactive systems. Starting with an abstract specification, step-by-step more concrete specifications are constructed by refining
actions into concrete processes. Thus the complexity of the design process is
reduced and furthermore verification can be facilitated [13]. Standard action
refinement uses a *strong* concept of inheritance of causally related abstract actions: if two actions sequentially follow each other on an abstract level, their
implementations (as described by the refinement) are also strictly sequential.
However, this has turned out to be too restrictive in practice: it might as well be
the case that some parts of the implementation of sequential actions overlap in
time. As proposed by [14], one way to overcome this problem is to combine action
refinement with a dependency relation on actions (in the sense of Mazurkiewicz
[16]): the ordering among abstract actions is only inherited to dependent parts
of their refinements. Dependency-based action refinement allows an overlapping
(concurrent execution) of *independent* parts of the implementation of sequential
abstract actions. Other approaches to design by refinement allowing such an
overlap, but not based on dependencies, can be found in [10, 21, 23].

In general, two approaches to the definition of an action refinement operator can be found in the literature. On the one hand, action refinement can be
defined as *syntactic substitution* on terms [1, 2, 18], on the other hand as *semantic substitution* in an appropriate denotational model [11, 3, 15, 8, 25, 3].

---

* The research reported in this paper was partially supported by the Human Capital
and Mobility Cooperation Network "EXPRESS" (Expressiveness of Languages for
Concurrency).

For dependency-based refinement, only the latter approach has been followed yet [14]. In general, the two approaches do not coincide (for a comparison see [12]). *Operational semantics* for action refinement, being consistent with a denotational model, have been defined in [6, 9, 22], however all of them generate transition systems with an enhanced labelling of transition (not just plain actions), adding e.g. causes or event names. The reason for this is essentially that plain transition systems can in general only be distinguished up to *bisimulation* and bisimulation is not a congruence for standard action refinement [7]. Bisimulation over augmented transition systems (as used for giving operational semantics to action refinement) coincides with equivalences (ST-bisimulation, history preserving bisimulation, event isomorphism) which are invariant under refinement.

In this paper we extend a process algebra with action dependencies (as presented in [24]) with an operator for action refinement and develop an operational semantics for it. The basic idea is to reduce refinement to *weak sequential composition* (which is dependency-based sequential composition allowing an overlapping of independent parts of the operands): if the abstract system can perform a sequence of action $a_1 a_2 \ldots$ then the refinement can perform $r(a_1) \cdot r(a_2) \cdot \ldots$ ($r(a)$ is the refinement of an action a according to a refinement function r, $\cdot$ denotes weak sequential composition). This allows independent parts of refinements of sequentially ordered actions to be executed concurrently. The transition relation of the resulting labelled transition system is solely labelled with actions.

The operational semantics is shown to be consistent with the denotational semantics of dependency-based action refinement as introduced in [26]. We furthermore show that bisimulation is a congruence for action refinement. Finally we illustrate the practical applicability of our approach by an example.

2 Definitions

We start with a brief repetition of the operational and denotational semantics for the process algebra **L** as developed in [24]. Thereafter, **L** will be extended by an action refinement operator.

We assume a distributed alphabet $\langle \mathbf{A}, I \rangle$, where $\mathbf{A}$ is a set of actions and $I \subseteq \mathbf{A} \times \mathbf{A}$ is a symmetric and irreflexive *independence relation*. $\mathbf{A}$ is ranged over by a, b, c. We denote $a \, D \, b$ iff $\neg(a \, I \, b)$, and $[A]_D := \{b \, D \, a \mid a \in A\}$. D is called the dependency relation. Intuitively, actions are dependent if they share some common resource, which can for instance be a shared variable, a database item or a printer. We also use a set of process variables $\mathbf{X}$ to allow for the definition of infinite behaviour. The language **L** is generated by the following productions:

$$B ::= \mathbf{0}_P \mid a \mid X \mid B \cdot B \mid B + B \mid B \parallel_A B \mid \mu X.B$$

where $a \in \mathbf{A}$, $A, P \subseteq \mathbf{A}$ and $X \in \mathbf{X}$. A term is called *closed* iff all variables are in the scope of a fixpoint operator. In the sequel we only consider closed terms. We let $B\{B'/X\}$ stand for the substitution of all occurrences of the variable X in B

by B'. Furthermore, we denote by $\alpha(B)$ the set of all actions which syntactically occur in B.

The operators have the following informal meaning: $\mathbf{0}_P$ describes an empty process (below we comment on the set P), a a process executing the action a, $\cdot$ is *weak* sequential composition, $+$ denotes a CCS-like choice (with a non-standard interaction with sequential composition), $\|_A$ a TCSP-like parallel composition with synchronisation on actions in A and $\mu X.B$ defines recursion. We only consider *guarded* recursion, where a term B is considered guarded if all variables X only occur in subterms of the form $a \cdot X$ with $[a]_D = \mathbf{A}$.

The information about the dependencies of actions is exploited to give a semantics to terms which ensures that independent actions never have to wait for one another to proceed. This results in a *weak* form of sequential composition which allows actions from the second operand that are independent of the first to be executed immediately. To capture the intended behaviour of weak sequential composition, a particular notion of *termination* is needed. It is not sufficient anymore to solely distinguish between successful termination and deadlock; instead, a process is said to be terminated *wrt. a set of actions P* (intuitively if it does not want to execute anything on which actions out of P depend). In a sequential composition, the second component can perform an action a if the first component is terminated wrt. a (e.g. $a \cdot b$, $a \, I \, b$, can execute b since a is terminated wrt. b). This notion of termination (called *permission*) is also reflected in the constants representing empty processes: for any $P \subseteq \mathbf{A}$, $\mathbf{0}_P$ is an empty process terminated wrt. P (thus $\mathbf{0}_{\{a\}} \cdot a$ allows execution of a). We let $\mathbf{1} := \mathbf{0}_\mathbf{A}$ stand for the completely terminated process.

In connection with weak sequential composition, the semantics of choice is also non-standard. Choices can be resolved by actions not participating in the choice but sequentially following it. As an example take $B = (a+b)\cdot c$, $a \, D \, c$, $b \, I \, c$. Since b and c are independent, c can also be executed first; afterwards, however, a is not possible anymore, since otherwise the specified order between dependent actions, a and c, is not met.

In Table 1 we give a structured operational semantics for $\mathbf{L}$. Besides a transition relation ($\rightarrow$) we also include a *permission relation* ($\cdots\rightarrow$) to describe the new form of termination. In contrast to the usual termination predicate, permission is a binary relation, since permission may change processes due to the resolution of choices (see above). This is reflected by the permission rule for choices: if only one operand of the choice permits an action, the term changes with the permission. The standard semantic model of labelled transition systems can be extended to permissions.

Definition 1. A *transition-permission-system* (tps for short) is a tuple $T = \langle S, \rightarrow, \cdots\rightarrow \rangle$ with
- S a set of *states*,
- $\rightarrow \subseteq S \times \mathbf{A} \times S$ a *transition relation*,
- $\cdots\rightarrow \subseteq S \times \mathbf{A} \times S$ a *permission relation*.

In particular, the rules in Table 1 define relations $\rightarrow$ and $\cdots\rightarrow$ over $\mathbf{L}$ that give rise to a transition-permission system $(\mathbf{L}, \rightarrow, \cdots\rightarrow)$.

$$
\begin{array}{ccc}
 & a \in P & a\, I\, b \\[4pt]
\hline
\end{array}
$$

$$0_P \ ..\xrightarrow{a} 0_P \qquad\qquad b \ ..\xrightarrow{a} b \qquad\qquad a \xrightarrow{a} 1$$

$$\frac{B_1 \ ..\xrightarrow{a} B' \quad B_2 \ ..\not\xrightarrow{a}}{B_1 + B_2 \ ..\xrightarrow{a} B'} \qquad \frac{B_2 \ ..\xrightarrow{a} B' \quad B_1 \ ..\not\xrightarrow{a}}{B_1 + B_2 \ ..\xrightarrow{a} B'} \qquad \frac{B_1 \xrightarrow{a} B'}{B_1 + B_2 \xrightarrow{a} B'} \qquad \frac{B_2 \xrightarrow{a} B'}{B_1 + B_2 \xrightarrow{a} B'}$$

$$\frac{B_1 \ ..\xrightarrow{a} B_1' \quad B_2 \ ..\xrightarrow{a} B_2'}{B_1 \cdot B_2 \ ..\xrightarrow{a} B_1' \cdot B_2'} \qquad \frac{B_1 \xrightarrow{a} B_1'}{B_1 \cdot B_2 \xrightarrow{a} B_1' \cdot B_2} \qquad \frac{B_1 \ ..\xrightarrow{a} B_1' \quad B_2 \xrightarrow{a} B_2'}{B_1 \cdot B_2 \xrightarrow{a} B_1' \cdot B_2'}$$

$$\frac{B_1 \ ..\xrightarrow{a} B_1' \quad B_2 \ ..\xrightarrow{a} B_2'}{B_1 \|_A B_2 \ ..\xrightarrow{a} B_1' \|_A B_2'} \qquad \frac{B_1 \xrightarrow{a} B_1', a \notin A}{B_1 \|_A B_2 \xrightarrow{a} B_1' \|_A B_2} \qquad \frac{B_2 \xrightarrow{a} B_2', a \notin A}{B_1 \|_A B_2 \xrightarrow{a} B_1 \|_A B_2'}$$

$$\frac{B_1 \xrightarrow{a} B_1' \quad B_2 \xrightarrow{a} B_2', a \in A}{B_1 \|_A B_2 \xrightarrow{a} B_1' \|_A B_2'} \qquad \frac{B\{\mu X.B/X\} \ ..\xrightarrow{a} B'}{\mu X.B \ ..\xrightarrow{a} B'} \qquad \frac{B\{\mu X.B/X\} \xrightarrow{a} B'}{\mu X.B \xrightarrow{a} B'}$$

Table 1. Operational semantics for **L**

In a transition-permission system T, a state s' is called *reachable* from a state s if $(s, s') \in (\bigcup_a \xrightarrow{a} \cup \ ..\xrightarrow{a})^*$. Furthermore, s is called *fully terminated* if $s \ ..\xrightarrow{a} s$ for all $a \in \mathbf{A}$, and *terminating* if for all s' reachable from s, there is a fully terminated s'' reachable from s'. The usual notion of (strong) bisimilarity of [19, 17] on labelled transition systems can be adapted to our setting.

Definition 2. Let $T = \langle S, \rightarrow, \ ..\rightarrow \rangle$ be a tps. *Bisimilariy over* T is the largest symmetrical relation $\sim \subseteq S \times S$ such that whenever $s_1 \sim s_2$ then
1. $s_1 \xrightarrow{a} s_1'$ implies $\exists s_2' : s_2 \xrightarrow{a} s_2'$ and $s_1' \sim s_2'$;
2. $s_1 \ ..\xrightarrow{a} s_1'$ implies $\exists s_2' : s_2 \ ..\xrightarrow{a} s_2'$ and $s_1' \sim s_2'$.

Denotational semantics. We now recall the event-based denotational model for **L** developed in [26], with a slight deviation to smoothen the definition of action refinement, later on. In this model, a process is represented by its set of maximal runs (this is the deviation from [26], where all runs where included, not just the maximal ones). Each run consists of a set of *events* that have happened (where an event corresponds to the execution of an action), together with their causal ordering and a labelling function indicating which actions have been executed. Furthermore, each run contains *termination information*, in the form of a set containing the actions with respect to which the run is terminated. The causal ordering has to be consistent with the dependency relation: only dependent actions may (and must) be ordered. For the construction of runs we assume a universe of events $\mathbf{E}$, ranged over by d, e.

Definition 3. A run is a tuple $u = \langle E, <, \ell, \checkmark \rangle$ where
- $E \subseteq \mathbf{E}$ is a (finite or infinite) set of events;
- $< \subseteq E \times E$ is an acyclic ordering on E;

- $\ell\colon E \to \mathbf{A}$ is a labelling function, such that $\ell(d)$ D $\ell(e)$ iff $d \le e$ or $e \le d$;
- $\checkmark \subseteq \mathbf{A}$ is a *termination set*.

The class of runs is denoted $\mathbf{R}$. We denote the elements of a run u by E_u, $<_u$, ℓ_u and $\checkmark_u$. $\le$ will denote the transitive closure of $<$. The independence relation I is extended to E_u through ℓ_u such that d D_u e iff $\ell_u(d)$ D $\ell_u(e)$. A run can be depicted by a graph; e.g., $\boxed{{}_1a\to{}_2b}$, $\mathbf{A}$ denotes a run consisting of two ordered events, 1 and 2 labelled with a and b, respectively, and termination set is $\mathbf{A}$. A run u is called a *prefix* of a run v if the following holds:
$$u \sqsubseteq v :\Leftrightarrow (E_u \subseteq E_v) \wedge (<_u \;=\; <_v \cap (E_u \times E_v)) \wedge \checkmark_u = \checkmark_v \setminus [\ell_v(E_v \setminus E_u)]_D \;.$$
If $\mathcal{P} \subseteq \mathbf{R}$, we denote $E_{\mathcal{P}} = \bigcup_{u \in \mathcal{P}} E_u$ and $\ell_{\mathcal{P}} = \bigcup_{u \in \mathcal{P}} \ell_u$. $\mathcal{P}$ is called *labelling consistent* if $\ell_{\mathcal{P}}$ is a function. Our denotational models will be nonempty, labelling consistent sets of runs, called *families of runs*. The class of families of runs is denoted $\mathbf{M}$. They are an extension of the *families of posets* proposed in [20]. Only the $\sqsubseteq$-maximal elements of a family of runs are considered significant. This is expressed by taking two models, $\mathcal{P}$ and $\mathcal{Q}$, as equivalent if there is a bijection $\phi\colon E_{\mathcal{P}} \to E_{\mathcal{Q}}$ such that $\phi^*(v) \in \max_{\sqsubseteq} \mathcal{Q}$ for all $u \in \max_{\sqsubseteq} \mathcal{P}$ (where ϕ^* denotes the natural pointwise extension of ϕ to $\mathbf{R}$).

In [26] we have presented a denotational semantics for $\mathbf{L}$ that can easily be adapted to the above families of runs, giving rise to a mapping $[\![\cdot]\!]\colon \mathbf{L} \to \mathbf{M}$. Due to lack of space, we omit the model constructions for the standard operators. As an example, assume $\mathbf{A} = \{a, b, c\}$ with a D c, a D b and b I c. A family of runs modelling the term $(a + b) \cdot (b \|_{\varnothing} c)$ is given by $\boxed{\begin{smallmatrix} & & {}_2b \\ & \nearrow & \\ {}_0a & \!\!\to\!\! & {}_3c \end{smallmatrix}}$, $\mathbf{A}$ and $\boxed{\begin{smallmatrix} {}_1b \to {}_2b \\ {}_3c \end{smallmatrix}}$, $\mathbf{A}$. Without proof, we state the following property:

Proposition 4. *B is terminating iff $\checkmark_u = \mathbf{A}$ for all $u \in \max_{\sqsubseteq}[\![B]\!]$.*

Families of runs distinguish concurrent from interleaving behaviour; moreover, the branching structure of the behaviour can be reconstructed via the names of events. Thus, the model is quite expressive; much more so, in fact, than necessary for our present purposes. In this paper, we interpret families of runs up to bisimulation, by regarding them as transition-permission systems in the sense of Definition 1:

(1) A transition corresponds to the occurrence of an event; the label of the transition is the event label. For an event to occur it must be *enabled* in a run, i.e., have no causal predecessors; afterwards, it can be discarded. Hence, if $e \in \min E_u$, the *remainder* of u after e is given by
$$u \setminus e := \langle E \setminus \{e\}, < \cap ((E \setminus \{e\}) \times (E \setminus \{e\})), \ell \restriction (E \setminus \{e\}), \checkmark_u \rangle$$

(2) A permission corresponds to the occurrence of a *future* action a. For any given run u, such a future action a can only be allowed if it is independent of all actions of u and u is terminated w.r.t. a. We say that u *permits* a iff $a \in \checkmark_u$ and $\forall e \in E_u. \ell_u(e)$ I a. The following relations are considered to hold iff the right hand sides are nonempty:
$$\mathcal{P} \xrightarrow{\ell_{\mathcal{P}}(e)} \{u \setminus e \mid u \in \mathcal{P}, e \in \min E_u\}$$
$$\mathcal{P} \overset{a}{\cdots\!\!\to} \{u \in \mathcal{P} \mid u \text{ permits } a\}$$

This gives rise to a transition-permission system $(\mathbf{M}, \rightarrow, \cdots \rightarrow)$. We then have the following consistency result (cf. [24]):

Theorem 5. *For arbitrary terms $B \in \mathbf{L}$, $B \sim [\![B]\!]$ (in the transition-permission system $\mathbf{L} \cup \mathbf{M}$).*

3 Action Refinement

We now extend our language with the *action refinement* operator discussed in the introduction. Syntactically, the extension is given by the term $B[r]$, where r stands for a *refinement function* $r \colon \mathbf{A} \to \mathbf{L}$. In contrast to standard action refinement in causality-based models [11], in our setting, the inheritance of abstract orderings by the concrete actions of the refinement is driven by the dependencies between the latter. Therefore, the refinement of ordered abstract actions may result in sets of events which partially overlap in their execution. This is demonstrated in the following example.

Example 1. Let $B = a \cdot b$ with $a\ D\ b$; hence $[\![B]\!] = \{\boxed{{}_1a \to {}_2b}, \mathbf{A}\}$. Let $r \colon a \mapsto a_1 \cdot a_2$ with $a_1\ D\ a_2$, and $b \mapsto b_1 \cdot b_2$ with $b_1\ D\ b_2$, such that $a_1\ D\ b_1$ and $a_2\ D\ b_2$ but $a_2\ I\ b_1$ and $a_1\ I\ b_2$. The only allowed execution of $B[r]$ by standard refinement would be $a_1 a_2 b_1 b_2$, where the entire refinement of b has to wait for a to complete. With dependency-based refinement we get the following run, which also allows an overlapping execution $a_1 b_1 a_2 b_2$:

$$\boxed{\begin{array}{ccc} {}_{(1,1)}a_1 & \to & {}_{(1,2)}a_2 \\ \downarrow & & \downarrow \\ {}_{(2,1)}b_1 & \to & {}_{(2,2)}b_2 \end{array}}, \mathbf{A}$$

In the following we extend both the operational and the denotational semantics of $\mathbf{L}$ to capture this type of dependency-based refinement. To achive this, we have to impose two restrictions on refinement functions:

(1) All images have to be *terminating* (see Section 2). This is a quite natural restriction, given the fact that abstractly, an action is atomic, i.e., cannot deadlock during its execution.

(2) The refinement has to *preserve dependency and independency*. This property is called *D-consistency* below. Preservation of dependency, in our case, means that if $a\ D\ b$ then a also has to depend on the *initial* actions of $r(b)$. Preservation of independency means that if $a\ I\ b$ then a is independent of *all* actions of $r(b)$. Formally, r is called terminating if $r(a)$ is terminating for all $a \in \mathbf{A}$, and D-consistent if

$$a\ D\ b \Longrightarrow r(a) \overset{b}{\cdots\!\!\not\rightarrow} \ \wedge \ \forall r(a) \overset{a'}{\rightarrow} : a'\ D\ b$$
$$a\ I\ b \Longrightarrow \forall a' \in \alpha(r(a)) : a'\ I\ b$$

Operational semantics. We develop SOS rules for refinement. With respect to the transitions of $B[r]$, it is clear that we need at least a rule of the following form: from $B \overset{a}{\rightarrow} B'$ and $r(a) \overset{b}{\rightarrow} C$, conclude $B[r] \overset{b}{\rightarrow} B''$ for some term B''. The interesting part is the choice of B''. It should capture the following points: (1) The refinement of a should be able to proceed, i.e., B'' should somehow contain

C; (2) B'' should resolve all the choices in B in the same way as in B'; (3) B'' should be able to start the refinements of all a-independent actions allowed by B' and (4) B'' should still contain the function r. A rule which captures all these aspects is:

$$\frac{B \xrightarrow{a} B', \; r(a) \xrightarrow{b} C}{B[r] \xrightarrow{b} C \cdot B'[r]} \tag{1}$$

Hence, B'' equals the sequential composition of C with $B'[r]$. To come back to Example 1, the overlapping execution of the refinement of a and b is thus derivable:

$$B[r] \xrightarrow{a_1} a_2 \cdot b[r] \xrightarrow{b_1} a_2 \cdot b_2 \cdot 1[r] \xrightarrow{a_2} 1 \cdot b_2 \cdot 1[r] \xrightarrow{b_2} 1 \cdot 1 \cdot 1[r]$$

The rule for permissions is straightforward, and reflects the intuition behind D-consistency: If the abstract system B permits an action a, then the refined system permits it as well.

$$\frac{B \mathrel{..\xrightarrow{a}} B'}{B[r] \mathrel{..\xrightarrow{a}} B'[r]} \tag{2}$$

The operational semantics of L-plus-refinement, therefore, is determined by Table 1 augmented by Equations (1) and (2). It is noteworthy that these operational refinement rules are simpler by far than the ones obtained in other approaches, in particular [9] and [22]. For instance, we no not rely on auxiliary operators of any kind.

An immediate question concerns the congruence of our semantic equivalence, bisimulation. This is proved by showing that our operational rules obey a certain format, which must at least allow negative premises. The format we choose is GSOS [4]. In order to apply this to our setting, with two kinds of transition relations, we can extend $\mathbf{A}$ by a set $\overline{\mathbf{A}}$ and define $\mathrel{..\xrightarrow{a}}$ to be $\xrightarrow{\bar{a}}$.

Theorem 6. $\sim$ *is a congruence for refinement.*

It is clear that the congruence property does not depend on the termination or D-consistency of r, since these requirements are not expressed in the operational rules in any way. On the other hand, for refinement functions that are not D-consistent, the operationally derived behaviour may deviate from the expected.

Example 2. Consider $B = a \cdot b$ with $a \, D \, b$, and a D-inconsistent refinement in which some initial action b_1 from $r(b)$ is independent of the refinement of a; e.g., $r : a \mapsto a', b \mapsto b_1 \cdot b_2$ such that $a' \, I \, b_1$ and $a' \, D \, b_2$. Intuitively, since ab_1b_2 is an execution of $B[r]$ and b_1 is independent of a, $B[r]$ should also be able to start with b_1. However, this cannot be inferred operationally: the only initial action of B is a, and therefore $B[r]$ cannot start with an action not coming from $r(a)$.

We will strengthen and formalise the concept of "intuitive correctness" for the operational semantics by investigating a denotational characterisation of the refinement operator, and showing that the resulting models coincide, at least for D-consistent refinement functions.

Denotational semantics. For the denotational semantics of $B[r]$, we first derive a semantic function $[\![r]\!]: \mathbf{A} \to \mathbf{M}$ according to $[\![r]\!]: a \mapsto [\![r(a)]\!]$, and then show how to apply such a function to an arbitrary model $\mathcal{P} \in \mathbf{M}$. The latter concerns a pointwise extension of the refinements of single runs.

- Given a model $\mathcal{P} \in \mathbf{M}$ and a semantic refinement function $\mathcal{R}: \mathbf{A} \to \mathbf{M}$, a *witness* is a function $w: E_{\mathcal{P}} \to \mathbf{R}$ such that $w(e) \in \mathcal{R}(\ell_{\mathcal{P}}(e))$ and $\checkmark_{w(e)} = \mathbf{A}$ for all $e \in E_{\mathcal{P}}$.
- Given a run $u \in \mathcal{P}$ and witness $w: E_{\mathcal{P}} \to \mathbf{R}$, the refinement of u by w replaces all $e \in E_u$ by their w-images, and orders the resulting events, *insofar they are dependent*, according to the ordering of u. This is defined formally below.
- The refinement of $\mathcal{P}$ by $\mathcal{R}$ is defined as the set of all $\mathcal{P}$-runs refined by all $\mathcal{R}$-witnesses: $\mathcal{R}(\mathcal{P}) = \{w(u) \mid w \text{ is a } \mathcal{R}\text{-witness}, u \in \mathcal{P}\}$.

Definition 7. Let u be a run and $w: E \to \mathbf{R}$ a function with $E_u \subseteq E$. The *refinement of u by w*, denoted $w(u)$, is defined as:

- $E_{w(u)} := \bigcup_{e \in E_u} \{e\} \times E_{w(e)}$;
- $(d, d') <_{w(u)} (e, e') :\Leftrightarrow (d = e \wedge d' <_{w(e)} e') \vee (d <_u e \wedge d' D_{w(u)} e')$;
- $\ell_{w(u)}: (e, e') \mapsto \ell_{w(e)}(e')$ for all $(e, e') \in E_{w(u)}$;
- $\checkmark_{w(u)} := \checkmark_u$.

We can now extend the semantic mapping $[\![\cdot]\!]$ to the language with refinement, through the rule $[\![B[r]]\!] := [\![r]\!]([\![B]\!])$. The consistency result of Theorem 5 can now be extended to the full language.

Theorem 8. *For arbitrary B and terminating, D-consistent r, $B[r] \sim [\![B[r]]\!]$.*

For the proof see [27]. Here, the requirement of D-consistency is crucial: The denotational semantics can describe the intuitively expected behaviour of Example 2. This shows that the operational and denotational interpretations may differ outside the class of D-consistent refinements. Moreover, for D-inconsistent refinements, bisimulation may fail to be a congruence. For more details see [27].

4 Example: Data base access

In this section we apply our theory to a small example inspired by Brinksma, Jonsson and Orava [5]. The example concerns a distributed data base that can be queried and updated. We assume that there are only two possible data, which we denote 1 and 2. The data base specification is modelled by the transition system $Data_S$ in the following figure:

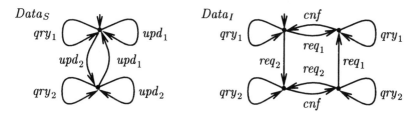

The problem considered in the paper is to change the interface of the data base, so that updating consists not of a single action but of two separate stages, in which the update is *requested* and *confirmed*, respectively. In our setting, this can be expressed by a refinement function $r: upd_i \mapsto req_i; cnf$. Moreover, it is required that in the meantime (between request and confirmation), querying the data base should still be possible. This results in the behaviour $Data_I$ above.

In our approach, this implementation can be obtained algebraically through an application of the refinement operator. The overlap between qry_i and cnf is obtained by setting the dependencies appropriately: $qry_i \; D \; req_j$ but $qry_i \; I \; cnf$.

$$Data_S = \mu D_1. \; qry_1 \cdot D_1 + upd_1 \cdot D_1 + upd_2 \cdot \mu D_2. \; qry_2 \cdot D_2 + upd_1 \cdot D_1 + upd_2 \cdot D_2$$
$$Data'_S = \mu D_2. \; qry_2 \cdot D_2 + upd_2 \cdot D_2 + upd_1 \cdot \mu D_1. \; qry_1 \cdot D_1 + upd_2 \cdot D_2 + upd_1 \cdot D_1$$

The operational behaviour of $Data_I = Data_S[r]$ is given by the left hand transition system in the following figure. The right hand system shows the case where $qry_i \; D \; cnf$ instead, in which case the next query must wait for the second phase of the updating to finish.

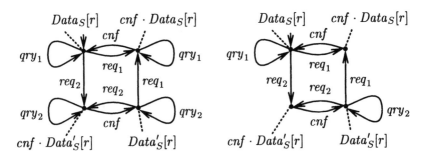

References

1. L. Aceto and M. Hennessy. Towards action-refinement in process algebras. *Information and Computation*, 103(2):204–269, 1993.
2. L. Aceto and M. Hennessy. Adding action refinement to a finite process algebra. *Information and Computation*, 115(2):179–247, 1994.
3. E. Best, R. Devillers, and J. Esparza. General refinement and recursion operators for the Petri box calculus. In Enjalbert, Finkel, and Wagner, eds., *STACS 93*, vol. 665 of *LNCS*, pp. 130–140. Springer, 1993.
4. B. Bloom, S. Istrail, and A. R. Meyer. Bisimulation can't be traced. *Journal of the ACM*, 42(1):232–268, Jan. 1995.
5. E. Brinksma, B. Jonsson, and F. Orava. Refining interfaces of communicating systems. In Abramsky and Maibaum, eds., *TAPSOFT '91, Volume 2*, vol. 494 of *LNCS*, pp. 297–312. Springer, 1991.
6. N. Busi, R. van Glabbeek, and R. Gorrieri. Axiomatising ST bisimulation equivalence. In Olderog, ed., *Programming Concepts, Methods and Calculi*, vol. A–56 of *IFIP Transactions*, pp. 169–188. IFIP, 1994.
7. L. Castellano, G. De Michelis, and L. Pomello. Concurrency vs. interleaving: An instructive example. *Bull. Eur. Ass. Theoret. Comput. Sci.*, 31:12–15, 1987. Note.
8. P. Darondeau and P. Degano. Refinement of actions in event structures and causal trees. *Theoretical Computer Science*, 118:21–48, 1993.

9. P. Degano and R. Gorrieri. A causal operational semantics of action refinement. *Information and Computation*, 122(1):97–119, 1995.

10. P. Degano, R. Gorrieri, and G. Rosolini. A categorical view of process refinement. In de Bakker, de Roever, and Rozenberg, eds., *Semantics: Foundations and Applications*, vol. 666 of *LNCS*, pp. 138–153. Springer, 1992.

11. R. van Glabbeek and U. Goltz. Equivalences and refinement. In Guessarian, ed., *18éme Ecole de Printemps d'Informatique Théorique Semantique du Parallelisme*, vol. 469 of *LNCS*, 1990.

12. U. Goltz, R. Gorrieri, and A. Rensink. Comparing syntactic and semantic action refinement. *Information and Computation*, 125(2):118–143, 1996.

13. M. Huhn. Action refinement and property inheritance in systems of sequential agents. In Montanari and Sassone, eds., *Concur'96*, vol. 1119 of *LNCS*, pp. 639–654. Springer, 1996.

14. W. Janssen, M. Poel, and J. Zwiers. Actions systems and action refinement in the development of parallel systems. In Baeten and Groote, eds., *Concur '91*, vol. 527 of *LNCS*, pp. 298–316. Springer, 1991.

15. L. Jategaonkar and A. Meyer. Testing equivalences for Petri nets with action refinement. In Cleaveland, ed., *Concur '92*, vol. 630 of *LNCS*, pp. 17–31. Springer, 1992.

16. A. Mazurkiewicz. Basic notions of trace theory. In de Bakker, de Roever, and Rozenberg, eds., *Linear Time, Branching Time and Partial Order in Logics and Models for Concurrency*, vol. 354 of *LNCS*, pp. 285–363. Springer, 1989.

17. R. Milner. *Communication and Concurrency*. Prentice-Hall, 1989.

18. M. Nielsen, U. Engberg, and K. G. Larsen. Fully abstract models for a process language with refinement. In de Bakker, de Roever, and Rozenberg, eds., *Linear Time, Branching Time and Partial Order in Logics and Models for Concurrency*, vol. 354 of *LNCS*, pp. 523–549. Springer, 1989.

19. D. Park. Concurrency and automata on infinite sequences. In Deussen, ed., *Proceedings 5th GI Conference*, vol. 104 of *LNCS*, pp. 167–183. Springer, 1981.

20. A. Rensink. Posets for configurations! In Cleaveland, ed., *Concur '92*, vol. 630 of *LNCS*, pp. 269–285. Springer, 1992.

21. A. Rensink. Methodological aspects of action refinement. In Olderog, ed., *Programming concepts, methods and calculi*, vol. A-56 of *IFIP Transactions*. IFIP, 1994.

22. A. Rensink. An event-based SOS for a language with refinement. In Desel, ed., *Structures in Concurrency Theory*, Workshops in Computing, pp. 294–309. Springer, 1995.

23. A. Rensink and R. Gorrieri. Action refinement as an implementation relation. In Bidoit and Dauchet, eds., *TAPSOFT '97: Theory and Practice of Software Development*, vol. 1214 of *LNCS*, pp. 772–786. Springer, 1997.

24. A. Rensink and H. Wehrheim. Weak sequential composition in process algebras. In Jonsson and Parrow, eds., *Concur '94: Concurrency Theory*, vol. 836 of *LNCS*, pp. 226–241. Springer, 1994.

25. W. Vogler. Failure semantics based on interval semiwords is a congruence for refinement. *Distributed Computing*, 4:139–162, 1991.

26. H. Wehrheim. Parametric action refinement. In Olderog, ed., *IFIP Transactions: Programming Concepts, Methods and Calculi*, pp. 247–266. Elsevier, 1994.

27. H. Wehrheim. *Specifying Reactive Systems with Action Dependencies: Modelling and Hierarchical Design*. PhD thesis, University of Hildesheim, 1996.

A Hierarchy for $(1, +k)$-Branching Programs with Respect to k

P. Savický, S. Žák [*]

Institute of Computer Science
Academy of Sciences of the Czech Republic
Pod vodárenskou věží 2
182 07 Praha 8
Czech Republic

Abstract. Branching programs (b. p.'s) or decision diagrams are a general graph-based model of sequential computation. The b. p.'s of polynomial size are a nonuniform counterpart of LOG. Lower bounds for different kinds of restricted b. p.'s are intensively investigated. An important restriction are so called k-b. p.'s, where each computation reads each input bit at most k times. Although, for more restricted syntactic k-b.p.'s, exponential lower bounds are proven and there is a series of exponential lower bounds for 1-b. p.'s, this is not true for general (nonsyntactic) k-b.p.'s, even for $k = 2$. Therefore, so called $(1, +k)$-b. p.'s are investigated.

For some explicit functions, exponential lower bounds for $(1, +k)$-b. p.'s are known. Investigating the syntactic $(1, +k)$-b. p.'s, Sieling has found functions $f_{n,k}$ which are polynomially easy for syntactic $(1, +k)$-b. p.'s, but exponentially hard for syntactic $(1, +(k-1))$-b. p.'s. In the present paper, a similar hierarchy with respect to k is proven for general (nonsyntactic) $(1, +k)$-b. p.'s.

1 Introduction

A branching program (b. p.) is a computation model for representing the Boolean functions. The input of a branching program is a vector consisting of the values of n Boolean variables. The branching program itself is a directed acyclic graph with one source. The out-degree of each node is at most 2. Every branching node, i.e. a node of out-degree 2, is labeled by an input variable and one of its out-going edges is labeled by 0, the other one by 1. The sinks (out-degree 0) are labeled by 0 and 1. A branching program determines a Boolean function as follows. The computation starts at the source. If a node of out-degree 1 is reached, the computation follows the unique edge leaving the node. In each branching node, the variable assigned to the node is tested and the out-going edge labeled by the actual value of the variable is chosen. Finally, a sink is reached. Its label determines the value of the function for the given input. By the size of a branching program we mean the number of its nodes.

The branching programs are a model of the configuration space of Turing machines where each node corresponds to a configuration. Thus the polynomial size b. p.'s represent a nonuniform variant of LOG. Hence, a superpolynomial lower bound on b. p.'s for a Boolean function computable within polynomial time would imply $P \neq LOG$.

In order to investigate the computing power of branching programs, restricted models were suggested. An important restriction are so called read-once branching programs (1-b. p.'s), where the restriction is such that during each computation on any input each variable is tested at most once.

The first exponential lower bounds for 1-b. p.'s were [18] and [19]. These results were improved in [5]. These three bounds are of magnitude $2^{\Omega(\sqrt{n})}$. A lower bound of magnitude $2^{\Omega n/\log n}$ was presented in [7]. The first lower bound of magnitude $2^{\Omega(n)}$, where n is the input size was $2^{n/c}$ lower bound for a large c for the function "parity of the number of triangles in a graph", see [2]. The constant c in this lower bound was improved, see [17], to a lower bound of magnitude $2^{n/2000}$ for the same function. For a different function, a lower bound $2^{n-o(n)}$ was proved in [14]. In [11], a lower bound $2^{c\sqrt{n}}$ is proved for multiplication.

Several generalizations of 1-b. p.'s are investigated. Recently, the most powerful among them are so called k-b. p.'s, where each computation is allowed to test each variable at most k times. Since no

---

[*] The research of both authors was supported by GA of the Czech Republic, grant No. 201/95/0976.

superpolynomial lower bounds even for 2-b. p.'s are known, even more restricted b. p.'s are investigated. Namely, so called $(1, +k)$-b. p.'s, where for every input, there are at most k variables that are tested during the computation more than once. If, moreover, the variables with repeated tests may be read at most two times, we obtain a class that contain 1-b. p.'s and is contained in 2-b. p.'s. For $(1, +k)$-b. p.'s, an exponential lower bound for k up to n^α for some fixed positive α may be found in [8], [13] and [20]. The results of [8] and [13] hold even for α arbitrarily close to 1. A superpolynomial lower bound for $k = o(n/\log n)$ is proved in [8].

The two restrictions mentioned above, namely k-b. p.'s and $(1, +k)$-b. p.'s can be made even stronger, if the restriction of repeated tests is applied not only to valid computation paths, but to every possible path from the source to a sink in the b. p., including inconsistent paths. In this way, we obtain so called syntactic k-b. p.'s and syntactic $(1, +k)$-b. p.'s. Note that each 1-b. p. is a syntactic 1-b. p., while for k-b. p.'s, $k \geq 2$ this is not true in general.

For syntactic k-b. p.'s, exponential lower bounds are known, see [3], [6], [9], for some $k = \Omega(\log n)$. The result of [3] and [6] hold even for nondeterministic k-b. p.'s. An exponential separation of syntactic k-b. p.'s and syntactic k^2-b. p.'s for any constant k was presented in [10]. For syntactic $(1, +k)$-b. p.'s, a strict hierarchy according to k was proved in [15] and [16] for k at most roughly $n^{1/3}$.

In the present paper, we generalize the last mentioned result in such a way that it holds also for (nonsyntactic) b. p.'s for $k \leq 1/2 \cdot n^{1/8} \log^{-1/4} n$. Namely, we present functions $f_{n,k}$ which are polynomially easy for $(1, +k)$-b. p.'s, but exponentially hard for $(1, +(k-1))$-b. p.'s.

In order to prove the hierarchy result, an exponential lower bound for $(1, +k)$-b. p.'s is proved. In comparison to [8], [13] and [20] the exponential lower bound is reached in a smaller range of k. On the other hand, while all the three mentioned results use in fact the same method, the present paper is based on a different method. The method of [8], [13] and [20] may be applied only to functions f satisfying the following requirement or its dual. If $f(x) = f(y) = 1$, then either $x = y$ or the Hamming distance of x and y is at least n^ε for some positive ε. The function used for the lower bound in the present paper does not have this property.

The structure of the paper is as follows. In Section 2, a function $f_{n,k}$ of n variables is defined and some its properties are proved. In Section 3 an exponential lower bound for $f_{n,k}$ in $(1, +(k-1))$-b. p.'s is proven using a theorem the proof of which is presented in Section 5. In Section 4, we present a polynomial size $(1, +k)$-b. p. for $f_{n,k}$ and summarize the main result.

2 The function and its basic properties

In this section, we define the function $f_{n,k}$ and prove some properties of the function. Informally, the function is defined as follows. The n variables are divided into k blocks of length m. For every $j = 1, 2, \ldots, k$, a weighted sum of the bits of block j determines an index i_j of some of the input bits. Then, the value of the function is the parity of the bits determined by i_j for $j = 1, 2, \ldots, k$. The exact definition of $f_{n,k}$ requires some technical notation.

For every natural number n, let $p(n)$ be the smallest prime greater than n. Consider the set $\{1, 2, \ldots, n\}$ as a subset of $Z_{p(n)}$, the field of the residue classes modulo $p(n)$. Then, for every $t \in Z_{p(n)}$, let $\omega(t) = t$, if $t \in \{1, 2, \ldots, n\}$ and $\omega(t) = 1$ otherwise.

Definition 1. For every $t = (t_1, t_2, \ldots, t_k) \in \{1, 2, \ldots, n\}^k$ and every $x \in \{0, 1\}^n$, let $Par(x, t) = x_{t_1} \oplus x_{t_2} \oplus \ldots \oplus x_{t_k}$.

Definition 2. Let k divide n and let $m = n/k$. Then, let $\psi_{n,k} : \{0, 1\}^n \to \{1, 2, \ldots, n\}^k$ be defined as follows. For every x let $\psi_{n,k}(x) =_{\text{def}} (t_1, t_2, \ldots, t_k)$, where for every $j = 1, 2, \ldots, k$,

$$t_j = \omega\left(\sum_{i=1}^{m} i\, x_{(j-1)m+i}\right),$$

where the sum is evaluated in $Z_{p(n)}$. Moreover, let $f_{n,k}(x) =_{\text{def}} Par(x, \psi_{n,k}(x))$.

In order to prove some required properties of $\psi_{n,k}$, we shall use the following theorem originally proved in [4]. A different proof of this theorem may be found in [1].

Theorem 3. (Dias da Silva and Hamidoune) *Let p be a prime and let h_1 and h_2 be integers. Moreover, let $h_2 \le h_1 \le p$ and let $A \subseteq Z_p$ such that $|A| = h_1$. Let A' be the set of all sums of h_2 distinct elements of A. Then, $|A'| \ge \min(p, h_2(h_1 - h_2) + 1)$.*

Corollary 4. *Let $\varepsilon > 0$ be fixed. Then, for every n large enough, the following is true. If $A \subseteq Z_{p(n)}$ and $|A| \ge (2 + \varepsilon)\sqrt{n}$, then, for every $t \in Z_{p(n)}$, there is a subset $B \subseteq A$ such that the sum of the elements of B is equal to t.*

Proof. Let $h_2 = \lfloor (1 + \varepsilon/2)\sqrt{n} \rfloor$ and $h_1 = 2h_2$. Choose any subset C of A of size h_1. By Theorem 3, there is at least $\min(p(n), h_2^2 + 1)$ different sums of h_2 distinct elements of C. We have $h_2^2 + 1 \ge (1 + \varepsilon)n - O(\sqrt{n})$. Since $p(n) = n + o(n)$, see [12], we have $\min(p(n), h_2^2 + 1) = p(n)$. Hence, for every $t \in Z_{p(n)}$, there is a set $B \subset A$ of h_2 elements adding up to t.

Lemma 5. *For any fixed $\varepsilon > 0$, for every n large enough and for every k as above, the following is true. If r is an integer such that $n > r \ge k((2 + \varepsilon)\sqrt{n} + 1)$, and if at most $n - r$ of the variables of the function $Par(x, \psi_{n,k}(x))$ are set to some constants, the restricted function is still not a constant function.*

The proof is omitted.

Lemma 6. *If at most $n/k - 3\sqrt{n}$ of the variables of the function $\psi_{n,k}$ are set to some constants, the restricted function still satisfies the following. For every choice of $1 \le i_1, i_2, \dots, i_k \le n$, there is a setting of the free variables such that the value of $\psi_{n,k}$ is equal to $(i_1, i_2, \dots, i_k)$.*

Proof. In each block, at least $3\sqrt{n}$ variables are free. Using Corollary 4, each entry of $\psi_{n,k}(x)$ may be set to any value from $\{1, 2, \dots, n\}$ independently of the other entries.

3   The lower bound

In this section, we prove an exponential lower bound for $f_{n,k}$ in $(1, +(k-1))$-b. p.'s, if k is not too large. There are several possibilities how to bound the number of repeated tests in a path. We use the following definition, i.e., we count only the number of different variables involved in the repeated tests, not the number of these tests.

Definition 7. Let P be a b. p. For every input x, let $R(x)$ be the set of indices of input bits that are read more than once during the computation for x. The b. p. P is called a $(1, +k)$-b. p., if for every x, $|R(x)| \le k$.

For a path α, let $\pi(\alpha)$ be the set of variables tested in α.

Definition 8. Let S be some set of paths in a b. p. going from a node u to a node v. Then the number

$$\left| \bigcup_{\alpha \in S} \pi(\alpha) \right| - \min_{\alpha \in S} |\pi(\alpha)|$$

will be called the *fluctuation* of S. If both occurrences of $\pi(\alpha)$ in the expression above are replaced by $\pi(\alpha) \cap I$, where I is a set of variables, we call the resulting number the fluctuation of S relative to I.

We say that an edge (u, v) is a *test* of a variable x_i, if u is of degree 2 and x_i is the label of u.

Definition 9. We say that a branching program P is (p, r)-*well-behaved*, if it satisfies the following three conditions:
(i) Every path from the source of P to a sink contains at least p tests.
(ii) The first p tests on any path starting in the source of P test p different variables.
(iii) If w is any node of P and S is the set of all paths with p tests leading from the source to w, then the fluctuation of S is less than r.

One of the key steps of the proof is the following theorem. Its proof may be found in Section 5.

Theorem 10. *Let n, k, p, r be integers such that $r < n$, $2kp \leq n - r$. Let f be a Boolean function of n variables such that any setting of at most $n - r$ variables to constants still leads to a nonconstant function. Let P be a $(1, +(k - 1))$-b. p. computing f. Then, there is a subprogram P' of P arising from P by setting at most $(2k - 1)p$ variables to constants that is (p, r)-well-behaved.*

To prove the lower bound, we shall combine Theorem 10 with the following Theorem 11 implicitly used already in [15]. If v_j is a vector, let $v_{j,i}$ be its i-th coordinate.

Theorem 11. *Let k, p and m be integers and let k divide p. For $j = 1, 2, \ldots, m$, let $v_j \in \{0, 1\}^I$, where I is some index-set of size p. Assume that for every k-tuple $i_1, i_2, \ldots, i_k \in I$, there is a function $\phi : \{0, 1\}^I \to \{0, 1\}$ such that*
(1) ϕ is computable by a decision tree of depth at most $k - 1$.
(2) For every $j = 1, 2, \ldots, m$ we have

$$\phi(v_j) = v_{j,i_1} \oplus v_{j,i_2} \oplus \ldots \oplus v_{j,i_k}.$$

Then, we have $m \leq 2^{p(1-1/k)}$.

The proof is omitted.
Now, we can state and prove the lower bound result.

Theorem 12. *Let n, k be integers, let k divide n and let $k \leq \sqrt{n}/3$. Then, every $(1, +(k - 1))$-b. p. computing $f_{n,k}$ has size at least*

$$2^{\left(\frac{n}{2k^3} - \frac{3\sqrt{n}}{2k^2} - 3k\sqrt{n} \log n - 1 \right)}.$$

Proof. First, let us introduce an auxiliary notation. For partial inputs $u_1, u_2, \ldots, u_s$ specifying disjoint sets of bits, let $[u_1, u_2, \ldots, u_s]$ denote the (partial) input specifying all the bits specified in some of u_j in the same way as in corresponding u_j.

Let $r = k(\lfloor 3\sqrt{n} \rfloor - 1)$, $q = n/k - 3\sqrt{n}$ and $p = \lfloor q/(2k^2) \rfloor k$. We have $2kp \leq q \leq n - r$. By Lemma 5, setting of at most $n - r$ variables in $Par(x, \psi_{n,k}(x)) = f_{n,k}$ leads to a nonconstant function. Hence, the function $Par(x, \psi_{n,k}(x))$ satisfies the assumption of Theorem 10 for our choice of k, p and r. Let P be a $(1, +(k - 1))$-b. p. of size c computing $f_{n,k}$. Consider the subprogram P' of P guaranteed by Theorem 10. Let u be the partial input with at most $(2k - 1)p$ fixed variables which yields P' and let w_1 be the source of P'. We have that P' is a $(1, +(k - 1))$-b. p. of size at most c computing the restriction of $Par(x, \psi_{n,k}(x))$ according to u.

Let w_2 be the node of P' such that the number of paths starting at w_1, ending in w_2 and containing p tests is maximal. There is at least $2^p/c$ such paths. Call the set of these paths $\mathcal{P}_1$. Each path tests some set of variables. Since P' is (p, r)-well-behaved, the fluctuation of $\mathcal{P}_1$ is less than r and hence, there are at most $\binom{p+r}{r}$ of different sets of variables tested along individual paths from $\mathcal{P}_1$. Let $\mathcal{P}_2$ be some of the largest subsets of $\mathcal{P}_1$ of paths testing exactly the same set of variables. Then, we have

$$|\mathcal{P}_2| \geq \frac{2^p}{c\binom{p+r}{r}}.$$

Each path in $\mathcal{P}_2$ together with u determines a partial input. For every partial input, it is possible to evaluate its contribution to the k entries of the value of $\psi_{n,k}$. By this, we mean the sums from Definition 2 restricted to bits with the value fixed by u and the given path from $\mathcal{P}_2$. The number of possible contributions is at most n^k. Hence, there is a subset $\mathcal{P}_3$ of $\mathcal{P}_2$ of paths with the same contributions and such that its size $m =_{\text{def}} |\mathcal{P}_3|$ satisfies

$$m \geq \frac{2^p}{c\binom{p+r}{r}n^k}. \tag{1}$$

Let $v_1, v_2, \ldots, v_m$ be the list of elements of $\mathcal{P}_3$ and let I be the set of indices of variables set to a constant by inputs v_j. By construction of $\mathcal{P}_2$, $|I| = p$. We are going to verify that the inputs $v_1, v_2, \ldots, v_m$ satisfy the assumption of Theorem 11.

Let us fix some $i_1, i_2, \ldots, i_k \in I$. Let x be a partial input such that $[u, v_1, x]$ is a total input satisfying $\psi_{n,k}([u, v_1, x]) = (i_1, i_2, \ldots, i_k)$. Such an x exists, since the number of bits fixed by $[u, v_1]$ is not larger than $2kp \leq q \leq n/k - 3\sqrt{n}$ and therefore we may apply Lemma 6.

Since all the partial inputs $v_1, v_2, \ldots, v_m$ have the same contributions to the sums in the Definition 2, $\psi_{n,k}([u, v_j, x]) = (i_1, i_2, \ldots, i_k)$ for all $j = 1, 2, \ldots, m$.

Consider the restriction P'' of P' according to the values of input bits from the input x. The only free input bits of P'' are the bits from I. For every v_j, the computation of P'' computes $f_{n,k}([u, v_j, x]) = Par([u, v_j, x], (i_1, i_2, \ldots, i_k)) = v_{j,i_1} \oplus v_{j,i_2} \oplus \ldots \oplus v_{j,i_k}$. Moreover, for every v_j, the computation of P'' reads all the bits from I, then it reaches the node w_2 and in the rest of the computation, at most $k - 1$ variables with indices in I are read. (Since P'' is also a $(1, +(k-1))$-b. p.)

Consider the subprogram of P'' starting in w_2 and let P''' be the decision tree obtained from this subprogram as follows. First, we expand the subprogram starting at w_2 into a tree. In the second step, we delete all edges of the tree that are not visited by any computation starting from w_2 for some of v_j. After this, some of the nodes of the tree might have out-degree 1. In the last step, every such node is deleted and the edge leading to it is redirect to the single successor of the considered node.

Every leaf of P''' is reached by a computation for some v_j, otherwise some of the edges of the path leading to the leaf would have been deleted. Hence, each path of P''' tests at most $k - 1$ variables, since it tests a subset of the set of variables read by some computation of P'' after the node w_2 for some of v_j.

Let ϕ be the function computed by P'''. Clearly, ϕ satisfies the assumption (1) of Theorem 11.

By construction of P''', P'' and P''' are equivalent on inputs v_j. Thus, for each $j = 1, 2 \ldots, m$, we have $\phi(v_j) = f_{n,k}([u, v_j, x]) = v_{j,i_1} \oplus v_{j,i_2} \oplus \ldots \oplus v_{j,i_k}$. Hence, ϕ satisfies also the assumption (2) of Theorem 11.

These arguments work for every k-tuple of indices from I. Hence, Theorem 11 implies $m \leq 2^{p(1-1/k)}$.

Together with (1), this implies

$$c \geq \frac{2^{p/k}}{\binom{p+r}{r} n^k}.$$

Since $\binom{p+r}{r} \leq n^r$, we have

$$c \geq 2^{p/k - r \log n - k \log n}. \tag{2}$$

The theorem now follows by substitution of the chosen values of p, r and k into the last estimate.

4 The hierarchy

We shall prove an upper bound for the function $f_{n,k}$ on $(1, +k)$-b. p.'s. Together with the lower bound from the previous section, it gives that $(1, +k)$-b. p.'s are more powerful than $(1, +(k-1))$-b. p.'s.

Theorem 13. Let $k = k(n) \leq 1/2 \cdot n^{1/8} \log^{-1/4} n$. Then, for every n large enough, we have (i) There is a $(1, +k)$-b. p. computing $f_{n,k}$ of size $O(n^2)$.
(ii) Every $(1, +(k-1))$-b. p. computing $f_{n,k}$ has size at least $2^{\Omega(n/k^3)}$.

Proof. Let us start with (i). We shall construct a $(1, +k)$-b.p. P computing $f_{n,k}$. Consider the input bits in the input x divided into k groups in the same way as in the definition of $\psi_{n,k}$. Let $\psi_{n,k}(x) = (i_1, i_2, \ldots, i_k)$. In order to describe P, we shall describe for every $j = 1, 2, \ldots, k$ a b. p. P_j computing $x_{i_1} \oplus x_{i_2} \oplus \ldots \oplus x_{i_j}$. Then, P is P_k.

Let $x_1, x_2, \ldots, x_m$ be the bits in the first group. The b. p. P_1 is leveled and it reads the bits in the first group in the natural ordering. For simplicity, assume that each level consists of $p(n)$ vertices corresponding to the residue classes mod $p(n)$. The computation starts in level 0 in the node corresponding to 0. After reading x_j, the computation reaches the j-th level in the node corresponding to the residue class $x_1 + 2x_2 + 3x_3 + \ldots + jx_j \mod p(n)$. For each $j = 0, 1, \ldots, m - 1$ and each node

w at level j, this determines the two nodes at level $j + 1$, where the edges from w lead to. Consider the node corresponding to $t \in Z_{p(n)}$ at level m. In this node, the variable $w(t)$ is tested and its value is the output of P_1.

The b. p. just described computes x_{i_1}, since the computation reaches the m-th level in the node corresponding to $x_1 + 2x_2 + 3x_3 + \ldots + mx_m \mod p(n)$ and by definition of $\psi_{n,k}$, we have $i_1 = w(x_1 + 2x_2 + 3x_3 + \ldots + mx_m)$.

Now, assume, P_j is constructed. In order to construct P_{j+1}, append to each of the two sinks of P_j a b. p., computing x_{i_j} in a way similar to the computation of x_{i_1} in P_1. We obtain a b. p. with four sinks corresponding to the four possible values of $x_{i_1} \oplus x_{i_2} \oplus \ldots \oplus x_{i_j}$ and $x_{i_{j+1}}$. Now, P_{j+1} is obtained by joining the sinks with the same value of $x_{i_1} \oplus x_{i_2} \oplus \ldots \oplus x_{i_{j+1}}$.

Note that, P_1 has size at most $p(n)n/k$. Moreover, for each $j = 2, 3, \ldots, k$, the b. p. P_j contains at most $2p(n)n/k$ additional nodes w.r.t. P_{j-1}. Hence, P_k is of size at most $2p(n)n = O(n^2)$.

In order to prove (ii), note that, if $k = k(n) \leq 1/2 \cdot n^{1/8} \log^{-1/4} n$, then

$$\frac{1}{2} \cdot \frac{n}{2k^3} \geq \frac{3\sqrt{n}}{2k^2} + 3k\sqrt{n} \log n + 1 .$$

Using this, Theorem 12 implies (ii).

5 Proof of Theorem 10

Let us start the proof by the following. Let an edge (u, v) be a test of a variable x_i. The Boolean value labeling this edge is called the value required by this test. Two tests of the same variable are consistent, if they require the same value.

First, we shall construct a sequence $v_0, v_1, \ldots, v_t$ of nodes of P, where v_0 is the source and v_t is some of the sinks and a sequence $T_1, T_2, \ldots, T_t$, where T_i is a set of some paths from v_{i-1} to v_i. The proof is then finished by showing that at least one of the nodes $v_0, v_1, \ldots, v_{2k-1}$ may be chosen as the source of the required (p, r)-well-behaved subprogram.

The two sequences are constructed by a process starting with v_0 being the source of P and with an empty sequence of sets. The process will be described in steps. In step j, we start with some sequence $v_0, v_1, \ldots, v_{j-1}$ of nodes and a sequence of sets $T_1, T_2, \ldots, T_{j-1}$, we add a new node v_j, a new set T_j and possibly modify the sets T_i for $i = 1, \ldots, j - 1$. The process stops, when v_j becomes a sink of P and we set $t = j$. In each step of the process, to each of the sets T_i a type A, B or C is assigned. The type is assigned when the set is created and it may be modified, if the set is changed at some later step.

Definition 14. Let $T_1, T_2, \ldots, T_j$ be the sets constructed at some step of the process. Let t_1, t_2 be tests of the same variable contained in $T_1, T_2, \ldots, T_j$. Then, we say that the test t_1 *preceeds* the test t_2, if either for some i both t_1 and t_2 are in some path $\alpha \in T_i$ and t_1 preceeds t_2 in α or t_1 is contained in T_{i_1} and t_2 in T_{i_2} and $i_1 < i_2$. Moreover, we say that the *rank* of a test t is h, if h is the maximum integer, for which there are tests $t_1, t_2, \ldots, t_h$ of the same variable, such that $t = t_h$ and for all $i = 2, 3, \ldots, h$, t_{i-1} preceeds t_i. A test is called a *repeated* test, if its rank is at least 2.

In each step of the process, we require that the sequence $T_1, T_2, \ldots, T_j$ is consistent. By this, we mean the following. If $\alpha_i \in T_i$ is any choice of one path from each of the sets, the concatenation of α_i is a consistent path in P. We will require even stronger structural property expressed as the conjunction of Requirement 1 and 2.

Requirement 1 Let x_i be any variable contained in a repeated test in $T_1, T_2, \ldots, T_j$. Then

(i) all its tests in $T_1, T_2, \ldots, T_j$ are consistent,

(ii) there is exactly one test of x_i of rank 1, say t_1,

(iii) there is exactly one test of x_i of rank 2, say t_2.

(iv) If $j_1 \leq j_2$ are such that t_1 is contained in $\alpha_1 \in T_{j_1}$ and t_2 is contained in $\alpha_2 \in T_{j_2}$, then $|T_{j_1}| = |T_{j_2}| = 1$ and t_2 is the last test of α_2.

Note that if Requirement 1 is satisfied, there may be inconsistent tests of some variable in $T_1, T_2, \ldots, T_j$, if all have rank 1. In particular, all these tests have to be contained in the same set T_{j_1} for some $j_1 = 1, 2, \ldots, j$.

Requirement 2 (i) If T_i is assigned type A, then it contains no test of rank 2.
(ii) If T_i is assigned type B, then it contains no test of rank 2 and, moreover, it contains exactly one path.
(iii) If T_i is assigned type C, then it contains exactly one path, this path contains exactly one test of rank 2 and this test is the last test of the path.

The procedure of creating sets T_j and the assignment of types will be such that in each step of the process, Requirements 1 and 2 will be satisfied.
Let a sequence $T_1, T_2, \ldots, T_{j-1}$ satisfying Requirements 1 and 2 be given. Note that, at the beginning of the process, i.e. if $j = 1$, the sequence of sets is empty and, hence, it satisfies both Requirements 1 and 2. Let us describe the procedure of creating the set T_j. For any path α starting at v_{j-1} calculate the ranks of tests in α according to the sequence $T_1, T_2, \ldots, T_{j-1}, \{\alpha\}$ of j sets. Any path α starting in v_{j-1} is called *good*, if there is no test of rank 2 in α.
Let α be a good path starting at v_{j-1}. Note that if some variable has a test of rank 1 in α, then there is no other test of the same variable in α and also no test of the variable in $T_1, T_2, \ldots, T_{j-1}$. This implies that, if some variable has a test of rank at least 3, then also the first test of this variable in α has rank at least 3. Hence, the variable is repeated already in $T_1, T_2, \ldots, T_{j-1}$ and all occurrences of this variable in $T_1, T_2, \ldots, T_{j-1}$ are consistent. Consequently, every test of such a variable in α is either consistent with all its preceeding tests in $T_1, T_2, \ldots, T_{j-1}$ or with none of them.

Definition 15. (i) A good path α is called *consistent*, if all tests in α of any variable that is repeated in $T_1, T_2, \ldots, T_{j-1}$ are consistent with all the tests of the same variable in $T_1, T_2, \ldots, T_{j-1}$.
(ii) A good path is called *maximal* good path, if it leads to a sink or to a node labeled by a variable x_i, such that adding a test of x_i to the path creates a test of rank 2.

If a consistent good path is not maximal, then it leads to a node, such that the variable x_i tested in it is either repeated in $T_1, T_2, \ldots, T_{j-1}$ or has no test there. In the former case, one of the edges leaving the node forms a consistent prolongation of the path. In the latter case, both edges leaving the node lead to a consistent prolongation. Hence, every consistent good path is a prefix of a consistent good path that is moreover a maximal good path.
We shall distinguish the following three cases. It is easy to see that if Case 1 does not occur, then at least one of Cases 2 or 3 occurs. If Case 2 and 3 occur simultaneously, Case 2 has higher priority. Consider all consistent maximal good paths starting at v_{j-1}.

Case 1: Every such path contains at least p tests of rank 1.
Case 2: Among these paths, there is a consistent good path α containing $< p$ tests of rank 1 and leading to a sink.
Case 3: Among these paths, there is a consistent good path α containing $< p$ tests of rank 1 and leading to a node w in which a variable x_i is tested, such that adding a test of x_i to the end of α produces a test of rank 2.

Now, we describe the procedure of creating the set T_j, the assignment of type to this set and the possible modifications in the previous sets.
In **Case 1**, let S be the set of all consistent good paths containing exactly p tests of rank 1 and such that the last test of the path has rank 1. Note that no path of S is a prefix of another. If a consistent good path reaches a node adding a test of rank 1, then a consistent good path may continue along both edges leaving the node. Hence, S consists of 2^p paths.
For every node u, let S_u be the subset of paths from S leading to the node u. Now, choose u so that the fluctuation of S_u relative to the variables in tests of rank 1 be maximal. Then, $v_j = u$, $T_j = S_u$ and its type is chosen to be A.

In **Case 2**, $T_j = \{\alpha\}$ and it will be considered of type B. In this situation, the process stops and t is set to j.

In **Case 3**, either there is some test of x_i (mentioned in Case 3 above) of rank 1 in $T_1, T_2, \ldots, T_{j-1}$ (Subcase 3a) or there is some test of x_i of rank 1 in α (Subcase 3b). Assume, Case 2 does not occur. Then, the two subcases are handled as follows:

Subcase 3a. Let j_1 be such that x_i has a test in T_{j_1}. If the type assigned to T_{j_1} is A, we choose a path β containing x_i in T_{j_1}, change the set T_{j_1} to $\{\beta\}$ and its type is changed to B. After this change, the test of x_i in β is the unique test of x_i in $T_1, T_2, \ldots, T_{j-1}$. Among the two edges leaving w, we choose the edge consistent with the test of x_i in β. Let α' be the path consisting of α and the chosen edge. Finally, let $T_j = \{\alpha'\}$ and let its type be C.

Subcase 3b. Let α' be the path consisting of α and the edge leaving w, which is consistent with the test of x_i already contained in α. Then, let $T_j = \{\alpha'\}$ and let its type be C.

It is easy to verify that in each case, the new sequence of sets $T_1, T_2, \ldots, T_j$ satisfies Requirements 1 and 2.

According to the description of the process, some tests contained in the new set T_j may be later deleted, if Subcase 3a occurs. Note, however, that if some test is not deleted until the end of the process, then, if its rank was 1, 2 or 3, it does not change and if it was more than 3, it is still at least 3 at the end of the process.

Assume, the process just described stopped with the sequence $T_1, T_2, \ldots, T_t$. Concatenation of any choice of $\alpha_i \in T_i$ for $i = 1, 2, \ldots, t$ forms a valid computation testing at most tp different variables and ending in a sink. By the assumptions of the theorem, setting of at most $n - r$ variables does not lead to constant subfunction. Hence, we have $tp > n - r \geq 2kp$ and so, $t \geq 2k + 1$.

The number of sets in the sequence $T_1, T_2, \ldots, T_t$ of type A, B or C will be denoted a, b and c respectively. The type B may be assigned to a set in $T_1, T_2, \ldots, T_t$ only in Case 2 and in Subcase 3a. Case 2 may occur only as the last step of the process and in Subcase 3a the new set is assigned type C. Hence, $c + 1 \geq b$.

Assume for a moment that there is no set of type A among the first $2k$ sets in the sequence. Then, $b + c = 2k$. Hence, we have $2c \geq 2k - 1$ and so $c \geq k$. Since c is the number of sets of type C, there is at least c tests of rank 2 occurring in sets T_j of size 1. Hence, these tests are repeated tests of different variables in any computation which may be created from $T_1, T_2, \ldots, T_t$. It is a contradiction, since P is a $(1, +(k-1))$-b. p.

Hence, there is a set T_j of type A among the first $2k$ sets of $T_1, T_2, \ldots, T_t$. Let T_j be the first of such sets. All the sets $T_1, T_2, \ldots, T_{j-1}$ are of type B or C and hence contain exactly one path. Let β be their concatenation. Note that β contains at most $(2k - 1)p$ tests of different variables. If we set the variables occurring in β to the values required in β, we obtain a subprogram P' of P, in which v_{j-1} is the source.

Those variables tested in T_j that are not fixed by β have no occurrence in $T_{j+1}, \ldots, T_t$. Hence, if the fluctuation of T_j relative to these variables is at least r, it is possible to choose $\alpha_i \in T_i$ for all $i \geq j$ so that the path $\beta \alpha_j \ldots \alpha_t$ contains no test of at least r variables. This is a contradiction with the assumptions of the theorem. Hence, T_j has fluctuation less than r relative to the variables not fixed by β.

Let $T'_1, T'_2, \ldots, T'_{j-1}$ be the sequence $T_1, T_2, \ldots, T_{j-1}$ at the beginning of the step j. Since $T_i \subset T'_i$, we have that β is a concatenation of some paths chosen from $T'_1, T'_2, \ldots, T'_{j-1}$. In the following, we shall derive the properties of tests in paths in P' using the properties of these tests in the context from the step j, in which T_j was created. In particular, this means that the rank of tests in the paths starting in v_{j-1} is calculated according to $T'_1, T'_2, \ldots, T'_{j-1}$. Also the notions maximal and consistent good path are considered according to the sequence $T'_1, T'_2, \ldots, T'_{j-1}$. If we consider a test in this way, we say that we consider the test in the original context. As a shorthand, the original rank refers to the rank considered in the original context.

Now, we are going to prove that P' is (p, r)-well-behaved. When created, the set T_j was assigned type A, otherwise, it cannot have type A at the end of the process. This may happen only in Case 1. Hence, the consistent maximal good paths in P starting at v_{j-1} satisfy the requirements of Case 1 in the original context.

Consider a path α' in P' from the source to a sink. Let α denote α' considered in the original context. Let γ be the longest prefix of α that is good. It may be α itself or it may end in a node w such that adding to γ the edge from w which is contained in α would create a test of rank 2. In this latter case, adding any edge from w to γ leads to a test of rank 2. Hence, in both cases, γ is a maximal good path in P. Moreover, γ is consistent. It follows by the requirement of Case 1 that γ contains at least p tests of rank 1. All these tests are not influenced by the setting according β. This implies condition (i) of the definition of (p, r)-well-behaved b. p.

Claim. In each path in P', all of the first at most p tests have original rank 1.

In order to prove the claim, let α' be any path in P' containing at least one test of original rank 2. Let γ' be the prefix of α' ending just before the first of such tests. Let γ denotes γ' considered in the original context. It is a good path and any prolongation of γ by one edge contains a test of rank 2. Hence, γ is a consistent maximal good path in P. Thus, it contains at least p tests of rank 1. These tests are not influenced by the setting according to β. Hence, γ' contains at least p tests of original rank 1 and no test of original rank 2. The claim follows.

The claim implies that no two of the first at most p tests in any path in P' starting in the source may contain the same variable. This implies (ii) of the definition of (p, r)-well-behaved b. p. Moreover, the claim also implies that the fluctuation of any set of paths with p tests in P' is the same as the fluctuation of the set relative to tests of original rank 1. Together with the definition of sets S and S_u in the description of Case 1, this implies that in every S_u the fluctuation is at most the fluctuation of T_j, which is by some previous paragraph at most r. This implies (iii) of the definition of (p, r)-well-behaved b. p.

Acknowledgement The authors are grateful to Endre Szemeredi for directing their attention to [1] and [4].

References

1. N. Alon, M. B. Nathanson and I. Z. Ruzsa, The polynomial method and restricted sums of congruence classes, *J. Number Theory*, to appear.
2. L. Babai, P. Hajnal, E. Szemeredi and G. Turan, A lower bound for read-once-only branching programs, *Journal of Computer and Systems Sciences*, vol. 35 (1987), 153–162.
3. A. Borodin, A.Razborov and R. Smolensky, On Lower Bounds for Read-k-times Branching Programs, *Computational Complexity* 3 (1993) 1 – 18.
4. J. A. Dias da Silva and Y. O. Hamidoune, Cyclic spaces for Grassmann derivatives and additive theory, *Bull. London Math. Soc.*, 26 (1994), 140–146.
5. P. E. Dunne, Lower bounds on the complexity of one–time–only branching programs, In *Proceedings of the FCT, Lecture Notes in Computer Science*, 199 (1985), 90–99.
6. S. Jukna, A Note on Read-k-times Branching Programs, *RAIRO Theoretical Informatics and Applications*, vol. 29, Nr. 1 (1995), pp. 75–83.
7. S. Jukna, Entropy of Contact Circuits and Lower Bounds on Their Complexity, *Theoretical Computer Science*, 57 (1988), pp. 113–129.
8. S. Jukna, A. A. Razborov, Neither Reading Few Bits Twice nor Reading Illegally Helps Much, TR96-037, ECCC, Trier.
9. E. A. Okolnishnikova, Lower bounds for branching programs computing characteristic functions of binary codes (in Russian), *Metody diskretnogo Analiza*, 51 (1991), 61–83.
10. E. A. Okolnishnikova, Comparing the complexity of binary k-programs (in Russian), Diskretnyj analiz i issledovanije operacij, 1995. Vol. 2, No. 4. pp. 54-73.
11. S. J. Ponzio, A lower bound for integer multiplication with read-once branching programs, *Proceedings of 27's Annual ACM Symposium on the Theory of Computing*, Las Vegas, 1995, pp. 130–139.
12. K. Prachar, *Distribution of Prime Numbers* (in German), *Primzahlverteilung*, Springer-Verlag, Berlin–Göttingen–Heidelberg, 1957.

13. P. Savický, S. Žák, A Lower Bound on Branching Programs Reading Some Bits Twice, to appear in TCS.

14. P. Savický, S. Žák, A Large Lower bound for 1-branching programs, TR96-036, ECCC, Trier.

15. D. Sieling, New Lower Bounds and Hierarchy Results for Restricted Branching Programs, TR 494, 1993, Univ. Dortmund, to appear in *J. of Computer and System Sciences*.

16. D. Sieling and I. Wegener, New Lower bounds and hierarchy results for Restricted Branching Programs, in *Proc. of Workshop on Graph-Theoretic Concepts in Computer Science WG '94*, Lecture Notes in Computer Science Vol. 903 (Springer,Berlin, 1994) 359 – 370.

17. J. Simon, M. Szegedy, A New Lower Bound Theorem for Read Only Once Branching Programs and its Applications, *Advances in Computational Complexity Theory* (J. Cai, editor), DIMACS Series, Vol. 13, AMS (1993) pp. 183–193.

18. I. Wegener, On the Complexity of Branching Programs and Decision Trees for Clique Functions, *JACM* 35 (1988) 461 – 471.

19. S. Žák, An Exponential Lower Bound for One-time-only Branching Programs, in *Proc. MFCS'84*, Lecture Notes in Computer Science Vol. 176 (Springer, Berlin, 1984) 562 – 566.

20. S. Žák, A superpolynomial lower bound for $(1, +k(n))$- branching programs, in *Proc. MFCS'95*, Lecture Notes in Computer Science Vol. 969 (Springer, Berlin, 1995) 319 – 325.

Routing with Finite Speeds of Memory and Network

Jop F. Sibeyn*

Abstract

On practical parallel computers, the time for routing a distribution of sufficiently large packets can be approximated by $\max\{T_f, T_b\}$. Here T_f is proportional to the maximum number of bytes a PU sends and receives, and T_b is proportional to the maximum number of bytes a connection in the network has to transfer. We show that several important routing patterns can be performed by a sequence of balanced all-to-all routings and analyze how to optimally perform these under the above cost-model. We concentrate on dimension-order routing on meshes, and assume that the routing pattern must be decomposed into a sequence of permutations. The developed strategy has been implemented on the Intel Paragon. In comparison with the trivial strategy, in which PU_i routes to $PU_{(i+t) \bmod P}$ in permutation t, $1 \leq t < P$, one gains between 10 and 20%.

1 Introduction

1.1 Importance of Balanced All-to-All Routing

On parallel computers, communication is essential: the processing units, PUs, need to exchange information packets to coordinate their activity (assuming that there is no common memory). This communication is performed over the connections of an interconnection network.

Many communication patterns have been considered, but there is one with an outstanding importance: *balanced all-to-all routing* (also called "balanced all-to-all personalized communication" [6]). In a balanced all-to-all routing on a network with P PUs, each PU PU_i, $0 \leq i < P$, initially holds P packets, $p_{i,j}$, $0 \leq j < P$ of size h. The task is to route these packets such that finally each PU PU_i, $0 \leq i < P$, holds all packets $p_{j,i}$, $0 \leq j < P$. In the remainder we will speak of *bata routing* and *bata distributions*.

The extraordinary importance of bata routing goes back on its role in BSP computations and in the simulation of PRAM algorithms. Suppose e.g. that each of the PUs simulates the operations of a large number H of PRAM PUs, and that one request has to be routed for each PU. Then on the average H/P requests have to be routed to each PU. If the pattern is more or less random, then the actual number hardly ever deviates from this by more than $7 \cdot \sqrt{H/P}$ (as follows from the Chernoff bounds). For $H/P \geq 10,000$, this is only a small fraction.

However, bata routing also has great importance for irregular routing problems. In the following we distinguish *permutation routing*, in which every PU is the source and destination of exactly one packet, and patterns where the PUs route to many destinations.

Permutation Routing. For most instances of permutation routing in practice, it will be best to rely on the hardware based router, and to send the packets directly. Sometimes, however, this may work very badly. For example, in the case of transpositions [1, 3], trivial routing leads to heavily congested connections. For some regular cases, one can design optimal schedules by hand, but this is tedious. Therefore, it is desirable to have a general purpose router, which routes all distributions within a minimal guaranteed amount of time. There are two basic approaches:

---

*Max-Planck-Institut für Informatik, Im Stadtwald, 66123 Saarbrücken, Germany. E-mail: jopsi@mpi-sb.mpg.de. WWW: http://www.mpi-sb.mpg.de/~jopsi/index.html.

1. Route all packets to intermediate destinations given by a globally chosen random permutation. Route all packets to their real destinations.

2. Divide all packets in P equal pieces. Route one piece from each PU to each PU. Route all pieces to their destination PUs.

Approach 1 is a variant of the algorithm by Valiant and Brebner [7]. Approach 2 has the advantage that no permutation has to be chosen, and that the routing is fixed and perfectly balanced, and can be performed in an optimal precomputed way. For packets that are so large, that the overhead for sending them away does not play a role, it will be faster than Approach 1.

General Distributions. For arbitrary distributions, there is not much to say without knowing the topology of the network. But, for the special case, that every PU sends and receives in total the same amount H of data, there are several approaches. Trivially sending all packets directly to their destinations may not work good in this case: if all packets are sent at the same time, the network may be flooded, and its performance may decrease. Therefore it is desirable to operate in rounds, such that in every round a permutation has to be routed. With more than one destination per PU and packets of different lengths, this is almost impossible to schedule efficiently. The same problem arises when one tries to generalize Approach 1.

A generalization of Approach 2 is far better. PU_i sends a fraction $1/P$ of $p_{i,j}$ to each PU; and all the data it sends to intermediate destination PU_j are bundled in one packet of length H/P. Thus, we find back a bata routing. The only difference with the application to permutation routing is, that now information about the lengths of the constituting parts must be sent along.

From the above we conclude that bata routing is the most suitable routing operation for all cases that each PU sends and receives approximately the same amount, even without the condition that all packets $p_{i,j}$ have the same length.

1.2 Approaches for All-to-All Routing

Theoretical Cost Model. The most common theoretical cost model, is to assume that in every step every PU can transfer one packet to each of its neighbors. Under this model, it is trivial to design optimal schedules for many regular interconnection networks, such as meshes and hypercubes. Here a d-dimensional $n \times \cdots \times n$ mesh consists of $P = n^d$ PUs, arranged in a grid in which adjacent PUs are connected. On meshes, packets can be routed to their destinations by rearranging them dimension by dimension. Because such an algorithm utilizes only the connections along one dimension at a time, d of them can be overlapped without delay. This leads to an algorithm running in exactly $h \cdot n^{d+1}/4$ steps. This is optimal, because $h \cdot n^{2 \cdot d}/4$ packets have to traverse in one direction over the n^{d-1} connections in a bisection of the network.

This Paper. The above given cost model means an oversimplification in most practical cases. We define a cost model that involves two parameters (measuring the capacity of the connections and the speed of the memory), which gives an accurate description for the case that the packet size h is sufficiently large. Under this model trivial strategies do not fully exploit the available hardware. We present a non-trivial bata-routing strategy for meshes, that assures optimal performance under our cost model. A strong feature of our strategy is that it assures optimality, whatever the basic parameters (speed of the memory and the connections) of the parallel computer are. Because there are no disadvantages connected to its use, it can be applied always. The only thing that depends on the parameters is the amount by which it is better than the default strategy, ranging from zero to almost 50%. On meshes of dimension two and higher, one may expect to gain at least 33% in most cases.

Previous Work. Recently bata routing has also been addressed by Stricker and Hardwick [6] (they also give references to more related literature). Their goals are the same, but their approach is very different. Their paper is mainly practical and focuses on improvements for bata routing on the Cray T3E. Our approach is mainly theoretical, and finally we show that it also makes sense in the context of the Intel Paragon. These are not two sides of the same page. We believe that our results may well outlive the Paragon. Much closer to our investigations is the work of Hambrusch e.a. [2] (based on Scott [4]). The considered problem is the same, and there is also an overlap in the solution. Nevertheless our paper still contains several interesting features:

- Applying a practically motivated cost-model, we give a quantitative analysis of the expected differences of several strategies for bata routing.

- Our approach is optimal for all side-lengths of the meshes.

In [2] many details fail. No motivating cost-model is provided, and therefore the chosen approach cannot be justified theoretically. Especially, the gain that can be expected cannot be quantified. Most important for practical applications, however, is that the schedules in [2] are not optimal for odd side-lengths n. Particularly for $n = 4 \cdot i + 1$, for some i, the difference in the routing time may be considerable (an example is given at the end of Section 5).

Contents. In the following section we define our cost model, motivate the decomposition into permutations and give a lower bound. Then, in Section 3 we analyze the trivial strategy. Hereafter we present the optimal strategy for one-dimensional processor arrays, and a generalization for higher dimensional meshes. Finally we describe our experiments on the Paragon. Due to a lack of space many important details had to be omitted. These are provided in the full version.

2 Preliminaries

Cost Model. In this paper we consider a cost model that gives a more accurate description of the routing time on practical systems than the store-and-forward model described above.

For the time to route a packet distribution, there are three factors that may be of importance: start-up time, T_s, feeding time, T_f, and bisection time, T_b. These are given by

$$T_s = a_s \cdot c_s, \qquad\qquad T_f = a_f \cdot c_f, \qquad\qquad T_b = a_b \cdot c_b.$$

Here c_s, c_f and c_b are constants. a_s gives the maximum number of packets a PU has to send away, a_f, the maximum sum of the sizes of the packets a PU has to send and receive and a_b the maximum number of bytes any connections has to transfer in one direction.

T_f depends on the speed of the memory and T_b on the capacity of the network. If the packets are large enough (how large, depends on the relative value of c_s), then the start-up time can be neglected (for small packets on the contrary, the start-up time determines the routing time). In that case, a reasonable approximation of the routing time for a distribution D is given by

$$T_{\text{route}}(D) = \max\{T_f, T_b\}. \tag{1}$$

Once again, this is only a simplified theoretical approximation of the cost in practice. On Parallel computers, the Intel Paragon for example, (1) does not hold exactly due to unexpected effects. For example, packets may obstruct each other in a fairly unpredictable way. To some extend the routing time is given by the sum of T_f and T_b.

Decomposition into Permutations. For bata routing, there is a trivial strategy: just send all packets away, relying on the hardware based router of the system. There are several reasons why this is not a good idea, and why it is preferable to decompose the routing into P permutations:

- In order to obtain efficient routing, a PU must issue a `receive` for every packet that is sent to it before it actually receives the packet. There may be upper-limits on the number of pending `receive`s, and handling them may become expensive if there are too many.

- In practice the throughput of a network is highest if it is not oversaturated. Thus, one should only route packets at the same time if it can be expected that this is faster than when they are routed separately.

- If only one packet of size h is sent at a time, we only need h buffer space. Sending more packets means that a larger part of the memory has to be reserved for buffers.

Thus, the all-to-all pattern should be split-up in several simpler patterns. The question is how many. Routing less than one permutation at a time possibly leaves much of the hardware unused. Routing more than one permutation at a time is not necessary, because from the following it will follow that such a decomposition in permutations can be constructed, that routing the permutations one-by-one is at worst as expensive as routing all of them at the same time.

Lower Bound. For comparison purposes we need a lower bound. A simple argument gives

Lemma 1 *On a d-dimensional mesh with side lengths $n_1 \leq n_2 \leq \cdots \leq n_d$ and $P = \prod_{j=1}^{d} n_j$ PUs in total, bata routing with packets of size h and costs according to (1), takes at least*

$$T_{min}(n_1, \ldots, n_d, h) = h \cdot P \cdot \max\{2 \cdot c_f, n_d/4 \cdot c_b\}.$$

In the remainder of this paper we will try to find an optimal decomposition of a bata distribution into P permutations. Here optimality is defined with respect to the cost model of (1).

3 Trivial Strategy

In [5] we applied the following trivial strategy:

for $t = 1$ **to** $P - 1$ **do**
    for all $0 < i < P$ **do in parallel**
        Route $p_{i,(i+t) \bmod P}$ from PU_i to $PU_{(i+t) \bmod P}$.

One-Dimensional Meshes. We analyze the trivial strategy for one-dimensional meshes. Let h be the size of the packets in bytes, and let a_b give the number of bytes a connection has to transfer in a single direction.

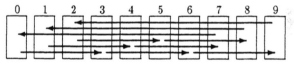

Figure 1: The routing pattern in Phase 3 of the trivial strategy on a one-dimensional processor array with 10 PUs. The connections in the center have to transfer three packets in each direction.

Lemma 2 *Consider a one-dimensional processor array with P PUs. In Phase t, $1 \leq t \leq P$, of the trivial strategy $a_b = h \cdot \min\{t, P - t\}$.*

Corollary 1 *Applying the trivial strategy for bata routing with packets of size h on a one-dimensional processor array with P PUs takes*

$$T_{triv}(P, h) = h \cdot \sum_{t=1}^{P-1} \max\{2 \cdot c_f, \min\{t, P - t\} \cdot c_b\}.$$

Evaluation and comparison with the lower bound of Lemma 1 shows that

Theorem 1 *On a one-dimensional processor array with P PUs, the trivial strategy is optimal, when either $2 \cdot c_f \leq c_b$, or $2 \cdot c_f \geq \lfloor P/2 \rfloor \cdot c_b$.*

Lemma 3 *On a one-dimensional processor array with P PUs, the time-consumption of the trivial strategy exceeds the lower bound by a factor $5/4$ for $2 \cdot c_f = P/4 \cdot c_b$.*

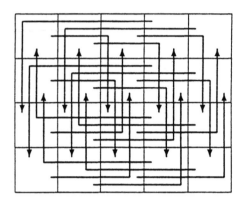

Figure 2: The routing pattern during Phase 12 of the trivial strategy on a 5 × 4 mesh.

Higher Dimensional Meshes. We consider how the trivial strategy performs for higher dimensional meshes. The PUs are indexed by the standard row-major indexing (this means that on an $n_1 \times \cdots \times n_d$ mesh, the PU at position $(x_1, \ldots, x_d)$ has index $\sum_{i=1}^{d} x_i \cdot \prod_{j=1}^{i-1} n_j$). On a two-dimensional mesh, the routing is a composition of two one-dimensional routings, see Figure 2 for an example. From this it follows easily that

Lemma 4 *Consider a d-dimensional mesh with side lengths $n_1, n_2, \ldots, n_d$ and $P = \prod_{j=1}^{d} n_d$, in Phase t, $1 \le t \le P$, of the trivial strategy*

$$a_b = h \cdot \max_{j=1}^{d} \{\min\{t_j, n_j - t_j\}\},$$

where the t_j are the numbers such that $\sum_{j=1}^{d}(t_j \cdot \prod_{i=1}^{j-1} n_i) = t$.

The maximum deviation from the lower bound increases with the dimension, and the trivial algorithm is not optimal anymore for weak networks:

Lemma 5 *On a two-dimensional $n \times n$ processor array, the time-consumption of the trivial strategy exceeds the lower bound by a factor of $4/3$ or more for all $2 \cdot c_f \le n/4 \cdot c_b$. For $2 \cdot c_f = n/4 \cdot c_b$, this factor equals $1^5/_{12}$.*

Proof: For $2 \cdot c_f = n/4 \cdot c_b$, both terms of the lower bound in Lemma 1 are equal: $h \cdot P \cdot n/4 \cdot c_b$. The time consumption during any phase can be computed by combining (1) and Lemma 4. Most practical is to compute for given t_2, the sum over the t_1, and then to sum over the t_2. Here t_1 and t_2 are defined to be the numbers such that $t_2 \cdot n + t_1 = t$. $\square$

Lemma 6 *On a d-dimensional $n \times \cdots \times n$ mesh, the deviation from the lower bound converges to a factor 2 in the limit for $d \to \infty$, for all $2 \cdot c_f \le n/4 \cdot c_b$.*

Proof: For $d \to \infty$, the average value over all $1 \le t < P$ of $\max_{j=1}^{d}\{\min\{t_j, n - t_j\}\}$ converges to $n/2$. Thus, the total time consumption converges to $P \cdot h \cdot n/2 \cdot c_b$, which is twice the lower-bound when $2 \cdot c_f \le n/4 \cdot c_b$. $\square$

Need for Refined Strategy. A powerful network is expensive. Thus, a manufacturer will carefully choose its parameters, and certainly not over-dimension it. For example, on the Paragon $c_f \simeq 1.1 \cdot 10^{-8}$ and $c_b \simeq 8.3 \cdot 10^{-9}$. This means that on partitions with maximal side-length at most 6, bisection effects are hardly noticeable, but for larger partitions, the performance of the trivial strategy starts to degrade. Generally, we may assume that for many practically relevant cases the trivial strategy performs far from optimal. For those cases, we present a refined strategy in the following sections.

4 Optimal Strategy for One-Dimensional Arrays

For a one-dimensional processor array with P PUs we must decompose the bata distribution into P permutations.[1] We will construct these permutations such that in each of them the number of packets going over any particular connection is the same. Namely, except for rounding, precisely the number of packets that must go over this connection divided by P.

4.1 Guiding Example

The case $P = 8$ gives an excellent example of what we intend to do in general.

from	to 0	1	2	3	4	5	6	7
0	0	1	2	3	4	5	6	7
1	7	0	1	2	3	4	5	6
2	6	7	0	1	2	3	4	5
3	5	6	7	0	1	2	3	4
4	4	5	6	7	0	1	2	3
5	3	4	5	6	7	0	1	2
6	2	3	4	5	6	7	0	1
7	1	2	3	4	5	6	7	0

from	to 0	1	2	3	4	5	6	7
0	0	1	2	3	4	5	6	7
1	4	6	7	5	0	1	2	3
2	5	2	3	1	7	4	0	6
3	6	7	0	4	2	3	5	1
4	3	5	6	2	1	7	4	0
5	7	0	5	6	3	2	1	4
6	2	4	1	7	6	0	3	5
7	1	3	4	0	5	6	7	2

Table 1: **Left:** The trivial decomposition of an all-to-all routing on a one-dimensional processor array of length eight into eight permutations. **Right:** An optimal decomposition of an all-to-all routing on a one-dimensional processor array of length eight into eight permutations.

The trivial schedule is represented in Table 1 on the left. The entries in the table indicate the phase in which a packet is routed from the PU indicated in the first row, to the PU indicated in the first column. For example, we see that PU_4 routes in Phase 6 to PU_2. That each PU sends and receives only one packet at a time, implies that all numbers should occur only once in every row and column, respectively.

From Table 1 we can immediately see the unbalanced character of the trivial strategy: in the indicated top-right quarter, there are four 4-s, and no 0-s. Without knowing the structure, this tells us that during Phase 4, four packets have to be routed over the bisection of the network, the connection from PU_3 to PU_4, and no packet during Phase 0.

Now consider our alternative in Table 1 on the right. Here all numbers $0, 1, \ldots, 7$ occur exactly twice in the 16 positions of the upper-right corner, indicating that in every Phase exactly two packets have to go over the bisection. Of course the bisection is not the only connection that may be heavily loaded, but it is the most critical one. For the other leftwards connections Table 1 indicates all critical sections. It is easy to check that in none of the seven indicated rectangles a number occurs more than twice. Because we are considering permutations, the rightwards connections are as heavily loaded as the leftwards connections.

What does this mean? Let us consider the cost of bata routing with the decomposition on the right side of Table 1. In each of the eight permutations a PU sends and receives at most one packet. The connections have to transfer at most two packets in each direction. Thus, the total cost is given by

$$T_{\text{opt}}(8, h) = 8 \cdot \max\{2 \cdot c_f, 2 \cdot c_b\} \cdot h.$$

This just matches the lower bound of Lemma 1. Thus, a single balanced decomposition gives optimal routing for *all* c_f and c_b.

---

[1] Actually, it is slightly faster to decompose into $P - 1$ permutations and the identity, but with an eye on our generalizations to higher dimensional meshes in Section 5, it is better to work with P permutations from the start.

4.2 Balanced Decomposition

Inspired by the above example for the case $P = 8$, we formulate our goal:

Goal 1 *For routing on a one-dimensional processor array with P PUs, decompose an all-to-all distribution into P permutations, such that in none of them a connection has to transfer more than $\lceil P/4 \rceil$ packets.*

We will say that the permutations are *balanced*. This is indeed what we are looking for:

Theorem 2 *If bata routing on a one-dimensional processor array with P PUs is performed by a sequence of P balanced permutations, then the routing is optimal except for rounding errors.*

Proof: $\qquad\qquad T_{\text{opt}}(P, h) \leq P \cdot \max\{2 \cdot c_f, \lceil P/4 \rceil \cdot c_b\} \cdot h.$ $\qquad\qquad\square$

It turns out to be really easy to construct balanced decompositions. For $P \leq 14$, we have constructed them by hand. Two examples are given in Table 2. Actually, the constructed decompositions are better than required. Not only is, for routing on an array of length P, the load on any connection bounded by $\lceil P/4 \rceil$ in every permutation, but also is the sum of the maximum loads over all P permutations exactly $\lfloor P/2 \rfloor \cdot \lceil P/2 \rceil$, as it should be.

		P	$=$	7		
0	1	2	3	4	5	6
3	2	4	0	6	1	5
1	3	5	4	0	6	2
4	5	3	6	1	2	0
6	0	1	2	5	4	3
2	4	6	5	3	0	1
5	6	0	1	2	3	4

				P	$=$	9		
0	1	2	3	4	5	6	7	8
6	4	5	1	3	7	0	8	2
5	3	7	8	2	1	4	0	6
1	7	3	6	8	0	2	5	4
2	6	1	0	7	3	8	4	5
7	5	6	4	0	8	1	2	3
3	0	8	2	1	4	5	6	7
4	8	0	5	6	2	7	3	1
8	2	4	7	5	6	3	1	0

Table 2: Optimal decompositions for one-dimensional processor arrays of length 7 and 9.

For larger P, one should have an algorithm that constructs decompositions. The whole problem can be transformed into a maximum-flow problem, but a heuristic with some backtracking works much faster. We suggest to fill the rectangles indicated in Figure 3 as follows:

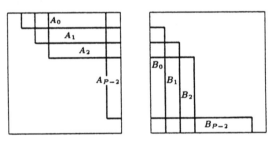

Table 3: The rectangles A_j (right) and B_j (left) that are used in DECOMPOSE.

Algorithm DECOMPOSE

1. **for** $j = 0$ **to** $P - 2$ **do**
   Fill A_j up with $P - j - 1$ numbers such that afterwards
   all numbers occur either $\lfloor \#A_j/P \rfloor$ or $\lceil \#A_j/P \rceil$ times.

2. **for** $j = 0$ **to** $P - 2$ **do**
   Fill B_j up with the $P - j - 1$ numbers from $A_j - B_j$,
   such that each number occurs at most once in every row.

3. **for** $j = 0$ **to** $P - 1$ **do**
   Determine the number that does not yet occur in Row j
   and Column j and write it on the diagonal position.

Here $\#A_j$ denotes the area of A_j. In Step 1, $\#A_j = (j + 1) \cdot (P - j - 1)$. One should choose an arrangement such that the numbers that come to stand on the left fringe of A_j (those positions that do not belong to A_{j+1}) are among the numbers in A_j with maximum frequency. In Step 2, by $A_j - B_j$ we mean the difference of two multisets. One should first place the numbers that have the smallest number of rows they can be allocated to, and place them as high as possible. One may have to do some back-tracking, but our experience shows that apparently the conditions are so loose that this happens rarely. Once all steps have been performed as specified, a balanced decomposition has been constructed: because of the specification in Step 1, in none of the rectangles A_j a number occurs more than $\lceil \#A_j/P \rceil$ times. As $\#A_j < P^2/4$, we have $\lceil \#A_j/P \rceil \leq \lceil P/4 \rceil$, as required.

5 Higher Dimensional Meshes

The extension to higher dimensional meshes is surprisingly simple. Consider an $n_1 \times n_2 \times \cdots \times n_d$ mesh, and let $P = \prod_{j=1}^{d} n_j$. Then we perform

1. Initially, each PU_i, $0 \leq i < P$, determines the numbers i_j, $1 \leq j \leq d$ such that $\sum_{j=1}^{d} (i_j \cdot \prod_{k=1}^{j-1} n_k) = i$: it determines its position along each of the coordinate axes.

2. Then each PU performs P routing phases. During some Phase $t = \sum_{j=1}^{d} (t_j \cdot \prod_{k=1}^{j-1} n_k)$, $0 \leq t < P$, PU_i, $0 \leq i < P$, determines the numbers i'_j, $1 \leq j \leq d$, where i'_j is the index of the PU to which PU_{i_j} would send during Phase t_j of bata routing on a one-dimensional processor array with n_j PUs; computes $i' = \sum_{j=1}^{d} (i'_j \cdot \prod_{i=1}^{j-1} n_i)$; and sends $p_{i,i'}$ to $PU_{i'}$.

Setting $i'_j = (i_j + t_j) \mod n_j$, we find back the trivial strategy (though for that case the formulation is unnecessarily complicated). Look at Figure 2, we see that in Phase 12, in all rows the routing is as in Phase 2 of the trivial strategy on a one-dimensional array with 5 PUs. In all columns the routing is as on a one-dimensional array with 4 PUs. Now again, the routings in all rows and columns are the same, but different, better than before.

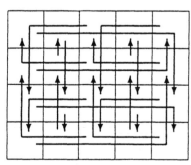

Figure 3: The routing pattern during Phase 12 of the optimal strategy on a 5×4 mesh.

In Figure 3 we give the example that corresponds to the one in Figure 2. To compute this we have to know how the routing on one-dimensional arrays with 4 and 5 PUs is performed. But, for Phase 12, we must only know the routing in Phase 2 for both networks:

$$P = 4, \text{ Phase } 2 : \quad (0, 1, 2, 3) \quad \mapsto \quad (2, 0, 3, 1),$$
$$P = 5, \text{ Phase } 2 : \quad (0, 1, 2, 3, 4) \quad \mapsto \quad (2, 1, 4, 3, 0).$$

Now we come to the main theorem of this paper:

Theorem 3 *Consider bata routing on an $n_1 \times n_2 \times \cdots \times n_d$ mesh. If we have balanced decompositions of all-to-all distributions for all n_j, $1 \leq j \leq d$, then the routing with the above scheme is optimal with respect to (1) except for rounding errors.*

Proof: The scheme is such that the routing in each one-dimensional subarray of size n_j is precisely the routing according to the balanced decomposition for a one-dimensional array of size n_j. This means that none of the connections in this subarray has to transfer more than $\lceil n_j/4 \rceil$ packets. Thus, assuming that $n_d \geq n_j$, for all $1 \leq j \leq d$, we conclude that globally no connection has to transfer more than $\lceil n_d/4 \rceil$ packets. Hence, each of the P phases costs $h \cdot \max\{2 \cdot c_f, \lceil n_d/4 \rceil \cdot c_b\}$ time. Multiplying by P gives the lower-bound of Lemma 1, except for the rounding-up. □

The decompositions in [2] may require one extra routing phase. For example, routing on a one-dimensional processor array with 5 PUs is performed by a sequence of six partial permutations. This means that for routing on a 5×5 mesh, 36 operations instead of 25 are needed. Thus, for relatively large c_f the schedule of [2] is almost 50% slower than ours.

6 Experiments

Description. We have written a simple program for the Intel Paragon, that performs the trivial strategy and the optimal strategy. For the optimal strategy we have utilized the constructed balanced decompositions. given in the appendix. In order to increase the accuracy of the measurements, every measurement was preceded by an identical dummy round. The measurements were made for 100 repeated routing with $h = 10^6$ bytes. This is quite extreme, but it exposes most clearly the difference between the strategies.

The Paragon is a mesh connected parallel computer. The access to the memory can be performed at about 90 MB / s, while the connections can transfer about 120 MB / s bidirectionally:

$$c_f \simeq 1.1 \cdot 10^{-8} \qquad\qquad c_b \simeq 8.3 \cdot 10^{-9}.$$

For the considered large packets, start-up time can be neglected. The Paragon which we have access to, has size 10×14, minus two PUs in the lower-right corner. We have tested a large number (about 80) different choices of height and width.

One-Dimensional Arrays. We tested one-dimensional arrays of length P up to 14. We found that for $P \leq 8$, the trivial strategy was a few percent faster. This will be caused by the fact that the trivial strategy consists of $P - 1$ sending rounds only. Then the optimal strategy becomes clearly better. For $P = 13$, it is 20% faster. Results are given in Table 4.

P	8	9	10	11	12	13	14
$Rate_{triv}$	90	90	86	88	94	85	90
$Rate_{opt}$	90	98	99	100	101	103	97
Difference in %	-2	8	16	14	14	20	14

Table 4: Results for routing on one-dimensional arrays. $Rate_{triv}$ and $Rate_{opt}$ indicate the number of MB / s that are transferred over the bisection.

We see that as soon as the bisection gets critical, for $P \geq 9$, the optimal strategy gets more out-of the bisection than the trivial one. The difference corresponds quite well to the theoretical possibilities. Further improvements can be achieved by minimizing the length of the cycles inside the permutations (a permutation that consists of $P/2$ exchanges is considerably faster than one cycle of length P). As this only has a positive effect for routing on one-dimensional arrays, we did not optimize for this criterion.

n	1	2	3	4	5	6	8	10
$Rate_{triv}$	85	79	79	75	77	73	69	65
$Rate_{opt}$	103	99	95	96	96	84	81	75
Difference in %	20	26	26	28	25	15	17	15

Table 5: Results for routing on $n \times 13$ partitions.

Two-Dimensional Arrays. For two-dimensional partitions, it becomes clearer that our simple cost-estimate does not reflect all the machine-specific details. Theoretically, the transfer rate of the bisection should be the same for a 4×13 partition as for a 1×13 partition: on the horizontal connections no congestion is to be expected. But, in practice this is not exactly true. It appears that packets that get blocked in their move along the columns, continue to occupy connections along the rows, which in turn may block packets along the columns. This is a serious effect, leading to a considerable deterioration of the performance.

The largest difference between the strategies occurs for a 8×9 partition: 29% (transfer-rates of 67 and 86 MB / s, respectively). Complete results for partitions of height 13 are given in Table 5. These are representative. Generally, for larger partitions the difference appears to stabilize around 15%. This is half of what could be expected. The following explanation seems plausible. In our analysis, we consider congestion in one direction to be the same as congestion in two directions. In practice, congestion in two directions is worse (but not twice as bad) as congestion in one direction. Thus, we have over-estimated our possible gain.

7 Conclusion

Under a praxis-inspired, but universal cost model we have designed an optimal routing strategy. The strategy assures optimality, independently of the parameters of the hardware. For two-dimensional meshes, theoretically, the improvement over a trivial strategy is at least 33% for the most relevant parameters. The strategies have been implemented on the Paragon. In practice the optimal strategy is indeed considerably better than the trivial strategy, but the difference is smaller than predicted. This is mainly due to the fact that our cost-model does not comprise all features of the actual costs on the Paragon. As the goal of this paper was to present a machine-independent approach we have not tried to optimize the performance on the Paragon. Our approach might be an excellent starting point for machine-specific optimizations.

Acknowledgement. This paper profited from discussions with Uli Meyer. The implementations were performed on the Paragon at the KFA in Jülich.

References

[1] Ding, K.-S., C.-T. Ho, J.-J. Tsay, 'Matrix Transpose on Meshes with Wormhole and XY Routing,' *Proc. 6th Symposium on Parallel and Distributed Processing*, pp. 656–663, IEEE, 1994.

[2] Hambrusch, S.E., F. Hameed, A. Khokhar, 'Communication Operations on Coarse-Grained Architectures,' *Parallel Computing*, 21, pp. 731–751, 1995.

[3] Kaufmann, M., U. Meyer, J.F. Sibeyn, 'Matrix Transpose on Meshes: Theory and Practice,' *Proc. 11th International Parallel Processing Symposium*, IEEE, 1997, to appear.

[4] Scott, D.S., 'Efficient All-to-All Communication Patterns in Hypercube and Mesh Topologies,' *Proc. 6th Distributed Memory Computing Conference*, pp. 398–403, 1991.

[5] Sibeyn, J.F., F. Guillaume, T. Seidel, 'Practical Parallel List Ranking,' *Proc. 4th Symposium on Solving Irregularly Structured Problems in Parallel*, LNCS, Springer-Verlag, 1997. To appear.

[6] Stricker, T.M., J.C. Hardwick, 'From AAPC Algorithms to High Performance Permutation Routing and Sorting,' *Proc. 8th Symp. on Parallel Algorithms and Architectures*, pp. 200–203, ACM, 1996.

[7] Valiant, L.G., G.J. Brebner, 'Universal Schemes for Parallel Communication,' *Proc. 13th Symposium on Theory of Computing*, pp. 263–277, ACM, 1981.

Queries and Algorithms
Computable by Polynomial Time
Existential Reflective Machines*

Extended Abstract

Jerzy Tyszkiewicz

Mathematische Grundlagen der Informatik,
RWTH Aachen, Ahornstraße 55,
D-52074 Aachen, Germany.
e-mail jurek@informatik.rwth-aachen.de

Abstract. We consider two kinds of reflective relational machines: the usual ones, which use first order queries, and existential reflective machines, which use only first order existential queries. We compare these two computation models. We build on already existing results for standard relational machines, obtained by Abiteboul, Papadimitriou and Vianu [1], so we prove only results for existential machines.

First we show that for both standard and existential reflective machines the set of polynomial time computable Boolean queries consists precisely of all $\mathcal{PSPACE}$ computable queries.

Then we go farther and compare which classes of algorithms both kinds of machines represent. Unless $\mathcal{PSPACE} = \mathcal{P}^{\mathcal{NP}}$, there are $\mathcal{PSPACE}$ queries which cannot be computed by polynomial time existential reflective machines which use only polynomial amount of relational memory, while it is possible for standard reflective machines. We conclude that existential reflective machines, being equivalent in computational power to unrestricted machines, implement substantially worse algorithms than the latter.

Concerning deciding $\mathcal{P}$ classes of structures, every fixpoint query can be evaluated by a polynomial time unrestricted reflective machine using constant number of variables, while existential reflective machines need $\lfloor n/2 \rfloor$ variables to implement the graph connectivity query. So again the algorithms represented by existential reflective machines are worse.

1   Introduction

1.1   Class of queries vs. class of algorithms

Many formalisms for representing queries and query languages have been considered and compared with each other. Two most prominent examples are the

---

* This research has been supported by a Polish KBN grant 8 T11C 002 11 and by the German Science Foundation DFG.

relational algebra and the predicate calculus, which I prefer to call first order logic. It is one of the early discovered and still most important facts of the database theory that these two formalisms express precisely the same class of queries. And this is typically the first question one asks when comparing two such formalisms: what are the classes of queries they represent?

However, every formalism comes typically with its own "intended" semantics, and therefore represents not only a class of queries, but a class of algorithms for computing queries, as well. Here the relational algebra and first order logic are not identical any more. As it is well-known, for the purpose of query optimization algebra behaves much better. So it seems that it is somehow easier to represent efficient algorithms in the algebra. There are even much more drastic examples of similar differences. One of them is the result of Suciu and Paredaens [6] that any algorithm for transitive closure query written in certain complex object algebra must use the powerset operator, and thus requires exponential time and space in the intended semantics. Another example are two kinds of structural recursion of Suciu and Wong [7], which, if added to the relational algebra for complex objects, represent precisely the same class of queries, but one of them appears to represent much better algorithms than the other does.

Thus we see that the issues of representing a class of queries and a class of algorithms, which compute these queries, can be different. In this paper we want to discuss yet another example of the mismatch class of queries vs. class of algorithms.

1.2 Reflective machines

Description of the machines. We define now the *reflective relational machines* in the sense of Abiteboul, Papadimitriou and Vianu [1], which is in turn an extension of the model of *loosely coupled relational machines* of Abiteboul and Vianu [2].

The set Databases of databases is understood in this paper to be identical with the set of all finite relational structures, written typically as $\mathbb{A}, \mathbb{B}, \ldots$; the corresponding Latin letters $A, B, \ldots$ stand for the universe sets of the structures, which are always initial segments of $\mathbb{N}$.

Let L be a query language, i.e., a recursive set of (partial) recursive functions $p :$ Databases $\rightarrow$ Databases such that for any isomorphism $\alpha : \mathbb{A} \rightarrow \mathbb{B}$ of two databases holds $\alpha(p(\mathbb{A})) = p(\alpha(\mathbb{A}))$. This condition is called *genericity*.

A *reflective relational L-machine* consists of a standard deterministic Turing machine component, including finite control and several work tapes. In addition it has a *relational store*, consisting of infinitely many relations R_i, $i = 1, 2, \ldots$ of variable arities over some fixed finite set. These relations are provided as an input for the machine. One of the work tapes is distinguished to be the *L-query tape*, on which the machine can write an arbitrary L-query of any finite subsignature of $R_1, R_2, \ldots$, followed by $\#$ and a natural number. Upon entering a special state ex the query is evaluated in one step on the actual contents of the store, and the result is placed in the relation whose number is the second part of contents of the query tape. In case the second number is 0, the query on the tape should be a sentence. Then the output (*true* or *false*), which is obtained analogously in one step, is used to determine the next state of the machine, along with the symbols seen by heads on the standard tapes. Creation and executing queries at runtime is called *reflection*, and hence the name of the machines.

The initial configuration of the machine is as usually, where the relational store contains the input: it can be an arbitrary database $\mathbb{A}$ of a fixed signature $(R_1, \ldots, R_m)$. The remaining relations $R_{m+1}, R_{m+2}, \ldots$ are set initially empty and will play the rôle of work registers. If a machine uses only Boolean queries, we call it *update-free*.

We can now easily define, what it means, that a reflective *L*-machine M computes a total function f : Databases $\rightarrow$ N or a total query p : Databases $\rightarrow$ Databases, where, recall, databases are identified with finite relational structures.

The following proposition follows directly from the definition of reflective machines and from the fact that queries are (by definition) generic.

Proposition 1. *If a function* f : Databases $\rightarrow \{0,1\}^*$ *is computable by an L-reflective machine and* $\mathbb{B} \cong \mathbb{A}$, *then* $f(\mathbb{A}) = f(\mathbb{B})$.

In particular, if a class $X \subseteq$ Databases *can be decided by an L-reflective machine, then it is isomorphism-closed, i.e., if* $\mathbb{A} \in X$ *and* $\mathbb{B} \cong \mathbb{A}$, *then* $\mathbb{B} \in X$. $\square$

The condition appearing in the proposition above is a very important one, therefore we introduce a special terminology for it, in a slightly more general setting.

Definition 2. A subset X of Databases is said to be a *class of structures* if it is closed under isomorphisms.

A class X is said to *be in a complexity class* $\mathcal{C}$ iff there exists a Turing machine, which decides X, given codes of structures as inputs. Those codes are defined to be concatenations of the rows of the adjacency matrix for graphs, and similarly for structures of other signatures. This means, that the machine must give the same answer for all possible codes of isomorphic structures.

Motivation. The authors of [1] considered reflective machines as a model of a massively parallel computing device. We believe, however, that it is even a better model of a situation encountered in databases accessed through a network, where users use reflection to compensate for the limitations of the query language—they compute queries not supported in the system by submitting sequences of simpler queries and analyzing their results. In this scenario varying the query language used by machines is quite natural, because it reflects the diversity of database systems accessible over networks. Let us consider the following example: we want to determine the author's Erdős number, i.e., the distance from Paul Erdős to J.T. in the graph consisting of all ever active mathematicians as vertices, where the edge relation is "x and y have published a joint paper". There are sources of all the necessary information, which can be accessed in the WWW, like the databases of *Zentralblatt für Mathematik* and *Mathematical Reviews*. But they do not allow formulating recursive queries, like the Erdős number query. The only way to circumvent this difficulty is to evaluate a large number of simple queries in the database, and determine the answer from their results. This is precisely the kind of computations modeled by reflective machines. Further argumentation can be found in author's [8,9].

We believe this model of computation is of practical relevance, because, no matter whether owners of Internet accessible databases wish it or not, their users will practice reflective querying to compute queries which are not provided in the system—and most likely they already do it.

Cost functions. There are two natural cost functions for algorithms implemented on reflective machines: one of them assumes evaluating queries at unit cost, and reflects the egotistic point of view of the remote user accessing the database. The other takes the cost of query evaluation into account and essentially reflects the point of view of the database server. We call this complexity the server complexity, in contrast to the former, which will be called the user complexity. Finally, in the paper we consider the intrinsic complexity of queries that are computed. In this situation we mean the complexity of recognizing classes of structures by standard Turing machines, as in Definition 2.

In most of the situations we speak just about complexity, when the context uniquely determines which one of the three above ones we mean.

Machine classes to be compared. In this paper we consider mainly two query languages: first order logic FO and semi-positive existential first order logic EFO. The first query language is widely known, the latter is a fragment of existential first order logic, in which only input relations $R_1, \ldots, R_m$ can be negated in queries, but they cannot be updated. This applies to the equality relation, as well. All remaining relations can be updated, but must appear unnegated in queries.

We call the reflective relational FO-machines (*standard*) *reflective machines*, and the reflective relational EFO-machines *existential reflective machines.*

Our two formalisms to be compared are reflective relational machines and existential reflective machines, which have polynomial time user complexity. We believe these class corresponds reasonably well to the sequences of queries that might be issued in practice. Then we look at what happens on the database server side, i.e., what is the server complexity of algorithms represented by such polynomial time reflective querying.

2 The results

2.1 Classes of computable queries

First we state the results obtained by Abiteboul, Papadimitriou and Vianu [1], which tell us about the abilities of standard reflective machines.

Theorem 3. *1. Every $\mathcal{PSPACE}$ class of finite structures can be decided by an update-free polynomial time reflective machine.*

2. *Every polynomial time reflective machine decides a $\mathcal{PSPACE}$ class of structures.* □

Our first result compares the power of existential reflective machines with their standard counterparts in absence of additional restrictions.

Theorem 4. *Polynomial time existential reflective machines can decide precisely all $\mathcal{PSPACE}$ classes of finite structures.*

Proof. One direction of the theorem follows trivially from Theorem 3, since an existential reflective machine is a standard reflective machine, as well, so the set of structures it accepts must be in $\mathcal{PSPACE}$.

In the proof of the other direction we show that for every polynomial time update-free standard reflective machine there is a polynomial time existential reflective machine, which recognizes the same class of structures. And our only real concern is how to simulate evaluation of an FO sentence using EFO queries.

We do it as follows: whenever the machine to be simulated attempts to evaluate an FO sentence, the simulating machine rewrites it first to the prenex normal form $Q_t x_t \ \ldots \ Q_2 x_2 \ Q_1 x_1 \ \varphi(x_1, x_2, \ldots, x_t)$ with $Q_i \in \{\forall, \exists\}$, and such that all negations in the quantifier-free matrix φ are applied to atoms. (Note the reversed indexing of quantifiers.) Since the simulated machine is update-free, these negations are thus applied to input relations and equalities, only. Initially our machine goes on asking Boolean queries $\exists x_1 \ldots \exists x_n \bigwedge_{i \neq j} x_i \neq x_j$ for $n = 1, 2, \ldots$, until it finds the first value of n for which the answer is *false*, determining thereby the cardinality of its input $\mathbb{A}$. Now the simulating machine runs the following algorithm, in which relations P_i of the relational store satisfy $\mathrm{arity}(P_i) = i$. In particular, P_0 is a Boolean flag.

Algorithm A.

input $Q_t x_t \ \ldots \ Q_2 x_2 \ Q_1 x_1 \ \varphi(x_1, x_2, \ldots, x_t)$;
       update P_t with $\varphi(x_1, x_2, \ldots, x_t)$;
       for $i = 1$ **to** t **do**
          if $Q_i = \forall$

           then update P_{t-i} with $\exists y_1 \ldots \exists y_n \left(\begin{array}{c} \bigwedge\limits_{j \neq k} y_j \neq y_k \\ \bigwedge\limits_{j=1}^{n} P_{t-i+1}(\frac{y_j}{x_i}, x_{i+1}, \ldots, x_t) \end{array} \right)$
           else update P_{t-i} with $\exists x_i P_{t-i+1}(x_i, x_{i+1}, \ldots, x_t)$
          fi
      od;
output P_0.

The idea behind the algorithm is indeed quite simple. If we can address each element of the universe with a separate existentially quantified variable, we can transform every FO sentence into an existential one, replacing universal quantifiers by conjunctions over all elements of the structure. This procedure causes an exponential blow-up of the size of the sentence. To avoid this we save the intermediate results in this process in the relational store and then call them in the next step of the transformation under the name of the store relation. This prevents the syntax from blowing up, at the cost of creating temporary results of superpolynomial size, because a relation of arity s requires space n^s to be stored, and we cannot bound s by any constant. Note a clear indication of a conflict of interests between the user and the server.   □

2.2   Algorithms for $\mathcal{PSPACE}$ queries

On the face of the problem, polynomial time existential machines and polynomial time standard reflective machines are equivalent, because they compute precisely the same queries. But the situation is more interesting, as the following result shows.

Theorem 5. *1. Every $\mathcal{P}^{\mathcal{NP}}$ class of finite structures can be decided by an update-free polynomial time existential reflective machine.*
  2. Every polynomial time existential reflective machine which updates only relations of bounded arity decides a $\mathcal{P}^{\mathcal{NP}}$ set of structures.

Proof. The proof of Item 1 is presented for directed graphs. It can be easily extended to arbitrary relational structures. We begin with a purely complexity-theoretic consideration. Let a finite directed graph $\mathbb{G}$ be encoded by a binary string code($\mathbb{G}$), resulting from concatenating rows of the adjacency matrix of $\mathbb{G}$, exactly as in Definition 2. Now let canon($\mathbb{G}$) = $\max_{\mathbb{G}'}\{\text{code}(\mathbb{G}') \mid \mathbb{G}' \cong \mathbb{G}\}$, where we take max with respect to the lexicographic ordering. canon($\mathbb{G}$) is thus a binary string of length n^2, determining the *isomorphism type* of $\mathbb{G}$.

Let L_{canon} be the following language:

$$L_{\text{canon}} = \{\text{code}(\mathbb{G})\#w \mid w \in \{0,1\}^*, \exists \mathbb{G}' \cong \mathbb{G}\ \exists v \in \{0,1\}^*\ \text{code}(\mathbb{G}') = w1v\}.$$

It is immediate from the definition that $L_{\text{canon}} \in \mathcal{NP}$. Now consider the following algorithm:

Algorithm B.

input code($\mathbb{G}$);
$\quad w := \epsilon$;
$\quad\quad$ **for** $i = 0$ **to** $|\mathbb{G}|^2$ **do**
$\quad\quad\quad$ **if** code($\mathbb{G}$)$\#w \in L_{\text{canon}}$ **then** $w := w1$ **else** $w := w0$ **fi**
$\quad\quad$ **od**;
output w.

It is easy to see, that this algorithm computes the function code($\mathbb{G}$) $\mapsto$ canon($\mathbb{G}$). This confirms the known fact that graph canonization is indeed in $\mathcal{P}^{\mathcal{NP}} = \Delta_2^p$, what has been first proven by Blass and Gurevich [3].

Now we are turning to the construction of an update-free existential reflective machine M, which will decide a given class of structures $X \in \mathcal{P}^{\mathcal{NP}}$.

In the initial phase, exactly as in the previous proof, our machine determines the cardinality n of its input structure $\mathbb{G}$. The next task is to implement Algorithm B on an existential update-free reflective machine, in order to show that the machine we construct will be able to determine $\mathrm{canon}(\mathbb{G})$, where $\mathbb{G}$ is its own relational input. In order to achieve this, it suffices to show how our machine can determine the membership of $\mathrm{code}(\mathbb{G})\#w$ in L_{canon}, for $w = w_1 \ldots w_{|w|}$. This is done by the Boolean query

$$
\exists x_1 \ldots \exists x_n \left(\begin{array}{l} \bigwedge\limits_{i \neq j} x_i \neq x_j \\[4pt] \bigwedge\limits_{\substack{1 \leqslant i \leqslant |w| \\ w_i = 1}} R(x_{i \ (\mathrm{mod}\ n)}, x_{\lfloor i/n \rfloor}) \\[4pt] \bigwedge\limits_{\substack{1 \leqslant i \leqslant |w| \\ w_i = 0}} \neg R(x_{i \ (\mathrm{mod}\ n)}, x_{\lfloor i/n \rfloor}) \\[4pt] R(x_{|w|+1 \ (\mathrm{mod}\ n)}, x_{\lfloor (|w|+1)/n \rfloor}) \end{array} \right).
$$

The sentence above asserts that there is an ordering of the elements of the universe (according to the indices of variables) such that under this ordering the concatenation of rows of the adjacency matrix of $\mathbb{G}$ agrees with w and the next digit following the copy of w in this concatenation is 1.

Since the class X we want to decide is in $\mathcal{P}^{\mathcal{NP}}$, there is a polynomial time Turing machine M with an $\mathcal{NP}$ oracle, which decides whether $\mathbb{G} \in X$, given $\mathrm{canon}(\mathbb{G})$ as input. After two previous steps our machine has already determined $\mathrm{canon}(\mathbb{G})$. Now it can simulate the behavior of M in the standard way, provided that we can show how to implement calls to the oracle by using Boolean queries to the relational store.

Because the machine is able to compute polynomial time reductions on its tapes, it is enough to show that it can solve some fixed $\mathcal{NP}$-complete problem by calls to the oracle. We choose the Hamiltonian path problem. Let n^k be the bound of the size of graphs $\mathbb{G} = \langle V, E \rangle$ which have to be tested for Hamiltonicity.

Let x, y denote vectors of variables of length k. Their equality and inequality should be understood vector-like, e.g., $x \neq y$ as $\bigvee_{i=1}^{k}(x)_i \neq (y)_i$. Now the test is performed by the following query.

$$
\exists x_1 \ldots \exists x_{|V|} \exists y_1 \ldots \exists y_{|V|} \left(\begin{array}{l} \bigwedge\limits_{i \neq j} x_i \neq x_j \\[4pt] \bigwedge\limits_{i \neq j} y_i \neq y_j \\[4pt] \bigwedge\limits_{i} \bigvee\limits_{j} x_i = y_j \\[4pt] \bigwedge\limits_{i < |V|} \bigvee\limits_{(j,k) \in E} y_i = x_j \wedge y_{i+1} = x_k \end{array} \right).
$$

The idea behind the above sentence is that elements witnessing $x_1, \ldots, x_{|V|}$ constitute the universe of $\mathbb{G}$ (first condition), the y's are a permutation of x's (second and third condition). The last condition expresses that under this permutation there is an edge between every two consecutive vertices.

The proof of Item 2 is omitted due to space limitations. □

Corollary 6.

$$server\text{-}\mathcal{PSPACE} = \mathcal{PSPACE} \iff \mathcal{P}^{\mathcal{NP}} = \mathcal{PSPACE}$$

holds in the world of polynomial user-time existential reflective machines. □

The conclusion is that existential reflective machines, despite the fact that they are of the same expressive power as their unrestricted counterparts, implement substantially worse algorithms.

2.3 Algorithms for $\mathcal{P}$ queries

Concerning recognizing classes of structures which are in $\mathcal{P}$, the situation is even more puzzling. To discuss this issue, we introduce a definition.

Definition 7 (Gurevich [5]). A complexity class $\mathcal{C}$ of classes of structures is said to be *recursively presentable* iff there is a recursively enumerable set T of Turing machines which obey the resource bounds of $\mathcal{C}$ and such that any class $X \in \mathcal{C}$ is recognized by at least one machine from T.

Another way to express it is to say that *there is a logic for* $\mathcal{C}$, or that there is a logic which *captures* $\mathcal{C}$.

The positive side of the problem is as follows.

Theorem 8. *For any recursively presentable class $\mathcal{C} \subseteq \mathcal{P}$ there exists an optimization technique of EFO queries such that all classes of structures in $\mathcal{C}$ can be recognized by polynomial server-time algorithms on existential reflective machines.*

Proof. Omitted. □

As a corollary, several well-known recursively presentable subclasses of $\mathcal{P}$, like fixpoint logic or $\mathcal{P}$ on ordered structures, admit polynomial server-time algorithms on existential reflexive machines.

On the other hand, one immediately asks if $\mathcal{P}$ itself is recursively presentable, repeating the question of Gurevich [5]:

Problem. *Is there a logic for $\mathcal{P}$?*

In fact, already Gurevich introducing his problem conjectured the answer be "no". On the other hand, proving the Gurevich Conjecture in affirmative would immediately imply separation of $\mathcal{P}$ from $\mathcal{NP}$, because there is are logics for $\mathcal{NP}$, e.g., according to Fagin's Theorem $\Sigma_1^1 = \mathcal{NP}$, see [4]. So there seems to be not much hope for polynomial server-time algorithms for all $\mathcal{P}$ classes of structures.

Being unable to determine the time consumption of both kinds of reflective machines we wan to compare, we use a finer measure of the quality of algorithms, the number of variables they need. We show now that some apparently very simple polynomial time queries can be implemented only by relatively complex algorithms in the model of existential reflective machines, unlike in the unrestricted model, for which the following holds.

Theorem 9 (Abiteboul and Vianu [2]). *Every query definable by a formula of fixpoint logic can be computed by a polynomial time reflective machine which uses exactly as many variables as there are in the defining formula.* □

This should be contrasted with the following result for existential reflective machines, which holds for all such machines, not only for the polynomial time ones.

Theorem 10. *Any existential reflective machine determining whether its input graph is connected must use $\lfloor n/2 \rfloor$ variables for inputs of size n.*

Proof (Sketch). Consider the following two graphs $\mathbb{G}_m$ and $\mathbb{H}_m$. $\mathbb{G}_m$ is a directed tree of depth 2, whose root is of out-degree m, all children of root are of out-degree 1, and grandchildren of root are leaves. $\mathbb{H}_m$ is $\mathbb{G}_m$ plus an isolated vertex.

$\mathbb{G}_m$ is connected and $\mathbb{H}_m$ is not, so the existential reflective machine which decides the class of connected graphs must accept $\mathbb{G}_m$ and reject $\mathbb{H}_m$. However, $\mathbb{G}_m$ and $\mathbb{H}_m$ satisfy precisely the same sentences of EFO with up to m variables. Consequently for odd $n = 2m + 1$ the machine must use a query with at least $m + 1$ variables in $\mathbb{G}_m$, and for even $n = 2m + 2$ the machine must use the same query in $\mathbb{H}_m$. The hypothesis of the theorem follows.

$\mathbb{G}_m \subseteq \mathbb{H}_m$, so every EFO sentence satisfied in $\mathbb{G}_m$ is satisfied in $\mathbb{H}_m$, as well. We have to show, that any EFO sentence with at most m variables satisfied in $\mathbb{H}_m$ is satisfied in $\mathbb{G}_m$, as well. The complete proof of this claim requires introducing an appropriate pebble game and cannot be fully presented here due to space limitations. Just to have the intuitive idea how the proof goes on, note that with m variables one cannot say in EFO that there are m elements in the middle layer of $\mathbb{H}$, keeping the isolated vertex at the same time. Therefore we never need more than $m - 1$ elements of the middle layer of $\mathbb{G}$ to satisfy any EFO sentence true in $\mathbb{H}$, which speaks about the isolated vertex, either. But then a vertex in the lower layer of $\mathbb{G}$ connected to an unused vertex from the middle layer can serve as a counterpart of the isolated vertex in $\mathbb{H}$. □

Consequently, a specific optimization technique is essentially needed if one wishes that the server-time complexity of the connectivity query remained polynomial. Note that such a method exists, according to Theorem 8. But under the naive evaluation method it would require time about $n^{\lfloor n/2 \rfloor}$ to be evaluated. To the contrary, no such technique is needed for unrestricted reflective machines, as Theorem 9 asserts. For such implementations even the naive evaluation technique for first order queries yields a polynomial server-time algorithm. Note that the connectivity query is definable even in Datalog with 3 variables, or, equivalently, in positive existential fixpoint logic with 3 variables. And nevertheless it requires a linear number of variables to be evaluated by existential reflexive machines.

Thus we have again found an evidence, that existential reflective machines implement much worse algorithms that the unrestricted reflective machines do.

2.4 Conclusion

Existential reflective machines compute in polynomial user-time precisely the same set of queries as their full first order counterparts.

On the other hand, the quality of algorithms for these queries represented by existential machines is worse on very complicated problems, such as those which are (widely believed to exist) in $\mathcal{PSPACE} \setminus \mathcal{P^{NP}}$, and on some very simple polynomial time queries, like graph connectivity, as well.

This confirms that we have indeed found a new example of a mismatch between the class of computable queries and the class of algorithms computing these queries, represented in a common formalism.

Acknowledgments. The author is very indebted to Martin Grohe, who suggested him better examples for the proof of Theorem 10. My own construction sufficed only for $(1 - o(1)) \log n$ variables lower bound.

References

1. S. Abiteboul, C. Papadimitriou and V. Vianu, The power of reflective relational machines, in: *Proc. 9th Symposium on Logic in Computer Science*, 1994, pp. 230–240.
2. S. Abiteboul and V. Vianu, Generic computation and its complexity, in: *Proc. ACM SIGACT Symp. on the Theory of Computing*, 1991, pp. 209–219.
3. A. Blass and Y. Gurevich, Equivalence relations, invariants and normal forms, *SIAM Journal on Computing* **123**(1995), pp. 172–184.
4. R. Fagin, Generalized first-order spectra and polynomial-time recognizable sets, in: R. M. Karp, ed., *Complexity of Computation, SIAM-AMS Proc.* **7**(1974), pp. 43–73.
5. Y. Gurevich, Logic and the challenge of computer science, in: E. Börger (ed.), *Current Trends in Theoretical Computer Science*, Computer Science Press 1988, pp. 1–57.
6. D. Suciu and J. Paredaens, Any algorithm in the complex object algebra needs exponential space to compute transitive closure, in: *Proc. 13th ACM Symposium on Principles of Database Systems*, 1994, pp. 201–209.
7. D. Suciu and L. Wong, On two forms of structural recursion, in: *G. Gottlob, M.Y. Vardi (eds.), Proc. ICDT'95*, Lecture Notes in Computer Science No. 893, Springer Verlag 1995, pp. 111–124.
8. J. Tyszkiewicz, Fine hierarchies of generic computation and their complexity, submitted. Preliminary version appeared as [9].
9. J. Tyszkiewicz, Fine hierarchies of generic computation, in: *F. Afrati and Ph. Kolaitis (eds.), Proc. ICDT'97*, Lecture Notes in Computer Science No. 1186, Springer Verlag 1997, pp. 125–139.

Partial Order Semantics and Read Arcs

Walter Vogler *

Inst. f. Informatik, Universität Augsburg, Germany

Abstract

We study a new partial order semantics of Petri nets with read arcs, where read arcs model reading without consuming, which is often more adequate than the destructive-read-and-rewrite modelled in ordinary nets. As basic observations we take ST-traces, which are sequences of transition starts and ends. We define processes of our nets and derive two partial orders modelling causality and start precedence. These partial orders are related to observations and systems states as in the ordinary approach the single partial order of a process is related to firing sequences and reachable markings. Our approach also supports a new view of concurrency as captured by steps.

1 Introduction

Describing the runs of a concurrent system by sequences of actions ignores the possible concurrency of these actions, which can be important e.g. for judging the temporal efficiency of the system. Alternatively, one can take step sequences, where a step consists of simultaneous actions, or partial orders to describe runs. We will use safe Petri nets to model concurrent systems; for these models, the most prominent partial order semantics are so-called processes. A process of a net N is essentially a very simple net consisting of events (transition firings in N) and conditions (tokens in N produced during the run); the process gives a partial order on these events and conditions.

The beauty of the approach is that operationally defined entities of N can now be derived order-theoretically: Each linearization of the events is a firing sequence of N, and vice versa, each firing sequence of N is a linearization of a unique process. We can view the process as a run and these linearizations as observations of the run; essentially by Szpilrajn's Theorem, we can reconstruct the partial order of the events simply as intersection of the total orders given by all these observations. Unordered conditions are coexisting tokens, and each

---

*Work on this paper was partially supported by the DFG (Project 'Halbordnungstesten'). Author's address: Inst. f. Informatik, Uni. Augsburg, D-86135 Augsburg, Germany. email: vogler@informatik.uni-augsburg.de

slice (maximal set of unordered conditions) is a reachable marking; each reachable marking is a slice of some process and each step a set of unordered events.

Recently, Petri nets with read arcs have found considerable interest [CH93, JK95, MR95, BG95, BP96]; read arcs – as the lines from s in Figure 1 – describe reading without consuming, e.g. reading in a database; consequently, a and b in N_1 can occur concurrently. In ordinary nets, loops (arcs from a to s and from s to a and similarly for b) would be used instead, which describe a destructive-read-and-rewrite and do not allow concurrency; this is certainly often not adequate. [MR95, JK95, BP96] define processes of nets with read arcs and generalize some of the results listed above, taking step sequences as observations. Whereas in Figure 1 [MR95, BP96] allow a step $\{a, b\}$ only for N_1, [JK95] allows this step also for N_2 and N_3; [JK95] views these nets as translations from nets with inhibitor arcs where these steps are intuitively reasonable if a and b start together and end some time later. For N_2 and N_3 themselves, this intuition does not seem very convincing. Also, an undesirable effect is that in N_3 the step reaches a marking that is not reachable by firing sequences. (Correspondingly, [JK95] allows more processes than [MR95, BP96].)

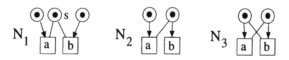

Figure 1

The purpose of the present paper is a partial order semantics under the assumption that activities have durations; consequently, observations of runs are ST-traces [Gla90, Vog92] where we see transitions start and then end. The respective states are ST-markings consisting of marked places and currently firing transitions; hence, ST-markings treat places and transitions on an equal footing just as nets themselves do.

When a starts in N_2, b remains enabled and can start during the occurrence of a; thus, a and b overlap in time and $\{a, b\}$ is observably a step; note that for a and b both to occur, a *has to start* before b. This view allows more concurrency than that of [MR95, BP96]. In fact, in the latter approach each net with read arcs can be translated to an ordinary net with the same partial order semantics. Such a translation does not exist for N_2 in our setting; $\{a, b\}$ is a step of N_2 but ba is not a firing sequence; this is impossible for ordinary nets. Hence, read arcs really make a difference in our approach, see also [Vog97]. On the other hand, our approach is a conservative extension of the ordinary setting since steps only reach markings that are also reachable by firing sequences.

Our processes are the same as those in [MR95], but the relational structures we derive from them are new; our spc-structures have two partial orders $\prec$ and $\sqsubset$ modelling causality and start precedence: $e \prec f$ means that e necessarily ends before f starts, while $e \sqsubset f$ means that e necessarily starts before f starts – compare a and b in N_2 above.

In Section 2, we define ST-traces, firing and step sequences for nets with read arcs and relate them to each other. Section 3 studies spc-structures and shows a suitable analogue of Szpilrajn's Theorem: each spc-structure is (in a way) the intersection of its so-called ST-linearizations. (Other generalizations of Szpilrajn's Theorem can be found in [JK93], but these cannot be applied here.) In Section 4, we define processes and the spc-structures they induce, and we show: Each order-theoretically derived ST-linearization of a process of some net N is an ST-trace of N; each cut (maximal unordered set of events and conditions) is an ST-marking reached along such an ST-trace. Vice versa, for each ST-trace of N we can construct a unique corresponding process, each reachable ST-marking is a cut of some process. For ordinary nets without read arcs our results are also of interest, since they give a generalization from firing (or step) sequences to ST-traces.

For the results on cuts, it is important that the spc-structures are defined on events and conditions. [JK95] also derives from a process a relational structure with two relations, but these are only defined on events, and they aim at step sequences such that the ST-traces of a net cannot be obtained.

Proofs and a more detailed comparison with existing approaches can be found in the full version under http://www.informatik.uni-augsburg.de/techreport/.

2 Petri nets, read arcs, steps and ST-traces

A *net* $N = (S, T, W, R, M_N)$ consists of finite disjoint sets S of *places* and T of *transitions*, the (ordinary) *arcs* $W \subseteq S \times T \cup T \times S$ (which all have weight 1), the set of *read arcs* $R \subseteq S \times T$, and the *initial marking* $M_N : S \to \{0, 1\}$; we always assume $(R \cup R^{-1}) \cap W = \emptyset$. When we introduce a net N, then this implicitly introduces its components $S, T, W, \ldots$ and similarly for other tuples. The tuple (S, T, W, R) is called a *net graph*. A net is called *ordinary*, if $R = \emptyset$.

For each $x \in S \cup T$, the *preset* of x is ${}^\bullet x = \{y \mid (y, x) \in W\}$, the *read set* of x is $\hat{x} = \{y \mid (y, x) \in R \cup R^{-1}\}$, and the *postset* of x is $x^\bullet = \{y \mid (x, y) \in W\}$. These notions are extended to sets, e.g. ${}^\bullet X = \bigcup_{x \in X} {}^\bullet x$. If $x \in {}^\bullet y \cap y^\bullet$, then x and y form a *loop*. A *marking* is a function $S \to \mathbb{N}_0$. We identify sets and characteristic functions, which map the elements of the sets to 1 and are 0 everywhere else; hence, we can e.g. add a marking and a postset of a transition.

In the basic firing rule, we simply regard read arcs as loops. A transition t is *enabled* under a marking M, denoted by $M[t\rangle$, if ${}^\bullet t \cup \hat{t} \leq M$. If $M[t\rangle$ and $M' = M + t^\bullet - {}^\bullet t$, we write $M[t\rangle M'$ and say that t can *fire* under M yielding the marking M'. This definition of enabling and firing can be extended to sequences as usual. If w is enabled under M_N, it is called a *firing sequence*. A marking M is called *reachable* if $\exists w \in T^* : M_N[w\rangle M$. The net is *safe* if $M(s) \leq 1$ for all places s and reachable markings M.

General assumption All nets considered in this paper are safe and T-restricted, i.e. each transition has a nonempty preset and a nonempty postset.

For a finite set X, $X^\pm$ denotes the union of two disjoint copies of X; for $x \in X$, the copies of x are x^+, the *start* of x, and x^-, the *end* of x. A sequence over $X^\pm$ is *closed*, if it contains each x^+ as often as the respective x^-. An *ST-sequence* over X is a sequence containing each x^+ once and each x^- at most once and only after the corresponding x^+.

$Q = (M, C)$ is an *ST-marking* of a net N, if M is a marking of N and $C \subseteq T$; C is the set of *currently firing* transitions. A transition start t^+ is *enabled* under Q, $Q[t^+\rangle$, if $M[t\rangle$; a transition end t^- is *enabled* under Q, $Q[t^-\rangle$, if $t \in C$. *Firing* yields a *follower ST-marking* given by $Q[t^+\rangle(M - {}^\bullet t, C \cup \{t\})$, thus the start t^+ checks the enabledness of t and consumes the input, and by $Q[t^-\rangle(M + t^\bullet, C - \{t\})$, i.e. t^- produces the output. We extend this definition to sequences, and if $Q_N = (M_N, \emptyset)[w\rangle Q$, then w is an *ST-trace* and Q a *reachable* ST-marking of N. The following results show in particular that ST-traces are a fairly conservative, refined version of firing sequences:

Prop. 1 *i) A marking M is reachable iff $(M, \emptyset)$ is a reachable ST-marking. Then $M[t_1 \ldots t_n\rangle M'$ iff $(M, \emptyset)[t_1^+ t_1^- \ldots t_n^+ t_n^-\rangle(M', \emptyset)$.*

ii) If (M, C) is a reachable ST-marking, then $(M, C)[t^+\rangle$ implies $t \notin C$. If w is an ST-trace, then t^+ and t^- occur alternatingly in w starting with t^+.

iii) If w' is obtained from an ST-trace w by moving some t^- to an earlier position that is still after the preceding t^+, then w' is an ST-trace as well and reaches the same ST-marking.

While the definitions of firing sequence and ST-trace are quite unquestionable, there are at least two different definitions of a step for nets with read arcs, and we define a third one by generalizing the observation made in the introduction for N_2. Recall that a *linearization* of a set X (or a partial order $\prec$ on X) is a sequence containing each element of X once (such that x occurs before y whenever $x \prec y$).

Def. 2 A transition t can *fire concurrently* to $G \subseteq T$ under a marking M, if $(M - {}^\bullet G)[t\rangle$. A set G with $\emptyset \neq G \subseteq T$ is a *step enabled under M* if for some linearization $t_1 \ldots t_n$ of G t_i can fire concurrently to $\{t_1, \ldots, t_{i-1}\}$ under M for $i = 1, \ldots, n$; $t_1 \ldots t_n$ is a *generation ordering* for G under M. The marking M' reached by firing G is $M - {}^\bullet G + G^\bullet$ and we write $M[G\rangle M'$; this is generalized to *step sequences* as usual. □

Steps are suitably defined as sets, since $t \notin G$ whenever t can fire concurrently to G under a reachable marking. We have already argued that, with our step definition, nets with read arcs cannot be simulated by ordinary nets – in contrast with results in [CH93, MR95]. That our definition is nevertheless a conservative extension is demonstrated in the following results. The following theorem also shows that steps and hence step sequences can be seen as special ST-traces. Also, steps are sets of transitions that can appear as currently firing (and thus concurrent) transitions in reachable ST-markings.

Theorem 3 *i) Let $t_1 \ldots t_n$ be a linearization of $G \neq \emptyset$ and M a reachable marking; G is a step under M with generation ordering $t_1 \ldots t_n$ iff $(M, \emptyset)[t_1^+ \ldots t_n^+\rangle$.*

In this case, $M[G\rangle M'$ implies $(M, \emptyset)[t_1^+ \ldots t_n^+\rangle(M - {}^\bullet G, G)[t_1^- \ldots t_n^-\rangle(M', \emptyset)$ and $M[t_1 \ldots t_n\rangle M'$. The markings reachable by step sequences are exactly the reachable markings.

ii) If G is a step under M, then there exists a reachable ST-marking (M', G).

The next theorem describes how a generation ordering for a step can be found; $\circ$ denotes composition of relations – e.g. $a(R^{-1}\circ W)b$ in N_2 in Figure 1.

Theorem 4 Let M be a reachable marking of a net N and $\emptyset \neq G \subseteq T$.

G is a step under M if and only if: ${}^\bullet G \cup \hat{G} \subseteq M$, the transitions in G have pairwise disjoint presets and the relation $R^{-1}\circ W$ is acyclic on G, i.e. $(R^{-1}\circ W)^+$ is irreflexive and thus a partial order. If G is a step, then the linearizations of $(R^{-1}\circ W)^+$ on G are exactly the generation orderings.

Corollary 5 Let M be a reachable marking of a net N. If G is a step under M and $\emptyset \neq G' \subseteq G$, then G' is a step under M. If N is an ordinary net and $\emptyset \neq G \subseteq T$, then G is a step under M iff for all $t, t' \in G$ we have $M[t\rangle$ and $t \neq t' \Rightarrow {}^\bullet t \cap {}^\bullet t' = \emptyset$; in this case, $M[w\rangle$ for each linearization w of G.

3 Structures for causality and start precedences

Usually, a partial order description of a system run is a set of events (and possibly conditions) ordered by some partial order $\prec$, modelling causality; i.e. for events e and f $e \prec f$ means that e necessarily ends before f starts. As argued in Section 1, we also have to consider for some events e and f that e necessarily *starts* before f starts; we will write $e \sqsubset f$ in this case. Obviously, $\sqsubset$ should also be a partial order and both relations should satisfy the requirements of the following definition. We write $e \sqsubseteq f$ whenever $e \sqsubset f \vee e = f$, $e \, co_\prec \, f$ when neither $e \prec f$ nor $f \prec e$; a set is a $co_\prec$-set if its elements are pairwise $co_\prec$.

Def. 6 An *spc-order* $p = (E, \prec, \sqsubset)$ consists of a finite set E, whose elements we call *events* in this section, and two partial orders $\prec$ and $\sqsubset$ on E such that $\prec \subseteq \sqsubset$ and $\prec \circ \sqsubset \subseteq \prec$ (i.e. $e \prec f \sqsubset g$ implies $e \prec g$ or equivalently $\prec \circ \sqsubseteq = \prec$).

An *spc-structure* $p = (E, \prec, \sqsubset, l)$ consists of an spc-order $(E, \prec, \sqsubset)$ and a *labelling* $l : E \to X$, where $e \, co_\prec \, f \wedge e \neq f$ implies $l(e) \neq l(f)$ for all $e, f \in E$. By this label requirement, the events with a given label x are totally ordered by $\prec$ and we can speak of the i-th event with label x; p is *canonical*, if $E \subseteq X \times \mathbb{N}$ and each $(x, i) \in E$ is the i-th event with label x. □

The above definition of an spc-order can be simplified since the other conditions imply that $\prec$ is a partial order. Obviously, each spc-structure is isomorphic to a canonical spc-structure, i.e. we can restrict attention to canonical spc-structures whenever this seems to be an advantage.

Graphically, we present an spc-order by connecting events e and f by an arrow if $e \prec f$ and by a dashed arrow if $e \sqsubset f$. (For spc-structures, we replace the events of E by their labels.) Implied arrows are often omitted, in particular we never draw an ordinary *and* a dashed arrow from e to f. If the arrows form an acyclic directed graph, then the drawing represents an spc-order:

Prop. 7 *Let E be a finite set with binary relations $\prec \subseteq \sqsubset$ such that $\sqsubset$ is acyclic. Then $p' = (E, \prec', \sqsubset')$ is an spc-order, where $\sqsubset'$ is $\sqsubset^+$ and $\prec'$ is $\prec \circ \sqsubset'$. For all spc-orders p'' with $\prec \subseteq \prec''$ and $\sqsubset \subseteq \sqsubset''$, we have $\prec' \subseteq \prec''$ and $\sqsubset' \subseteq \sqsubset''$.*

The proposition says that p' contains just all the orderings implied by the arrows in $\prec$ and $\sqsubset - \prec$; we call p' the spc-order *induced* by $\prec$ and R for any relation R with $\sqsubset = R \cup \prec$.

From a partial order, we can derive its augmentations (or extensions) to total orders; total orders obviously represent sequences and vice versa; the derived sequences are the linearizations. Similarly, one can order-theoretically define the derived step-sequences. In the case of spc-orders, we will analogously define which spc-orders correspond to sequences, step-sequences and ST-sequences; then, from a given spc-order, we can again derive sequences etc. order-theoretically as augmentations. First, we identify the spc-orders that correspond to sequences, step sequences and – more or less – to ST-sequences.

Def. 8 Let p be an spc-order. Then, p is an *spc-sequence*, if $\prec$ is a total order; the linearization w of $\prec$ is the *corresponding sequence*. If $co_\prec$ is an equivalence relation, p is an *spc-step-sequence*; the obvious sequence w of the equivalence classes ordered according to $\prec$ is the *corresponding step-sequence*. For an spc-structure p, p is analogously an *spc-trace* or an *spc-step-trace*; replacing each $e \in E$ by its label in the respective w gives the *corresponding trace* or *step-trace* with w as *underlying* sequence or step-sequence.

If $\sqsubset$ is total, p is an *interval-spc-order*. An ST-sequence w over E is a *corresponding ST-sequence* if for all e, f: e^+ occurs before f^+ if and only if $e \sqsubset f$; e^- occurs before f^+ if and only if $e \prec f$. As above, we derive from this the definitions of *interval-spc-structure*, *corresponding ST-trace* and *underlying* ST-sequence. □

The definitions of the first part are straightforward generalizations from partial orders. Since the labelling l of an spc-structure is injective on the equivalence classes of $co_\prec$ by the label requirement, the corresponding step trace is a sequence of sets (and not multisets). The second part needs more explanations. A partial order $\prec$ on E is an *interval order*, if for all $e, e', f, f' \in E$ we have: if $e \prec e'$ and $f \prec f'$, then $e \prec f'$ or $f \prec e'$; in this case, we can associate each $e \in E$ with an interval of real numbers such that $e \prec f$ iff the interval of e lies completely before that of f; a basic reference for interval orders is [Fis85, Chapter 2]. The following result explains the name interval-spc-order.

Prop. 9 *If p is an interval-spc-order, then $\prec$ is an interval order.*

Different from sequences and step-sequences, an interval-spc-order does not have a unique corresponding ST-sequence, but a set of such sequences. The next result shows that these sequences coincide up to simple modifications; each of the sequences allows to reconstruct the interval-spc-order, i.e. an interval-spc-order is a simple abstraction of an ST-sequence.

Prop. 10 *Let p be an interval-spc-order and I the set of its corresponding ST-sequences. Then there exists a closed w in I, and I is the set of sequences v that can be obtained from w by repeatedly replacing some $e^- f^-$ by $f^- e^-$ and deleting some e^- at the end of w.*

For an arbitrary $v \in I$, put $e \prec' f$ if e^- occurs in v before f^+ and put $e \sqsubset' f$ if e^+ occurs in v before f^+. Then $\prec = \prec'$ and $\sqsubset = \sqsubset'$.

This result carries over to interval-spc-structures, since by the following lemma we can determine from a corresponding ST-trace the underlying ST-sequence.

Lemma 11 *Let p be a canonical interval-spc-structure and w be a corresponding ST-trace, x a label. Then, x^+ and x^- alternate in w starting with x^+, and the i-th x^+ and x^- correspond to $(x, i)^+$ and $(x, i)^-$ in the underlying ST-sequence.*

Corollary 12 *Let p be a canonical interval-spc-structure and I the set of its corresponding ST-traces. Then there exists a closed w in I, and I is the set of sequences v that can be obtained from w by repeatedly replacing some $x^- y^-$ by $y^- x^-$ and deleting some x^- at the end of w.*

For an arbitrary $v \in I$, put $(x, i) \prec' (y, j)$ if the i-th x^- occurs in v before the j-th y^+ and put $(x, i) \sqsubset' (y, j)$ if the i-th x^+ occurs in v before the j-th y^+. Then $\prec = \prec' \wedge \sqsubset = \sqsubset'$.

Observe that the abstraction from ST-traces made by interval-spc-structures is compatible with the application to nets: if w in 12 is an ST-trace of a net, then the modifications v are also ST-traces of the net by Proposition 1 – independently of the net. Now we will describe how we can order-theoretically derive sequences etc. from an spc-order.

Def. 13 An spc-order $p' = (E, \prec', \sqsubset')$ is an *augmentation* of an spc-order p, if $\prec \subseteq \prec'$ and $\sqsubset \subseteq \sqsubset'$. If p' is additionally an spc-sequence, an spc-step-sequence or an interval-spc-order, then it is called a *linear*, *step* or *interval augmentation*.

A *linearization* of p is the corresponding sequence of a linear augmentation of p. Analogously, a *step linearization* and an *ST-linearization* correspond to a step and an interval augmentation of p.

This definition carries over to spc-structures. Linearizations etc. are analogously defined as corresponding traces, step traces and ST-traces with underlying sequences as in Definition 8. □

We will use spc-structures to model system runs; an ST-trace is an observation and, as we have seen, an interval-spc-structure is a moderate abstraction of an observation. By definition, such abstract observations can be derived order-theoretically from a run: they are the interval augmentations. We now show how to read off the ST-linearizations etc. directly, demonstrating how $\prec$ and $\sqsubset$ describe relationships between starts and ends.

Theorem 14 *w is an ST-linearization of an spc-order p iff it is an ST-sequence over E where e^+ occurs before f^+ if $e \sqsubset f$ and e^- occurs before f^+ if $e \prec f$;*

w is a linearization of p iff it is a linearization of $(E, \emptyset)$ such that $e \sqsubset f$ implies that e occurs before f;

w is a step linearization of p iff it is a sequence of sets forming a partition of E where: $e \prec f$ implies that the set containing e occurs before the set containing f; $e \sqsubset f$ implies that the set with e does not occur later than the set with f.

Theorem 14 also tells us how to read off the ST-linearizations etc. of an spc-structure p: we simply read off the ST-linearizations of the spc-order $(E, \prec, \sqsubset)$ and apply the labelling. The next theorem implies that ST-linearizations are all we need – here, we state this for spc-structures only.

Theorem 15 *Let p be an spc-structure, $x_i \in X$ and $\emptyset \neq X_i \subseteq X$ for $i = 1, \ldots, n$. $x_1 \ldots x_n$ is a linearization of p iff $x_1^+ x_1^- \ldots x_n^+ x_n^-$ is an ST-linearization of p. $X_1 \ldots X_n$ is a step linearization of p iff for all i $X_i = \{x_{i1}, \ldots, x_{im_i}\}$ for some indexing and $x_{11}^+ \ldots x_{1m_1}^+ x_{11}^- \ldots x_{1m_1}^- \ldots x_{n1}^- \ldots x_{nm_n}^-$ is an ST-linearization of p.*

Observe that Theorem 15 fits Proposition 1 i) and Theorem 3 i): if we have an spc-structure p and a net N such that all ST-linearizations of p are ST-traces of N, then all linearizations (step linearizations) of p are firing sequences (step sequences) of N; vice versa, if we can find for each ST-trace w of N an spc-structure p of a certain type such that w is an ST-linearization of p, then we can also find for each firing sequence or step sequence w of N an spc-structure p of this type such that w is a linearization or step linearization of p. Hence, if we want to study the behaviour of nets using spc-structures, we can restrict attention to ST-traces. The main result of this section shows that we can reconstruct a run from the set of its abstract observations.

Theorem 16 *Let p be an spc-order or -structure and I the set of its interval augmentations. Then $\prec = \bigcap_{p' \in I} \prec'$ and $\sqsubset = \bigcap_{p' \in I} \sqsubset'$.*

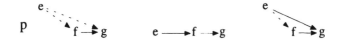

Figure 2

Proposition 10 and Corollary 12 show that we can reconstruct an interval-spc-order or -structure (up to isomorphism) from any of its ST-sequences or -traces. By Theorem 16, we can reconstruct an spc-order or -structure from its interval augmentations and, hence, (up to isomorphism) from its ST-linearizations. These results do not hold for step sequences. Figure 2 shows on the right an spc-step-sequence, where we cannot derive $e \sqsubset f$ from its corresponding step sequence $\binom{e}{f}g$. The spc-order p on the left cannot be reconstructed from its two step augmentations – which are also shown –, because we cannot derive that $\neg e \prec g$.

If we are only interested in step sequences, it is irrelevant whether $e \prec g$ or only $e \sqsubset g$. But if we are interested in the durations of events and runs, this difference is important: assume e.g. that e has duration 3 in p and f and g have duration 1; then e can start time 1 before f and carry on in parallel with g, such that the whole run p takes time 3. If we had $e \prec g$, the whole run would take at least time 4. Temporal efficiency of ordinary nets (where events have durations), partial order semantics and ST-traces have been explored in [Vog95].

4 Processes of nets with read arcs

A process is essentially a so-called occurrence net describing one run of another net N. Transitions of occurrence nets are called events and model the firings of transitions of N, places are called conditions and model tokens. We will extend the definition of processes to nets with read arcs, essentially following [MR95]. For a partial order $\prec$, we put $min_\prec(Y) = \{y \in Y \mid x \prec y \text{ for no } x \in Y\}$.

Def. 17 For a T-restricted net graph $O = (B, E, F, A)$, we define two relations on $B \cup E$: $\sqsubset$ is $(F \cup A \cup (A^{-1} \circ F))^+$ and $\prec$ is $F \circ \sqsubset$. O is an *occurrence net* if $|^\bullet b|, |b^\bullet| \leq 1$ for all $b \in B$ and $F \cup A \cup (A^{-1} \circ F)$ is acyclic, i.e. $\sqsubset$ is a partial order.

We denote the spc-order $(B \cup E, \prec, \sqsubset)$ induced by F and $A \cup (A^{-1} \circ F)$ according to Proposition 7 by $spc(O)$ and $min_\prec(B \cup E)$ by $^\bullet O$. We call the places $b \in B$ *conditions*, the transitions $e \in E$ *events*. □

As usual, a token is produced and consumed by (at most) one event; in this sense, conditions are unbranched in an occurrence net, but additionally they may be incident to some read arcs. If xFy, then x starts and ends firing or holding before y starts; if bAe, then e reads b, i.e. the holding of b starts before the firing e starts. As discussed in the introduction, bAe and bFf enforce that e starts before f. Thus, $\sqsubset$ intuitively gives an ordering of starts and should be acyclic, and $spc(O)$ should model the necessary relations between starts and ends of conditions and events in the run described by O; in our graphical notation for spc-orders, F gives the ordinary arrows while $A \cup (A^{-1} \circ F)$ gives the dashed arrows.

As in the classical setting, a process of a net N is an occurrence net O whose events correspond to transitions of N and whose conditions to places of N; $^\bullet O$ corresponds to the initial marking of N, F and A to the arcs and read arcs of N.

Def. 18 A *process* $\pi = (O, l)$ of a net N consists of an occurrence net O and a *labelling* $l : B \cup E \to S \cup T$ such that

i) $l(B) \subseteq S$, $l(E) \subseteq T$ ii) l is injective on $^\bullet O$ and $l(^\bullet O) = M_N$

iii) $\forall e \in E$: l is injective on $^\bullet e$, $e^\bullet$ and $\hat{e}$; $l(^\bullet e) = {}^\bullet l(e)$, $l(e^\bullet) = l(e)^\bullet$, $l(\hat{e}) = \widehat{l(e)}$.

We put $^\bullet \pi = {}^\bullet O$ and call $fspc(\pi) = (B \cup E, \prec, \sqsubset, l)$ the *full spc-structure* of π and its restriction $spc(\pi)$ to E (in all components) the *spc-structure* of π. An *ST-linearization* of π is an ST-linearization of $spc(\pi)$.

A *cut* of π is a maximal $co_\prec$-set of $fspc(\pi)$; a *slice* is a cut $D \subseteq B$. A cut D *corresponds* to the ST-marking $(\{l(b) \mid b \in B \cap D\}, \{l(e) \mid e \in E \cap D\})$ and a slice D *corresponds* to the marking $l(D)$. □

The first main result of this section shows that the order-theoretically derived ST-linearizations, (step) linearizations, cuts and slices of a process are ST-traces, firing (or step) sequences, reachable ST-markings and markings, which are behaviourally defined. One shows the 'ST-statements' first, the other statements then follow with Theorem 15, Proposition 1 and Theorem 3.

Theorem 19 *Let π be a process of some net N. Then all ST-linearizations of π are ST-traces of N, all (step) linearizations are firing (or step) sequences of N. The labelling l is injective on all cuts. Cuts correspond exactly to those ST-markings that can be reached along ST-linearizations of π, slices correspond exactly to those markings that can be reached along (step) linearizations of π.*

All the behaviourally defined entities can also be derived order-theoretically:

Theorem 20 *Let N be a net. For each ST-trace or firing or step sequence v of N, there is a unique process π of N which has v as ST-linearization or (step) linearization. For each reachable ST-marking Q or marking M of N, there is a process π of N with a cut or slice that corresponds to Q or M.*

Corollary 21 *For a step G under a reachable marking, there exists a process π and a $co_\prec$-set $E' \subseteq E$, such that l is injective on E' and $l(E') = G$.*

References

[BG95] N. Busi and R. Gorrieri. A Petri net semantics for π-calculus. In L. Insup and S. Smolka, editors, *CONCUR 95*, LNCS 962, 145–159. Springer, 1995.

[BP96] N. Busi and M. Pinna. Non-sequential semantics for contextual P/T-nets. In J. Billington and W. Reisig, editors, *Applications and Theory of Petri Nets 1996*, LNCS 1091, 113–132. Springer, 1996.

[CH93] S. Christensen and N.D. Hansen. Coloured Petri nets extended with place capacities, test arcs, and inhibitor arcs. In M. Ajmone-Marsan, editor, *Applications and Theory of Petri Nets 1993*, LNCS 691, 186–205. Springer, 1993.

[Fis85] P.C. Fishburn. *Interval Orders and Interval Graphs*. J. Wiley, 1985.

[Gla90] R.J. v. Glabbeek. The refinement theorem for ST-bisimulation semantics. In M. Broy and C.B. Jones, editors, *Programming Concepts and Methods, Proc. IFIP Working Conference*, 27–52. Elsevier Science Pub. (North-Holland), 1990.

[JK93] R. Janicki and M. Koutny. Order structures and generalisations of szpilrajn's theorem. In R. Shyamasundar, editor, *Found. Software Techn. and Theor. Comp. Sci. '93*, LNCS 761, 348–357. Springer, 1993.

[JK95] R. Janicki and M. Koutny. Semantics of inhibitor nets. *Information and Computation*, 123:1–16, 1995.

[MR95] U. Montanari, F. Rossi. Contextual nets. *Acta Informatica*, 32:545–596, 1995.

[Vog92] W. Vogler. *Modular Construction and Partial Order Semantics of Petri Nets*. Lect. Notes Comp. Sci. 625. Springer, 1992.

[Vog95] W. Vogler. Timed testing of concurrent systems. *Information and Computation*, 121:149–171, 1995.

[Vog97] W. Vogler. Efficiency of asynchronous systems and read arcs in Petri nets. To appear in ICALP 97, LNCS. Springer, 1997.

Author Index

Lecture Notes in Computer Science

For information about Vols. 1–1229

please contact your bookseller or Springer-Verlag

Vol. 1267: E. Biham (Ed.), Fast Software Encryption. Proceedings, 1997. VIII, 289 pages. 1997.

Vol. 1268: W. Kluge (Ed.), Implementation of Functional Languages. Proceedings, 1996. XI, 284 pages. 1997.

Vol. 1269: J. Rolim (Ed.), Randomization and Approximation Techniques in Computer Science. Proceedings, 1997. VIII, 227 pages. 1997.

Vol. 1270: V. Varadharajan, J. Pieprzyk, Y. Mu (Eds.), Information Security and Privacy. Proceedings, 1997. XI, 337 pages. 1997.

Vol. 1271: C. Small, P. Douglas, R. Johnson, P. King, N. Martin (Eds.), Advances in Databases. Proceedings, 1997. XI, 233 pages. 1997.

Vol. 1272: F. Dehne, A. Rau-Chaplin, J.-R. Sack, R. Tamassia (Eds.), Algorithms and Data Structures. Proceedings, 1997. X, 476 pages. 1997.

Vol. 1273: P. Antsaklis, W. Kohn, A. Nerode, S. Sastry (Eds.), Hybrid Systems IV. X, 405 pages. 1997.

Vol. 1274: T. Masuda, Y. Masunaga, M. Tsukamoto (Eds.), Worldwide Computing and Its Applications. Proceedings, 1997. XVI, 443 pages. 1997.

Vol. 1275: E.L. Gunter, A. Felty (Eds.), Theorem Proving in Higher Order Logics. Proceedings, 1997. VIII, 339 pages. 1997.

Vol. 1276: T. Jiang, D.T. Lee (Eds.), Computing and Combinatorics. Proceedings, 1997. XI, 522 pages. 1997.

Vol. 1277: V. Malyshkin (Ed.), Parallel Computing Technologies. Proceedings, 1997. XII, 455 pages. 1997.

Vol. 1278: R. Hofestädt, T. Lengauer, M. Löffler, D. Schomburg (Eds.), Bioinformatics. Proceedings, 1996. XI, 222 pages. 1997.

Vol. 1279: B. S. Chlebus, L. Czaja (Eds.), Fundamentals of Computation Theory. Proceedings, 1997. XI, 475 pages. 1997.

Vol. 1280: X. Liu, P. Cohen, M. Berthold (Eds.), Advances in Intelligent Data Analysis. Proceedings, 1997. XII, 621 pages. 1997.

Vol. 1281: M. Abadi, T. Ito (Eds.), Theoretical Aspects of Computer Software. Proceedings, 1997. XI, 639 pages. 1997.

Vol. 1282: D. Garlan, D. Le Métayer (Eds.), Coordination Languages and Models. Proceedings, 1997. X, 435 pages. 1997.

Vol. 1283: M. Müller-Olm, Modular Compiler Verification. XV, 250 pages. 1997.

Vol. 1284: R. Burkard, G. Woeginger (Eds.), Algorithms — ESA '97. Proceedings, 1997. XI, 515 pages. 1997.

Vol. 1285: X. Jao, J.-H. Kim, T. Furuhashi (Eds.), Simulated Evolution and Learning. Proceedings, 1996. VIII, 231 pages. 1997. (Subseries LNAI).

Vol. 1286: C. Zhang, D. Lukose (Eds.), Multi-Agent Systems. Proceedings, 1996. VII, 195 pages. 1997. (Subseries LNAI).

Vol. 1287: T. Kropf (Ed.), Formal Hardware Verification. XII, 367 pages. 1997.

Vol. 1288: M. Schneider, Spatial Data Types for Database Systems. XIII, 275 pages. 1997.

Vol. 1289: G. Gottlob, A. Leitsch, D. Mundici (Eds.), Computational Logic and Proof Theory. Proceedings, 1997. VIII, 348 pages. 1997.

Vol. 1290: E. Moggi, G. Rosolini (Eds.), Category Theory and Computer Science. Proceedings, 1997. VII, 313 pages. 1997.

Vol. 1292: H. Glaser, P. Hartel, H. Kuchen (Eds.), Programming Languages: Implementations, Logigs, and Programs. Proceedings, 1997. XI, 425 pages. 1997.

Vol. 1294: B.S. Kaliski Jr. (Ed.), Advances in Cryptology — CRYPTO '97. Proceedings, 1997. XII, 539 pages. 1997.

Vol. 1295: I. Prívara, P. Ružička (Eds.), Mathematical Foundations of Computer Science 1997. Proceedings, 1997. X, 519 pages. 1997.

Vol. 1296: G. Sommer, K. Daniilidis, J. Pauli (Eds.), Computer Analysis of Images and Patterns. Proceedings, 1997. XIII, 737 pages. 1997.

Vol. 1297: N. Lavrač, S. Džeroski (Eds.), Inductive Logic Programming. Proceedings, 1997. VIII, 309 pages. 1997. (Subseries LNAI).

Vol. 1298: M. Hanus, J. Heering, K. Meinke (Eds.), Algebraic and Logic Programming. Proceedings, 1997. X, 286 pages. 1997.

Vol. 1299: M.T. Pazienza (Ed.), Information Extraction. Proceedings, 1997. IX, 213 pages. 1997. (Subseries LNAI).

Vol. 1300: C. Lengauer, M. Griebl, S. Gorlatch (Eds.), Euro-Par'97 Parallel Processing. Proceedings, 1997. XXX, 1379 pages. 1997.

Vol. 1302: P. Van Hentenryck (Ed.), Static Analysis. Proceedings, 1997. X, 413 pages. 1997.

Vol. 1303: G. Brewka, C. Habel, B. Nebel (Eds.), KI-97: Advances in Artificial Intelligence. Proceedings, 1997. XI, 413 pages. 1997. (Subseries LNAI).

Vol. 1304: W. Luk, P.Y.K. Cheung, M. Glesner (Eds.), Field-Programmable Logic and Applications. Proceedings, 1997. XI, 503 pages. 1997.

Vol. 1305: D. Corne, J.L. Shapiro (Eds.), Evolutionary Computing. Proceedings, 1997. X, 313 pages. 1997.

Vol. 1308: A. Hameurlain, A M. Tjoa (Eds.), Database and Expert Systems Applications. Proceedings, 1997. XVII, 688 pages. 1997.

Vol. 1310: A. Del Bimbo (Ed.), Image Analysis and Processing. Proceedings, 1997. Volume I. XXI, 722 pages. 1997.

Vol. 1311: A. Del Bimbo (Ed.), Image Analysis and Processing. Proceedings, 1997. Volume II. XXII, 794 pages. 1997.

Vol. 1312: A. Geppert, M. Berndtsson (Eds.), Rules in Database Systems. Proceedings, 1997. VII, 213 pages. 1997.

Vol. 1314: S. Muggleton (Ed.), Inductive Logic Programming. Proceedings, 1996. VIII, 397 pages. 1997. (Subseries LNAI).

Vol. 1315: G. Sommer, J.J. Koenderink (Eds.), Algebraic Frames for the Perception-Action Cycle. Proceedings, 1997. VIII, 395 pages. 1997.